MEMS

Applications

Mechanical Engineering Series
Frank Kreith and Roop Mahajan - Series Editors

The MEMS Handbook

Second Edition

MEMS

Applications

Edited by
Mohamed Gad-el-Hak

CRC Press
Taylor & Francis Group
Boca Raton   London   New York

CRC Press is an imprint of the
Taylor & Francis Group, an **informa** business

A TAYLOR & FRANCIS BOOK

Foreground: A 24-layer rotary varactor fabricated in nickel using the Electrochemical Fabrication (EFAB®) technology. See Chapter 6, *MEMS: Design and Fabrication*, for details of the EFAB® technology. Scanning electron micrograph courtesy of Adam L. Cohen, Microfabrica Incorporated (www.microfabrica.com), U.S.A.

Background: A two-layer surface macromachined, vibrating gyroscope. The overall size of the integrated circuitry is 4.5 × 4.5 mm. Sandia National Laboratories' emblem in the lower right-hand corner is 700 microns wide. The four silver rectangles in the center are the gyroscope's proof masses, each 240 × 310 × 2.25 microns. See Chapter 4, *MEMS: Applications* (0-8493-9139-3), for design and fabrication details. Photograph courtesy of Andrew D. Oliver, Sandia National Laboratories.

CRC Press
Taylor & Francis Group
6000 Broken Sound Parkway NW, Suite 300
Boca Raton, FL 33487-2742

First issued in paperback 2019

© 2006 by Taylor & Francis Group, LLC
CRC Press is an imprint of Taylor & Francis Group, an Informa business

No claim to original U.S. Government works

ISBN-13: 978-0-8493-9139-2 (hbk)
ISBN-13: 978-0-367-39164-5 (pbk)
ISBN 13: 978-0-8493-2106-1 (set)

Library of Congress Cataloging-in-Publication Data

MEMS : applications / edited by Mohamed Gad-el-Hak.
    p. cm. -- (Mechanical engineering series)
  Includes bibliographical references and index.
  ISBN 0-8493-9139-3 (alk. paper)
    1. Microelectromechanical systems. 2. Detectors. 3. Microactuators. 4. Robots. I. Gad-el-Hak, Mohamed, 1945- II. Mechanical engineering series (Boca Raton Fla.)

TK7875.M423 2005
621.381--dc22   2005051409

Library of Congress Card Number 2005051409

Visit the Taylor & Francis Web site at
http://www.taylorandfrancis.com

and the CRC Press Web site at
http://www.crcpress.com

Preface

In a little time I felt something alive moving on my left leg, which advancing gently forward over my breast, came almost up to my chin; when bending my eyes downward as much as I could, I perceived it to be a human creature not six inches high, with a bow and arrow in his hands, and a quiver at his back. ... I had the fortune to break the strings, and wrench out the pegs that fastened my left arm to the ground; for, by lifting it up to my face, I discovered the methods they had taken to bind me, and at the same time with a violent pull, which gave me excessive pain, I a little loosened the strings that tied down my hair on the left side, so that I was just able to turn my head about two inches. ... These people are most excellent mathematicians, and arrived to a great perfection in mechanics by the countenance and encouragement of the emperor, who is a renowned patron of learning. This prince has several machines fixed on wheels, for the carriage of trees and other great weights.

(From *Gulliver's Travels—A Voyage to Lilliput*, by Jonathan Swift, 1726.)

In the Nevada desert, an experiment has gone horribly wrong. A cloud of nanoparticles — micro-robots — has escaped from the laboratory. This cloud is self-sustaining and self-reproducing. It is intelligent and learns from experience. For all practical purposes, it is alive.

It has been programmed as a predator. It is evolving swiftly, becoming more deadly with each passing hour.

Every attempt to destroy it has failed.

And we are the prey.

(From Michael Crichton's techno-thriller *Prey*, HarperCollins Publishers, 2002.)

Almost three centuries apart, the imaginative novelists quoted above contemplated the astonishing, at times frightening possibilities of living beings much bigger or much smaller than us. In 1959, the physicist Richard Feynman envisioned the fabrication of machines much smaller than their makers. The length scale of man, at slightly more than 10^0 m, amazingly fits right in the middle of the smallest subatomic particle, which is approximately 10^{-26} m, and the extent of the observable universe, which is of the order of 10^{26} m. Toolmaking has always differentiated our species from all others on Earth. Close to 400,000 years ago, archaic *Homo sapiens* carved aerodynamically correct wooden spears. Man builds things consistent with his size, typically in the range of two orders of magnitude larger or smaller than himself. But humans have always striven to explore, build, and control the extremes of length and time scales. In the voyages to Lilliput and Brobdingnag in *Gulliver's Travels*, Jonathan Swift speculates on the remarkable possibilities which diminution or magnification of physical dimensions provides. The Great Pyramid of Khufu was originally 147 m high when completed around 2600 B.C., while the Empire State Building constructed in 1931 is presently 449 m high. At the other end of the spectrum of manmade artifacts, a dime is slightly less than 2 cm in diameter. Watchmakers have practiced the art of miniaturization since the 13th century. The invention of the microscope in the 17th century opened the way for direct observation of microbes and plant and animal cells. Smaller things were manmade in the latter half of the 20th century. The transistor in today's integrated circuits has a size of 0.18 micron in production and approaches 10 nanometers in research laboratories.

Microelectromechanical systems (MEMS) refer to devices that have characteristic length of less than 1 mm but more than 1 micron, that combine electrical and mechanical components, and that are fabricated using integrated circuit batch-processing technologies. Current manufacturing techniques for MEMS include surface silicon micromachining; bulk silicon micromachining; lithography, electrodeposition, and plastic molding; and electrodischarge machining. The multidisciplinary field has witnessed explosive growth during the last decade and the technology is progressing at a rate that far exceeds that of our understanding of the physics involved. Electrostatic, magnetic, electromagnetic, pneumatic and thermal actuators, motors, valves, gears, cantilevers, diaphragms, and tweezers of less than 100 micron size have been fabricated. These have been used as sensors for pressure, temperature, mass flow, velocity, sound and chemical composition, as actuators for linear and angular motions, and as simple components for complex systems such as robots, lab-on-a-chip, micro heat engines and micro heat pumps. The lab-on-a-chip in particular is promising to automate biology and chemistry to the same extent the integrated circuit has allowed large-scale automation of computation. Global funding for micro- and nanotechnology research and development quintupled from $432 million in 1997 to $2.2 billion in 2002. In 2004, the U.S. National Nanotechnology Initiative had a budget of close to $1 billion, and the worldwide investment in nanotechnology exceeded $3.5 billion. In 10 to 15 years, it is estimated that micro- and nanotechnology markets will represent $340 billion per year in materials, $300 billion per year in electronics, and $180 billion per year in pharmaceuticals.

The three-book *MEMS set* covers several aspects of microelectromechanical systems, or more broadly, the art and science of electromechanical miniaturization. MEMS design, fabrication, and application as well as the physical modeling of their materials, transport phenomena, and operations are all discussed. Chapters on the electrical, structural, fluidic, transport and control aspects of MEMS are included in the books. Other chapters cover existing and potential applications of microdevices in a variety of fields, including instrumentation and distributed control. Up-to-date new chapters in the areas of microscale hydrodynamics, lattice Boltzmann simulations, polymeric-based sensors and actuators, diagnostic tools, microactuators, nonlinear electrokinetic devices, and molecular self-assembly are included in the three books constituting the second edition of *The MEMS Handbook*. The 16 chapters in *MEMS: Introduction and Fundamentals* provide background and physical considerations, the 14 chapters in *MEMS: Design and Fabrication* discuss the design and fabrication of microdevices, and the 15 chapters in *MEMS: Applications* review some of the applications of microsensors and microactuators.

There are a total of 45 chapters written by the world's foremost authorities in this multidisciplinary subject. The 71 contributing authors come from Canada, China (Hong Kong), India, Israel, Italy, Korea, Sweden, Taiwan, and the United States, and are affiliated with academia, government, and industry. Without compromising rigorousness, the present text is designed for maximum readability by a broad audience having engineering or science background. As expected when several authors are involved, and despite the editor's best effort, the chapters of each book vary in length, depth, breadth, and writing style. These books should be useful as references to scientists and engineers already experienced in the field or as primers to researchers and graduate students just getting started in the art and science of electromechanical miniaturization. The Editor-in-Chief is very grateful to all the contributing authors for their dedication to this endeavor and selfless, generous giving of their time with no material reward other than the knowledge that their hard work may one day make the difference in someone else's life. The talent, enthusiasm, and indefatigability of Taylor & Francis Group's Cindy Renee Carelli (acquisition editor), Jessica Vakili (production coordinator), N. S. Pandian and the rest of the editorial team at Macmillan India Limited, Mimi Williams and Tao Woolfe (project editors) were highly contagious and percolated throughout the entire endeavor.

Mohamed Gad-el-Hak

Editor-in-Chief

Mohamed Gad-el-Hak received his B.Sc. (summa cum laude) in mechanical engineering from Ain Shams University in 1966 and his Ph.D. in fluid mechanics from the Johns Hopkins University in 1973, where he worked with Professor Stanley Corrsin. Gad-el-Hak has since taught and conducted research at the University of Southern California, University of Virginia, University of Notre Dame, Institut National Polytechnique de Grenoble, Université de Poitiers, Friedrich-Alexander-Universität Erlangen-Nürnberg, Technische Universität München, and Technische Universität Berlin, and has lectured extensively at seminars in the United States and overseas. Dr. Gad-el-Hak is currently the Inez Caudill Eminent Professor of Biomedical Engineering and chair of mechanical engineering at Virginia Commonwealth University in Richmond. Prior to his Notre Dame appointment as professor of aerospace and mechanical engineering, Gad-el-Hak was senior research scientist and program manager at Flow Research Company in Seattle, Washington, where he managed a variety of aerodynamic and hydrodynamic research projects.

Professor Gad-el-Hak is world renowned for advancing several novel diagnostic tools for turbulent flows, including the laser-induced fluorescence (LIF) technique for flow visualization; for discovering the efficient mechanism via which a turbulent region rapidly grows by destabilizing a surrounding laminar flow; for conducting the seminal experiments which detailed the fluid–compliant surface interactions in turbulent boundary layers; for introducing the concept of targeted control to achieve drag reduction, lift enhancement and mixing augmentation in wall-bounded flows; and for developing a novel viscous pump suited for microelectromechanical systems (MEMS) applications. Gad-el-Hak's work on Reynolds number effects in turbulent boundary layers, published in 1994, marked a significant paradigm shift in the subject. His 1999 paper on the fluid mechanics of microdevices established the fledgling field on firm physical grounds and is one of the most cited articles of the 1990s.

Gad-el-Hak holds two patents: one for a drag-reducing method for airplanes and underwater vehicles and the other for a lift-control device for delta wings. Dr. Gad-el-Hak has published over 450 articles, authored/edited 14 books and conference proceedings, and presented 250 invited lectures in the basic and applied research areas of isotropic turbulence, boundary layer flows, stratified flows, fluid–structure interactions, compliant coatings, unsteady aerodynamics, biological flows, non-Newtonian fluids, hard and soft computing including genetic algorithms, flow control, and microelectromechanical systems. Gad-el-Hak's papers have been cited well over 1000 times in the technical literature. He is the author of the book *"Flow Control: Passive, Active, and Reactive Flow Management,"* and editor of the books *"Frontiers in Experimental Fluid Mechanics," "Advances in Fluid Mechanics Measurements," "Flow Control: Fundamentals and Practices," "The MEMS Handbook,"* and *"Transition and Turbulence Control."*

Professor Gad-el-Hak is a fellow of the American Academy of Mechanics, a fellow and life member of the American Physical Society, a fellow of the American Society of Mechanical Engineers, an associate fellow of the American Institute of Aeronautics and Astronautics, and a member of the European Mechanics

Society. He has recently been inducted as an eminent engineer in Tau Beta Pi, an honorary member in Sigma Gamma Tau and Pi Tau Sigma, and a member-at-large in Sigma Xi. From 1988 to 1991, Dr. Gad-el-Hak served as Associate Editor for *AIAA Journal*. He is currently serving as Editor-in-Chief for *e-MicroNano.com*, Associate Editor for *Applied Mechanics Reviews* and *e-Fluids*, as well as Contributing Editor for Springer-Verlag's *Lecture Notes in Engineering* and *Lecture Notes in Physics*, for McGraw-Hill's Year Book of Science and Technology, and for CRC Press' *Mechanical Engineering Series*.

Dr. Gad-el-Hak serves as consultant to the governments of Egypt, France, Germany, Italy, Poland, Singapore, Sweden, United Kingdom and the United States, the United Nations, and numerous industrial organizations. Professor Gad-el-Hak has been a member of several advisory panels for DOD, DOE, NASA and NSF. During the 1991/1992 academic year, he was a visiting professor at Institut de Mécanique de Grenoble, France. During the summers of 1993, 1994 and 1997, Dr. Gad-el-Hak was, respectively, a distinguished faculty fellow at Naval Undersea Warfare Center, Newport, Rhode Island, a visiting exceptional professor at Université de Poitiers, France, and a Gastwissenschaftler (guest scientist) at Forschungszentrum Rossendorf, Dresden, Germany. In 1998, Professor Gad-el-Hak was named the Fourteenth ASME Freeman Scholar. In 1999, Gad-el-Hak was awarded the prestigious Alexander von Humboldt Prize — Germany's highest research award for senior U.S. scientists and scholars in all disciplines — as well as the Japanese Government Research Award for Foreign Scholars. In 2002, Gad-el-Hak was named ASME Distinguished Lecturer, as well as inducted into the Johns Hopkins University Society of Scholars.

Contributors

Yuxing Ben
Department of Mathematics
Massachusetts Institute of
    Technology
Cambridge, Massachusetts, U.S.A.

Paul L. Bergstrom
Department of Electrical and
    Computer Engineering
Michigan Technological University
Houghton, Michigan, U.S.A.

Alberto Borboni
Dipartimento di Ingegneria
    Meccanica
Università degli studi di Brescia
Brescia, Italy

Hsueh-Chia Chang
Department of Chemical and
    Biomolecular Engineering
University of Notre Dame
Notre Dame, Indiana, U.S.A.

Haecheon Choi
School of Mechanical and
    Aerospace Engineering
Seoul National University
Seoul, Republic of Korea

Thorbjörn Ebefors
SILEX Microsystems AB
Jarfalla, Sweden

Mohamed Gad-el-Hak
Department of Mechanical
    Engineering
Virginia Commonwealth University
Richmond, Virginia, U.S.A.

Yogesh B. Gianchandani
Department of Electrical
    Engineering and Computer Science
University of Michigan
Ann Arbor, Michigan, U.S.A.

Gary G. Li
Freescale Semiconductor
    Incorporated
Tempe, Arizona, U.S.A.

Lennart Löfdahl
Thermo and Fluid Dynamics
Chalmers University of Technology
Göteborg, Sweden

E. Phillip Muntz
University of Southern California
Department of Aerospace and
    Mechanical Engineering
Los Angeles, California, U.S.A.

Ahmed Naguib
Department of Mechanical
    Engineering
Michigan State University
East Lansing, Michigan, U.S.A.

Andrew D. Oliver
Principal Member of the
    Technical Staff
Advanced Microsystems Packaging
Sandia National Laboratories
Albuquerque, New Mexico, U.S.A.

Jae-Sung Park
Department of Electrical and
    Computer Engineering
University of Wisconsin—Madison
Madison, Wisconsin, U.S.A.

G. P. Peterson
Rensselaer Polytechnic Institute
Troy, New York, U.S.A.

David W. Plummer
Sandia National Laboratories
Albuquerque, New Mexico, U.S.A.

Choondal B. Sobhan
Department of Mechanical
    Engineering
National Institute of Technology
Calicut, Kerala, India

Göran Stemme
Department of Signals, Sensors and
    Systems
School of Electrical Engineering
Royal Institute of Technology
Stockholm, Sweden

Melissa L. Trombley
Department of Electrical and
    Computer Engineering
Michigan Technological University
Houghton, Michigan, U.S.A.

Fan-Gang Tseng
Department of Engineering and
    System Science
National Tsing Hua University
Hsinchu, Taiwan, Republic of China

Stephen E. Vargo
Siimpel Corporation
Arcadia, California, U.S.A.

Chester G. Wilson
Institute for Micromanufacturing
Louisiana Tech University
Ruston, Los Angeles, U.S.A.

Marcus Young
University of Southern California
Department of Aerospace and
    Mechanical Engineering
Los Angeles, California, U.S.A.

Yitshak Zohar
Department of Aerospace and
    Mechanical Engineering
University of Arizona
Tucson, Arizona, U.S.A.

Table of Contents

The farther backward you can look,
the farther forward you are likely to see.

(Sir Winston Leonard Spencer Churchill, 1874–1965)

Janus, Roman god of
gates, doorways and all
beginnings, gazing both
forward and backward.

As for the future, your task is not to foresee, but to enable it.

(Antoine-Marie-Roger de Saint-Exupéry, 1900–1944,
in Citadelle [*The Wisdom of the Sands*])

1

Introduction

Mohamed Gad-el-Hak
Virginia Commonwealth University

How many times when you are working on something frustratingly tiny, like your wife's wrist watch, have you said to yourself, "If I could only train an ant to do this!" What I would like to suggest is the possibility of training an ant to train a mite to do this. What are the possibilities of small but movable machines? They may or may not be useful, but they surely would be fun to make.

(From the talk "There's Plenty of Room at the Bottom," delivered by Richard P. Feynman at the annual meeting of the American Physical Society, Pasadena, California, December 1959.)

Toolmaking has always differentiated our species from all others on Earth. Aerodynamically correct wooden spears were carved by archaic *Homo sapiens* close to 400,000 years ago. Man builds things consistent with his size, typically in the range of two orders of magnitude larger or smaller than himself, as indicated in Figure 1.1. Though the extremes of length-scale are outside the range of this figure, man, at slightly more than 10^0 m, amazingly fits right in the middle of the smallest subatomic particle, which is

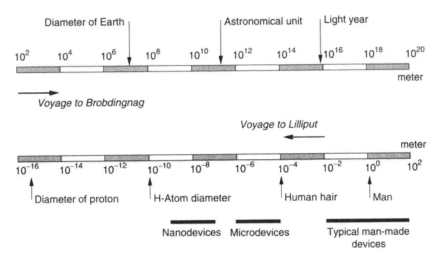

FIGURE 1.1 Scale of things, in meters. Lower scale continues in the upper bar from left to right. One meter is 10^6 microns, 10^9 nanometers, or 10^{10} Angstroms.

approximately 10^{-26} m, and the extent of the observable universe, which is of the order of 10^{26} m (15 billion light years); neither geocentric nor heliocentric, but rather egocentric universe. But humans have always striven to explore, build, and control the extremes of length and time scales. In the voyages to Lilliput and Brobdingnag of *Gulliver's Travels*, Jonathan Swift (1726) speculates on the remarkable possibilities which diminution or magnification of physical dimensions provides.[1] The Great Pyramid of Khufu was originally 147 m high when completed around 2600 B.C., while the Empire State Building constructed in 1931 is presently — after the addition of a television antenna mast in 1950 — 449 m high. At the other end of the spectrum of manmade artifacts, a dime is slightly less than 2 cm in diameter. Watchmakers have practiced the art of miniaturization since the 13th century. The invention of the microscope in the 17th century opened the way for direct observation of microbes and plant and animal cells. Smaller things were man-made in the latter half of the 20th century. The transistor — invented in 1947 — in today's integrated circuits has a size[2] of 0.18 micron (180 nanometers) in production and approaches 10 nm in research laboratories using electron beams. But what about the miniaturization of mechanical parts — machines — envisioned by Feynman (1961) in his legendary speech quoted above?

Manufacturing processes that can create extremely small machines have been developed in recent years (Angell et al., 1983; Gabriel et al., 1988, 1992; O'Connor, 1992; Gravesen et al., 1993; Bryzek et al., 1994; Gabriel, 1995; Ashley, 1996; Ho and Tai, 1996, 1998; Hogan, 1996; Ouellette, 1996, 2003; Paula, 1996; Robinson et al., 1996a, 1996b; Tien, 1997; Amato, 1998; Busch-Vishniac, 1998; Kovacs, 1998; Knight, 1999; Epstein, 2000; O'Connor and Hutchinson, 2000; Goldin et al., 2000; Chalmers, 2001; Tang and Lee, 2001; Nguyen and Wereley, 2002; Karniadakis and Beskok, 2002; Madou, 2002; DeGaspari, 2003; Ehrenman, 2004; Sharke, 2004; Stone et al., 2004; Squires and Quake, 2005). Electrostatic, magnetic, electromagnetic, pneumatic and thermal actuators, motors, valves, gears, cantilevers, diaphragms, and tweezers of less than 100 μm size have been fabricated. These have been used as sensors for pressure, temperature, mass flow, velocity, sound, and chemical composition, as actuators for linear and angular motions, and as simple components for complex systems, such as lab-on-a-chip, robots, micro-heat-engines and micro heat pumps (Lipkin, 1993; Garcia and Sniegowski, 1993, 1995; Sniegowski and Garcia, 1996; Epstein and Senturia, 1997; Epstein et al., 1997; Pekola et al., 2004; Squires and Quake, 2005).

Microelectromechanical systems (MEMS) refer to devices that have characteristic length of less than 1 mm but more than 1 micron, that combine electrical and mechanical components, and that are fabricated using integrated circuit batch-processing technologies. The books by Kovacs (1998) and Madou (2002) provide excellent sources for microfabrication technology. Current manufacturing techniques for MEMS include surface silicon micromachining; bulk silicon micromachining; lithography, electrodeposition, and plastic molding (or, in its original German, *Lithographie Galvanoformung Abformung, LIGA*); and electrodischarge machining (EDM). As indicated in Figure 1.1, MEMS are more than four orders of magnitude larger than the diameter of the hydrogen atom, but about four orders of magnitude smaller than the traditional manmade artifacts. Microdevices can have characteristic lengths smaller than the diameter of a human hair. Nanodevices (some say NEMS) further push the envelope of electromechanical miniaturization (Roco, 2001; Lemay et al., 2001; Feder, 2004).

The famed physicist Richard P. Feynman delivered a mere two, albeit profound, lectures[3] on electromechanical miniaturization: "There's Plenty of Room at the Bottom," quoted above, and "Infinitesimal Machinery," presented at the Jet Propulsion Laboratory on February 23, 1983. He could not see a lot of use for micromachines, lamenting in 1959 that "(small but movable machines) may or may not be useful, but they surely would be fun to make," and 24 years later said, "There is no use for these machines, so I still don't

---

[1] *Gulliver's Travels* were originally designed to form part of a satire on the abuse of human learning. At the heart of the story is a radical critique of human nature in which subtle ironic techniques work to part the reader from any comfortable preconceptions and challenge him to rethink from first principles his notions of man.

[2] The smallest feature on a microchip is defined by its smallest linewidth, which in turn is related to the wavelength of light employed in the basic lithographic process used to create the chip.

[3] Both talks have been reprinted in the *Journal of Microelectromechanical Systems*, vol. 1, no. 1, pp. 60–66, 1992, and vol. 2, no. 1, pp. 4–14, 1993.

understand why I'm fascinated by the question of making small machines with movable and controllable parts." Despite Feynman's demurring regarding the usefulness of small machines, MEMS are finding increased applications in a variety of industrial and medical fields with a potential worldwide market in the billions of dollars.

Accelerometers for automobile airbags, keyless entry systems, dense arrays of micromirrors for high-definition optical displays, scanning electron microscope tips to image single atoms, micro heat exchangers for cooling of electronic circuits, reactors for separating biological cells, blood analyzers, and pressure sensors for catheter tips are but a few of the current usages. Microducts are used in infrared detectors, diode lasers, miniature gas chromatographs, and high-frequency fluidic control systems. Micropumps are used for ink jet printing, environmental testing, and electronic cooling. Potential medical applications for small pumps include controlled delivery and monitoring of minute amount of medication, manufacturing of nanoliters of chemicals, and development of artificial pancreas. The much sought-after lab-on-a-chip is promising to automate biology and chemistry to the same extent the integrated circuit has allowed large-scale automation of computation. Global funding for micro- and nanotechnology research and development quintupled from $432 million in 1997 to $2.2 billion in 2002. In 2004, the U.S. National Nanotechnology Initiative had a budget of close to $1 billion, and the worldwide investment in nanotechnology exceeded $3.5 billion. In 10 to 15 years, it is estimated that micro- and nanotechnology markets will represent $340 billion per year in materials, $300 billion per year in electronics, and $180 billion per year in pharmaceuticals.

The multidisciplinary field has witnessed explosive growth during the past decade. Several new journals are dedicated to the science and technology of MEMS; for example *Journal of Microelectromechanical Systems, Journal of Micromechanics and Microengineering, Microscale Thermophysical Engineering, Microfluidics and Nanofluidics Journal, Nanotechnology Journal*, and *Journal of Nanoscience and Nanotechnology*. Numerous professional meetings are devoted to micromachines; for example Solid-State Sensor and Actuator Workshop, International Conference on Solid-State Sensors and Actuators (Transducers), Micro Electro Mechanical Systems Workshop, Micro Total Analysis Systems, and Eurosensors. Several web portals are dedicated to micro- and nanotechnology; for example, <http://www.smalltimes.com>, <http://www.emicronano.com>, <http://www.nanotechweb.org/>, and <http://www.peterindia.net/NanoTechnologyResources.html>.

The three-book *MEMS set* covers several aspects of microelectromechanical systems, or more broadly, the art and science of electromechanical miniaturization. MEMS design, fabrication, and application as well as the physical modeling of their materials, transport phenomena, and operations are all discussed. Chapters on the electrical, structural, fluidic, transport and control aspects of MEMS are included in the books. Other chapters cover existing and potential applications of microdevices in a variety of fields, including instrumentation and distributed control. Up-to-date new chapters in the areas of microscale hydrodynamics, lattice Boltzmann simulations, polymeric-based sensors and actuators, diagnostic tools, microactuators, nonlinear electrokinetic devices, and molecular self-assembly are included in the three books constituting the second edition of *The MEMS Handbook*. The 16 chapters in *MEMS: Introduction and Fundamentals* provide background and physical considerations, the 14 chapters in *MEMS: Design and Fabrication* discuss the design and fabrication of microdevices, and the 15 chapters in *MEMS: Applications* review some of the applications of microsensors and microactuators.

There are a total of 45 chapters written by the world's foremost authorities in this multidisciplinary subject. The 71 contributing authors come from Canada, China (Hong Kong), India, Israel, Italy, Korea, Sweden, Taiwan, and the United States, and are affiliated with academia, government, and industry. Without compromising rigorousness, the present text is designed for maximum readability by a broad audience having engineering or science background. As expected when several authors are involved, and despite the editor's best effort, the chapters of each book vary in length, depth, breadth, and writing style. The nature of the books — being handbooks and not encyclopedias — and the size limitation dictate the noninclusion of several important topics in the MEMS area of research and development.

Our objective is to provide a current overview of the fledgling discipline and its future developments for the benefit of working professionals and researchers. The three books will be useful guides and references to the explosive literature on MEMS and should provide the definitive word for the fundamentals and applications of microfabrication and microdevices. Glancing at each table of contents, the reader may rightly sense an overemphasis on the physics of microdevices. This is consistent with the strong conviction of the Editor-in-Chief that the MEMS technology is moving too fast relative to our understanding of the unconventional physics involved. This technology can certainly benefit from a solid foundation of the underlying fundamentals. If the physics is better understood, less expensive, and more efficient, microdevices can be designed, built, and operated for a variety of existing and yet-to-be-dreamed applications. Consistent with this philosophy, chapters on control theory, distributed control, and soft computing are included as the backbone of the futuristic idea of using colossal numbers of microsensors and microactuators in reactive control strategies aimed at taming turbulent flows to achieve substantial energy savings and performance improvements of vehicles and other manmade devices.

I shall leave you now for the many wonders of the small world you are about to encounter when navigating through the various chapters of these volumes. May your voyage to Lilliput be as exhilarating, enchanting, and enlightening as Lemuel Gulliver's travels into "Several Remote Nations of the World." *Hekinah degul!* Jonathan Swift may not have been a good biologist and his scaling laws were not as good as those of William Trimmer (see Chapter 2 of *MEMS: Introduction and Fundamentals*), but Swift most certainly was a magnificent storyteller. *Hnuy illa nyha majah Yahoo!*

References

Amato, I. (1998) "Formenting a Revolution, in Miniature," *Science* 282, no. 5388, 16 October, pp. 402–405.

Angell, J.B., Terry, S.C., and Barth, P.W. (1983) "Silicon Micromechanical Devices," *Faraday Transactions I* 68, pp. 744–748.

Ashley, S. (1996) "Getting a Microgrip in the Operating Room," *Mech. Eng.* 118, September, pp. 91–93.

Bryzek, J., Peterson, K., and McCulley, W. (1994) "Micromachines on the March," *IEEE Spectrum* 31, May, pp. 20–31.

Busch-Vishniac, I.J. (1998) "Trends in Electromechanical Transduction," *Phys. Today* 51, July, pp. 28–34.

Chalmers, P. (2001) "Relay Races," *Mech. Eng.* 123, January, pp. 66–68.

DeGaspari, J. (2003) "Mixing It Up," *Mech. Eng.* 125, August, pp. 34–38.

Ehrenman, G. (2004) "Shrinking the Lab Down to Size," *Mech. Eng.* 126, May, pp. 26–29.

Epstein, A.H. (2000) "The Inevitability of Small," *Aerospace Am.* 38, March, pp. 30–37.

Epstein, A.H., and Senturia, S.D. (1997) "Macro Power from Micro Machinery," *Science* 276, 23 May, p. 1211.

Epstein, A.H., Senturia, S.D., Al-Midani, O., Anathasuresh, G., Ayon, A., Breuer, K., Chen, K.-S., Ehrich, F.F., Esteve, E., Frechette, L., Gauba, G., Ghodssi, R., Groshenry, C., Jacobson, S.A., Kerrebrock, J.L., Lang, J.H., Lin, C.-C., London, A., Lopata, J., Mehra, A., Mur Miranda, J.O., Nagle, S., Orr, D.J., Piekos, E., Schmidt, M.A., Shirley, G., Spearing, S.M., Tan, C.S., Tzeng, Y.-S., and Waitz, I.A. (1997) "Micro-Heat Engines, Gas Turbines, and Rocket Engines — The MIT Microengine Project," AIAA Paper No. 97-1773, AIAA, Reston, Virginia.

Feder, T. (2004) "Scholars Probe Nanotechnology's Promise and Its Potential Problems," *Phys. Today* 57, June, pp. 30–33.

Feynman, R.P. (1961) "There's Plenty of Room at the Bottom," in *Miniaturization*, H.D. Gilbert, ed., pp. 282–296, Reinhold Publishing, New York.

Gabriel, K.J. (1995) "Engineering Microscopic Machines," *Sci. Am.* 260, September, pp. 150–153.

Gabriel, K.J., Jarvis, J., and Trimmer, W., eds. (1988) *Small Machines, Large Opportunities: A Report on the Emerging Field of Microdynamics, National Science Foundation*, published by AT&T Bell Laboratories, Murray Hill, New Jersey.

Gabriel, K.J., Tabata, O., Shimaoka, K., Sugiyama, S., and Fujita, H. (1992) "Surface-Normal Electrostatic/Pneumatic Actuator," in *Proc. IEEE Micro Electro Mechanical Systems '92*, pp. 128–131, 4–7 February, Travemünde, Germany.

Garcia, E.J., and Sniegowski, J.J. (1993) "The Design and Modelling of a Comb-Drive-Based Microengine for Mechanism Drive Applications," in *Proc. Seventh International Conference on Solid-State Sensors and Actuators (Transducers '93)*, pp. 763–766, Yokohama, Japan, 7–10 June.

Garcia, E.J., and Sniegowski, J.J. (1995) "Surface Micromachined Microengine," *Sensor. Actuator. A* **48**, pp. 203–214.

Goldin, D.S., Venneri, S.L., and Noor, A.K. (2000) "The Great out of the Small," *Mech. Eng.* **122**, November, pp. 70–79.

Gravesen, P., Branebjerg, J., and Jensen, O.S. (1993) "Microfluidics — A Review," *J. Micromech. Microeng.* **3**, pp. 168–182.

Ho, C.-M., and Tai, Y.-C. (1996) "Review: MEMS and Its Applications for Flow Control," *J. Fluids Eng.* **118**, pp. 437–447.

Ho, C.-M., and Tai, Y.-C. (1998) "Micro–Electro–Mechanical Systems (MEMS) and Fluid Flows," *Annu. Rev. Fluid Mech.* **30**, pp. 579–612.

Hogan, H. (1996) "Invasion of the Micromachines," *New Sci.* **29**, June, pp. 28–33.

Karniadakis, G.E., and Beskok A. (2002) *Microflows: Fundamentals and Simulation*, Springer-Verlag, New York.

Knight, J. (1999) "Dust Mite's Dilemma," *New Sci.* **162**, no. 2180, 29 May, pp. 40–43.

Kovacs, G.T.A. (1998) *Micromachined Transducers Sourcebook*, McGraw-Hill, New York.

Lemay, S.G., Janssen, J.W., van den Hout, M., Mooij, M., Bronikowski, M.J., Willis, P.A., Smalley, R.E., Kouwenhoven, L.P., and Dekker, C. (2001) "Two-Dimensional Imaging of Electronic Wavefunctions in Carbon Nanotubes," *Nature* **412**, 9 August, pp. 617–620.

Lipkin, R. (1993) "Micro Steam Engine Makes Forceful Debut," *Sci. News* **144**, September, p. 197.

Madou, M. (2002) *Fundamentals of Microfabrication*, second edition, CRC Press, Boca Raton, Florida.

Nguyen, N.-T., and Wereley, S.T. (2002) *Fundamentals and Applications of Microfluidics*, Artech House, Norwood, Massachusetts.

O'Connor, L. (1992) "MEMS: Micromechanical Systems," *Mech. Eng.* **114**, February, pp. 40–47.

O'Connor, L., and Hutchinson, H. (2000) "Skyscrapers in a Microworld," *Mech. Eng.* **122**, March, pp. 64–67.

Ouellette, J. (1996) "MEMS: Mega Promise for Micro Devices," *Mech. Eng.* **118**, October, pp. 64–68.

Ouellette, J. (2003) "A New Wave of Microfluidic Devices," *Ind. Phys.* **9**, no. 4, pp. 14–17.

Paula, G. (1996) "MEMS Sensors Branch Out," *Aerospace Am.* **34**, September, pp. 26–32.

Pekola, J., Schoelkopf, R., and Ullom, J. (2004) "Cryogenics on a Chip," *Phys. Today* **57**, May, pp. 41–47.

Robinson, E.Y., Helvajian, H., and Jansen, S.W. (1996a) "Small and Smaller: The World of MNT," *Aerospace Am.* **34**, September, pp. 26–32.

Robinson, E.Y., Helvajian, H., and Jansen, S.W. (1996b) "Big Benefits from Tiny Technologies," *Aerospace Am.* **34**, October, pp. 38–43.

Roco, M.C. (2001) "A Frontier for Engineering," *Mech. Eng.* **123**, January, pp. 52–55.

Sharke, P. (2004) "Water, Paper, Glass," *Mech. Eng.* **126**, May, pp. 30–32.

Sniegowski, J.J., and Garcia, E.J. (1996) "Surface Micromachined Gear Trains Driven by an On-Chip Electrostatic Microengine," *IEEE Electron Device Lett.* **17**, July, p. 366.

Squires, T.M., and Quake, S.R. (2005) "Microfluidics: Fluid Physics at the Nanoliter Scale," *Rev. Mod. Phys.* **77**, pp. 977–1026.

Stone, H.A., Stroock, A.D., and Ajdari, A. (2004) "Engineering Flows in Small Devices: Microfluidics Toward a Lab-on-a-Chip," *Annu. Rev. Fluid Mech.* **36**, pp. 381–411.

Swift, J. (1726) *Gulliver's Travels*, 1840 reprinting of *Lemuel Gulliver's Travels into Several Remote Nations of the World*, Hayward & Moore, London, Great Britain.

Tang, W.C., and Lee, A.P. (2001) "Military Applications of Microsystems," *Ind. Phys.* **7**, February, pp. 26–29.

Tien, N.C. (1997) "Silicon Micromachined Thermal Sensors and Actuators," *Microscale Thermophys. Eng.* **1**, pp. 275–292.

2

Inertial Sensors

Paul L. Bergstrom and
Melissa L. Trombley
Michigan Technological University

Gary G. Li
Freescale Semiconductor Incorporated

2.1   Introduction

Inertial sensors are designed to convert, or transduce, a physical phenomenon into a measurable signal. This physical phenomenon is an inertial force. Often this force is transduced into a linearly scaled voltage output with a specified sensitivity. The methodologies utilized for macroscopic inertial sensors can and have been utilized for micromachined sensors in many applications. It is worth considering what factors have led to the introduction of micromachined inertial sensors. As will be demonstrated in this chapter, differences in linear and angular sensor application requirements impact the choice of micromachining technology, transducer design, and system architecture. The system requirements often delineate micromachining technology options very clearly, although most sensing mechanisms and micromachining technologies have been applied to inertial sensors. First, the chapter will address design parameters for linear inertial sensors, or accelerometers. Technologies applied to accelerometers will demonstrate the major physical mechanisms implemented in sensing inertial displacement. Next, design parameters specific to rotational inertial rate sensors, also called angular rate sensors or gyroscopes, will be presented. An overview of several recognized microsystem fabrication processes is also given with discussion of how technology can influence system and sensor, or transducer, design.

2.2   Applications of Inertial Sensors

Three primary areas that often are considered in micromachined device applications include packaged volume or size, system cost, and performance. Often, these three drivers cannot be met in a single technology choice. Packaged volume or overall system size is usually an easy goal for micromachined inertial sensors versus their macroscopic counterparts. Micromachining technologies are capable of reducing the sensor element and electronics board components to the scale of one integrated or two co-packaged chips, in small plastic or ceramic packages. Figure 2.1 shows two such examples: (a) an integrated accelerometer technology produced by Analog Devices, Inc., and (b) a stacked co-packaged accelerometer produced by Freescale Semiconductor, Inc. shown here for a quad flat no-lead (QFN) package.

System cost is also an important goal for micromachined inertial sensors. Because of their technological relation to the microelectronics industry, micromachined sensors can be batch fabricated, sharing process cost over large volumes of sensors, and reducing the overall process constraints significantly. While many individual processes may be significantly more expensive than their macroscopic counterparts, because the benefits of scale can be applied, the impact is greatly reduced.

Meeting a targeted device performance with sufficient profitability requires improvements in production costs per unit. One factor in this cost improvement is die area utilization. Maximizing device sensitivity per unit area minimizes die area. Current sensor device requirements for occupant safety systems have allowed the incorporation of surface micromachined technologies in early generation technologies. Surface micromachining technologies use the successive deposition of sacrificial and structural layers to produce an anchored, yet freestanding device typically made of polycrystalline silicon in structural thicknesses less than three microns [Ristic et al., 1992]. These technologies have been successful for current design requirements, but are being replaced by technologies that demonstrate application flexibility and improved die area utilization.

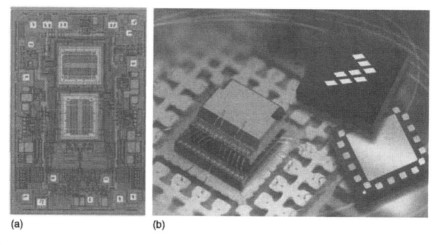

(a)   (b)

FIGURE 2.1   (See color insert following page 2-12.) Examples of two high-volume accelerometer products. Example (a) is the top view micrograph of the Analog Devices, Inc. ADXL250 two-axis lateral monolithically-integrated accelerometer. Example (b) is a perspective view of the Freescale Semiconductor, Inc. wafer-scale packaged accelerometer and control chips stack-mounted on a lead frame prior to plastic injection molding. (Photos courtesy of Analog Devices, Inc. and Freescale Semiconductor, Inc.)

Micromachined sensors can suffer in overall system sensitivity and full-scale range compared to macroscopic counterparts. The tradeoff between system size and cost and performance is often directly coupled, although notable exceptions exist to that rule. Micromachined technologies have succeeded in applications driven by size and cost with modest performance requirements. One such application is the automobile.

Automotive applications motivate most of the micromachined inertial sensor technology development efforts. Automotive inertial sensor applications are listed in Table 2.1 [MacDonald, 1990]. Accelerometers are one of the largest volume micromachined products; they are used in airbag control systems in automobile occupant restraining systems. Devices designed for airbag systems typically require inertial sensitivities of $20\,g$ to $100\,g$ full-scale for front impact airbags and $100\,g$ to $250\,g$ full-scale for side impact airbags, where one g represents the acceleration due to earth's gravity. Single-axis inertial sensors are also used in vehicle dynamics for active suspension systems with typical required inertial sensitivities from $0.5\,g$ to $10\,g$. Future occupant safety systems are beginning to require more sensors to tailor a system response to the conditions of the crash event. Crash variables can include impact location, occupant position and weight, use of seat belts, and crash severity. A future occupant safety system may use many multi-axis transducers distributed around the automobile to determine whether an airbag should be deployed and at what rate.

Vehicle dynamics and occupant safety systems are increasing in complexity and capability. Encompassing active suspensions, traction control, rollover safety systems, low-g accelerometers, yaw rate sensors, and tilt rate sensors are employed and tied into engine, steering, and antilock braking systems to return control of the vehicle to the driver in an out-of-control situation. Combined with front and side impact airbag systems, the system will determine the severity of the event and deploy seat belt pretensioners, side bags, head bags, window bags, and interaction airbags between occupants as necessary. Versions of these systems are being introduced on more and more vehicles today. In the future, many of these systems will be merged, providing greater system capability and complexity.

Angular inertial rate sensor technologies, encompassing pitch, roll, and yaw rate sensing, require significantly higher effective sensitivities than the analogous accelerometer. Such devices typically exhibit milli-g to micro-g resolution in order to produce stable measurements of rotational inertia with less than one degree per minute drift. Design considerations for such devices encompass the same micromachining technologies but add significant device and system complexity to achieve stable and reliable results.

TABLE 2.1   Inertial Sensor Applications in the Automobile

	Range	Application Comments
	$\pm 1\,g$	Anti-lock braking (ABS), traction control systems (TCS), virtual reality (VR),
	$\pm 2\,g$	Vertical body motion
	$\pm 50\,g$	Front air bag deployment, wheel motion
	± 100–$250\,g$	Side (B-pillar) air bag deployment
	± 100–$250°/s$	Roll and yaw rate for safety and stability control
Resolution	<0.1%	Full-scale, all applications
Linearity	<1%	Full-scale, all applications
Output noise	<0.005–0.05% FS/$\sqrt{\text{Hz}}$	Full-scale signal, all applications
Offset drift	<1 g/s	Accelerometer
	<0.1°/s/s	Gyroscope
Temperature range	−40 to 85°C	Operational conditions
	−55 to 125°C	Storage conditions
Cross-axis sensitivity	<1 to 3%	Application dependent
Frequency response	DC to 1–5 kHz	Airbag deployment
	DC to 10–100 Hz	Gyroscopes and 1–2 g accelerometers
Shock survivability	>500 g	Powered all axes
	>1500 g	Unpowered all axes

Outside the transportation and vehicular marketplace there are many applications for sub-*g* inertial sensor products including virtual reality systems, intelligent toys, industrial motion control, hard drive head protection systems, video camera image stabilization systems, shipping damage detectors, robotic warehouse operations, GPS receivers and inertial navigation systems.

2.3   Basic Acceleration Concepts

A wide range of platforms exists for micromechanical inertial sensors, and designers choose and optimize a particular style based on reference frame and the quantity being measured. The reference frame is primarily a function of the sensor application and depends upon whether the measurement is linear or rotational. The linear sensor is generally defined in Cartesian coordinates and measures the kinematic force due to a linear acceleration as shown in Figure 2.2. The angular rate sensor can be defined in either a cylindrical or a Cartesian space and measures the angular velocity of a rotation about its primary axis. This angular rate measurement is usually due to the coupled Coriolis force on a rotating or vibrating body.

Linear acceleration a can be defined as

$$\mathbf{a} = \frac{d^2\mathbf{r}}{dt^2} = \frac{d\mathbf{v}}{dt} \tag{2.1}$$

where r denotes linear displacement in meters (m) and v is the linear velocity in meters per second (m/s). The equation is written in vector notation to indicate that while in most systems only one axis of motion is allowed (where the scalar, *x*, would replace r), off axis interactions need to be considered in complex system design.

Angular rate sensors measure angular velocity, ω, which is defined as

$$\omega = \frac{d\theta}{dt} \tag{2.2}$$

where θ is the angular displacement in radians, and the angular velocity, ω, is measured in radians per second (rad/s). Angular acceleration, α, may also be measured and is defined as

$$\alpha = \frac{d^2\theta}{dt^2} = \frac{d\omega}{dt} \tag{2.3}$$

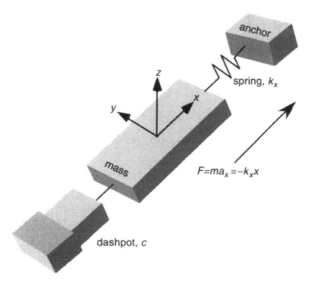

FIGURE 2.2   Cartesian reference frame for linear accelerometers. The figure shows an *X*-axis device for reference, including anchor, spring, seismic mass, and dashpot.

To measure the angular velocity of a moving body, the Coriolis Effect is most often utilized in micro-machined *vibratory* sensors, although rotating micromachined devices have been demonstrated [Shearwood et al., 1999]. If a proof-mass is driven into oscillation in one axis, rotation of the reference frame will displace the oscillating mass into a second orthogonal axis. This coupling is the Coriolis Effect and is given by Equation 2.4:

$$F_c = 2m\mathbf{v} \times \Omega \qquad (2.4)$$

which is a force produced when a vibratory mass, *m*, moving at a velocity, **v**, is placed in a rotation Ω. This Coriolis-induced force is orthogonal to the vibratory motion as defined by the vector cross product. Nonidealities or limitations in the design, fabrication, and operation of angular rate sensors will generate coupling terms that confound the orthogonal Coriolis force measured in these sensors, and require clever and complex solutions in the device and system architectures and fabrication technologies [Painter and Shkel, 2003].

2.4 Linear Inertial Sensor Parameters

Linear inertial sensors typically consist of four components: a seismic mass, also called a proof-mass; a suspension in the form of one or more elastic springs; a dashpot to provide motion stabilization; and a method by which the displacement of the seismic mass is measured. The mass is used to generate an inertial force due to an acceleration or deceleration event; the elastic spring will mechanically support the proof-mass and restore the mass to its neutral position after the acceleration is removed. The dashpot is usually the volume of air, or controlled ambient, captured inside the package or cavity surrounding the device; it is designed to control the motion of the seismic mass in order to obtain favorable frequency response characteristics. The sense methodology converts the mechanical displacement to an electrical output. Linear devices are classified either as in-plane (often denoted *X-axis* or *X*-lateral) and out-of-plane (or *Z*-axis). The choice of axis is primarily driven by the application. Front airbag systems require lateral sensing, whereas side and satellite airbag sensors are often mounted vertically and call for *Z*-axis sensing. While these two types of devices are similar in concept and operate much in the same manner, *X*-axis and *Z*-axis devices have very different designs and often bear little physical resemblance. Increasingly, designers and manufacturers are looking toward multiple axis inertial sensing in the same package or assembly and are considering more complicated and elaborate electromechanical structures to achieve the design goals.

The successful implementation of microfabrication technology to produce an inertial sensor requires more than the micromachining technology alone. The tradeoffs between transducer and circuitry define the system approach and often demonstrate the complexity of the systems these micromechanical devices are intended to replace. All micromachining technologies exhibit limitations that must be accounted for in the design of the overall system. The designs of the transducer for a given technology and of the overall system are interwoven. The transducer design and technology capability must first be addressed. The integration of the transducer function at the system level also defines the partitioning and complexity of the technology.

The manufacturability of a transducer structure is of paramount importance and should be vigorously considered during the design stage. In theory, one may design a sensor structure to be as sensitive as desired, but if the structure cannot be manufactured in a robust manner, the effort is wasted. An oversized proof-mass or a soft spring may dramatically increase the probability of stiction resulting in yield loss, for example (see section 2.7 on Micromachining Technology Manufacturing Issues). In this regard, the design must follow certain design rules, which will differ from one technology to another, in order to improve the yield of manufacturing.

2.4.1 Converting Acceleration to Force: The Seismic Mass

The application of an inertial load exerts a force on the proof-mass, which is translated into a displacement by the elastic spring. The simple force equation for a static load is shown in Equation 2.5 — where *m* is the proof mass (kg) and **a** is the static or steady-state inertial acceleration (m/s^2) — is given by

FIGURE 2.3 Scanning electron micrograph of an ultra-high-resolution X-lateral accelerometer with micro-g resolution. (Photo courtesy of K. Najafi, University of Michigan.)

$$F = ma = \frac{d^2x}{dt^2} \qquad (2.5)$$

The accelerometer design needs to precisely control the displacement of the seismic mass. The structure should be sufficiently massive and rigid to act in a well-behaved manner. In general, it should have at least an order of magnitude greater stiffness than that of the elastic spring in the axis of sensitivity. For a lateral sensor, the design of a sufficiently massive proof-mass benefits from thicker structures. For a given transducer area, the mass increases linearly with thickness, and the stiffness out-of-plane increases by the third power. A good example of a massive proof-mass in a lateral (X-axis) accelerometer is shown in Figure 2.3. Najafi et al. (2003) demonstrated this micro-g resolution lateral inertial sensor along with a Z-axis accelerometer and vibratory ring gyroscopes.

In most applications, sufficient proof-mass stiffness is not difficult to achieve. However, design considerations may be impacted by other system requirements. For example, in lateral capacitive structures, the incorporation of interdigitated sense fingers to the periphery of the proof-mass can complicate the proof-mass behavior. These sense fingers are often only a few times stiffer than the elastic springs and introduce a non-ideal behavior to the sensitivity for large inertial loads. Minimizing this impact is often a significant design tradeoff between device area and nominal capacitance for a device. Again, the thickness of the sense fingers improves the stiffness of a beam design and increases nominal capacitance per finger.

The size and stiffness of the mass can also be affected by technological constraints such as the nature of the sacrificial release etch. Since the proof-mass is typically by far the largest feature in an inertial sensor, the release of the proof-mass from the substrate by a chemical etchant may require a substantial duration to complete the lateral dissolution of the underlying sacrificial layer [Monk et al., 1994]. Designers often incorporate a series of *etch holes* through the structure to expedite the release, though this may require increasing the size of the mass in order to maintain transducer performance. These sacrificial etch holes are shown in Figure 2.4, in the proof-mass of the Freescale Semiconductor, Inc. Z-axis inertial sensor design. Various chemical mixtures have been developed specifically to improve the rate and quality of inertial sensor release [Williams et al., 2003; Overstolz et al., 2004].

2.4.2 Converting Force to Displacement: The Elastic Spring

The elastic spring is required to provide kinematic displacement of the proof-mass in the axis of sensitivity; this will produce a suitable sense signal while being sufficiently rigid in other axes to eliminate cross-axis

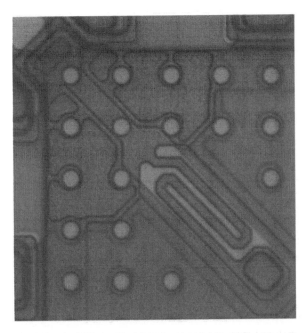

FIGURE 2.4   (See color insert following page 2-12.) Top view micrograph of a Z-axis accelerometer quadrant showing a folded spring and sacrificial etch holes designed into the proof-mass structure. (Photo courtesy Freescale Semiconductor, Inc.)

sensitivity. The force obtained from an applied acceleration produces an opposite restoring force according to Equation 2.6, where K is the elastic spring constant tensor (N/m) and x is the spatial displacement of the mass with respect to the reference frame generated by the inertial load:

$$\mathbf{F} = -K\mathbf{x}. \tag{2.6}$$

In the ideal case, the proof-mass displacement is well controlled and in one axis only. The ratio of the proof-mass to spring constant defines the sensitivity of the system to inertial loads.

The spring constant associated with the proof-mass suspension is governed by its geometry and material properties. An initial approximation for one dimension can be made using the standard equation for the deflection of an anchored cantilever beam under a point force applied at the opposite end:

$$k = \frac{Ehw^3}{4l^3}. \tag{2.7}$$

E is the modulus of elasticity (Pa) for the spring material and h, w, and l are the thickness, width, and length (m), respectively, of the beam. Micromachined springs are often folded or bent around a radius to improve the performance of the overall device; consequently, spring constant estimates require combinations of individual segment values [Boser and Howe, 1996]. Bent beams are often used to relieve residual intrinsic stress in the micromachined material and stabilize device parameters across wafers and production lots. Folded beams perform this function as well as reduce the device topology for the overall structure. They also improve the sensitivity of the inertial sensor to package stress by reducing the spacing between anchored points of the springs around the periphery of the sensor structure, as shown in Figure 2.4. It should be noted that Equation 2.7 is intended only to serve as an estimate for the purpose of initial design; computer-based simulation through finite element modeling or other means is highly recommended.

For a lateral accelerometer, increasing the out-of-plane stiffness proves to be one of the major challenges for many micromachining technologies. Out-of-plane stiffness strongly impacts drop shock immunity and cross axis sensitivity. Increasing the thickness of the beam increases this value by the third power, while the spring constant in the lateral axis of sensitivity only increases linearly. The aspect ratio, or ratio of the height to width of the spring, impacts the relative difference in stiffness both in- and out-of-plane. Since the spring constant in the lateral axis of sensitivity is defined by the chosen geometry of the device, simply increasing its geometrical length or reducing the beam width can compensate for the increase in spring constant of a beam. While this reduces the out-of-plane value somewhat, it has a considerably smaller linear impact compared to the cubic increase in beam stiffness due to the increase in thickness.

2.4.3   Device Damping: The Dashpot

While the proof-mass and elastic spring can be designed for a static condition, the design of the dashpot should provide an optimal dynamic damping coefficient through squeeze-film damping by the proper choice of the sensor geometry and packaging pressures. The extent of this effect is defined by the aspect ratio of the space between the plates and the ambient pressure. A large area-to-gap ratio results in significantly higher squeeze numbers, resulting in greater damping as described by Starr (1990).

Micromachined inertial sensor devices are often operated in an isolated environment filled with nitrogen or other types of gas such that the gas functions as a working fluid and dissipates energy. A gas film between two closely spaced parallel plates oscillating in normal relative motion generates a force — due to compression and internal friction — that opposes the motion of the plates. The damping due to such a force (related to energy loss in the system) is referred to as *squeeze film damping*. In other cases, two closely spaced parallel plates oscillate in a direction parallel to each other, and this damping generated by a gas film is referred to as *shear damping*. Under the small motion assumption in one axis, the flow-induced force is linearly proportional to the displacement and velocity of the moving plates. A single-axis dynamic model of an accelerometer is shown in Equation 2.8. The coefficient of the velocity is the damping coefficient, c,

$$m\frac{d^2\Delta x}{dt^2} + c\frac{d\Delta x}{dt} + k_x\Delta x = -ma_x \tag{2.8}$$

The solution to the differential equation demonstrates the fundamental resonant mode for the primary axis of motion shown. More generally, inertial sensors have multiple on- and off-axis resonant modes to be controlled, but the resonance in the axis of sensitivity is the primary effect with the greatest impact on device operation. The magnitude of the resonance peak is determined by the magnitude of the damping coefficient, c. This is the major contributing factor in the stability of the system and it demonstrates the tradeoff between spring constant and seismic mass. Typically, the design intent is to push the resonant modes far beyond the typical frequencies of interest for operation. A seismic mass that is too large or a spring constant that is too small drops the magnitude of the fundamental resonant frequency and reduces the system margin for operation.

For a given inertial sensor design, the magnitude of the damping is impacted by the aspect ratio of the spaces surrounding the seismic mass, the packaged pressure, and other internal material losses in the system. Figure 2.5 shows how packaged pressure influences the magnitude of the resonant peak for a surface micromachined lateral polysilicon accelerometer. By reducing the magnitude of the resonance peak using squeeze film damping at higher pressures, the motion of the seismic mass is brought under better control. In this case, for a given aspect ratio, increasing the pressure increases the damping coefficient, c, by approximately a factor of two. Alternatively, increasing the aspect ratio has a significant impact on the damping coefficient, c, for a given ambient pressure. In the case of high-pressure packaging, improving the aspect ratio of a design can minimize technology complexity by reducing the required packaging pressure to manageable and controllable values.

While the unit "N/m" (Newton per meter) for spring constant is well understood, the unit "kg/s" (kilogram per second) for damping coefficient is somewhat abstract, and it is often difficult to grasp its magnitude. Consequently, for a given mass–spring–dashpot oscillator, a non-dimensional damping ratio, ξ, is often used in practice. An under damped system corresponds to $\xi < 1$, a critically damped system corresponds to $\xi = 1$, and an over damped system corresponds to $\xi > 1$. To correlate the dimensional

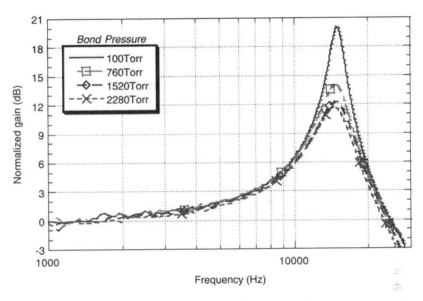

FIGURE 2.5 Normalized mechanical response over frequency for a 3 μm-thick lateral accelerometer capped at various pressures.

damping c with the non-dimensional damping ξ, consider the following governing equation of motion in one axis for a forced simple harmonic oscillator:

$$m\ddot{x} + c\dot{x} + kx = f_0 \sin \omega t, \tag{2.9}$$

where m is mass, c is damping coefficient, and k is spring constant. Introducing a non-dimensional damping, ξ, such that

$$\xi = \frac{c}{2}\sqrt{\frac{m}{k}} \equiv \frac{c}{2m\omega_n} \tag{2.10}$$

where $\omega_n = \sqrt{k/m}$ is the system's natural frequency. With ξ and ω_n defined, Equation 2.9 can be written into a standard form of

$$\ddot{x} + 2\xi\omega_n\dot{x} + \omega_n^2 x = \frac{f_0}{m}\sin \omega t \tag{2.11}$$

The steady state solution to Equation 2.11 is therefore

$$x(t) = \frac{1}{\sqrt{\left[1 - \left(\frac{\omega}{\omega_n}\right)^2\right]^2 + \left[2\xi\frac{\omega}{\omega_n}\right]^2}} \times \frac{f_0}{k}\sin(\omega t - \phi) \equiv \bar{x}\sin(\omega t - \phi) \tag{2.12}$$

where $\bar{x}$ is amplitude of the response and ϕ is phase difference. The damping ratio ξ has a large influence on the amplitude in the frequency region near resonance. For $\frac{\omega}{\omega_n} \ll 1$, the amplitude of response is simply the static displacement f_0/k. Near resonance, the amplitude of the response may be amplified greatly for small values of ξ. Although the critical damping condition of $\xi = 1$ would eliminate the vibration altogether, the maximum flatness of the system response can be achieved at a damping of $\xi = 0.65$. This can be clearly seen in Figure 2.6.

Another term related to damping is the quality factor Q. In forced vibration, Q is a measure of the sharpness of resonance. It is defined as a ratio of the system's natural frequency ω_n over a bandwidth

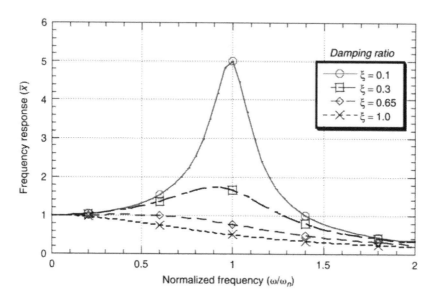

FIGURE 2.6 (See color insert following page 2-12.) The frequency response $\bar{x}$ versus normalized frequency ratio ω/ω_n.

$\omega_2 - \omega_1$, where ω_1 and ω_2 are two frequencies at which the response, $\bar{x}$, is 0.707 of its peak value. Manipulating Equation 2.12, it can be shown that, for small damping,

$$Q = \frac{\omega_n}{\omega_2 - \omega_1} = \frac{1}{\sqrt{1 + 2\xi} - \sqrt{1 - 2\xi}} \approx \frac{1}{2\xi} \qquad (2.13)$$

Therefore, at low damping, the quality factor Q is approximately inversely proportional to damping ξ. A quality factor of 5 means a damping ratio of 0.1.

To eliminate resonance in a transducer, a damping of $\xi > 0.65$ is required [Blech, 1983]. Both ambient pressure and the transducer structure influence the damping. To illustrate this, consider a sensing structure modeled for a 3 μm polysilicon technology. The simulated damping at 1 atm is about 0.11. If a thicker sensing structure were used, the damping would increase in an approximately quadratic manner with thickness, t. At $t = 10$ μm, the damping ratio is 1.1, an increase by a factor of 10. If the spring stiffness were kept at constant while increasing the thickness, t, the pace of damping increase would go up faster. At $t = 10$ μm, the damping is more than 2 and becomes over-damped.

If damping is too low in a micromachined lateral accelerometer, the severe degree of resonance of the accelerometer, upon an impact of external force, may produce a large signal that overloads the control circuitry resulting in system failure. High damping (near critical) is generally desired for accelerometers. Angular rate sensors, on the other hand, require low damping in order to achieve sufficient sensitivity of the system under a given driving force and in certain applications. Therefore, in designing a MEMS device, the consideration of damping must be taken into account at the earliest stage.

In general, the capping pressure for a micromachined system is below or much below the atmospheric pressure. As pressure decreases, the mean free path of the gas molecules (nitrogen for example) increases. When the mean free path is comparable to the air gap between two plates, one may no longer be able to treat the gas as continuum. Therefore, an effective viscosity coefficient is introduced such that governing equations of motion for fluid at relatively high pressures can still be used to treat fluid motion at low pressures where the mean free path is comparable or even larger than the air gap of the plates.

Based on earlier lubrication theory [Burgdorfer, 1959; Blech, 1983], Veijola et al. (1995, 1997, 1998) and Andrews et al. (1993, 1995) conducted extensive studies on squeeze film damping. They showed that the viscous damping effect of the air film dominates at low frequencies or squeeze numbers, but the flow-induced spring becomes more prominent at high frequencies or squeeze numbers. For systems where only squeeze film damping is present and for small plate oscillations, experimental studies [Veijola et al., 1995;

Andrews et al., 1993] for large square parallel plates are in general agreement with theory. In particular, the study carried out by Corman et al. (1997) is especially important for systems exposed to very low ambient pressures. For bulk micromachined resonator structures in silicon, they presented a squeeze film damping comparison between theory and experiments for pressures as low as 0.1 mbar. The agreement between theory and experiment is considered to be reasonable. By employing the technique of Green's function solution, Darling et al. (1998) also conducted a theoretical study on squeeze film damping for different boundary conditions. The resulting models are computationally compact, and thus applicable for dynamic system simulation purposes. Additional studies by van Kampen and Wolffenbuttel (1998), Reuther et al. (1996), and Pan et al. (1998) also dealt with squeeze film damping with emphasis on numerical aspects of simulations. From these studies, one may get a comprehensive understanding of the micromachined systems and their dynamic behavior and gain insight into how to fine-tune the design parameters to achieve higher sensitivity and better overall performance.

It is also worth mentioning the experimental work by Kim et al. (1999) and Gudeman et al. (1998) on MEMS damping. Kim and coworkers investigated the squeeze film damping for a variety of perforated structures by varying the size and number of perforations. Through finite element analysis, they found that the model underestimated the squeeze film damping by as much as 66% of the experimental values. Using a doubly supported MEMS ribbon of a grating light valve device, Gudeman and coworkers were able to characterize the damping by introducing a concept of "damping time." They found a simple linear relationship between the damping time of the ribbons and the gas viscosity when corrected for rarefaction effects.

All the above-mentioned literature is on squeeze film damping for parallel plates oscillating in the normal direction. There are relatively few studies on shear damping in lateral accelerometers. A study by Cho et al. (1994) investigated viscous damping for laterally oscillating microstructures. It was found that Stokes-type fluid motion models viscous damping more accurately than a Couette-type flow field. The theoretical damping was also compared with experimental data, and a discrepancy of about 20% still remains between the theoretically estimated (from Stokes-type model) and the measured damping.

2.4.4 Mechanical to Electrical Transduction: The Sensing Method

Many viable approaches have been implemented to measure changes in linear velocity. The capability of any technology needs to be tempered by the cost and market focus required by the application. Many sensing methodologies have been successfully demonstrated for inertial sensors. Piezoresistive, resonant frequency modulation, capacitive, floating gate FET sensing, strain FET sensing, and tunneling-based sensing will be discussed briefly below. Other variations continue to be demonstrated [Noell et al., 2002]. A study of the materials used in micromachining can be found in Part 2 of this Handbook. Detailed explanations of the electronic properties of materials utilized for sensing methodologies can be found in references such as Seeger's *Semiconductor Physics: An Introduction* (1985) and others.

Piezoresistive sensing has been successfully demonstrated in bulk micromachined and single-crystal inertial sensors [Roylance and Angell, 1979; Partridge et al., 1998]. This sense method has been utilized for many years in pressure sensor structures quite successfully, demonstrating very sensitive devices within marketable costs [Andersson et al., 1999; Yoshii et al., 1997]. This technique is sensitive to temperature variations but can be compensated electronically, as Lee et al. (2004) demonstrated. Yoshii et al. (1997) and Ding et al. (1999) have demonstrated monolithic integration of circuitry with piezoresistive elements. Thermal sensitivity, junction noise, and junction leakage are issues with piezoresistive sensing that require compensation for highly sensitive systems.

Resonant frequency shifts in a structure caused by inertial forces have been applied to some of the most sensitive and highest performing inertial sensor products on the market today [Madni et al., 2003; Barbour and Schmidt, 2001]. Resonant-beam tuning-fork style inertial sensors are in production for many high-end applications. Zook et al. (1999) and others have demonstrated resonant systems providing greater than 100 g full-scale range with milli-g resolution. These techniques are sensitive to temperature and generally require sensitive and complex control circuitry to keep the transducers resonating at a controlled magnitude.

Of all the alternatives, capacitive sensing has the broadest application to current inertial sensor products and for ultra high performance accelerometer systems [Chae and Najafi, 2004]. There are several reasons why capacitive sensing enjoys broad interest. The methodology is inherently temperature insensitive and appropriate precautions in transducer and circuit design result in nearly zero temperature coefficients of offset and sensitivity [Lee and Wise, 1981]. One can implement scaling the method to suit different sensing ranges by scaling the device capacitances to provide larger output signals for small inertial loads. The technique can be implemented in a wide variety of micromechanical processes ranging from bulk micromachining to surface micromachining [Hermann et al., 1995; Delapierre, 1999]. All axes of motion can be sensed capacitively. Figure 2.7 shows a quadrant of the Freescale Semiconductor, Inc. *X*-lateral inertial sensor utilizing interdigitated capacitive comb sensing and folded beam suspension. CMOS control circuitry is especially well suited to measure capacitances, leading to broader application of this technique implemented with switched capacitor sense circuitry.

Kniffin et al. (1998) and others have demonstrated floating gate FET structures to measure inertial forces. The technique allows a direct voltage transduction from the inertial force via a floating gate FET structure. The technique is complicated in such a structure, however, by the sensitivity of the air gap to variations in work function and the difficulty in providing a stable bias condition for the FET device. Variations in the packaged environment can result in quite large offsets in the response, which has made this methodology difficult to implement industrially.

The use of FET-based strain sensors has not been broadly studied for micromachined inertial sensors. Strain measurements in packaging development for large-scale CMOS circuits has been studied extensively [Jaeger et al., 1997] and has been extended to inertial sensing providing direct voltage sensitivity to inertial forces as demonstrated by Haronian (1999). Concerns remain regarding the manufacturability and stability of a FET device at the base of a micromachined strain gauge beam. However, this technique holds promise for direct integration of inertial sensing devices with CMOS.

Devices based on electron tunneling can provide extreme sensitivity to displacement. Because tunneling current is so strongly dependent on the space between the cathode and anode in the system, closed loop operation is the most common configuration considered. Recent efforts to produce tunneling-based inertial sensors can be found in Rockstad et al. (1995), Wang et al. (1996), and Yeh and Najafi (1997). Significant challenges remain for this methodology, however. Drift and issues related to the long-term stability of the

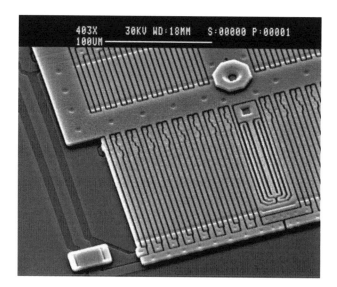

FIGURE 2.7 Perspective view scanning electron micrograph (SEM) of a *X*-lateral accelerometer quadrant showing interdigitated differential capacitive measurement, a folded spring design, and over-travel stops for the proof-mass structure. (Photo courtesy Freescale Semiconductor, Inc.)

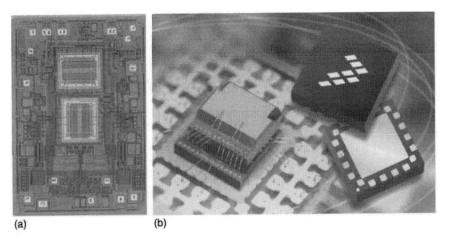

(a)   (b)

COLOR FIGURE 2.1   Examples of two high-volume accelerometer products. Example (a) is the top view micrograph of the Analog Devices, Inc. ADXL250 two-axis lateral monolithically-integrated accelerometer. Example (b) is a perspective view of the Freescale Semiconductor, Inc. wafer-scale packaged accelerometer and control chips stack-mounted on a lead frame prior to plastic injection molding. (Photos courtesy of Analog Devices, Inc. and Freescale Semiconductor, Inc.)

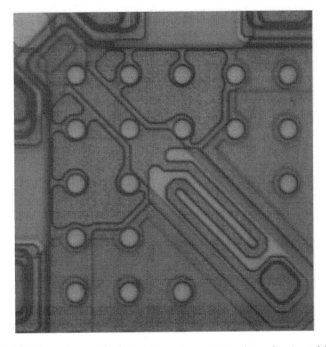

COLOR FIGURE 2.4   Top view micrograph of a Z-axis accelerometer quadrant showing a folded spring and sacrificial etch holes designed into the proof-mass structure. (Photo courtesy Freescale Semiconductor, Inc.)

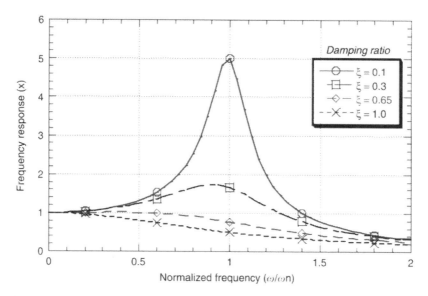

COLOR FIGURE 2.6 The frequency response $\bar{x}$ versus normalized frequency ratio ω/ω_n.

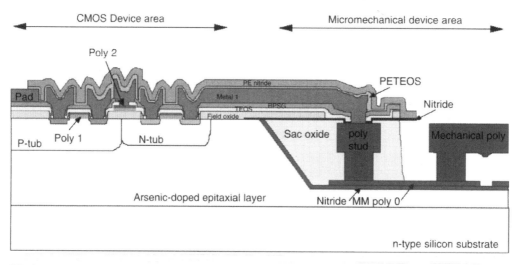

COLOR FIGURE 2.13 Cross-sectional diagram of the IMEMS process developed at Sandia National Laboratories demonstrating the transducer formed in a recessed moat and sealed prior to the commencement of the high density CMOS process. (Photo courtesy Sandia National Laboratories.)

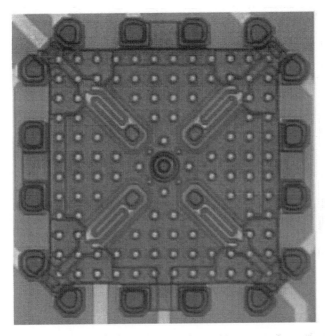

COLOR FIGURE 2.14   Top view micrograph of a Z-axis capacitive accelerometer in three polysilicon layers. The design allows for high inertial sensitivity with a low temperature sensitivity. (Photo courtesy Freescale Semiconductor, Inc.)

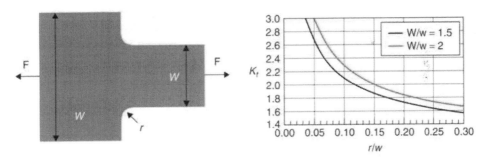

COLOR FIGURE 4.10   Stress concentrations for a flat plate loaded axially with two different widths and fillet radius r. The maximum stress is located around the fillets.

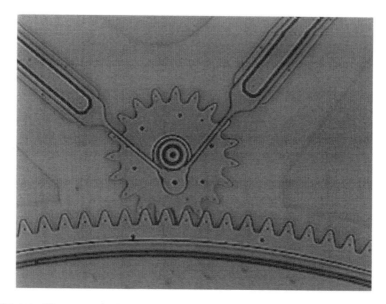

COLOR FIGURE 4.32 The gear teeth of the small gear are wedged underneath the teeth of the large diameter gear. In this case, gear misalignment is about 2.5 mm in the vertical direction.

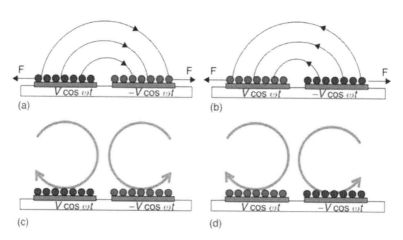

COLOR FIGURE 9.4 Schematic illustration of the capacitive charging: (a) and (b) demonstrate the electric field, and F represents time averaged Maxwell force; (c) and (d) demonstrate the flow profile.

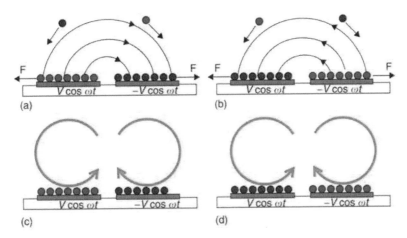

COLOR FIGURE 9.5 Schematic illustration of the Faradaic charging: (a) and (b) on the left, anions are driven to the same electrode surface where cations are produced by a Faradaic anodic reaction during the half-cycle when the electrode potential is positive; (c) and (d) the flow directions are opposite to those in Figure 9.4.

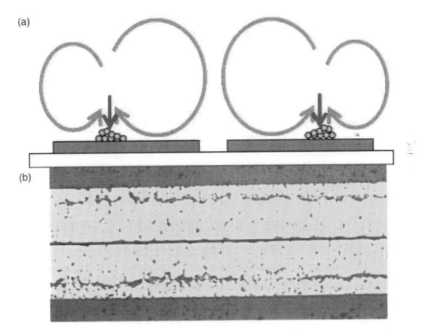

COLOR FIGURE 9.7 Particle focusing lines along the stagnation points for capacitive charging. The vertical force toward the electrode is a weak DEP or gravitational force. The circulation is opposite for Faradaic charging. An actual image of the assembled particles is shown below.

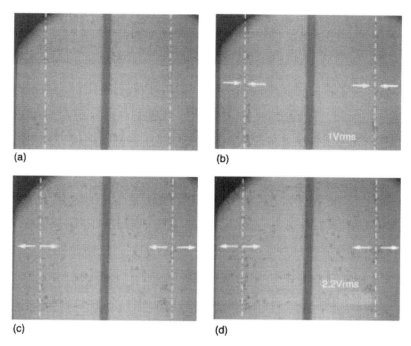

(a)   (b)

(c)   (d)

COLOR FIGURE 9.8   The writing and erasure processes for Au electrodes at $\omega = 100\,\text{Hz}$. The frames are taken at 0 s, 5 s, 10 s, and 15 s after the field is turned on. The initial voltage is 1.0 Vrms and is increased to 2.2 Vrms at 7.0 s. Particles on the electrode in the first two frames (a) and (b) move in directions consistent with electro-osmotic flow due to capacitive charging and assemble into lines. They are erased by Faradaic charging in the next two frames (c) and (d). The arrows demonstrate the direction of particle motion. The dashed lines are located at the theoretical $L/\sqrt{2}$.

COLOR FIGURE 9.9   Bacteria trapping by AC electroosmotic flow.

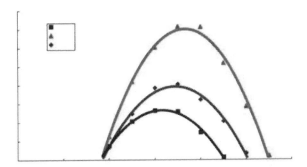

COLOR FIGURE 10.11   Bubble volume variation versus time for three different heater designs under same heat flux of 1.2 GW/m^2, courtesy Yang, et al. (2004).

COLOR FIGURE 11.7   Silicon wafer into which an array of micro heat pipes has been fabricated.

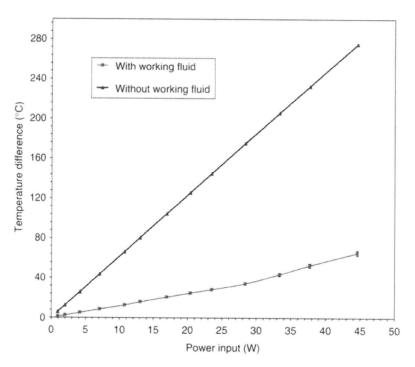

COLOR FIGURE 11.10 Temperature difference of micro heat pipe arrays with or without working fluid. (Reprinted with permission from Wang, Y., Ma, H.B., and Peterson, G.P. (2001) "Investigation of the Temperature Distributions on Radiator Fins with Micro Heat Pipes," *AIAA J. Thermophysics and Heat Transfer* 15(1), pp. 42–49.)

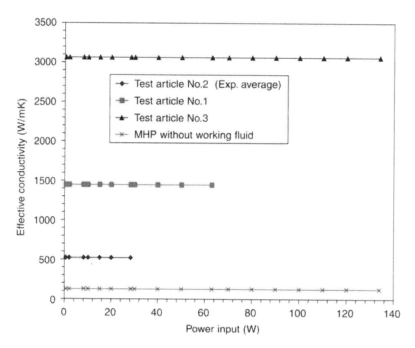

COLOR FIGURE 11.11 Effective thermal conductivity of micro heat pipe arrays. (Reprinted with permission from Wang, Y., Ma, H.B., and Peterson, G.P. (2001) "Investigation of the Temperature Distributions on Radiator Fins with Micro Heat Pipes," *AIAA J. Thermophysics and Heat Transfer* 15(1), pp. 42–49.)

tunneling tip, as well as low frequency noise sources, remain [Grade et al., 1996]. Recently, Shashkin et al. (2004) proposed that Fowler–Nordheim tunneling-based inertial sensing could provide a more stable alternative using parallel electrodes resulting in high sensitivity. As tunneling-based technologies expand in application, researchers will find solutions to mitigate the current limitations of the methodology.

2.5   Rotational Inertial Sensor Parameters

Linear and rotational inertial sensors have much in common; for example, both exhibit a structure comprising a specific mass as well as a flexible means by which this structure is anchored to the substrate, and both types of sensors are often manufactured through the same or similar technologies. Unlike a linear inertial sensor, however, the transducer of an angular rate sensor needs to be driven into oscillation in order to generate a measurable signal (in most cases). This requirement comes from the coupling of vibratory motion by the Coriolis Effect to produce a positional shift sufficient for sensing. The requirement adds both transducer and circuit complexity to the system. Upon a rotation of the transducer about its sense axis, a Coriolis force is generated in the presence of a rotational velocity of the reference frame, which in turn drives the transducer structure orthogonally as given in Equation 2.4. This means that a minimum of two orthonormal axes of motion is required in order to suitably measure the small Coriolis force exerted on a resonating proofmass during rotation. Rollover sensors typically resonate in plane and measure normal to the surface. Axes of sensitivity for gyroscopic sensors are shown in Figure 2.8. The scalar governing equation of motion for a gyroscopic device with a resonating mass in the Y-axis, rotated about the Z-axis is given by Equation 2.14,

$$\frac{d^2x}{dt^2} + 2\xi\omega_n\frac{dx}{dt} + \omega_n^2 x = 2\Omega_z\frac{dy}{dt} \tag{2.14}$$

where Ω_z is the rate of rotation and y is linear velocity of the structure due to the drive. One may make an analogy between rotational and linear sensors if the Coriolis term ($2\,\Omega_z\,dy/dt$) is considered an acceleration. According to a typical automotive spec where the full range of angular velocity is 100 deg/sec an equivalent acceleration, a, is given by Equation 2.15,

$$a = 3.5\frac{dy}{dt} \tag{2.15}$$

In general, the driving frequency is near resonance and the vibration amplitude of the transducer structure is about 1 µm. Assuming a natural frequency of 10 kHz, the resulting Coriolis acceleration of Equation 2.15 has a value of 0.022 mg, demonstrating that this force-induced acceleration is very small.

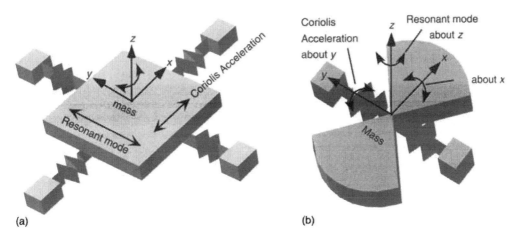

(a)   (b)

FIGURE 2.8   Reference frames for rotational gyroscopes based on the Coriolis effect showing axes of sensing for (a) yaw and (b) roll applications.

For most applications, a single axis angular rotation measurement is required. Such a single axis rate sensor can be built by sensing induced displacement from an oscillating rotor or from a linearly oscillating structure. Although these two types of rate sensor designs appear to be very different, the operation principles are the same. In both cases, when the reference frame (or device substrate) experiences a rotation along the input axis, the oscillating mass (either translational or rotary), in a direction perpendicular to the input axis (referred to as the drive axis), would induce a Coriolis force or torque in a direction perpendicular to both the input axis and the drive axis. With the amplitude of the drive oscillation fixed and controlled, the amplitude of the sensing oscillation is proportional to the rate of rotation of the mounting foundation. Feng and Gore (2004) show a mathematical model for the dynamic behavior of vibratory gyroscopes.

Because the coupling of the Coriolis Effect is orthogonal to the vibratory motion in a micromachined device, two degrees of mechanical freedom are required. One degree of freedom is utilized for the excitation of the vibratory motion, and the second degree of freedom orthogonal to the first is required for sensing. This requirement couples tightly into the technology choice for rotational inertial sensors, because the axis of sensitivity defines which mechanical degrees of freedom are required to sense it. For example, a very thick high aspect ratio technology — as is possible with direct wafer bonded structures — might not be the most suitable for a device that is required to move out of the plane of the wafer. However, as with their linear counterparts, most technologies and sensing methodologies have been applied to vibratory sensors with new combinations of methodologies always under consideration.

Putty and Najafi (1994) provide a discussion of the varieties of rotational inertial sensors, including vibrating prismatic beams [Greiff et al., 1991], tuning fork designs [Voss et al., 1997; Hiller et al., 1998], coupled accelerometers [Lutz et al., 1997; Kobayashi et al., 1999; Park et al., 1999], and vibrating shells [Putty and Najafi, 1994; McNie et al., 1999]. As illustrated in Figure 2.9, He and Najafi (2002) demonstrate an all-silicon vibrating ring gyroscope with very good performance. Multiple-axis systems have also been demonstrated [Juneau et al., 1997; Fujita et al., 1997]. In all cases, the vibrating structure is displaced orthogonally to the direction of the vibrating motion. This configuration can lead to system errors related to the transducer structure and the electronics. The primary error related to the transducer is called *quadrature error* and is discussed in the next sub-section.

As an alternative to single proof-mass designs, a concept involving two coupled oscillating masses has emerged, with one mass for driving and one mass for sensing. One of the first such designs is documented by Hsu et al. (1999), who used an outer ring as the drive mass and an inner disk as the sense mass. The driving mass is actuated by a set of rotary comb structures and oscillates about the Z-axis (or the vertical axis). The sensing disk is anchored to the substrate in such a way that the stiffness about the Z-axis is significantly greater

FIGURE 2.9   Perspective view scanning electron micrograph of a single-crystalline silicon vibratory ring gyroscope. (Photo courtesy K. Najafi, University of Michigan.)

than stiffness about the other axes. The outer ring and the inner disk are connected by a set of flexible beams or linkages. When there is no input of angular rotation, the oscillation of the drive mass about the Z-axis has virtually no impact on the motion of the sense disk. When the device experiences a rotation about either the X- or Y-axis, the Coriolis force-induced torque drives the inner disk into a rocking motion about the Y- or X-axis. Electrode pads underneath the disk measure the variation of capacitance, which is proportional to the input angular rate. Another advantage of this two-mass design is that the dual proof-mass structure permits the ring and the disk to be excited independently so that each can be dynamically compensated for manufacturing non-uniformity.

Several other vibrational devices have been demonstrated also involving two mutually perpendicular oscillating masses [Kobayashi et al., 1999; Park et al., 1999]. In these designs, the drive mass is forced to oscillate along the Y-axis by comb actuators and the sense mass is forced to oscillate along the X-axis by a Coriolis force. The magnitude of this Coriolis force is proportional to the input angular rate along the Z-axis. The angular rate of rotation is measured by detecting changes in capacitance with interdigitated comb structures attached to the sense mass. The linkage between the drive and sense masses is designed in such a way that the Coriolis force is transferred from the drive mass to the sense mass in an efficient way; yet the feedback from the motion of the sense mass to the drive mass is kept to a minimum by frequency matching. Acar and Shkel (2003) proposed a variation on this scheme using four masses in a decoupled mode between a two-degree of freedom drive oscillator and a two-degree of freedom sense oscillator to further reduce offsets and improve the performance.

2.5.1 Design Considerations: Quadrature Error and Coupled Sensitivity

Most of the earlier sensor designs involve a single proof-mass for both driving and sensing. The proof-mass is supported by a set of multiple slender beam linkages, usually made of the same semiconductor materials as the proof-mass, to allow for movement in two mutually perpendicular axes. A major drawback in a single proof-mass design is the cross-axis coupling between the drive axis and the sense axis, a phenomenon commonly referred to as quadrature error. This coupling can be attributed to defects or small non-orthogonalities in the mechanical structure. Because the sense displacement is a minute fraction of the typical drive displacement, small structural defects can generate large quadrature errors in the system. Quadrature is compensated for by enhanced structural design, as will be demonstrated in the examples below, as well as by the generation of quadrature canceling force feedback of position in the control electronics [Geen, 1998]. Fortunately, the quadrature error coupled from the drive vibration is 90 degrees out-of-phase with respect to Coriolis-induced vibration and can be phase-discriminated to a large degree in the control circuitry at the expense of additional control circuit complexity. However, the continued increasing complexity in the structural design in modern micromachined gyroscopes indicates that quadrature error cancellation cannot be completely resolved by the control circuit.

The sensitivity requirements for rotational inertial sensors far exceed those for most linear inertial systems, both in terms of the transducer design and the circuit complexity. In vibrational systems, structural sensitivity and absolute stability in the control electronics are required to accurately measure rotational rate. Because the magnitude of the driven vibration is directly proportional to the magnitude of the Coriolis-induced output displacement, the structure and electronics are designed to maximize the coupling and stability of the magnitude. Structurally, the driven oscillations can have large displacements, on the order of microns or even tens of microns in some cases. The devices are also commonly operated in a near-vacuum environment to minimize the impact of mechanical damping on the structure to maximize the resonant response, or the Q, of the system. Electronically, precise control of the driven vibration amplitude is paramount. Phase discriminating circuitry such as phase-locked-loop (PLL) control is used to drive the device displacement at or near resonance to maximize the displacement while precise amplitude control is maintained. Phase discrimination and synchronous phase demodulators are also required to sense the Coriolis force displacement and cancel quadrature effects [Geen, 1998; Kobayashi et al., 1999]. As with accelerometer systems, the sense circuitry can be operated in open loop or closed loop force feedback configurations to sense displacement with the system tradeoffs discussed in a later section.

2.6   Micromachining Technologies for Inertial Sensing

Micromachining technology, implemented to produce the transducer device, is coupled to the physical principle used to sense the inertial displacement. Comprehensive details regarding these technological developments are described in Section II of this Handbook. Bulk silicon micromachined technologies were first implemented for inertial sensors. However, polysilicon-based surface micromachined technologies dominate the current marketplace for micromachined inertial sensors. The trend is to use higher aspect ratio "surface" micromachined technologies to produce inertial sensors.

Surface micromachined capacitive inertial sensors were broadly demonstrated commercially as a result of the collaboration between Analog Devices, Inc. and the University of California-Berkeley in the introduction of the Analog Devices, Inc. iMEMS™ BiCMOS integrated surface micromachined accelerometer process technology, as shown in Figure 2.1(a). The technology embeds a 2 µm micromechanical polysilicon layer into a BiCMOS process flow [Chau et al., 1996]. Application of this process has more recently been expanded to gyroscopes, with the ADXRS150 utilizing a 4 µm structure [Lewis et al., 2003]. Sandia National Laboratories [Smith et al., 1995], Motorola, Inc., Sensor Products Division [Ristic et al., 1992] (now Freescale Semiconductor, Inc.), and Siemens [Hierold et al., 1996], among others, have all demonstrated industrial surface micromachined inertial sensor technologies. Limitations to surface micromachining are primarily related to the technological challenges in producing low-stress, high aspect ratio structures that have demonstrated benefits for sensitivity, mechanical damping properties, and insensitivity to off-axis motion.

Epitaxially deposited polysilicon eliminates the aspect ratio limitations of the standard LPCVD polysilicon deposition typically used in surface micromachining. This technology, sometimes referred to as "epipoly" technology, also allows the monolithic integration of CMOS or BiCMOS circuitry with higher aspect ratio capacitive transducers, typically on 10–12 µm-thick epitaxial layers [Kirsten et al., 1995; Offenberg et al., 1995; Geiger et al., 1999; Reichenbach et al., 2003; Baschirotto et al., 2003]. Epitaxial deposition of silicon is cost-competitive for micromachining to thicknesses of 50 µm. These high aspect ratio transducer structures are relatively insensitive to out-of-plane motion and provide suitable mechanical damping at reasonable packaging pressures. This material has desirable film properties with nearly immeasurable intrinsic stress and a high deposition rate [Gennissen and French, 1996].

With a reasonably flexible interconnect scheme, epipoly technologies have demonstrated monolithic integration with CMOS and BiCMOS circuitry as well as device thicknesses ranging from 8 µm to over 50 µm. Challenges for this technology include that co-deposited epitaxial silicon and polycrystalline silicon have different deposition rates that complicate fabrication. High temperature polycrystalline films typical of epitaxial deposition also suffer from severe surface roughness and very large semi-conical crystalline grains. Solutions to many of these issues have been documented by various sources [Kirsten et al., 1995; Gennissen and French, 1996; Bergstrom et al., 1999].

Direct wafer bond (DWB) technology has long demonstrated the successful incorporation of thick capacitively- or piezoresistively-sensed inertial sensor structures. Recent advances in this technique have demonstrated improved device interconnect through the use of a silicon-on-insulator (SOI) handle wafer with defined interconnect [Ishihara et al., 1999]. Piezoresistive elements have also been incorporated on the sidewalls of very high aspect ratio DWB structures to provide transducers with both piezoresistive and capacitive sensing mechanisms [Partridge et al., 1998]. DWB transducer technologies provide great process and device flexibility. Very high aspect ratio structures are possible, approaching bulk-wafer thicknesses if necessary, providing excellent out-of-plane insensitivity and mechanical damping properties. Monolithic integration with CMOS is also possible. The technology requires significant process capability to successfully produce DWB structures at high yield.

As SOI microelectronic device technologies gain popularity for high performance mainstream CMOS process technologies, the substrate material required for micromachining becomes cost competitive with alternative transducer technology approaches, making SOI more appealing for inertial sensing applications. SOI technology, as a descendant of DWB technology, provides technological flexibility with desirable device properties, including the out-of-plane insensitivity and high damping associated with high aspect ratio structures. SOI technology provides the advantage of single-crystal silicon sensor structures with very well behaved mechanical properties and extraordinary flexibility for device thickness, as with DWB technologies. Thicknesses can range from submicron to hundreds of microns for structural layers. Unlike DWB, SOI technologies often lack the flexibility of pre-bond processing of the handle and active wafers to form microcavities or buried contact layers that are often implemented in DWB technologies. Another technology hurdle has been the choice of methodology to minimize parasitics to the handle wafer. Even so, SOI has demonstrated a significant increase in its popularity as a micromachining substrate [Delapierre, 1999; Lemkin, Juneau et al., 1999; Park et al., 1999; McNie et al., 1999; Noworolski and Judy, 1999; Lehto, 1999; Usenko and Carr, 1999]. Lemkin and Boser (1999) demonstrated the monolithic integration of SOI inertial sensors with CMOS. While technological hurdles still need to be overcome for broad industrialization of SOI MEMS devices, the technology holds great promise for a broad technological platform with few limitations. Macdonald and Zhang (1993) at Cornell University developed a process known as SCREAM (Single Crystal Reactive Etching and Metallization), which produces SOI-like high aspect ratio single crystal silicon transducers using a two-stage dry etching technique on a bulk silicon substrate. This technology had been limited by the difficulty in electrically isolating the transducer structure from the surrounding substrate. Sridhar et al. (1999) demonstrate that this technology can now produce fully isolated high aspect ratio transducers in bulk silicon substrates. With fully isolated structures, this technology can produce 20:1 aspect ratio devices for thicknesses to 50 μm and can be monolithically integrated with circuitry. Xie et al. (2000) and Yan et al. (2004) have also demonstrated a related two-stage release methodology on capacitive inertial devices formed in a CMOS integrated technology. This technology shows promise for full integration of high aspect ratio lateral inertial structures with CMOS. Also, Haronian (1999) demonstrated an integrated FET readout for an inertial mass released using the SCREAM process.

The development of metal micromachined structures by electroforming has demonstrated 300 μm thick nickel structures with submicron gaps formed using LIGA (Lithographie, Galvanoformung, Abformung) techniques [Ehrfeld et al., 1987]. LIGA-like process techniques using reasonably high-resolution thick UV photoresist processes, have resulted in inertial sensor development in nickel, permalloy, and gold post-CMOS micromachining [Putty and Najafi, 1994; Wycisk et al., 1999]. Very thick structures are possible using these techniques with effective "buried" contacts to the underlying circuitry in the substrate. As a single-layer process addition, the technology adds minimal cost to the overall sensing system. High aspect ratio structures are possible. However, the material properties of electroformed materials are difficult to stabilize and can be prone to creep.

Traditionally a bulk micromachined technology, the application of deep anisotropic etching of (110)-oriented silicon wafers has produced novel inertial sensors with very high aspect ratios. Aspect ratios up to 200:1 have been demonstrated using this technique, although the practical application of the technology may limit the maximum aspect ratio to below 100:1 [Hölke and Henderson, 1999]. This technology offers an elegant solution providing extremely high aspect ratios compared to anisotropic dry etching techniques. The technology is somewhat limited in its application flexibility because the deep trenches are crystallographically defined by the intersection of the (111) planes with the surface of the wafer and must be arranged in parallelograms in the etch mask. Circular and truly orthogonal structures are not easily configured for this technology.

2.7 Micromachining Technology Manufacturing Issues

The manufacturability of a transducer structure should be considered as important as the performance of the device. In theory, a sensor structure may be designed as sensitive as desired, but if the structure cannot be manufactured in a robust manner, the effort is futile. Issues such as release or in-use stiction, stability of material properties, and the control of critical processes in the manufacture of inertial sensors should be investigated and understood. The impact of high aspect ratio technologies creates new challenges in controlling and maintaining processes.

2.7.1   Stiction

Stiction is a term used in micromachining to describe two conditions: release and in-use stiction. Release stiction is the irreversible latching of some part of the moveable structure in the device caused during the release etch and drying processes. In-use stiction is the irreversible latching of the moveable structure during device operation. All high aspect ratio technologies require a step to release the moveable structure from the supporting substrate or sidewalls at some point in the process flow. This process is not always, but is most often a wet etch of a dielectric layer, using a solution containing hydrofluoric acid and water. Release stiction typically occurs during the drying step following a wet solution process as the surface tension forces in the liquid draw the micromachined structure into intimate contact with adjacent surfaces. The close contact and typically hydrated surfaces result in van der Waals attraction along the smooth parallel surfaces, bonding the layers to each other [Mastrangelo and Hsu, 1993]. Too large a proof-mass or too soft a spring may dramatically increase the probability of stiction resulting in yield loss.

There are many techniques employed to reduce or eliminate release stiction. Supercritical CO_2 drying processes avoid surface tension forces completely and often result in very good stiction yields. However, this technique has been difficult to implement in industrial process conditions. Surface modifications, often based on fluorinated polymer coatings, have been used to reduce surface tension forces on the micromachined structure during release and drying with some success. As hydrophobic materials, these monolayer coatings require significant surface treatment and have not found broad industrial utilization yet. Other techniques have been employed with some success, all at the cost of additional process complexity and structural compromises. The latest trend has been to utilize dry release processing, often related to a DRIE-last process flow for a high aspect ratio device that is exposed through the substrate [Amini et al., 2004]. However, the problem with stiction yield loss is increased with aspect ratio because the surface tension forces act over a larger area. High aspect ratio structures must be designed with care to minimize the complications from release stiction.

In-use stiction issues also increase with the aspect ratio of a device. For capacitive accelerometers, the proof-mass closing in on an actuated electrode can cause electrostatic latching if the electrostatic force becomes larger than the elastic spring's restoring force. This condition is called *pull-in* or *electrostatic latching* and is design dependent. High aspect ratio designs result in more capacitive coupling force for a given device topography and can be more prone to latching. Many devices are designed with over-travel stops to reduce the risk from this compromising situation.

2.7.2   Material Stability

While providing design performance and off axis stiffness, high aspect ratio devices remain sensitive to the stability of material properties. This is particularly important for polycrystalline silicon devices, since the deposition process can result in variations in the average intrinsic stress of a film as well as generate stress gradients throughout the sensor layer. However, all associated materials result in significant impacts on the device performance and repeatability. Stability and uniformity of backside film stacks, plasma-assisted deposited dielectric films, and even the proximity of metallizations in the front-end process can impact the uniform and controlled behavior of a device.

2.7.3   High Aspect Ratio Structures

As was mentioned previously, there are many advantages to using thicker structures for both linear and rotational inertial sensors. However, high aspect ratio silicon structures require low-stress structural layers as well as deep etching capability. The former issue is not of concern for bulk micromachined devices, but if the structure is to be formed from a deposited film, there are trade-offs among the various deposition methods and conditions in order to obtain a uniform, smooth layer free of stress-induced curvature, particularly if a reasonably high deposition rate is desired. Many of the common challenges associated with surface micromachining are exacerbated as the thickness of the structural layer is increased. As previously stated, epitaxial deposition of polysilicon structures results in very low-stress films with a high deposition rate, but additional measures are often required to reduce surface roughness. Another solution involves the deposition of polysilicon into trench-based forms [Chae et al., 2004]. Both of these methods typically involve high-temperature processing, which may impose restrictions on fabrication sequencing, as will be discussed in a later section.

Deep etching for high aspect ratio structures has taken on several different forms based on the desired shape and uniformity of the resulting trenches. In the case of (110) silicon technologies, a wet anisotropic etchant utilizes the etch rates along different crystallographic planes in the device material to control the profile of the trenches formed. Control of such processing requires accurate alignment of the etch mask to the crystallographic planes in order to successfully control the aspect ratio to a designed parameter [Hölke and Henderson, 1999]. The technique is also sensitive to impurities in the crystal. In most high aspect ratio technologies, however, a deep dry reactive ion etch (DRIE) of the trenches forms the structure. Deep trench etching has been implemented using various techniques, but a process pioneered by Bosch [Laerme et al., 1999] in which the film is cycled between modes of reactive etching and sidewall passivation has demonstrated a clear predominance as an alternative. Control of the etch properties and profile is the most significant challenge for high aspect ratio technologies. Many potential process conditions can degrade the etch profile or complicate the uniformity of the process for across-wafer and wafer-to-wafer variations in the process. These variations in profile and width strongly impact the design parameters such as spring constant and damping, etc.

2.7.4   Inertial Sensor Packaging

Package interactions are just as critical as device technology choices and often contribute significant performance shifts from package to package [Li et al., 1998]. Micromachined inertial sensors, while robust on a micro scale, are fragile at the assembly scale and easily damaged and often require two levels of packaging: (1) wafer level packaging, which is usually hermetic to provide damping control and to protect the MEMS devices from the subsequent assembly operations; and (2) conventional electronic packaging of die-bonding, wire-bonding, and molding to provide a housing for handling, mounting, and board level interconnection. The package must fulfill several basic functions: (1) to provide electrical connections and isolation, (2) to dissipate heat through thermal conduction, and (3) to provide mechanical support and isolate stress. An industrially relevant packaging process must be stable, robust, and easily automated, and must take testability into account.

Wafer level packaging techniques include silicon-to-glass anodic bonding [Dokmeci et al., 1997], thermocompression bonding using glass frit [Audet et al., 1997] or eutectic [Wolffenbuttel and Wise, 1994; Cheng et al., 2000], direct wafer bonding [Huff et al., 1991], and monolithic capping technologies [Burns et al., 1995] utilizing one or more wafers. These techniques allow the transducer device to be sealed at the wafer level to protect the movable components from damage during assembly. The wafer level package also provides the sensor with a controlled ambient to preserve the damping characteristics of the proof-mass. Figure 2.1(b) shows a Freescale Semiconductor, Inc. accelerometer die with a wafer level cap in silicon mounted on top of the co-packaged CMOS control IC.

The unique challenge of sensor packaging is that in addition to providing a mounting foundation to a PC board, one must control stresses that are induced by mismatch in the thermal expansion coefficients of the materials used to fabricate the package and the external thermal loading of the package. These stresses must be kept at a level low enough to avoid impact to the sensor or control circuitry performance. An example of this challenge is illustrated in Figure 2.10, adapted from Li et al. (1998), where external package-induced stresses on an accelerometer die produced a 0.15 μm curvature out-of-plane for the die from center to edge, resulting in a device offset that would not be present for the die in wafer form. For capacitive devices capable of resolving displacements at the nanometer or sub-nanometer scale, excessive curvature due to stress on the die at a late point in the assembly can be catastrophic. In general, different MEMS devices have different stress tolerance levels. Therefore, each MEMS package must be uniquely designed and evaluated to meet special requirements [Dickerson and Ward, 1997; Tang et al., 1997].

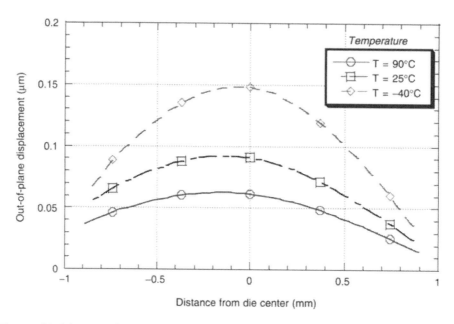

FIGURE 2.10   Die deformation due to a chip-packaging scheme demonstrating the impact of packaging technologies on the industrialization microelectromechanical systems. (Adapted from Li, G.X., Bergstrom, P.L., Ger, M.–L., Foerstner, J., Schmiesing, J.E., Shemansky, F.A., Mahadevan, D., and Shah, M.K. [1998] "Low Stress Packaging of a Micro-Machined Accelerometer," *Proc. 3rd International Symposium on Electronic Package Technology*, pp. 553–62.)

2.7.5   Impact Dynamics

Micromachined devices may demonstrate weaknesses not typically found in macroscopic sensor systems that occur during the system assembly of the micromachined inertial sensor with other electronic components. A common test to determine how robust a sensor is to system assembly is the drop test. When a packaged device is dropped from a tabletop to a hard surface floor, both the package and the microstructure undergo sudden changes in their respective velocity. Assuming that there is no energy loss during the flight of the drop, both the package and the microstructure would have a downward velocity of $v = \sqrt{2gh}$ immediately prior to impact. After impact, the package may be stuck to the ground or may bounce back with a smaller velocity. The motion of the package can be calculated by ignoring the influence of the microstructure, which is at least five orders of magnitude smaller than the mass of the package itself [Li and Shemansky, 2000]. The motion of the microstructure as a result of the impact is governed by a second order ordinary differential equation of a standard form, as in Equation 2.11 [Meirovitch, 1975]. Equation 2.17 gives the maximum displacement,

$$z_{max} = \sqrt{\frac{2mgh}{k}}\, d_0(\xi, r) \tag{2.16}$$

where $d_0(\xi, r)$ is a unit-less scaling function only of the damping ratio, ξ, and the elasticity of the collision as defined by a restitution coefficient, r, with $0 \leqslant r \leqslant 1$ [Li and Shemansky, 2000]. Knowing the seismic-mass travel as a result of a mechanical drop, the equivalent g load on the structure can be determined in a similar fashion and is graphically shown in Figure 2.11 for a specific device application.

At a one-meter inelastic ($r = 0$) drop, the impact experienced by a device is similar to a situation where it is subjected to an acceleration greater than 20,000 g. At a damping ratio of 1.5 (over damped), the equivalent acceleration is 14,000 g. These values of acceleration would be doubled if the package impact with the floor is elastic ($r = 1$). Nevertheless, it is clear that the drop-induced acceleration is large, much larger than one normally expects to encounter during normal operation.

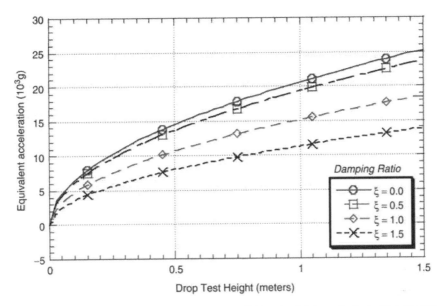

FIGURE 2.11 Calculation of the equivalent *g* load on an accelerometer due to an inelastic drop shock for several damping ratios. The device conditions are: $m = 0.6\,\mu g$, $k = 5\,N/m$, $r = 0$.

The analysis on a lumped mass and spring model helps to provide a picture regarding the magnitude of proof-mass travel and the large *g*-force induced by a drop. A MEMS structure, however, has distributed properties such as mass and stiffness. Therefore, it is sometimes necessary to carry the analysis one step further to take the flexibility of the proof-mass into account. This is especially important for those lateral sensors and actuators of comb-type designs where the conductive movable component is large in lateral dimensions and is susceptible to bending. The maximum possible displacement in a spring-supported structure is the sum of the travel as a lumped mass and the bending caused by dropping.

A typical lateral accelerometer is basically comprised of an array of sensing and actuating electrodes, a central spine plate, and multiple supporting springs. In such a design, both the movable and stationary sensing electrodes can be modeled as cantilevers such that when subjected to acceleration, the fixed end follows the motion of the attachment and the free end deflects. As for the central spine plate, it can be modeled either as a hinged–hinged beam or a clamped–clamped beam depending on the position of supporting springs.

Per Li and Shemansky (2000), the maximum displacement, z_{max}, for a 2 µm thick polysilicon cantilever is graphically shown in Figure 2.12, where $h = 1.2\,m$, $E = 161,000\,MPa$, $I = 4/3\,\mu m^4$, $r = 0.5$. Assuming there is no damping, a 100 µm cantilever would bend approximately 8.5 µm near its free end, and the bending would be 19 µm for a beam 150 µm long. When damping is included, the beam deflection becomes smaller as indicated by the dashed and dotted curves in Figure 2.12. The damping ratio for a nominal lateral accelerometer of 2 µm thick polysilicon and 1.5 µm finger gap is approximately 0.1. Therefore, depending on finger dimension, the drop induced bending could be very excessive and cause structural damage.

2.8 System Issues for Inertial Sensors

The partitioning of the transduction and controlled output of an inertial sensing system has led to many variations on what inertial sensor technology should look like. The coupling between the micromachined structure and the microsystem, including the method of control and the choice of interface electronics, is very close. A change in an aspect of one strongly impacts the requirements for the other. The motivations and potential for combining the integrated circuit with the transducer and the impact of the control circuit architecture on the overall system will demonstrate the many system design tradeoffs required to produce a complete inertial sensing system.

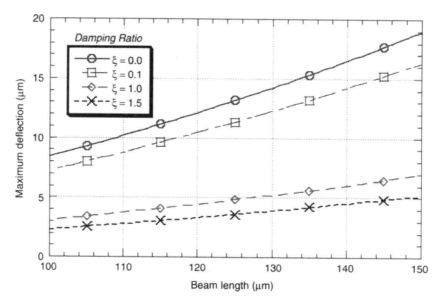

FIGURE 2.12   Calculation of the maximum beam displacement, z_{max}, for a 2 µm thick cantilever beam with different damping ratios and lengths (Li and Shanasky).

2.8.1   System Partitioning: One-Chip or Multi-Chip

The sense methodologies and transducer technologies discussed above have all been demonstrated both as multi-component and as monolithically integrated systems. One-chip versus package-level multi-chip integration of transducer and circuitry has generated many passionate discussions regarding the viability of each approach. System cost and capability are the two primary motivations for the choice to monolithically integrate or to co-package components.

Silicon die area utilization, process complexity, wafer scale testing requirements, and packaging costs all significantly contribute to the overall production costs for any sensor system. Some multi-component systems have mitigated process complexity by isolating transducer and integrated circuit processes and co-packaging the unique system pieces. System requirements have been met by this multi-chip methodology. Front-end silicon process complexity and cost have been traded for back-end testing and packaging costs, which make up a large fraction of the overall product costs. There is no clear application boundary for multi-chip system partitioning compared to single-chip integration. Industrialized sensor products have demonstrated two-chip solutions successfully implemented for even the most complex and demanding system requirements in military and aerospace applications [Delapierre, 1999; Chae et al., 2004].

Monolithically integrated transducer technologies have advantages in approaching the fundamental sensing performance limitations for many applications by mitigating inter-chip parasitics associated with multichip integration. Trade-offs in system costs are moving toward the incorporation of additional front-end complexity to design for testability and mitigate back-end testing and packaging costs. An additional benefit to monolithic integration of transducer and circuitry is the minimization of silicon die area. Die area may become *the* driving motivation to monolithic integration for future sensor applications. The drive to smaller outline package surface mount technologies also requires a smaller silicon footprint in the package, again leading towards integration.

There are motivating reasons to consider monolithic integration of transducer technologies with circuitry for system performance and system cost. Each application should be reviewed carefully to provide system requirements with the least expensive, high performance technology. With future applications demanding greater performance in an increasingly small package, integratable transducer technologies are prudent to consider for future technology applications. In monolithically integrated technology, the

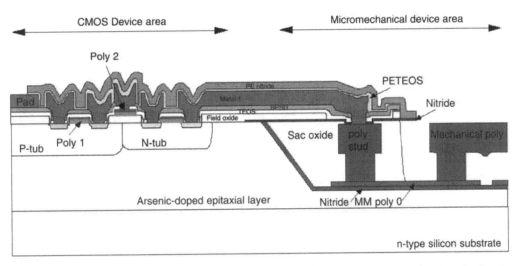

FIGURE 2.13 (See color insert following page 2-12.) Cross-sectional diagram of the IMEMS process developed at Sandia National Laboratories demonstrating the transducer formed in a recessed moat and sealed prior to the commencement of the high density CMOS process. (Photo courtesy Sandia National Laboratories.)

method of integration defines the manufacturability, process cost, testability, and the ability to integrate the technology. This decision includes both the integration method and the choice of integrated circuit technology and should fully address the system requirements.

2.8.2 Sensor Integration Approaches

Three methodologies classify the integration of a micromachined element into an integrated circuit process: transducer first, transducer middle or interleaved, and transducer last integrations. All have demonstrated monolithically integrated inertial sensor technologies. All have noted strengths and weaknesses. The *transducer first* integration merges the process that creates the transducer element prior to a standard integrated circuit process. A notable example of a transducer first integration is the Buried Transducer Process, now know as Sandia's Integrated MicroElectroMechanical Systems technology, or IMEMS, demonstrated by Smith et al. (1995) at Sandia National Laboratories. The transducer is formed in a recessed region in the field of the wafer and sealed in a stack of dielectric films prior to the beginning of a CMOS process flow. This dielectrically sealed moat is planarized using chemical mechanical polishing (CMP) prior to the start of the CMOS flow. Electrical interconnections are formed as part of the CMOS process, with the transducer released as a final step in the process flow. A cross-sectional diagram is shown in Figure 2.13. The benefit of this approach is that the transducer is largely decoupled from the remainder of the integrated circuit process and can be transported between integrated circuit processes. This allows the use of standard integrated circuit process steps, with minimal impact to the integrated circuit processing steps and only a limited impact on thermal budget, allowing the optimization of the transducer without impacting the standard integrated circuit process. The challenge of this versatile approach lies in the fact that all of the integrated circuit processing steps are still ahead, increasing the risk of contamination.

The *transducer middle* integration merges the process that created the transducer element with the standard integrated circuit process by reusing and minimally adjusting and inserting processing steps for the formation of the transducer. A well-known example of this technique is the Analog Devices, Inc. iMEMS™ process [Chau et al., 1996]. This technology merges a 2 μm-thick micromechanical polysilicon layer for the transducer with a high-density mixed-signal BiCMOS process technology, interleaving the transducer process with the integrated circuit flow. An example of this technology is shown in Figure 2.1(a). The benefit of this approach lies in the reuse of existing integrated circuit process steps used in the creation of the transducer, which has the potential to minimize the number of additional process steps required to implement this approach. The risk for this approach lies in its specialization. Since the transducer is merged into an integrated circuit process, a unique process flow is generated for the specific system. A merged process will impact the integrated circuit device parameters as well as the transducer parameters. The reuse of existing process steps complicates the optimization of the transducer as well as the integrated circuit.

The *transducer last* integration merges the transducer process with the standard integrated circuit process by inserting the transducer formation at the end of the integrated circuit processing steps. This integration can take on many forms, including the approach taken by Xie et al. (2000), in which they used layers and structures formed during the CMOS process to define post-processed regions for subsequent deep reactive ion etching of high aspect ratio structures. A more flexible approach uses low temperature post-processing of structural layers to integrate micromechanical structures with circuitry [Franke et al., 2000; Honer and Kovacs, 2000; Xia et al., 2004; Wu et al., 2004]. The benefit of this approach lies in the reuse of a minimal set of existing integrated circuit back-end processing steps used in the creation of the transducer, thereby having the potential to minimize the number of additional process steps. Since the transducer is formed after the completion of all the standard integrated circuit device formation steps, the impact on integrated circuit device parametrics is defined primarily by the circuitry's exposure to the maximum temperature of film deposition or the plasma processing required to form the transducer. In addition, there is some measure of transportability of the transducer process between existing integrated circuit platforms. The difficulty with this method lies in the potential complexity of the integration approach and its impact on the isolation and interconnect of the transducer. Since the integrated circuit processing has been completed before the formation of the transducer, the patterning of the transducer around the existing interconnect scheme becomes difficult, and the release etch in the absence of an adequate etch stop can be a real challenge. Solutions to that challenge have been proposed through dry release [Xie et al., 2000] or CMOS-benign release in hydrogen peroxide [Franke et al., 2000]. While transportability between different integrated circuit platforms is possible, there will be a degree of optimization required, which will be specific to the platform used.

The introduction of monolithically integrated inertial sensor technologies requires clever system design to understand the impact of the device behavior in the system. Improvements in test methodology and the potential to isolate transducer and integrated circuit behavior are required. In particular, the suitable calibration of self-test forces and the decoupling of drive and sense responses is even more important for the active transducers required for rotational inertia sensing.

2.8.3   System Methodologies: Open or Closed Loop Control

Sensing methodologies utilized in inertial sensing generally fall under two categories: open loop or closed loop control architectures. Suffice it to say that there are motivations to pursue both open loop and closed loop force-feedback control for inertial sensing systems. Open loop control systems, straightforwardly, measure changes in the sense signal, whether it is a change in piezoresistance, or capacitance, or other, as a result of the inertial load displacing the seismic mass from its zero state position. These signals are typically amplified, compensated, filtered, buffered, and output as control variables either as analog voltages or digital control signals to the larger system. Open loop control schemes tend to be relatively immune to small production variations in the transducer element, are inherently stable systems relying on no feedback signals, provide ratiometric output signals, and, perhaps most importantly, are often smaller in die area than their closed loop counterparts.

Closed loop control schemes rely on feedback to control the position of the seismic mass via a force feedback, or force rebalancing, at its rest position. The force feedback required is proportional to the magnitude of the inertial load. There is great potential in this methodology. This force feedback defines the sensitivity of the system and acts as an electrostatic spring force, which is added in Equation 2.6. It also contributes in Equation 2.8, impacting the dynamic behavior of the system and modifying the damping conditions. The most prevalent system configuration utilizing force rebalancing is in capacitive inertial devices. A notable example of an inertial sensing system utilizing force rebalancing is the Analog Devices, Inc. ADXL50, which is a lateral 50g accelerometer product as described in Goodenough (1991) and Sherman et al. (1992). Closed loop force feedback systems have the potential for very high sensitivity and have been implemented in gyroscopic systems due to the minute forces and displacements generated by the Coriolis effect [Putty and Najafi, 1994; Greiff and Boxenhorn, 1995].

2.8.4 System Example: Freescale Semiconductor Two-Chip X-Axis Accelerometer System

Freescale Semiconductor's 40-*g* X-lateral accelerometer can be used as an example of a high-volume product that will demonstrate the design tradeoffs and broad classes of issues required to produce inertial sensors. The packaged system, as shown in perspective in Figure 2.1(b) on a leadframe prior to final assembly, is comprised of a two-chip co-packaged system in a dual in-line plastic package. The sensor die, capped with a hermetically sealed silicon cap, is wire bonded to the adjacent integrated circuit control die. Low-stress adhesive and coating materials are used to minimize mechanical coupling of the sensor die to the metal leadframe and the plastic injection molding materials and to minimize system offsets.

The transducer, a portion of which is shown in a perspective view SEM image in Figure 2.7, is produced using surface micromachining in polysilicon to form a suspended polysilicon lateral double-sided capacitive structure. The central proof-mass plate is configured to produce an over damped Z-axis response at a given capping pressure. The X-axis damping is defined by the aspect ratio between the suspended sense electrodes and the fixed capacitor plates forming the lateral capacitive structure. The folded-beam spring design minimizes coupling to external mechanical strains on the chip, controls the X-axis spring constant to the designed parameter, maximizes the Y-axis spring constants to minimize off-axis coupling, and provides the electrical connection to the center electrode for the lateral capacitive measurement. The Z-axis motion in this device is constrained by a series of motion limiters, limiting the out-of-plane motion to a designed tolerance. The double-sided capacitor structure is formed by the definition of a plurality of adjacent left and right electrodes to the electrodes formed on the proof-mass in a configuration often described as a capacitive "comb" structure. This comb structure, through the multiplication of the small capacitive coupling in each left, center, and right set of electrodes provides sufficient capacitive coupling to suitably sense the inertial deflection of the proof-mass center electrode with respect to the left and right differential electrodes.

This device demonstrates a clear system partitioning. Minimizing process complexity, the two-chip approach allows the potential maximization of process capability for both sensor and control circuitry for the given application. Alternatively, fabrication process and assembly costs are minimized for the performance required by the application by isolating the micromechanical and circuit elements. This technique has been implemented in many other accelerometer and gyroscopic systems over the past decade. Figure 2.14 shows a Freescale Semiconductor 40-*g* Z-axis accelerometer implemented in a similar two-chip system configuration to the X-lateral accelerometer as described in Ristic et al. (1992). Other inertial sensor systems have been demonstrated as two chip approaches with vastly different system requirements [Delapierre, 1999; Yazde and Najafi, 1997; Ayazi and Najafi, 1998; Greiff et al., 1991; Hiller et al., 1998]. Figure 2.15 shows a high sensitivity Z-axis two-chip capacitive inertial sensor developed at the University of Michigan. Najafi et al. (2003) states a capacitance sensitivity of 5.6 pF/*g* and a noise floor for this device of 1.08 µg/√Hz due to its massive proof-mass and compliant springs.

Two-chip methodologies do have some limitations. Inter-chip parasitic capacitances pre-load the control circuitry with a non-sensitive capacitance; this requires the control system to discriminate small capacitance changes in a total capacitance several times larger than the capacitance of the accelerometer device alone. This inter-chip parasitic capacitance can be on the order of two to five times the transducer nominal capacitance, requiring clever circuit techniques to minimize the impact of the parasitic coupling. Other capacitive sensor systems have implemented single-chip integration to eliminate the inter-chip parasitic coupling at the expense of increased process complexity.

2.8.5 System Example: Michigan Vibratory Ring Gyroscope

As an example of a rotational inertial sensor, a study by Ayazi and Najafi (1998) presented a detailed analysis on the design and scaling limits of vibrating ring gyroscopes and their implementation using a

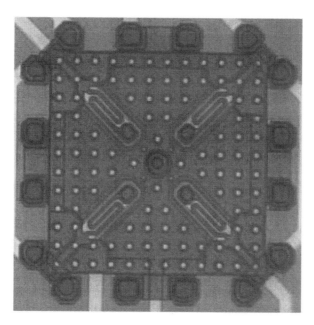

FIGURE 2.14   (See color insert following page 2-12.) Top view micrograph of a Z-axis capacitive accelerometer in three polysilicon layers. The design allows for high inertial sensitivity with a low temperature sensitivity. (Photo courtesy Freescale Semiconductor, Inc.)

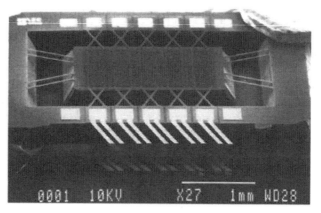

FIGURE 2.15   Perspective view scanning electron micrograph of a high resolution Z-axis capacitive accelerometer incorporating a bulk silicon proof-mass with trench-embedded polysilicon sense electrodes. (Photo courtesy K. Najafi, University of Michigan.)

combined bulk and surface micromachining technology. A high aspect ratio p^{++}/polysilicon trench and refill fabrication technology was used to realize the 30–40 μm thick polysilicon ring structure with 0.9 μm ring-to-electrode gap spacing, as shown in Figure 2.16. The theoretical analysis of the ring gyroscope shows that several orders of magnitude improvement in performance can be achieved through materials development and design. By taking advantage of the high quality factor of polysilicon, submicron ring-to-electrode gap spacing, high aspect ratio polysilicon ring structure produced using deep dry etching, and the all-silicon feature of this technology, a tactical grade vibrating ring gyroscope with random walk as small as 0.05 deg/√Hz was realized. Ayazi et al. (2000) enhanced this technology with an even higher aspect ratio 60 μm thick ring. This device demonstrated a random walk of 0.04 deg/√Hz with

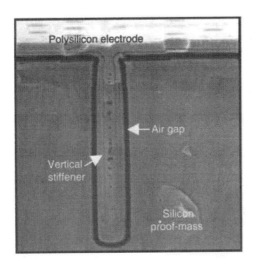

FIGURE 2.16 Cross sectional scanning electron micrograph of the vertical stiffener in the polysilicon trench-filled inertial sensor structures. (Photo courtesy K. Najafi, University of Michigan.)

a theoretical Brownian noise floor of 0.01 deg/$\sqrt{Hz}$. The single crystal silicon vibratory ring gyro shown in Figure 2.9 demonstrated a 10°/hr/Hz resolution [Najafi et al., 2003].

2.9 Concluding Remarks

Inertial sensors, both linear and rotational, have seen broad commercial and industrial application of micromachining technologies driving system cost, size, and performance. Many of the early automotive applications, driven initially by cost and package size with modest performance requirements, resulted in a niche product field in which micromachined accelerometers could successfully compete with macroscopic competitors. As the technology field has matured and broadened, increasing performance expectations in what could be considered *traditional* micromachined product areas will push the technology frontiers much harder. Such traditional areas might include front airbag crash sensing and the future utilization of distributed sensor systems in automobiles for systems like stability control, ride control, future generation occupant safety systems, rollover, and a cadre of potential new applications. Not only will micromachined inertial sensors need to produce the low cost, small size parts they have demonstrated so successfully, but they also will need to demonstrate significant system performance gains at the modest costs and packages in order to continue to flourish. There is certainly room for the next generation of technology, device, and system designers to creatively demonstrate exciting, challenging, and expanding applications for micromachined inertial sensing.

References

Acar, C., and Shkel, A.M. (2003) "Nonresonant Micromachined Gyroscopes with Structural Mode-Decoupling," *IEEE Sensors J.*, 3(4), pp. 497–506.

Amini, B.V., Pourkamali, S., and Ayazi, F. (2004) "A High Resolution, Stictionless, CMOS Compatible SOI Accelerometer with a Low Noise, Low Power, 0.25μm CMOS Interface," *Digest IEEE Conference on Micro and Nano Electro Mechanical Systems (MEMS 2004)*, pp. 572–75, January 2004.

Andersson, G.I., Hedenstierna, N., Svensson, P., and Pettersson, H. (1999) "A Novel Bulk Gyroscope," *Technical Digest IEEE International Conference on Solid-State Sensors and Actuators (Transducers '99)*, pp. 902–05, IEEE, New York.

Andrews, M.K., and Harris, P.D. (1995) "Damping and Gas Viscosity Measurements using a Microstructure," *Sensors Actuators A*, 49, pp. 103–08.

Andrews, M.K., Harris, I., and Turner, G. (1993) "A Comparison of Squeeze Film Theory with Measurements on a Microstructure," *Sensors Actuators A*, **36**, pp. 79–87.

Audet, S.A., Edenfeld, K.M., and Bergstrom, P.L. (1997) "Motorola Wafer-Level Packaging for Integrated Sensors," *Micromachine Devices* **2**, pp. 1–3.

Ayazi, F., and Najafi, K. (1998) "Design and Fabrication of High-Performance Polysilicon Vibrating Ring Gyroscope," *Technical Digest IEEE International Micro Electro Mechanical Systems Workshop (MEMS '98)*, pp. 621–26, IEEE, New York.

Ayazi, F., Chen, H.H., Kocer, F., He, G., and Najafi, K. (2000) "A High Aspect-Ratio Polysilicon Vibrating Ring Gyroscope," *Technical Digest Solid-State Sensor and Actuator Workshop*, Hilton Head, SC, pp. 289–92, Transducers Research Foundation, Cleveland, OH.

Barbour, N., and Schmidt, G. (2001) "Inertial Sensor technology trends," IEEE Sensors Journal, pp. 332–39, IEEE, New York.

Baschirotto, A., Gola, A., Chiesa, E., Lasalandra, E., Pasolini, F., Tronconi, M., and Ungaretti, T. (2003) "A $\pm$1-g Dual-Axis Linear Accelerometer in a Standard 0.5-μm CMOS Technology for High-Sensitivity Applications," *IEEE J. Solid-State Circuits*, **38**(7), pp. 1292–97.

Bergstrom, P.L., Bosch, D.R., and Averett, G. (1999) "Investigation of Thick Polysilicon Processing for MEMS Transducer Fabrication," *Proc. 1999 SPIE Symposium on Micromachining and Microfabrication: Materials and Device Characterization in Micromachining II*, SPIE 3875, pp. 87–96.

Blech, J.J. (1983) "On Isothermal Squeeze Films," *J. Lubr. Technol.*, **105**, pp. 615–20.

Boser, B.E., and Howe, R.T. (1996) "Surface Micromachined Accelerometers," *IEEE J. Solid-State Circuits*, **31**(3), pp. 366–75.

Burgdorfer, A. (1959) "The Influence of the Molecular Mean Free Path on the Performance of Hydrodynamic Gas Lubricated Bearings," *J. Basic Eng.*, March 1959, pp. 94–100.

Burns, D.W., Zook, J.D., Horning, R.D., Herb, W.R., and Guckel, H. (1995) "Sealed-Cavity Resonant Microbeam Pressure Sensor," *Sensors Actuators A*, **48**, pp. 179–86.

Chae, J., Kulah, H., and Najafi, K. (2004) "An *In-Plane* High-Sensitivity, Low-Noise Micro-g Silicon Accelerometer with CMOS Readout Circuitry," *IEEE J. Microelectromech. Syst.*, **13**(4), pp. 628–35.

Chau, K.H.–L., Lewis, S.R., Zhao, Y., Howe, R.T., Bart, S.F., and Marcheselli, R.G. (1996) "An Integrated Force-Balanced Capacitive Accelerometer for Low–g Applications," *Sensors Actuators A*, **54**, pp. 472–76.

Cheng, Y.T., Lin, L., and Najafi, K. (2000) "Localized Silicon Fusion and Eutectic Bonding for MEMS Fabrication and Packaging," *J. Microelectromech. Syst.*, **9**, pp. 3–8.

Cho, Y.–H., Pisano, A.P., and Howe, R.T. (1994) "Viscous Damping Model for Laterally Oscillating Microstructures," *J. Microelectromech. Syst.* **3**, pp. 81–87.

Corman, T., Enoksson, P., and Stemme, G. (1997) "Gas Damping of Electrostatically Excited Resonators," *Sensors Actuators A*, **61**, pp. 249–55.

Darling, R.B., Hivick, C., and Xu, J. (1998) "Compact Analytical Modeling of Squeeze Film Damping with Arbitrary Venting Conditions using a Green's Function Approach," *Sensors Actuators A*, **70**, pp. 32–41.

Delapierre, G. (1999) "MEMS and Microsensors: From Laboratory to Industry," *Technical Digest IEEE International Conference on Solid-State Sensors and Actuators (Transducers '99)*, pp. 6–11, IEEE, New York.

Dickerson, T., and Ward, M. (1997) "Low Deformation and Stress Packaging of Micro-Machined Devices," *IEE Colloquium on Assembly and Connections in Microsystems*, pp. 7/1–73, IEE, London.

Ding, X., Czarnocki, W., Schuster, J.P., and Roeckner, B. (1999) "DSP-Based CMOS Monolithic Pressure Sensor for High Volume Manufacturing," *Technical Digest IEEE International Conference on Solid-State Sensors and Actuators (Transducers '99)*, pp. 362–65, IEEE, New York.

Dokmeci, M.R., von Arx, J.A., and Najafi, K. (1997) "Accelerated Testing of Anodically Bonded Glass-Silicon Packages in Salt Water," *Technical Digest International Conference on Solid-State Sensors and Actuators (Transducers '97)*, pp. 283–86, IEEE, New York.

Ehrfeld, W., Bley, B., Götz, F., Hagmann, P., Maner, A., Mohr, J., Moser, H.O., Münchmeyer, D., Schelb, W., Schmidt, D., and Becker, E.W. (1987) "Fabrication of Microstructures Using the LIGA Process," *Technical Digest 1987 IEEE Micro Robots and Teleoperators Workshop*, pp. 11/1–11/11, IEEE, New York.

Feng, Z.C., and Gore, K. (2004) "Dynamic Characteristics of Vibratory Gyroscopes," *IEEE Sensors J.*, 4(1), pp. 80–84.

Franke, A.E., Jiao, Y., Wu, M.T., King, T.-J., and Howe, R.T. (2000) "Post-CMOS Modular Integration of Poly-SiGe Microstructures," *Technical Digest Solid-State Sensor and Actuator Workshop*, Hilton Head, SC, pp. 18–21, Transducers Research Foundation, Cleveland, OH.

Fujita, T., Mizuno, T., Kenny, R., Maenaka, K., and Maeda, M. (1997) "Two-Dimensional Micromachined Gyroscope," *Technical Digest IEEE International Conference on Solid-State Sensors and Actuators (Transducers '97)*, pp. 887–90, IEEE, New York.

Geen, J. (1998) "A Path to Low Cost Gyroscopy," *Technical Digest Solid-State Sensor and Actuator Workshop*, Hilton Head, SC, pp. 51–54, Transducers Research Foundation, Cleveland, OH.

Geiger, W., Merz, J., Fischer, T., Folkmer, B., Sandmaier, H., and Lang, W. (1999) "The Silicon Angular Rate Sensor System MARS–RR," *Technical Digest IEEE International Conference on Solid-State Sensors and Actuators (Transducers '99)*, pp. 1578–81, IEEE, New York.

Gennissen, P.T.J., and French, P.J. (1996) "Applications of Bipolar Compatible Epitaxial Polysilicon," *Proc. SPIE*, 2882, pp. 59–65.

Goodenough, F. (1991) "Airbags Boom when IC Accelerometer Sees 50 *g*," *Electron. Design*, 39, pp. 45–6, 49–51, 54, 56.

Grade, J. Barzilai, A., Reynolds, J.K., Liu, C.H., Partridge, A., Kenny, T.W., VanZandt, T.R., Miller, L.M., and Podosek, J.A. (1996) "Progress in Tunnel Sensors," *Technical Digest 1996 Solid State Sensor and Actuator Workshop*, Hilton Head, SC, pp. 72–75, Transducers Research Foundation, Cleveland, OH.

Greiff, P., and Boxenhorn, B. (1995) *Micromechanical Gyroscopic Transducer with Improved Drive and Sense Capabilities*, US Patent #5,408,877, April 1995.

Greiff, P., Boxenhorn, B., King, T., and Niles, L. (1991) "Silicon Monolithic Micromechanical Gyroscope," *Technical Digest IEEE International Conference on Solid-State Sensors and Actuators (Transducers '91)*, pp. 966–68, IEEE, New York.

Gudeman, C.S., Staker, B., and Daneman, M. (1998) "Squeeze Film Damping of Doubly Supported Ribbons in Noble Gas Atmospheres [MEMS Light Valve]," *Technical Digest 1998 Solid-State Sensor and Actuator Workshop*, Hilton Head, SC, pp. 288–91, Transducers Research Foundation, Cleveland, OH.

Haronian, D. (1999) "Direct Integration (DI) of Solid State Stress Sensors with Single Crystal Micro-Electro-Mechanical Systems for Integrated Displacement Sensing," *Technical Digest IEEE Workshop on Micro Electro Mechanical Systems (MEMS '99)*, pp. 88–93, IEEE, New York.

He, G., and Najafi, K. (2002) "A Single-Crystal Silicon Vibrating Ring Gyroscope," *Digest IEEE Conference on Micro and Nano Electro Mechanical Systems (MEMS 2002)*, pp. 718–21.

Hermann, J., Bourgeois, C., Porret, F., and Kloeck, B. (1995) "Capacitive Silicon Differential Pressure Sensor," *Technical Digest IEEE International Conference on Solid-State Sensors and Actuators (Transducers '95)*, pp. 620–623, IEEE, New York.

Hierold, C., Hildebrant, A., Näher, U., Scheiter, T., Mensching, B., Steger, M., and Tielert, R. (1996) "A Pure CMOS Micromachined Integrated Accelerometer," *Technical Digest IEEE Workshop on Micro Electro Mechanical Systems (MEMS '96)*, pp. 174–79, IEEE, New York.

Hiller, K., Wiemer, M., Billep, D., Huhnerfurst, A., Gessner, T., Pyrko, B., Breng, U., Zimmermann, S., and Gutmann, W. (1998) "A New Bulk Micromachined Gyroscope with Vibration Enhancement Coupled Resonators," *Technical Digest 6th International Conference on Micro Electro, Opto, Mechanical Systems and Components (Micro Systems Technologies '98)*, pp. 115–20, VDE Verlag, Berlin.

Hölke, A., and Henderson, H.T. (1999) "Ultra-Deep Anisotropic Etching of (110) Silicon," *J. Micromech. Microeng.*, 9, pp. 51–57.

Honer, K.A., and Kovacs, G.T.A. (2000) "Sputtered Silicon for Integrated MEMS Applications," *Technical Digest Solid-State Sensor and Actuator Workshop*, Hilton Head, SC, pp. 308–311, Transducer Research Foundation, Cleveland, OH.

Hsu, Y., Reeds, J.W., and Saunders, C.H. (1999) *Multi-Element Micro-Gyro*, US Patent #5,955,668, September 1999.

Huff, M.A., Nikolich, A.D., and Schmidt, M.A. (1991) "A Threshold Pressure Switch Utilizing Plastic Deformation of Silicon," *Technical Digest IEEE International Conference on Solid-State Sensors and Actuators (Transducers '91)*, pp. 177–80, IEEE, New York.

Ishihara, K., Yung, C.–F., Ayón, A.A., and Schmidt, M.A. (1999) "An Inertial Sensor Technology Using DRIE and Wafer Bonding with Enhanced Interconnect Capability," *Technical Digest IEEE International Conference on Solid-State Sensors and Actuators (Transducers '99)*, pp. 254–57, IEEE, New York.

Jaeger, R.C., Suhling, J.C., Liechti, K.M., and Liu, S. (1997) "Advances in Stress Test Chips," *Technical Digest ASME 1997 International Mechanical Engineering Congress and Exposition*: Applications of Experimental Mechanics to Electronic Packaging, pp. 1–5, ASME, New York.

Juneau, T., Pisano, A.P., and Smith, J.H. (1997) "Dual Axis Operation of a Micromachined Rate Gyroscope," *Technical Digest IEEE International Conference on Solid-State Sensors and Actuators (Transducers '97)*, pp. 883–86, IEEE, New York.

Kim, E.S., Cho, Y.H., and Kim, M.U. (1999) "Effect of Holes and Edges on the Squeeze Film Damping of Perforated Micromechanical Structures," *Technical Digest IEEE Workshop on Micro Electro Mechanical Systems (MEMS '99)*, pp. 296–301, IEEE, New York.

Kirsten, M., Wenk, B., Ericson, F., Schweitz, J.Å., Riethmüller, W., and Lange, P. (1995) "Deposition of Thick Doped Polysilicon Films with Low Stress in an Epitaxial Reactor for Surface Micromachining Applications," *Thin Solid Films*, 259, pp. 181–87.

Kniffin, M.L., Wiegele, T.G., Masquelier, M.P., Fu, H., and Whitfield, J.D. (1998) "Modeling and Characterization of an Integrated FET Accelerometer," *Technical Digest IEE International Conference on Modeling and Simulation of Microsystems, Semiconductors, Sensors and Actuators*, pp. 546–51, Computational Publications, Cambridge, MA.

Kobayashi, S., Hara, T., Oguchi, T., Asaji, Y., Yaji, K., and Ohwada, K. (1999) "Double-Frame Silicon Gyroscope Packaged Under Low Pressure by Wafer Bonding," *Technical Digest IEEE International Conference on Solid-State Sensors and Actuators (Transducers '99)*, pp. 910–13, IEEE, New York.

Laerme, F., Schilp, A., Funk, K., and Offenberg, M. (1999) "Bosch Deep Silicon Etching: Improving Uniformity and Etch Rate for Advanced MEMS Applications," *Technical Digest IEEE International Conference on Micro Electro Mechanical Systems (MEMS '99)*, pp. 211–16, IEEE, New York.

Lee, K.I., Takao, H., Sawada, K., and Ishida, M. (2004) "Analysis and Experimental Verification of Thermal Drift in a Constant Temperature Control Type Three-Axis Accelerometer for High Temperatures with a Novel Composition of Wheatstone Bridge," *Digest IEEE Conference on Micro and Nano Electro Mechanical Systems (MEMS 2004)*, pp. 241–44.

Lee, K.W., and Wise, K.D. (1981) "Accurate Simulation of High-Performance Silicon Pressure Sensors," *Technical Digest IEEE International Electron Devices Meeting (IEDM '81)*, pp. 471–474, IEEE, New York.

Lehto, A. (1999) "SOI Microsensors and MEMS," *Proc. Silicon-on-Insulator Technology and Devices IX*, ECS Proc. vol. 99–3, pp. 11–24, ECS, Pennington, NJ.

Lemkin, M., and Boser, B.E. (1999) "A Three-Axis Micromachined Accelerometer.

Lemkin, M., Juneau, T.N., Clark, W.A., Roessig, T.A., and Brosnihan, T.J. (1999) "A Low-Noise Digital Accelerometer Using Integrated SOI–MEMS Technology," *Technical Digest IEEE International Conference on Solid-State Sensors and Actuators (Transducers '99)*, pp. 1294–97, IEEE, New York.

Lewis, S., Alie, S., Brosnihan, T., Core, C., Core, T., Howe, R., Geen, J., Hollocher, D., Judy, M., Memishian, J., Nunan, K., Paine, R., Sherman, S., Tsang, B., and Wachtmann, B. (2003) "Integrated Sensor and Electronics Processing for >10^8 "iMEMS" Inertial Measurement Unit Components," *Proc. Int. Electron. Devices Meet.*, pp. 949–52.

Li, G.X., and Shemansky, F.A. (2000) "Drop Test and Analysis on Micro-Machined Structures," *Sensors Actuators A*, 85, pp. 280–86.

Li, G.X., Bergstrom, P.L., Ger, M.–L., Foerstner, J., Schmiesing, J.E., Shemansky, F.A., Mahadevan, D., and Shah, M.K. (1998) "Low Stress Packaging of a Micro-Machined Accelerometer," *Proc. 3rd* International Symposium on Electronic Package Technology (ISEPT '98), pp. 553–62.

Lutz, M., Golderer, W., Gerstenmeier, J., Marek, J., Maihöfer, B., Mahler, S., Münzel, H., and Bischof, U. (1997) "A Precision Yaw Rate Sensor in Silicon Micromachining," *Technical Digest IEEE International Conference on Solid-State Sensors and Actuators (Transducers '97)*, pp. 847–50, IEEE, New York.

MacDonald, G.A. (1990) "A Review of Low Cost Accelerometers for Vehicle Dynamics," *Sensors Actuators A*, 21, pp. 303–07.

MacDonald, N.C., and Zhang, Z.L. (1993) *RIE Process for Fabricating Submicron, Silicon Electromechanical Structures*, US Patent #5,198,390, March 1993.

Madni, A.M., Costlow, L.E., and Knowles, S.J. (2003) "Common Design Techniques for BEI GyroChip Quartz Rate Sensors for Both Automotive and Aerospace/Defense Markets," *IEEE Sensors J.*, 3(5), pp. 569–78.

Mastrangelo, C.H., and Hsu, C.H. (1993) "Mechanical Stability and Adhesion of Microstructures Under Capillary Forces. II. Experiments," *J. Microelectromech. Syst.*, 2, pp. 44–55.

McNie, E., Burdess, J.S., Harris, A.J., Hedley, J., and Young, M. (1999) "High Aspect Ratio Ring Gyroscopes Fabricated in [100] Silicon On Insulator (SOI) Material," *Technical Digest IEEE International Conference on Solid-State Sensors and Actuators (Transducers '99)*, pp. 1590–93, IEEE, New York.

Meirovitch, L. (1975) *Elements of Vibration Analysis*, McGraw–Hill: New York.

Monk, D.J., Soane, D.S., and Howe, R.T. (1994) "Hydrofluoric Acid Etching of Silicon Dioxide Sacrificial Layers. I. Experimental Observations," *J. Electrochem. Soc.*, 141(1), pp. 264–69.

Najafi, K., Chae, J., Kulah, H., and He, G. (2003) "Micromachined Silicon Accelerometers and Gyroscopes," *IEEE International Robotics and Systems Conference (IROS)*, Las Vegas, NV.

Noell, W., Clerc, P.-A., Dellmann, L., Guldimann, B., Herzig, H.-P., Manzardo, O., Marxer, C.R., Weible, K.J., Dändliker, R., and de Rooij, N. (2002) "Applications of SOI-based Optical MEMS," *IEEE J. Select. Top. Quant. Electron.*, 8(1), pp. 148–54.

Noworolski, J.M., and Judy, M. (1999) "VHARM: Sub-Micrometer Electrostatic MEMS," *Technical Digest IEEE International Conference on Solid-State Sensors and Actuators (Transducers '99)*, pp. 1482–85, IEEE, New York.

Offenberg, M., Lärmer, F., Elsner, B., Münzel, H., and Riethmüller, W. (1995) "Novel Process for a Monolithic Integrated Accelerometer," *Technical Digest IEEE International Conference on Solid-State Sensors and Actuators (Transducers '95)*, pp. 589–92, IEEE, New York.

Overstolz, T., Clerc, P.A., Noell, W., Zickar, M., and de Rooij, N.F. (2004) "A Clean Wafer-Scale Chip-Release Process without Dicing Based on Vapor Phase Etching," *Digest IEEE Conference on Micro and Nano Electro Mechanical Systems (MEMS 2004)*, pp. 717–20.

Painter, C.C., and Shkel, A.M. (2003) "Active Structural Error Suppression in MEMS Vibratory Rate Integrating Gyroscopes," *IEEE Sensors J.*, 3(5), pp. 595–606.

Pan, F., Kubby, J., Peeters, E., Tran, A.T., and Mukherjee, S. (1998) "Squeeze Film Damping on the Dynamic Response of a MEMS Torsion Mirror," *J. Micromech. Microeng.*, 8, pp. 200–08.

Park, Y., Jeong, H.S., An, S., Shin, S.H., and Lee, C.W. (1999) "Lateral Gyroscope Suspended by Two Gimbals Through High Aspect Ratio ICP Etching," *Technical Digest IEEE International Conference on Solid-State Sensors and Actuators (Transducers '99)*, pp. 972–75, IEEE, New York.

Partridge, A., Reynolds, J.K., Chui, B.W., Chow, E.M., Fitzgerald, A.M., Zhang, L., Cooper, S.R., Kenny, T.W., and Maluf, N.I. (1998) "A High Performance Planar Piezoresistive Accelerometer," *Technical Digest Solid-State Sensor and Actuator Workshop*, Hilton Head, SC, pp. 59–64, Transducers Research Foundation, Cleveland, OH.

Putty, M., and Najafi, K. (1994) "A Micromachined Vibrating Ring Gyroscope," *Technical Digest Solid State Sensor and Actuator Workshop*, Hilton Head, SC, pp. 213–20, Transducers Research Foundation, Cleveland, OH.

Reichenbach, R., Schubert, D., and Gerlach, G. (2003) "Micromechanical Triaxial Acceleration Sensor for Automotive Applications," *Digest International Conference on Solid-State Sensors, Actuators, and Microsystems (Transducers '03)*, pp. 77–80. Boston.

Reuther, H.M., Weinmann, M., Fischer, M., von Munch, W., and Assmus, F. (1996) "Modeling Electrostatically Deflectable Microstructures and Air Damping Effects," *Sensors Mater.*, 8, pp. 251–69.

Ristic, Lj., Gutteridge, R., Dunn, B., Mietus, D., and Bennett P. (1992) "Surface Micromachined Polysilicon Accelerometer," *Technical Digest IEEE Solid-State Sensor and Actuator Workshop*, Hilton Head, SC, pp. 118–21, IEEE, New York.

Rockstad, H.K., Reynolds, J.K., Tang, T.K., Kenny, T.W., Kaiser, W.J., and Gabrielson, T.B. (1995) "A Miniature, High-Sensitivity, Electron Tunneling Accelerometer," *Technical Digest IEEE International Conference on Solid-State Sensors and Actuators (Transducer '95)* 2, pp. 675–78, IEEE, New York.

Roylance, L.M., and Angell, J.B. (1979) "A Batch-Fabricated Silicon Accelerometer," *IEEE Trans. Electron Devices* 26, pp. 1911–17.

Seeger, K. (1985) *Semiconductor Physics: An Introduction*, Springer-Verlag: New York.

Shashkin, V.I., Vostokov, N.V., Vopilkin, E.A., Klimov, A.Y., Volgunov, D.G., Rogov, V.V., and Lazarev, S.G. (2004) "High-Sensitivity Accelerometer Based on Cold Emission Principle," *IEEE Sensors J.*, 4(2), pp. 211–15.

Shearwood, C., Ho, K.Y., and Gong, H.Q. (1999) "Testing of a Micro-Rotating Gyroscope," *Technical Digest IEEE International Conference on Solid-State Sensors and Actuators (Transducers '99)*, pp. 984–87, IEEE, New York.

Sherman, S.J., Tsang, W.K., Core, T.A., Payne, R.S., Quinn, D.E., Chau, K.H.-L., Farash, J.A., and Baum, S.K. (1992) "A Low Cost Monolithic Accelerometer; Product/Technology Update," *Technical Digest IEEE International Electron Devices Meeting*, pp. 501–04, IEEE, New York.

Smith, J., Montague, S., and Sniegowski, J. (1995) "Material and Processing Issues for the Monolithic Integration of Microelectronics with Surface-Micromachined Polysilicon Sensors and Actuators," *SPIE*, 2639, pp. 64–73.

Sridhar, U., Lau, C.H., Miao, Y.B., Tan, K.S., Foo, P.D., Liu, L.J., Sooriakumar, K., Loh, Y.H., John, P., Lay Har, A.T., Austin, A., Lai, C.C., and Bergstrom, J. (1999) "Single Crystal Silicon Microstructures Using Trench Isolation," *Technical Digest IEEE International Conference on Solid-State Sensors and Actuators (Transducers '99)*, pp. 258–61, IEEE, New York.

Starr, J.B. (1990) "Squeeze-Film Damping in Solid-State Accelerometers," *Technical Digest IEEE Solid-State Sensor and Actuator Workshop*, Hilton Head, SC, pp. 44–47, IEEE, New York.

Tang, T.K., Gutierrez, R.C., Stell, C.B., Vorperian, V., Arakaki, G.A., Rice, J.T., Li, W.J., Chakraborty, I., Shcheglov, K., Wilcox, J.Z., and Kaiser, W.J. (1997) "A Packaged Silicon MEMS Vibratory Gyroscope for Microspacecraft," *Technical Digest IEEE International Workshop on Micro Electro Mechanical Systems (MEMS '97)*, Nagoya, Japan, pp. 500–05, IEEE, New York.

Usenko, Y., and Carr, W.N. (1999) "SOI Technology for MEMS Applications," *Proc. Silicon-on-Insulator Technology and Devices IX*, ECS Proc. 99–3, pp. 347–52, ECS, Pennington, NJ.

van Kampen, R.P., and Wolffenbuttel, R.F. (1998) "Modeling the Mechanical Behavior of Bulk-Micro-Machined Silicon Accelerometers," *Sensors Actuators A*, 4, pp. 137–50.

Veijola, T., and Ryhanen, T. (1995) "Model of Capacitive Micromechanical Accelerometer including Effect of Squeezed Gas Film," *Technical Digest 1995 IEEE Symposium on Circuits and Systems*, pp. 664–67, IEEE, New York.

Veijola, T., Kuisma, H., and Lahdenperä, J. (1997) "Model for Gas Film Damping in a Silicon Accelerometer," *Technical Digest IEEE International Conference on Solid-State Sensors and Actuators (Transducers '97)* 2, pp. 1097–1100, IEEE, New York.

Veijola, T., Kuisma, H., and Lahdenperä, J. (1998) "The Influence of Gas–Surface Interaction on Gas–Film Damping in a Silicon Accelerometer," *Sensors Actuators A*, 66, pp. 83–92.

Voss, R., Bauer, K., Ficker, W., Gleissner, T., Kupke, W., Rose, M., Sassen, S., Schalk, J., Seidel, H., and Stenzel, E. (1997) "Silicon Angular Rate Sensor for Automotive Applications with Piezoelectric Drive and Piezoresistive Read-Out," *Technical Digest IEEE International Conference on Solid-State Sensors and Actuators (Transducers '97)*, pp. 879–82, IEEE, New York.

Wang, J., McClelland, B., Zavracky, P.M., Hartley, F., and Dolgin, B. (1996) "Design, Fabrication and Measurement of a Tunneling Tip Accelerometer," *Technical Digest 1996 Solid State*

Sensor and Actuator Workshop, Hilton Head, SC, pp. 68–71, Transducers Research Foundation, Cleveland, OH.

Williams, K.R., Gupta. K., and Wasilik, M. (2003) "Etch Rates for Micromachining Processing-Part II," *J. Microelectromech. Syst.*, 12(6), pp. 761–78.

Wolffenbuttel, R.F., and Wise, K.D. (1994) "Low-Temperature Silicon Wafer-to-Wafer Bonding using Gold at Eutectic Temperature," *Sensors Actuators A*, 43, pp. 223–29.

Wu, J., Fedder, G.K., and Carley, L.R. (2004) "A Low-Noise Low-Offset Capacitive Sensing Amplifier for a 50-μg/$\sqrt{Hz}$ Monolithic CMOS MEMS Accelerometer," *IEEE J. Solid-State Circuits*, 39(5), pp. 722–30.

Wycisk, M., Tönnesen, T., Binder, J., Michaelis, S., and Timme, H.–J. (1999) "Low Cost Post-CMOS Integration of Electroplated Microstructures for Inertial Sensing," *Technical Digest IEEE International Conference on Solid-State Sensors and Actuators (Transducers '99)*, pp. 1424–27, IEEE, New York.

Xia, H., Yang, Y., and Bergstrom, P.L. (2004) "Low Temperatures Silicon Films Deposition by Pulsed Cathodic Arc Process for Microsystem Technology," *Symposium on Amorphous and Nanocrystalline Silicon Science and Technology (808), Materials Research Society Meeting*, pp. 371–76, San Francisco, CA.

Xie, H., and Fedder, G.K. (2003) "Fabrication, Characterization, and Analysis of a DRIE CMOS-MEMS Gyroscope," *IEEE Sensors J.*, 3(5), pp. 622–31.

Xie, H., Erdmann, L., Zhu, X., Gabriel, K.J., and Fedder, G.K. (2000) "Post-CMOS Processing for High-Aspect-Ratio Integrated Silicon Microstructures," *Technical Digest 2000 Solid State Sensor and Actuator Workshop*, Hilton Head, SC, pp. 77–80, Transducers Research Foundation, Cleveland, OH.

Yan, G., Wang, C., Zhang, R., Chen, Z., Liu, X., and Wang, Y.Y. (2004) "Integrated Bulk-Micromachined Gyroscope Using Deep Trench Isolation Technology," *Digest IEEE Conference on Micro and Nano Electro Mechanical Systems (MEMS 2004)*, pp. 605–08.

Yazde, N., and Najafi, K. (1997) "An All-Silicon Single-Wafer Fabrication Technology for Precision Micro-accelerometers," *Technical Digest IEEE International Conference on Solid-State Sensors and Actuators (Transducers '97)*, pp. 1181–1184, IEEE New York.

Yeh, C., and Najafi, K. (1997) "Micromachined Tunneling Accelerometer with a Low-Voltage CMOS Interface Circuit," *Technical Digest IEEE International Conference on Solid-State Sensors and Actuators (Transducers '97)*, pp. 1213–1216, IEEE, New York.

Yoshii, Y., Nakajo, A., Abe, H., Ninomiya, K., Miyashita, H., Sakurai, N., Kosuge, M., and Hao, S. (1997) "1 Chip Integrated Software Calibrated CMOS Pressure Sensor with MCU, A/D Converter, D/A Converter, Digital Communication Port, Signal Conditioning Circuit and Temperature Sensor," *Technical Digest IEEE International Conference on Solid-State Sensors and Actuators (Transducers '97)*, pp. 1485–88, IEEE, New York.

Zook, J.D., Herb, W.R., Bassett, C.J., Stark, T., Schoess, J.N., and Wilson, M.L. (1999) "Fiber-optic Vibration Sensor Based on Frequency Modulation of Light-Excited Oscillators," *Technical Digest IEEE International Conference on Solid-State Sensors and Actuators (Transducers '99)*, pp. 1306–09, IEEE, New York.

3

Micromachined Pressure Sensors: Devices, Interface Circuits, and Performance Limits

Yogesh B. Gianchandani
University of Michigan

Chester G. Wilson
Louisiana Tech University

Jae-Sung Park
Shriner's Institute, Boston

3.1 Introduction

Pressure sensors represent one of the greatest successes of micromachining technology. In the past four decades, this field has produced both commercially available and research-oriented devices for a variety of automotive, biomedical, and industrial applications. Automotive applications include pressure sensors for the engine manifold, fuel lines, exhaust gases, tires, seats, and other uses. Biomedical applications that have been proposed or developed include implantable devices for measuring ocular, cranial, or bowel pressure, and devices in catheters that can aid procedures such as angioplasty. Many industrial applications exist that relate to monitoring manufacturing processes. In the semiconductor sector, for example, process steps such as plasma etching or deposition and chemical vapor deposition are very sensitive to operating pressures.

In the long history of using micromachining technology for pressure sensors, device designs have evolved as the technology has progressed, allowing pressure sensors to serve as technology demonstration vehicles [Wise, 1994]. A number of sensing approaches that offer different relative merits have evolved, and there has been a steady march toward improving performance parameters such as sensitivity, resolution, and dynamic range. Although multiple options exist, silicon has been a popular choice for the structural material of micromachined pressure sensors partly because its material properties are adequate, and there is significant manufacturing capacity and know-how that can be borrowed from the integrated circuit industry. The primary focus in this chapter is on schemes that use silicon as the structural material. The chapter is divided into six sections. The first section introduces structural and performance concepts that are common to a number of micromachined pressure sensors. The second and third sections focus in some detail on piezoresistive and capacitive pick-off schemes for detecting pressure. These two schemes form the basis of the vast majority of micromachined pressure sensors available commercially and studied by the MEMS research community. Fabrication, packaging, and calibration issues related to these devices are also addressed in these sections. The fourth section describes servo-controlled pressure sensors, which represent an emerging theme in research publications. The fifth section surveys alternative approaches and transduction schemes that may be suitable for selected applications. It includes a few schemes that have been explored with non-micromachined apparatus, but may be suitable for miniaturization in the future. The sixth section concludes the chapter.

3.2   Device Structure and Performance Measures

The essential feature of most micromachined pressure sensors is an edge-supported diaphragm that deflects in response to a transverse pressure differential across it. This deformation is typically detected by measuring the stresses in the diaphragm, or by measuring the displacement of the diaphragm. An example of the former approach is the *piezoresistive pick-off*, in which resistors are formed at specific locations of the diaphragm to measure the stress. An example of the latter approach is the *capacitive pick-off*, in which an electrode is located on a substrate some distance below the diaphragm to capacitively measure its displacement. The choice of silicon as a structural material is amenable to both approaches because it has a relatively large piezoresistive coefficient and because it can serve as an electrode for a capacitor as well.

3.2.1   Pressure on a Diaphragm

The deflection of a diaphragm and the stresses associated with it can be calculated analytically in many cases. It is generally worthwhile to make some simplifying assumptions regarding the dimensions and boundary conditions. One approach is to assume that the edges are simply supported. This is a reasonable approximation if the thickness of the diaphragm, h, is much smaller than its radius, a. This condition prevents transverse displacement of the neutral surface at the perimeter, while allowing rotational and longitudinal displacement. Mathematically, it permits the second derivative of the deflection to be zero at the edge of the diaphragm. However, the preferred assumption is that the edges of the diaphragm are rigidly affixed (built-in) to the support around its perimeter. Under this assumption the stress on the lower surface of a circular diaphragm can be expressed in polar coordinates by the equations:

$$\sigma_r = \frac{3 \cdot \Delta P}{8h^2} [a^2(1 + v) - r^2(3 + v)] \tag{3.1}$$

$$\sigma_t = \frac{3 \cdot \Delta P}{8h^2} [a^2(1 + v) - r^2(1 + 3v)] \tag{3.2}$$

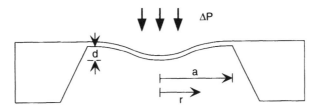

FIGURE 3.1 Deflection of a diaphragm under applied pressure.

where the former denotes the radial component and the latter the tangential component, *a* and *h* are the radius and thickness of the diaphragm, *r* is the radial co-ordinate, ΔP is the pressure applied to the upper surface of the diaphragm, and *v* is Poisson's ratio (Figure 3.1) [Timoshenko and Woinowsky-Krieger, 1959; Samaun et al., 1973; Middleoek and Audet, 1994]. In the (100) plane of silicon, Poisson's ratio shows four-fold symmetry, and varies from 0.066 in the [011] direction to 0.28 in the [001] direction [Evans and Evans, 1965' Madou, 1997]. These equations indicate that both components of stress vary from the same tensile maximum at the center of the diaphragm to different compressive maxima at its periphery. Both components are zero at separate values of *r* between zero and *a*. In general, piezoresistors located at the points of highest compressive and tensile stress will provide the largest responses. The deflection of a circular diaphragm under the stated assumptions is given by:

$$d = \frac{3 \cdot \Delta P(1 - v^2)(a^2 - r^2)^2}{16Eh^3} \tag{3.3}$$

where *E* is the Young's modulus of the structural material. This is valid for a thin diaphragm with simply supported edges, assuming a small defection.

Like Poisson's ratio, the Young's modulus for silicon shows four-fold symmetry in the (100) plane, varying from 168 GPa in the [011] direction to 129.5 GPa in the [100] direction [Greenwood, 1988; Madou, 1997]. When polycrystalline silicon (polysilicon) is used as the structural material, the composite effect of grains of varying size and crystalline orientation can cause substantial variations. It is important to note that additional variations in mechanical properties may arise from crystal defects caused by doping and other disruptions of the lattice. Equation 3.3 indicates that the maximum deflection of a diaphragm is at its center, which comes as no surprise. More importantly, it is dependent on the radius to the fourth power, and on the thickness to the third power, making it extremely sensitive to inadvertent variations in these dimensions. This can be of some consequence in controlling the sensitivity of capacitive pressure sensors.

3.2.2 Square Diaphragm

For pressure sensors that are micromachined from bulk Si, it is common to use anisotropic wet etchants that are selective to crystallographic planes, which results in square diaphragms. The deflection of a square diaphragm with built-in edges can be related to applied pressure by the following expression:

$$\Delta P = \frac{Eh^4}{(1 - v^2)a^4} \left[4.20\frac{w_c}{h} + 1.58\frac{w_c^3}{h^3} \right] \tag{3.4}$$

where *a* is half the length of one side of the diaphragm [Chau and Wise, 1987]. This equation provides a reasonable approximation of the maximum deflection over a wide range of pressures, and is not limited to small deflections. The first term within this equation dominates for small deflections, for which $w_c < h$, whereas the second term dominates for large deflections. For very large deflections, it approaches the deflection predicted for flexible membranes with a 13% error.

3.2.3 Residual Stress

It should be noted that the analysis presented above assumes that the residual stress in the diaphragm is negligibly small. Although mathematically convenient, this is often not the case. In reality, a tensile stress of 5–50 MPa is not uncommon. This may significantly reduce the sensitivity of certain designs, particularly if the diaphragm is very thin. Following the treatment in Chau and Wise (1987) for a small deflection in a circular diaphragm with built-in edges, the governing differential equation is:

$$\frac{d^2\phi}{dr^2} + \frac{1}{r}\frac{d\phi}{dr} - \left(\frac{\sigma_i h}{D} + \frac{1}{r^2}\right)\phi = -\frac{\Delta P \cdot r}{2D} \tag{3.5}$$

where σ_i is the intrinsic or residual stress in the undeflected diaphragm, $D = Eh^3/[12(1 - v^2)]$, and $\phi = -dw/dr$ is the slope of the deformed diaphragm. The solution this differential equation provides w:

$$w = \frac{\Delta P \cdot a^4\left[I_0\left(\frac{kr}{a}\right) - I_0(k)\right]}{2k^3 I_1(k)D} + \frac{\Delta P \cdot a^2(a^2 - r^2)}{4k^2 D} \tag{3.6}$$

in tension, and

$$w = \frac{\Delta P \cdot a^4\left[J_0\left(\frac{kr}{a}\right) - J_0(k)\right]}{2k^3 J_1(k)D} + \frac{\Delta P \cdot a^2(a^2 - r^2)}{4k^2 D} \tag{3.7}$$

in compression. In these expressions, J_n and I_n are the Bessel function and the modified Bessel function of the first kind of order n, respectively. The term k is given by:

$$k^2 = \frac{|\sigma_i| a^2 h}{D} = \frac{12(1 - v^2)|\sigma_i| a^2}{Eh^2} \tag{3.8}$$

The maximum deflection (at the center of the diaphragm), normalized to the deflection in the absence of residual stress, is provided by:

$$w_c' = \frac{16[2 - 2I_0(k) + kI_1(k)]}{k^3 I_1(k)} \tag{3.9}$$

in tension, and

$$w_c' = \frac{16[2 - 2J_0(k) + kJ_1(k)]}{k^3 J_1(k)} \tag{3.10}$$

in compression. It is instructive to evaluate the dependence of this normalized deflection to the dimensionless intrinsic stress, which is defined by:

$$\sigma_i' = \frac{(1 - v^2)\sigma_i a^2}{Eh^2} \tag{3.11}$$

As shown in Figure 3.2, residual stress can have a tremendous impact on deflection: a tensile dimensionless stress of 1.3 diminishes the center deflection by 50%. For tensile (positive) values of σ_i' exceeding 10, the center deflection (not normalized) can be approximated as for a *membrane*:

$$w_c = \frac{\Delta P \cdot a^2}{4\sigma_i h} \tag{3.12}$$

Returning to Figure 3.2, it is evident that the deflection can be increased by compressive stress. However, even relatively small values of compressive stress can result in buckling, so it is not generally perceived as a feature that can be reliably exploited.

3.2.4 Composite Diaphragms

In micromachined pressure sensors, it is often the case that the diaphragm is fabricated not from a single material but from composite layers. For example, a silicon membrane can be covered by a layer of SiO_2

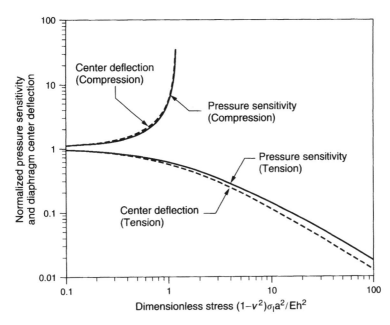

FIGURE 3.2 Normalized deflection of a circular diaphragm as a function of dimensionless stress. (Reprinted with permission from Chau, H., and Wise, K. [1987a] "Noise Due to Brownian Motion in Ultrasensitive Solid-State Pressure Sensors," *IEEE Transactions on Electron Devices* **34**, pp. 859–865.)

or Si_xN_y for electrical isolation. In general, these films can be of comparable thickness, and have values of Young's modulus and residual stress that are significantly different. The residual stress of a composite membrane is given by:

$$\sigma_c t_c = \sum_m \sigma_m t_m \tag{3.13}$$

where σ_c and t_c denote the composite stress and thickness, respectively, while σ_m and t_m denote the stress and thickness of individual films. Furthermore, if the Poisson's ratio of all the layers in the membrane is comparable, the following approximation may be used for the Young's modulus:

$$E_c t_c = \sum_m E_m t_m \tag{3.14}$$

where the suffixes have the same meaning as in the preceding equation.

3.2.5   Categories and Units

Pressure sensors are typically divided into three categories: absolute, gauge, and differential (relative) pressure sensors. Absolute pressure sensors provide an output referenced to vacuum, and often accomplish this by vacuum sealing a cavity underneath the diaphragm. The output of a gauge pressure sensor is referenced to atmospheric pressure. A differential pressure sensor compares the pressure at two input ports, which typically transfer the pressure to different sides of the diaphragm.

A number of different units are used to denote pressure, which can lead to some confusion when comparing performance ratings. One atmosphere of pressure is equivalent to 14.696 pounds per square inch (psi), 101.33 kPa, 1.0133 bar (or centimeters of H_2O at 4°C), and 760 Torr (or millimeters of Hg at 0°C).

3.2.6   Performance Criteria

The performance criteria of primary interest in pressure sensors are sensitivity, dynamic range, full-scale output, linearity, and the temperature coefficients of sensitivity and offset. These characteristics depend on the device geometry, the mechanical and thermal properties of the structural and packaging materials, and selected sensing scheme. Sensitivity is defined as a normalized signal change per unit pressure change to reference signal:

$$S = \frac{1}{\theta} \frac{\partial \theta}{\partial P} \tag{3.15}$$

where θ is output signal and $\partial \theta$ is the change in this pressure due to the applied pressure ∂P. Dynamic range is the pressure range over which the sensor can provide a meaningful output. It may be limited by the saturation of the transduced output signal such as the piezoresistance or capacitance. It also may be limited by yield and failure of the pressure diaphragm. The full-scale output (FSO) of a pressure sensor is simply the algebraic difference in the end points of the output. Linearity refers to the proximity of the device response to a specified straight line. It is the maximum separation between the output and the line, expressed as a percentage of FSO. Generally, capacitive pressure sensors provide highly non-linear outputs, and piezoresistive pressure sensors provide fairly linear output.

The temperature sensitivity of a pressure sensor is an important performance metric. The definition of temperature coefficient of sensitivity (TCS) is:

$$\text{TCS} = \frac{1}{S} \frac{\partial S}{\partial T} \tag{3.16}$$

where S is sensitivity. Another important parameter is the temperature coefficient of offset (TCO). The offset of a pressure sensor is the value of the output signal at a reference pressure, such as when $\Delta P = 0$. Consequently, the TCO is:

$$\text{TCO} = \frac{1}{\theta_0} \frac{\partial \theta_0}{\partial T} \tag{3.17}$$

where θ_0 is offset, and T is temperature. Thermal stresses caused by differences in expansion coefficients between the diaphragm and the substrate or packaging materials are some of the many possible contributors to these temperature coefficients.

3.3 Piezoresistive Pressure Sensors

The majority of commercially available micromachined pressure sensors are bulk micromachined piezoresistive devices. These devices are etched from single crystal silicon wafers, which have relatively well-controlled mechanical properties. The diaphragm can be formed by etching the back of a <100> oriented Si wafer with an anisotropic wet etchant such as potassium hydroxide (KOH). An electrochemical etch-stop, dopant-selective etch-stop, or a layer of buried oxide can be used to terminate the etch and control the thickness of the unetched diaphragm. This diaphragm is supported at its perimeter by a portion of the wafer that was not exposed to the etchant and remains at full thickness (Figure 3.1). The piezoresistors are fashioned by selectively doping portions of the diaphragm to form junction-isolated resistors. Although this form of isolation permits significant leakage current at elevated temperatures and the resistors present sheet resistance per unit length that depends on the local bias across the isolation diode, it allows the designer to exploit the substantial piezoresistive coefficient of silicon and locate the resistors at the points of maximum stress on the diaphragm.

Surface micromachined piezoresistive pressure sensors have also been reported. Sugiyama et al. (1991) used silicon nitride as the structural material for the diaphragm. Polycrystalline silicon (polysilicon) was used both as a sacrificial material and to form the piezoresistors. This approach permits the fabrication of small devices with high packing density. However, the maximum deflection of the diaphragm is limited to the thickness of the sacrificial layer, and can constrain the dynamic range. In Guckel et al. (1986), (reprinted in *Microsensors* (1990)) polysilicon was used to form both the diaphragm and the piezoresistors.

3.3.1 Design Equations

In an anisotropic material such as single crystal silicon, resistivity is defined by a tensor that relates the three directional components of the electric field to the three directional components of current flow. In general, the tensor has nine elements expressed in a 3 $\times$ 3 matrix, but they reduce to six independent values from symmetry considerations:

$$\begin{bmatrix} \varepsilon_1 \\ \varepsilon_2 \\ \varepsilon_3 \end{bmatrix} = \begin{bmatrix} \rho_1 & \rho_6 & \rho_5 \\ \rho_6 & \rho_2 & \rho_4 \\ \rho_5 & \rho_4 & \rho_3 \end{bmatrix} \begin{bmatrix} j_1 \\ j_2 \\ j_3 \end{bmatrix} \tag{3.18}$$

where ε_i and j_i represent electric field and current density components, and ρ_i represent resistivity components. Following the treatment in Kloeck and de Rooij (1994) and Middleoek and Audet (1994), if the Cartesian axes are aligned to the <100> axes in a cubic crystal structure such as silicon, $\rho_1, \rho_2,$ and ρ_3 will be equal because they all represent resistance along the <100> axes, and are denoted by ρ. The remaining components of the resistivity matrix, which represent cross-axis resistivities, will be zero because unstressed silicon is electrically isotropic. When stress is applied to silicon, the components in the resistivity matrix change. The change in each of the six independent components, $\Delta\rho_i$, will be related to all the stress components. The stress can always be decomposed into three normal components (σ_i), and three shear components (τ_i), as shown in Figure 3.3. The change in the six components of the resistivity matrix (expressed as a fraction of the unstressed resistivity ρ) can then be related to the six stress components by a 36-element tensor. However, due to symmetry conditions, this tensor is populated by only three nonzero components, as shown:

$$\begin{bmatrix} \rho_1 \\ \rho_2 \\ \rho_3 \\ \rho_4 \\ \rho_5 \\ \rho_6 \end{bmatrix} = \begin{bmatrix} \rho \\ \rho \\ \rho \\ 0 \\ 0 \\ 0 \end{bmatrix} + \begin{bmatrix} \Delta\rho_1 \\ \Delta\rho_2 \\ \Delta\rho_3 \\ \Delta\rho_4 \\ \Delta\rho_5 \\ \Delta\rho_6 \end{bmatrix}; \frac{1}{\rho}\begin{bmatrix} \Delta\rho_1 \\ \Delta\rho_2 \\ \Delta\rho_3 \\ \Delta\rho_4 \\ \Delta\rho_5 \\ \Delta\rho_6 \end{bmatrix} = \begin{bmatrix} \pi_{11} & \pi_{12} & \pi_{12} & 0 & 0 & 0 \\ \pi_{12} & \pi_{11} & \pi_{12} & 0 & 0 & 0 \\ \pi_{12} & \pi_{12} & \pi_{11} & 0 & 0 & 0 \\ 0 & 0 & 0 & \pi_{44} & 0 & 0 \\ 0 & 0 & 0 & 0 & \pi_{44} & 0 \\ 0 & 0 & 0 & 0 & 0 & \pi_{44} \end{bmatrix} \begin{bmatrix} \sigma_1 \\ \sigma_2 \\ \sigma_3 \\ \tau_1 \\ \tau_2 \\ \tau_3 \end{bmatrix} \tag{3.19}$$

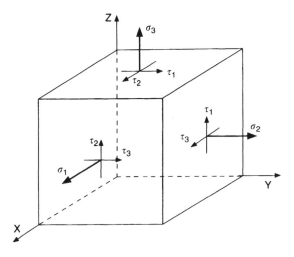

FIGURE 3.3   Definition of normal and shear stresses.

where the π_{ij} coefficients, which have units of Pa^{-1}, may be either positive or negative, and are sensitive to doping type, doping level, and operating temperature. It is evident that π_{11} relates the resistivity in any direction to stress in the same direction, whereas π_{12} and π_{44} are cross-terms.

Equation 3.19 was derived in the context of a coordinate system aligned to the <100> axes, and is not always convenient to apply. A preferred representation is to express the fractional change in an arbitrarily oriented diffused resistor by:

$$\frac{\Delta R}{R} = \pi_l \sigma_l + \pi_t \sigma_t \tag{3.20}$$

where π_l and σ_l are the piezoresistive coefficient and stress parallel to the direction of current flow in the resistor (i.e., parallel to its length), and π_t and σ_t are the values in the transverse direction. The piezoresistive coefficients referenced to the direction of the resistor may be obtained from those referenced to the <100> axes in Equation 3.19 by using a transformation of the coordinate system. It can then be stated that:

$$\pi_l = \pi_{11} + 2(\pi_{44} + \pi_{12} - \pi_{11})(l_1^2 m_1^2 + l_1^2 n_1^2 + n_1^2 m_1^2) \tag{3.21}$$

$$\pi_t = \pi_{12} - (\pi_{44} + \pi_{12} - \pi_{11})(l_1^2 l_2^2 + m_1^2 m_2^2 + n_1^2 n_2^2) \tag{3.22}$$

where l_1, m_1, and n_1 are the direction cosines (with respect to the crystallographic axes) of a unit length vector, which is parallel to the current flow in the resistor whereas l_2, m_2, and n_2 are those for a unit length vector perpendicular to the resistor. Thus, $l_i^2 + m_i^2 + n_i^2 = 1$. As an example, for the <111> direction, in which projections to all three crystallographic axes are equal, $l_i^2 = m_i^2 = n_i^2 = 1/3$.

A sample set of piezoresistive coefficients for Si is listed in Table 3.1. It is evident that π_{44} dominates for p-type Si, with a value that is more than 20 times larger than the other coefficients. By using the dominant coefficient and neglecting the smaller ones, Equations 3.21 and 3.22 can be further simplified. It should be noted, however, that the piezoresistive coefficient can vary significantly with doping level and operating temperature of the resistor. A convenient way in which to represent the changes is to normalize the piezoresistive coefficient to a value obtained at room temperature for weakly doped silicon [Kanda, 1982]:

$$\pi(N,T) = P(N,T)\, \pi_{ref.} \tag{3.23}$$

Figure 3.4 shows the variation of parameter P for p-type and n-type Si, as N, the doping concentration, and T, the temperature, are varied.

Figure 3.5 plots the longitudinal and transverse piezoresistive coefficients for resistor orientations on the surface of a (100) silicon wafer. Note that each figure is split into two halves, showing π_l and π_t simultaneously for p-type Si in one case and n-type Si in the other case. Each curve would be reflected in the horizontal axis if drawn individually. Also note that for p-type Si, both π_l and π_t peak along <110>, whereas for n-type Si, they peak along <100>. Since anisotropic wet etchants make trenches aligned to <110> on these wafer surfaces, p-type piezoresistors, which can be conveniently aligned parallel or perpendicular to the etched pits, are favored.

Consider two p-type resistors aligned to the <110> axes and near the perimeter of a circular diaphragm on a silicon wafer: assume that one resistor is parallel to the radius of the diaphragm, whereas the other is perpendicular to it. Using the equations presented previously, it can be shown that as pressure is applied, the fractional change in these resistors is equal and opposite:

$$\left(\frac{\Delta R}{R}\right)_{ra} = -\left(\frac{\Delta R}{R}\right)_{ta} = -\Delta P \cdot \frac{3\pi_{44} a^2 (1 - \nu)}{8h^2} \tag{3.24}$$

where the subscripts denote radial and tangentially oriented resistors. The complementary change in these resistors is well-suited to a bridge-type arrangement for readout, as shown in Figure 3.6.

TABLE 3.1 A Sample of Room Temperature Piezoresistive Coefficients in Si in 10^{-11} Pa^{-1}. (Reprinted with Permission from Smith, C.S. [1954] "Piezoresistance Effect in Germanium and Silicon," *Physical Review* **94**, pp. 42–49

Resistivity	π_{11}	π_{12}	π_{44}
7.8 Ω cm, p-type	6.6	−1.1	138.1
11.7 Ω cm, n-type	−102.2	53.4	−13.6

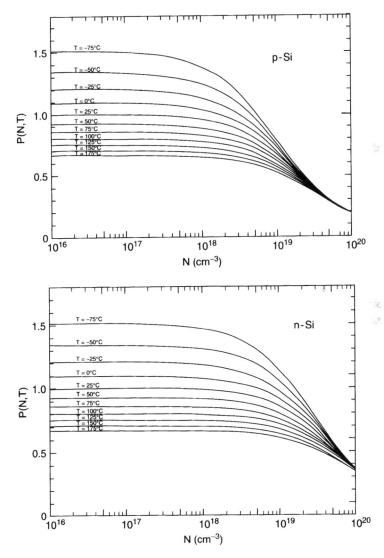

FIGURE 3.4 Variation of piezoresistive coefficient for n-type and p-type Si. (Reprinted with permission from Kanda, Y. [1982] "A Graphical Representation of the Piezoresistive Coefficients in Si," *IEEE Transactions on Electron Devices* **29**, pp. 64–70.)

(The bridge-type readout arrangement is suitable for square diaphragms as well.) The output voltage in this case is given by:

$$\Delta V_0 = V_s \frac{\Delta R}{R} \qquad (3.25)$$

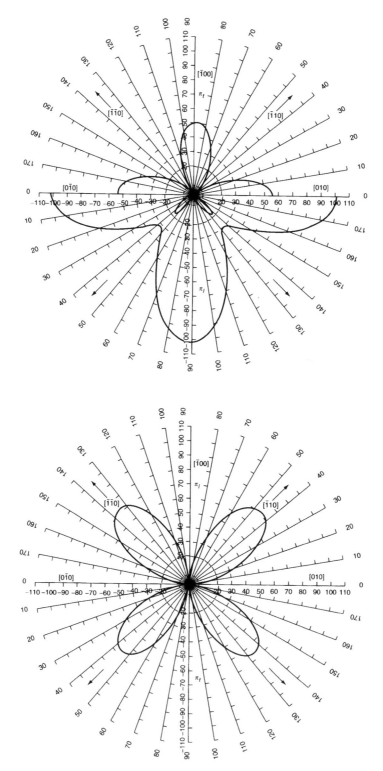

FIGURE 3.5 Longitudinal and transverse piezoresistive coefficients for n-type (upper) and p-type (lower) resistors on the surface of a (100) Si wafer. (Reprinted with permission from Kanda, Y. [1982] "A Graphical Representation of the Piezoresistive Coefficients in Si," *IEEE Transactions on Electron Devices* **29**, pp. 64–70.)

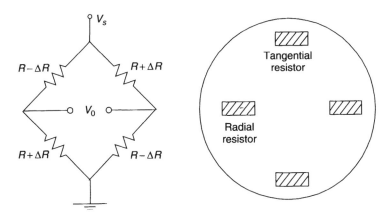

FIGURE 3.6   Schematic representation of Wheatstone bridge configuration (left) and placement of radial and tangential strain sensors on a circular diaphragm (right).

Since the output is proportional to the supply voltage V_s, the output voltage of piezoresistive pressure sensors is generally presented as a fraction of the supply voltage per unit change in pressure. It is proportional to a^2/h^2, and is typically on the order of 100 ppm/Torr. The maximum fractional change in a piezoresistor is in the order of 1% to 2%. It is evident from Equations 3.24 and 3.25 that the temperature coefficient of sensitivity, which is the fractional change in the sensitivity per unit change in temperature, is primarily governed by the temperature coefficient of π_{44}. A typical value for this is in the range of 1000–5000 ppm/K. While this is large, its dependence on π makes it relatively repeatable and predictable, and permits it to be compensated.

A valuable feature of the resistor bridge is the relatively low impedance that it presents. This permits the remainder of the sensing circuit to be located at some distance from the diaphragm, without deleterious effects from parasitic capacitance that may be incorporated. This stands in contrast to the output from capacitive pickoff pressure sensors, for which the high output impedance creates significant challenges.

3.3.2   Scaling

The resistors present a scaling limitation for the pressure sensors. As the length of a resistor is decreased, the resistance decreases and the power consumption rises, which is not favorable. As the width is decreased, minute variations that may occur because of non-ideal lithography or other processing limitations will have a more significant impact on the resistance. These issues constrain how small a resistor can be made. As the size of the diaphragm is reduced, the resistors will span a larger area between its perimeter and the center. Since the maximum stresses occur at these locations, a resistor that extends between them will be subject to stress averaging, and the sensitivity of the readout will be compromised. In addition, if the nominal values of the resistors vary, the two legs of the bridge become unbalanced, and the circuit presents a non-zero signal even when the diaphragm is undeflected. This offset varies with temperature and cannot be easily compensated because it is in general unsystematic.

3.3.3   Noise

There are three general sources of noise that must be evaluated for piezoresistive pressure sensors, including mechanical vibration of the diaphragm, electrical noise from the piezoresistors, and electrical noise from the interface circuit. Thermal energy in the form of Brownian motion of the gas molecules surrounding the diaphragm causes variations in its deflection that can be treated as though caused by an equivalent pressure. The treatment in Chau and Wise (1987) and Chau and Wise (1987a) provides a solution for rarefied gas environments in which the mean free path between collisions of the gas molecules is much larger than the dimensions of the diaphragm. For a circular diaphragm with built-in edges, the noise equivalent mean square pressure at low frequencies (i.e. in a bandwidth limited by the pressure sensor), is given by:

$$\overline{p_n^2} = \alpha \sqrt{\frac{32kT}{\pi}} \frac{\left(\sqrt{m_1}P_1 + \sqrt{m_2}P_2\right)B}{a^2} \tag{3.26}$$

where B is the system bandwidth, T is absolute temperature, k is Boltzmann's constant, m_1 and m_2 are the masses of gas molecules, P_1 and P_2, are the respective pressures on the two sides of the diaphragm, and α is 1.7. It is worth noting that this treatment of intrinsic diaphragm noise is for relatively low pressures, at which the levels gas noise is relatively small. It ignores the thermal vibrations within the structural material because these are typically even smaller. However, this is not necessarily true in very high vacuum environments, which may permit noise from the thermal vibrations of the structural materials to dominate.

At higher pressures, for which viscous damping dominates, the mean square noise pressure on the surface of the diaphragm is given by:

$$\overline{p_n^2} = 4kTR_a \tag{3.27}$$

where R_a is the mechanical resistance (damping) coefficient per unit active area, i.e., R/A, where R is the total damping coefficient and A is the area of the diaphragm [Gabrielson, 1993]. The term R may also be stated as $m\omega_0/Q$, in which m is the effective mass per unit area of the diaphragm, ω_0 is the resonant frequency, and Q is the mechanical quality factor of the diaphragm resonance.

Squeeze film damping between the diaphragm and any opposing surface can be a very significant contributor to sensor noise. Since this is more important in capacitive pressure sensors because of the relatively small gap between the diaphragm and the sensing counter-electrode, it is discussed in the next section.

In addition, thermally generated acoustic waves also deserve attention at higher pressures. The approximate two-sided power spectrum is given by Chau and Wise (1987):

$$S_a(f) = \frac{2\pi kT\rho f^2}{c} \tag{3.28}$$

where S_a is the input noise pressure, ρ is the density, f is the frequency, and c is the speed of sound in the fluid surrounding the pressure sensor. This is evidently significant at higher frequencies, and is typically significant above 10 KHz.

The electrical noise from the piezoresistor, identified above as the second of three components, is also Brownian in origin. The equivalent noise pressure from this source can be presented as:

$$\overline{p_n^2} = \frac{4kTRB}{(V_s S_p)^2} \tag{3.29}$$

where V_s is the voltage of the power supply used for the resistor bridge, and S_p is the sensitivity of the piezoresistive pressure sensor [Chau and Wise, 1987]. It was noted following Equation 3.25 that S_p is proportional to (a^2/h^2), which makes this component of mean square noise pressure proportional to (h^4/a^4).

In many cases, the dominant noise source is not the sensor itself but the interface circuit. If ΔV_{min} represents the minimum voltage difference that can be resolved by the interface circuit, then the pressure resolution limit becomes:

$$\Gamma_{ckt} = \frac{\Delta V_{min}}{(V_s S_p)} \tag{3.30}$$

which is proportional to (h^2/a^2) [Chau and Wise, 1987].

In comparing the noise sources identified for piezoresistive pressure sensors, the following examples are useful [Chau and Wise, 1987]. For a small device with diaphragm length of 100 μm and thickness 1 μm, the Brownian noise in a 100 Hz bandwidth and 760 Torr on both sides of the diaphragm is $<10^{-5}$ Torr.

The piezoresistor noise, assuming a $2\,k\Omega$ bridge resistance and a 5 V supply, is about 0.11 mTorr, and the circuit noise is 1.5 mTorr assuming that the minimum resolvable voltage is $0.5\,\mu V$. In contrast, a 1 mm long diaphragm of $1\,\mu m$ thickness has $4 \times 10^{-6}\,mTorr$ Brownian noise, $1.1 \times 10^{-3}\,mTorr$ piezoresistor noise, and $9.5 \times 10^{-3}\,mTorr$ circuit noise. In all of these calculations, the component identified in Equation 3.27 is ignored.

3.3.4 Interface Circuits for Piezoresistive Pressure Sensors

An instrumentation amplifier is a natural choice for interfacing with a piezoresistive full-bridge. In such an arrangement, the differential output labeled V_0 in Figure 3.6 would be connected to the two input terminals labeled v_1 and v_2 in Figure 3.7. The instrumentation amplifier has two stages [Sedra and Smith, 1998]. In the first stage, which is formed by operational amplifiers A_1 and A_2, the virtual short circuits between the two inputs of each of these amplifiers force the current flow in the resistor R_1 to be $(v_1 - v_2)/R_1$. Since the input impedance of the op amps is large, the same current also flows through the two resistors labeled R_2, causing the voltage difference between the outputs of A_1 and A_2 to be:

$$v_{o1} - v_{o2} = \left(1 + \frac{2R_2}{R_1}\right)(v_1 - v_2) \tag{3.31}$$

Thus, the gain provided by this first stage can be changed by varying the value of R_1. The second stage is simply a difference amplifier formed by op amp A_3 and the surrounding resistors. This causes the output voltage to be:

$$v_0 = \left(-\frac{R_4}{R_3}\right)(v_{o1} - v_{o2}) \tag{3.32}$$

A separate gain stage can be added to the circuit past the instrumentation amplifier. This may also be used to subtract the temperature dependence of the signal that can be generated as reference voltage from the top of a current-driven bridge.

Spencer et al. (1985) described another type of circuit that can be used to read out a resistor bridge, which is a voltage controlled duty-cycle oscillator (VCDCO). As shown in Figure 3.8, this circuit incorporates a cross-coupled multi-vibrator formed essentially by Q_1, Q_2, and the surrounding passive elements. The input voltage, v_{IN}, which is provided by the output of the resistor bridge, determines the ratio of the currents in the emitter-coupled stage formed by Q_3 and Q_4:

$$\frac{I_{C3}}{I_{C4}} = \exp\left(\frac{qv_{IN}}{kT}\right) \tag{3.33}$$

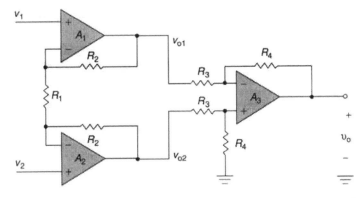

FIGURE 3.7 Schematic of an instrumentation amplifier used to read out a resistor bridge circuit (Reprinted with permission from Sedra, A.S., and Smith, K.C. [1998] *Microelectronic Circuits*, 4th ed., Oxford University Press, Oxford.)

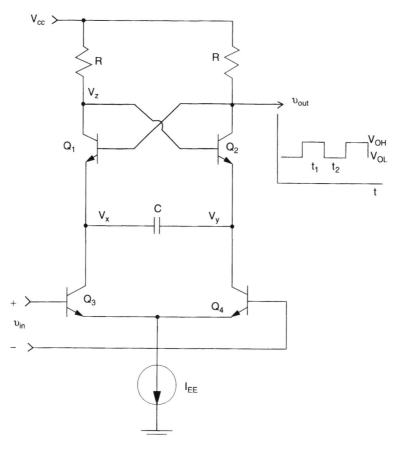

FIGURE 3.8   Schematic of a duty cycle oscillator used to read out resistance-based sensors (Reprinted with permission from Spencer, R.R., Fleischer, B.M., Barth, P.W., and Angell, J.B. (1985) "The Voltage Controlled Duty-Cycle Oscillator: Basis for a New A-to-D Conversion Technique," *Rec. of the Third International Conference On Solid-State Sensors and Actuators*, pp. 49–52.)

The output of the circuit, as illustrated in Figure 3.8, is a square wave, which shows a dependence on the input voltage in both the duty cycle and period. To understand the operation of the circuit and obtain a simple quantitative model for its behavior, one may begin with the assumption that the output has just switched low, causing Q_1 to turn off. In terms of the timing diagram, this is the beginning of t_2. The entire tail current I_{EE}, then, is provided by Q_2, and neglects the base current of Q_2, $v_z \approx V_{CC}$, which causes v_y to eventually settle at approximately $V_{CC} - 0.7\,\mathrm{V}$; i.e., below v_z by the drop across the base-to-emitter diode of Q_2. This condition also sets v_{OUT} at $V_{CC} - R.I_{EE}$. With Q_1 turned off, Q_3 continues to extract I_{C3} from C until v_x drops below v_{OUT} by 0.7 V, which causes Q_1 to turn on, drop v_z, and thereby turn off Q_2. This in turn causes v_{OUT} to rise, ending the time period t_2. Thus, at the end of t_2, the voltage across C, $v_x - v_y \approx (V_{CC} - R.I_{EE} - 0.7) - (V_{CC} - 0.7\,\mathrm{V}) = -R.I_{EE}$. Similarly, at the end of t_1, this is $+R.I_{EE}$. The change in voltage across C, in either t_1 or t_2, is therefore $2R.I_{EE}$. The resulting change in charge must be balanced by the current flow in the given time interval, thus:

$$t_1 \approx \frac{2RCI_{EE}}{I_{C4}}; \, t_2 \approx \frac{2RCI_{EE}}{I_{C3}} \qquad (3.34)$$

Since $I_{C3} + I_{C4} = I_{EE}$, using equations (3.33) and (3.34), it can be shown that:

$$t_1 \approx 2RC\left\{ 1 + \exp\left(\frac{qv_{IN}}{kT} \right) \right\}, \text{ and} \qquad (3.35)$$

$$v_{IN} = \left(\frac{kT}{q}\right)\ln\left(\frac{t_1}{t_2}\right) \tag{3.36}$$

This equation can be used to determine the input voltage from a measurement of t_1 and t_2. An important result is that any ratio that involves t_1 and t_2 is independent of all circuit variables; drifts in their values over time will not affect the accuracy and resolution of the readout. An exception to this immunity is the input offset of Q_3 and Q_4, which can affect the signal. It is also worth noting that the output is pseudo-digital, in that it is discretized in voltage, although not in time. However, the time durations can be easily clocked, making the circuit conducive to analog-to-digital conversion of the output signal.

3.3.5 Calibration and Compensation

Regardless of the choice of interface circuit, it is generally necessary to calibrate the overall output. For the circuit in Figure 3.8, the input voltage can be expressed as the sum of two components: one which carries the pressure information as represented in Equation 3.25, and a non-ideal offset term which was previously ignored:

$$v_{IN} = \left(\frac{\Delta R}{R}\right)V_s + v_{OFF} = \Delta P \cdot SV_s + v_{OFF} \tag{3.37}$$

where S is the sensitivity of the pressure sensor, V_s is the supply to the piezoresistor bridge, and v_{OFF} is the offset voltage presented not only by the resistor bridge, but also incorporates the offset due to mismatch in Q_3 and Q_4 in Figure 3.8. The sensitivity and offset voltage are both functions of temperature, with linearized dependences described in Equations 3.16 and 3.17 as TCS and TCO, respectively. Using (3.26) and (3.37), it can be stated that the pressure across the diaphragm is given by:

$$\Delta P = \left(\frac{kT}{qSV_s}\right)\ln\left(\frac{t_1}{t_2}\right) - \left(\frac{v_{OFF}}{SV_s}\right) \tag{3.38}$$

Use of this equation requires the determination of S, v_{OFF}, and their temperature coefficients. The minimum calibration, therefore, requires the application of two known pressures at two known temperatures. Using this method, 10-bit resolution has been achieved [Spencer et al., 1985]. The conversion time was 3 ms, and the LSB was $26\,\mu V$ with an input noise of $690\,\mu V_{RMS}$ over a 100 KHz measurement bandwidth. The linear proportionality of v_{IN} to $ln(t_1/t_2)$, as anticipated in Equation 3.28, held within $\pm 5\,\mu V$ from $-20\,mV$ to $+20\,mV$.

Like the compensation of temperature dependencies, non-linearity in sensor response (at a single temperature) can also be compensated. Linearity is expressed as a percentage of the full-scale output. Without compensation, for piezoresistive pressure sensors this is typically around 0.5% (8 bits). As will be discussed later, it can be substantially poorer for capacitive sensors because of their inherently non-linear response. Digital compensation will be addressed within that context.

3.4 Capacitive Pressure Sensors

Capacitive pressure sensors were developed in the late 1970s and early 1980s. The flexible diaphragm in these devices serves as one electrode of a capacitor, whereas the other electrode is located on a substrate beneath it. As the diaphragm deflects in response to applied pressure, the average gap between the electrodes changes, leading to a change in the capacitance. The concept is illustrated in Figure 3.9.

Although many fabrication schemes can be conceived, Chau and Wise (1988) described an attractive bulk micromachining approach. Areas of a silicon wafer were first recessed, leaving plateaus to serve as anchors for the diaphragm. Boron diffusion was then used to define an etch stop in the regions that would eventually form the structure. The top surface of the silicon wafer was then anodically bonded to a glass wafer that had been inlaid with a thin film of metal, which served as the stationary electrode and provided lead transfer to circuitry. The undoped silicon was finally dissolved in a dopant-selective etchant such as

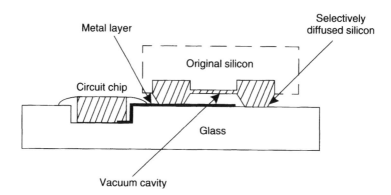

FIGURE 3.9 Schematic of a pressure sensor fabricated on a glass substrate using a dopant-selective etch stop. (Reprinted with permission from Chau, H., and Wise, K. [1988] "An Ultra-Miniature Solid-State Pressure Sensor for Cardiovascular Catheter," *IEEE Transactions on Electron Devices* 35(12), pp. 2355–2362.)

ethylene diamine pyrocatechol (EDP). In order to maintain a low profile and small size, the interface circuit was hybrid-packaged into a recess in the same glass substrate, creating a transducer that was small enough to be located within a cardiovascular catheter of 0.5 mm outer diameter. Using a 2 μm thick, 560 × 280 μm² diaphragm with a 2 μm capacitor gap, and a circuit chip 350 μm wide, 1.4 mm long, and 100 μm thick, pressure resolution <2 mmHg was achieved [Ji et al., 1992].

The capacitive sensing scheme circumvents some of the limitations of piezoresistive sensing. For example, since resistors do not have to be fabricated on the diaphragm, scaling down the device dimensions is easier because concerns about stress averaging and resistor tolerance are eliminated. In addition, the largest contributor to the temperature coefficient of sensitivity — the variation of π_{44} — is also eliminated. The full-scale output swing can be 100% or more, in comparison to about 2% for piezoresistive sensing. There is virtually no power consumption in the sense element as the DC current component is zero. However, capacitive sensing presents other limitations: (1) the capacitance changes non-linearly with diaphragm displacement and applied pressure; (2) even though the fractional change in the sense capacitance may be large, the absolute change is small and considerable caution must be exercised in designing the sense circuit; (3) the output impedance of the device is large, which affects the interface circuit design; and (4) the parasitic capacitance between the interface circuit and the device output can have a significant negative impact on the readout, which means that the circuit must be placed in close proximity to the device in a hybrid or monolithic implementation. An additional concern is related to lead transfer and packaging. In the case of absolute pressure sensors the cavity beneath the diaphragm must be sealed in vacuum. Transferring the signal at the counter electrode out of the cavity in a manner that retains the hermetic seal can present a substantial manufacturing challenge.

3.4.1 Design Equations

The capacitance between two parallel electrodes can be expressed as:

$$C = \frac{\varepsilon_0 \varepsilon_r A}{d} \tag{3.39}$$

where ε_0, ε_r, A, and d are permittivity of free space (8.854×10^{-14} F/cm), relative dielectric constant of material between the plates (which is unity for air), effective electrode area, and gap between the plates. Since the gap is not uniform when the diaphragm deflects, finite element analysis is commonly used to compute the response of a capacitive pressure sensor. However, it can be shown that for deflections that are small compared to the thickness of the diaphragm, the sensitivity, which is defined as the fractional

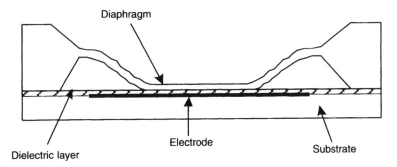

FIGURE 3.10   A touch-mode capacitive pressure sensor.

change in the capacitance per unit change in pressure, has the following proportionality [Chau and Wise, 1987]:

$$S_{cap} = \frac{\Delta C}{C \cdot \Delta P} = \frac{1}{PAd} \int_A w\, dA = 0.0746 \frac{1 - v^2}{E} \frac{a^4}{h^3 d} \tag{3.40}$$

where d is the nominal gap between the diaphragm and the electrode, and the remaining variables represent the same quantities as in section 3.2. Comparing to Equation 3.24, which provides the fractional resistance change in a piezoresistive device, it is evident that capacitive devices have an increased dependence on a/h, the ratio of the radius to the thickness of the diaphragm. In addition, they are dependent on d, the capacitive gap. The sensitivity of capacitive pressure sensors is generally in the range of 1000 ppm/Torr, which is about 10× greater than piezoresistive pressure sensors. This is an important advantage, but it carries a penalty for linearity and dynamic range.

The dynamic range of a capacitive pressure sensor is conventionally perceived to be limited by the "full-scale" deflection, which is defined to exist at the point that $\Delta C = C_0$, i.e., the point at which the device capacitance doubles (in the absence of parasitics). However, in reality the capacitance keeps rising as the pressure increases even beyond the threshold at which the diaphragm touches the substrate beneath it. As the pressure increases, the contact area grows, extending the useful operating range of the device (Figure 3.10). As long as the electrode is electrically isolated from the diaphragm, such as by an intervening thin film of dielectric material, the sensor can continue to function. This mode of operation is called "touch mode" [Cho et al., 1992; Ko et al., 1996]. The concept of touch-mode operation exploits the fact that in a capacitive pressure sensor, the substrate provides a natural over-pressure stop that can delay or prevent rupture of the diaphragm. This is a feature lacking in most piezoresistive pressure sensors. Since most of the diaphragm may, in fact, be in contact with the substrate in this mode, the output capacitance is relatively large. For example, a 4 μm thick p^{++} silicon diaphragm of 1500 × 447 μm^2 area and a nominal capacitive gap of 10.4 μm was experimentally shown to have a dynamic range of 120 psi in touch mode, whereas the conventional range was only 50 psi [Ko et al., 1996].

3.4.2   Residual Stress

The impact of residual stress upon diaphragm deflection was addressed in this chapter, and it was noted that stress has greater impact on thin diaphragms. To calculate the sensitivity of a capacitive pressure sensor with a diaphragm that is under stress, the capacitance at any pressure must be calculated by using Equation 3.6 for the deflection w in Equation 3.40. This leads to the following expressions for normalized sensitivity (i.e., sensitivity in the presence of stress divided by sensitivity in the absence of stress) [Chau and Wise, 1987]:

$$S'_{cap} = 192 \left[\frac{1}{k^4} + \frac{1}{8k^2} - \frac{1}{2k^3} \frac{I_0(k)}{I_1(k)} \right] \qquad \text{in tension, and} \tag{3.41}$$

$$S'_{cap} = 192\left[\frac{1}{k^4} - \frac{1}{8k^2} - \frac{1}{2k^3} \frac{J_0(k)}{J_1(k)} \right] \quad \text{in compression,} \qquad (3.42)$$

where the variables are defined as for Equation 3.6. These equations are plotted in Figure 3.2, and as expected, the lines are very close to those of the normalized deflection at the center of the diaphragm, as expressed by Equations 3.9 and 3.10. For tensile diaphragms, when the dimensionless intrinsic stress expressed by Equation 3.11 is greater than about 10, the sensitivity (not normalized) can be approximated as:

$$S_{cap} = \frac{a^2}{8\sigma_i hd} \quad \text{for } \sigma'_i \geq 10(\text{tensile}) \qquad (3.43)$$

In a large, thin diaphragm, the sensitivity can easily be degraded by one or two orders of magnitude by a few tens of MPa of residual tension.

3.4.3 Expansion Mismatch

One cause of temperature coefficients in capacitive pressure sensors is expansion mismatch between the substrate and diaphragm, which can be minimized by the proper choice of materials and fabrication sequence. For example, when a glass substrate is used, it is important to select one that is expansion-matched to the silicon. In general, the expansion coefficient of materials changes with temperature, i.e., the expansion is not linearly related to temperature, as shown in Figure 3.11 [Greenwood, 1988]. Hoya™ SD-2 is designed to be very closely matched to silicon, but the more readily available #7740 Pyrex™ glass provides acceptable performance for many cases. Naturally, the best results can be expected when the substrate material is the same as the diaphragm material. Anodic bonding of the silicon structure to a silicon substrate wafer that was coated with 2.5–5 µm of glass has been shown to produce devices with low TCO

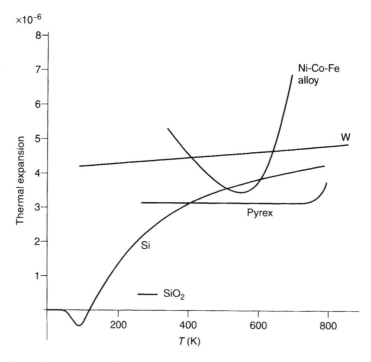

FIGURE 3.11   Thermal expansion coefficients of various materials as a function of temperature. (Reprinted with permission from Greenwood, J.C. [1988] "Silicon in Mechanical Sensors," *J. Phys. E, Sci. Instrum.*, **21**, pp. 1114–1128.)

[Hanneborg and Ohlckers, 1990]. The typical TCO of silicon pressure sensors that use anodically bonded glass substrates is <100 ppm/K. Since the sensitivity is in the range of 1000 ppm/Torr, the TCO can be stated as 0.1 Torr/K.

The expansion mismatch between the structural materials used for the diaphragm and substrate has an impact not only on the operation of the device, but can also affect the manufacturing yield because of the shear stresses produced in during the bonding process. If it is assumed for simplicity that CTE of Si is independent of temperature, the shear stress generated on a bonding rim around the perimeter of a circular area is:

$$\tau_{gs} = \frac{t_s^2}{W} \frac{(\alpha_g - \alpha_s)\Delta T}{\left(\dfrac{1 - v_s}{E_s} t_s + \dfrac{1 - v_g}{E_g} t_g \right)} \tag{3.44}$$

where ΔT is the difference between the bonding temperature and operating temperatures, α, E, v, t_s, t_g and W denote the expansion coefficient, Young's modulus, Poisson's ratio, thickness of the Si wafer, thickness of the glass wafer, and width of bonding ring between silicon and glass. If the shear force is sustained by a small area, this can lead to failure manifested as cracks in the glass as the wafers are cooled after anodic bonding [Park and Gianchandani, 2003]. However, it is evident from this equation that reducing the Si wafer thickness relative to glass can alleviate this problem. It is convenient to chemically thin the Si wafer prior to bonding to release stress. For example, if it is assumed that $\alpha_s = 2.3$ ppm/K, $\alpha_g = 1.3$ ppm/K, $\Delta T = 300$°C, $E_s = 160$ GPa, $E_g = 73$ GPa, $t_g = 600$ μm, $v = 0.23$ for both Si and glass and $w = 20$ μm, the shear stress reduces from >160 MPa for a 300 μm thick Si wafer to <30 MPa for a 100 μm thick Si wafer.

3.4.4 Trapped Gas Effects in Absolute Pressure Sensors

For absolute pressure sensors, the primary source of temperature sensitivity can be expansion of gas entrapped in the sealed reference cavity. This gas may come from an imperfect vacuum ambient during the sealing. The pressure due to this gas changes as the diaphragm is deflected and the cavity volume changed. This can lead to erroneous readings and a loss of sensitivity [Ji et al., 1992]. If the gas is un-reactive, this component can be estimated by the ideal gas equation [Park and Gianchandani, 2003]. After anodic bonding and dissolution, as the device is returned to room temperature, the contraction of the gas causes a diaphragm deflection. At equilibrium, the pressure in the cavity is:

$$P_1 = \frac{mRT_1}{V_0 - \Delta V} \approx \frac{mRT_1}{V_0 - \dfrac{a^6(P_a - P_1)\pi}{192D}} \tag{3.45}$$

where P_1 is pressure, P_a is the atmospheric or applied external pressure, V is volume of the vacuum cavity, m is entrapped air mass, R is universal gas constant for air (0.287 kN m/kg·K), and D is flexural rigidity — simply $Eh^3/[12(1 - v^2)]$ for most cases, T is absolute temperature, the subscript 0 indicates thermodynamic state at bonding, and subscript 1 indicates the thermodynamic state in the cavity at room temperature after the dissolution process. The volume change in the cavity after cooling down is ΔV:

$$\Delta V = 2\pi \int_0^a rw(r)dr \approx 2\pi \int_0^a \frac{P_a - P_1}{64D} r(a^2 - r^2)^2 \, dr = \frac{a^6(P_a - P_1)\pi}{192D} \tag{3.46}$$

where $w(r)$ is deflection of the circular diaphragm. To obtain this, it is assumed that the deflection is small, the boundary condition is clamped, and axisymmetric bending theory can be applied. Comparison with FEA shows a discrepancy of about 7%. The net pressure across the diaphragm can be estimated from the preceding equations.

Another source of entrapped gas is out-diffusion from the cavity walls after sealing, which is more difficult to quantify [Henmi et al., 1994]. To some extent, this can be blocked by using a thin film metal coating as a diffusion barrier [Cheng et al., 2001].

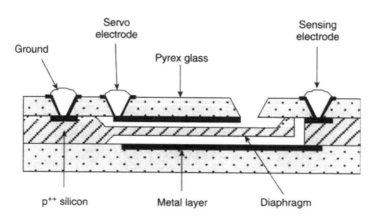

FIGURE 3.12   Lead transfer in an electrostatically servo-controlled pressure sensor. (Reprinted with permission from Wang, Y., and Esashi, M. [1997] "A Novel Electrostatic Servo Capacitive Vacuum Sensor," *Proc., IEEE International Conference on Solid-State Sensors and Actuators [Transducers '97]*, 2, pp. 1457–1460.)

Various sealing methods have been developed based on the principle of creating a cavity by etching a sub-surface sacrificial layer through a narrow access hole and then sealing off the access by depositing a thin film at a low pressure [Murakami, 1989]. The reactive sealing method reported in Guckel et al. (1987) and Guckel et al. (1990) provides some of the lowest sealed cavity pressures. In this approach, chemical vapor deposited polysilicon is used as the sealant. A subsequent thermal anneal causes the trapped gas to react with the interior walls of the cavity, leaving a high vacuum. Although appropriate for surface micro-machined cavities, the thermal budget of the reactive sealing process that uses polysilicon is too high for devices that use glass substrates. A similar effect can be achieved with metal films called non-evaporable getters (NEG), which are suitable for use with glass wafers. Henmi et al. (1994) used a NEG that was a Ni/Cr ribbon covered with a mixture of porous Ti and Zr–V–Fe alloy. To achieve the best results it was initially heated to 300°C to desorb gas, next it was sealed within the reference cavity, and finally heated to 400°C to be activated. The final cavity pressure achieved was reported to be $<10^{-5}$ Torr.

3.4.5   Lead Transfer

For capacitive pressure sensors, lead transfer to the sense electrode located within the sealed cavity has been a persistent challenge. For the type of device shown in Figure 3.9, in which the anodic bond must form a hermetic seal over the metal lead patterned on the glass substrate, the metal layer thickness may not exceed 50 nm (unless, of course, inlaid into the substrate) [Lee and Wise, 1982]. An addition, a thin film dielectric must be used to separate the silicon anchor from the lead. Epoxy is sometimes applied at the end of the fabrication sequence to strengthen the seal over the lead. However, this is not a favored solution, and alternatives have been developed. In one approach, a hole is etched or drilled in the substrate or in a rigid portion of the microstructure next to the flexible diaphragm, the lead is transferred through it, and then it is sealed with epoxy or metals [Wang and Esashi, 1997; Giachino et al., 1981, Peters in [Gia 81]]. Figure 3.12 illustrates the approach used by Wang and Esashi (1997). In another approach, sub-surface polysilicon leads are used with the help of chemical–mechanical polishing to achieve a planar bonding surface that allows a hermetic seal (Figure 3.13). This can be done at the wafer scale but results in substantial parasitic capacitance, which may influence the choice of interface circuit [Chavan and Wise, 1997].

3.4.6   Design Variations

A number of variations to the conventional choices for structures and materials have been reported. These include ultra-thin dielectric diaphragms, bossed or corrugated diaphragms, double-walled diaphragms with embedded rigid electrodes, and structures with sense electrodes located external to the sealed cavity.

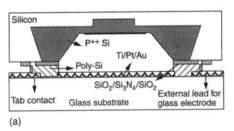
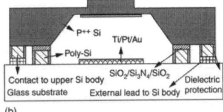

FIGURE 3.13 Sealed lead transfer for a capacitive pressure sensor using a sub-surface polysilicon layer. (Reprinted with permission from Chavan, A.V., and Wise, K.D. [1997] "A Batch-Processed Vacuum-Sealed Capacitive Pressure Sensor," *International Conference on Solid-State Sensors and Actuators [Transducers]*, pp. 1449–1451.)

For achieving very high sensitivities, Equation 3.40 suggests that the area of the diaphragm should be increased and the thickness should be decreased. However, a larger and thinner diaphragm develops high stress under applied pressure, which may cause device failure. Dielectric materials compatible with silicon processing have been explored as alternative structural materials for pressure sensors. In particular, stress-compensated silicon nitride has been the focus of some interest. It is both chemically and mechanically robust, with a yield strength twice that of silicon. High sensitivities were reached by using a large and thin silicon nitride diaphragm of 2 mm diameter and 0.3 μm thickness [Zhang and Wise, 1994]. Silicon nitride is under high tensile residual stress after LPCVD deposition, which reduces the diaphragm deflection. To compensate this residual stress, a silicon dioxide layer, which contributes compressive stress, was used between two silicon nitride layers (60 nm/180 nm/60 nm). To compensate the non-linear behavior due to large deflections, a 3 μm thick boss (p^{++} Si) was located in the center of the diaphragm. Its diameter was 60% of the diaphragm; this percentage is generally a good compromise: a greater percentage lowers sensitivity, a smaller percentage does not provide the fullest benefit toward linearity. The sensitivity was 10,000 ppm/Pa (5 fF/mTorr), which is 10× greater than the typical value for capacitive devices. The minimum resolution pressure was 0.1 mTorr. The TCO and TCS were, respectively, 910 ppm/K and −2900 ppm/K. The dynamic range was >130 Pa.

Introducing corrugation into a pressure sensor diaphragm allows a longer linear travel, and can contribute toward larger dynamic range. Corrugations can be created by wet or dry etching. Ding (1992), and Zhang and Wise (1994a) present bossed and corrugated diaphragms under tensile, neutral, and compressive stress. Both papers report a square root dependence of deflection versus pressure for unbossed corrugated pressure sensors. Residual stress in the diaphragm can result in bending when corrugations are present even in the absence of an applied pressure. The use of a boss in the center has been shown to considerably improve this, but can cause a deflection in the opposite direction.

Wang et al. (2000) reported a differential capacitive pressure sensor consisting of a double-layer diaphragm with an embedded electrode. As shown in Figure 3.14, one capacitor is formed between the upper diaphragm and middle electrode, and another between the electrode and lower diaphragm, serving as a capacitive half-bridge. The diaphragms are mechanically connected to each other, and move in the same direction, which is dependent on the pressure difference between the upper and lower surfaces of the entire structure. The advantage of this pressure sensor is that the sense cavity is sealed, but still can generate a signal based on differential pressure. In the reported effort, a diaphragm size of 150 × 150 μm², thickness of 2 μm, and capacitor gaps of nominally 1 μm were used. With one side of the pressure sensor at atmosphere, the nominal capacitance change is 86 fF for a differential pressure change from −80 kPa to 80 kPa.

Most masking steps in the fabrication of pressure sensors are consumed on electrical lead transfer from the inside of the vacuum cavity to the outside as shown in the previous devices. A pressure sensor that eliminates the problem of the sealed lead transfer by locating the pick-off capacitance outside the sealed cavity is illustrated in Figure 3.15 [Park and Gianchandani, 2000]. A skirt-shaped electrode extends outward from the periphery of the vacuum-sealed cavity, and serves as the element which deflects under pressure. The stationary electrode is metal patterned on the substrate below this skirt. As the external pressure

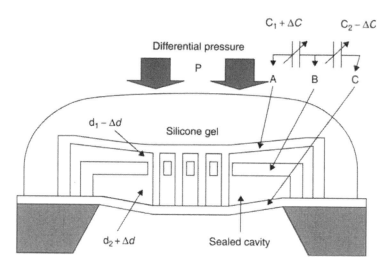

FIGURE 3.14 Differential capacitive pressure sensor with double-layer diaphragm and embedded rigid electrode. (Reprinted with permission from Wang, C.C., Gogoi, B.P., Monk, D.J., and Mastrangelo, C.H. [2000] "Contamination Insensitive Differential Capacitive Pressure Sensors," *Proc., IEEE International Conference on Microelectromechanical Systems*, pp. 551–555.)

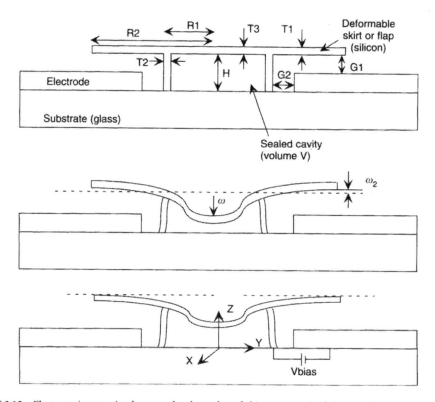

FIGURE 3.15 Electrostatic attraction between the electrode and skirt opposes the deflection due to external pressure.

increases, the center of the diaphragm deflects downwards, and the periphery of the skirt rises, reducing the pick-off capacitance. This deflection continues monotonically as the external pressure increases beyond the value at which the center diaphragm touches the substrate, so this device can be operated in touch mode for expanded dynamic range. To fabricate this, a silicon wafer was first dry etched to the

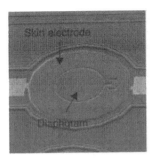 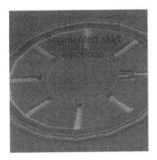 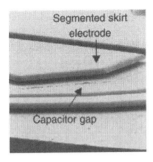

FIGURE 3.16   SEM of fabricated devices from Park and Gianchandani (2003). (a) Device with continuous skirt electrode. (b) Device with segmented skirt electrode. (c) Gap between skirt electrode and substrate. The metal thin film on the substrate and the dummy anchor are also visible.

desired height of the cavity, and then selectively diffused with boron to define the radius of the pressure sensor. The depth of the boron diffusion determined the eventual thickness of the structural layer. The silicon wafer was then flipped over and anodically bonded to a glass wafer that had been inlaid with a Ti/Pt metal that serves as interconnect and provides the bond pads. The undoped Si was finally dissolved in EDP, leaving the pressure sensor on the glass substrate. Fabricated devices are shown in Figure 3.16. Numerical modeling indicates that for a device with diaphragm radius of 500 μm and total radius of 1 mm, cylinder height of 30 μm, uniform wall thickness of 5 μm, and nominal sense capacitor gap of 5 μm, the nominal capacitance is 3.86 pF and the sensitivity is about −2900 ppm/kPa in non-touch mode, but drops to −270 ppm/kPa in touch mode.

Mastrangelo (1995) described a single crystal silicon pressure diaphragm using epitaxial deposition. The advantage of this kind of sensor is that it does not require bonding on glass as most p^{++} silicon pressure sensors do, and the diaphragm stress condition and material properties are predictable, as single crystal silicon provides more consistent performance than polysilicon.

3.4.7   Digital Compensation

One of the drawbacks of capacitive pressure sensors is their inherent non-linearity. Piezoresistive pressure sensors offer better linearity without compensation, but these too need improvement for many applications. Digital compensation involves the correction of the non-linearity in software of the digital interface (or computer) by using polynomials or look-up tables. It is attractive because it is relatively easy to implement and reconfigure. Crary et al. (1990) described case studies involving the use of polynomials for compensation of both capacitive and piezoresistive pressure sensors. In this effort, bi-variate compensation was used to correct for changes in temperature and non-linearity of the response to pressure. First, the output (response surface) of a pressure sensor was measured at a variety of fixed temperatures and pressures as determined by other sensors, which were used as references. By averaging over a number of measurements at each such point, it was determined that the deviation of the sensor response was 13 bits, which could be regarded as the noise level. The compensation formula was then determined by using stepwise regression to fit the response to a polynomial of the form $P = \Sigma_{m,n} R^m T^n$, where P is the actual pressure, R is the sensor response, T is the temperature, and m and n are integers from 0–5. For a commercial piezoresistive sensor, it was found that the standard error (taken as 1/4 of the 95% confidence interval) was 4.6 bits using just the linear terms in the polynomial; but it improved to 9.7 bits with the inclusion of higher-order terms which resulted in a polynomial of 7 terms. For a capacitive pressure sensor, a similar effort resulted in standard errors of 4.6 and 7.5 bits, respectively.

3.4.8   Noise

Some of the noise sources in capacitive pressure sensors are similar to those presented previously in this chapter for piezoresistive devices, but there are also some important differences. Equation 3.26 still holds for the noise from a circular diaphragm with built-in edges at low frequencies and low pressure, but within it α is now 1.2 instead of 1.7. The noise pressure due to viscous damping at higher pressures is obtained from Equation 3.27, as before. However, squeeze film damping in the small gap that generally exists between the deflecting diaphragm and counter electrode can be a major contributor to diaphragm noise [Gabrielson, 1993]:

$$\overline{p_n^2} = 4kTR_{sq} \approx 4kT(R_{film} || R_{perf}), \text{ where} \tag{3.47}$$

$$R_{film} = \frac{3\mu A^2}{2\pi d^2} \left[\frac{N\,s}{m} \right], \text{ and} \tag{3.48}$$

$$R_{perf} = \frac{12\,\mu A^2 \left[\dfrac{F}{2} - \dfrac{F^2}{8} - \dfrac{\ln F}{4} - \dfrac{3}{8} \right]}{N\pi d^3} \tag{3.49}$$

In these equations, d is the average gap between the diaphragm and electrode, and μ is the viscosity of the fluid between the plates (18×10^{-6} kg/m $\cdot$ s for air at 20°C). R_{sq} is the damping coefficient due to squeeze film damping. It can be approximated by the parallel combination (i.e., the product divided by the sum) of damping coefficients R_{film} and R_{perf}. The former, Equation 3.48, is for damping between two unperforated parallel plates; whereas the latter, Equation 3.49, is for damping between the diaphragm and a perforated plate with N perforations and the fraction of open area in the plate represented by F. The former equation accounts for fluid flows only around the perimeter of the plate, whereas the latter one accounts for flows only through the perforations.

As for piezoresistive pressure sensors, acoustic radiation can be described by Equation 3.49.

The electrical noise from the sensing piezoresistor is replaced by electrical noise from the sensing capacitance. This is given by Chau and Wise (1987):

$$\overline{p_n^2} = \frac{kT(C_0 + C_{stray})}{(V_s C_0 S_{cap})^2} \tag{3.50}$$

where C_0 is the sense capacitance and C_{stray} is the parasitic capacitance in parallel with it. This mean square pressure is proportional to $h^6 g^3/a^{10}$, i.e., it increases very rapidly as the sensor is scaled down in radius. It should be noted that Equation 3.50 provides the broadband noise, which is diminished by the bandwidth of the system in most cases.

In order to measure the readout capacitance in a pressure sensor, it is generally necessary to apply a voltage bias between the electrode and the diaphragm. This bias may also be used to electrostatically pre-deflect the diaphragm, and increase the nominal capacitance. However, electrical noise in the applied bias will result in the application of instantaneous electrostatic forces, which can induce noise vibrations in the diaphragm. Since this noise is dependent not only on the electrical source but also on the damped temporal response of the diaphragm, it is not easily expressed by a closed-form equation.

The ability of the interface circuit to detect small changes in capacitance can impose an important practical limitation of the usefulness of a pressure sensor. If ΔC_{min} is the minimum resolvable capacitance change for any given circuit, then the pressure resolution limit becomes:

$$\Gamma_{ckt} = \frac{\Delta C_{min}}{(C_0 S_{cap})} \tag{3.51}$$

which is proportional to $(h^3 g^2/a^6)$ [Chau and Wise, 1987].

3.4.9 Interface Circuits

There are several types of circuits that have been developed for interfacing with capacitive sensors. The two most commonly used with pressure sensors are switched-capacitor charge amplifiers and capacitance-to-frequency converters.

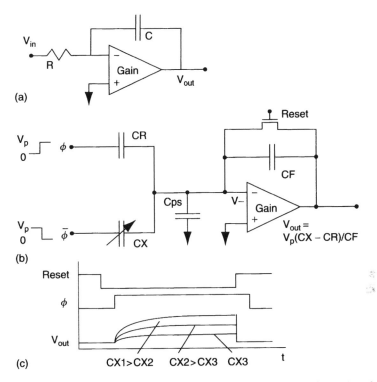

FIGURE 3.17   (a) A simple integrator using an operational amplifier; (b) A switched capacitor charge integrator. (Reprinted with permission from Park, Y.E., and Wise, K.D. [1983] "An MOS Switched Capacitor Readout Amplifier for Capacitive Pressure Sensors," *Record of the IEEE Custom IC Conference*); (c) A timing diagram for control and output signals used in (b).

Switched-capacitor charge amplifiers generally offer high sensitivity and immunity from parasitic capacitance. They are based upon charge integrators. A charge integrator can be made using operational amplifiers, as illustrated in Figure 3.17a. For this, the output voltage is given nominally by:

$$V_{out} = -\frac{1}{RC} \int_{t=0}^{t} v_i(t)dt \qquad (3.52)$$

The sensor interface (Figure 3.17b) utilizes a two-phase clock to compare the variable sense capacitor (C_X) to a reference capacitor (C_R) [Park and Wise, 1983]. A reset pulse initially nulls the output (Figure 3.17c). Subsequently, when the clock switches, a charge proportional to the clock amplitude and the difference between the reference and variable capacitor is forced into the feedback capacitor C_F, producing an output voltage given by:

$$V_{out} = \frac{V_P(C_X - C_R)}{C_F} \qquad (3.53)$$

Since this scheme measures the difference between two capacitors, it is unaffected by parasitic capacitance (C_P) values that are common to both.

One concern related to the switched capacitor interface circuit is the switching noise caused by the downward transition of the reset signal. Because the upward transition occurs after the output is sampled, and causes the output to be nulled, it is not critical. The difficulty with the downward transition is that since it is rapid, the charge stored in C_{gs}, the capacitance between the gate and source of the reset transistor, cannot immediately change in response to it, and this is reflected as a spike in the output signal. One solution is to introduce a dummy transistor as shown in Figure 3.18. This transistor is sized to have a net capacitance C_{dummy}, that is the same as C_{gs}. Its source and drain are shorted, so that it does not interfere with the signal transmission. It is switched by a complement of the reset line, so that the downward spike

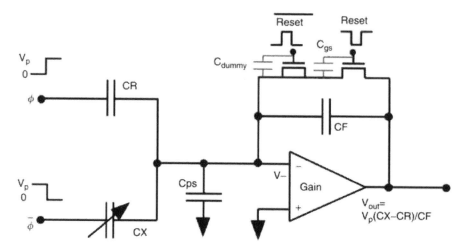

FIGURE 3.18   Cancellation of reset switching noise by using a dummy transistor in a switched capacitor interface circuit.

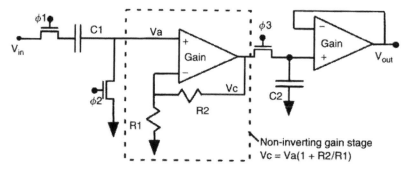

FIGURE 3.19   Correlated double-sampling used to reduce noise, particularly in conjunction with the switched capacitor charge amplifier interface circuit.

on C_{gs} appears as an upward spike across C_{dummy}, and they effectively cancel each other out. If properly designed, the dummy transistor can reduce the reset noise by an order of magnitude.

It is evident from the preceding discussion that despite the general simplicity and robustness of the switched capacitor charge amplifier, it is (as any other interface circuit) susceptible to multiple sources of electronic noise. Switched capacitor circuits, in general, suffer from switching noise that appears as spikes in the voltage waveform at pulse edges (i.e., clock noise). One of the approaches used to combat this is correlated double sampling (CDS), which is also commonly used as a signal processing method in CCD image sensors [Hynecek, 1992]. This technique samples the input to the signal processing circuit block at two different points in time, and produces a signal that is proportional to the difference between them. If the input is switched between a reference null value and the actual signal value, noise components that are common to both intervals will be automatically subtracted away. In this sense, the noise that can be eliminated must be correlated between the two sample times. Generally, noise due to the power supply, clock noise, and even kT/C noise can be cancelled in this manner. A simplified representation of the CDS circuit is shown in Figure 3.19. It basically consists of a non-inverting amplifier of gain A, a unity gain buffer, three sampling switches, and two storage capacitors. Initially, switches $\phi 1$ and $\phi 2$ are pulsed on, while $\phi 3$ is not. This charges up capacitor $C1$ to a reference value V_{s1}, while V_{out} is held at ground. The noise voltage is then stored in $C1$. In the second sample period, $\phi 1$ and $\phi 3$ are pulsed on, while $\phi 2$ is not. If the input is stable at V_{s2} during this period, then V_{out} settles at $A(V_{s2} - V_{s1})$, and the noise components that are common to both sampling periods are cancelled out.

Interface circuits that convert capacitance to frequency are attractive because they offer pseudo-discretized outputs for which exact values are not important and the useful information is carried in the time between signal transitions. This was true of the oscillator for piezoresistive sensors previously described in this chapter, but may certainly be used for capacitive interfaces as well. A simple capacitance-to-frequency converter is illustrated in Figure 3.20(a) [Ko, 1986]. This uses two op amps: one in the integrator, which serves as the variable delay element; and the other in the inverting amplifier, which serves as a comparator and switches between high and low saturated output values. The weakness of this approach is that it is susceptible to parasitic capacitances and resistances, temperature drifts, and other sources of variation in the nominal oscillation frequency. A differential approach that compares the measured capacitance to a reference value is generally more robust. Figure 3.20(b) shows an oscillator interface circuit that is switched between the sensor and a reference capacitor by a periodic waveform [Wise and Najafi, 1989]. Within each such sampling period, the output oscillates at a frequency that is related to the capacitance, which presents a variable loading to one of the elements. The circuit is basically a Schmitt oscillator with an element that has hysteresis in its transfer function. The timing diagram in Figure 3.20(b) explains how the circuit operates. Assume that V_a is descending. As it passes the transition level V_{IL}, V_b switches low, causing V_c to switch high. This causes V_a to start ascending. As it passes the transition level V_{IH}, the various outputs switch to the opposite state. Thus, if I_L is the current available to discharge the capacitance CX of the sensor during the descent of V_a, then the duration of the descent is nominally $t_1 = CX(V_{IH}-V_{IL})/I_L$. Similarly, if I_H is the current available for charging up CX, then the duration of the ascent is $t_2 = CX(V_{IH}-V_{IL})/I_H$. The period if the waveform is $T = t_1 + t_2$.

Figure 3.20(c) shows how a Schmitt trigger, the element that provides hysteresis in the oscillator discussed above, can be implemented using an operational amplifier. The input voltage to this circuit is compared to a reference value, which is simply generated by a voltage divider connected to the output. The output generally remains saturated at the upper or lower limit of the dynamic range, thereby providing two reference values. The transfer characteristics of the circuit are shown in the same figure. Figure 3.20(d) shows a CMOS implementation of the Schmitt trigger. In this, transistors M3 and M6 are small, and oppose the switching of the ouput as it transitions from low to high and high to low, respectively. It should be noted that the circuits shown in Figures 3.20(c) and (d) are *inverting* stages, while the one used in Figure 3.20(b) is *non-inverting*.

Despite their advantages of simplicity and accuracy, capacitance-to-frequency converters are not without compromise, and represent a rather slow method of reading out the signal. For example, to achieve 10-bit accuracy it is necessary to count 2^{11} (i.e., 2048) pulses. (The last bit provides a resolution of only $\pm 1/2$ of the least significant bit.) For a differential measurement, the circuit would have to switch between the sensor and the reference, taking twice as long.

One of the most highly integrated pressure sensors that has been reported is described in Chavan and Wise (2000). It offered 15-bit resolution over a 300 Torr range, which was achieved by using an on-chip 3-bit multiplexing circuit to select between five diaphragms that were all on the same device. Each of the diaphragms used a center boss. One diaphragm covered the full dynamic range with reduced sensitivity, and its output was used to select between the others for high-resolution output over a narrow range, which was tailored by varying the overall diameters from 1000 µm–1100 µm. The cavities were sealed in vacuum, which caused a diaphragm deflection of about 9.8 µm at atmospheric pressure, leaving a capacitive gap of 0.5 µm between the sense electrodes and resulting in a nominal capacitance of 8–10 pF. The circuit (Figure 3.21) incorporated a buffered analog output with a three-stage programmable switched capacitor amplifier using polysilicon capacitors for reference. The front end was a differential charge integrator with a folded cascode amplifier. Correlated double-sampling was used to lower the 1/f noise and the amplifier offset. The circuit also offered separate calibration and operation modes. The fabrication was accomplished using a 20-mask process using one silicon wafer and a glass substrate: 15 for the on-chip p-well circuitry, three for the transducers, and two for the glass processing. The circuitry was fabricated in recessed cavities 2.5 µm deep, and the transducers in cavities 8 µm deep, allowing the bonding anchors to be the highest points on the silicon wafer. Then a 2 µm thick level of polysilicon was deposited for electrical feed-throughs, followed by chemical mechanical polishing to make the device flat for bonding to the glass wafer. Unprotected silicon was then dissolved to expose the diaphragm.

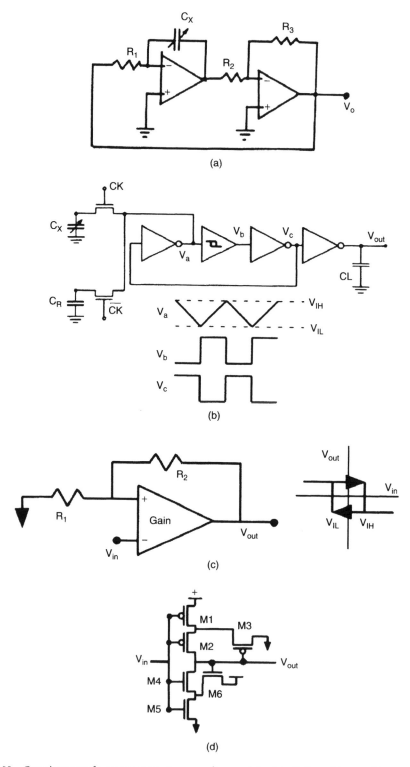

FIGURE 3.20 Capacitance-to-frequency converters are often used for interfacing with capacitive sensors: (a) an implementation using op amps [Ko, 1986]; (b) an implementation that incorporates a reference capacitor, permitting differential measurements; (c) an inverting Schmitt trigger implemented with an operational amplifier; (d) a CMOS implementation of an inverting Schmitt trigger.

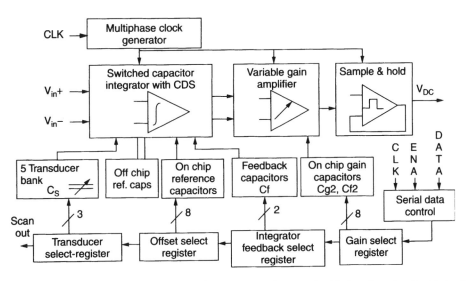

FIGURE 3.21 System architecture for a multi-element sensor integrated with interface circuitry that achieves 15-bit resolution. (Reprinted with permission from Chavan, A.V., and Wise, K.D. [2000] "A Monolithic Fully Integrated Vacuum Sealed CMOS Pressure Sensor," *Proc., IEEE International Conference on Microelectromechanical Systems*, pp. 341–346.)

3.5 Servo-Controlled Pressure Sensors

When a measured signal from a pressure change is used for feedback control to restore the signal to its reference, the pressure sensor is called a *closed-loop* or *servo-controlled* pressure sensor. Capacitive pressure sensors offer many attractive advantages such as high sensitivity, low temperature coefficients, and low power consumption. However, they tend to compromise linearity and dynamic range. Servo-controlled operation offers a solution to this problem. This concept has been widely used in micromachined inertial sensors to increase sensitivity, linearity, and dynamic range. To implement this concept, an actuator is necessary to drive the pressure sensor. The most common choice for this purpose is an electrostatic actuator. This tends to be a natural choice, particularly if a capacitive pick-off is already used in the device. However, it faces two constraints. The first is that the applied voltages provide only attractive forces, which adds complexity to the structural design and device fabrication. The second is that for voltages smaller than the instability limit (i.e., the "pull-in" voltage at which the a diaphragm would collapse to the actuation electrode), the electrostatic pressure is relatively small. The solutions that have been developed are described in the following case studies.

In the structure shown in Figure 3.12, the Si diaphragm is suspended between two glass wafers on which the sensing and feedback electrodes are located [Wang and Esashi, 1997]. A perforation in one glass wafer provides access to applied pressure. Lead transfer for the lower electrode is through the silicon wafer: this provides access to all the electrodes from the upper surface of the device. This device was fabricated by bulk silicon micromachining and required two glass wafers and one Si wafer. For 10 Torr (1.33 kPa) applied pressure, 70 V was applied to balance the electrostatic actuator.

A surface micromachined device that provided low voltage operation was presented by Gogoi and Mastrangelo (1999). The electrode area was designed to be 100× larger than the pressure diaphragm area to reduce the restoring voltage. The device has a 20 μm × 20 μm diaphragm, and 250 μm × 250 μm foot print area. The diaphragm and the feedback electrodes were made from polysilicon. The fabrication process required 15 masks and the device was self-encapsulated, which can greatly simplify packaging. For 100 kPa applied pressure, the actuation voltage was less than 12 V.

To avoid structural complications and the high operating voltage over a wide pressure range, another closed loop pressure sensor has been devised [Park and Gianchandani, 2001; Park and Gianchandani,

2003]. The diaphragm in this case extends to a skirt electrode, and both are attached to a narrow sidewall that acts as a flexural hinge (Figure 3.15). This structure can be used in open-loop mode, i.e., without servo-controlled operation. The flexible diaphragm that deforms in response to applied pressure is not directly used for sensing. Instead, it is attached to a skirt-like external electrode that deflects *upward* as the diaphragm deflects *downward*. This movement is detected capacitively by a counter electrode located under it. For closed-loop operation, it permits the deflection of the skirt to be balanced by a voltage bias on the sense electrode without an extra structural layer on the top of diaphragm. Moreover, the concentric layout of the sealed cavity and the skirt permits the electrode to naturally occupy a larger area than the diaphragm, as preferred for electrostatic feedback. It is notable that in this feedback scheme, the skirt and not necessarily the diaphragm, is restored to its reference position. Low operating voltages are possible, because there is no need to completely balance the barometric force and the electrode is much more flexible than diaphragm.

The system-level representation of this pressure sensor is shown in Figure 3.22, and the parameters used are defined in Table 3.2. The deflection of the diaphragm in response to applied barometric force causes a deflection of the skirt electrode and leads to a change in the readout capacitance. Based on this capacitance change, an equivalent barometric force can be defined as acting on the skirt electrode:

$$F_d = Z_d P \tag{3.54}$$

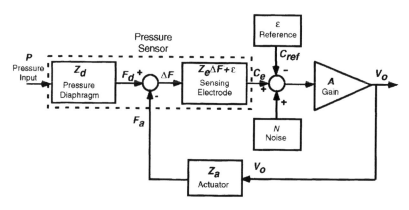

FIGURE 3.22 Block diagram for a servo-controlled capacitive pressure sensor [Park and Gianchandani, 2003].

TABLE 3.2 Definition of Variables Used in the System Representation of a Servo-Controlled Pressure Sensor

Variable	Physical Meaning
A [V/F]	Circuit gain from capacitance to voltage
C_e [F]	Capacitance output of pressure sensor
C_{ref} [F]	Reference capacitance
ε [F]	Capacitance offset at zero pressure
F_a [N]	Electrostatic force generated in feedback
F_d [N]	Mechanical force on sense electrode due to P
ΔF [N]	Net force on electrode
N [F]	Capacitance measurement uncertainty
P [N/m²]	Barometric pressure
V_o [V]	Output voltage
Z_a [N/V]	Ratio of F_a to V_o
Z_d [m²]	Ratio of F_d to P
Z_e [F/N]	Capacitance change per unit force on sense electrode

An electrostatic force, F_a, is applied by biasing the lower electrode to restore the capacitance. This is non-linearly related to the applied voltage and capacitive gap. The total equivalent force applied on the electrode after feedback is:

$$\Delta F = F_d - F_a \qquad (3.55)$$

An expression for F_a can now be derived. In open-loop operation:

$$C_e = Z_e \Delta F + \varepsilon \qquad (3.56)$$

$$V_o = A(C_e - C_{ref}) \qquad (3.57)$$

C_e can be provided by substituting (3.55) into (3.56):

$$C_e = Z_e(F_d - F_a) + \varepsilon \qquad (3.58)$$

Substituting (3.58) into (3.57), V_o can be provided by:

$$V_o = A[Z_e(F_d - F_a) + \varepsilon - C_{ref}] \qquad (3.59)$$

$$V_o = AZ_e(F_d - F_a) \quad \text{if } C_{ref} = \varepsilon \qquad (3.60)$$

This signal is fed back to the electrostatic actuator, which generates force F_a:

$$F_a = Z_a V_o \qquad (3.61)$$

Substituting (3.54) and (3.61) into (3.60), the output of the servo-controlled system is:

$$V_o = AZ_e(Z_d P - Z_a V_o) \qquad (3.62)$$

$$V_o = \frac{Z_e A}{1 + Z_e Z_a A} Z_d P \qquad (3.63)$$

$$V_o = \frac{1}{Z_a} Z_d P \quad \text{as } Z_e A \to \infty \qquad (3.64)$$

The electrostatic force can be obtained by substituting (3.63) into (3.61):

$$F_a = Z_a \frac{Z_e A}{1 + Z_e Z_a A} Z_d P \qquad (3.65)$$

where Z_e is capacitance change per unit pressure, Z_a is the electrostatic pressure per unit applied voltage, and A is the transformation from capacitance to voltage. Therefore, the net force applied on the diaphragm is:

$$\Delta F = Z_d P - F_a = Z_d P - Z_a \frac{Z_e A}{1 + Z_e Z_a A} Z_d P \qquad (3.66)$$

As amplifier gain increases, the steady state equivalent force on the diaphragm is zero:

$$\Delta F \to 0 \quad \text{as } Z_e A \to \infty \qquad (3.67)$$

The bias applied to the lower electrode also serves as the output signal:

$$V_o = \frac{Z_e A}{1 + Z_e Z_a A} Z_d P \to \frac{Z_d}{Z_a} P \quad \text{as } A \to \infty \qquad (3.68)$$

TABLE 3.3 Device dimensions and measurement results from Park and Gianchandani (2003), which are in terms of gauge pressure. C_0 is the capacitance at zero gauge pressure. *Linearity for DEV1 was measured over 0–250 kPa. For DEV2, DEV3, and DEV4, linearities were measured over 0–100 kPa. (Reprinted with Permission from Park, J.-S., and Gianchandani, Y.B. [2003] "A Servo-Controlled Capacitive Pressure Sensor Using a Capped-Cylinder Structure Microfabricated by a Three-Mask Process," *IEEE/ASME Journal of Microelectromechanical Systems* 12[2], pp. 209–220.)

DEVICE			DEV1	DEV2	DEV3	DEV4 Segmented Skirt
T1 = T3 (µm)			9.7	12	12	12
T2 (µm)			18	18	20	20
H = G1 (µm)			10	15	15	15
R1 (µm)			500	500	500	500
R2 (µm)			1000	1000	1000	1000
Bonding Temperature (°C)			550	550	450	450
Unpackaged	Open Loop	C_0 (pF)	2.910	1.562	1.684	1.514
		Sensitivity (ppm/kPa)	−408	−154	−124	−136
		Sensitivity (fF/kPa)	−1.21	−0.24	−0.21	−0.21
		Linearity (%)	<3.1	<4.5	<14.6	<6.5
	Closed Loop	Sensivity (V/kPa)	0.516	1.436	1.208	1.642
		Linearity* (%) (>20 kPa)	<3.2	<2.6	<2.6	<1.8
Packaged	Open Loop	C_0 (pF)			2.882	2.646
		Sensitivity (ppm/kPa)			−137	−164
		$\Delta C/\Delta P$ (fF/kPa)			−0.39	−0.44
		Linearity* (%)			<11.0	<4.7
	Closed Loop	Sensitivity (V/kPa)			0.748	0.819
		Linearity (%) (>20 kPa)			<1.4	<1.8

For the predictive modeling of this device, finite element analysis is preferred because the simultaneous flexing of the diaphragm, skirt electrode, and sidewall are not easily expressed in closed-form analytical manner. In general, it is necessary to use coupled-field analysis to model this device because both pressure and electrostatic forces must be simulated.

Experimentally, a number of devices with 2 mm overall diameter, but varying in other dimensions, were fabricated and tested. An alternative design in which skirt electrode was divided into segments to relieve residual stress provided 10%–20% more open-loop sensitivity for similar overall dimensions. The device dimensions and performance results are summarized in Table 3.3. In addition, Figures 3.23 and 3.24, respectively, provide examples of the open-loop and servo-controlled responses of some of the fabricated devices.

Some additional points are worth noting about the operation of this type of device. First, the operating voltages required for servo-controlled operation are dependent on the dimensional parameters. For example, for DEV1 (Table 3.3), the voltage bias needed to achieve servo-controlled operation over a dynamic range of 0–100 kPa was 31.2–73.2 V, with the capacitance held at 2.91 pF. This can be reduced by increasing the diameter of the skirt. A superior approach is to reduce the capacitive gap between the skirt and the counter-electrode, not by reducing the height of the cavity, which could compromise the flexibility of the hinge offered by the sidewall, but by raising the surface of the counter-electrode by electroplating it up to a greater thickness. However, reducing the capacitive gap would increase the non-linearity of the open-loop response.

Second, the sensitivity of the closed-loop response can be electrically changed by changing the set-point of the nominal capacitance. For DEV1, if the nominal capacitance was set at 2.96 pF instead of 2.91 pF,

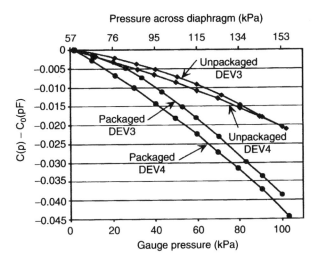

FIGURE 3.23   Measured response of DEV3 and DEV4 in open-loop operation before and after packaging. [Park and Gianchandani, 2003].

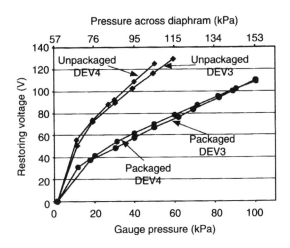

FIGURE 3.24   Measured response of DEV3 and DEV4 in servo-controlled operation before and after packaging. [Park and Gianchandani, 2003].

the 0–100 kPa dynamic range required a bias variation over only 53.9–82.3 V, and provided a sensitivity of 0.284 V/kPa. This could be implemented by biasing the Si structure at −68 V and varying the electrode bias over ±15 V, allowing standard (low-voltage) electronics to be used for the servo-controlled operation.

Third, a bias voltage may also be used to tune the open-loop response of the device. For DEV1, the average open-loop sensitivity over a 0–105 kPa dynamic range increased by 33% (from −328 ppm/kPa to −437 ppm/kPa) in the presence of a 65 V bias. The nominal capacitance increased from 2.91 pF to 3.03 pF.

Fourth, packaging can be a significant challenge for a device in which the sense gap is external to the sealed cavity. For the devices reported in Gogoi and Mastrangelo (1999), all the electrodes were located within the sealed cavity, so this problem did not arise. Park and Gianchandani (2003) addressed the challenge by encasing the device in a metal package that had a flexible diaphragm and an inert liquid (Fluorinert75™) filler to transmit the pressure from this diaphragm to the sensor. The dielectric constant of the liquid makes the readout capacitance larger. Since open-loop sensitivity is defined in fractional (normalized) terms, it does not change. However, with respect to servo-controlled operation, the factors Z_a (electrostatic pressure/voltage) and Z_e (capacitance change/barometric pressure change), increase linearly with the dielectric constant of the liquid, which is about 1.7. This leads to a proportional reduction in the voltages required for servo-controlled operation of the device. The liquid ambient also helps to reduce mechanical vibrations of the skirt electrode, and can help to reduce noise. Thus, the liquid filler can have many benefits, although it is not without drawbacks. Most notably, the relatively high coefficient of thermal expansion of this liquid, at 1380 ppm/K, is about 100× higher than that of the metal housing. Unless calibrated and compensated, this can lead to problems.

3.5.1 Measurement Uncertainty and Noise

In the system-level representation of servo-controlled operation shown in Figure 3.22, the uncertainty or noise in the capacitance measurement is indicated by N, and the output voltage due to it is:

$$\Delta V_o = \frac{A}{1 + Z_e Z_a A} N \tag{3.69}$$

Dividing Equation 3.68 by Equation 3.69, the resulting signal-to-noise ratio is independent of the elements Z_a and A, which are related to the electrostatic feedback force. Servo-controlled operation does not change the nominal value of the signal-to-noise ratio of the transducer. However, Equation 3.68 shows that for large A, the uncertainty in the readout is reduced by factor of $1/(Z_e Z_a)$. A liquid medium high dielectric constant is clearly an asset in this case, since Z_e and Z_a both increase. Even though the sensitivity of the pressure sensor is nominally un-changed by the relative dielectric constant of the liquid ambient inside the package, the resolution — defined as $\Delta C_{min}/(C \cdot S)$, where ΔC_{min} is the minimum detectable capacitance change and S is the sensitivity — is increased.

3.6 Other Approaches

Several pressure sensor concepts that do not rely on capacitive and piezoresistive sensing have been developed. The motivation to pursue alternative approaches is typically driven by a unique application, or by stringent requirements in a certain performance category. Resonant beam devices offer particularly high resolution and a pseudo-digital output. Tunneling devices offer high sensitivity, optical devices are ideal for harsh environments, and Pirani devices are well-suited for high vacuum applications.

3.6.1 Resonant Beam Pick-Off

Frequency measurements of resonating micromachined components can be considerably more reliable and accurate than the measurement of a resistor or capacitor. In addition, since the output is naturally quantized, interfacing to a digital system can be easier. Two categories of devices using resonant sensing have been reported in the past: strain-based and damping-based. The former category uses a strain-sensitive resonating beam as a substitute for a piezoresistor within a deflecting diaphragm. The typical strain-based resonating beam pressure sensor will incorporate end-clamped beams to detect the strain in a diaphragm, which can be related to applied pressure. The resonating beam may be driven by electrostatic, magnetic, or even optical stimulus. The resonant frequency can be measured by these transduction methods or by implanted piezoresistors on the beam itself. The damping-based resonating device will frequently be a single cantilevered paddle or a lateral-comb resonator. Gaseous damping mechanisms vary with the ratio of the device size to the mean free path of molecules in the ambient. As the mean free path varies with pressure, so does the damping; this information is extracted by measuring the amplitude, phase, or quality factor of the resonator.

Ikeda et al. (1990) described one of the first commercially available strain-based pressure sensors to use resonant beams embedded within a diaphragm. Not only was this one of the first devices of its kind to be reported, but it also achieved very high performance, with accuracies better than 0.01% of the full-scale range. In order to provide a differential signal, this device utilized two resonating beams: one in the center of the diaphragm and one at the perimeter (Figure 3.25[a]). The beams were vacuum-encapsulated

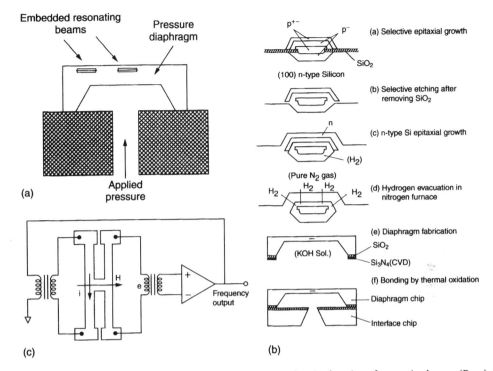

FIGURE 3.25 (a) Schematic of a resonating beam pressure sensor showing location of resonating beams. (Reprinted with permission from Ikeda, K., Kuwayama, H., Kobayashi, T., Watanabe, T., Nishikawa, T., Yoshida, T., and Harada, K. [1990] "Silicon Pressure Sensor Integrates Resonant Strain Gauge on Diaphragm," *Sensors and Actuators* A21, pp. 146–150.); (b) fabrication process; (c) interface to the pressure sensor.

by a fabrication process that used four levels of epitaxial growth (Figure 3.25[b]). The cavities in which the beams were encapsulated were sealed at a pressure of 1 mTorr, which was made possible by low pressure of the epitaxial deposition by which the cavities were sealed. This allowed high Q factors for the beams to exceed 50,000. The temperature sensitivity of the device was limited by the thermal expansion of silicon, with a coefficient of -40 ppm/K. In operation, a beam was excited into resonance by driving an AC current through the resonating beam, and applying a DC magnetic field (Figure 3.25[c]). The current was controlled by active feedback circuitry to keep the oscillation amplitude constant. Smaller resonance amplitude helped minimize second order non-linearities in the vibration, in particular hysteresis of the frequency response. The high accuracy that was a result of this design addressed the aerospace market, where the sensor has been commercialized.

Another resonating beam pressure sensor used a thin single crystal silicon membrane [Petersen et al., 1991]. This was high temperature fusion bonded to a second wafer, which formed the square diaphragm of the pressure sensor. A single resonant beam was fabricated by etching two slots through the membrane. This beam was excited to resonance electrostatically by a voltage on the order of a few millivolts.

Burns et al. (1994) reported a resonant beam sensor that utilized a modulated laser beam as the driving mechanism for the resonator. This structure was developed with a single resonator on the membrane. However, a reference resonator can be located off the diaphragm to provide a differential output. The beam could also be driven electrostatically and was optimized to reduce secondary vibration modes. The sensitivity was 3880 Hz/psi with a dynamic range of 0–10 psi. The dynamic range was adjusted by varying diaphragm thickness, which allowed the same masks to be used for different models. The temperature stability was 0.01 ppm/K. The device was operated with off-chip interface circuitry that also provided compensation for improving linearity. The final noise variance was 0.2–0.4 Hz. The ratios of microbeam length-to-diaphragm diameter were selected to optimize frequency sensitivity.

Stemme and Stemme (1990) described an interesting dual-diaphragm cavity structure. The device consisted of two diaphragms bonded together at the circumference, forming a pressure sensitive capsule that could provide an absolute pressure measurement (Figure 3.26[a]). This structure was vibrated in a torsional mode. Changes in external pressure varied the torsional stiffness, and thereby the frequency of oscillation. A hollow channel, extending from the perimeter of the structure to the sealed cavity between the diaphragms, permitted this device to be used for differential measurements. The geometry provided a very high quality factor, Q, reaching 80,000 in vacuum, and around 2400 in air. This performance was achieved partially by perforating the corners of the structure — where the motion has the largest amplitude — in order to reduce damping (Figure 3.26[b]). The structure was electrostatically oscillated and the displacement was measured by interferometry with a HeNe laser. Typical pressure sensitivities were from 14%/bar and 19%/barr for the absolute and differential modes, respectively, at atmosphere. Typical resonant frequencies were 15–19 KHz. The temperature coefficients of sensitivity were $+60$ ppm/K and -16 ppm/K for the absolute and differential devices, respectively. The higher temperature coefficient of the absolute pressure sensor was principally the result of trapped gas during the bonding process.

Damping-based resonant devices rely on the variation of the dissipative forces with pressure. Interaction between a solid structure and the ambient gas is most commonly described by the Knudsen number, K_n. This is the dimensionless ratio of the molecular mean free path to the dimension of the moving mass. The free molecular flow region is defined $K_n > 10$, whereas the viscous flow region is defined as $K_n < 10$. In the former, the drag force is proportional to the velocity of the mass; whereas in the latter, it is linearly proportional to both velocity and acceleration [Kawamura et al., 1987; Landau and Lifschitz, 1970]. This causes the amplitude, phase, and quality factor of an excited oscillating beam to depend on the ambient pressure.

Kawamura et al. (1987) described a pressure sensor based on damping measurements. A micromachined silicon paddle, formed by KOH etching, varied 1.8 mm–9 mm in width, 1.2 mm–6 mm in length, and 0.040 mm–0.35 mm in thickness. Its oscillation amplitude decreased with increasing pressure beginning at a critical point, which varied with the paddle size, ranging from 10^{-2} to 10^5 Pa (Figure 3.27). Calibration of the amplitude and phase delay permitted pressures within this range to be measured.

Another device relying on the pressure dependence of the kinetic properties of gases was described by Andrews et al. (1992). In this work, a rectangular cavity with a boss was fabricated in silicon, and a paddle fabricated over the cavity. The paddle was electrostatically resonated. The resonant frequency was a function of the paddle dimensions and the pressure of the trapped gas in the cavity. As the pressure increased, the gas in the cavity provided an increasing stiffness component, increasing the resonance frequency. Pressures from 10^{-2} Pa to atmosphere were measured. As this device benefits from squeeze-film damping, the Q was relatively low. At atmospheric pressure, the Q was in the range of 10–20, and 92% of the spring stiffness was contributed by the trapped air, with only 8% from the structure. Despite the low Q, the device had very stable characteristics: a pressure resolution of 1 part in 30,000 was achieved at atmosphere. Interestingly, the frequency of the output was not a function of gas species, showing that all gas compressions were isothermal and could be approximated as ideal. However, gas viscosity had an effect on the Q of the device, and viscosity is species dependent, so the device could also be used to discriminate gas types. One advantage of all paddle-type pressure sensors is that because there is no loading on any diaphragm, the long-term expected creep is minimal.

3.6.2 Tunneling Pick-Off Pressure Sensors

Capacitive pressure sensors require the fabrication of very thin and large-area diaphragms in order to provide high sensitivity and facilitate the measurement of small (average) deflections. However, tunneling current established between an electrode and the diaphragm can be used to measure nanometer scale deflections. This is because the tunneling current is an exponential function of the separation distance between the tunneling tip and counter-electrode [Wickramasinghe, 1989]. A critical advantage offered by this transduction scheme is that it opens up the possibility of using smaller and thicker diaphragms than allowed for capacitive devices of similar sensitivity.

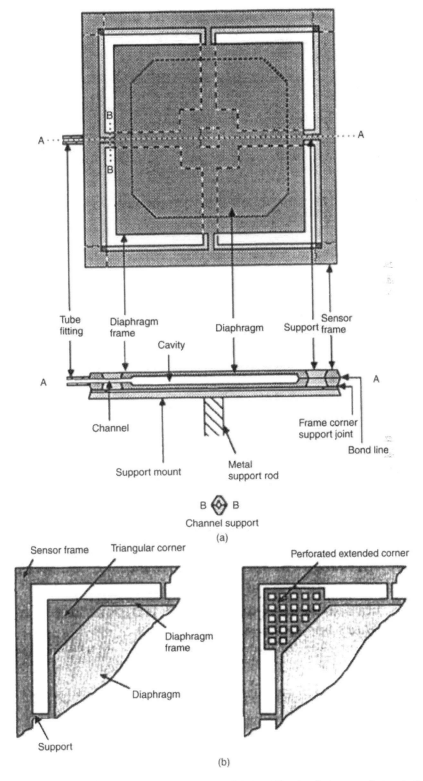

FIGURE 3.26 (a) A dual-diaphragm pressure sensor using variation of torsional resonance frequency for sensing. (Reprinted with permission from Stemme, E., and Stemme, G. [1990] "A Balanced Resonant Pressure Sensor," *Sensors and Actuators* **A21–A23**, pp. 336–341.) (b) Perforated corners reduce damping and increase quality factor.

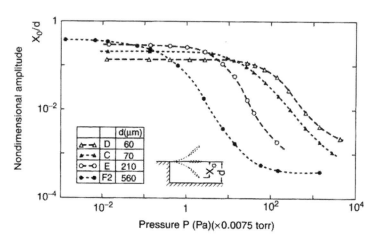

FIGURE 3.27 Normalized variation of vibration amplitude with ambient pressure. (Reprinted with permission from Kawamura, Y., Sato, K., Terasawa, T., and Tanaka, S. [1987] "Si Cantilever–Oscillator as a Vacuum Sensor," *International Conference on Solid-State Sensors and Actuators (Transducers)*, pp. 283–286.)

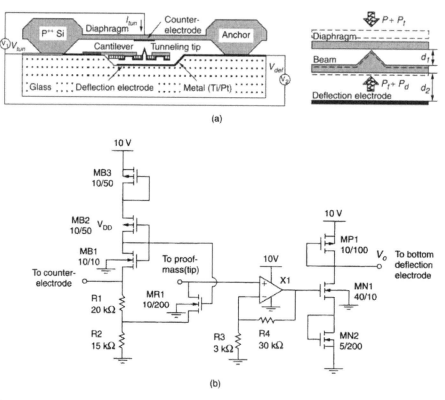

FIGURE 3.28 (a) Device structure and (b) an interface circuit for a pressure sensor with tunneling current pickoff. (Reprinted with permission from Chingwen, Y., and Najafi, K. [1998] "CMOS Interface Circuitry for a Low Voltage Micromachined Tunneling Accelerometer," *Journal of Microelectromechanical Systems*, March, pp. 6–9.)

Yeh and Najafi (1994) reported on a pressure sensor using tunneling current. This device used two levels of micromachined silicon above a patterned glass substrate (Figure 3.28). The first silicon layer was used to form a cantilever beam with a tunneling electrode, while the second formed a diaphragm above the tip. The glass substrate had a patterned metal electrode to electrostatically deflect the cantilever beam away from the diaphragm, with the intent to maintaining a constant separation in the tunneling gap as the diaphragm deflected under applied external pressure.

The small tunneling current that exists through a single tip can be an asset for low-power operation, but can also raise concerns about noise. In contrast to scanning tunneling microscopes, which are intended to provide very high spatial resolution in lateral scans, the pressure sensor does not require any rastering motion of the tunneling tip. Therefore, a number of tips can be self-selected for tunneling from the nano-scale topographical variations in a mesa-shaped tunneling electrode. As these tips wear, others take their place. The operating current is summed over many tips operating in parallel.

While this type of pressure sensor may be operated in an open-loop mode, in which the tunneling current is monitored as a function of applied pressure with fixed bias voltages on the tunneling electrodes as well as the deflection electrode, the dynamic range is only about 10 mTorr because of the exponential dependence of the current upon the tunneling gap. An increase of 10 V in either the deflection voltage or the tunneling voltage produces two orders of magnitude change in the tunneling current. In this mode, the device can perhaps be used as a threshold pressure sensor. The more conventional applications, which require a wider dynamic range and linear response, can be satisfied by servo-controlled operation.

Several interface circuits for the servo-controlled operation of tunneling pressure sensors are presented in [Yeh and Najafi, 1998]. The schematic of one that was fabricated and tested is illustrated in Figure 3.28(b). Transistor MR1, a long NMOS, was biased in the linear region. As the tunneling current traveled through MR1, its drain-to-source voltage increased in a roughly linear manner. This was amplified by 10× with the following non-inverting op amp stage, and then the signal was inverted and buffered by the final gain stage. The output signal, V_o, was used to modulate the deflection electrode, which regulates the tunneling gap by deflecting a proof-mass on a membrane. The final stage was designed to have a high output impedance to prevent over-current should the cantilever and the deflection electrode come in contact. The circuit was operated at 10 V. In an alternative implementation, MR1 can be replaced by a diode, then the current is exponential with the voltage across the diode, as is the tunneling current. This provides a deflection voltage that is linear with the tunneling gap.

3.6.3   Optical Pick-Off Pressure Sensors

Optical devices offer particular advantages in speed and remote sensing, and may be utilized more in the future for specific niche applications. As the sensors can be addressed and read with light, they are potentially useful for hazardous environments. In areas that are flammable, remote, or corrosive, it may not be practical to run electrical cabling.

The measurement of high-speed turbulent air-flow on aeronautical surfaces requires remote sensing, and good spatial and temporal resolution. Miller et al. (1997) described an approach to correlate pressure fluctuations in the ambient to mechanical vibrations of the surface. An array of pressure-sensitive cavities, designed to act as interferometric Fabry–Perot etalons, was used in conjunction with an external near-infrared laser (Figure 3.29). This scheme eliminated electrical connections to the devices, which could affect the pressure distribution and the air flow patterns. The devices were fabricated from 0.55 mm thick silicon-on-insulator wafers, with a 55 μm thick upper layer, and a 0.4 μm thick buried SiO_2 layer. The etalon cavities were formed by patterning and wet-etching the front surface down to the oxide layer using KOH. The buried oxide ensured an even depth, and a smooth, reflective trench surface. A second SOI wafer was then bonded to it face-to-face. The exposed back of the second wafer was then patterned and etched, forming a 55 μm thick diaphragm. A half-wave coating of SiO_2 was deposited on the diaphragm. The fabricated etalons were 1 mm square with an internal pressure of 0.8 atm. The array was interrogated with a 1.5 μm wavelength laser, and interference patterns, corresponding to quarter wavelength deformations, were observed with a video camera.

A number of other device concepts using optical sensing have also been proposed. In one approach, the device was located at the end of an optical fiber, and included a rigid substrate, followed by a cavity at a reference pressure, followed by a flexible membrane [Greywall, 1997]. The device acted as a Fabry–Perot interferometer. A parameter, β, was defined that was proportional to the square of the radius

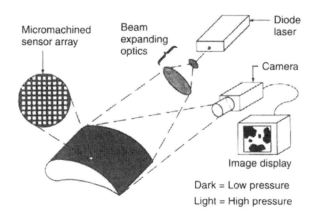

FIGURE 3.29   Optical measurement of pressure on a wing surface using interferometric micromachined devices. (Reprinted with permission from Miller, M.F., Allen, M.G., Arkilic, E., Breuer, K.S., and Schmidt, M.A. [1997] "Fabry–Perot Pressure Sensor Arrays for Imaging Surface Pressure Distributions," *International Conference on Solid-State Sensors and Actuators [Transducers],* pp. 1469–1472.)

and inversely proportional to membrane stress, thickness, and to cavity length. The larger the β, the larger the membrane displacement, and therefore, the larger the variation of the reflectivity with applied pressure. As a consequence, larger β improved sensitivity, but reduced dynamic range. In this device, the optical intensity was the same for several different values of pressure, as the length of the reference cavity cycled through multiples of quarter wavelengths as the membrane was deflected. The intensity was also affected by stray light leaking into the system, the surface roughness of the device, and the index of refraction of the medium being measured.

An *in-line* Fabry–Perot device was developed for sensing pressure during the process of drilling for oil wells [Maron, 2000]. The fiber optic waveguide had two partially transparent mirrors internal to the fiber. These served as an interferometer for light traveling along the fiber, which was stretched across the interior of a metal package. Variations in pressure along the length of the fiber modulated the interferometer and could be detected optically. The device was able to measure peak pressures in the range of 15,000–30,000 Torr, with resolution on the order of 300 Torr.

Kao and Taylor (1996) describe a device that is conceptually a hybrid between the approaches described in Greywall (1997) and Maron (2000). A fiber with internal mirrors (once again forming a Fabry–Perot cavity) was attached under tension between two ends of a cylindrical metal housing. The housing, which was sealed at a reference pressure, had a flexible diaphragm at one end. As the external pressure increased, the diaphragm was deformed, changing the longitudinal tension on the fiber, and consequently the optical length between the two mirrors. A device with 50.8 μm membrane thickness, a 3.05 mm diameter, and a fiber length of 1.9 cm was able to measure pressure from 1–100 Torr with a resolution of 0.4 Torr.

There are a few limited applications where the pressures are high enough to essentially preclude mechanical sensors. One application would be in a diamond anvil, where high pressure physics experiments are conducted, and pressures can reach the megabar range. These pressures are usually achieved in very small volumes, and it is impractical to utilize any feedthroughs, so remote sensing is a must. A common sensor for this is a ruby crystal [Trzeciakowski et al., 1992]. The crystal is illuminated with a laser, and the very narrow florescent atomic excitation lines due to the chromium impurity shift to lower energies with pressure. The resolution is around 1–2 kbar, which is adequate for the megabar range, but inadequate for lower ones. Measuring pressure from the wavelength of atomic excitations offers advantages over intensity measurements, as roughness and reflection coefficients are no longer an issue. Spectral lines shift with pressure in many materials, but effect is limited to fairly low <77 K temperatures due to phonon interaction.

A similar device has been made from compound semiconductor multiple quantum–well (MQW) structures using 10 or more alternating GaAs and AlGaAs layers of 10 nm thickness to confine electrons

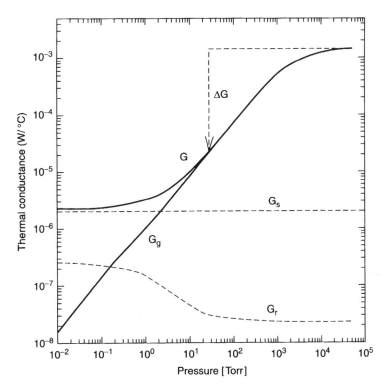

FIGURE 3.30 Thermal conductance as a function of pressure. (Reprinted with permission from Chou, B.C.S., and Shie, J.S. [1997] "An Innovative Pirani Pressure Sensor," *International Conference on Solid-State Sensors and Actuators [Transducers]*, pp. 1465–1468.)

and holes as excitons (i.e., loosely bound electron–hole pairs) [Trzeciakowski et al., 1992]. The material was placed in the pressure chamber and externally stimulated with an off-the-shelf laser. Optical emission can be excited with Ar, Kr, or HeNe gas lasers. The spectral lines due to exciton emission shift with applied pressure. The absorption of optical energy and attenuation through the multiple layers broadened the characteristic emissions. Better results were found in wider GaAs wells, with higher aluminum concentration in the AlGaAs regions. The heterostructure sample showed about 20 times the sensitivity of a ruby crystal with 0.02 kbar accuracy over a 40 kbar range. The temperature coefficient was about 40 bar/K.

3.6.4 Pirani Sensors

Pirani sensors are vacuum sensors that utilize temperature sensitive resistors as heating and sensing elements to measure pressure-dependent heat conduction across an air gap. This concept is widely used in conventional sensors, and micromachined versions have been reported as well [Chou and Shie, 1997; Stark et al., 2003]. Figure 3.30 illustrates thermal conductance as a function of pressure for air. In this figure, the total thermal conductance, G, is an S-curve; G_s is the solid heat transfer-through leads, and various contact points; G_r is radiative thermal conduction; and G_g is the gas heat conduction. Gas heat conduction dominates radiative heat conduction above 1 Torr, and increases linearly with pressure (on a log-log plot), up to a saturation point. The device is most useful over the range in which G is dominated by G_g. This dynamic range can be increased by minimizing the air gap across which the thermal conductance is measured, that is, by increasing the pressure at which the conduction saturates. It is also helpful to suppress the parasitic conduction paths: increase G_g relative to the competing heat loss mechanisms. Chou and Shie (1997) described a $50 \times 50 \, \mu m^2$ membrane with PECVD low stress silicon nitride supported a

100 nm platinum film acting as a 200 Ω temperature sensitive resistor. A sacrificial layer of polysilicon was used to suspend this structure 0.3 μm above the substrate. An identical platinum resistor located on the substrate served as a reference. A bridge-type readout scheme was used to measure the temperature-dependent resistivity as a function of pressure. The small air gap allowed an operating region from 10^{-6} to 10^4 Torr, extending the upper limit well above atmospheric pressure.

3.7 Conclusions

This chapter has provided a perspective of breadth and depth of the developments in pressure sensors, which represent one of the richest applications of micromachining technology to date. Although most of the commercially available options use piezoresistive sense schemes, capacitive sensing has been a favorite research theme for a number of years, and other options have also emerged. The information presented on noise limits and interface circuits can be extended to other devices and applications. Novel structures have been proposed to address special needs such as on-chip encapsulation and feedback control. Highly sensitive devices have been developed using thin film dielectric diaphragms and by using tunneling current for sensing. Optical sensing appears to be particularly promising for applications in harsh environments. There is little doubt that this standard-bearer for the MEMS field will continue to evolve with exciting new structures and technologies in the future.

References

Andrews, M.K., Turner, G.C., Harris, P.D., and Harris, I.M. (1993) "A Resonant Pressure Sensor Based on a Squeezed Film of Gas," *Sensors and Actuators A*, pp. 219–226.

Burns, D.W., Zook, J.D., Horning, R.D., Herb, W.R., and Guckel, H.G. (1994) "A Digital Pressure Sensor Based on Resonant Microbeams," *Solid State Sensor and Actuator Workshop*, pp. 221–224.

Chau, H., and Wise, K. (1987) "Scaling Limits in Batch-Fabricated Silicon Pressure Sensors," *IEEE Trans. Electron. Devices*, **34**, pp. 850–858.

Chau, H., and Wise, K. (1988) "An Ultra-Miniature Solid-State Pressure Sensor for Cardiovascular Catheter," *IEEE Trans. Electron. Devices*, 35(12), pp. 2355–2362.

Chau, H., and Wise, K. (1987a) "Noise Due to Brownian Motion in Ultrasensitive Solid-State Pressure Sensors," *IEEE Trans. Electron. Devices*, **34**, pp. 859–865.

Chavan, A.V., and Wise, K.D. (2000) "A Monolithic Fully Integrated Vacuum Sealed CMOS Pressure Sensor," *Proc., IEEE International Conference on Microelectromechanical Systems*, pp. 341–346.

Chavan, A.V., and Wise, K.D. (1997) "A Batch-Processed Vacuum-Sealed Capacitive Pressure Sensor," *International Conference on Solid-State Sensors and Actuators (Transducers)*, pp. 1449–1451.

Cheng, Y.T., Hsu, W.T., Lin, L. Nguyen, C.T., and Najafi, K. (2001) "Vacuum Packaging Technology Using Localized Aluminum/Silicon–to–Glass Bonding," *Proc., IEEE International Conference on Microelectromechanical Systems*, pp. 18–21.

Cho, S.T., Najafi, K. Lowman, C.L., and Wise, K.D. (1992) "An Ultrasensitive Silicon Pressure-Sensor Based Microflow Sensor," *IEEE Electron Devices*, **ED-39** pp. 825–835.

Chou, B.C.S., and Shie, J.S. (1997) "An Innovative Pirani Pressure Sensor," *International Conference on Solid-State Sensors and Actuators (Transducers)*, pp. 1465–1468.

Crary, S., Baer, W.G., Cowles, J.C., and Wise, K.D. (1990) "Digital Compensation of High Performance Silicon Pressure Transducers," *Sensors Actuators A*, pp. 70–72.

Ding, X. (1997) "Behavior and Application of Silicon Diaphragms with a Boss and Corrugations," *International Conference on Solid-State Sensors and Actuators (Transducer)*, pp. 166–169.

Evans, J.J., and Evans, R.A. (1965) *J. Appl. Phys.*, **36**, pp. 153–156.

Gabrielson, T.B. (1993) "Mechanical–Thermal Noise in Micromachined Acoustic and Vibration Sensors," *IEEE Trans. Electron. Devices*, 40(5), pp. 903–909.

Giachino, J.M., Haeberle, R.J., and Crow, J.W. U.S. Patents #4261086 (April, 1981) and #4386453 (June, 1983). Gogoi, B., and Mastrangelo, C.H. (1999) "A Low Voltage Force-Balanced Pressure Sensor with Hermetically Sealed Servomechanism," *Proc., IEEE International Conference on Microelectro-mechanical Systems*, pp. 493–498.

Greenwood, J.C. (1988) "Silicon in Mechanical Sensors," *J. Phys. E, Sci. Instrum.*, 21, pp. 1114–1128.

Greywall, D.S. (1997) "Micromechanical Light Modulators, Pressure Gauges, and Thermometers Attached to Optical Fibers" *J. Micromech. Microeng.*, 4, pp. 343–352.

Guckel, H., Burns, D.W., and Rutigliano, C.T. (1986) "Design and Construction Techniques for Planar Polysilicon Pressure Transducers with Piezoresistive Read-Out," *Solid State Sensor and Actuator Workshop*.

Guckel, H., Burns, D.W., Rutigliano, C.R., Showers, D.K., and Uglow, J. (1987) "Fine Grained Polysilicon and its Application to Planar Pressure Transducers," *International Conference on Solid-State Sensors and Actuators (Transducers)*, pp. 277–282.

Guckel, H. Sniegowski, J.J., and Christenson, T.R. "Construction and Performance Characteristics of Polysilicon Resonating Beam Force Transducers," in *Integrated Micro-Motions: Micromachining, Control, and Applications: A Collection of Contributions Based on Lectures Presented at the Third Toyota Conference*, Harashima, F., ed., Elsevier, Amsterdam, 1990, pp. 393–404.

Hanneborg, A., and Ohlckers, P. (1990) "A Capacitive Silicon Pressure Sensor with Low TCO and High Long-Term Stability," *Sensors Actuators A*, 21–23, pp. 151–154.

Henmi, H., Shoji, S., Shoji, Y., Yoshmi, K., and Esashi, M. (1994) "Vacuum Packaging for Microsensors by Glass-Silicon Anodic Bonding," *Sensors Actuators A*, 43, pp. 243–248.

Hynecek, J. (1992) "Theoretical Analysis and Optimization of CDS Signal Processing Methods for CCD Image Sensors," *IEEE Trans. Electron. Devices*, 39 (11), pp. 2497–2507.

Ikeda, K., Kuwayama, H., Kobayashi, T., Watanabe, T., Nishikawa, T., Yoshida, T., and Harada, K. (1990) "Silicon Pressure Sensor Integrates Resonant Strain Gauge on Diaphragm," *Sensors Actuators A*, 21, pp. 146–150.

Ji, J., Cho, S.T., Zhang, Y., and Najafi, K. (1992) "An Ultraminiature CMOS Pressure Sensor for a Multiplexed Cardiovascular Catheter," *IEEE Trans. Electron. Devices*, 39, pp. 2260–2267.

Kanda, Y. (1982) "A Graphical Representation of the Piezoresistive Coefficients in Si," *IEEE Trans. Electron. Devices*, 29, pp. 64–70.

Kao, T.W., and Taylor, H.F. (1996) "High-Sensitivity Intrinsic Fiber-Optic Fabry–Perot Pressure Sensor," *Opt. Lett.*, 21, pp. 615–618.

Kawamura, Y., Sato, K., Terasawa, T., and Tanaka, S. (1987) "Si Cantilever-Oscillator as a Vacuum Sensor," *International Conference on Solid-State Sensors and Actuators (Transducers)*, pp. 283–286.

Kloeck, B., and de Rooij, N.F. "Mechanical Sensors," in *Semiconductor Sensors*, Sze, S.M., John Wiley, ed., New York, 1994.

Ko, W.H. (1986) "Solid-State Capacitive Pressure Transducers," *Sensors Actuators*, 10, pp. 303–320.

Ko, W.H., Wang, Q., and Wang, Y. (1996) "Touch Mode Pressure Sensors for Industrial Applications," *Solid State Sensor and Actuator Workshop*, pp. 244–248.

Landau, L.D., and Lifschitz, E.M. *Fluid Mechanics*, Pergamon Press, Oxford, 1970.

Lee, Y.S., and Wise, K.D. (1982) "A Batch-Fabricated Silicon Capacitive Pressure Transducer with Low Temperature Sensitivity," *IEEE Trans. Electron. Devices*, 29, pp. 42–48.

Madou, M. *Fundamentals of Microfabrication*, CRC Press, Boca Raton, FL, 1997, pp. 460–464.

Maron, R.J. "High Sensitivity Fiber Optic Pressure Sensor For Use In Harsh Environments," U.S. Patent #6016702, January 25, 2000.

Mastrangelo, C.H. "Method for Producing a Silicon-on-Insulator Capacitive Surface Micromachined Absolute Pressure Sensor," U.S. Patent #5470797, November 28, 1995.

Microsensors, Muller, R.S., Howe, R.T., Senturia, S.D., Smith, R.L., and White, R.M., et al., eds., IEEE Press, Washington, 1990, pp. 329–333.

Middleoek, S., and Audet, S.A. *Silicon Sensors*, Delft University of Technology, The Netherlands, 1994.

Miller, M.F., Allen, M.G., Arkilic, E., Breuer, K.S., and Schmidt, M.A. (1997) "Fabry–Perot Pressure Sensor Arrays for Imaging Surface Pressure Distributions," *International Conference on Solid-State Sensors and Actuators (Transducers)*, pp. 1469–1472.

Muller, R.S., Howe, R.T., Smith, R.L., and White, R.M., et al., eds., *Microsensors*, IEEE Press, Washington, 1990, pp. 350–351.

Murakami, K. "Pressure Transducer and Method Fabricating Same," U.S. Patent #4838088, June 13, 1989.

Park, J.-S., and Gianchandani, Y.B. (2000) "A Capacitive Absolute-Pressure Sensor with External Pick-off Electrodes," *J. Micromech. Microeng.*, 10, pp. 528–533.

Park, J.-S., and Gianchandani, Y.B. (2001) "A Servo-Controlled Capacitive Pressure Sensor with a Three-Mask Fabrication Sequence," *International Conference on Solid-State Sensors and Actuators (Transducers)* 1, pp. 506–509.

Park, J.-S., and Gianchandani, Y.B. (2003) "A Servo-Controlled Capacitive Pressure Sensor Using a Capped-Cylinder Structure Microfabricated by a Three-Mask Process," *IEEE/ASME J. Microelectromech. Syst.*, 12(2), pp. 209–220.

Park, Y.E., and Wise, K.D. (1983) "An MOS Switched Capacitor Readout Amplifier for Capacitive Pressure Sensors," *Record of the IEEE Custom IC Conference.*

Peters, A.J., and Marks, E.A., U.S. Patent #4586109 (April, 1986).

Petersen, K., Pourahmadi, F., Brown, J., Parsons, P., Skinner, M., and Tudor, J. (1991) "Resonant Beam Pressure Sensor Fabricated with Silicon Fusion Bonding," *Solid State Sensor and Actuator Workshop*, pp. 664–667.

Samaun, Wise, K.D., and Angell, J.B. (1973) "An IC Piezoresistive Pressure Sensor for Biomedical Instrumentation," *IEEE Trans. Biomed. Eng.*, 20, pp. 101–109.

Sedra, A.S., and Smith, K.C. (1998) *Microelectronic Circuits*, 4th ed., Oxford University Press, Oxford.

Smith, C.S. (1954) "Piezoresistance Effect in Germanium and Silicon," *Phys. Rev.*, 94, pp. 42–49.

Spencer, R.R., Fleischer, B.M., Barth, P.W., and Angell, J.B. (1985) "The Voltage Controlled Duty-Cycle Oscillator: Basis for a New A-to-D Conversion Technique," *Rec. of the Third International Conference On Solid-State Sensors and Actuators*, pp. 49–52, re-printed in *Microsensors*, Muller, R.S., Howe, R.T., Senturia, S.D., Smith, R.L., and White, R.M., et al., eds., IEEE Press, Washington, 1991.

Stark, B.H., Mei, Y., Zhang, C., and Najafi, K. (2003) "A Doubly Anchored Surface Micromachined Pirani Gauge for Vacuum Package Characterization" *Proc., IEEE International Conference on Microelectromechanical Systems*, pp. 506–509.

Stemme, E., and Stemme, G. (1990) "A Balanced Resonant Pressure Sensor," *Sensors Actuators A*, 21–23, pp. 336–341.

Sugiyama, S., Shimaoka, K., and Tabata, O. (1991) "Surface Micromachines Micro-Diaphragm Pressure Sensors," *International Conference on Solid-State Sensors and Actuators (Transducers)*, pp. 188–191.

Timoshenko, S., and Woinowsky-Krieger, S. *Theory of Plates and Shells*, McGraw-Hill, New York, 1959.

Trzeciakowski, W., Perlin, P., Teisseyre, H., Mendonca, C.A., Micovic, M., Ciepielewski, P., and Kaminska, E. (1992) "Optical Pressure Sensors Based on Semiconductor Quantum Wells, *Sensors Actuators A*, 32, pp. 632–638.

Wang, C.C., Gogoi, B.P., Monk, D.J., and Mastrangelo, C.H. (2000) "Contamination Insensitive Differential Capacitive Pressure Sensors," *Proc., IEEE International Conference on Microelectromechanical Systems*, pp. 551–555.

Wang, Y., and Esashi, M. (1997) "A Novel Electrostatic Servo Capacitive Vacuum Sensor," *Proc., IEEE International Conference on Solid-State Sensors and Actuators (Transducers '97)*, pp. 1457–1460.

Wickramasinghe, H.K. (1989) "Scanned-Probe Microscopes," *Sci. Am.*, Oct., pp. 98–105.

Wise, K.D., and Najafi, K. "VLSI Sensors in Medicine," in *VLSI in Medicine*, Einspruch, N.G., and Gold, R.D. eds., Academic Press, San Diego, CA, 1989.

Wise, K.D. (1994) "On Metrics for MEMS," *MEMS Newslett.*, 1(1).

Yeh, C., and Najafi, K. (1994) "Bulk Silicon Tunneling Based Pressure Sensors" *Solid State Sensor and Actuator Workshop*, pp. 201–204.

Yeh, C., and Najafi, K. (1998) "CMOS Interface Circuitry for a Low Voltage Micromachined Tunneling Accelerometer," *J. Microelectromech. Syst.*, March, pp. 6–9.

Zhang, Y., and Wise, K.D. (1994) "An Ultra-Sensitive Capacitive Pressure Sensor with a Bossed Dielectric Diaphragm," *Solid State Sensor and Actuator Workshop*, pp. 205–208.

Zhang, Y., and Wise, K.D. (1994) "Performance of Non-Planar Silicon Diaphragms Under Large Deflections," *J. Microelectromech. Syst.*, pp. 59–68.

4

Surface Micromachined Devices

Andrew D. Oliver and
David W. Plummer
Sandia National Laboratories

4.1 Introduction

One of the great promises of MEMS and surface micromachining is the small size of the devices that can be fabricated. From Jonathan Swift's stories about the Lilliputians in 1726, to the dollhouses and model trains that fascinate both children and adults in the present, humans are fascinated by small things. The advantages of smaller machines are sometimes very important: in aviation and space applications, a decrease in size and weight corresponds to an increased range or a reduction in the amount of fuel required for a given mission. The advantages of the different physical effects at the microscale level are less obvious. Smaller scales mean that surface micromachined devices are more resistant to shock and vibration than macrosized machines, because the component strength decreases as the square of the dimensions while the mass and inertia decrease as the cube of the dimensions. Another difference is that surface forces, such as van der Waals forces and electrostatic attraction, are much more important at the microscale than at the macroscale, while volume forces such as gravity are much less important. Reduced assembly costs are another advantage of surface micromachined devices. Surface micromachining allows the creation of machines that are assembled at the same time as their constituent components. Instead of using skilled workers to assemble intricate mechanisms by hand or investing in complicated machinery, the assembly is done as a batch process during the integrated-circuit-derived fabrication process. Preassembly imposes certain limitations on the designer, such as the inability to build devices with as-fabricated stored mechanical energy or preload. Instead, structures such as springs must have energy added to them. Preassembly also requires structures that operate out of the plane of the substrate to be erected prior to use.

Perhaps the greatest disadvantage of surface micromachined devices is the limit on the number and type of layers available to a designer. Usually only one structural material with a restricted number of layers (usually two or three) is available to the designer. The thickness of the layers is restricted by the limited deposition rate of equipment and stress in the thin films. The uninterrupted span of a structure in the plane of the substrate is limited by the speed of the release etch. The materials used in surface micromachines may also have residual stress or stress gradients. Having only one structural material also leads to tribology issues. The friction and wear created by like materials rubbing on each other is a large problem in some surface micromachines.

The intent of this chapter is to provide theoretical and practical information to designers of surface micromachined mechanisms and devices. The design of surface micromachined mechanisms has much in common with the design of macrosized mechanisms, but there are also differences. For example, the function of four bar linkages is similar regardless of the size, but the joints and connections between the linkages differ at the microscale. Rotating joints are commonly used in macrosized mechanisms, and flexible joints are more common in surface micromachined mechanisms. The design of surface micromachined mechanisms requires a good understanding of the fabrication process as well as an appreciation of the unique advantages of designing in the microdomain.

This chapter covers some of the strengths and weaknesses of surface micromachines and some of the basics of the mechanics of materials. It also discusses the actuation and packaging of surface micromachines. The chapter concludes by describing some failure modes in surfaced micromachined mechanisms.

4.2 Material Properties and Geometric Considerations

Loads and forces acting on a body or structure will cause it to undergo a change in shape. Most engineering materials will return to their original shape after the load is removed, providing that the load is not too large. If the body returns to its original shape after the load is removed, it has experienced elastic deformation. If the load exceeds the elastic limit, the body will not completely return to its original shape after the load is removed and if the body retains some permanent change in shape, it has experienced plastic or nonelastic deformation.

Most mechanisms are designed so that the expected loads cause only elastic deformation: plastic deformation only occurs during failure. The designer must be able to quantify the threshold between elastic and plastic deformation. The equations typically used in design are based on certain assumptions about the material and its geometric configuration. The theory of elasticity contains expressions relating deformation and load that do not require simplifying assumptions, but those are usually too mathematically complex for designers to use.

One assumption made in the derivation of simple design equations is that the material is isotropic. An isotropic material has elastic properties that are the same in all directions. The designer is cautioned about blindly accepting the assumption of isotropy. Single-crystal silicon [Worthman and Evans, 1965], individual crystals of polysilicon [Madou, 2002], and electroformed materials in the LIGA process [Sharpe et al., 1997] are known to show nonisotropic behavior.

A second common assumption is homogeneity. A homogeneous body is one that has the same properties throughout its entire extent. Most materials consist of an aggregate of very small crystals spread throughout the body. When a cross section of the body contains a large number of crystals, the effects of forces and loads are dispersed and shared so the assumption of homogeneity is fulfilled. With polycrystalline surface micromachining techniques, mechanism dimensions can approach the size scale of the crystals. In those cases, the critical cross section may contain only a few crystals. The effect of the load cannot be averaged and nonhomogeneity occurs. With the cautions described above, the equations to be presented are a good starting point for surface micromachine designers.

4.3 Stress and Strain

When a body supports a load, the material is under stress. Stress is a measure of the intensity of the body's reaction to the load and is a field quantity. It is measured as the force per unit of exposed cross-sectional area. In SI units, stress is measured in Pascals (Newtons per square meter).

In many instances, the distribution of stress across the area is ignored so that only the average stress is considered. This is especially true in the cases of axial loads along slender bodies that cause only lengthening or shortening. The average stress, σ, is equal to the load divided by the cross-sectional area:

$$\sigma = \frac{F}{A} \tag{4.1}$$

Tension is stress that tends to lengthen a body and compression is stress that tends to shorten a body.

The deformation, δ, of a body under load is dependent on the size and shape of the body. The amount of deformation normalized by the dimensions of the body, L, is called strain, ε. Strain is represented mathematically as:

$$\varepsilon = \frac{\delta}{L} \tag{4.2}$$

4.3.1 Young's Modulus

In a body undergoing elastic deformation in the normal direction, the ratio of stress to strain is known as the Young's modulus, E, which is also called the modulus of elasticity:

$$E = \frac{\sigma}{\varepsilon} \tag{4.3}$$

If the body is undergoing shear, the corresponding proportionality constant is G, the modulus of rigidity or the shear modulus of elasticity. Both of these proportionality constants have units of Pascals (Pa). These material properties describe the stiffness of the material. A compliant material, such as rubber, has a small Young's modulus (0.5 GPa); a hard material, such as stainless steel (e.g. ASTM 18–8), has a large Young's modulus (190 GPa); and a single-crystal diamond has a Young's modulus of 1035 GPa.

4.3.2 Poisson's Ratio

Poisson's ratio, υ, describes the ratio of the transverse strain to the axial strain when a body is subjected to an axial load:

$$\upsilon = -\frac{\varepsilon_{transverse}}{\varepsilon_{axial}} \tag{4.4}$$

This can be visualized by thinking about pieces of rubber: if a rubber band is stretched (a positive axial strain), then the rubber band becomes narrower, which is a negative transverse strain. Conversely, if a pencil eraser is compressed, then it becomes wider. In most cases, υ has a value between 0.2 and 0.5. A value of 0.5 corresponds to a conservation of volume.

In situations where the shear modulus is not known, it can be estimated using E and υ in the following relationship:

$$E = 2G(1 + \upsilon) \tag{4.5}$$

if the materials are assumed to be isotropic.

4.3.3 Contact Stresses

When two bodies are pressed together, the point or line of contact expands to area contact. As the surfaces move together, their progress is interrupted when a single point from one body, an asperity, contacts an asperity from the other body. These asperities attempt to carry the entire contact force, but quickly yield and plastically deform. This allows the bodies to move closer together until more asperities touch. This process of contact and yield continues until there is enough total contact area to support the force at the yield strength of the material. These regions of contact, called *a-spots*, typically form a circular pattern known as the Hertzian stress circle.

If the contacting surfaces have a radius of curvature, the radius of the resulting Hertzian stress circle can be calculated using the equation originally developed by Hertz:

$$a = \sqrt[3]{\frac{3F}{4} \times \frac{(1 - v_1^2)/E_1 + (1 - v_2^2)/E_2}{(1/r_1 + 1/r_2)}} \tag{4.6}$$

where F is the force of contact, v is Poisson's ratio, E is Young's modulus, and r is the radii of curvature for the two contacting surfaces [Johnson, 1985]. For the case of a curved surface contacting a plane, the radius of curvature of the plane is equal to infinity. It is good design practice to ensure that the apparent contact area (determined by the nominal dimensions) is larger than the Hertzian contact area. A more detailed description can be found in Roark and Young (1989).

4.3.4 Stress in Films and Stress Gradients

One of the important nonidealities in designing surface micromachined mechanisms is as fabricated stress in the thin films. The stress has different sources and effects depending on the application and the geometries involved. One of the sources of stress in thin films is the difference in the thermal expansion coefficients between the film and the substrate, especially if the film was deposited at an elevated temperature. Another source of stress is deposition of films away from equilibrium conditions such as in chemical vapor deposition. An example of stress in thin films is thermally grown silicon dioxide (SiO_2) that is in compression because the SiO_2 molecules are larger than the lattice spacing of the host silicon atoms.

There are also many effects of stresses in thin films. Excessive stress can result in cracking or delamination of thin films. Stress in films can also be relieved as strain or deformed microstructures. For example, a compressive stress will buckle a clamped–clamped beam if the stress is above a certain level. This will be discussed later in the section on buckling. Stress will also change the compliance, spring constant, and resonant frequency of beams. Chapter 6 of the text by Hsu (2004) includes a description of some test structures for measuring stress in thin films. A simulation of the change in the resonant frequency of a beam as a function of compressive stress is shown in Figure 4.1. Control of residual stress is important in resonant devices, for example a resonant clamped–clamped beam resonator used as an oscillator, which is shown in Figure 4.2. A good summary of resonant devices for wireless applications can be found in Nguyen et al. (1998).

One method of relieving stresses in thin-film mechanical elements is the use of folded flexures. Folded flexures relieve axial stress because each flexure is free to expand or contract in the axial direction. This eliminates most of the stress caused by the fabrication process and by thermal expansion mismatches between the flexures and the substrate. Crab-leg flexures and meandering flexures also relieve some of the residual stress. All three flexure types are shown in Figure 4.3.

Another nonideality of surface micromachines are stress gradients in the films. Stress gradients are variations in the stress of the film through the thickness of the film and cause cantilevers to either curl away from or toward the substrate. This can be partially compensated for by folding the flexures. Stress gradients will also warp thin plates or gears into dome shapes and cause cantilever beams to bend [Fan et al., 1990].

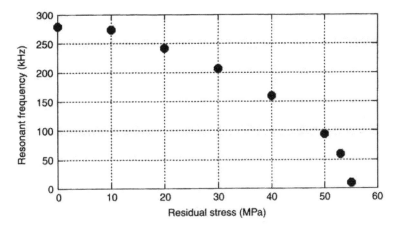

FIGURE 4.1 The resonate frequency for a clamped–clamped beam 300 μm long by 30 μm wide by 3 μm thick under various amounts of residual compressive stress. (Courtesy of Fernando Bitsie, Sandia National Laboratories.)

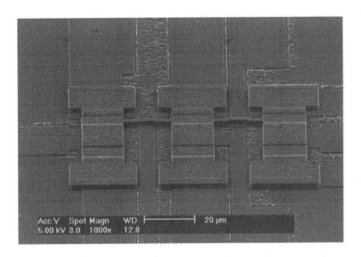

FIGURE 4.2 Clamped–clamped beams used as a micromechanical filter. (Photograph courtesy of M. Abdelmoneum, University of Michigan.)

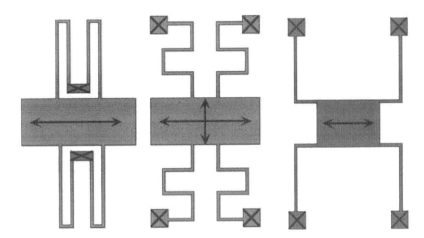

FIGURE 4.3 Three flexure designs. From left to right: folded flexures, meandering flexures, and crab-leg flexures.

4.3.5   Wear

One of the limitations of surface micromachined devices is wear. This is not surprising as these mechanisms can have touching, rubbing, or impacting surfaces which are both typically constructed of the same material. This is different than conventional machines, which commonly use dissimilar materials to reduce wear. Also, surface micromachines typically run without liquid lubrication. Tanner (2000) and Dugger et al. (1999) have conducted a great deal of research into the wear of polysilicon surface micromachines. They report that the wear of polysilicon surface micromachines has similarities to the wear of macroscopic mechanisms. Friction and wear have been correlated with the normal forces on a surface: the greater the force, the higher the wear rate. Also, the wear rate was found to increase at operating frequencies above the resonant frequency. It has been theorized that wear is caused by adhesion and stick-slip behavior between two moving surfaces. The accumulation of wear debris that results in three-body wear (the two surfaces and the wear debris) has also been observed in surface micromachined mechanisms. Figure 4.4 shows the effects of extreme, accelerated wear on a surface micromachined pin joint and a hub. Another source of information on the tribology of surface micromachined devices is Chapter 16 of the *Handbook of Micro/Nano Tribology, 2nd ed.*, by Bhushan (1999). For increased reliability, many designers of surface micromachines attempt to minimize or eliminate rubbing and contacting surfaces.

4.3.6   Stiction

Potentially, one of the most destructive influences on the performance of surface micromachined mechanisms is stiction, which is the unintended adhesion of surface micromachined parts to each other or to the substrate. It is a combination of various effects that conspire to pull flat, compliant MEMS structures together and to create permanent bonds between them. Contributing factors to stiction include very flat (nanometer-scale roughness) surfaces, large surface areas, compliant structures, electrostatic attraction, humidity, and hydrophilic surfaces. There are two major categories of stiction: stiction due to the release process and in-use stiction. The main cause of stiction during the release process is capillary forces of the liquids in the release process that pull the parts together. Release-stiction is alleviated with various hydrophobic coatings, and using critical point, or sublimation drying processes. A cause of in-use stiction is electrostatic attraction. This often involves insulating materials, such as silicon nitride, that trap charges. The trapped charge causes an electrostatic force in which the magnitude can shift in an unpredictable manner. Other causes of in-use stiction are van der Waals forces, hydrogen bonding, and the condensation of water, which results in capillary forces.

Forces due to stiction are not typically measured in Newtons or Pascals, but rather in terms of work of adhesion Γ, J/m^2 An equation for the stiction force due to van der Waals forces is:

$$\Gamma = \frac{A}{12\pi x^2} \tag{4.7}$$

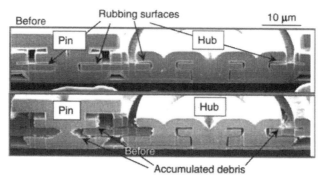

FIGURE 4.4   A pin joint and hub joint before and after accelerated wear. (Courtesy of D. Tanner, Sandia National Laboratories.)

where A, the Hamaker constant, is about the 5×10^{-20} J, for a silicon oxide or hydrocarbon surface and x is the average separation. The adhesive force due to stiction is the derivative of the adhesive energy and has units of N/m²:

$$\frac{d\Gamma}{dx} = -\frac{A}{6\pi x^3} \tag{4.8}$$

Equation 4.7 has values in the micro Joule per square meter range for separations of 5–10 nm. If water or other liquid causes capillary forces, the value of Γ will be in the milli Joule per square meter range. The mechanism designer can decrease stiction by reducing the surface area of the parts, for example by using a spoked gear instead of a solid gear. Another way of reducing the surface area is to use dimples, which are minimum feature-sized protrusions on the bottom of a surface micromachined part. Dimples greatly reduce the surface area in contact with the substrate. It is also important to design structures that are stiff in the out-of-plane direction because it increases the amount of force required to pull the structure down to the substrate. The designer should also minimize the area of any contacting surfaces. The best method of avoiding the problem with dielectric layers accumulating stray charges is to use a conductive ground plane to cover any exposed insulating layers and to connect every object to an electrical ground. If this is not possible (for instance, with a rotating gear), at a minimum, the hub should be tied to the ground. Packaging is another area where the mechanism designer can combat stiction. If water vapor in the package condenses on the surface micromachine, the water will cause capillary adhesive forces similar to those found in the release process. Therefore, a package with a dewpoint well below the minimum storage or operating temperature of the mechanism is essential.

4.4 Machine Design

The fundamental difference between structures and machines is that machines have intended degrees of freedom — usually one. Micromachines can be further divided into two classes by the element used to allow the motion. Elastic-mode machines use compliance elements, such as springs and flexures, to allow motion. Rigid-body-mode (or gross-motion) machines permit motion through joints and bearings. Gross-motion machines allow parts to accumulate motion. As an example of gross motion, a gear spinning on a hub accumulates angular displacement. Elastic-mode machines force the machine parts to oscillate about a fixed point or axis.

An exciting trend in surface micromachined mechanisms has been the emphasis on using only elastic mode elements. These devices have advantages that include no contacting surfaces, no wear, no rubbing surfaces, and reduced stiction. They are also well suited to materials such as polysilicon or silicon that can tolerate large strains without fracturing or plastically deforming. One drawback is that compliant mechanisms often operate past the linear region and thus they require the use of elliptic integrals or numerical analysis in the design process. Also, compliant mechanisms need to be combined with other types of mechanisms such as rotary joints in order for them to accumulate displacement. By themselves, they are only oscillatory. An excellent reference on the theory and design of compliant mechanisms, including their application to surface micromachines, is *Compliant Mechanisms* by Howell.

4.4.1 Compliance Elements: Columns, Beams, and Flexures

Compliance elements or springs allow machine parts to move. Springs take many forms, but they are primarily realized in microsystems as beams or flexures. Elastic-mode machines are usually focused on resonating devices. In some cases, the spring element also serves as the resonating mass and the spring must be evaluated using the beam equations from continuum mechanics. If the spring supports another element whose mass is significantly larger, then the spring can be treated as a lumped element described by a parameter called the spring constant or spring rate. The spring rate is the ratio of applied force to deflection:

$$k = \frac{F}{\delta} \tag{4.9}$$

where k is the spring constant, F is the applied force and δ is the resulting deflection.

FIGURE 4.5   Illustration showing the elongation of a column. L is the original length of the column, and δ is the change in the length of the column.

4.4.1.1   Columns

A column is a structural member that supports pure tension or compression. Columns are loaded along their longitudinal axis and are generally stiff in comparison to beams and flexures. Columns are not frequently used by themselves but more often as part of more complex structures so they are important to understand. If loaded in compression beyond a critical value, columns fail by buckling. Buckling is a large lateral deformation caused by an axial load. The force F on a column creates an elongation, δ, shown in Figure 4.5: this relationship is derived from Equations 4.1 to 4.3:

$$\frac{F}{A} = E\frac{\delta}{L} \tag{4.10}$$

where E is Young's modulus, A is the cross-sectional area and L is the length of the column. The spring rate for columns is:

$$k = \frac{EA}{L} \tag{4.11}$$

4.4.1.2   Beams

Beams are structures characterized by their ability to support bending and they provide a linear degree of freedom. Moments and transverse loads applied to beams deform them into curves, as opposed to axial loads that only make them longer or shorter. As the loaded beam takes on a curved shape, one side of the beam becomes longer while the opposite side is shortened. The stretched surface of the beam is in tension and the shortened surface of the beam is in compression. A plane through the geometric center of the beam, called the neutral axis, does not change in length at all and is unstressed as a result of bending.

For simple cases, the maximum stress in a beam occurs at the surface and is independent of the material composing the beam. The magnitude of the stress of a beam bent into a curved shape by either a moment or a transverse load is:

$$\sigma = \frac{Mc}{I} \tag{4.12}$$

where M is the moment causing the curvature, c is the distance from the neutral axis to the surface of the beam (usually half the thickness), and I is the area moment of inertia of the beam's cross section.

Moment of inertia is a measure of the dispersion of area about the centroid. In an analogy to statistics, the centroid is directly analogous to the mean, and the moment of inertia is analogous to the standard deviation. There are different types of *moment of inertia*. Moment of inertia is calculated when the axis of rotation is in the same plane as the cross section. The *polar moment of inertia* is calculated for cases when the axis of rotation is normal to the cross section. Expressions for moment of inertia for common geometric shapes are available in most mechanical engineering handbooks, as well as mechanics of materials textbooks. The moments of inertia for the two common shapes encountered in surface micromachined

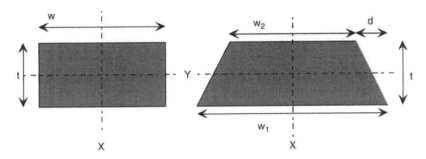

FIGURE 4.6   Moment of inertia for two common beam shapes.

mechanisms are illustrated in Figure 4.6. The lines through the cross sections in Figure 4.6 show the axis of rotation for the curving beam. The moment of inertia for a rectangular beam with a rotation about the y-axis of rotation is:

$$I_Y = \frac{wt^3}{12}$$

(4.13)

and for rotation about the x-axis:

$$I_X = \frac{tw^3}{12}$$

(4.14)

where w and t are the two dimensions shown in Figure 4.6. For a trapezoidal beam about the y-axis the moment of inertia is:

$$I = \frac{t^3(w_1^2 + 4w_1w_2 + w_2^2)}{36(w_1 + w_2)}$$

(4.15)

and the moment of inertia about the x-axis is:

$$I = \frac{t}{36(w_1 + w_2)}\left[w_1^4 + w_2^4 + 2w_1w_2(w_2^2 + w_1^2) - d(w_1^3 + 3w_1^2w_2 - 3w_1w_2^2 - w_2^3) \right. \\ \left. + d^2(w_1^2 + 4w_1w_2 + w_2^2)\right]$$

(4.16)

The polar moment of inertia for torsional loads is defined as:

$$J = I_X + I_Y$$

(4.17)

Example

At a given cross section in a beam, the moment acting on it is 2 nN-m. The beam is rectangular with a thickness of 2 μm and a width of 30 μm. What is the maximum stress in the beam if the beam is bending away from the substrate?

The stress caused by bending in the beam is $\sigma = Mc/I$, where c is the distance from the neutral axis to the surface of interest. Because we want the maximum stress, c is half the thickness. The moment of inertia must be calculated from Equation 4.13, the expression for rectangular cross sections.

$$I = \frac{30\ \mu m \times (2\ \mu m)^3}{12} = 2 \times 10^{-23}\ m^4$$

The resulting maximum stress at the cross section is:

$$\sigma = \frac{2 \times 10^{-9}\ \text{Nm} \times 1 \times 10^{-6}\ m}{2 \times 10^{-23}\ m^4} = 10^8\ \text{N/m}^2 = 100\ \text{MPa}$$

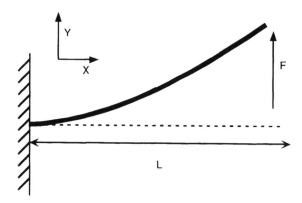

FIGURE 4.7 Cantilever beam spring. The deformation of the beam is due to the load *F* and can be calculated using Equations 4.19 and 4.20.

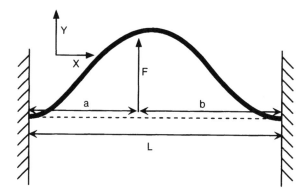

FIGURE 4.8 Deflection of a fixed-beam spring due to a point load. The applied load is *F*, the load's location is shown by *a* and *b* on the x-axis, and the spring deflects in the direction of the y-axis.

Note that in this example it was not necessary to know the length of the beam or the location of the applied loads. However, this information is required to calculate the moment.

Beams can be categorized by their supports. A beam held rigid on one end and free to move on the other is called a *cantilever* beam. A beam held rigid at both ends is called a *fixed* beam. For the same dimensions, cantilever beams are more flexible than fixed beams, so cantilevers have lower natural frequencies and larger deflections. Drawings of a cantilever beam spring and a fixed beam spring are shown in Figures 4.7 and 4.8, respectively.

The equations for deflection of cantilever beams and fixed beams are relatively straightforward and can be applied accurately in most cases. In situations of complex geometry or intricate loading or if more precision is required, alternate methods, such as finite element analysis, should be employed.

4.4.1.3 Stress Concentration

Stress equations are developed for homogeneous bodies with mathematically simple geometries. In real devices, the geometry is usually complex and this can lead to regions of highly localized stresses not predicted by the simple equations. Stress concentrations are significant for ductile materials only when the loads are repeated. For the case of static loading, stress concentration is important only for materials that are both brittle and relatively homogeneous. These amplifications of stress must be accounted for in design. Stress concentration is an empirical contrivance for design engineers because stress concentration factors better predict the stress than the nominal stress equations. More importantly, information about stress concentration can be used as design guidance to avoid performance complications or failures.

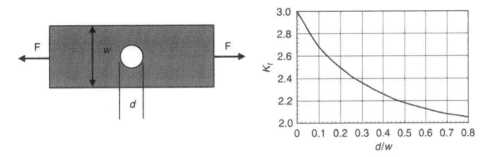

FIGURE 4.9   Stress concentrations for a flat plate loaded axially with a circular hole. The maximum stress is located around the hole.

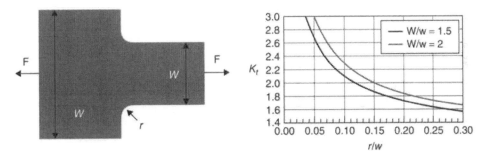

FIGURE 4.10   (See color insert following page 2-12.) Stress concentrations for a flat plate loaded axially with two different widths and fillet radius *r*. The maximum stress is located around the fillets.

Because geometries occur in an infinite variety of ways, it is not possible to find an empirical stress concentration that applies for every case. Thus, engineering judgment with considerable estimation is essential.

Etch release holes, an accommodation to the fabrication process, serve as stress concentrations in plates. Sharp corners in Manhattan (right-angle) geometries also create stress concentrations. The stress concentration factor, K_t, is used to adjust the stress calculated from the standard equations, such as Equations 4.1 and 4.11:

$$\sigma_{CONCENTRATION} = K_t \sigma_{NOMINAL} \tag{4.18}$$

For flat plates loaded axially with a circular hole in them, the nominal stress is given by:

$$\sigma_{NOM} = \frac{F}{t(w - d)} \tag{4.19}$$

where *t* is the thickness, *w* is the width of the plate, and *d* is the diameter of the hole. The stress concentration K_t is shown in Figure 4.9 along with a drawing of the geometry. For square etch release holes, refer to Figure 4.9 and Equation 4.19, with the length of the sides of the square opening being equal to the diameter of the circle.

For flat plates with two different widths, the nominal stress is given by Equation 4.20. Figure 4.10 illustrates a drawing of the geometry and a graph of the stress concentration. Note that the stress concentration factor depends on the radius of the fillet that is denoted as *r*. For surface-micromachines designed with a Manhattan geometry, *r* is nominally zero but the fabrication process will create small non-zero radii in the corners. The process engineer should be able to provide a reasonable estimate for *r*.

$$\sigma_{NOM} = \frac{F}{wt} \tag{4.20}$$

4.4.1.4   Cantilever Beam Springs

The deformation y at any arbitrary point x along a cantilever beam, illustrated in Figure 4.7, can be calculated by:

$$y = \frac{F}{6EI}(3Lx^2 - x^3) \tag{4.21}$$

where L is the length of the beam, E is Young's modulus of the beam material and F is the applied load. The maximum deflection occurs at the end of the beam and can be calculated by:

$$y_{MAX} = \frac{FL^3}{3EI} \tag{4.22}$$

The maximum bending stress in a cantilever beam occurs at the attachment or fixed end where the internal moment is FL. For a rectangular cross section of thickness t and width w, the maximum stress can be calculated with Equation 4.12 to be:

$$\sigma_{MAX} = \frac{6FL}{wt^2} \tag{4.23}$$

The fundamental resonant frequency as given in Hertz of a simple cantilever beam is:

$$f_1 = \frac{3.52}{2\pi}\sqrt{\frac{EI}{\rho AL^4}} \tag{4.24}$$

where E is Young's modulus, A is the cross-sectional area and ρ is the density (mass per unit volume).

Example

A cantilever beam fabricated in polycrystalline silicon with a rectangular cross section is electrostatically pulled down to the substrate by a small electrode at the end of the beam. The beam is $50\,\mu m$ wide, $6\,\mu m$ thick, and $200\,\mu m$ long. The gap between the bottom of the beam and the electrode is $2\,\mu m$. What is the required electrostatic force? If the beam were suddenly released, at what frequency would it vibrate?

The equation for cantilever beam deflection can be solved for force:

$$F = \frac{3EIy}{L^3} \tag{4.25}$$

A reasonable value for the Young's modulus for polycrystalline silicon is $155\,GPa$ or $0.155\,N/\mu m^2$. The area moment of inertia for a rectangular cross section is $I = wt^3/12 = (50\,\mu m)(6\,\mu m)^3/12 = 900\,\mu m^4$. So,

$$F = \frac{3(0.155\,N/\mu m^2)(900\,\mu m^4)(2\,\mu m)}{(200\,\mu m)^3} = 105\,\mu N$$

A reasonable value for the density of solid polycrystalline silicon is $2.33 \times 10^{-15}\,kg/\mu m^3$. The cross-sectional area of the beam is $A = wt = (50\,\mu m)(6\,\mu m) = 300\,\mu m^2$:

$$f = \frac{3.52}{2\pi}\sqrt{\frac{(0.155\,N/\mu m^3)(900\,\mu m^4)(10^6\,\mu/m)}{[2.33(10^{-15})kg/\mu m^3](50\,\mu m \times 6\,\mu m)(200\,\mu m)^4}} = 198\,kHz$$

4.4.1.5   Fixed Beam Springs

The deflection at any point $0 < x < a$ along a fixed beam, shown in Figure 4.8, can be calculated using the following equation:

$$y = \frac{Fb^2x^2[3aL - (3a + b)x]}{6L^3EI} \tag{4.26}$$

This equation is based on a point-applied load somewhere along the length of the beam. The deflection caused by a distributed load is somewhat different. Deflection can be estimated by assuming the entire load is applied at the center with some error. An exact solution can be found in texts on mechanics of materials such as Timoshenko (1958).

The fundamental frequency for a fixed beam is:

$$f_1 = \frac{22.4}{2\pi} \sqrt{\frac{EI}{\rho A L^4}} \qquad (4.27)$$

The form of this equation is similar Equation 4.24 for cantilever beams. Note that fixing the other end of the beam increases the fundamental frequency by more than a factor of six.

Example

Assuming the same parameters as the cantilever beam example, except that it is fixed on both ends, calculate the force applied at the center necessary to pull it down to the substrate and the resulting frequency of vibration when released.

For this case, $x = a = b = L/2$, so the equation for deflection reduces to:

$$y_{MAX} = \frac{FL^3}{192\,EI} \qquad (4.28)$$

and solving for load:

$$F = \frac{192\,EIy}{L^3} = \frac{192(0.155\,\text{N/}\mu\text{m}^2)(900\,\mu\text{m}^4)(2\,\mu\text{m})}{(200\,\mu\text{m})^3} = 6700\,\mu\text{N}$$

$$f = \frac{22.4}{2\pi} \sqrt{\frac{(0.155\,\text{N/}\mu\text{m}^2)(900\,\mu\text{m}^4)(10^6\,\mu\text{m/m})}{[2.33(10^{-15})\,\text{kg/}\mu\text{m}^3](6\,\mu\text{m} \times 50\,\mu\text{m})(200\,\mu\text{m})^4}} = 1.26\,\text{MHz}$$

4.4.1.6 Flexures

Flexures are typically used when a rotational degree of freedom is desired. As opposed to beams where the applied moments cause curvature, torque applied to flexures causes twisting. The ends of the flexure will rotate relative to one another through some angle θ. The resultant twisting creates shear stress in the element. Strictly speaking, the only cross section for which the simple torsion equation applies is circular. Circular cross sections do not appear frequently in micromachine design; rather, rectangular cross sections are the norm. The simple equation can be adapted for use with little problem.

The shear stress in a circular bar arising from an applied torque is:

$$\tau = \frac{Tc}{J} \qquad (4.29)$$

where τ is the shear stress, T is the applied torque, c is the distance from the central axis to the point where the stress is desired, and J is the polar area moment of inertia.

The equation above must be modified to account for rectangular cross sections violating a basic assumption used in deriving the equation above. More precise analysis suggests the following equation:

$$\tau = \frac{9T}{2wt^2} \qquad (4.30)$$

where w is the width (larger dimension) of the flexure and t is the thickness (smaller dimension). The deflection created by the applied torque is:

$$\theta = \frac{TL}{JG} \qquad (4.31)$$

where L is the length of the flexure, G is the shear modulus and J is the polar moment of inertia.

Example

A surface micromachined device made of aluminum contains a 150 μm square plate and is supported by 6 μm-wide torsional flexures forming an axis of rotation through the plate's midpoint. The plate is 4 μm thick and the flexures are 20 μm long. If one edge is pulled down to the substrate 3 μm from the bottom of the plate, what is the resulting shear stress in the flexure? When the force is released, at what frequency will the system vibrate? Use a modulus of rigidity of 26 GPa and a density of 2660 kg/m³.

Assuming that the plate does not significantly deform, the flexures are required to rotate through an angle of 2.3° for the plate to touch the substrate. This angle can be estimated by:

$$\theta = \tan^{-1}\left(\frac{3\,\mu m}{(150\,\mu m/2)}\right) = 2.3° = 0.040 \text{ rad}$$

From Equation 4.17 the polar area moment of inertia for a rectangular section is:

$$J = I_x + I_y = \frac{wt}{12}(w^2 + t^2) = \frac{(6\,\mu m)(4\,\mu m)}{12}\left[(6\,\mu m)^2 + (4\,\mu m)^2\right] = 104\,\mu m^4$$

The torque required to rotate one flexure through the angle θ is:

$$T = \frac{JG\theta}{L} = \frac{(104\,\mu m^4)(0.0772\,N/\mu m^2)(0.04 \text{ rad})}{20\,\mu m} = 1.6 \times 10^{-2}\,\mu Nm$$

This is the torque per flexure. The total torque on the plate is twice that value because there are two flexures. If it could be applied at the edge, the value of the force needed to create this much rotation is:

$$F = 2\frac{T}{r} = \frac{1.6 \times 10^{-2}\,N \cdot \mu m}{50\,\mu m} = 6.4 \times 10^{-4}\,N$$

The resulting shear stress in each flexure is:

$$\tau = \frac{9T}{2wt^2} = \frac{9 \times 1.6 \times 10^{-2}\,N \cdot \mu m}{2(6\,\mu m)(4\,\mu m)^2} = 7.5 \times 10^{-4}\,N/\mu m^2 = 0.75\,GPa$$

In a dynamics context, the system can be a modeled as a rigid body vibrating about a spring. The differential equation of motion for a simple mass–spring system and the associated natural frequency are:

$$i\ddot{\varphi} + K\varphi = 0 \tag{4.32}$$

$$f_n = \frac{1}{2\pi}\sqrt{\frac{K}{i}} \tag{4.33}$$

where i is the mass moment of inertia (different than area moment of inertia used to calculate stress) and K is the torsional spring rate. The spring rate of the flexures is:

$$K = \frac{T}{\theta} = \frac{JG}{L} = \frac{(104\,\mu m^4)(0.026\,N/\mu m^2)}{20\,\mu m} = 0.14\,\mu Nm/rad$$

The mass of the plate is:

$$m = \rho wlt = [2.71 \times 10^{-18}\,kg/\mu m^3](150\,\mu m)^2(4\,\mu m) = 2.44 \times 10^{-10}\,kg$$

The mass moment of inertia for a thin rectangular plate vibrating about its center is:

$$i = \frac{1}{12}mw^2 = \frac{[2.44 \times 10^{-10}\,kg](150\,\mu m)^2}{12} = 4.5 \times 10^{-7}\,kg \cdot \mu m^2$$

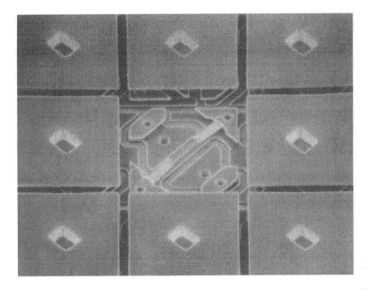

FIGURE 4.11 One pixel of the Texas Instruments surface micromachined DMD device. Each pixel is $16\,\mu m \times 16\,\mu m$. (Photograph courtesy of Texas Instruments.)

The natural frequency is:

$$f_n = \frac{1}{2\pi} \sqrt{\frac{(0.14\,\mu N \cdot m)}{4.57 \times 10^{-19}\,kg \cdot m^2}} = 87\,kHz$$

The device in this problem is similar to the Texas Instruments DMD device, which is shown in Figure 4.11.

4.4.1.7 Springs in Combinations

Springs are energy storage elements analogous to electrical capacitors. The laws for calculating equivalent spring rates are the same as those for calculating equivalent capacitance. Springs are in parallel if they undergo the same deflection and they are in series if they support the same loading. The equivalent spring rate for springs in parallel is:

$$k_{eq} = k_1 + k_2 + k_3 + \cdots \tag{4.34}$$

where springs in parallel experience the same deflection δ.

The equivalent spring rate for spring in series is:

$$\frac{1}{k_{eq}} = \frac{1}{k_1} + \frac{1}{k_2} + \frac{1}{k_3} + \cdots \tag{4.35}$$

where springs in series can be identified because each spring experiences the same force. Springs in parallel and series are shown in Figures 4.12 and 4.13, respectively.

Whether the design goal is to tailor the performance or to conserve space, it is sometimes desirable to use springs in combinations. Figure 4.14 shows an oscillating structure supported by a complex suspension system. The suspension is comprised of a parallel combination of four spring systems each of which is a series combination of three individual beam and column elements. In the example shown in Figure 4.14, the U-shaped springs are all attached to the substrate and to the resonator shuttle. When the shuttle moves, each of the four U-shaped springs deflects the same amount — the motion of the shuttle.

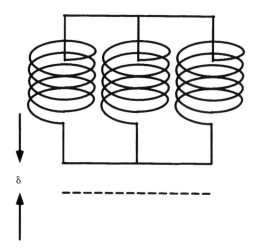

FIGURE 4.12   Springs in parallel.

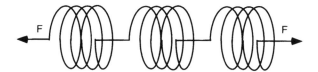

FIGURE 4.13   Springs in series.

There are four different paths from the shuttle to the ground, which is another indication that the springs are in parallel.

Closer inspection of a single U-shaped spring reveals three beams in series. Because the beams form a single path between the shuttle and ground, they must carry the same load. A free-body diagram shows that the load from one beam is transmitted fully to the next until the load to transmitted to ground.

Example

Calculate the equivalent spring rate of the U-shaped spring shown in Figure 4.15, which is a simplification of one of the springs shown in Figure 4.14.

The spring system consists of three springs in series: two cantilever beam springs connected by one column. It is reasonable to treat spring 2 simply as a column even though it supports reaction moments from springs 1 and 3, because none of the resulting bend in spring 2 contributes to the motion of the shuttle. The springs are clearly in series, as the load from one spring is transmitted fully to the next and each spring is allowed to deflect whatever amount is required to support the load.

For springs in series, the equivalent spring rate is given by:

$$\frac{1}{k_{eq}} = \frac{1}{k_1} + \frac{1}{k_2} + \frac{1}{k_3} \tag{4.35}$$

For surface micromachines, all the springs are fabricated of the same material and thus have the same material properties. Springs 1 and 3 are cantilever beams and their rates, respectively, are:

$$k_1 = \frac{3EI_1}{L_1^3} \text{ and } k_3 = \frac{3EI_3}{L_3^3}$$

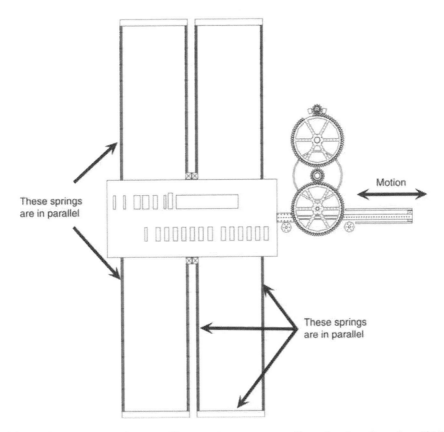

FIGURE 4.14 Surface micromachined oscillating mechanism supported by springs in series and parallel. This drawing is a modification of one by G.E. Vernon and M.A. Polosky.

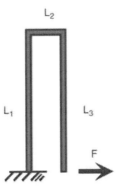

FIGURE 4.15 U-shaped spring comprised of three beam springs connected in series and parallel.

From Equation 4.11, the rate for the column, spring 2, is:

$$k_2 = \frac{EA_2}{L_2}$$

The equivalent rate for the three-spring combination is:

$$\frac{1}{k_{eq}} = \frac{1}{k_1} + \frac{1}{k_2} + \frac{1}{k_3} = \frac{L_1^3}{3EI_1} + \frac{L_2}{EA_2} + \frac{L_3^3}{3EI_3}$$

4.4.1.8 Buckling

When a short column is loaded in compression, the average compressive stress is calculated by simply dividing the load by the cross-sectional area as in Equation 4.1. However, when the column is long and slender, the situation is complicated by the possibility of static instability, also known as lateral buckling.

When loaded below the buckling threshold, a column under a compressive load will get shorter while remaining essentially straight. After the buckling threshold is exceeded, the column deflects normally to the axis of the column and the stress increases rapidly. These stresses can cause failure by rupture.

The phenomenon of buckling is much different than that of bending. A beam will begin to bend as soon as any moment (bending load) is present. In contrast, a column will not exhibit any lateral deflection until the critical or buckling load is reached. Above the critical load, additional loading causes large increases in lateral deflection. Because of the tensile properties of polysilicon, it is possible to design a column to buckle without exceeding the yield strength. If yield does not occur, the column will return to its original straight position after the load is removed. If designed appropriately, a buckling column can be used as an out-of-plane flexure [Garcia 1998].

Buckling is a complex nonlinear problem and predicting the shape of the buckled structure is beyond the scope of this text. It is, however, covered in several references including Timoshenko and Gere (1961), Brush and Almroth (1975), Hutchinson and Koiter (1970), and Fang and Wickert (1994). However, the equation for predicting the onset of buckling is relatively straightforward. For a column with one end fixed and the other free to move in any direction (i.e., a cantilever), the critical load to cause buckling is:

$$F_{CR} = \frac{\pi^2 EI}{4L^2} \tag{4.36}$$

It is important to remember that the critical load for buckling cannot exceed the maximum force supported by the material. In other words, if the load needed to cause buckling is larger than the load needed to exceed the compressive strength, the column will fail by rupture before buckling.

4.4.1.9 Hinges and Hubs

A different type of rigid-body-mode machine is a vertical axis hinge. These hinges are used in surface micromachining to create three-dimensional structures out of a two-dimensional surface micromachining process. In the surface micromachining process, hinges are constructed by stapling one layer of polysilicon over another layer of polysilicon with a sacrificial layer between them. A cross section of a simple hinge is illustrated in Figure 4.16. A more complex hinge is shown in Figure 4.17. Hinges have some

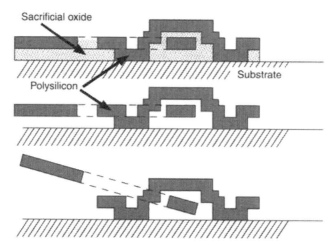

FIGURE 4.16 Cross section of a polysilicon mirror hinge. The top figure shows the device before release, the middle drawing is after release, and the bottom depicts the actuated device.

advantages over flexible joints. One advantage is that no stress is transmitted to the hinged part so greater angles of rotation are possible. Also, the performance of hinged structures is not influenced by the thickness of the material and deformation of the machine is not required. The limitations of hinged structures are that the sacrificial layers must be thick enough for the hinge pin to rotate and at least two released layers of material are required. Hinges are also susceptible to problems with friction, wear, and the associated reliability problems. The same limitations and advantages are true for structures that rotate parallel to the plane of the substrate such as gears and wheels. A cross section of a simple hub and gear is shown in Figure 4.18 and a complex hub structure is shown in Figure 4.4.

4.4.1.10 Actuators

There are several different actuation techniques for surface micromachine mechanisms. Actuators for all types of MEMS devices are covered in Chapter 5, but the most common actuator types for surface micromachines — electrostatic and thermal — are also covered here. Electrostatic actuators harness the attractive Coulomb force between charged bodies. For a constant voltage between two parallel plates, the energy stored between the plates is given by:

$$W = \frac{\varepsilon_0 \varepsilon_r A V^2}{2y} \tag{4.37}$$

where ε_0 is the dielectric constant of free space (8.854×10^{-12} F/m), ε_r is the relative dielectric constant, for air it is 1.0, A is the area, V is the voltage, and y is the distance. The force between the plates is attractive and given by:

$$F = -\frac{dW}{dy} = \frac{\varepsilon_0 \varepsilon_r A V^2}{y^2} \tag{4.38}$$

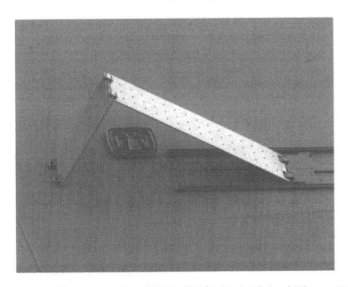

FIGURE 4.17 Hinged polysilicon micromirror fabricated in the Sandia National Laboratories SUMMiT™ process. (Photograph courtesy of Sandia National Laboratories.)

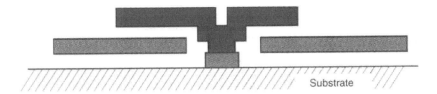

FIGURE 4.18 Cross section of a simple hub and gear fabricated in a two level surface micromachining process.

Note that this force is dependent on $1/y^2$. Therefore, the Coulomb force is very strong at small gaps, but drops off rapidly as the gap increases. If the plates are not fully engaged, as in Figure 4.19, there will be tangential as well as normal forces. The equation for the tangential motion is obtained by modifying Equation 4.37 by substituting the area term, A, with the product of the lateral dimensions z and x:

$$W = \frac{\varepsilon_0 \varepsilon_r A V^2}{2y} = \frac{\varepsilon_0 \varepsilon_r xz V^2}{2y} \tag{4.39}$$

The derivative of energy with respect to position is force.

$$F = \frac{dW}{dx} = \frac{\varepsilon_0 \varepsilon_r z V^2}{2y} \tag{4.40}$$

Note that this force is not dependent on the lateral position x. Comb-drive actuators utilize these tangential forces with banks of comb fingers. The surface micromachined comb-drive in Figure 4.20 typically operates at voltages of 90 V and has output forces of around $10\,\mu N$. Comb-drives can operate at speeds up to 10 s of kHz and consume only the power necessary to charge and discharge their capacitive plates.

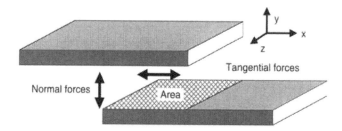

FIGURE 4.19   Illustration of normal and tangential forces in an electrostatic actuator.

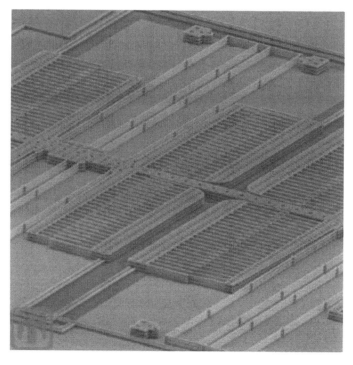

FIGURE 4.20   Electrostatic comb-drive actuator fabricated in the SUMMiT V™ process at Sandia National Laboratories.

Because the output force of an electrostatic comb-drive is proportional to the square of the applied voltage, these devices are often operated at higher voltages than most analog and digital integrated circuits. This has complicated their implementation into systems. It should be noted that Equations 4.37 through 4.40 are for electrostatic actuators connected to a power supply and in constant voltage mode.

Example

The plate in Figure 4.19 is 50 μm long, 6 μm thick, and 3 μm wide. It has a neighboring electrode that is 1 μm away. If 80 V are applied between the electrode and its identical neighbor and fringing is ignored, what are the forces in the normal and tangential directions if the combs are aligned in the width dimension and overlap by 40 μm and 10 μm in the length dimension? How much does the force change in both directions if the distance is reduced to 0.1 μm? If there are 30 electrodes at 80 V interlaced with 31 electrodes at 0 V as in Figure 4.20 with a separation of 1 μm between the electrodes what is the net tangential force? Assume that $\varepsilon_r = 1.0$.

Parallel plate electrostatic forces for the 40 μm of overlap case are:

$$F_{normal} = \frac{\varepsilon_0 \varepsilon_r x z V^2}{y^2} = \frac{8.854 \times 10^{-12} \, F/m \times 1 \times 40 \times 10^{-6} \, m \times 6 \times 10^{-6} \, m \times (80 \, V)^2}{(1 \times 10^{-6} \, m)^2} = 14 \times 10^{-6} \, N$$

and for the 10 μm case:

$$F_{normal} = \frac{\varepsilon_0 \varepsilon_r x z V^2}{y^2} = \frac{8.854 \times 10^{-12} \, F/m \times 1 \times 10 \times 10^{-6} \, m \times 6 \times 10^{-6} \, m \times (80 \, V)^2}{(1 \times 10^{-6} \, m)^2} = 4.4 \times 10^{-6} \, N$$

For the tangential force, the lateral dimension, x, is not included in Equation 4.40 so both calculations yield the same result.

$$F_{tangential} = \frac{dW}{dx} = \frac{\varepsilon_0 \varepsilon_r z V^2}{2y} = \frac{8.854 \times 10^{-12} \, F/m \times 1 \times 40 \times 10^{-6} m \times (80 V)^2}{2 \times 1 \times 10^{-6} m} = 1.7 \times 10^{-7} \, N$$

The reader should note that the parallel plate force is much higher than tangential force. If the separation is reduced to 0.1 μm, then the normal force is increased by a factor of 100 to 440 μN and the tangential force is increased by a factor of 10 to 1.7 μN. For 30 energized electrodes and 31 non-energized electrodes, the tangential force is multiplied by the number of energized electrodes and then doubled to account for both faces of the electrode:

$$F_{30 \, fingers} = 170 \, nN \times 30 \times 2 = 10 \, \mu N$$

Note that the parallel plate force of one electrode engaged 40 μm is still higher than the tangential force of 30 electrodes. However, the tangential force comb-drive has a force that is independent of position, while the parallel plate case falls off rapidly with distance. However, the high forces at small separations makes parallel plate actuators useful for electrical contact switches.

Thermal actuators have higher forces (hundreds of μN up to a few mN) and operate at lower voltages (1 V to 15 V) than electrostatic actuators. They operate by passing current through a thermally isolated actuator. The actuator increases in temperature through resistive heating and expands, thus moving the load. One type of thermal actuator has two beams that expand different amounts relative to each other [Guckel et al, 1992; Comtois, 1998]. These pseudo bimorph or differential actuators use a wide beam and a narrow beam that are electrically resistors in series and mechanically flexures in parallel. Because the narrow beam has a higher resistance than the wide beam, it expands more and bends the actuator in an arc around the anchors.

Another type of thermal actuator uses two beams that are at a shallow angle. Bent beam thermal actuators generally have strokes of between 5 μm and 50 μm [Que et al., 2001; Cragun and Howell, 1999]. Unlike the pseudo bimorph devices, they move in a straight line instead of an arc. Like the pseudo bimorph actuators, they have an output force that falls off quickly with displacement. Both types of thermal actuators are shown in Figure 4.21.

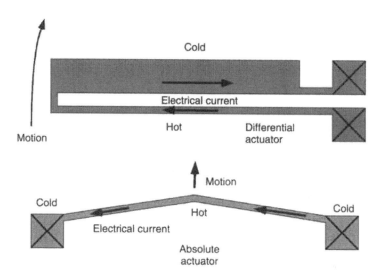

FIGURE 4.21   Drawings of absolute and differential thermal actuators.

4.5   Packaging

This section covers only aspects of packaging that are especially relevant to surface micromachines. One of the challenges of packaging surface micromachines is that the packaging is application specific and varies greatly between different types of devices. This is one reason that packaging tends to be the most expensive part of surface micromachined devices. Therefore it is very important that designers of surface micromachines understand packaging and collaborate with engineers specializing in packaging while designing their device. Some of the main purposes of packages for surface micromachines include electrical and mechanical connections to the next assembly and protection from the environment.

The mechanical attachment between a die and a package can be achieved in several different ways. The main criteria for choosing a die-attach method include: the temperature used during the die-attach process; the amount of stress the die-attach process induces on the die; the electrical and thermal properties of the die-attach; the preparation of the die for the die-attach process; and the amount of outgassing that is emitted by the die-attach. Die-attach methods that do not outgas include silver-filled glasses and eutectics (gold–silicon). Both of these induce a large amount of stress onto the die as well as requiring temperatures of around 400°C. Epoxy-based die-attaches use much lower temperatures (up to 150°C) and induce lower stress on the die than eutectics or silver-filled glasses. The low stress means that they can be used for larger die. However, depending on the type of epoxy, they do outgas water and corrosive chemicals such as ammonia. As an alternative, flip-chip processes combine the process of die-attach and electrical interconnection by mounting the die upside-down on solder balls. A note of caution: surface micromachines are strongly influenced by coatings and contamination or removal of these coatings, which can happen during temperature cycling, is detrimental to the micromachine.

Electrical interconnect to surface micromachines is typically accomplished by wire bonding. Wirebonders use a combination of heat, force, and ultrasonic energy to weld a wire (normally aluminum or gold) to a bondpad on the surface micromachined die. The other end of the wire is welded to the package. Although, wirebonds in high volume production have been done on less than 50 μm centers, for low volume applications it is easier if the bondpads are fairly large. Metal coated bond pads that are 125 μm or larger on a side with 250 μm center-to-center spacing allow reworking of the wire bonds and for non-automated wire bonding equipment to be used. These numbers are on the conservative side but could be followed in the absence of process specific information. Figure 4.22 shows an Analog Devices ADXL50 surface micromachined accelerometer with its wirebonds and epoxy die-attach.

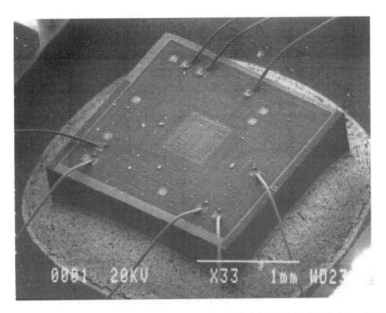

FIGURE 4.22 Analog Devices accelerometer after the package lid has been removed. The epoxy die-attach and wire bonds are clearly visible. This device had its package lid removed at Sandia National Laboratories. (Photograph courtesy of Jon Custer of Sandia National Laboratories.)

While electrical and mechanical connections to surface micromachines are similar to other types of micromachines and integrated circuits, the mechanical protection aspects of the packaging can be quite different. The package must protect the surface micromachine from handling by people or machines, from particles and dust that might mechanically interfere with the device, and from water vapor that can induce stiction. Packages for some resonant devices must maintain a vacuum and all packages must keep out dust and soot in the air. However, sensor packages must allow the MEMS device to interact with its environment. Because surface micromachines are very delicate and fragile after release, even a careful packaging process can damage a released die. Therefore, one of the trends in packaging is to encapsulate and mechanically protect the devices as early as possible in the manufacturing process. Henry Guckel at the University of Wisconsin was one of the first developers of an integrated encapsulation technique [Guckel, 1991]. In this design, the surface micromachine was covered by an additional layer of structural material during the fabrication process. This last structural layer completely encapsulates the surface micromachine with the exception of a hole used to permit removal of the sacrificial layer by the release etchant. After the release etch, the hole is sealed with materials ranging from LPCVD films, to sputtered films, to solders. There is a good summary of this and other sealing techniques in Hsu (2004).

Another method of encapsulating the device is to use wafer bonding. In this technique a cap wafer is bonded to the device wafer forming a protective cover. One common method involves anodic bonding of a glass cover wafer over the released surface micromachines. A second involves the use of intermediate layers such as glass frit, silicon gold eutectic, and aluminum. The wafer bonding techniques are more independent of the fabrication process than the wafer level deposition processes. Because of the lack of stiction forces between the cap and the substrate and because film stresses in the cap are not a problem, bonded caps can be used for larger devices than deposited caps. A package formed by Corning 7740 glass (pyrex) bonded to a surface micromachine is shown in Figure 4.23.

Conventional packaging such as ceramic or metal packages also protect surface micromachines from the environment. In this case, the package provides electrical and mechanical interconnections to the next assembly as well as mechanical protection. These types of packages tend to be more expensive than plastic packages, which can be used if encapsulation is done on the wafer level. These packages are typically sealed using either a welding operation that keeps the surface micromachine at room temperature or a

FIGURE 4.23   A surface micromachine encapsulated by a piece of glass. Comb-drives can be seen through the right side of the mechanically machined cap. This process can be accomplished either on a die or wafer basis. (Photograph courtesy of A. Oliver, Sandia National Laboratories).

belt sealing operation that elevates the temperature of the entire assembly to several hundred degrees Centigrade.

4.6   Applications

The applications section of this chapter will present some surface micromachined mechanisms and discuss them with regard to some of the mechanical concepts discussed earlier. The chapter concludes with some design rules and lessons learned in the design of surface micromachined devices.

4.6.1   Countermeshing Gear Discriminator

One example of a surface micromachined mechanism is the countermeshing gear discriminator that was invented by Polosky et al. (1998). This device has two large wheels with coded gear teeth. Counter-rotation pawls restrain each wheel so that it can rotate counterclockwise and is prevented from rotating clockwise. The wheels have three levels of teeth that are designed so they will interfere if the wheels are rotated in the incorrect sequence. Only the correct sequence of drive signals will allow the wheels to rotate and open an optical shutter. If mechanical interference occurs, the mechanism is immobilized in the counterclockwise direction by the interfering gear teeth and by the counter-rotation pawls in the clockwise direction. A drawing of the device and a close-up of the teeth are shown in Figures 4.24 and 4.25, respectively. The wheels have three levels of intermeshing gear teeth that will allow only one sequence of rotations out of the more than the 16 million that are possible. Because the gear teeth on one level are not intended to interfere with gear teeth on another level and because the actuators must remain meshed with the code wheels, the vertical displacement of the code wheels must be restricted. This was accomplished with dimples on the underside of the coded wheels that limited the vertical displacement to 0.5 µm. Warpage of the large 1.9-mm-diameter coded wheels is reduced by adding ribs with an additional layer of polysilicon.

The large coded wheels are prevented from rotating backward by the counter-rotation pawls. These devices must be compliant in one direction and capable of preventing rotation in the other direction. The next example discusses counter-rotation pawls.

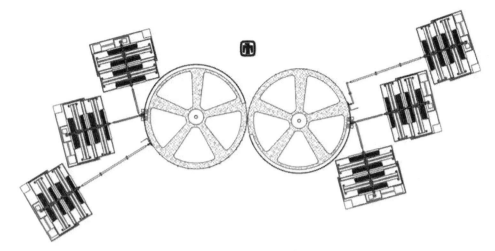

FIGURE 4.24   The countermeshing gear discriminator. The two code wheels are the large gears with five spokes in the center of the drawing; the counter-rotation pawls are connected to the comb-drives; and the long beams in the upper right and lower left portion of the photograph. (Drawing courtesy of M.A. Polosky, Sandia National Laboratories.)

FIGURE 4.25   Teeth in the countermeshing gear discriminator. The gear tooth on the left is on the top level of polysilicon and the gear tooth on the right is on the bottom layer. If the gears do not tilt or warp, the teeth should pass without interfering with each other. (Photograph courtesy of Sandia National Laboratories.)

Example

Figure 4.26 shows a counter-rotation pawl. The spring is 180 μm long from the anchor to a stop (labeled as *l*) and 20 μm long from the stop to the end of the flexible portion (denoted as *a*), with a width of 2 μm and a thickness of 3 μm. The Young's modulus is 155 GPa. Assume that the tooth on the free end of

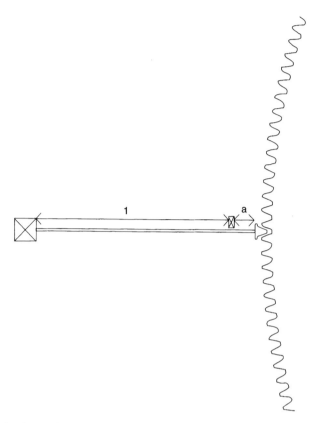

FIGURE 4.26 Example of a simple counter-rotation pawl. The stop is assumed to have a width of 0.

the beam does not affect the stiffness and that the width of the stop is negligible. Find the spring constant of the pawl if the gear is rotated in the counterclockwise direction. Comment on the spring constant if the gear is rotated in the clockwise direction.

In the counterclockwise direction, the spring constant k is:

$$k = \frac{Ew^3t}{4L^3} = 0.12 \text{ N/m}$$

using a length $l + a$ of 200 μm. In the clockwise direction, it is tempting to redefine the spring length as 20 μm. The resulting spring constant is 116 N/m. Unfortunately, this is an oversimplification because the beam will deform around the stop. The exact equation is in Timoshenko's *Strength of Materials* [Timoshenko, 1958] and in Equation 4.41:

$$k = \frac{1}{\left(\dfrac{a^3}{3EI} + \dfrac{a^2l}{4EI} \right)} \tag{4.41}$$

This equation results in a spring constant of 15 N/m, which is still very stiff but not as stiff as the results of the oversimplified calculation.

4.6.2 Microengine

One important element of many polysilicon mechanism designs is the *microengine*. This device, described by Garcia and Sniegowski (1995) and shown in Figures 4.27 to 4.29, uses an electrostatic comb-drive

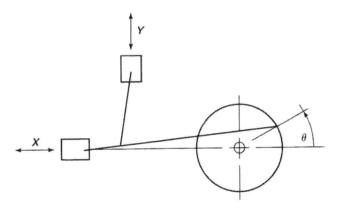

FIGURE 4.27 Mechanical representation of a microengine.

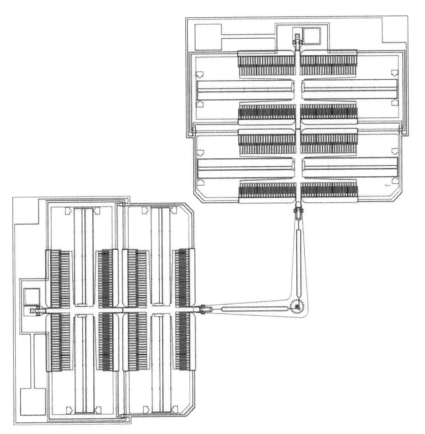

FIGURE 4.28 Drawing of a microengine. The actuator measures 2.2 mm × 2.2 mm and produces approximately 55 pN-m of torque.

connected to a pinion gear by a slider-crank mechanism with a second comb-drive to move the pinion past the top and bottom dead center. Two comb-drives are necessary because the torque on a pinion produced by a single actuator has a dependence on angle and is given by the following equation:

$$Torque = F_0 r |\sin(\theta)| \tag{4.42}$$

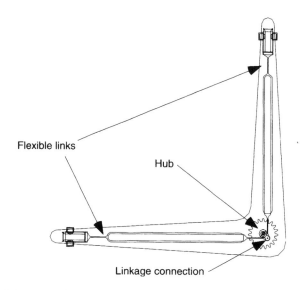

FIGURE 4.29   Detail of Figure 4.28 showing the thin linkages that connect the comb-drives to the gear.

where F_0 is the force of the comb-drive, θ describes the angle between hub of the output gear and the linkage, and r is the distance between the hub and the linkage. The torque produced by the "Y" actuator has a similar dependence on angle. The inertia of the rotating pinion gear is not great enough to rotate the gear past the top and bottom dead center. One important feature of this device is that the conversion from linear motion to rotary motion requires the beams between the actuator and the driven gear to bend. The bending is permitted by a polysilicon linkage that is 40 μm in length and 1.5 μm in width with a thickness of 2.5 μm.

Example

The comb-drive labeled "X" in Figure 4.27 has an actuator arm that must bend 17 μm in the lateral direction as the gear rotates from 0° to 90°. The gear is connected to the comb-drive via a 50-μm-wide beam that is 500 μm long in series with a thin flexible link that is 1.5 μm wide and 45 μm long. The thin link is connected to the comb-drive actuator and both beams have a Young's modulus of 155 GPa. Given that both beams are 2.5 μm thick, approximately how much force does it take to bend to the flexible linkage to rotate the gear from q = 0 to q = 90° if friction and surface forces are neglected? A drawing of this linkage is shown in Figure 4.29.

Assume that each segment is a cantilever beam spring that has one end fixed while the other end is undergoing a small deflection. For each beam:

$$y_{max} = \frac{PL^3}{3EI} \tag{4.43}$$

and

$$k = \frac{P}{y_{MAX}} = \frac{3EI}{L^3} \tag{4.44}$$

Recall that for rectangular cross sections:

$$I = \frac{tw^3}{12} \tag{4.14}$$

Using substitution, the equivalent spring constant for the long beam is:

$$k = \frac{Etw^3}{4L^3} = \frac{155 \text{ GPa} \times (2.5 \,\mu\text{m})(50 \,\mu\text{m})^3}{4 \times (500 \,\mu\text{m})^3} = 97 \text{ N/m}$$

and for the short beam:

$$k = \frac{Etw^3}{4L^3} = \frac{155\,\text{GPa} \times (2.5\,\mu\text{m})(1.5\,\mu\text{m})^3}{4 \times (45\,\mu\text{m})^3} = 3.6\,\text{N/m}$$

By Equation 4.35, the equivalent spring rate is:

$$\frac{1}{k_{eq}} = \frac{1}{k_1} + \frac{1}{k_2} = \frac{1}{97\,\text{N/m}} + \frac{1}{3.6\,\text{N/m}} = 3.5\,\text{N/m}$$

We can employ some simple trigonometry and calculus to determine the needed deflection of the flexure. From the section on cantilever beam springs:

$$y = \frac{P}{6EI}\left(3Lx^2 - x^3\right) \tag{4.21}$$

Because the majority of the bending occurs in the thin flexible link the desired slope of the flexible link at its end is:

$$\frac{dy}{dx} = \frac{17\,\mu\text{m}}{545\,\mu\text{m}} = \frac{P}{6EI}\left(6Lx - 3x^2\right) \tag{4.45}$$

For the case of the small flexible link, I is equal to:

$$I = \frac{tw^3}{12} = \frac{2.5\,\mu\text{m} \times (1.5\,\mu\text{m})^3}{12} = 7 \times 10^{-25}\,\text{m}^4$$

Because $x = L = 45\,\mu\text{m}$, P can be calculated as:

$$P = \frac{17\,\mu\text{m}}{545\,\mu\text{m}} \times 6 \times 155\,\text{GPa} \times 7.0 \times 10^{-25}\,\text{m}^4 \times \frac{1}{\left(6 \times (45\,\mu\text{m})^2 - 3 \times (45\,\mu\text{m})^2\right)} = 3.4\,\mu\text{N}$$

4.6.3 Micro-Flex Mirror

The Micro-Flex mirror is a device that deforms out of plane through buckling [Garcia, 1998]. It can be built in any surface micromachining process that has at least one level of released structural material. This device can be configured as a mirror, an optical shutter, a valve, or any structure that requires a plate or beam to move out of the plane of fabrication. A scanning electron microscope (SEM) photograph of the device is shown in Figure 4.30. The mirror consists of a long flexible beam connected to a plate that in turn is connected to two anchors via two additional flexible beams. The device operates by buckling. When a force is placed on the long flexible beam in the direction of the anchors, the structure is placed under a compressive force. When the force exceeds the critical value described by Equation 4.36, the structure buckles. Because the long flexible beam is larger in the direction parallel to the plane of the substrate than it is in the direction away from the substrate, it preferentially buckles out of the plane of the substrate. Also, because the plate and the two anchor beams are wider than the long flexible beam, the majority of the bending occurs in the long flexible beam and not in the plate or the anchor beams. The next example discusses the buckling criteria.

Example

Determine how much axial force is needed for a micromachined polysilicon mirror to buckle given the following dimensions. The main beam is $300\,\mu\text{m}$ long, $4\,\mu\text{m}$ wide, and $1\,\mu\text{m}$ thick with a Young's modulus of $155\,\text{GPa}$. Assume that the beam can be simplified to be a cantilever. Neglect the buckling of the anchors and the mirror.

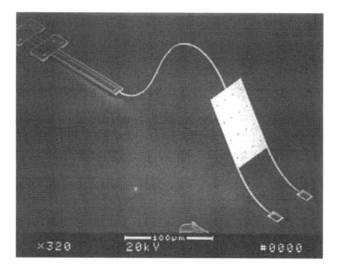

FIGURE 4.30 A flexible pop-up mirror that operates via buckling. In this photograph the buckling is out of the plane of the substrate. (Photograph courtesy of E.J. Garcia of Sandia National Laboratories.) A further description of this device can be found in Garcia, E.J. (1998) "Micro-Flex Mirror and Instability Actuation Technique," in *Proc. 1998 IEEE Conf. on Micro Electro Mechanical Systems* (MEMS '98), pp. 470–474.

The moment of inertia for this situation is:

$$I = \frac{wt^3}{12} = \frac{4 \times 10^{-6} \text{ m} \times (1 \times 10^{-6} \text{ m})^3}{12} = 3.3 \times 10^{-25} \text{ m}^4$$

From the mechanics of materials section we know that the minimum force to buckle the beam is:

$$F_{cr} = \frac{\pi^2 EI}{4L^2} \tag{4.36}$$

Substituting this into the equation for force and using 155 GPa as the value of Young's modulus we have the following:

$$F_{cr} = \frac{\pi^2 \times 155 \times 10^9 \text{ N/m}^2 \times 3.3 \times 10^{-25} \text{ m}^4}{4 \times (300 \times 10^{-6} \text{ m})^2} = 1.4 \,\mu\text{N}$$

As the example shows, it is necessary to have a great deal of force in order to buckle the flexible mirror. In the original paper by Garcia (1998), a transmission (shown in Figure 4.31) was used to increase the force on the mirror. The transmission traded displacement for force to ensure that the mirror buckled. The amount of deflection after buckling is a nonlinear dynamics problem and beyond the scope of this book. For those readers interested in the subject, good references include the work of Timoshenko and Gere (1961), Brush and Almroth (1975), and Hutchinson and Koiter (1970). Fang and Wickert (1994) have examined the subject for micromachined beams. One interesting aspect of buckling is that it is impossible to determine if the beam will initially buckle toward or away from the substrate. If the beam buckles away from the substrate, it will continue to deflect away from the substrate. If it initially buckles toward the substrate, it will contact the substrate and further compression of the beam will result in the structure buckling away from the substrate.

Example

For the mechanism shown in Figure 4.31, how much mechanical force is made available by the transmission if the microengine output gear has a diameter of 50 μm and an applied torque of 50 pN-m, the large spoked wheel has a diameter of 1600 μm, and the pin joint that links the spoked wheel to the mirror

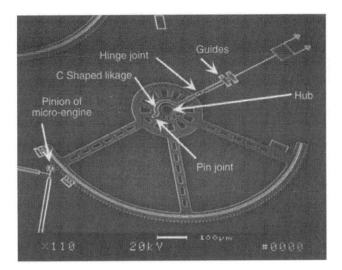

FIGURE 4.31 A flexible mirror in its initial state and a large gear with spokes. The spoked wheel acts as a transmission and is used to gain enough mechanical advantage to buckle the beam. Two important features are the hinge joint and guides which convert the off-axis motion of the "c-shaped" linkage to motion that is aligned with the mirror. The c-shaped linkage was designed to avoid fabricating the linkage above the rotating joint and thus causing fabrication problems. (Photograph courtesy of E.J. Garcia, Sandia National Laboratories.)

is 100 μm from the hub? Determine the maximum force in the direction of the mirror at the pin joint, neglecting friction, if there is a gap of 0.5 μm between the linkage and the pin joint.

The torque available at the pin joint is:

$$torque = 50\,pNm \times \frac{1600\,\mu m}{50\,\mu m} = 1600\,pNm$$

and the radial force at the pin joint is:

$$F_{radial} = \frac{1600\,pNm}{100\,\mu m} = 16\,\mu N$$

If there was no gap in the pin joint the force at the pin joint would be given by:

$$F_{mirror} = \frac{F_{radial}}{\sin(\theta)} \qquad (4.47)$$

where θ is the angle between a line connecting the pin joint and the mirror and a line connecting the pin joint and the hub. However, there is a gap or slop in the pin joint. $\sin(\theta)$ is given by:

$$\sin(\theta) = \frac{0.5\,\mu m}{100\,\mu m} = 0.005$$

If the motion of the spoked wheel is assumed to be in a straight line instead of an arc. The resulting output force is:

$$F_{mirror} = \frac{F_{radial}}{\sin(\theta)} = \frac{16\,\mu N}{0.005} = 3.2\,mN$$

This example is taken after Garcia (1998), who included a more detailed derivation that accounts for the rotation of a "c"-shaped linkage.

Example

The large forces produced by the linkage system in the previous example are important for overcoming stiction. Calculate the adhesive force due to van der Waals forces for a 150 μm × 150 μm mirror if the surface of the mirror is separated from the substrate by 8 nm and the surface under the mirror is silicon dioxide with a Hamaker constant of 5×10^{-20} J. Compare this result to a mirror that has four 2 μm by 2 μm dimples that are also 8 nm from the substrate.

From Equation 4.8:

$$\frac{d\Gamma}{dx} = -\frac{A}{6\pi x^3} = -\frac{5 \times 10^{-20} \text{ Nm}}{6\pi \times (8 \times 10^{-9} \text{ m})^3} = 5180 \frac{\text{N}}{\text{m}^2}$$

For a mirror that is 150 μm × 150 μm the force required is:

$$F = \frac{d\Gamma}{dx} \times area = 5180 \frac{\text{N}}{\text{m}^2} \times (150 \times 10^{-6} \text{ m})^2 = 120 \,\mu\text{N}$$

For a mirror with dimples:

$$F = \frac{d\Gamma}{dx} \times area = 5180 \frac{\text{N}}{\text{m}^2} \times 4 \times (2 \,\mu\text{m})^2 = 83 \text{ nN}$$

Note that the adhesive forces due to stiction are much smaller when dimples are used.

4.7   Failure Mechanisms in MEMS

For most practioners, the field of MEMS and surface micromachines has a steep learning curve. Often the learning occurs through repeated iterations of the same design, but this can be very time consuming and expensive because the time between design completion and testing is usually measured in months and the price per fabrication run is many thousands of dollars. This section describes some failures in surface micromachined mechanisms. The hope is that the reader will gain a deeper appreciation for the complexities of surface micromachined mechanism design and learn about some of the pitfalls.

4.7.1   Vertical Play and Mechanical Interference in Out-of-Plane Structures

Surface micromachined parts typically have a thickness that is very small in relationship to their width or breadth. In the out-of-plane direction, the thickness is limited to a few micrometers due to the limited deposition rates of deposition systems and the stresses in the deposited films. In the plane of the substrate, structures can be millimeters across. These factors typically lead to surface micromachined structures that have a very small aspect ratio as well as stiffness issues in the out-of-plane direction due to the limited thickness of the parts. The result is that designers of surface micromachines need to design structures in three dimensions and account for potential movements out of the plane of the substrate. A potential problem occurs when two gears fabricated in the same structural layer of polysilicon fail to mesh because one or both of the gears are tilted. Another is when structures moving above or below another structure mechanically interfere with each other when it was intended for them to not touch each other. Both of these instances will be examined separately.

An example of the out-of-plane movement of gears is illustrated in Figure 4.32. In this instance, the driven gear in the top of the figure has been wedged underneath the large load gear at the bottom of the photograph. The way to prevent this situation is to understand the forces that create the out-of-plane motion and to reduce or restrain them. One way of reducing the relative vertical motion of meshed gears is to increase the ratio of the radius of the hub to the radius of the gear. Mathematically, the maximum displacement of the outside edge of a gear from its position parallel to the plane of the substrate is:

$$Y_{max} = R \times Y/X \qquad (4.48)$$

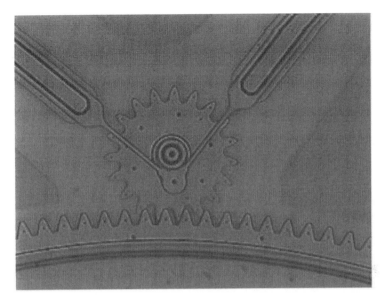

FIGURE 4.32 (See color insert following page 2-12.) The gear teeth of the small gear are wedged underneath the teeth of the large diameter gear. In this case, gear misalignment is about 2.5 mm in the vertical direction.

where the radius of the hub is X, the vertical play in the hub is Y, the radius of the gear is R, and the maximum vertical displacement is Y_{max}. The increase in hub diameter, while it can increase the amount of adhesive forces in the hub, does not increase the amount of frictional forces in the rotating or sliding structure.

The vertical displacement of a gear can also be limited by contact with the substrate. One way of limiting this is to place dimples on the underside of the gear. Dimples are commercially available in several surface micromachining processes including the MUMPS process offered by MEMSCAP and the SUMMiT™ process offered by Sandia National Laboratories. The dimples are generally spaced well apart from each other, depending on the stiffness of the gear and the prevalence of stiction. They typically occupy less than 1% of the surface area.

Another method of restraining the vertical motion of surface micromachined parts involves the use of clips above or below the gear or other moving structure as shown in Figure 4.33. Generally three or four clips are used, which is enough to restrain the moving structure while limiting the increase in the amount of frictional forces on the gear. Dimples on the bottom of the clips can further reduce the amount of play. Of course, all of these techniques can be combined to reduce vertical play.

Another type of reliability problem is the tendency for mechanical systems to increase their level of entropy and become disconnected or disengaged. The following example illustrates the point. A pin and socket joint was the connection between an actuator and an optical shutter. A cross section of the joint is shown in Figure 4.34. Unfortunately, the side walls of the socket are slightly sloped and the arm is not restrained in the vertical direction. In Figure 4.35, the force between a pin and the optical shutter was relieved, not by the horizontal movement of the shutter, but by the vertical movement of the pin, which slid up the sloped side walls of the shutter and out of the socket. The proper way to design this joint would be to use an additional layer of polysilicon below the shutter to restrain the linkage in the vertical direction. An alternative approach would be to attach the linkage to the shutter and make the linkage compliant to compensate for the rotation of the shutter.

Unintended vertical motion of surface micromachined parts can be exacerbated by structures that are soft in the vertical direction. Figure 4.36 shows a beam with a dimple that is caught on a gear. There should have been no interaction between the two elements because the fabricated dimple should have cleared the gear by $0.5\,\mu m$. Unfortunately in this case, neither the gear nor the spring is adequately electrically grounded, which may have caused some electrostatic attractive force between the two. Stiction

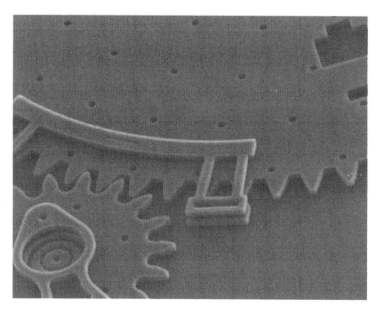

FIGURE 4.33   SEM of clip-over gear. The purpose of the clip is to limit the vertical displacement of the gear and to help ensure that the gear remains meshed. The line in the center of the clip is a dimple that further reduces the amount of vertical play in the gear. (Photograph courtesy of the Intelligent Micromachine Department at Sandia National Laboratories.)

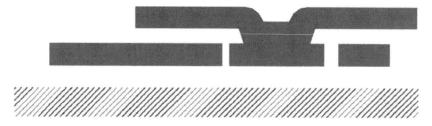

FIGURE 4.34   A cross section of a pin and socket joint.

FIGURE 4.35   A cross section of the pin and socket joint where the pin has moved out of the joint.

may also have contributed to the proximity of the two parts. The spring constant of this beam in the vertical direction is much too soft at 1.5×10^{-3} N/m. The solution to the problem is to make the surface micromachined part stiff in the vertical direction. As noted earlier in Equation 4.44:

$$k = \frac{Ew^3t}{4L^3} \tag{4.44}$$

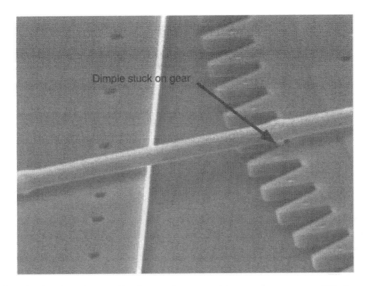

FIGURE 4.36 SEM of polysilicon beam with dimples that is caught on surface micromachined gear. The main problem with the beam is that it was designed to be compliant in the vertical direction. A secondary problem was that the dimple was small enough to be caught in the gear teeth.

for a rectangular cantilever beam. Therefore, to increase the stiffness in the vertical direction the thickness of the structure should be increased and the length should be decreased. If the spring constant is increased, it takes more energy to pull the structure down to the substrate. Another less effective option is to increase the width. A completely different but complementary approach would be to change the size of the dimple so that it is larger than the space between the gear teeth.

A source of unintended vertical motion in surface micromachined parts is electrostatic forces. Unintended electrostatic forces in surface micromachines have two basic causes. One is fixed charge buildup in dielectric layers. Charges in silicon nitride and silicon dioxide, especially silicon nitride, are not mobile and tend to attract other structures. This can be a great problem in accelerometers or gyroscopes because they are designed to be sensitive to the vertical displacement of the proof-mass. The best way to avoid the influences of trapped charge is to minimizing exposure of the dielectrics to radiation and energetic electrons. The other method is to shield the moveable mechanism parts from the charge buildup in the dielectric by using a conductive ground plane between the movable structure and the dielectrics.

Unintended electrostatic forces are also important when features and structures in a design are left electrically floating. In this situation, as in the dielectric charging case, the floating potentials can eventually cause attractive forces. The cure for this situation is to fix all potentials to a known level. For dielectrics, a grounded conductor should shield them from other structures and layers. For rotating gears or other structures that are not electrically connected to the substrate, the hubs or mechanical restraints should be connected to the ground. If the restraints and the moving component momentarily touch, the moving part can discharge through the support to the substrate. Electrostatic attraction of floating components can be a maddening problem, as the potentials on the various conductors and dielectrics can shift randomly and cause structures to be unexpectedly attracted to each other.

4.7.2 Electrical Failures

One problem that is unique to surface micromachines made of conductive materials is that of electrical shorting between different parts of the machine. One common situation occurs when the fingers of a

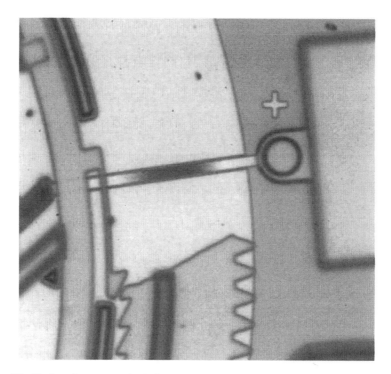

FIGURE 4.37   The blackened trace near the "+" sign was caused by a dimple on a grounded gear touching a high potential signal line. The color change is probably caused by roughing of the polysilicon surface. (Photograph courtesy of S. Barnes, Sandia National Laboratories.)

comb-drive touch. Another situation happens when the dimple of a moving structure such as a gear or a shutter comes in contact with an underlying voltage carrying trace. An example is shown in Figure 4.37. The problem is that bare polysilicon structures (and silicon–germanium) are neither good conductors nor good insulators even when coated by a thin anti-stiction coating or native oxides. The native oxide coatings and anti-stiction coatings are prone to breakdown under the high voltages commonly used to power electrostatic actuators. Because insulating dielectrics are not normally used between layers, the designer must resort to minimizing the chances for contact between conductors at different potentials. One method of doing this is to make structures, especially comb-drives, stiff in all directions but the direction of intended motion. Another is to use stops or other mechanical restraints to prevent contact between conductors at different potentials.

Electrostatic discharge (ESD) is another failure mechanism in surface micromachined mechanisms. One common cause of ESD is static electricity from people from handling devices. While the momentary shock caused by walking across carpet and touching a metal object may seem harmless, ESD is a common cause of failure in integrated circuits and can also occur in surface micromachines. ESD events can be reduced by the consistent use of ground straps by personnel handling the devices, anti-static workplace furniture and tools, and ESD resistant packaging. Figure 4.38 shows an electrode on a comb-drive that was partially melted due to an ESD event.

4.7.3   Lithographic Variations

One problem common to all types of mechanism design is tolerances in the manufacturing process. This is an especially grave problem in surface micromachined mechanisms because the manufacturing tolerances approach the size of the parts. For example, assume a cantilever beam spring with a width of 2 μm

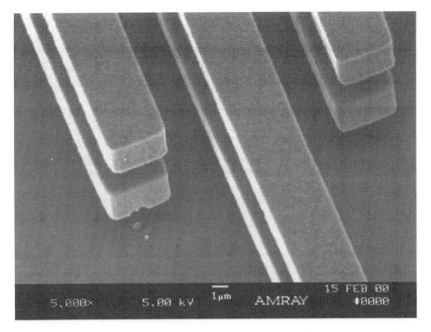

FIGURE 4.38   SEM of comb-drive electrode damaged by an ESD event. Note the damage to the left most comb and to the substrate underneath it. (Image courtesy of J.A. Walraven, Sandia National Laboratories.)

and a variation in the width of 0.2 μm. Cantilever beam springs have a spring constant that was given in Equation 4.44:

$$k = \frac{Ew^3t}{4L^3} \tag{4.44}$$

Due to the cubic relation between w and k, a small change in w will result in a large change in k. For example, if the line-width changes by 1%, the resulting spring constant changes by about 3%:

$$\Delta k = \left(\frac{\partial k}{\partial w}\right)\Delta w = \frac{3Ew^2}{4L^3}\Delta w \tag{4.49}$$

$$\frac{\Delta k}{k} = 3\frac{\Delta w}{w} \tag{4.50}$$

There are a few ways of minimizing the effects of features that vary in size. The first is not to design minimum-sized features. The line-width is not reduced as a fixed percentage of the feature size but instead all features are reduced by the same amount. For example, a 0.2-μm reduction in the size of a 5-μm-wide beam spring is only a 4% reduction in the width of the spring and results in approximately a 12% change in the spring-constant, while a 1-μm-wide spring reduced by the same amount changes the line-width by 20% and the spring constant by 49%.

The second method of minimizing the variation of mechanism performance due to line-width is to borrow techniques from integrated analog circuit layout techniques. A good introduction to this topic can be found in the text by Johns and Martin (1997). Probably the most applicable technique they discuss is to build dummy features adjacent to the intended mechanical features. Dummy features make processing boundary conditions (for example, in reactive ion etching or chemical mechanical polishing) similar for all parts of the mechanism and reduce the variation in feature size between mechanism parts. An example of this technique is shown in Figure 4.39, where dummy columns are placed adjacent to the column on the micro-flex mirror. This will help ensure that the column on the micro-flex mirror is etched

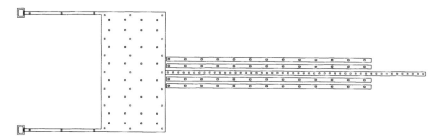

FIGURE 4.39 Dummy columns next to a column on a micro-flex mirror. The structures ensure that the micro-flex mirror is matched to other micro-flex mirrors on the same substrate and is exposed to similar processing conditions.

to a reproducible width and will be matched to other columns elsewhere on the chip. Care must be taken however, so that the dummy parts do not interfere with the mechanism parts.

4.8   Conclusion

This chapter has attempted to provide information to the designers of surface micromachines. It is intended only as a reference and is not intended to supercede information about a particular technology or application from local fabrication, packaging, testing, and reliability experts. The reader is strongly encouraged to talk to the local experts in these areas at his or her place of business or study. These people can save the designer time and money and greatly increase the chances of a successful design.

Acknowledgments

The authors would like to thank their co-workers at Sandia National Laboratories for their assistance. Many of the devices and concepts reported here are due to their efforts. Special thanks are due to Maarten de Boer, Chris Dyck, Jon Custer, Danelle Tanner, Michael Baker, Marc Polosky, Ernie Garcia, Robert Huber, Leslie Phinney, James Allen, and Fernando Bitsie for helpful discussions and suggestions. Sandia is a multiprogram laboratory operated by Sandia Corporation, a Lockheed Martin Company, for the United States Department of Energy's National Nuclear Security Administration under contract DE-AC04-94AL85000.

References

Bhushan, B., ed. (1999) *Handbook of Micro/Nano Tribology, 2nd ed.*, CRC Press, Boca Raton, FL.

Brush, D.O., and Almroth, B.O. (1975) *Buckling of Bars, Plates, and Shells*, McGraw-Hill, New York.

Chironis, N.P., and Sclater, N., eds. (1996) *Mechanisms and Mechanical Devices Sourcebook, 2nd ed.*, McGraw-Hill, New York.

Comtois, J.H., Michalicek, M.A., and Barron, C.C. (Oct. 1998) "Electrothermal Actuators Fabricated in Four-Level Planarized Surface Micromachined Polycrystalline Silicon," *Sensors Actuators A, 70*, pp. 23–31.

Cragun, R., and Howell, L.L. (Nov. 1999) "Linear Thermomechanical Microactuators," *Microelectromechanical Systems (MEMS)*, at the 1999 ASME International Mechanical Engineering Congress and Exposition, pp. 181–188.

Dugger, M.T., Poulter, G.A., and Ohlhausen, J.A. (1999) "Surface Passivation for Reduced Friction and Wear in Surface Micromachined Devices," in *Proc. Fall MRS Symp.*, Boston, MA.

Fan, L.S., Muller, R.S., Yun, W., Howe, R.T., and Huang, J. (1990) "Spiral Microstructures for the Measurement of Average Strain Gradients in Thin Films," in *Proc. 1990 IEEE Conf. on Micro Electro Mechanical Systems* (MEMS '90), pp. 177–181.

Fang, W., and Wickert, J.A. (1994) "Post-Buckling of Micromachined Beams," *J. Micromech. Microeng.*, 4(3), pp. 116–122.

Garcia, E.J. (1998) "Micro-Flex Mirror and Instability Actuation Technique," in *Proc. 1998 IEEE Conf. on Micro Electro Mechanical Systems* (MEMS '98), pp. 470–474.

Garcia, E.J., and Sniegowski, J.J. (1995) "Surface Micromachined Microengine as the Driver for Micromechanical Gears," in *Digest International Conference on Solid-State Sensors and Actuators*, Stockholm, Sweden, pp. 365–368.

Guckel, H., Klein, J., Christenson, T., Skrobis, K., Laudon, M., and Lovell, E.G. (1992) "Thermo-Magnetic Metal Flexure Actuators," in *Proc. Solid State Sensor and Actuator Workshop*, Hilton Head, pp. 73–75.

Guckel. H. (1991) "Surface Micromachined Pressure Transducers," *Sensors Actuators, A Phys.*, 28, pp. 132–146.

Howell, L.L. (2001) *Compliant Mechanisms*, John Wiley & Sons, New York.

Hsu, T. (2004) *MEMS Packaging*, INSPEC, The Institution of Electrical Engineers, London, United Kingdom.

Hutchinson, J.W., and Koiter, W.T. (1970) "Postbuckling Theory," *Appl. Mech. Rev.*, 23, pp. 1353–1366.

Johns, D.A., and Martin, K. (1997) *Analog Integrated Circuit Design*, John Wiley & Sons, New York.

Johnson, K.L. (1985) *Contact Mechanics*, Cambridge University Press, Cambridge, U.K.

Kovacs, G.T. (1998) *Micromachined Transducers Sourcebook*, WCB/McGraw-Hill, Boston.

Madou, M. (2002) *Fundamentals of Microfabrication: The Science of Miniaturization*, CRC Press, Boca Raton, FL.

Nguyen, C.T.-C., Katehi, L.P.B., and Rebeiz, G.M. (Aug., 1998) "Micromachined Devices for Wireless Communications," in *Proc. IEEE.*, 86, pp. 1756–1768.

Polosky, M.A., Garcia, E.J., and Allen, J.J. (1998) "Surface Micromachined Counter-Meshing Gears Discrimination Device," in *SPIE 5th Annual International Symposium on Smart Structures and Materials*, San Diego, CA.

Que, L., Park, J.S., and Gianchandani, Y.B. (June, 2001) "Bent-Beam Electrothermal Actuators Part I: Single Beam and Cascaded Devices," *J. Microelectromech. Syst.*, 10, pp. 255–262.

Rai-Choudhury, P. ed. (2000) *MEMS and MOEMS: Technology and Applications*, SPIE, Bellingham, Washington.

Roark, R., and Young, W. (1989) *Formulas for Stress and Strain*, McGraw-Hill, New York.

Senturia, S.D. (2001) *Microsystem Design*, Kluwer Academic Publishers, Norwell, MA.

Sharpe, W.N., Jr., LaVan, D.A., and Edwards, R.L. (1997) "Mechanical Properties of LIGA-Deposited Nickel for MEMS Transducers," in *Proc. 1997 International Conference on Sensors and Actuators*, Chicago pp. 607–610.

Shigley, J.E., and Mischke, C.R. (1989) *Mechanical Engineering Design, 5th ed.*, McGraw-Hill, New York.

Tanner, D.M. (Invited Keynote) (2000) *Proc. 22nd Int. Conf. on Microelectronics*, Nis, Yugoslavia, pp. 97–104.

Timoshenko, S.P. (1958) *Strength of Materials*, Van Nostrand, New York.

Timoshenko, S.P., and Gere, S.P. (1961) *Theory of Elastic Stability*, McGraw-Hill, New York.

Worthman, J.J., and Evans, R.A. (1965) "Young's Modulus, Shear Modulus and Poisson's Ratio in Silicon and Germanium," *J. Appl. Phys.*, 36, pp. 153–156.

Further Reading

Among the several excellent sources of information on surface micromachining and surface micro-machined devices are the recommended books by Madou (2002), Kovacs (1998), Senturia (2001), and Rai-Choudhury (2000). There are many conferences in the field including those organized by IEEE, SPIE, and ASME. Some of the heavily read journals in the field of MEMS include *Sensors and Actuators* and the *Journal of Microelectromechanical Systems*. For information on foundry services for surface micromachines, both MEMSCAP and Sandia provide information about their processes. The Internet address for MEMSCAP is www.memsrus.com; for Sandia, the address is www.mems.sandia.gov. Good references on mechanical engineering and mechanism design include *Mechanical Engineering Design* [Shigley and Mischke, 1989], *Mechanisms and Mechanical Devices Sourcebook* [Chironis and Sclater, 1996], *Microsystem Design* [Senturia, 2001], and *Compliant Mechanisms* [Howell, 2001]. References on packaging include *MEMS Packaging* Hsu (2004). Information on the testing of surface micromachines is in *MEMS Packaging* [Hsu, 2004] and *MEMS and MOEMS: Technology and Applications* [Rai-Choudhury, 2000].

5

Microactuators

Alberto Borboni
Università degli studi di Brescia

5.1 Introduction

This chapter will provide the definition of a microactuator within the context of a micromachine. A micromachine is a system that uses a small control energy to cause an observable (or controllable) perturbation to the environment. This chapter will address only machines able to generate an observable mechanical perturbation to the environment: that is, a machine able to generate a perturbation on environmental mechanical properties such as position, velocity, acceleration, force, pressure, and work. These types of machines are defined as "mechanical machines." Within the group of mechanical machines, we are specifically interested in machines able to generate microperturbations to the environment. Microperturbation is defined as a perturbation in the environmental mechanical properties that are recorded in micro units, or some multiples of micro units, defined in terms of SI units (for example, a 1 micron change of the position of an object or a generation of a 1 micronewton force on an object). Figure 5.1 illustrates how mechanical micromachines fit into the scheme of machines and micromachines.

From a functional standpoint, a mechanical machine is a system composed of an actuator, a transmission, and a user. The actuator generates the mechanical work; the transmission transforms this work and connects the actuator with the user; and the user acts directly on the environment (Figure 5.2).

A microactuator can have macro dimensions to generate a microenvironmental perturbation; however, for our purposes, we will consider actuators with at least one dimension in the order of microns (less than 1 mm), so we are referring to micromechanical actuators in the strictest sense (Figure 5.3).

In this chapter, the term microactuator will be used to describe a micromechanical actuator of less than 1 mm or almost in the strictest sense (a very small micromechanical actuator larger than 1 mm).

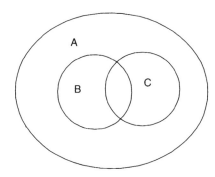

FIGURE 5.1   Definition of a mechanical micromachine. A is the set of all machines; B is the set of mechanical machines; and C is the set of micromachines. The intersection of B and C is the set of mechanical micromachines.

FIGURE 5.2   Functional definition of a mechanical machine. A is the actuator, T is the transmission, U is the user, e is an energy, m is a mechanical function, u is a usable mechanical function, and p is an environmental perturbation.

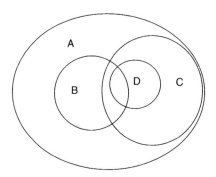

FIGURE 5.3   Definition of a micromechanical actuator in strict sense. A is the set of all the actuators; B is the set of mechanical actuators; C is the set of microactuators; and D is the set of microactuators in the strictest sense. The set of micromechanical actuators in the strictest sense is the intersection of B, C, and D.

5.1.1   Functional Characteristics and Choosing Criteria

The functional characteristics of an actuator are those of its mechanical output. These can be described by the relationship between generated torque and angular speed (if the actuator produces a rotary motion) or by the relationship between generated force and linear speed (if the actuator produces a linear motion). For this discussion, we will assume a rotary motion. If the motor is operating under steady-state conditions (input power, load to overcome, and the environment are constant) then the relationship between generated torque and angular speed can be graphed by a line on the torque-speed plane. This relationship is called the actuator characteristic (Figure 5.4).

On a bilogarithmic scale, it is possible to define three groups of ideal actuators (see Figure 5.5): torque generators, described by horizontal straight lines (where T is constant); speed generators, described by vertical straight lines (where ω is constant); and, power generators, described by oblique straight curves (where $T \cdot \omega$ is constant).

Usually real actuators differ from ideal actuators; however, in some aspects of its actuator characteristics, ideal actuators can approximate real actuators. Usually the actuator output can be manipulated with

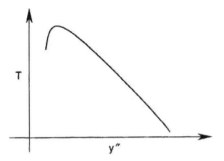

FIGURE 5.4 Mechanical characteristics of an actuator. *T* is the generated torque and ω is the generated angular speed.

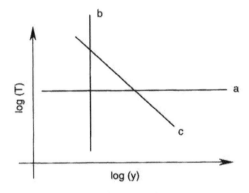

FIGURE 5.5 Ideal actuator: *a* is a torque generator, *b* is a speed generator, and *c* is a power generator.

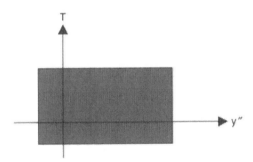

FIGURE 5.6 Field of the actuator characteristics.

a control variable, allowing, the actuator characteristic to be changed. Therefore, the set of all the actuator characteristics can be described by an extended portion of the T–ω plane (Figure 5.6).

Further, the actuator is also able to provide power for a limited amount of time, outside the standard field of the actuator characteristics, creating a joint field or an overloaded field (Figure 5.7).

This chapter has addressed the description of an actuator in terms of its mechanical output. But it is not possible to correctly select the most suitable actuator for a specific application, until one has information on the machine's purpose. Referring to Figure 5.2, this is represented by the user motion and the user load. The description of the desired mechanical user input is similar to the mechanical actuator output, the two related variables being the user torque (or force) and the user angular speed (or linear speed). To compare the desired user motion and load with the available actuator, the same reference should be

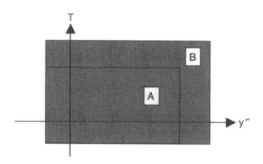

FIGURE 5.7   Fields of the actuator characteristics. *A* is the nominal field and *B* is the overloaded field.

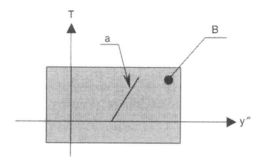

FIGURE 5.8   Comparing desired actuator with available actuator: *a* is the desired actuator characteristic and *B* is the field of the available actuator characteristics; *a* is within the field *B*, so the available actuator can be used for our needs.

used. Referring to Figure 5.2, to transform the actuator output to the user input, we use the transmission function:

$$T_u = f(T, x) \quad \omega_u = g(\omega, x) \tag{5.1}$$

where T_u and ω_u are, respectively, the user torque and speed, while f and g are the direct transmission functions, which can be dependent upon many variables and parameters synthesized into vector x. To transform the desired user input into the desired actuator output, we will use the inverse transmission function:

$$T_d = f^{inv}(T_{du}, x) \quad \omega_d = g^{inv}(\omega_{du}, x) \tag{5.2}$$

where T_{du} and ω_{du} are, respectively, the desired user torque and speed; T_d and ω_d are, respectively, the desired actuator torque and speed; and f^{inv} and g^{inv} are the inverse transmission functions.

Now, we can compare the desired actuator characteristic (T_d vs. ω_d) with the available actuator characteristic (T vs. ω). If the desired actuator characteristic is within the available actuator field of characteristics, we can use this available actuator for our needs (Figure 5.8).

If there is more than one desired actuator characteristic, the entire desired field should be compared with the available actuator characteristics field (see Figure 5.9).

The optimal selection of an actuator should take into account the transmission; usually they are selected simultaneously. Some suggestions for this can be found in Histand and Alciatore (1999); Groβ et al. (2000); Legnani, Adamini and Tiboni (2002). The optimal selection of a microactuator is a very difficult task. Many parameters should be taken in account, such as environmental conditions. Thus, big approximations are frequently acceptable.

There are two additional problems that complicate the correct selection of an actuator. First, the microactuators are not always well characterized from a mechanical viewpoint; therefore a precise selection is

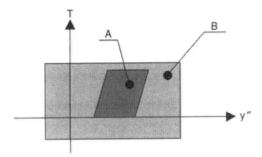

FIGURE 5.9 Comparing desired actuator with available actuator. *A* is the field of the desired actuator characteristics and *B* is the field of the available actuator characteristics. *A* is within field *B*, so the available actuator can be used for our needs.

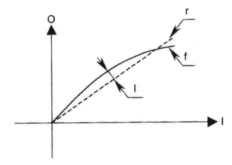

FIGURE 5.10 Definition of linearity. *O* is the actuator output, *I* is the actuator input, *r* is the reference line, *f* is the input-output function, and *l* is a measure of the linearity.

sometimes impossible. The second problem is a factor of the microtransmission. A scale factor can introduce undesired dynamical phenomena, such as high friction. An elastic joint can be used to avoid complex connections, however it exhibits a nonlinear dynamic behavior. For these reasons, some specialized optimization algorithms can be applied via a custom software program to achieve an optimal selection. These selection techniques should take into account many other properties of the considered actuators. One of the most interesting characteristics is the linearity. Linearity refers to the input–output function of the actuator, where the input is a physical property of the input energy (current, tension, temperature, etc.) and the output is a physical property of the mechanical output energy (position, speed, acceleration, power, etc.). The maximum difference between a reference linear line and the actuator output is a measure of the linearity (Figure 5.10). In fact, a high linearity is synonymous of a simple relationship between input and output and it implies facility of commanding the actuator.

An actuator is used to generate a motion, but how much of the commanded motion is similar to the generated motion? The answer can be found in the concept of "positioning precision," which is defined by three subconcepts: accuracy, repeatability, and resolution (Figure 5.11). Accuracy is the distance between the average reached position and the target position. Repeatability is the average distance between the different reached positions and the same target position. And resolution is the minimum incremental motion of the actuator.

The dynamics of the actuator can be influenced by much undesired phenomena, including hysteresis, threshold fluctuations, noise, and backlash; any of which can reduce the performance of the system. Hysteresis is the difference in the actuator output when it is reached from two opposite directions (Figure 5.12). The threshold is the smallest initial increment of the input able to generate an actuator output. Noise is a measure of the fluctuations in the output with zero input and backlash is the lost motion after reversing direction.

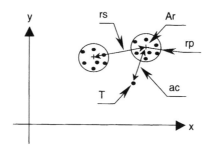

FIGURE 5.11   Definition of planar positioning precision. *x* and *y* are two spatial coordinates, *T* is the target position, *a* is a measure of the accuracy, *Ar* is the average reached position, *rs* is a measure of the resolution, and *rp* is a measure of the repeatability.

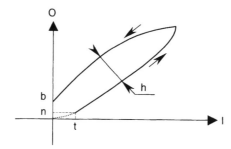

FIGURE 5.12   Definition of hysteresis, threshold, noise, and backlash. *O* is the actuator output, *I* is the actuator input, *t* is a measure of the threshold, *h* is a measure of the maximum hysteresis, *b* is a measure of the backlash, and *n* is a measure of the noise.

When the microactuator is composed of two or more parts mutually connected by mechanical joints, additional mechanical properties of this connection should be taken into account. In particular, the connection can be affected by compliance in terms of mechanical backlash, level of stiffness, friction, or mechanical hysteresis. This compliance can dramatically transform the dynamics of the actuator. Mechanical backlash is the jarring reaction or striking back of a mechanical part in the actuator when the motion is not uniform or when a sudden force is applied. The level of stiffness is the ability to convert a force into a displacement without any deformations. Friction is the resistance that any body meets when moving over another body. And the mechanical hysteresis is the difference in the actuator displacement when it is reached from two opposite directions.

The particular properties of the actuator can be dependent on some parameters or can be related to each other. For example, if the static behavior of the actuator is considered (the force-displacement curve), one can observe that the load is able to change this curve: the stiffness is dependent upon the applied load. Similarly the sensitivity of the actuator, that is, the ratio of change in the actuator output to an incremental change in its input, is usually a function of temperature. To describe these characteristics, many general or specialized dynamical models were developed, some of which will be expanded upon in this chapter. It is significant to note how the introduction of "smart material" into the development of actuation technology has changed the way of thinking about the actuators, not only during the design stage, but also during the performance evaluation stage. The use of smart materials can reduce the number of bodies forming the actuator to one or two, minimizing the problem of connections. The number of design possibilities is significantly expanded by the introduction of smart materials, due to their ability to generate different shapes and movements. For these reasons, the development of a microactuator needs to remain "open-minded" and the evaluation of its performances should take new factors into consideration. Some of these promising smart materials are listed in Table 5.1 and some of their characteristics are evaluated to understand their degree of usability.

TABLE 5.1 Properties of Some Actuators Made with Smart Materials

Properties	Electrostrictive	Elec.Rheological	Magnetostrictive	Ni-Ti SMA	Piezo Ceramic
Cost	Moderate	Moderate	Moderate	Low	Moderate
Maturity	Moderate	Moderate	Moderate	Fair	Good
Networkable	Yes	Yes	Yes	Yes	Yes
Embedability	Good	Fair	Good	High	High

TABLE 5.2 Classification of Microactuators Based on the Output Motion

Type of Embedding	Generated Motion
Actuator embedded in the machine	Complex three-dimensional Complex bi-dimensional motion Linear motion Rotational motion
Modular actuator	Linear motion Rotational motion
Special un-modular and un-embedded actuator	Complex three-dimensional Complex bi-dimensional motion

TABLE 5.3 Classification of Microactuators Based on Integration-digitalization Level

Integration of the Design	Digitalization of Mechanical Output
Single component	Analogical output
Serial integration	Analogical force and possible digital stroke
Parallel integration	Possible digital force and analogical stroke
Serial and parallel integration	Possible digital force and possible analogical stroke

Undoubtedly, many microactuators are present on the market and in research laboratories, therefore a classification would be useful to frame this chapter into a schematic knowledge system. A first definition of microactuators can be based on their mechanical output, particularly on the generated motion. If the actuator is embedded within the other components of the machine, it is difficult to distinguish the actuator from the machine, so the complete machine can be thought of as the actuator. This embedded actuator is able to generate complex motions. If the actuator is only connected to the transmission of the machine, it usually is able to generate only simple motions. This second choice is usually selected to increase the modularity of the system. A scheme of the classification, based on generated motion, is shown in Table 5.2.

Some actuation technologies allow for the integration of many identical microactuators (or nanoactuators) that combined to form a unique actuator (or microactuator). Every integrated microactuator is able to exert a force and to generate a simple motion (a linear motion for this example). If the microactuators are integrated in a parallel configuration, the force exerted by the new actuator is the sum of the forces exerted by every single microactuator. The stroke of the new actuator is the same stroke of each single microactuator (Mavroidis, 2002). If the microactuators are integrated in a serial configuration, the force exerted by the new actuator is the same force exerted by each single microactuator and the stroke of the new actuator is the sum of the strokes of each single microactuator. If each integrated microactuator is unable to exert any force or its maximum exertable force and if it is unable to generate any motion or the total stroke motion, then the resulting actuator is able to exert a digitalized force or a digitalized stroke, depending on the integration configuration. Therefore, it is possible to define an actuator based on the digitalization level of the mechanical output or on the integration level of the design (see Table 5.3).

Finally, a classification of microactuators can be based on the type of input energy (Table 5.4). The most important are electrical, fluidic, thermal, chemical, optical, and acoustic.

TABLE 5.4   Classification of Microactuators Based on the Input Energy

Input Energy	Physical Class	Actuator
Electrical	Electric and magnetic field	Electrostatic
		Electromagnetic
	Molecular forces	Piezoelectric
		Piezoceramic
		Piezopolymeric
		Magnetostrictive
		Electrostrictive
		Magnetorheological
		Electrorheological
Fluidic	Pneumatic	High pressure
		Low pressure
	Hydraulic	Hydraulic
Thermal	Thermal expansion	Bimetallic
		Thermal
		Polymer gels
	Shape memory effect	Shape memory alloys
		Shape memory polymers
Chemical	Electrolytic	Electrochemical
	Explosive	Pyrotechnical
Optical	Photomechanical	Photomechanical
		Polymer gels
Acoustic	Induced vibration	Vibrating

This chapter will adopt a classification of microactuators based on the form of the input energy and will deal with only some of the microactuators listed in Table 5.4. In particular, piezoelectric, electromagnetic, shape memory alloys, and electrostatic microactuators will be discussed. Some information will also be provided on polymeric, electrorheological, SMA polymeric, and chemical microactuators.

5.2   Piezoelectric Actuators

The discovery of the piezoelectric phenomena is due to Pierre and Jacque Curie, who experimentally demonstrated the connection between crystallographic structure and macroscopic piezoelectric phenomena and published their results in 1880. Their first results were only on the direct piezoelectric effect (from mechanical energy to electric energy). The next year, Lippman theoretically demonstrated the existence of an inverse piezoelectric effect (from electric energy to mechanical energy). The Curie brothers then gave value to Lippman's theory with new experimental data, opening the way to piezoelectric actuators. After some tenacious theoretical and experimental work in the scientific community, Voigt synthesized all the knowledge in the field using a properly tensorial approach, and in 1910 published a comprehensive study on piezoelectricity. The first application of a piezoelectric system was a sensor (direct effect), a submarine ultrasonic detector, developed by Lengevin and French in 1917. Between the first and second World Wars many applications of natural piezoelectric crystals appeared, the most important being ultrasonic transducers, microphones, accelerometers, bender element actuators, signal filters, and phonograph pick-ups. During World War II, the research was stimulated in the United States, Japan, and Soviet Union, resulting in the discovery of piezoelectric properties of piezoceramic materials exhibiting dielectric constants up to 100 times higher than common cut crystals. The research on new piezoelectric materials continued during the second half of the twentieth century with the development of barium titanate and lead zirconate titanate piezoceramics. Knowledge was also gained on the mechanisms of piezoelectricity and on the doping possibilities of piezoceramics to improve their properties. These new results allowed high performance and low cost applications and the exploitation of a new design approach (piezocomposite structures, polymeric materials, new geometries, etc.) to develop new classes of sensors and, especially new classes of actuators.

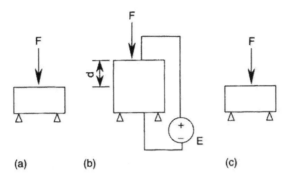

FIGURE 5.13 Inverse piezoelectric effect. An external force *F* is applied to a piezoelectric parallelepiped as in configuration (a). When an electric tension generator gives power to the actuator, it results in a displacement *d*, as in configuration (b). If the electric power is disconnected, the piezoelectric parallelepiped returns to its initial condition, as in configuration (c).

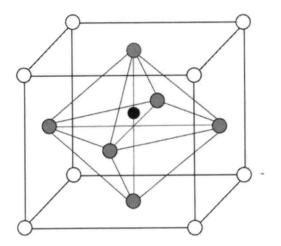

FIGURE 5.14 Structure of a PZT cell under the Curie temperature. White particles are large divalent metal ions, gray particles are oxygen ions, and the black particle is a tetravalent metal ion.

5.2.1 Properties of Piezoelectric Materials

A piezoelectric material is characterized by the ability to convert electrical power to mechanical power (inverse piezoelectric effect) by a crystallographic deformation. When piezoelectric crystals are polarized by an electric tension on two opposite surfaces, they change their structure causing an elongation or a shortening, according to the electric field polarity. The electric charge is converted to a mechanical strain, enabling a relative movement between two material points on the actuator. If an external force or moment is applied to one of the two selected points, opposing a resistance to the movement, this "conceptual actuator" is able to win the force or moment, resulting in a mechanical power generation (Figure 5.13). The most frequently used piezoelectric materials are piezoceramics such as PZT, a polycrystalline ferroelectric material with a tetragonal-rhombahedral structure. These materials are generally composed of large divalent metal ions; such as lead; tetravalent metal ions, such as titanium or zirconium (Figure 5.14); and oxygen ions. Under the Curie temperature, these materials exhibit a structure without a center of symmetry. When the piezoceramics are exposed to temperatures higher than Curie point, they transform their structure, becoming symmetric and loosing their piezoelectric ability.

Common piezoelectric materials are piezoceramics such as lead zirconate titanate (PZT) and piezoelectric polymers such as polyvinylidene fluoride (PVDF). To improve the performance of piezoceramics, the research proposed new formulations, PZN–PT and PMN–PT. These new formulations extend the strain from the 0.1%–0.2% (for PZT) to 1% (for the new formulations) and are able to generate a power density five times higher than that of PZT. Piezoelectric polymers are usually configured in film structures and exhibit high voltage limits, but have low stiffness and electromechanical coupling coefficients. Piezoceramics are much stiffer and have larger electromechanical coupling coefficients; therefore polymers are not usually chosen as actuators.

We can use the constitutive equations to describe the relationship between the electric field and the mechanical strain in the piezoelectric media:

$$\begin{aligned} S_{ij} &= s_{ijkl}^E T_{kl} + \underline{d}_{ijk} E_k \\ D_i &= d_{ijk} T_{jk} + \varepsilon_{ij}^T E_j \end{aligned}, \quad i, j, k, l \in 1, 2, 3 \tag{5.3}$$

where D is the three-dimensional vector of the electric displacement, E is the three-dimensional vector of the electric field density, S is the second order tensor of mechanical strain, T is the second order tensor of mechanical stress, s^E is the fourth order tensor of the elastic compliance, ε^T is the second order tensor of the permeability, and d is the third order tensor of the piezoelectric strain. Note, through a transposition, $\underline{d}$ is related to d as that can be observed in the commonly used matrix form of the constitutive equation:

$$\begin{aligned} S_I &= s_{I,J}^E T_J + d_{I,j} E_j \\ D_i &= d_{i,J} T_J + \varepsilon_{i,j}^T E_j \end{aligned}, \quad i, j \in 1, 2, 3; I, J \in 1, 2 \dots 6 \tag{5.4}$$

The first version of the constitutive equation is useful to explicate the dimensions of the tensors, while the second is more concise. There are many ways of writing these equations: another interesting version (Equation 5.5) gives the strain in terms of stress and electric displacement, introducing the voltage matrix g and the β matrix:

$$\begin{aligned} S_I &= s_{I,J}^D T_J + g_{I,j} E_j \\ E_i &= -g_{i,j} T_J + \beta_{i,j}^T D_j \end{aligned}, \quad i, j \in 1, 2, 3; I, J \in 1, 2 \dots 6 \tag{5.5}$$

If the tensors/vectors D, S, E, and T are rearranged into nine-dimensional column vectors, the constitutive equation can then take the form of Equation 5.6, Equation 5.7, Equation 5.8, or Equation 5.9, according to the selection of dependent and independent variables:

$$\begin{cases} D = \varepsilon^T E + d : T \\ S = d^t E + s^E : T \end{cases} \tag{5.6}$$

$$\begin{cases} D = \varepsilon^S E + e : S \\ T = -e^t E + c^E : S \end{cases} \tag{5.7}$$

$$\begin{cases} E = \beta^T D - g : T \\ S = g^t D + s^D : T \end{cases} \tag{5.8}$$

$$\begin{cases} E = \beta^S D - h : S \\ T = -h^t D + c^D : S \end{cases} \tag{5.9}$$

The above constitutive equations exhibit linear relationships between the applied field and the resulting strain. As an example, we can consider the tensor d. The experimental values of its components are obtained by an approximation, as it depends upon the strain and the applied electric field. This approximation consists of the hypothesis of low variation of applied voltage and the resulting strain. If the considered region is out of the field of linearity, then new values should be used to estimate the tensor d (the constitutive equations are linear, but the value of d is different in each small considered region). Or, a unique nonlinear constitutive equation could be used (d is no more constant but is a function of S and

E, resulting in a theoretically correct but really complex approach). Another consideration is based on the "aging effect" of piezoceramic materials represented by a logarithmic decay of their properties with time. Therefore, over time, a new value of *d* should be estimated to obtain a correct model for the piezoceramic material.

Considering linear constitutive equations (for each small region of the considered variables) in combination with the hypothesis of low strain, we can write:

$$\nabla_s v = \frac{\partial S}{\partial t}, \; with \;\; \nabla_s(\mathrm{o}) \equiv \frac{1}{2}\left(\nabla(\mathrm{o}) + \nabla(\mathrm{o}^T)\right) \tag{5.10}$$

where *v* is the speed of a basic element of piezoceramic. Using Equation 5.10, the equation of motion (Equation 5.11), the Maxwell equations (Equation 5.12), and the previously mentioned constitutive Equations 5.7 and 5.8 (the latter is used only to find the time derivative of *D*), we can obtain the general Christoffel equations of motion (Equation 5.13).

$$\nabla \mathrm{o} T = \rho \frac{\partial v}{\partial t} - F \tag{5.11}$$

where *ρ* is the density of the material and *F* is the resulting internal force reduced to a surface force (with the divergence theorem).

$$\begin{cases} -\nabla \times E = \mu_0 \dfrac{\partial H}{\partial t} \\[2mm] \nabla H = \dfrac{\partial D}{\partial t} + J \end{cases} \tag{5.12}$$

where *μ*₀ is the permeability constant, *H* is the electromagnetic induction, and *J* is current density.

$$\begin{cases} \nabla \mathrm{o} c^E : \nabla_s v = \rho \dfrac{\partial^2 v}{\partial t^2} - \dfrac{\partial F}{\partial t} + \nabla \mathrm{o} e \dfrac{\partial E}{\partial t} \\[3mm] -\nabla \times \nabla \times E = \mu_0 \varepsilon^S \dfrac{\partial^2 E}{\partial t^2} + \mu_0 e : \nabla_s \dfrac{\partial v}{\partial t} + \mu_0 \dfrac{\partial J}{\partial t} \end{cases} \tag{5.13}$$

In order to obtain a simple set of equations from Equation 5.13, we can neglect the presence of force, current density, and the rotational term of *E*. The Fourier theorem allows us to transform, under reasonable hypotheses, a periodic function (or in general a function defined into a finite time frame) into a sum of trigonometric functions such that we can consider only a single wave propagating through the media. The usual geometries of piezoelectric actuators are planar, therefore we will consider only planar waves (Equation 5.14). Under these simplifications, we can obtain a simplified set of Christoffel equations (Equation 5.15).

$$f(r, t) = e^{j(\omega t - k l \mathrm{o} r)} \tag{5.14}$$

where *ω* is the angular pulsation, *I* is the direction of the wave, and the constant *k* should be determined.

$$\begin{cases} -k^2 \left(l_{iK} c^E_{KL} l_{Lj}\right) \cdot v_j + \rho \omega^2 v_i = -j\omega \cdot k^2 \left(l_{iK} e_{Kj} l\right) \cdot V \\[2mm] \omega^2 k^2 \left(l_i \varepsilon^S_{ij} l_j\right) \cdot V = -j\omega \cdot k^2 \left(l_i e_{iL} l_{Lj}\right) \cdot v_j \end{cases} \tag{5.15}$$

where *V* is the electric potential, while l_i, l_j, $l_{i,K}$ and l_{Lj} are the matrices of the directional cosines. The resulting equation can be solved to calculate the value of the potential energy to have a desired speed (under the limitations of the used technology). Knowing the model for the actuating principle, this equation can be implemented in a more complex model of a complete actuator or it can be used as a "metaphor" for the behavior of the actuator. The second approach allows the definition of a "virtual" piezoelectric object implementing, through a proper calibration, some correction of the piezoelectric matrix of the "real" piezoceramic to take into account some unconsidered phenomena.

5.2.2   Properties of Piezoelectric Actuators

In 2001, Niezrecki et al. in 2001 proposed a review of the state of the art of piezoelectric actuation. This section will use this scheme and provide some explanations of the most common actuation systems.

Piezoelectric actuators are composed of elementary PZT parts that can be divided into three categories (depending on the used piezoelectric relation) of axial actuators, transversal actuators, and flexural actuators (Figure 5.15). Axial and transversal actuators are characterized by greater stiffness, reduced stroke, and higher exertable forces, while flexural actuators can achieve larger strokes but exhibit lower stiffness.

Although we have shown in Figure 5.15 piezoelectric elementary parts with a parallelepiped shape, piezoelectric materials are produced in a wide range of forms using different production techniques—from simple forms, such as rectangular patches or thin disks, to custom very complex shapes. Because of the reduced displacements, piezoelectric materials are not usually used directly to generate a motion, but are connected to the user by a transmission element (Figure 5.2). Therefore, the piezoelectric actuators are not just "simple actuators," they are complete machines with an actuation system (the PZT element) and a transmission allowing the transformation of mechanical generated power in a desired form. In fact, the primary design parameters of a piezoelectric actuator (referring to the entire actuating machine, not only the elementary PZT part) include

- the functional parameters — displacement, force, and frequency — and
- the design parameters — size, weight, and electrical input power.

Underlining only the functional parameters, the generated mechanical power is essentially a trade off between these three parameters; the actuator architecture is devoted to increment one or two of these

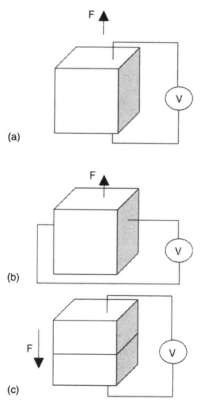

FIGURE 5.15   Elementary PZT part, where *F* is the exerted force and *V* is the electric potential difference between two faces. (a) is an axial actuator, (b) is a transversal actuator, and (c) is a flexural actuator.

parameters at the cost of the other parameters. Piezoelectric actuators are characterized by noticeable exerted forces and high frequencies, but also significantly reduced strokes; therefore the architecture designs aim to improve the stroke, reducing force or frequency. A distinction can then be made between

- force-leveraged actuators and
- frequency-leveraged actuators.

The leverage effect can be gained with an integrated architecture or with external mechanisms, so another distinction can be made between

- *internally leveraged* actuators and
- *externally leveraged* actuators.

The most common internally force-leveraged actuators include:

(1) Stack actuators
(2) Bender actuators
(3) Unimorph actuators and
(4) Building-block actuators.

The externally force-leveraged actuators can be subdivided as:

(1) Lever arm actuators
(2) Hydraulic amplified actuators
(3) Flextensional actuators and
(4) Special kinematics actuators.

The frequency-leveraged actuators can be, in general, led back to inchworm architecture.

Stack actuators consist of multiple layers of piezoceramics (Figure 5.16). Each layer is subjected to the same electrical potential difference (electrical parallel configuration), so the total stroke results is the sum of the stroke of each elementary layer, while the total exertable force is the force exerted by a single elementary layer. The leverage effect on the stroke is linearly proportional to the ratio between the elementary piezoelectric length and the actuator length.

The most common stack architectures can gain some microns stroke, exerting some kilonewtons forces, with about ten microseconds time responses.

Bender actuators consist of two or more layers of PZT materials subjected to electric potential differences, which induce opposite strain on the layers (Figure 5.17). The opposite strains cause a flexion of the bender, due to the induced internal moment in the structure. This architecture is able to generate an amplification of the stroke as a quadratic function of the length of the actuator, resulting in a stroke of

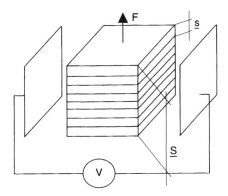

FIGURE 5.16 Stack actuator. *V* is the electrical potential difference applied to each piezoelectric element, *F* is the total exertable force. The stroke *s* of a single piezoelectric element is proportional to *s*, while the total stroke *S* is proportional to *S*.

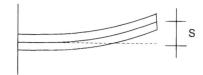

FIGURE 5.17 Bender actuator, where S is the total stroke.

more than one millimeter. Different configurations of bender actuators are available, such as end supported, cantilever, and many other configurations with different design tricks to improve stability or homogeneity of the movement.

Unimorph actuators are a special class of bender actuators, which are composed of a PZT layer and a non-active host. Two common unimorph architectures are: Rainbow, developed by Heartling (1994) and Thunder, developed at NASA Langley Research Center (Wise, 1998). These are characterized by a prestressed configuration. Being stackable, they are able to gain important strokes (some millimeters).

Building block actuators consist of various configurations characterized by the ability to combine the elementary blocks in series or parallel configurations to form an arrayed actuation system with improved stroke by series arrays and improved force by parallel arrays. There are various state-of-the-art elementary blocks available such as C-blocks, recurve actuators, and telescopic actuators.

The first class of externally leveraged actuators to be examined is the lever arm actuator class. Lever arm actuators are machines composed of an elementary actuator and a transmission able to amplify the stroke and reduce the generated force. The transmission utilized is a leverage mechanism or a multistage leverage system. To reduce design complexity, the leverage system is generally composed of two simple elements: a thin and flexible member (the fulcrum) and a thicker, more rigid, long element (the leverage arm). Another externally leveraged architecture is hydraulic amplification. In this configuration, a piezoelectric actuator moves a piston, which pumps a fluid into a tube moving another piston of a reduced section. The result is a very high stroke amplification (approximately 100 times); however, this amplification involves some problems due to the presence of fluids, microfluidic phenomena, and high frequency mechanical waves that are transmitted to the fluids. The third class of externally amplified actuators is flextensional actuators. This class is characterized by the presence of a flexible component with a proper shape, able to amplify displacement. It differs from the lever arm actuators approach, because of its closed-loop configuration, resulting in a higher stiffness but reduced amplification power. To increment the stroke amplification, this class of actuators can be used in a building-block architecture. A typical example is a stack of Moonie actuators. The research on actuation architecture is very dynamic. New design solutions emerge in literature and on the market frequently; therefore it would be improper to generalize these classifications based on only the three described classes of externally leveraged actuators: lever arm, hydraulic, and flextensional.

The final class of externally leveraged actuators uses the frequency leverage effect. These actuators are basically reducible to inchworm systems (Figure 5.18). In general, they are composed of three or more actuators, alternatively contracting, to simulate an inchworm movement. The resulting system is a very precise actuator, with very high stroke (more than 10 mm), but with a reduced natural frequency.

The behavior of PZT actuators can be affected by undesired physical phenomena such as hysteresis. In fact, hysteresis can account for as much as 30% of the full stroke of the actuator (Figure 5.19). An additional problem is the occurrence of spurious additional resonance frequencies under the natural frequencies. These additional frequencies introduce undesired vibrations, reducing positioning precision and the overall performance of the actuator. Furthermore, the depoling effect, which results in an undesired depolarization of artificially polarized materials, occurs when a too-high temperature of the PZT is gained, a too-large potential is imposed to the actuation system, or a too-high mechanical stress is applied. To avoid undesired phenomena, the actuator should be maintained within a proper range of temperatures, mechanical stresses, and electrical potential. A design able to counteract such undesired effects could be studied and a control system implemented. The piezoelectric effect could be implemented

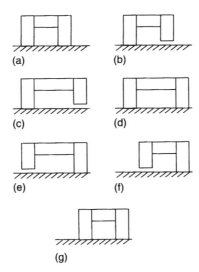

FIGURE 5.18 Sequential inchworm movement.

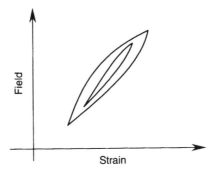

FIGURE 5.19 Hysteresis in PZT actuators.

to sense mechanical deformations. With a very compact design, a controlled electromechanical system can be developed. This is one of the many reasons piezoelectric actuators have become so successful.

5.3 Electromagnetic Actuators

5.3.1 Electromagnetic Phenomena

The research on electromagnetic phenomena and their ability to generate mechanical interactions is ancient. The first scientific results were from William Gilbert (1600), who, in 1600, published *De Magnete*, a treatise on the principal properties of a magnet — the presence of two poles and the attraction of opposite poles. In 1750, John Michell (1751) and then in 1785, Charles Coulomb (1785–1789) developed a quantitative model for these attraction forces discovered by Gilbert. In 1820, Oersted (1820) and independently, Biot and Savart (1820), discovered the mechanical interaction between an electric current and a magnet. In 1821, Faraday (1821) discovered the moment of the magnetic force. Ampere (1820) observed a magnetic equivalence of an electric circuit. In 1876, Rowland (1876) demonstrated that the magnetic effects due to moving electric charges are equivalent to the effects due to electric currents. In 1831, Michael Faraday (1832) and, independently, Joseph Henry (1831) discovered the possibility of generating an electric current with a variable magnetic field. In 1865, James Clerk Maxwell (1865) developed the first comprehensive theory on electromagnetic field, introducing the modern concepts of electromagnetic waves.

Though not considered so by his contemporaries, his theory was revolutionary, as it did not require the presence of a media, the ether, to propagate the electromagnetic field. Later, in 1887, Hertz (1887) experimentally demonstrated the existence of electromagnetic waves. This opened up the possibility of neglecting the ether and, in 1905, aided the formulation of the theory of relativity by Albert Einstein (1905). While electromagnetic physics was being studied, new advances in electromagnetic motion systems were developed. The earliest experiments were undertaken by M.H. Jacobi (1835) in 1834 (moving a boat). Though the first complete electric motor was built by Antonio Pacinotti (1865) in (1860). The first induction motor was invented and analysed by G. Ferraris (1888) in 1885 and later, independently, by N. Tesla (1888), who registered a patent in the United States in 1888. Many other macro electromagnetic motors were later developed and research in this field remained very active. Research in the field of electric microactuators started in 1960 with W. McLellan, who developed a 1/64th inch cubed micromotor in answer to a challenge by R. Feynman. Since then an indefinite number of inventions and prototypes have been presented to the scientific community, patented, and marketed. Therefore, outlining the history of microelectromagnetic actuators is an almost impossible task; however, by observing the new technologies produced, we are able to trace the key inventions and ideas to the formation of actual components.. For example, the isotropic and anisotropic etching techniques that were developed in the 1960s generated bulk micromachining in 1982; sacrificial layer techniques also developed in the 1960s generated surface micromachining in 1985. Some more recent technologies include silicon fusion bonding, LIGA technology, micro electro, and discharge machining.

5.3.2 Properties of Electromagnetic Actuators

Electromagnetic actuators can be classified according to four attributes: geometry, movement, stroke, and type of electromagnetic phenomena (Tables 5.5–5.8).

We will use the Lagrange equations of motion or Newtonian equations of motions to derive the models of each single type of actuator:

$$\frac{d}{dt}\left[\frac{\partial(\Gamma + D)}{\partial \dot{q}_i}\right] - \frac{\partial(\Gamma - \Pi)}{\partial q_i} = Q_i, \quad i = 1,2 \dots n \tag{5.16}$$

$$\sum_{j=l}^{m} F_j(t, r) = m\frac{d^2r}{dt^2}$$

TABLE 5.5   Classification of Electromagnetic Microactuators Based on Geometry

Geometry of the Actuator
Planar
Cylindrical
Spherical
Toroidal
Conical
Complex shape

TABLE 5.6   Classification of Electromagnetic Microactuators Based on Movement

Movement of the Actuator
Prismatic
Rotative
Complex movement

where the first and second equation represents, respectively, Lagrangian and Newtonian approach; the used symbols Γ, D, and Π denote, respectively, the kinetic, potential, and dissipated energies; while q_i and Q_i represent, respectively, the generalized coordinates and the generalized forces applied to the system. The Newtonian approach corresponds to the use of Newton's second law of motion, where F_j is a vectorial force applied to the system, r is the vector representing the position and the geometrical configuration of the system, and m is the mass of the system. The presence of rotary movement can imply the use of an analogous rotary version of Newton's second law. Multibody systems can be also considered with the use of a simple and concise matrix method. To develop MEMS models, Γ, D, and Π, in the Lagrangian approach, and F, in the Newtonian approach, should take account of mechanical and electrical terms.

We will first consider the Lagrangian approach as it is able to simply mix many forms of physical interactions without separating the model into many parts. If we consider elementary mechanical movement (prismatic and rotational) and elementary electric circuits, and then apply a lumped parameters model with a Lagrangian approach, we can propose the synthetic Table 5.9, to associate each elementary parameter with each kind of energy mentioned in (Equation 5.16) and select the correct generalized coordinates and forces.

The proposed table is mainly outlining many other forms of energy deemed worthy of consideration, as well as other basic or complex models to be taken into account. For instance, the effect of impacts due

TABLE 5.7 Classification of Electromagnetic Microactuators Based on the Stroke

Stroke of the Actuator
Limited stroke
Unlimited stroke

TABLE 5.8 Classification of Electromagnetic Microactuators Based on Electromagnetic Phenomena

Electromagnetic Phenomena of the Actuator
Direct current microactuators
Induction microactuators
Syncronous microactuators
Stepper microactuators

TABLE 5.9 Selection of Terms for Lagrange Equations of Motion

Generalized Elementary Movement	Terms of Lagrange Equation	Elementary Lumped Parameter
Prismatic mechanical movement	Kinetic energy	m: mass (prismatic inertial term)
	Potential energy	k: linear stiffness
	Dissipative energy	f: linear friction
	Generalized coordinate	x: linear coordinate along the trajectory
	Generalized force	F: applied force
Rotational mechanical movement	Kinetic energy	J: moment of inertia
	Potential energy	k: rotational stiffness
	Dissipative energy	f: rotational friction
	Generalized coordinate	t: rotational coordinate along the trajectory
	Generalized force	T: applied torque
Electric circuit	Kinetic energy	L: selfinductance
		M: mutual inductance
	Potential energy	C: capacitance
Thermal	Dissipative energy	R: resistance
	Generalized coordinate	q: electrical charge
	Generalized force	u: applied voltage

to backlash between mechanical components can be considered introducing stiffness function, which are governed by the distance between two consecutive components of the kinematical chain. The resulting elastic force can be evaluated with the product of the stiffness function and the distance between the two components. We consider this function as constant or polynomial function of the distance between the two components; however, it assumes the value of zero when the absolute distance between the two considered consecutive components is less than a half of the backlash:

$$k_{12}(x - y) \equiv \underline{k}_{12} + \frac{k_{12}}{2} \left(\frac{|x - y - g/2|}{x - y - g/2} - \frac{|x - y + g/2|}{x - y + g/2} \right) \qquad (5.17)$$

where k_{12} is the modified stiffness function, x is the generalized position of a mechanical component of the microactuator, y is the generalized position of the consecutive mechanical component, g is the backlash, and k_{12} is the real stiffness of the junction between the two adjacent components (Figure 5.20).

In a similar way, the backlash can also be considered in the definition of the dissipative forces between two adjacent components. The definition of the friction is a more complex problem, because of the presence of different type of frictions. Three important friction phenomena should be considered: Coulomb friction, viscous friction, and static friction:

$$F \equiv \left(f_v + f_s \exp\left(-\varphi|\dot{\xi}|\right) + f_c|\dot{\xi} \right) \frac{|\dot{\xi}|}{\dot{\xi}} \qquad (5.18)$$

where f_v, f_s, f_c and φ are optimal parameters, while $\xi \equiv x - y$ is the relative position of two adjacent mechanical components. The presence of backlash introduces a discontinuity in the frictional parameters (Equation 5.19), and the situation can radically increment the complexity of the frictional model.

$$F \equiv \left\{ f_v(1 + \Delta_1\delta(\xi)) + f_s(1 + \Delta_2\delta(\xi))\exp\left[-\varphi(1 + \Delta_3\delta(\xi))|\dot{\xi}|\right] + f_c(1 + \Delta_4\delta(\xi))|\dot{\xi}| \right\} \frac{|\dot{\xi}|}{\dot{\xi}},$$

$$\delta(\xi) \equiv \frac{1}{2} \left(\frac{|\xi - g/2|}{\xi - g/2} - \frac{|\xi + g/2|}{\xi + g/2} \right) \qquad (5.19)$$

where $\Delta_1, \Delta_2, \Delta_3$, and Δ_4 are optimal parameters, which take into account the variation of frictional effects in the "backlash-disconnected" joint. Many other observations can also be introduced to model a particular phenomenon such as the hysteretic behavior of general mechanical systems, but these remarks will be considered only for shape memory microactuators. An example for electrical parameters is the dependence of resistivity based on temperature (Equation 5.20). In fact, a fluctuation in the temperature range due to unexpected phenomena and changes in internal temperature caused by electrical power feeding and Joule effect affect the resistivity of a microelectromechanical system, and its related resistance, thus altering its functionality.

$$\rho = \rho_0 \left[1 + \sum_{i=1}^{n} \alpha_i (T - T_0)^i \right] \qquad (5.20)$$

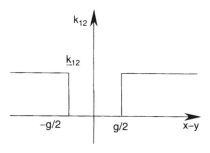

FIGURE 5.20   Modified stiffness function to take account of backlash.

where ρ is the resistivity, ρ_0 is the reference resistivity, α_i is the temperature dependence coefficient, T is the temperature of the resistor, and T_0 is the reference temperature. Other examples of non-linearity in the electrical parameters can be observed in non-linear components, such as diodes.

Regarding the general microelectromechanical model theory, a suitable approach allows the modeling of electro-mechanical systems using a lumped parameter approach. If designers consider the general relations between magnetic, electric, and mechanical fields, and referring to the Maxwell's equations, they could eventually apply a finite element approach to improve the model's power:

$$\nabla \circ E = 0$$

$$\nabla \circ B = 0$$

$$\nabla \times E = -\frac{\partial B}{\partial t} \tag{5.21}$$

$$\nabla \times B = \mu \varepsilon \frac{\partial E}{\partial t}$$

5.3.3 DC Mini- and Microactuators

Because of thermal problems related with brush friction, DC microactuators are not the most viable choice within the class of electromagnetic microactuators. To develop a complete model of the actuator, one can select a cylindrical geometry, having a rotational movement and an unlimited stroke. One can also consider a single conductive coil, which is turned on a cylindrical rotor, able to rotate around its axis. The coil is also embedded in a magnetic field B and it is fed with an electric current as in Figure 5.21. Due to the action of the magnetic field on the electric current, the coil is affected by the mechanical couple:

$$\tau = i \cdot A \times B \tag{5.22}$$

where τ is the mechanically generated couple, A is the area vector perpendicular to the coil, and B is the magnetic field intensity. We can now consider a set of n coils equally spaced on the rotor, which are affected by a set of p magnetic fields perpendicular to the rotor axis. The same current i is in all the coils, and the electric potential of the rotor is applied to the coils by brushes, which are used to change the direction of the current every half turn of the rotor. The rotor then, is subject to the force computed by:

$$\tau(\theta) = i \cdot |A| \cdot |B| \cdot \sum_{j=0}^{n-1} \sum_{k=0}^{p-1} \left| \cos\left(\theta + \pi \cdot \frac{j \cdot p + k \cdot n}{p \cdot n}\right) \right| \tag{5.23}$$

where θ is the angular coordinate which identifies the position of the rotor in respect to the microactuator's stator (Figure 5.22).

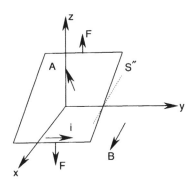

FIGURE 5.21 Coil with an area A, an electric current i and embedded in a magnetic field B.

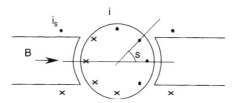

FIGURE 5.22   Scheme of a DC microactuator, where B is the magnetic field intensity, i is the rotor's current, and i_s is the stator's current used to generate the magnetic field.

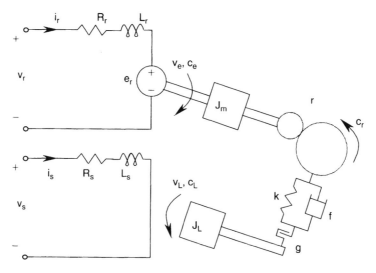

FIGURE 5.23   Circuit model of a DC microactuator. The subdivision takes place in three parts: (s) is the stator's electric circuit, (r) is the rotor's electric circuit, and (m) is the mechanical circuit of the rotor. (R) is the mechanical circuit of the rotor after the speed reduction stage integrated in the microactuator.

When the number of coils, and/or the number of magnetic fields, is sufficiently high, the mechanical torque expression can be further simplified:

$$\tau = i \cdot |A||B| \cdot n \cdot \frac{2p}{\pi} \tag{5.24}$$

Therefore, under the assumption that the magnetic field is generated by the stator's current, the magnetic torque can be expressed as the product of a constant term and two electric variables: the rotor's and the stator's currents:

$$\tau = k_s \cdot i \cdot i_s \tag{5.25}$$

To obtain the value of the angular speed of the rotor, we assume the temporary disregard of dissipations and other "parasitic phenomena" due to their effect, only before and after the transformation from electric to mechanical energy. For this reason, all the input electrical energy from the rotor is converted into mechanical energy and the angular speed of the rotor can be expressed:

$$\omega_m = e/\left(k_s i_s\right) \tag{5.26}$$

where e is the rotor's electrical potential. With the aid of the transformation equations (Equation 5.25–5.26), the microactuator behavior can be described from a standpoint of circuit's approach as in Figure 5.23.

The symbols mentioned in Figure 5.23 are summarized as follows:

(s) Stator's circuit
v_s is the stator's electric potential
i_s is the stator's current
R_s and L_s are the resistance and the inductance of the stator
(r) Rotor's circuit
v_r is the rotor's electric potential
i_r is the rotor's current
R_r and L_r are the resistance and the inductance of the rotor
e_r is the rotor's potential that linked with the rotor's current, is transformed in mechanical power which
    value is related with the stator's current and with the theoretical motion speed:

$$e_r = k_s i_s \dot{\alpha}_e \qquad (5.27)$$

where k_s is the electric–mechanic transformation constant.
(m) Mechanical stage
τ_e is the theoretic motor torque which depends upon the rotor's and stator's currents

$$\tau_e = k_s i_s i_r \qquad (5.28)$$

α_e is the theoretic position of the rotor
J_m is the motor's inertia
J_L is the load's inertia
f is the reduced friction coefficient of the system as described in Equation 5.19
k is the reduced stiffness coefficient of the system as described in Equation 5.17
g is the reduced backlash coefficient of the system
α_R is the theoretic position after the reduction stage (Equation 5.29)
α_L is the position of the load
τ_L is the applied torque of the load

$$\alpha_e = \alpha_r \cdot r \qquad (5.29)$$

where r (>1) is the mechanical reduction ratio. With a simple circuit analysis, a lumped parameter model
of the microactuator can be obtained (Equation 5.30):

$$
\begin{cases}
v_s = R_s i_s + L_s \dfrac{di_s}{dt} \\[2mm]
v_r = R_r i_r + L_r \dfrac{di_r}{dt} + e_r \\[2mm]
e_r = k_s i_s \dot{\alpha}_e \\[2mm]
\alpha_R = \alpha_c / r \\[2mm]
\tau_e = k_s i_s i_r \\[2mm]
\tau_R \equiv k(\alpha_R - \alpha_L) + f(\dot{\alpha}_R - \dot{\alpha}_L) \\[2mm]
\left(\tau_e - \dfrac{\tau_R}{r} \right) = J_m \ddot{\alpha}_e \\[2mm]
\tau_R - \tau_L = J_L \ddot{\alpha}_L
\end{cases}
\qquad (5.30)
$$

The result of the model in Equation 5.30 is a correct approximation of the non-linear model behavior of
the electromechanical microactuator after a proper definition of the considered parameters. If any of the

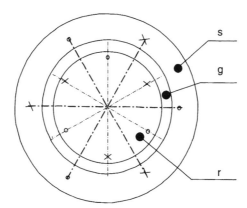

FIGURE 5.24 Physical scheme of rotary induction actuator. *s*, *r*, and *g* represent, respectively, the stator, the rotor, and the gap.

parameters of the model are not able to reach the desired predictive precision, more complex models could be considered in order to take into account neglected phenomena. Considering the introduction of fictitious mechanical compliances or electrical parasitic phenomena could be followed by a calibration procedure ruled directly by real experimental results or by a finite element model. Should a model not be implemented in a controlling algorithm, a finite element model can be directly used to provide good predictive result. Some useful finite element software are available on the market to achieve a proper model, thus creating new finite element software is not necessary.

5.3.4 Induction Mini- and Microactuators

Induction actuators consist of a mobile and a static part and the transformation from electric to mechanical energy due to the inductance of each part of the microactuator. This section will address a cylindrical rotary actuator with unlimited stroke and, to simplify the analysis, only a common configuration is considered: a three phase, two pole actuation system. Figure 5.24 shows a rotary induction motor. The mobile part (the rotor) has a cylindrical shape and is able to rotate around its axis; the static part (the stator) has the same axis as the rotor and is separated from it by an air gap. Both are composed of ferromagnetic material and incorporate lengthwise holes carrying conductive wires that are close to the air gap.

To develop a mathematical model of the dynamic behavior of the considered microactuator, it is necessary to define a reference system and an angular variable, which identify the position of the rotor (Figure 5.25). In particular where the variable θ_i defines the position of the i-th rotor's winding (r_i) in respect to the first stator's winding (s_1):

$$\theta_i = \theta + \frac{2\pi}{f}\left(i - 1\right) \tag{5.31}$$

where θ is indicated in Figure 5.25, as the angle between s_1 and r_1, and f is the number of phases. The dynamic model can be generalized as a multi-pole micromachine for which the ideal actuator speed can be calculated by:

$$\dot{\theta}_a = 2\dot{\theta}/p \tag{5.32}$$

where p is the (even) number of poles, θ is indicated in Figure 5.25, and θ_a is the ideal position of the actuator. Therefore, the dynamic electromagnetic behavior of the system can be described by the matrix equation:

$$\begin{cases} e = R \cdot i + \dot{\varphi} \\ \varphi = L \cdot i \end{cases} \tag{5.33}$$

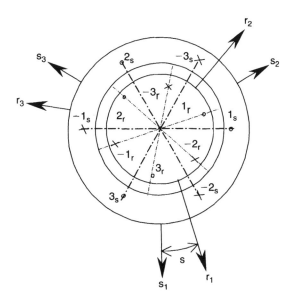

FIGURE 5.25 Definition of the position of the rotor's windings: i_r and j_s represent, respectively, the ith rotor's winding and the jth stator's winding; r_i and s_j represent, respectively the axis of the ith rotor's winding and the axis of the jth stator's winding, and θ is the angular position of the first rotor's winding in reference to the position of the first stator's winding.

where the following vectors are described in Equation 5.34: the vector e of electric potential, the vector e of electric currents, and the vector φ of magnetic fluxes; the matrix R of electric resistances is defined in Equation 5.35; and the matrix L of inductances is defined in Equation 5.36.

$$e = \begin{bmatrix} e_{s1} & e_{s2} & e_{s3} & e_{r1} & e_{r2} & e_{r3} \end{bmatrix}^T, \qquad i = \begin{bmatrix} i_{s1} & i_{s2} & i_{s3} & i_{r1} & i_{r2} & i_{r3} \end{bmatrix}^T,$$
$$\varphi = \begin{bmatrix} \varphi_{s1} & \varphi_{s2} & \varphi_{s3} & \varphi_{r1} & \varphi_{r2} & \varphi_{r3} \end{bmatrix}^T \tag{5.34}$$

where the index associated to each element of the three vectors refer to the stator/rotor winding shown in Figure 5.25.

$$R = \begin{bmatrix} I(3) \cdot R_s & 0(3) \\ 0(3) & I(3) \cdot R_r \end{bmatrix} \tag{5.35}$$

where R_s and R_r are, respectively, the stator and rotor winding resistances, while $I(3)$ and $0(3)$ are, respectively, the three-dimensional identity and zero matrixes.

$$L = \begin{bmatrix} L_{1,1} & L_{1,2} \\ L_{2,1} & L_{2,2} \end{bmatrix} \tag{5.36}$$

where the submatrixes $L_{1,1}$ and $L_{2,2}$, respectively, of statoric and rotoric inductances are defined in Equation 5.37, while the submatrixes $L_{1,2}$ and $L_{2,1}$ of mutual inductances are defined in Equation 5.38.

$$L_{1,1} = \begin{bmatrix} L_s & M_s & M_s \\ M_s & L_s & M_s \\ M_s & M_s & L_s \end{bmatrix}, L_{2,2} = \begin{bmatrix} L_r & M_r & M_r \\ M_r & L_r & M_r \\ M_r & M_r & L_r \end{bmatrix} \tag{5.37}$$

where L_s and L_r are the selfinductance of, respectively, each stator's and each rotor's winding, and M_s and M_r are the mutual inductance of, respectively, two stator's or two rotor's windings.

$$L_{1,2} = L_{2,1}^T = [l_{i,j}]_{i=1\ldots3, j=1\ldots3}, \quad l_{i,j} = M_{sr}\cos\left[\theta + \frac{2\pi}{3}(j - i)\right] \tag{5.38}$$

where M_{sr} is the mutual inductance between a stator's and a rotor's winding and θ is indicated in Figure 5.25. Coupled with the appropriate electric dynamics, and with consideration of speed reduction stage, the mechanical equilibrium equations (Equation 5.39) would develop in the complete micromechanical model of the system.

$$\begin{cases} (\tau_e - \dfrac{\tau_R}{r} = J_m \ddot{\theta} \\[2mm] \tau_R - \tau_L = J_L \ddot{\theta}_L \\[2mm] \tau_R \equiv k(\theta_R - \theta_L) + f(\dot{\theta}_R - \dot{\theta}_L) \end{cases} \tag{5.39}$$

where τ_e is the electric torque generated by the electromagnetic interaction (Equation 5.43), τ_R is torque transferred by the compliance of the mechanical stage, τ_L is the torque of the load; J_m and J_L are, respectively, the inertia of the motor and the inertia of the load; r (>1) is the mechanical reduction ratio; k and f are, respectively the stiffness (Equation 5.17) and the friction (Equation 5.19) coefficient of the reduction stage; θ, θ_R and θ_L are, respectively, the theoretic position of the rotor, the theoretic position after the reduction stage (Equation 5.40) and the position of the load.

$$\theta = \theta_R \cdot r \tag{5.40}$$

The computation of the electric torque τ_e is then related to the calculation of the generated power:

$$\begin{aligned} p &= e^T i \\[2mm] &= (R_i + \dot{\varphi})^T i \\[2mm] &= (i^T R^T i)_1 + (i^T \dot{L}^T i)_2 + (i^T \dot{L}^T i)_3 \end{aligned} \tag{5.41}$$

where the e and i are, respectively the vector of the electric potentials and the vector of the electric currents (Equation 5.34), R is the matrix of the electric resistance (Equation 5.35), φ is the vector of magnetic fluxes (Equation 5.34), L is the matrix of inductances (Equation 5.36). The three resulting terms of the power Equation 5.41 are, respectively, the electric power converted into heat, the electric power converted into mechanical power, and the variation of electromagnetic power of the system. The second term, the conversion of electric power into mechanical power, is used to obtain the electric torque τ_e; only the evaluation of the time derivative of the transposed matrix of the inductances is required:

$$\dot{L}^T = \begin{bmatrix} 0 & \dot{L}_{21} \\ \dot{L}_{12} & 0 \end{bmatrix}, L_{12} = [i_{i,j}]_{i=1\ldots3, j=1\ldots3}, i_{i,j} = -\dot{\theta} \cdot M_{sr} \sin\left[\theta + \frac{2\pi}{3}(j - i)\right] \tag{5.42}$$

Because the time derivative of the transposed matrix of the inductances is linearly dependent by the time derivative of the angular position of the rotor, the electric torque is easily found by:

$$\tau_e = \frac{i^T \dot{L}^T i}{\dot{\theta}} \tag{5.43}$$

5.3.5 Synchronous Mini- and Actuators

Synchronous microactuators consist of a mobile and a static part and the transformation from electric to mechanical energy is due to inductance of each part of the microactuator. The denomination of "synchronous machine" is due to the synchronization between the magnetic fields of the actual parts of which the microactuator is comprised. To simplify the studies, a cylindrical rotary actuator with unlimited stroke is taken into consideration and only a common configuration is analyzed: three-phase two-pole actuation system. In Figure 5.26 a rotary synchronous actuators is shown: the mobile part (rotor) can rotate around its axis, the static part (stator) has the same axis of the rotor, and it is separated from it by a gap. The stator is composed of ferromagnetic material and is furnished with liner conductive wires, while the composition of the rotor will be given in the following.

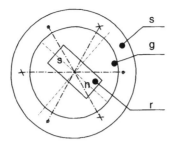

FIGURE 5.26 Physical scheme of rotary synchronous actuator: *s*, *r*, and *g* represent, respectively, the stator, the rotor, and the gap between them.

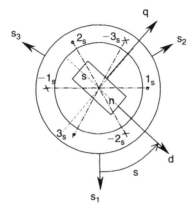

FIGURE 5.27 Definition of the rotor's position: i_s represents the ith stator's winding; s_i represents the axis of the ith stator's winding; *d* and *q* are, respectively, the direct and the quadrature magnetic axis, while θ represents the angular position of the direct magnetic axis in reference to the position of the first stator's winding.

To develop a mathematical model of the dynamic behavior of the microactuator under consideration, it is necessary to define a reference system and an angular variable that identifies the position of the rotor (Figure 5.27). In particular, the position of the rotor in respect to the i-th stator's winding (s_i) is indicated by the variable θ_i:

$$\theta_i = \theta + \frac{2\pi}{f}(i - 1) \tag{5.44}$$

where θ indicated in Figure 5.27, is the angle between s_1 and *d*, while *f* is the number of phases.

In mathematical terms, the dynamic electromagnetic behavior of the system can be described by the matrix equation 5.45, which takes the same form as the Equation 5.34.

$$\begin{cases} e = R \cdot i + \dot{\varphi} \\ \varphi = L \cdot i \end{cases}$$

$$e = \begin{bmatrix} e_{s1} & e_{s2} & e_{s3} & e_e & 0 & 0 \end{bmatrix}^T, \quad i = \begin{bmatrix} i_{s1} & i_{s2} & i_{s3} & i_e & i_D & i_Q \end{bmatrix}^T,$$

$$\varphi = \begin{bmatrix} \varphi_{s1} & \varphi_{s2} & \varphi_{s3} & \varphi_e & \varphi_D & \varphi_Q \end{bmatrix}^T \tag{5.45}$$

where the vector *e*, *i* and φ represent, respectively, the electric potential, the electric current and the magnetic flux, the matrix *R* of the electric resistance and the matrix *L* of the inductance are defined in Equation 5.46. The indices *s1*, *s2*, and *s3* associated with each element, refer to the stator's winding shown in Figure 5.27. The index *e* is associated with the excitation winding on the rotor (Figure 5.28), while the indices *D* and *Q* are, respectively, associated with the equivalent damping circuits of the direct and quadrature axis (Figure 5.29).

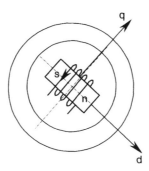

FIGURE 5.28   Excitation winding on the rotor.

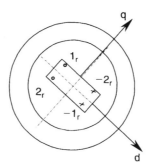

FIGURE 5.29   Equivalent damping windings on the rotor. The electromagnetic characteristics of the two rotor's windings are projected on the quadrature and direct axes to form two equivalent virtual windings.

$$R = Diag(R_s, R_s, R_s, R_e, R_D, R_Q), L = \begin{bmatrix} L_{1,1} & L_{1,2} \\ L_{2,1} & L_{2,2} \end{bmatrix} \tag{5.46}$$

where $Diag(xx)$ is the diagonal matrix, while R_s, R_e, R_D, and R_Q are, respectively, the resistance of the stator's winding, the excitation's winding, the direct axis winding, and the quadrature axis winding. The submatrixes $L_{1,1}$, $L_{2,2}$, $L_{1,2}$ and $L_{2,1}$ are defined in Equations 5.47–5.49.

$$L_{1,1} = [l_{i,j}]_{i=1\ldots3, j=1\ldots3}, l_{i,j} = \begin{cases} l_0 + L_{00} + L_2 \cos\left[2\theta - \dfrac{4\pi}{3}(i-1)\right], & i = j \\ -\dfrac{L_{00}}{2} + L_2 \cos\left[2\theta - \dfrac{4\pi}{3}(5-i-j)\right], & i \neq j \end{cases} \tag{5.47}$$

where l_0 is the statoric dispersion selfinductance, L_{00} is the net statoric selfinductance, and L_2 is the amplitude of the second Fourier term of each self-inductance of the stator.

$$L_{2,2} = \begin{bmatrix} L_e & L_{eD} & 0 \\ L_{eD} & L_D & 0 \\ 0 & 0 & L_Q \end{bmatrix} \tag{5.48}$$

where L_e, L_D and L_Q are, respectively, the excitation, the direct axis and the quadrature axis selfinductance, while L_{eD} is the mutual inductance between excitation and direct axis windings.

$$L_{1,2} = L_{2,1}^T = [l_{i,j}]_{i=1\ldots3, j=1\ldots3}, \quad l_{i,j} = \begin{cases} L_{mf} \cos(\theta_i), & j = 1 \\ L_{mD} \cos(\theta_i), & j = 2 \\ -L_{mQ} \sin(\theta_i) & j = 3 \end{cases} \tag{5.49}$$

where L_{mf}, L_{mD} and L_{mq} are the maximum values of the mutual inductance between the rotor's and stator's windings. The electric couple exerted by the actuator can be calculated, as in Equations 5.41 and 5.43, considering electric input power converted into mechanical power divided by the angular speed. The resulting real mechanical torque can be obtained with an equilibrium equation, like Equation 5.39, which takes into consideration the friction, stiffness, and backlash of the mechanical components.

5.4 Shape Memory Actuators

The first known study on shape memory effect is credited to Olander (1932) who observed that an object made of a Au–Cd alloy, if plastically deformed and then heated, is able to compensate the plastic deformation and to recover its original shape. In 1938, Greninger and Mooradian (1938), changing the temperature of a Cu–Zn alloy, observed the formation of a crystalline phase. In 1950, Chang and Read (1951) used an X-ray analysis to understand the phenomenon of crystalline phase formation in shape memory alloys and, in 1958, they showed the first shape memory Au–Cd actuator. An important milestone in the development of the shape memory alloys was reached by researchers at the Naval Ordanance Laboratory in 1961 lead by Buehler (1965), who observed the same shape memory effect in a Ni–Ti alloy. Ni–Ti alloys are cheaper, easier to work with, and less dangerous than Au–Cd alloys. Many applications appeared on the market during the 1970s: these were initially static in nature, but were later followed by dynamic examples, thus starting the era of smart technology in microactuators.

Control techniques were later developed to improve performance, with a particular focus on reducing cooling and heating time. Further data were acquired in reference to the properties of different shape memory alloys. These new results, along with the ability to mass-produce Ni–Ti microelements, will be a key to open the door to a much wider usage of smart materials. During the last two decades of the twentieth century, many advances have been made in the modeling of SMA behavior, including the control in regard to the change of shape; new design approaches to improve performances, particularly response-time and reliability; and allowing for low cost mass production. For instance, the response time of the system was significantly reduced by observing the relationship between the geometry of the system, its thermal inertia, and the use of particular material combinations to generate a Peltier cell.

The reliability of a SMA is based on the knowledge of the physical properties of the alloy: this knowledge is useful to prevent irreversible damages and to guarantee a repeatable behavior. The aims of preventing damage and gaining repeatable behavior, were attained through experimental works with new simple and complex models, varying design techniques, and the development of precise inputs and control of heating feeding. Advances in the control segment are some of the most important aspects of SMA systems in order to enhance repeatable high-speed and high-precision performance of such micro-SMA machines. Such advances were made possible through new forms of power, spurred by research in production technology. The consequential cost reduction of SMA objects followed shortly, thus allowing for the mass production of industrial and medical applications. Arguably a suitable choice in the marketplace, SMA systems provide new study opportunities placing alternate instruments in the designer's hands for new applications.

5.4.1 Properties of Shape Memory Alloys

A shape memory alloy that deforms at a low temperature will regain its original undeformed shape when heated to a higher temperature. This behavior is due to the thermoelastic martensitic phase transformation and its reversal (Bo and Lagoudas 1999a); the martensitic phase consists in a thick arrangement of crystallographic planes, characterized by a high relative mobility. Many alloys exhibit shape memory effect and the level of commercial interest of each and every alloy is correlated with its ability to easily recover its initial position or to exert a significantly high force. Based on these criteria the SMA transformation of a Ni–Ti alloy is an interesting consideration. When a Ni–Ti alloy is subject to an external force, the various planes will slide without a break in the crystallographic connections; the atoms of the structure are subject only to a reduced movement and through a subsequent heating, it will revert the structure to its initial position, resulting in the production of a significantly high mechanical force.

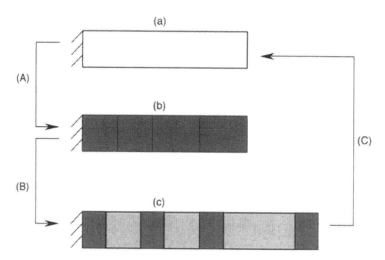

FIGURE 5.30 Shape memory effect.

The martensitic phase can appear in two different forms, depending on the history of the material. Twinned martensite is derived from the austenitic phase with cooling and it has a herringbone pattern, while detwinned martensite is due to a sliding of the crystallographic planes. A scheme of a simple cycle of transformations able to exhibit the shape memory effect is shown in Figure 5.30: a Ni–Ti bar with a fixed end and a free end without any external load is considered, apart from the thermomechanical loads imposed during the transformations (A), (B), and (C) to generate the shape memory effect. The first condition of the material (a) is the austenitic phase, after an external cooling (A), the material is converted into twinned martensite (b). From a macroscopic viewpoint, we cannot observe any deformation of the material or any external force exerted by the bar. Then a mechanical traction is applied to the bar (B) and some portions of the bar are converted, during its elongation, into detwinned martensite (c). Finally, after an external heating (C), the material is converted again into austenite and the bar recovers its original shape.

It should be observed that the martensitic generation process does not require an introduction of external substances and it depends only on the achievement of a critical temperature (M_s, or martensite start). The second important aspect of this transformation is the heat production, and finally an hysteretic phenomenon is observed when, at the same temperature, an austenitic phase and a martensitic phase coexist. The austenite-to-martensite transformation is subdivided in two subsequent stages: Bain strain and lattice-invariant shear (Figure 5.31). Bain strain generates, from austenite (a) and through a movement of the plane interface, a martensitic structure (m_b), then the lattice-invariant transformation can generate a reversible deformation with a twinning (m_{s1}) or a permanent deformation with a slip (m_{s2}). A reversible deformation is needed to exhibit a shape memory effect, so it is important to avoid slipped martensite (m_{s2}) and to generate twinned martensite (m_{s1}).

From a macroscopic viewpoint, the martensitic formation can be represented in the temperature diagram (Figure 5.32). Even though austenitic and martensitic macroscopic properties are slightly different, the martensitic fraction can be used as the dependent variable to define the behavior of the shape memory material. An interesting characteristic of this diagram is the hysteresis, which can occur between 20–40°C and is associated with the micro-frictions within the structure.

The classical shape memory effect consists in the cycle of Figure 5.30 and is usually called one-way shape memory effect (OWSM), because the shape transformation can be thermally commanded during the martensite-to-austenite conversion. This is because the material is able to "remember" only the austenitic shape. A more complete shape memory effect is the two-way memory effect (TWSM), which consists of the ability of recovering an austenitic shape as well as a martensitic shape. This second phenomenon results in reduced forces and deformations, so it is not very common in commercial applications.

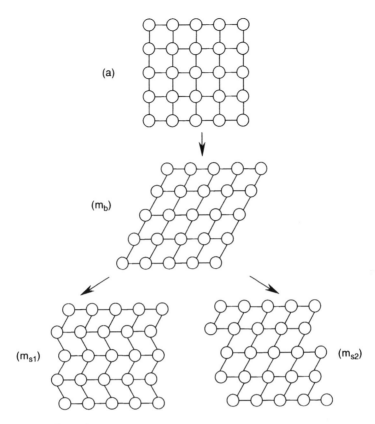

FIGURE 5.31 Martensite formation.

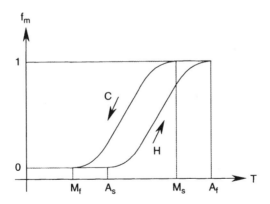

FIGURE 5.32 Martensitic fraction versus temperature. M_s, M_f, A_s and A_f are, respectively, martensite start, martensite finish, austenite start, and austenite finish critical temperatures. H and C are, respectively, heating and cooling transformations.

TWSM can be obtained from a shape memory alloy after a thermomechanical training, which generates a new microstructure (stress-biased martensite) and reduces the mechanical properties of the material.

The microscopic and macroscopic properties of the Ni–Ti alloys can be changed opportunely by varying the percentage of Ni or introducing additive components (Table 5.10). An increment in the percentage of Ni can reduce critical temperatures and increment the austenitic breaking stress. Fe and Cr are used to reduce critical temperatures. Cu reduces the hysteretic cycle and decreases the stress that is necessary to deform the martensite. O and C are usually avoided because they diminish the mechanical performance.

TABLE 5.10   Standard NiTi Alloys

Alloy code	A_s (°C)	A_f (°C)	Composition (%)
S	$-5 \div 15$	$10 \div 20$	~55.8 Ni
N	$-20 \div -5$	$0 \div 20$	~56.0 Ni
C	$-20 \div -5$	$0 \div 10$	~55.8 Ni, 0.25 Cr
B	$15 \div 45$	$20 \div 40$	~55.6 Ni
M	$45 \div 95$	$45 \div 95$	$55.1 \div 55.5$ Ni
H	>95	$95 \div 115$	<55.0 Ni

TABLE 5.11   Some Common Commercial SMA Components

Treatment	Description	Characteristics
Cold processed	The client execute thermal treatment	Do not exhibit SME
Straight annealed	Thermal treatment is executed by producer	Wire exhibiting SME
Flat annealed	Thermal treatment is executed by producer	Plate exhibiting SME
Special annealed	Thermal treatment is executed by producer	Special shape exhibiting SME

Some thermomechanical treatments are necessary to induce a proper shape memory effect, however, since shape memory alloys are commercialized after intermediate treatments, a classification such as in Table 5.11 can be used to select the correct SMA component.

To induce the shape memory effect, an austenitic shape must be "saved." This can be obtained through a thermal treatment consisting of a heating of the cold processed component in a hot mould to over 400°C for several minutes. Hardening will then result after a rapid cooling. The TWSM can also be obtained through a more complex approach consisting of the cyclic repetition of cold and hot deformations, which allows the "saving" of an austenitic shape as well as of a martensitic shape. Unfortunately, the TWSM process will lead to a degradation of mechanical performance. If compared with the OWSM, the recoverable deformation is only 2% (versus 5%–8%). In addition, its maximum stress is significantly reduced, SME disappears if the temperature of the material exceeds 250°C, and it is time-unstable; consequently the TWSM mechanical performance cannot be guaranteed after a high number of load cycles. It should be noted that not only Ni–Ti exhibit SMA; some Cu alloys can be used, such as CuZnAl or CuAlNi and others with Mn, as well as some Fe alloys (FePt, FePd, and FeNiCoTi). However, the description of their characteristics is beyond the scope of this chapter.

5.4.2   Thermoelectromechanical Models of SMA Fibers

The characteristics of SMA components can be estimated using different methods. Some of them include: differential scanning calorimetry (DSC), liquid bath, resistivity measure, cycle with constant applied load, and traction test. The DSC technique is able to measure absorbed and released heat in a SMA specimen during crystallographic transformations. Liquid bath is a liquid with a controllable temperature in which a SMA specimen is immersed and the imposed temperature is related to the macroscopic shape conversion. Resistivity measure is based on the change of resistivity during crystallographic transformation. An austenite–martensite–austenite cycle can be imposed with a constant mechanical load, allowing the measure of maximum and minimum deformations, which can be associated with critical points of the transformations. A traction test can be executed at a constant temperature to measure mechanical properties of the material and to relate them to the fixed temperature. The numerical characterization of the alloy can be used to properly set a thermoelectromechanical dynamic model, which allows the open-loop control of the SMA used as an actuator. This chapter will consider only dynamic-kinematic models of fibers.

Designing dynamic models of SMA fibers is an active field and a classification of the different approaches could be helpful in choosing the correct modelizing method. One classification can be based

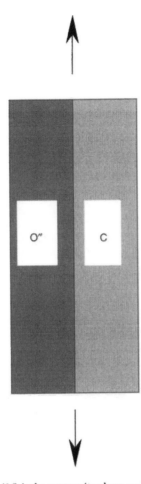

FIGURE 5.33 SMA Voigt model where (*M*) is the martensite phase, and (*A*) is the austenite phase.

on deriving principles such as empirical models, micromechanical models, and thermodynamical models. Some models are directly structured to describe the behavior of SMA fibers (a one-dimensional problem), while other models are developed for more complex problems (two- or three-dimensional problems) and then are reduced to describe only the characteristics of a one-dimensional problem. Dynamic SMA models are usually composed by a stress-strain relationship and a kinematic law. We can distinguish models where these two laws are inseparable, separable but coupled, or separable and uncoupled. The majority of the models are numerical because of the high non-linearity of the shape memory effect; however simplified analytic models do exist. Despite their imprecision, analytic models are very useful for a first dimensioning of a microactuator and in closed-loop applications. Many other distinctions can be considered such as the presence of martensite variants in 3-D models or the set of independent variables of the model. However, the proposed taxonomy can be a good starting point to deepen the world of shape memory models. This chapter will describe a famous numeric model that was started with a work of Tanaka (1986), first modified by Liang and Rogers (1990), and later by Brinson (1993), while Brailoyski et al. (1996) and Potapov and da Silva (2000) added some interesting simplified analytic models.

A simple micromechanical derivation of the first model was proposed by Brinson and Huang (1996). A parallel Voigt model of austenite and martensite phases is considered in a one-dimensional specimen, as in Figure 5.33. The specimen is subject to an external stress, so an elastic strain appears and, due to the considered Voigt model, the relations between austenite and martensite stress and strain are:

$$\sigma = (1 - \xi)\sigma_a + \xi\sigma_m \quad \varepsilon_m = \varepsilon_a$$
$$\sigma_a = E_a\varepsilon_a \quad \sigma_m = E_m\varepsilon_m \tag{5.50}$$

where σ is the stress of the specimen, ξ is the martensite fraction, σ_a and σ_m are, respectively, austenite and martensite stress, ε_a and ε_m are, respectively, austenite and martensite strain, while E_a and E_m are, respectively, austenite and martensite Young modulus. Relations in Equation 5.50 can be combined to obtain:

$$\sigma = [\xi \cdot E_m + (1 - \xi) \cdot E_a] \cdot \varepsilon_a \tag{5.51}$$

Now consider a temperature increment able to generate a phase transformation: it results in an adjustment of the SMA Voigt model (Figure 5.34) and the consequential total strain is the sum of the elastic strain and of the transformation strain, thus the stress-strain relation can be modified as:

$$\sigma = [\xi \cdot E_m + (1 - \xi) \cdot E_a] \cdot (\varepsilon - \varepsilon_L \cdot \xi_S) \tag{5.52}$$

where ε_L is the maximum residual strain of the transformation and ξ_S is the detwinned martensite fraction.

The stress–strain relation (Equation 5.52) must be coupled with a kinetics equation:

$$\begin{cases} \xi = \dfrac{1 - \xi_0}{2} \cos\left\{ \dfrac{\pi}{\sigma_S^{cr} - \sigma_f^{cr}} \left[\sigma - \sigma_f^{cr} - C_M(T - M_S) \right] \right\} \\ \text{with } T > M_S, \quad \sigma_S^{cr} + C_M(T - M_S) < \sigma < \sigma_f^{cr} + C_M(T - M_S) \end{cases} \tag{5.53}$$

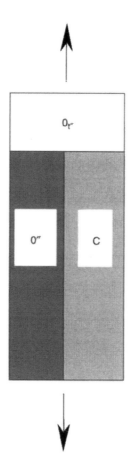

FIGURE 5.34 SMA Voigt model where (*M*) is the martensite phase, (*A*) is the austenite phase, and (*M_t*) is the transformed martensite.

where ξ_0 is the initial strain, M_s is the critical martensite start temperature, σ_s^{cr} and σ_f^{cr} are, respectively, the martensite start and finish critical stress, C_M is the slope of the border line of the detwinned martensite field in the stress-temperature diagram. A complete explanation of the model can be found in (Brinson, 1993).

5.4.3 Control Techniques

There are three main important control techniques based on three different physical phenomena: current control, resistance control, and Peltier effect control. There exist recent advances in the control of SMA actuators, especially of wire and rod actuators for linear movements, which will be discussed. Of particular emphasis is the use of some physical properties to achieve the desired behavior of the considered system. The scope is a brief mention of the control problems of SMA actuators, referring the reader to the specific bibliography for further detailed explanation. The first type of control approach for SMA actuators considered is current control. SMA fibers are electrical conductors having particular values of electrical resistance. Because of this resistance and because of the Joule effect, a current going through the wire is able to generate heat, and consequently the current can be used to control the temperature of the SMA wire.

Many authors reported studies of this type of SMA control: Bhattacharyya et al. (2000) considered a SMA wire subjected to an electric current, under no mechanical load, and under proper assumptions of the material's properties. Under their assumptions, the energetic balance can be formulated as in Equation 5.54 referring to Figure 5.35:

$$\frac{\partial}{\partial x}\left[K(\xi)\frac{\partial T}{\partial x}\right] + \rho_E(\xi)J(t)^2 - \frac{2h}{R}(T - T_{amb}) = C_V(\xi)\frac{\partial T}{\partial t} - H\frac{\partial \xi}{\partial t} \tag{5.54}$$

where ξ is martensite fraction (Bhattacharyya proposes a cooling and a heating law, to be associated to the thermodynamic energetic equation); K, ρ_E and C_V are, respectively, thermal conductivity, electrical resistivity and heat capacity; h is the convection coefficient (Bhattacharyya proposes a linear definition of these properties); H is the transformation latent heat; R is the cross-sectional radius of the wire; T and T_{amb} are, respectively, the temperature of the wire and of the environment; t is the time variable; and J is the current density, which is used to generate a forced heating of the wire. The boundary and the initial conditions of the Cauchy problem can be expressed by Equation 5.55, if there is a wire with a 2L length.

$$\begin{cases} T(x, t)|_{x=\pm L, \forall t \in \Re} = T_{amb} \\ T(x, t)|_{\forall x \in -L, L, t=0} = T_{amb}, \quad \xi(T(x, t))|_{\forall x \in -L, L, t=0} = \xi_0 \end{cases} \tag{5.55}$$

The second control scheme taken into consideration is based on the change of wire resistance during the phase transformation (Airoldi et al., 1995). When the wire is subjected to a generic thermal load and a constant mechanical load, a linear relationship is observed between the strain and the relative variation of the resistance of the wire (Wu et al., 2000). The slope of the relationship is dependent upon the specific

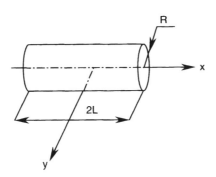

FIGURE 5.35 SMA wire.

transformation, and the relationship between the stress and the relative variation of resistance can be quite complex. This functional aspect can be used to control the movement of the microactuator with a simple control scheme. A quantitative work in this direction is reported in Sittner et al. (2000). The authors studied the behavior of a $Ni_{50}Ti_{45}Cu_5$ alloy and, utilizing the principles of Equation 5.56 during martensitic transformation and some other theoretical and experimental observations, they developed a new behavioral model of this alloy:

$$\frac{\Delta R}{R_0} = \frac{\Delta \rho}{\rho_0} + (1 + 2v)\varepsilon + \frac{C^R(T - T^R)}{\rho_0} \tag{5.56}$$

where R_0 and ρ_0 are, respectively, the resistance and resistivity in the austenitic phase at the reference temperature T^R; ΔR and $\Delta \rho$ are, respectively, the absolute change of resistance and resistivity during the transformation at the generic T temperature; ε is the resulting strain; and v and C^R are two material parameters. The authors, along with other researchers (De Araujo et al., 1999), observed a linear relation between relative resistivity and strain in martensitic and austenitic transformation:

$$\frac{\Delta \rho}{\rho_0} = K\varepsilon \tag{5.57}$$

where K represents the constant slopes associated with the shape memory transformations.

The third, and most promising, control scheme considered is based on the Peltier effect (Lagoudas and Kinra, 1993), which focuses on the ability to increment the working frequency of SMA microactuators. This special phenomenon is produced using a thermoelectric element, obtained by sandwiching a SMA layer between two semiconductor layers [a positively doped (P) and a negatively doped (N)]. Alternating the current direction, the SMA layer becomes the hot or the cold junction of a thermoelectric couple, because the Peltier effect generated by electric current causes a temperature differential at a junction of the dissimilar metals. The general thermal transfer model for this control scheme is proposed in (Bhattacharyya, Faulkner and Almaraj, 2000):

$$K_i\frac{\partial^2 T_i}{\partial x^2}(x, t) + \rho_i J^2(t) - H\frac{P_C}{A_C}\left(T_i(x, t) - T_0\right) = C_v^i\frac{\partial T_i}{\partial t}(x, t), \quad x \in I_i, t > 0, i \in \{P, S, N\}$$

$$I_P = \left]-L - \frac{d}{2}, -\frac{d}{2}\right[, I_S = \left]-\frac{d}{2}, \frac{d}{2}\right[, I_N = \left]\frac{d}{2}, L + \frac{d}{2}\right[\tag{5.58}$$

where P, S, and N are, respectively, a positively doped semiconductor, a SMA layer, and a negatively doped semiconductor (Figure 5.36); π is the electrical resistivity; J is the current density; C_v^i is the heat capacity per unit volume; T is the temperature with t as its time variable; P_C and A_C are the perimeter and the area of the cross-section; and H is the heat convection coefficient.

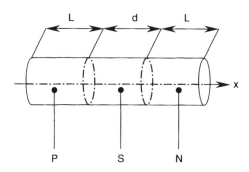

FIGURE 5.36 SMA Peltier actuator.

The interface conditions consist in the temperature continuity and the equal exchange of net heat flux across the interfaces:

$$T_S\left(-\frac{d}{2},t\right) = T_P\left(-\frac{d}{2},t\right), \; T_S\left(\frac{d}{2},t\right) = T_N\left(\frac{d}{2},t\right)$$

$$-K_S\frac{\partial T_S}{\partial x}\left(-\frac{d}{2},t\right) = -K_P\frac{\partial T_P}{\partial x}\left(-\frac{d}{2},t\right) + \alpha_P T_P\left(\frac{d}{2},t\right)J(t) \tag{5.59}$$

$$-K_S\frac{\partial T_S}{\partial x}\left(\frac{d}{2},t\right) = -K_N\frac{\partial T_N}{\partial x}\left(\frac{d}{2},t\right) + \alpha_N T_N\left(\frac{d}{2},t\right)J(t)$$

where α_N and α_P are the Seeback coefficients. The end boundary conditions are defined in Equation 5.60 and the initial conditions consist in the thermal equilibrium of the system at the T_0 environmental temperature.

$$T_P\left(-L-\frac{d}{2},t\right) = T_0, \quad T_N\left(L+\frac{d}{2},t\right) = T_0 \tag{5.60}$$

The solution of the differential problem provides a model to understand the behavior of the SMA system under the presence of the current density J.

Acknowledgments

Part of this research is financed by the Italian Ministry of Research, University and Instruction (MIUR), under the fund COFIN 2003 Project MiniPar, and part is financed by a grant of the University of Brescia on microactuators. The author would like to thank Riccardo Adamini and Enzo Locatelli for their fruitful collaboration; Mohamed Gad-el-Hak, Andrew Oliver and Mauro Ferrari for the discussion on the chapter scheme; Rodolfo Faglia for the discussion on the definition of microactuator; Vittorio Ferrari for the discussion on piezoelectric actuators; Giovanni Legnani for the discussion on electromagnetic actuators; and Roberto Roberti for the discussion on shape memory alloys.

References

Airoldi, G., Besseghini, S., and Riva, G. (1995) "Smart Behaviour in a CuZnAl Single Crystal Alloy," *Transactions of the International Conference on Martensitic Transformations, J. Phys. IV*, 5, Part 2, Session 4, Paper 12.

Ampere, A.M. (1820) "Conclusions D'un Mémoire sur l' Action Mutuelle de deux courans électriques, sur celle qui existe entre un courant électrique et un aimant, et celle de deux aimans l'un sur l'autre," *Journal de physique, de chimie, d'histoire naturelle et des arts* 91, Paris, pp. 76–78.

De Araujo, C.J., Morin, M., and Guenin, G. (1999) "Electro-Thermomechanical Behaviour of a Ti-45.0Ni-5.0Cu (at.%) Alloy During Shape Memory Cycling," *Mater. Sci. Eng. A*, 273–275, pp. 305–309.

Bhattacharyya, A., Faulkner, M.G., and Amalraj, J.J. (2000) "Finite Element Modeling of Cyclic Thermal Response of Shape Memory Alloy Wires with Variable Material Properties," *Comput. Mater. Sci.*, 17, pp. 93–104.

Biot, J.B., and Savart, F. (1820) "Expériences électromagnétiques sur la mesure de l'action exercée à distance sur une particule de magnétisme, par un fil conjonctif," *Journal de physique, de chimie, d'histoire naturelle et des arts*, Paris, pp. 151.

Bo, Z., and Lagoudas, D.C. (1999a) "Thermomechanical Modeling of Polycrystalline SMAs Under Cyclic Loading, Part I: Theoretical Derivations," *Int. J. Eng. Sci.*, 37, pp. 1089–1140.

Bo, Z., and Lagoudas, D.C. (1999b) "Thermomechanical Modeling of Polycrystalline SMAs Under Cyclic Loading, Part II: Material Characterization and Experimental Results for a Stable Transformation Cycle," *Int. J. Eng. Sci.*, 37, pp. 1141–1173.

Bo, Z., and Lagoudas, D.C. (1999c) "Thermomechanical Modeling of Polycrystalline SMAs Under Cyclic Loading, Part III: Evolution of a Plastic Strains and Two-way Shape Memory Effect," *Int. J. Eng. Sci.*, 37, pp. 1175–1203.

Bo, Z., and Lagoudas, D.C. (1999d) "Thermomechanical Modeling of Polycrystalline SMAs Under Cyclic Loading, Part IV: Modeling of Minor Hysteresis Loops," *Int. J. Eng. Sci.*, 37, pp. 1175–1203.

Brailovski, V., Trochu, F., and Daigneault, G. (1996) "Temporal Characteristics of Shape Memory Linear Actuators and Application to Circuit Breakers," *Mater. Design*, 17, pp. 151–158.

Brinson, L.C. (1993) "One Dimensional Constitutive Behavior of Shape Memory Alloys: Thermomechanical Derivation with Non-Constant Material Functions," *J. Intell. Mater. Syst. Struct.*, 4(2), pp. 229–242.

Brinson, L.C., and Huang, M.S. (1996) "Simplifications and Comparisons of Shape Memory Alloy Constitutive Models," *J. Intell. Mater. Syst. Struct.* 16, pp. 108–114.

Buehler, W.J., and Wiley, R.C. (1965) "Nickel-based Alloys. Technical Report," US Patent 3,174,851.

Chang, L.C., and Read, T.A. (1951) "Plastic Deformation and Diffusionless Phase Changes in Metals. The Gold-Cadmium Beta Phase" *Trans. AIME* 189, pp. 47–52.

Coulomb, C.A. (1785–1789) "Memoires sur l'electricite et le magnetisme," Memoires de l'Academie Royale des Science de Paris.

Einstein, A. (1905) "Zur Elektrodynamik bewegter Korper," Annalen der Physik 17, pp. 891–921.

Faraday, M. (1821) "On some new Electro-Magnetical Motions, and the Theory of Magnetism," *Quarterly Journal of Science* 12, pp. 75–96.

Faraday, M. (1832) "Experimental Researches in Electricity," Philosophical transactions of the Royal Society of London 122, pp. 125–194.

Ferraris, G. (1888) "Rotazioni elettrodinamiche prodotte per mezzo di Correnti alternate," Torino.

Gilbert, W. (1600) "Guilelmi Gilberti Colcestrensis De magnete, magneticisque corporibus, et de magnete tellure: physiologia nova, plurimus argumentis, experimentis demonstrata," Londini, excudebat P. Short.

Greninger, A.B., and Mooradian, V.G. (1938) "Strain Transformation in Metastable Copper-Zinc and Beta Copper-Tin Alloys," *Trans. AIME* 128, pp. 337–368.

Groβ, H., Hamann, J., and Wiegärtner, G. (2000) "Elektrische Vorschubantriebe in der Automatisierungstechnik: Grundlagen, Berechnung, Bemessung," Publicis MCD Corporate Publishing.

Heartling, G. (1994) "Rainbow Ceramics — A New Type of Ultra-High-Displacement Actuator," *Am. Ceram. Soc. Bull.*, 73, pp. 93–96.

Henry, J. (1831) "On Account of a Large Electro-Magnet, Made for the Laboratory of Yale College," *American Journal of Science and Arts* 20, pp. 201–203.

Hertz, H.R. (1887) "Uber sehr schnelle elektrische Schwingungen." Annalen der Physik und Chemie 31, pp. 421–448.

Histand, M.B., and Alciatore, D.G. (1999) *Introduction to Mechatronics and Measurement Systems*, McGraw-Hill, New York, pp. 333–337.

Jacobi, M.H. (1835) "Memoire sur l'application de l'Electromagnetisme au Mouvement des machines," Memoires de l'Academie de Saint Petersbourg.

Janocha, H. (1999) *Adaptronics and Smart Structures*, Springer-Verlag, Berlin.

Lagoudas, D.C., and Kinra, V.K. (1993) "Design of High Frequency SMA Actuators," *Disclosure of Invention Tamus 803*, TAMU, College Station, TX.

Legnani, G., Tiboni, M., and Adamini, R. (2002) "Meccanica Degli Azionamenti," Progetto Leonardo.

Liang, C., and Rogers, C.A. (1990) "One-Dimensional Thermomechanical Constitutive Relations for Shape Memory Materials," *J. Intell. Mater. Syst. Struct.*, 1(2), pp. 207–234.

Mavroidis, C. (2002) "Development of Advanced Actuators Using Shape Memory Alloys and Electrorheological Fluids," *Res. Nondestr. Eval.*, 14, pp. 1–32.

Maxwell, J.C. (1865) "A Dynamical Theory of the Electromagnetic Field," Philosophical Transactions of the Royal Society of London 155, pp. 459–512.

Michell, J. (1751) "A treatise of artificial magnets," Cambridge, printed by J. Bentham.

Niezrecki, C., Brei, B., Balakrishnan, S., and Moskalik, A. (2001) "Piezoelectric Actuation: State of the Art," *The Shock Vib. Dig.*, 33, pp. 269–280.

Oersted, H.C. (1820) "Experimenta circa effectum conflictus electrici in acum magneticam," Annales de chimie et de physique Paris, Tome IV, pp. 417–425.

Olander, A. (1932) "An Electrochemical Investigation of Solid Cadmium-Gold Alloys," *J. Am. Chem. Soc.* 56, pp. 3819–3833.

Pacinotti, A. (1865) "Descrizione di una macchinetta elettromagnetica," Nuovo Cimento XIX, p. 378.

Potapov, P.L., and da Silva, E.P. (2000) "Time Response of Shape Memory Alloy Actuators," *J. Intell. Mater. Syst. Struct.*, 11, pp. 125–134.

Proceedings of the IEEE Workshops on Micro Electro Mechanical Systems, (1987–1996).

Proceedings of the International Symposia on Micro Machine and Human Science, (1990–1995).

Rowland, H.A. (1876) "On a Magnetic Effect of Electric Connect," Physical papers of Henry Augustus Rowland, John Hopkins University.

Sittner, P., Vokoun, D., Dayananda, G.N., and Stalmans, R. (2000) "Recovery Stress Generation in Shape Memory $Ti_{50}Ni_{45}Cu_5$ Thin Wires," *Mater. Sci. Eng. A*, 286, pp. 298–311.

Special Issue on Micro-Machine System, (1991) *J. Robotics Mechatronics*.

Tanaka, (1986) "Thermomechanics of Transformation Pseudoelasticity and Shape Memory Effect in Alloys," *Int. J. Plast.*, 1, pp. 59–72.

Tesla, N. (1888) US Patents 381, 968-381, 969-382, 279-390, 415-390, 820.

Wise, S.A. (1998) "Displacement Properties of RAINBOW and THUNDER Piezoelectric Actuators," *Sensors Actuators A*, 69, pp. 33–38.

Wu, X.D., Fan, Y.Z., and Wu, J.S. (2000) "A Study on the Variations of the Electrical Resistance for NiTi Shape Memory Alloy Wires During the Thermo-Mechanical Loading," *Mater. Design*, 21, pp. 511–515.

6

Sensors and Actuators
for Turbulent Flows

Lennart Löfdahl
Chalmers University of Technology

Mohamed Gad-el-Hak
*Virginia Commonwealth
University*

6.1 Introduction

During the last decade MEMS devices have had major impacts on both industrial and medical applications. Examples of the former group are accelerometers for automobile airbags, scanning electron microscope tips to image single atoms, microheatexchangers for electronic cooling, and micromirrors used for light-beam steering. Micropumping is useful for ink-jet printing and cooling of electronic equipment. Existing and prospective medical applications include reactors for separating biological cells, controlled delivery and measurement of minute amount of medication, pressure sensors for catheter tips, and development of an artificial pancreas. The concept of MEMS includes a variety of devices, structures, and systems: this chapter will focus only on MEMS devices that generally can be categorized as microsensors and microactuators. Particularly, attention is directed toward microdevices used for turbulent-flow diagnosis and control.

MEMS are created using specialized techniques derived and developed from IC technology in a process often called micromachining. There exists today a vast variety of sensors and actuators and several associated technologies for their fabrication. However, three main technologies are usually distinguished when discussing micromachining: bulk micromachining, surface micromachining, and micromolding.

Bulk micromachining involves different techniques that use a simple, single-crystal, silicon wafer as structural material. Using anisotropic silicon etching and wafer bonding, three-dimensional structures such as pressure sensors, accelerometers, flow sensors, micropumps, and different resonators have been fabricated. This branch has been under development for more than twenty years and may now be considered as a well-established technology.

In the second group, surface micromachining, the silicon substrate is used as support material, and different thin films such as polysilicon, silicon dioxide, and silicon nitride provide sensing elements and electrical interconnections as well as structural, mask, and sacrificial layers. The basis of surface micromachining is sacrificial etching where free-standing, thin-film structures (polysilicon) are free-etched by lateral underlying sacrificial layer (silicon dioxide). Surface micromachining is very simple but powerful, and despite the two-dimensional nature of this technique, different complex structures such as pressure sensors, micromotors, and actuators have been fabricated.

The third group, micromolding (the LIGA technique), is more similar to conventional machining in concept. A metal mold is formed using lithographic techniques, which allow fine feature resolution. Typically, tall structures with submicrometer resolution are formed. Products created by micromolding techniques include thermally actuated microrelays, micromotors, and magnetic actuators. In a rapidly growing field like MEMS, numerous surveys on fabrication have been published a few of which include: Petersen (1982), Linder et al. (1992), Brysek et al. (1994), Diem et al. (1995), and Tien (1997). The books by Madou (2002) and Kovacs (1998) are valuable references, and the entire second part of this handbook focuses on MEMS design and fabrication. The emphasis in this chapter is on MEMS applications for turbulence measurements and flow control.

For more than 100 years, turbulence has been a challenge for scientists and engineers. Unfortunately, no simple solution to the "closure problem" of turbulence exists, so for the foreseeable future turbulence models will continue to play a crucial role in all engineering calculations. The modern development of turbulence models is basically directed towards applications to high-Reynolds-number flows ($Re > 10^6$). This development will be a joint effort between direct numerical simulations of the governing equations and advanced experiments. However, an "implicit closure problem" is inherent in the experiments, since an increase in Reynolds number will automatically generate smaller length-scales and shorter time-scales, which both in turn require small and fast sensors for a correct resolution of the flow field. MEMS offer a solution to this problem because sensors with length- and time-scales of the order of the relevant Kolmogorov microscales can now be fabricated. Additionally, these sensors are produced with high accuracy at relatively low cost per unit. For instance, MEMS pressure sensors can be used to determine fluctuating pressures beneath a turbulent boundary layer with a spatial resolution that is about one order of magnitude finer than what can be achieved with conventional transducers.

MEMS sensors can be closely spaced together on one chip, and such multi-sensor arrays are of significant interest when measuring correlations of fluctuating pressure and velocity, and, in particular, for their applications in aeroacoustics. Moreover, the low cost and energy consumption per unit device will play a key role when attempting to cover a large macroscopic surface with sensors to study coherent structures. More elaborate discussion on turbulence and the closure problem can be found in textbooks in the field [e.g., Tennekes and Lumley, 1972; Hinze 1975], or in surveys on turbulence modeling [e.g., Robinson, 1991; Speziale, 1991; Speziale et al., 1991; Hallbäck, 1993]. The role and importance of fluctuating pressures and velocities in aeroacoustics is well covered in the books by Goldstein (1976), Dowling and Ffowcs-Williams (1983), and Blake (1986).

Reactive flow control is another application where microdevices may play a crucial future role. MEMS sensors and actuators provide opportunities for targeting the small-scale coherent structures in macroscopic turbulent shear flows. Detecting and modulating these structures may be essential for a successful control of turbulent flows to achieve drag reduction, separation delay, and lift enhancement. To cover areas of significant spatial extension, many devices are needed requiring small-scale, low-cost, and low-energy-use components. In this context, the miniaturization, low-cost fabrication, and low-energy consumption of microsensors and microactuators are of utmost interest and promise a quantum leap in control-system performance. Combined with modern computer technologies, MEMS yields the essential matching of the length- and time-scales of the phenomena to be controlled. These issues and other aspects of flow control are well summarized in a number of reviews on reactive flow control [e.g. Wilkinson, 1990; Gad-el-Hak, 1994; Moin and Bewley, 1994], and in the books by Gad-el-Hak et al. (1998), Gad-el-Hak (2000) and Gad-el-Hak and Tsai (2005). The topic is also detailed in Chapters 13, 14 and 15 of the present handbook.

In this chapter, we focus on specific applications of MEMS in fluid dynamics, namely to measure turbulence and to reactively control fluid flows in general and turbulent flows in particular. To place these applications in perspective, we start by giving a brief description of MEMS fabrication; the next section is devoted to a brief general introduction to turbulence, discussion on tools necessary for the analysis of turbulent flows, and some fundamental findings made in turbulence. Specific attention is paid to small scales, which are of significant interest both in turbulence measurements and in reactive flow control, and we discuss the spatial- and temporal-resolution requirements. MEMS sensors for velocities, wall-shear stress, and pressure measurements are then discussed.

As compared to conventional technologies, an extremely small measuring volume can be achieved using MEMS-based velocity sensors. Most commonly, the velocity sensors are designed as hot-wires with the sensitive part made of polysilicon, but other principles are also available which will be discussed.

A significant parameter for control purposes is the fluctuating wall-shear stress, since it determines the individual processes transferring momentum to the wall. MEMS offers a unique possibility for direct as well as indirect measurements of this local flow quantity. Different design principles of conventional and MEMS-based wall-shear-stress sensors are discussed together with methods for calibrating those sensors.

The discussion of pressure sensors is focused on measurements of the fluctuating pressure field beneath turbulent boundary layers. Some basic design criteria are given for MEMS pressure sensors and advantages and drawbacks are elucidated. Significant quantities like rms-values, correlations, and advection velocities of pressure events obtained with MEMS sensors, yielding spatial resolution of 5–10 viscous units, are compared to conventional measurements.

In the last section, we address a real challenge and a necessity for reactive flow control, MEMS-based flow actuators. Our focus is on three-dimensional structures, and we discuss actuators working with the bi-layer effect as a principle. Electrostatic and magnetic actuators operating through external forces are also discussed together with actuators operating with mechanical folding. We also summarize the one-layer structure technology and discuss the out-of-plane rotation technology that has been made possible with this method. In connection with the actuator section, we discuss MEMS-fabricated devices such as pumps and turbines. Finally, the chapter is ended with reflections on the future possibilities that can be achieved in turbulence measurements and flow control by using MEMS technology.

6.2   MEMS Fabrication

6.2.1   Background

MEMS can be considered as a logical step in the silicon revolution, which took off when silicon microelectronics revolutionized the semiconductor and computer industries with the manufacturing of integrated circuits. An additional dimension is now being added by micromachines, because they allow the integrated circuit to break the confines of the electronic world and interact with the environment through sensors and actuators. It can be said that microelectromechanical systems will have in the near future the same impact on society and the economy as the IC has had since the early 1960s. The key element for the success of MEMS will be, as pointed out by Tien (1997), "the integration of electronics with mechanical components to create high-functionality, high-performance, low-cost, integrated microsystems." In other words, the material silicon and the MEMS fabrication processes are crucial to usher in a new era of micromachines.

Silicon is a well-characterized material. It is strong, being essentially similar to steel in modulus of elasticity, stronger than stainless steel in yield-strength, and exceeds aluminum in strength-to-weight ratio. Silicon has high thermal conductivity; low bulk expansion coefficient; and its electronic properties are well-defined and sensitive to stress, strain, temperature, and other environmental factors. In addition, the lack of hysteresis and the property of being communicative with electronic circuitry make silicon an almost perfect material for fabricating microsensors and microactuators for a broad variety of applications.

In MEMS fabrication, silicon can be chemically etched into various shapes, and associated thin-film materials such as polysilicon, silicon nitride, and aluminum can be micromachined in batches into a vast variety of mechanical shapes and configurations. Several technologies are available for MEMS fabrication, but three main technologies are usually distinguished: bulk micromachining, surface micromachining, and micromolding. An important characteristic of all micromachining techniques is that they can be complemented by standard IC batch-processing techniques such as ion implantation, photolithography, diffusion epitaxy, and thin-film deposition. This section will provide a background of the three main technologies from a user viewpoint. Readers who are interested in more comprehensive information on fabrication are referred to more elaborate work in the field [e.g. Petersen, 1982; O'Connor, 1992; Bryzek et al., 1994; Tien, 1997; Kovacs, 1998; Madou, 2002]. Part II of the present handbook focuses on MEMS design and fabrication.

6.2.2   Microfabrication

6.2.2.1   Bulk Micromachining

Bulk micromachining is the oldest technology for making MEMS. The technique has been used to fabricate sensors for about 20 years. The mechanical structures are created within the confines of a silicon wafer by selectively removing parts of the wafer material by using orientation-dependent etching of single-crystal silicon substrate. Etch-stopping techniques and masking films are crucial in the bulk micromachining process. The etching can be either isotropic, anisotropic, or a combination of both. In isotropic etching, the etch rate is identical in all directions, while in anisotropic etching the etch rate depends on the crystallographic orientation of the wafer. Two commonly used etchants are ethylene diamine and pyrocatechol (EDP), and an aqueous solution of potassium hydroxide (KOH). Because it is important to be able to stop the etching process at a precise location, etch-stopping techniques have been developed. One such method is based on the fact that heavily doped regions etch more slowly than un-doped regions; hence, by doping a portion of the material, the etch process can be made selective. Another technique for etch-stopping is electrochemical in nature and is based on the fact that etching stops upon encountering a region of different polarity in a biased *pn*–junction.

The following is a good illustration of the different steps in the bulk microfabrication process. Tien (1997) has summarized the processing steps necessary for micromachining a hole and a diaphragm in a wafer (Figure 6.1). Silicon nitride is used as an etch mask since it is not etched by either EDP or KOH. To stop the etch process at a specific location, and thereby form the diaphragm, a region heavily doped with boron is used. Holes and diaphragms, as shown in Figure 6.1, constitute the basis for many mechanical devices as for example pressure transducers which today are commercially available for measurements in the range of 60 Pa–68 MPa.

The fabrication of a pressure transducer is straightforward as has been summarized by Bryzek et al. (1994). As illustrated in Figure 6.2, the process starts with a silicon substrate that is polished on both sides. Boron-doped piezoresistors and both p^+ and n^+ enhancement regions are introduced by means of diffusion and ion implantation. Piezoresistors are the sensitive elements in pressure and acceleration sensors because their resistance varies with stress and temperature, the latter being the unwanted part of the signal if the objective is to measure force. A thin layer of deposited aluminum or other metal creates the ohmic contacts and connects the piezoresistors into a Wheatstone bridge. Finally, the device side of the wafer is protected and the back is patterned to allow formation of an anisotropically etched diaphragm. After stripping and cleaning, the wafer is anodically bonded to Pyrex® and finally diced.

Bulk micromachining is the most mature of the micromachining technologies and constitutes the base for many microdevices like silicon pressure sensors and silicon accelerometers. The fabrication process is straightforward and does not need much elaborate equipment, but the technique is afflicted with some severe limiting drawbacks. Since the geometries of the structures are restricted by the aspect ratio inherent in the fabrication method, the devices tend to be larger than those made with other micromachining technologies. As a consequence of this, expensive silicon "real state" is wasted. Another drawback is the

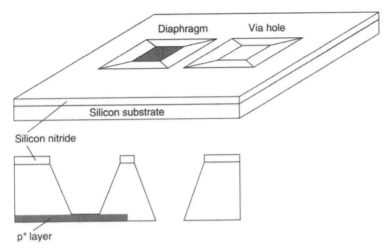

FIGURE 6.1 Bulk micromachined structures, diaphragm and via hole, in a silicon substrate. Depositioned silicon nitride is the mask for the wet etch and the doped silicon layer serves as an etch stop for the diaphragm formation. (Reprinted with permission from Tien, N.C. [1997] "Silicon Micromachined Thermal Sensors and Actuators," *Microscale Thermophysical Eng.* 1, pp. 275–292.)

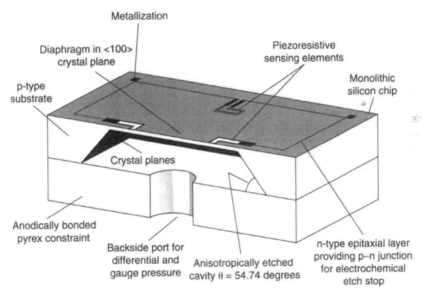

FIGURE 6.2 A bulk micromachined pressure sensor shown in cross-section. The sensor contains a thin silicon diaphragm formed by etching the silicon wafer with alkaline-hydroxide. The diaphragm deflection depends on pressure and is sensed by boron-doped piezoresistors. (Reprinted with permission from Bryzek, J., Peterson, K., and McCulley, W. [1994] "Micromachines on the March," *IEEE Spectrum* 31, May, pp. 20–31.)

use of alkaline etchants which unfortunately are not compatible with IC manufacturing. However, strategies to circumvent these drawbacks are available and details on such methods can be found in Bryzek et al. (1994) and Tien (1997).

6.2.2.2 Surface Micromachining

In contrast to bulk micromachining techniques, surface micromachining does not penetrate the wafer. Instead, thin-film materials are selectively added to and removed from the substrate during the processing.

Polysilicon is used as the mechanical material with sacrificial material like silicon dioxide sandwiched between layers of polysilicon. Both materials are commonly deposited using low-pressure chemical vapor deposition. Both wet and dry etching are essential and the sacrificial layers constitute the basis of surface micromachining.

To illustrate the processes needed in surface micromachining, a simplified fabrication of a polysilicon slider with a central rail has been summarized by Tien (1997), and the basic steps are illustrated in Figure 6.3. Two layers of structural polysilicon and sacrificial oxide are needed for this design, and Figure 6.3a illustrates the first sacrificial oxide layer and how the deposition and patterning of the first polysilicon layer have been completed. Figures 6.3b and 6.3c show the deposition of the second sacrificial oxide layer together with the free etching of the anchor openings through the oxide. The next step is the deposition and patterning of the second polysilicon layer, which is followed by the removal of the sacrificial oxide used to release the structure. More details including used etchants, sacrificial layers, and other "tricks" made in the fabrication process can be found in Tien (1997).

An essential advantage of surface micromachining is that there is no constraint on the miniaturization of the devices fabricated other than those raised by limitations in the lithography technology. Another important benefit is that structurally complex mechanical systems, including free-standing or moveable parts, can be created by stacking multiple layers of material. In addition, surface micromachining offers a high degree of compatibility with IC processing, an important trait assuming the future success of MEMS will be linked to the integration of electronics with mechanical systems. The main drawback of surface micromachining is that it is a thin-film technology that creates essentially two-dimensional structures. However, this has been circumvented by creative designs [see e.g. Pister et al., 1992; Tien et al., 1996a; 1996b].

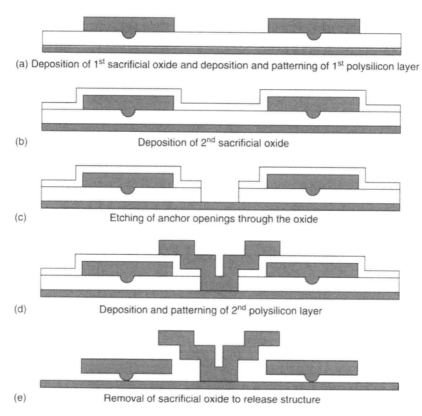

(a) Deposition of 1ˢᵗ sacrificial oxide and deposition and patterning of 1ˢᵗ polysilicon layer

(b)   Deposition of 2ⁿᵈ sacrificial oxide

(c)   Etching of anchor openings through the oxide

(d)   Deposition and patterning of 2ⁿᵈ polysilicon layer

(e)   Removal of sacrificial oxide to release structure

FIGURE 6.3 Polysilicon surface micromachining process for the fabrication of a slider with a central rail. (Reprinted with permission from Tien, N.C. [1997] "Silicon Micromachined Thermal Sensors and Actuators," *Microscale Thermophysical Engineering* 1, pp. 275–292.)

6.2.2.3 Micromolding

Although the micromolding technique is more similar to conventional machining in concept, it should be discussed in connection with micromachining because it is capable of producing minute devices using advanced IC lithography. In this group, the LIGA process method — introduced in the late 1980s by Ehrfeld et al. (1988) — is the most common. LIGA is a German acronym for "LIthographie Galvanoformoung Abformung." In English, LIGA is lithography, electroforming, and molding. The method basically relies on forming a metal mold using lithographic techniques. To form the mold, a thick layer of photoresist placed on top of a conductive substrate is exposed and developed using X-ray lithography. As illustrated in Figure 6.4, the metal is then electroplated from the substrate through the openings in the photoresist. After removing the photoresist, the metal mold can be used for pouring low-viscosity polymers such as polyimide, polyimethyl metacrylathe, and other plastic resins. After curing, the mold is removed leaving behind microreplicas of the original pattern. Products created by LIGA are three-dimensional and include for example, thermally actuated microrelays, micromotors, magnetic actuators, micro-optics, and micro-connectors, as well as a host of micromechanical components like joints, springs, bearings, and gears.

An extension of the LIGA process, the SLIGA technique, gains another degree of design freedom by combining LIGA with the use of sacrificial layers. Keller and Howe (1995) have presented the HEXSIL technique, which also includes a sacrificial layer and creates polysilicon components. A drawback of the micro-molding method is that the assembly of small parts and the integration of electronics with mechanical

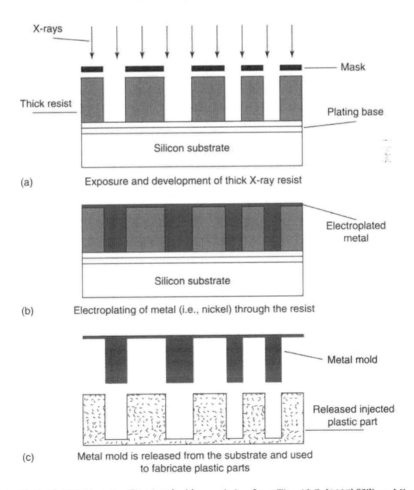

FIGURE 6.4 The basic LIGA process. (Reprinted with permission from Tien, N.C. [1997] "Silicon Micromachined Thermal Sensors and Actuators," *Microscale Thermophysical Eng.* 1, pp. 275–292.)

devices can be a real challenge. Additionally, the X-ray equipment needed for the fabrication is quite expensive.

To conclude this section, it is worth mentioning that much of what is known about the design of mechanical structures scales down to the microstructure level very nicely. However, the same cannot be said for the properties of materials moving from the bulk to the thin-film regimes. For instance, residual stresses within thin films can produce unwanted tension or compression within the microstructure. Microdefects can be ignored for thickness greater than 10 μm, but become important in the 1 μm range, which is typical for surface micromachining. Finally, microfriction, surface tension, and van der Waals forces can create undesired stiction or adhesion [see Israelachvili, 1991].

6.3   Turbulent Flows

For more than 100 years, turbulence has been a fascinating challenge to scientists in fluid mechanics. It is very easy to observe turbulent flows and to form a picture of its nature by looking at the plume of a smoke stack for instance. Such visualization shows clearly that the turbulent flow field contains numerous eddies of different size, orientation, and intensity. The largest eddies have a spatial extension of approximately the same size as the width of the flow field, while the smallest eddies are of the size where viscous effects become dominant and energy is transferred from kinetic to internal. To qualitatively analyze turbulent flow fields, the eddies are conveniently described by length, time, and velocity scales.

This section provides a general discussion on the classification of small and large length-scales and their importance in analyzing and modeling turbulent flows. We find that the width of the wavenumber spectrum is proportional to the Reynolds number in such a way that high *Re* generates smaller scales. Since turbulent flows are high-Reynolds-number flows, it is clear that a knowledge of scales and in particular, the small scales, is essential for the analysis and the modeling of turbulence. MEMS offers through miniaturization of sensors and actuators unique opportunities to resolve, as well as to target for control, the smallest scales even at high *Re*. The scale discussion here constitutes a cornerstone for the following sections, which will consider the use of MEMS sensors and actuators for measuring and controlling turbulence. For those readers new to turbulence, the section begins with a brief introduction to the subject, leading to simple ways for estimating typical scales, and sensor requirements for a particular flow field.

6.3.1   Definition of Turbulence

During the century in which turbulence has been formally studied, many different definitions have been contemplated. The first attempt to define turbulence was made in the late nineteenth century by Osborne Reynolds who simply stated that turbulence was a "sinuous motion." Later, more comprehensive and detailed definitions have been given, and each definition commonly has been associated with the current fashion of approaching the closure problem of turbulence. Hence, the definition by G. I. Taylor in the thirties had clear links to the statistical treatments of turbulence [Taylor, 1935], by Peter Bradshaw in the sixties to hot-wire measurements [Bradshaw, 1971], and by Marcel Lesieur in the late eighties to large-eddy and direct numerical simulations [Lesieur, 1991].

The most pragmatic definition is probably the one given by Tennekes and Lumley (1972), who provide not quite a definition, but instead seven characteristics of turbulence. It is stated that turbulence is irregular, or random, and this makes a deterministic approach impossible, so in the analysis one must rely on statistical methods. Diffusivity is another crucial feature of turbulence, which is important since it causes rapid mixing and increased rates of momentum, heat, and mass transfer. Turbulent flows occur always at high Reynolds numbers, which implies that they are always associated with small scales and complex interaction between the viscous and the nonlinear inertia terms in the equations of motion. All turbulent flows are three-dimensional and are characterized by high levels of vorticity fluctuations. The viscous shear stresses perform deformation work, which increases the internal energy at the expense of the kinetic energy, meaning that all turbulent flows are strongly dissipative. If no energy is supplied, turbulent flows eventually decay. Under ordinary circumstances turbulence is a continuum phenomenon, so turbulent flows obey the continuum hypothesis and the governing equations of fluid mechanics are applicable instantaneously. Even the smallest scales of a turbulent field are under normal conditions much larger than any molecular length-scale. Finally, turbulent flows are flows, which means that all turbulent flows are unique and no general solution to problems associated with turbulence is in sight. In spite of the latter statement, turbulent flows have many characteristics in common and this fact is exploited in the following subsection dealing with methods of analysis.

6.3.2 Methods for Analyzing Turbulence

Turbulence is one of the unsolved problems in classical physics, and it is still almost impossible to make accurate quantitative predictions for turbulent flows without relying heavily on empirical data. This is basically due to the fact that no methodology exists for obtaining stochastic solutions to the nonlinear partial differential equations describing the instantaneous three-dimensional flow. Moreover, statistical studies of the equations of motion always lead to a situation where there are more unknowns than equations, the closure problem of turbulence. This can easily be derived and is shown in most textbooks in the field [e.g. Tennekes and Lumley, 1972; Hinze, 1975; Pope, 2000]. Excluding direct numerical simulations of the governing equations which thus far have been used only for simple geometries and low Reynolds numbers, it can be stated that all computations, both scientific and engineering, of turbulent flows will even in the foreseeable future need experiment, modeling and analysis.

One powerful tool in the study of turbulent flows is dimensional analysis because it may be possible under certain conditions to argue that the structure of turbulence depends only on a few independent variables. Then, dimensional analysis dictates the relation between the dependent and independent variables, and the solution is known except for a numerical coefficient. An example where dimensional analysis has been successful is in the derivation of the region called the 'inertial subrange' in the turbulence kinetic energy spectrum. Here the slope obeys the so-called $-5/3$-law.

Since turbulent flows are characterized by high Reynolds numbers, it is reasonable to require that a description of turbulence should behave properly as the Reynolds number approaches infinity. This method of analysis is called asymptotic invariance, and has been successfully used in the development of the theory for turbulent boundary layers. In analyzing turbulence, the concept of local invariance or "self-preservation" is often invoked. This tool is powerful when the turbulent flow can be characterized as if it was controlled mainly by its immediate environment, and this situation occurs typically in the far downstream region of a wake, jet or free-shear layer. There, the time- and length-scales vary only slowly in the downstream direction, and if the turbulence time-scales are sufficiently small, it can be assumed that the flow has sufficient time to adjust to its gradually changing environment. The turbulence then is dynamically similar everywhere provided the average quantities are nondimensionalized with local length- and time-scales.

More details on the physics of turbulence can be found in classical textbooks in the field [Townsend, 1976; Monin and Yaglom, 1975; Tennekes and Lumley, 1972; Hinze, 1975]. There are also many good modern books available on the subject [McComb, 1990; Lesieur, 1991; Holmes et al., 1996; Pope, 2000]. The important point to the present chapter is that almost all methods for analyzing turbulence are heuristic and are not derived from first principles. Detailed measurements of flow quantities will continue therefore to be an essential component of the arsenal of attacks on the turbulence problem. In this context, MEMS-based sensors have widened the horizon of experiments and can be used for measuring turbulence reliably and inexpensively at high Reynolds numbers.

6.3.3 Scales

As mentioned, turbulence is a high-Reynolds-number phenomenon that is characterized by the existence of numerous length- and time-scales. The spatial extension of the length-scales is bounded from above by the dimensions of the flow field and from below by the diffusive and dissipative action of the molecular viscosity. If we limit our interest to shear flows, which are basically characterized by two large length-scales — one in the streamwise direction (the convective or longitudinal length-scale) and the other perpendicular to the flow direction (the diffusive or lateral length-scale) — we obtain a more well-defined problem. Moreover, at sufficiently high Reynolds numbers, the boundary layer approximation applies and it is assumed that there is a wide separation between the lateral and the longitudinal length-scales. This leads to some attractive simplifications in the equations of motion, for instance the elliptical Navier–Stokes equations are transferred to the parabolic boundary-layer equations [see Hinze, 1975]. So in this approximation, the lateral scale is approximately equal to the extension of the flow perpendicular to the flow direction (the boundary layer thickness), and the largest eddies have typically this spatial extension.

The large eddies are most energetic and play a crucial role both in the transport of momentum and contaminants. A constant energy supply is needed to maintain the turbulence, and this energy is extracted from the mean flow into the largest most energetic eddies. The lateral length-scale is also the relevant scale for analyzing this energy transfer. However, there is an energy destruction in the flow due to the action of the viscous forces (the dissipation), and other smaller length-scales are needed for the analysis of this process.

As the eddy size decreases, viscosity becomes a more significant parameter since one property of viscosity is its effectiveness in smoothing out velocity gradients. The viscous and the nonlinear terms in the momentum equation counteract each other in the generation of small-scale fluctuations. While the inertial terms try to produce smaller and smaller eddies, the viscous terms check this process and prevent the generation of infinitely small scales by dissipating the small-scale energy into heat. In the early 1940s, Kolmogorov (1941a; 1941b) developed the universal equilibrium theory. One cornerstone of this theory is that the small-scale motions are statistically independent of the relatively slower large-scale turbulence. An implication of this is that the turbulence at the small scales depends only on two parameters, namely the rate at which energy is supplied by the large-scale motion and the kinematic viscosity. In addition, it is assumed in the equilibrium theory that the rate of energy supply to the turbulence should be equal to the rate of dissipation. Hence, in the analysis of turbulence at small scales, the dissipation rate per unit mass ε is a relevant parameter together with the kinematic viscosity v. Kolmogorov (1941) used simple dimensional arguments to derive a length-scale, time-scale, and a velocity-scale relevant for the small-scale motion, respectively given by:

$$\eta = \left(\frac{v^3}{\varepsilon} \right)^{1/4} \tag{6.1}$$

$$\tau = \left(\frac{v}{\varepsilon} \right)^{1/2} \tag{6.2}$$

$$\upsilon = (v\varepsilon)^{1/4} \tag{6.3}$$

These scales are accordingly called the Kolmogorov microscales, or sometimes the inner scales of the flow. As they are obtained through a physical argument, these scales are the smallest scales that can exist in a turbulent flow and they are relevant for both free-shear and wall-bounded flows.

In boundary layers, the shear-layer thickness provides a measure of the largest eddies in the flow. The smallest scale in wall-bounded flows is the viscous wall unit, which will be shown here to be of the same order as the Kolmogorov length-scale. Viscous forces dominate over inertia in the near-wall region, and the characteristic scales are obtained from the magnitude of the mean vorticity in the region and its viscous diffusion away from the wall. Thus, the viscous time-scale t_v is given by the inverse of the mean wall vorticity:

$$t_v = \left[\frac{\partial \bar{U}}{\partial y} \bigg|_w \right]^{-1} \tag{6.4}$$

where $\bar{U}$ is the mean streamwise velocity. The viscous length-scale ℓ_v is determined by the characteristic distance by which the (spanwise) vorticity is diffused from the wall, and is thus given by:

$$\ell_v = \sqrt{vt_v} = \sqrt{\frac{v}{\partial \bar{U}/\partial y |_w}} \tag{6.5}$$

where v is the kinematic viscosity. The wall velocity-scale (so-called friction velocity, u_τ) follows directly from the above time- and length-scales:

$$u_\tau = \frac{\ell_v}{t_\tau} = \sqrt{v \frac{\partial U}{\partial y}\bigg|_w} = \sqrt{\frac{\tau_w}{\rho}} \tag{6.6}$$

where τ_w is the mean shear stress at the wall, and ρ is the fluid density. A wall unit implies scaling with the viscous scales, and the usual ( )$^+$ notation is used; for example, $y^+ = y/\ell_v = yu_\tau/v$. In the wall region, the characteristic length for the large eddies is y itself, while the Kolmogorov scale is related to the distance from the wall y as follows:

$$\eta^+ = \frac{\eta u_\tau}{v} \approx (\kappa y^+)^{1/4} \tag{6.7}$$

where κ is the von Karman constant (≈ 0.41). As y^+ changes in the range of 1–5 (the extent of the viscous sublayer), η changes from 0.8 to 1.2 wall units.

We now have access to scales for the largest and smallest eddies of a turbulent flow. To continue our analysis of the cascade energy process, it is necessary to find a connection between these diverse scales. One way of obtaining such a relation is to use the fact that at equilibrium, the amount of energy dissipating at high wavenumbers must equal the amount of energy drained from the mean flow into the energetic large-scale eddies at low wavenumbers. In the inertial region of the turbulence kinetic energy spectrum, the flow is almost independent of viscosity. Because since the same amount of energy dissipated at the high wavenumbers must pass this "inviscid" region, an inviscid relation for the total dissipation may be obtained by the following argument. The amount of kinetic energy per unit mass of an eddy with a wavenumber in the inertial sublayer is proportional to the square of a characteristic velocity for such an eddy, u^2. The rate of transfer of energy is assumed to be proportional to the reciprocal of one eddy turnover time, u/ℓ, where ℓ is a characteristic length of the inertial sublayer. Hence, the rate of energy that is supplied to the small-scale eddies via this particular wavenumber is of order u^3/ℓ, and this amount of energy must be equal to the energy dissipated at the highest wavenumber, expressed as:

$$\varepsilon \approx \frac{u^3}{\ell} \tag{6.8}$$

Note that this is an inviscid estimate of the dissipation since it is based on large-scale dynamics and does not either involve or contain viscosity. More comprehensive discussion of this issue can be found in Taylor (1935) and Tennekes and Lumley (1972). From an experimental perspective, this is a very important expression since it offers one way of estimating the Kolmogorov microscales from quantities measured in a much lower wavenumber range.

Since the Kolmogorov length- and time-scales are the smallest scales occurring in turbulent motion, a central question will be how small these scales can be without violating the continuum hypothesis. By looking at the governing equations, it can be concluded that high dissipation rates are usually associated with large velocities, and this situation is more likely to occur in gases than in liquids so it would be sufficient to show that for gas flows the smallest turbulence scales are normally much large than the molecular scales of motion. The relevant molecular length-scale is the mean free path, Λ, and the ratio between this length and the Kolmogorov length scale, η, is the microstructure Knudsen number and can be expressed as [Corrsin, 1959]:

$$Kn = \frac{\Lambda}{\eta} \approx \frac{Ma^{1/4}}{Re} \tag{6.9}$$

where the turbulence Reynolds number, Re, and the turbulence Mach number, Ma, are used as independent variables. It is obvious that a turbulent flow will interfere with the molecular motion only at high Mach number and low Reynolds number, and this is a very unusual situation occurring only in certain gaseous nebulae. (Note that in microduct flows and the like, the Re is usually too small for turbulence to even exist. So the issue of turbulence Knudsen number is mute in those circumstances even if rarefaction effects become strong.) Thus, under normal condition the turbulence Knudsen number falls in the group of continuum flows. It is noteworthy however, that measurements using extremely thin hot-wires, small MEMS sensors, or flows within narrow MEMS channels can generate values in the slip-flow regime and even beyond, and this implies that for instance the no-slip condition may be questioned.

6.3.4   Sensor Requirements

It is the ultimate goal of all measurements in turbulent flows to resolve both the largest and smallest eddies that occur in the flow. At the lower wavenumbers, the largest and most energetic eddies occur, and normally there are no problems associated with resolving these eddies. Basically, this is a question of having access to computers with sufficiently large memory for storing the amount of data that may be necessary to acquire from a large number of distributed probes, each collecting data for a time period long enough to reduce the statistical error to a prescribed level. However, at the other end of the spectrum, both the spatial and the temporal resolutions are crucial and this puts severe limitations on the sensors to be used. It is possible to obtain a relation between the small and large scales of the flow by substituting the inviscid estimate of the total dissipation rate, Equation 6.8, into the expressions for the Kolmogorov microscales, Equations 6.1–6.3:

$$\frac{\eta}{\ell} \approx \left(\frac{u\ell}{\nu}\right)^{3/4} = Re^{3/4} \tag{6.10}$$

$$\frac{\tau u}{\ell} \approx \left(\frac{u\ell}{\nu}\right)^{-1/2} = Re^{-1/2} \tag{6.11}$$

$$\frac{\upsilon}{u} \approx \left(\frac{u\ell}{\nu}\right)^{-1/4} = Re^{-1/4} \tag{6.12}$$

where Re is the Reynolds number based on the speed of the energy containing eddies u and their characteristic length ℓ. Since turbulence is a high-Reynolds-number phenomenon, these relations show that the small length-, time-, and velocity-scales are much less than those of the larger eddies, and that the separation in scales widens considerably as the Reynolds number increases. Moreover, this also implies that the assumptions made on the statistical independence and the dynamical equilibrium state of the small structures will be most relevant at high Reynolds numbers. Another conclusion from the above relations is that if two turbulent flow fields have the same spatial extension (i.e., same large-scale) but different Reynolds numbers, there would be an obvious difference in the small-scale structure in the two flows. The low-Reynolds-number flow would have a relatively coarse small-scale structure, while the high-Reynolds-number flow would have much finer small eddies.

To spatially resolve the smallest eddies, sensors are needed that are of approximately the same size as the Kolmogorov length-scale for the particular flow under consideration. This implies that as the Reynolds number increases smaller sensors are required. For instance, in the self-preserving region of a plane-cylinder wake at a modest Reynolds number, based on the cylinder diameter of 1840, the value of η varies in the range of 0.5–0.8 mm [Aronson and Löfdahl, 1994]. For this case, conventional hot-wires can be used for turbulence measurements. However, an increase in the Reynolds number by a factor of ten will yield Kolmogorov scales in the micrometer range and call for either extremely small conventional hot-wires or MEMS-based sensors. Another example of the Reynolds number effect on the requirement of small sensors is a simple two-dimensional, flat-plate boundary layer. At a momentum thickness Reynolds number of $Re_\theta = 4000$, the Kolmogorov length-scale is typically of the order of $50\,\mu m$, and in order to resolve these scales it is necessary to have access to sensors that have a characteristic active measuring length of the same spatial extension.

Severe errors will be introduced in the measurements by using a sensor that is too large because such a sensor will integrate the fluctuations due to the small eddies over its spatial extension and the energy content of these eddies will be interpreted by the sensor as an average "cooling." When measuring fluctuating quantities, this implies that these eddies are counted as part of the mean flow and their energy is "lost." The result will be a lower value of the turbulence parameter, which will wrongly be interpreted as a measured attenuation of the turbulence [Ligrani and Bradshaw, 1987]. However, since turbulence measurements deal with statistical values of fluctuating quantities, it may be possible to loosen the spatial constraint of having a sensor of the same size as η, to allow a sensor dimensions which are slightly larger than the Kolmogorov scale, say on the of order of η.

For boundary layers, the wall unit has been used to estimate the smallest necessary size of a sensor for accurately resolving the smallest eddies. For example, Keith et al. (1992) state that ten wall units or less is a relevant sensor dimension for resolving small-scale pressure fluctuations. Measurements of fluctuating velocity gradients, essential for estimating the total dissipation rate in turbulent flows, are another challenging task. Gad-el-Hak and Bandyopadhyay (1994) argue that turbulence measurements with probe lengths greater than the viscous sublayer thickness (about 5 wall units) are unreliable particularly near the surface. Many studies have been conducted on the spacing between sensors necessary to optimize the formed velocity gradients [Aronson et al., 1997, and references therein]. A general conclusion from both experiments and direct numerical simulations is that a sensor spacing of 3–5 Kolmogorov lengths is recommended. When designing arrays for correlation measurements, the spacing between the coherent structures will be the determining factor. For example, when studying the low-speed streaks in a turbulent boundary layer, several sensors must be situated along a lateral distance of 100 wall units, the average spanwise spacing between streaks. This requires quite small sensors, and many attempts have been made to meet these conditions with conventional sensor designs. However, in spite of the fact that conventional sensors like hot-wires have been fabricated in the micrometer size-range (for their diameter but not their length), they are usually hand-made, difficult to handle, and are too fragile: here the MEMS technology has really opened a door for new applications.

It is clear from the above that the spatial and temporal resolutions for any probe to be used to resolve high-Reynolds-number turbulent flows are extremely tight. For example, both the Kolmogorov scale and the viscous length-scale change from few microns at the typical field Reynolds number — based on the momentum thickness — of 10^6 to a couple of hundred microns at the typical laboratory Reynolds number of 10^3. MEMS sensors for pressure, velocity, temperature, and shear stress are at least one order of magnitude smaller than conventional sensors [Ho and Tai, 1998; Löfdahl et al., 1994a; 1994b]. Their small size improves both the spatial and temporal resolutions of the measurements, typically few microns and few microseconds, respectively. For example, a micro-hot-wire (called hot-point) has very small thermal inertia and the diaphragm of a micro-pressure-transducer has a correspondingly fast dynamic response. Moreover, the microsensors' extreme miniaturization and low energy consumption make them ideal for monitoring the flow state without appreciably affecting it. Lastly, literally hundreds of microsensors can be fabricated on the same silicon chip at a reasonable cost, making them well suited for distributed measurements and control. The UCLA/Caltech team [Ho and Tai, 1996; 1998, and references therein] has been very effective in developing many MEMS-based sensors and actuators for turbulence diagnosis and control.

The next section will focus attention on sensors used for measurements in turbulent flows. Specifically we discuss sensors for velocity, pressure, and wall-shear stress, quantities which so far have been difficult to measure and where the introduction of MEMS has created completely new perspective.

6.4   Sensors for Turbulence Measurements and Control

By definition, a transducer is a device that is actuated by power from one system and supplies power, usually in another form, to a second system. Hence, in an electromechanical transducer one connection to the environment typically conducts electrical energy and another conducts mechanical energy. Microphones, loudspeakers, strain gauges, and electric motors are examples of electromechanical transducers, which in turn may be categorized into sensors and actuators. A sensor is a device that responds to physical stimulus such as velocity, pressure, and temperature, and transmits a resulting impulse for either measurement or control purposes. The output of a sensor may depend on more than one variable. Ideally, the sensor monitors the state of a system without affecting it, while an actuator operates on the system the other way around: it imposes a state without regard to the system load.

Typically, a sensor converts the physical parameter to be measured into an electrical signal which is subsequently analyzed and interpreted. The studied physical parameters are usually classified into different groups, such as chemical, mechanical, and thermal signals. The transducer or sensor "helps" the electronic to "see," "hear," "smell," "taste," or "touch." In many applications, a sensor can be divided into a sensing part and a converting part. For instance, for a piezoresistive pressure sensor, the sensing part is the deflecting diaphragm and the converting part is the piezoresistor, which converts the deflection of the diaphragm into an electrical signal. It is generally more difficult to control a system than monitor it, and in the last decade it has been more challenging for scientists and engineers to design and build microsensors than microactuators. As a consequence, progress in sensors lags behind that in actuators.

6.4.1 Background

A typical MEMS sensor is well below 1 mm in size, and at least one order of magnitude smaller than traditional sensors used to measure instantaneous flow quantities such as velocity, pressure, and temperature. The small spatial extension implies that both the inertial mass and the thermal capacity are reduced, which makes these sensor suited for measurements of flow quantities in high-Reynolds-number turbulent flows where both high-frequency response and fine spatial resolution are essential. For instance, pressure and velocity sensors with diaphragm side length and wire length of less than 100 μm are in use today. MEMS sensors are not hand-made, but are produced by photolithographic methods. This implies that each unit is fabricated to extremely low tolerance and at low cost. (Normally the fabrication of a prototype sensor is very costly, but once the fabrication principle has been outlined the unit cost drops dramatically.) The latter trait makes it possible to use a large number of sensors to cover large areas and volumes of the flow field. This in turn makes it feasible to study coherent structures and to effectively execute reactive control for turbulent shear flows. An additional important advantage of microfabrication is that it enables packing of sensors in arrays on the same silicon chip. A major difficulty, however, is that the leads of each sensor have to be connected to an external signal-conditioning instrument, and the handling of numerous signal paths is tedious and occupies a large portion of the precious surface area of the chip. For example, Ho and Tai (1998) state that their array of wall-shear-stress sensors containing 85 elements occupies 1% of the area, while the leads take about 50% of the surface. However, current research attempts to solve this problem.

This section addresses MEMS-based sensors for use in turbulence measurements and reactive flow control applications. In particular, it reviews the state-of-the art of microsensors used to measure the instantaneous velocity, wall-shear stress, and pressure, which we deem as quantities of primary importance in turbulence diagnosis and control. For each group, we give a general background, design criteria, and calibration procedure, and provide examples of measurements conducted with MEMS-based sensors. When possible we compare the results to conventional measurements.

6.4.2 Velocity Sensors

6.4.2.1 Background

Turbulence is one of the unsolved problems of classical physics, and it is almost impossible to make predictions for turbulent flows without heavily relying on empirical data. Since turbulence obeys the continuum hypothesis, the governing equations are known and an analysis of these equations shows that mean velocities, higher-order moments of fluctuating velocities, and products of gradients of fluctuating velocities are needed for future development of turbulence models. To this end, thermal anemometers have been the most significant tool for measuring these quantities. The introduction of MEMS has extended the range of applicability of the thermal anemometer and has provided incentives for conducting new measurements in high-Reynolds-number flows. This progress is basically achieved by the increased spatial and temporal resolutions that are feasible through the miniaturization and formation of sensor arrays.

According to King (1914), the first experiments using thermal anemometers were conducted in 1902, but otherwise, the work of King (1915; 1916) on the design of hot-wires and on the theory of heat convection from cylinders is considered as the starting point for the era of thermal anemometry research. Early experiments were usually limited to measurements of mean velocities. In the late 1920s, however, the emphasis shifted toward measurements of fluctuating velocities [see Dryden and Kuethe, 1930]. Since then, numerous papers in the field have been published where all the measurements have been conducted using thermal anemometry with hot-wires or hot-films as sensing elements. Many researchers have made significant contributions to thermal anemometry and should be given credit in any complete review of the subject; however, this chapter will focus only on the advantages gained from MEMS for improving measurements with current thermal anemometry. For literature review on conventional hot-wires, the reader is referred to the survey papers by Comte-Bellot (1976), Freymuth (1983; 1992), and Fingerson and Freymuth (1996), as well as to the books by Hinze (1975), Lomas (1986) and Perry (1982).

In this subsection, we recall the principle of thermal anemometry for velocity measurements and summarize the characteristics of hot-wire sensors operated in constant-temperature circuit. Since the governing equations of thermal anemometry are the same whether the sensor is a conventional hot-wire or a MEMS-based probe, we discuss these equations for one, two, and three sensors, and remark on their applicability for MEMS. We provide an overview of current MEMS-based sensors used for velocity measurements and discuss the results of experiments conducted with these probes.

6.4.2.2 Thermal-Sensor Principle

The thermal anemometer can be regarded as a device used for measuring significant quantities for turbulence diagnosis and for reactive flow control. The sensors used are small, conducting elements which are heated by an electric current and cooled by the flow. All modes of heat transfer are present but forced convection is usually the dominate mode. From the temperature or rather resistance attained by the sensor, it is possible to gain the desired information on the instantaneous velocity vector. In order to thoroughly investigate a turbulent flow field, usually more than one sensor is needed and multi-sensor arrangements are commonly used in measurements forming the base for development of turbulence models.

The generally small sensor size yields a good spatial resolution and frequency response [Freymuth, 1977] making thermal anemometry especially suited for studying flow details in turbulent flows. A simple thermal anemometer is shown schematically in Figure 6.5. The minute resistor mounted between two prongs constitutes the sensing part and forms one arm in a Wheatstone bridge. For moderate temperature changes, the hot-wire resistance changes linearly with its temperature:

$$R = R_r[1 + \alpha(T_m - T_r)] \tag{6.13}$$

where R_r is the resistance at the reference temperature T_r, T_m is the mean sensor temperature along its length, and α is the temperature coefficient of resistance. The latter value is critical, since if the hot-wire sensor did not vary in resistance with temperature, there would be no signal from a thermal anemometer. The ambient fluid temperature T_a is often used as reference temperature T_r, and the value of α depends on the reference temperature used.

The Wheatstone bridge arrangement shown in Figure 6.5 is designed so that the resistance R_1 is large compared to that of the sensor. Then, the current I through the hot-wire is nearly constant. This implies, that any increase in heat transfer from the hot-wire to its surrounding will cause the sensor to cool, and this in turn will decrease the hot-wire resistance R (a decrease in the voltage E_{12} and a decrease in the amplifier output E). A decrease in heat transfer between sensor and fluid will have the opposite effect and create an increase in E. Without the feedback amplifier, the principle scheme shown in Figure 6.5 is basically an uncompensated, constant-current hot-wire anemometer, and this kind of system dominated in the infancy of the thermal anemometer era. Since then, advances have taken place in hot-wire fabrication, electronic control circuits, and data acquisition. The result is that the constant-current operation of thermal sensors has been largely replaced by the constant-temperature operation which offers much better stability and frequency response through high-gain feedback amplifiers. Stability criteria and techniques for checking the frequency response are now well understood for constant-temperature systems and the introduction of digital techniques have significantly expanded the capabilities for analyzing the resulting data. Today the nonlinear output is no longer a limitation; correlations, power spectra, and amplitude probability distributions are all readily obtainable.

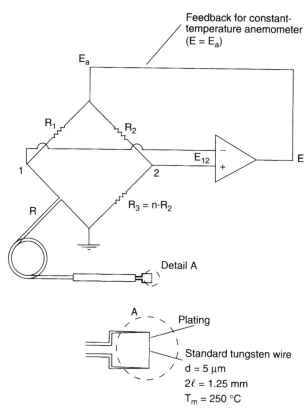

FIGURE 6.5 Basic elements of a hot-wire anemometer. (Reprinted with permission from Fingerson, L.M., and Freymuth, P. [1996] "Thermal Anemometers," in *Fluid Mechanics Measurements*, 2nd ed., R.J. Goldstein, ed., pp. 115–173, Taylor and Francis, Washington, D.C.)

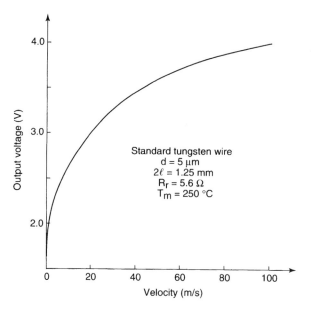

FIGURE 6.6 Typical calibration curve for a hot-wire anemometer. (Reprinted with permission from Fingerson, L.M., and Freymuth, P. [1996] "Thermal Anemometers," in *Fluid Mechanics Measurements*, 2nd ed., R.J. Goldstein, ed., pp. 115–173, Taylor and Francis, Washington, D.C.)

6.4.2.3 Calibration

A calibration curve is usually obtained by measuring a number of distinct velocities in a nearly inviscid flow, for example in the freestream outside a boundary layer. A Pitot tube is used as standard for the velocity, and at the same time the output voltage of the anemometer is recorded. A typical calibration curve for a single hot-wire is shown in Figure 6.6. This curve shows a nonlinear output with the velocity and a sensitivity decrease as the velocity increases. However, the sensitivity as a percentage of reading stays nearly constant which makes the hot-wire anemometer useful over a wide range of velocities. Under normal conditions the hot-wire is calibrated in a well-defined flow. Today the nonlinearity of the sensor characteristics is not a major constraint and many different transfer functions and cooling laws are available [Perry, 1982].

Once the calibration curve is obtained, good measurements in the unknown environment can be directly made provided that the fluid temperature, composition, and density are the same as those during the calibration. In addition, it is assumed that the turbulence intensity is small, say less than 15% of the mean velocity, and that the flow is incompressible. Once these conditions are fulfilled, the single hot-wire can be used to measure the streamwise mean and fluctuating velocities. Today, most modern laboratories use fast analog-to-digital converters (ADC) implying that all the above functions, including the linearization, are performed with high accuracy employing digital computers.

6.4.2.4 Hot-Wire Sensors

In the design of hot-wire sensors there are many compromises. For example, the length and diameter are an optimization between several conflicting criteria. Concerning the hot-wire length, a short sensor is desired to maximize the spatial resolution and minimize the aerodynamic stress. However, a long sensor is desired to minimize conduction losses to the support, provide more uniform temperature distribution, and reduce prong interference. Regarding the sensor diameter, a small diameter is desired to eliminate output noise due to separated flow around the sensor, to maximize the frequency response of the wire due to thermal inertia, to create higher heat transfer coefficient, to maximize the spatial resolution, and to improve the signal-to-noise ratio at high frequencies. On the other hand, the solidity of the wire is essential, and a large diameter is desirable to increase the probe strength and to reduce contamination effects due to particles in the fluid.

For research work, typical values of the traditional wire diameter are in the range of 2.5–5 μm, and the wire length-to-diameter ratios are usually in the range of $100 < \ell/d < 600$. The choice of wire material is also a compromise. Tungsten, platinum, and platinum–iridium are common hot-wire materials. Tungsten is desirable because of its high temperature coefficient of resistance and high strength. Platinum is available in small diameters, has good temperature coefficients of resistance, and does not oxidize. Platinum–iridium is a wire that does not oxidize, has good strength, but unfortunately has low temperature coefficient of resistance. The article by Fingerson and Freymuth (1996) has an elaborate section on wire materials and summarizes the properties of the most commonly used materials. An analytical relation between the flow velocity and anemometer output is desirable when physically interpreting the heat transfer process in complex flow situations not covered by the calibration and when correcting for various sources of error and approximations. For a constant-temperature anemometer, the heat transfer from the sensor to its environment, Q, can be expressed in terms of the anemometer output voltage E as follows:

$$Q = \frac{E^2 R}{(R + R_1)} = \phi + K \tag{6.14}$$

where R represents the resistance of the sensor only. The rate of heat transfer, Q, is equal to the electrical power input to the sensor and consists of two parts: convective heat transfer, ϕ, between the heated portion of the sensor and the flowing fluid, and conductive heat transfer, K, between the heated portion of the sensor and its support prongs. Our aim is now to establish a relation between the heat transfer, or electrical power input, and the flow velocity. Bremhorst and Gilmore (1976) concluded that the static- and dynamic-calibration coefficients agree to within 3% for the velocity range of 3–32 m/s. Based on this, and neglecting radiation, we can for the stationary case focus our interest on ϕ, which can be expressed as:

$$\phi = Nu \times 2\pi\ell\kappa_f(T_m - T_a) \tag{6.15}$$

where Nu is the Nusselt number, $(\equiv h_c d/\kappa_f)$, d is the sensor diameter, h_c is the coefficient of convective heat transfer, κ_f is the fluid thermal conductivity at $T_f = (T_m + T_a)/2$. T_m is the mean cylinder temperature, T_a is the ambient fluid temperature, and 2ℓ is the length of the sensitive portion of the wire.

The main problem now is to find a representative expression of Nusselt number in terms of fluid and sensor parameters. General expressions on this quantity can be found in Hinze (1975), Perry (1982) and Fingerson and Freymuth (1996). Fortunately, most applications permit a reduction of the parameters needed and a commonly used simplified expression is:

$$Nu = A + B\,Re^{0.5} \tag{6.16}$$

where A and B are constants determined through the calibration, and Re is the Reynolds number based on the wire diameter and the speed being measured. This is a very useful expression for ordinary measurements and its only drawback is that it does not represent the Nu number over a wide range of velocities. More elaborate expressions have, however, been proposed by, for example, Collis and Williams (1959):

$$Nu = (A + B\,Re^n)\left(1 + \frac{\alpha_T}{2}\right)^{0.17} \tag{6.17}$$

where α_T is the overheat ratio, and also by Kramers (1946):

$$Nu = 0.42\,Pr^{0.26} + 0.57\,Pr^{0.53}\,Re^{0.50} \tag{6.18}$$

where Pr is the Prandtl number. The latter expression covers the ranges $0.71 < Pr < 525$ and $2 < Nu < 20$, and can be used even when the fluid is water.

A discussion on parameters neglected in the general expression can be found in the references cited above. The ambient temperature, wire aspect-ratio, and conduction to wire supports have a large influence on the Nusselt number. Additionally, the Knudsen number (defined in the turbulence section) has been shown to be crucial in measurements using fine wires and small MEMS devices. In the next sub-subsection, we consider the important issue of the angular dependence of the heat transfer, and the equations necessary for measurement of one, two, and three velocity components.

6.4.2.5 Mean and Fluctuating Velocities

Figure 6.7 illustrates the instantaneous velocity components sensed by a hot-wire element. The simplest analytical expression for the angular sensitivity for an infinitely long wire is the so-called "cosine-law:"

$$U_{eff} = U\cos\alpha_1 \tag{6.19}$$

where U_{eff} is the effective cooling velocity past the sensor. This equation essentially states that the velocity component along the wire ($U\sin\alpha_1$) has no cooling effect, and that the sensor is rotationally symmetrical in both construction and response. However, the real hot-wire has a finite length and the velocity component parallel to the sensor must be considered. Therefore, Champagne (1965) suggested an extended expression for the effective velocity:

$$U_{eff} = U\sqrt{\cos^2\alpha_1 + k_T^2\sin^2\alpha_1} \tag{6.20}$$

where k_T is an empirically determined factor which varies as a function of both α_1 and the velocity. Champagne (1965) found that k_T decreases nearly linearly with ℓ/d from a value of $k_T = 0.2$ at $\ell/d = 200$, to zero at $\ell/d = 600$–800. Clearly, a large aspect-ratio sensor decreases the influence of the velocity component along the wire. Comte-Bellot et al. (1971) has shown that aerodynamic effects from both support needles and probe body must be considered, and Jörgensen (1971) has suggested a cooling law to take these effects into account:

$$U_{eff} = \sqrt{U_N^2 + k_T^2 U_T^2 + k_N^2 U_{BN}^2} \tag{6.21}$$

where U_{BN} is the velocity perpendicular to both sensor and supporting prongs. The value of k_N can range from 1.0 to 1.2, depending on the design of probe support and needles [Drubka et al., 1977]. It is important to note that neither k_T nor k_N can be considered as constants for all angles and velocities. The cooling laws are basically methods for calibrating a hot-wire. More elaborate computer-oriented cooling-law relations have been presented. For example, Lueptov et al. (1988) have suggested the so-called "look-up" table, where a direct transfer of voltages into velocities and angles of attack is conducted.

It is common practice to use a single hot-wire probe perpendicular to the flow direction to measure the time-averaged velocity, $\bar{U}$, and corresponding fluctuating velocity mean-square, $\overline{u^2}$. In Figure 6.7, the wire is oriented with the prongs in the streamwise direction so that $U_N = U_1$, $U_T = U_2$, and $U_{BN} = U_3$ where U_1, U_2, and U_3 are the orthogonal components of the velocity vector U. The effective instantaneous cooling velocity can then be expressed as:

$$U_{eff} = \sqrt{U_1^2 + k_T^2 U_2^2 + k_N^2 U_3^2} \tag{6.22}$$

If the mean flow is in the U_1 direction and if Reynolds decomposition is introduced, then the mean velocities $\bar{U}_2 = \bar{U}_3 = 0$, and the effective cooling velocity may be expressed as:

$$U_{eff} = \sqrt{(\bar{U}_1 + u_1)^2 + k_T^2 u_2^2 + k_N^2 u_3^2} \tag{6.23}$$

Since k_T is small and $k_N \approx 1$, we can write:

$$U_{eff} = \sqrt{(\bar{U}_1 + u_1)^2 + u_3^2} \tag{6.24}$$

If it is further assumed that u_3 is small, then:

$$\bar{U}_{eff} = \bar{U}_1 = \bar{U} \tag{6.25}$$

$$\sqrt{\overline{u_1^2}} = \sqrt{\overline{u^2}} \tag{6.26}$$

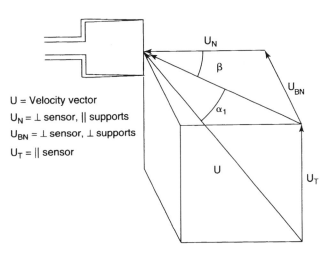

U = Velocity vector
U_N = ⊥ sensor, ∥ supports
U_{BN} = ⊥ sensor, ⊥ supports
U_T = ∥ sensor

FIGURE 6.7 Velocity components relative to a hot-wire sensor. (Reprinted with permission from Fingerson, L.M., and Freymuth, P. [1996] "Thermal Anemometers," in *Fluid Mechanics Measurements*, 2nd ed., R.J. Goldstein, ed., pp. 115–173, Taylor and Francis, Washington, D.C.)

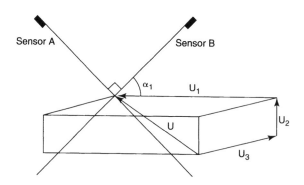

FIGURE 6.8 Configuration of a cross hot-wire sensor. (Reprinted with permission from Fingerson, L.M., and Freymuth, P. [1996] "Thermal Anemometers," in *Fluid Mechanics Measurements*, 2nd ed., R.J. Goldstein, ed., pp. 115–173, Taylor and Francis, Washington, D.C.)

Two velocity components can be measured with a cross hot-wire or by rotating a single probe and taking several sequential measurements [Fujita and Kovasznay, 1968]. Three velocity components can be measured with a triple hot-wire or by rotating a cross hot-wire. Figure 6.8 shows a cross wire with two sensors A and B, located so that they are sensitive to U_2. The effective cooling velocities of the two sensors can be expressed as:

$$U_{A,eff}^2 = (U_1 \cos \alpha_1 - U_2 \sin \alpha_1)^2 + k_T^2 (U_1 \cos \alpha_1 + U_2 \sin \alpha_1)^2 + k_N^2 U_3^2 \tag{6.27}$$

$$U_{B,eff}^2 = (U_1 \sin \alpha_1 + U_2 \cos \alpha_1)^2 + k_T^2 (U_1 \cos \alpha_1 - U_2 \sin \alpha_1)^2 + k_N^2 U_3^2 \tag{6.28}$$

If the coordinates are selected so that the mean velocity $\bar{u}_3 = 0$, and if it is assumed once again that the sensors are sufficiently long, then $k_T \rightarrow 0$, $k_N \rightarrow 1$, and Equations 6.27 and 6.28 are reduced to:

$$U_{A,eff}^2 = (U_1 \cos \alpha_1 - U_2 \sin \alpha_1)^2 + u_3^2 \tag{6.29}$$

$$U_{B,eff}^2 = (U_1 \sin \alpha_1 + U_2 \cos \alpha_1)^2 + u_3^2 \tag{6.30}$$

Assuming that u_3 is small and that the sensors are oriented so that $\alpha_1 = 45°$, the streamwise and normal velocities are then expressed as, respectively:

$$U_1 = \frac{1}{\sqrt{2}}(U_{A,eff} + U_{B,eff}) \tag{6.31}$$

$$U_2 = \frac{1}{\sqrt{2}}(U_{A,eff} - U_{B,eff}) \tag{6.32}$$

It is a common practice to orient the wire in the mean flow direction so that $\bar{U}_2 = 0$. In this case, the expressions for the mean velocity and Reynolds stresses are:

$$\bar{U} = \frac{1}{\sqrt{2}}(U_{A,eff} + U_{B,eff}) \tag{6.33}$$

$$\overline{u_1^2} = \frac{1}{2}(u_{A,eff} + u_{B,eff})^2 \tag{6.34}$$

$$\overline{u_2^2} = \frac{1}{2}(u_{A,eff} - u_{B,eff})^2 \tag{6.35}$$

$$\overline{u_1 u_2} = \frac{1}{2}\overline{(u_{A,eff} + u_{B,eff})(u_{A,eff} - u_{B,eff})} \tag{6.36}$$

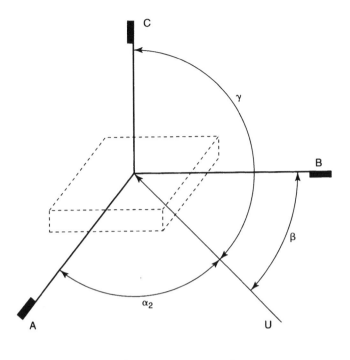

$$U^2_{A,eff} = U^2(\sin^2\alpha_2 + k^2\cos^2\alpha_2)$$

$$U^2_{B,eff} = U^2(\sin^2\beta + k^2\cos^2\beta)$$

$$U^2_{C,eff} = U^2(\sin^2\gamma + k^2\cos^2\gamma)$$

$$U^2 = \frac{U^2_A + U^2_B + U^2_C}{2 + k^2}$$

FIGURE 6.9 Direction sensitivity of a triple hot-wire sensor. (Reprinted with permission from Fingerson, L.M., and Freymuth, P. [1996] "Thermal Anemometers," in *Fluid Mechanics Measurements*, 2nd ed., R.J. Goldstein, ed., pp. 115–173, Taylor and Francis, Washington, D.C.)

Equations 6.34–6.36 are the fundamental relations used in most cross hot-wire measurements. As mentioned earlier, more information about the flow field can be obtained if all three velocity components are measured. This can for instance be done as shown schematically in Figure 6.9 by adding a third sensor C whose axis is at an angle α_3 to U_1 and in the $(U_1 U_3)$-plane. The effective cooling velocity of the third wire will then be:

$$U^2_{C,eff} = (U_1 \sin\alpha_3 + U_3 \cos\alpha_3)^2 + k^2_T(U_1 \cos\alpha_3 - U_3 \sin\alpha_3)^2 + k^2_N U^2_2 \tag{6.37}$$

Together with the equations for the cross hot-wire we have three equations and three unknowns, so the instantaneous velocity vector can be determined. Once the instantaneous components are available, all turbulence parameters can be calculated. Again, the equations can be simplified if the sensors are sufficiently long so that $k_T \to 0$ and $k_N \to 1$, and if the sensors are oriented so that $\overline{U_2} = \overline{U_3} = 0$, and $\alpha_1 = \alpha_2 = 45°$. The simplified equations are given below and were used by Fabris (1978) to make three-component velocity measurements:

$$U^2_{A,eff} = U^2(\sin^2\alpha_2 + k^2\cos^2\alpha_2) \tag{6.38}$$

$$U^2_{B,eff} = U^2(\sin^2\beta + k^2\cos^2\beta) \tag{6.39}$$

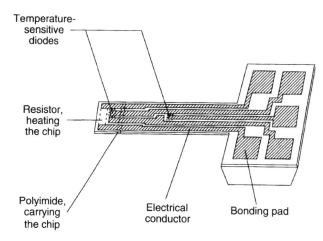

FIGURE 6.10   Schematic of a single-wire MEMS sensor consisting of three parts, the base plate 1.5 $\times$ 1.0 $\times$ 0.3 mm^3, the silicon beam 1.6 $\times$ 0.4 $\times$ 0.03 mm^3, and the chip 0.4 $\times$ 0.3 $\times$ 0.03 mm^3. (Reprinted with permission from Löfdahl, L., Stemme, G., and Johansson, B. [1992] "Silicon Based Flow Sensors for Mean Velocity and Turbulence Measurements," *Exp. Fluids* **12**, pp. 391–393.)

$$U^2_{C,eff} = U^2(\sin^2\gamma + k^2\cos^2\gamma) \tag{6.40}$$

$$U^2 = \frac{U^2_A + U^2_B + U^2_C}{2 + k^2} \tag{6.41}$$

All the equations developed in this section for velocity measurements using one, two, and three hot-wires are used in the same way when making velocity measurements using MEMS-based sensors. Few variants of the thermal principle for velocity measurements are used in microsensors, and these are discussed in the next subsection.

6.4.2.6   MEMS Velocity Sensors

Since turbulence is a high-Reynolds-number phenomenon with small length- and time-scales, it is generally preferred that sensors used to measure instantaneous quantities would have small physical size. Although it is true that conventional hot-wires with sensor diameter in the micrometer range have been reported in the literature, these wires are usually very fragile, expensive, and hand-made. Conventional velocity sensors are difficult to use and not feasible to employ in large numbers. MEMS sensors are not afflicted with these drawbacks, and as is clear from earlier discussion are well suited to fabricate in dense arrays at moderate cost.

Löfdahl et al. (1992) presented the first MEMS-based sensor for the determination of mean velocities and turbulence intensities. A schematic drawing of this sensor is depicted in Figure 6.10. Different over-heat ratios and length-to-width ratios of the sensor were used: 4/3, 5/1, and 10/1. The principle of operation is very similar to that used in conventional hot-wires, but some deviations are important to note. Instead of using a bridge balance, the operation relies on a voltage difference which is formed between two temperature-sensitive *pn*–junction diodes. The "hot" diode monitors the temperature of the chip, which is electrically heated by an integrated resistor and cooled by the flow. The "cold" diode, which is positioned on the beam as shown in Figure 6.10, adjusts the set overheat ratio relative to the ambient air temperature. Thus, the power dissipated in the heated resistor is a measure of the instantaneous flow velocity.

Both single- and double-chip sensors were fabricated by Löfdahl et al. (1992). Figure 6.11 shows the perpendicular arrangement of two heated sensors for measuring two velocity components. The performance of this micro-velocity-sensor was tested in a two-dimensional, flat-plate boundary layer at a Reynolds number based on distance from leading edge, of 4.2 $\times$ 10^6. Figure 6.12 compares the mean velocity profile in a turbulent boundary layer as measured with a conventional and MEMS-based sensor. The microsensor measured mean velocities with the same accuracy as a corresponding conventional hot-wire. Moreover,

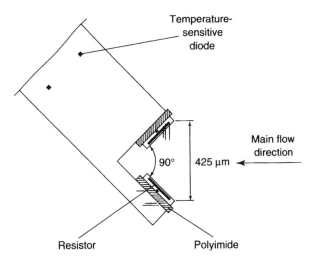

FIGURE 6.11 Double chip sensor for the determination of fluctuating velocity correlations. (Reprinted with permission from Löfdahl, L., Stemme, G., and Johansson, B. [1992] "Silicon Based Flow Sensors for Mean Velocity and Turbulence Measurements," *Exp. Fluids* 12, pp. 391–393.)

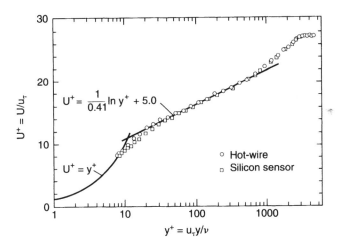

FIGURE 6.12 Typical mean velocity profile measured with a MEMS sensor and a conventional single hot-wire. Here, $U/u_\tau = f(u_\tau y/v)$, where u_τ is the friction velocity. (Reprinted with permission from Löfdahl, L., Stemme, G., and Johansson, B. [1992] "Silicon Based Flow Sensors for Mean Velocity and Turbulence Measurements," *Exp. Fluids* 12, pp. 391–393.)

it was also demonstrated that the microsensor had spatial and temporal resolutions that made it suited for turbulence measurements. Figure 6.13 shows the streamwise turbulence intensity and the Reynolds stress as measured by a conventional X-wire and by two silicon microsensors. The MEMS-based sensors operated with good resolution even when the temperature of the heated part was reduced considerably. A clear drawback of this micro-hot-film is the proximity of the heated part of the sensor to the surface of the chip, rendering the probe insensitive to changes in flow direction. This makes the silicon sensor unsuited for use in three-dimensional flows where the primary flow direction is not known a priori.

Jiang et al. (1994) presented a MEMS-based velocity sensor with the hot-wire free-standing in space without any nearby structures, so that cooling velocities can be determined in the same way as with a conventional hot-wire. Their sensor is shown in Figure 6.14. It has a polysilicon hot-element that is greatly reduced in size, typically about 0.5 μm thick, 1 μm wide, and 10–160 μm long. The dynamic performance

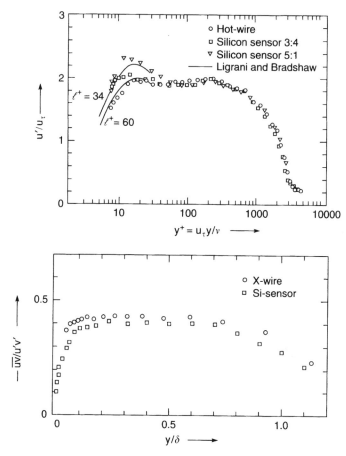

FIGURE 6.13   Streamwise velocity fluctuation (top plot), and turbulent shear stress (bottom plot) measured with MEMS sensors and conventional hot-wires. (Reprinted with permission from Löfdahl, L., Stemme, G., and Johansson, B. [1992] "Silicon Based Flow Sensors for Mean Velocity and Turbulence Measurements," *Exp. Fluids* **12**, pp. 391–393.)

and sensitivity of this sensor have been tested. A heating time of $2\,\mu s$ and a cooling time of $8\,\mu s$ for the $30\,\mu m$-long sensor in constant-current mode have been achieved. For constant-temperature operation, a time constant of $0.5\,\mu s$ for the $10\,\mu m$-long sensor has been recorded. The corresponding cut-off frequency is $1.4\,MHz$. The calibration curves of a $20\,\mu m$-long micro-hot-wire at two different angles are shown in Figure 6.15. The average sensitivity was found to be $20\,mV/m/s$ at an input current of $0.5\,mA$. No turbulence measurements have been reported using this sensor. It is noteworthy that the silicon hot-wires have a trapezoidal cross-section, which might cause severe uncontrolled errors in turbulence measurements.

A severe drawback of commercially available triple hot-wire probes is the large measuring volume, typically a sphere with a diameter of $3\,mm$. This is far too large to be acceptable for turbulence measurements at realistic Reynolds numbers. Ebefors et al. (1998) have presented a MEMS-based triple-hot-wire sensor, which is shown schematically in Figure 6.16. The X- and Y-hot-wires are located in the wafer plane while the third, Z-wire, is rotated out of the plane using a radial polyimide joint. The silicon chip size is $3.5 \times 3.0 \times 0.5\,mm^3$, and the three wires are each $500 \times 5 \times 2\,\mu m^3$. The sensor is based on the thermal anemometer principle, and the polyimide microjoint technique is used to create a well-controlled, out-of-plane rotation of the third wire. The time-constant of the hot-wire resistance change caused by heating without any flow was measured, and the cooling and heating time-constants were found to be 120 and $330\,\mu s$, respectively. By operating the sensor at constant temperature, these response times can be improved. Only some very preliminary velocity measurements have been conducted with this sensor in channel flow.

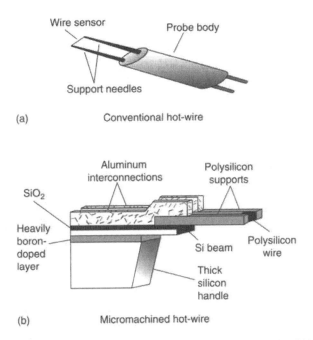

(a)   Conventional hot-wire

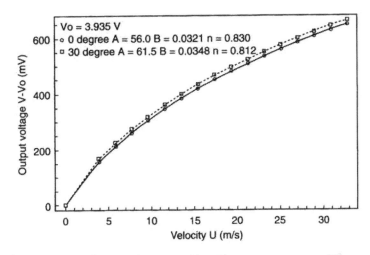

(b)   Micromachined hot-wire

FIGURE 6.14 Structure of hot-wire probes. (a) Conventional hot-wire; (b) Micromachined hot-wire. (Reprinted with permission from Jiang, F., Tai, Y.-C., Ho, C.-M., and Li, W. [1994] "A Micromachined Polysilicon Hot-Wire Anemometer," *Solid-State Sensor and Actuator Workshop*, Hilton Head, South Carolina.)

FIGURE 6.15 Calibration curves of a 20 μm-long, MEMS-based hot-wire sensor at two different angles. (Reprinted with permission from Jiang, F. Tai, Y.-C., Ho, C.-M., and Li, W. [1994] "A Micromachined Polysilicon Hot-Wire Anemometer," *Solid-State Sensor and Actuator Workshop*, Hilton Head, South Carolina.)

6.4.2.7 Outlook for Velocity Sensors

MEMS velocity sensors are superior to conventional hot-wires in the sense that they can be reduced in size to the order of microns. This is an important advantage, but for velocity measurements it is presumably the MEMS properties of accurate fabrication at low unit-cost that are even more relevant. For engineering purposes this means that velocity sensors, which may not necessarily be extremely small, can be used for controlling and optimizing system performance. Due to the low price per unit, numerous control points can be sensed, which is of utmost importance in controlling room ventilation, for instance.

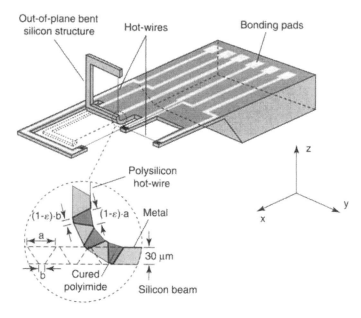

FIGURE 6.16   MEMS triple hot-wire. (Reprinted with permission from Ebefors, T., Kälvesten, E., and Stemme, G. [1998] "New Small Radius Joints Based on Thermal Shrinkage of Polyimide in V-Grooves for Robust Self-Assembly 3-D Microstructures," *J. Micromech., and Microeng.* 8, pp. 188–194.)

In turbulence research, small sensors are required for correct interpretation of the flow phenomena. However, here the most significant advantage of the MEMS sensors is that they can be spaced close together and fabricated in arrays for studies of coherent structures in the flow. In particular, further development of devices for measuring the instantaneous, three-dimensional velocity using for example the triple-hot-wire presented by Ebefors et al. (1998) would be of great interest. Further miniaturization of this sensor and positioning of each sensor in closely spaced arrays would offer the possibility of measuring flow events, both in the direction normal to the surface and in the spanwise direction. An integration of such sensor arrays with arrays of pressure transducers would be beneficial in instantaneous mapping of the flow field for studies of coherent structures. Combined with proper actuators, MEMS-based velocity sensors can be used for effective reactive control of difficult-to-tame turbulent shear flows.

6.4.3   Wall-Shear Stress Sensors

6.4.3.1   Background

From a scientific and engineering perspective, the wall-shear stress is an essential quantity to compute, measure, or infer in a wall-bounded turbulent flow. Time-averaged values of this quantity are indicative of the global state of the flow along a surface and can be used to determine body-averaged properties like the skin-friction drag. The time resolved part of the wall-shear stress is a measure of the unsteady structures in the flow which are responsible for the individual momentum transfer events and is an indicator of the coherent portion of the turbulence activities. Spatially distributed values of the instantaneous wall-shear stress can be used in a feedforward or feedback control-loop to effect beneficial changes in the boundary layer. A majority of wall-shear stress studies conducted rely on the premise that the mean velocity-gradient and the heat transfer rate near or at the wall are both proportional to the wall-shear stress. The former relationship was established already in the beginning of the 1920s by Stanton et al. (1920), and the latter a few years later by Leveque (1928).

During the last few decades, numerous experiments have been conducted where measurements of wall-shear stress have been the kernel. The success of those endeavors depends basically on the complexity of the

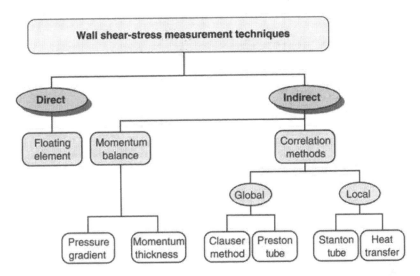

FIGURE 6.17 Classification of wall-shear stress measurements techniques. After Haritonidis (1989). [Haritonidis, J.H. (1989) "The Measurements of Wall-Shear Stress," in *Advances in Fluid Mechanics Measurements*, M. Gad-el-Hak, ed., pp. 229–261, Springer-Verlag, Berlin.]

flow, the geometry of the solid boundaries, and the limitations of the measuring device used. Some facts have been established like the wall-shear stress distribution along flat plates and simple bodies of revolution, but in general our knowledge of wall-shear stress (and in particular its fluctuating component) is limited. This is because it is both difficult and cumbersome to measure wall-shear stress. It is a parameter of small magnitude and some typical values should be kept in mind for the following discussion. A submarine cruising at 30 km/h has an estimated value of the shear stress of about 40 Pa, an aircraft flying at 420 km/h, 2 Pa, and a car moving at 100 km/h, 1 Pa. Such small forces per unit area put heavy demands on the resolution of the measuring devices used. Those estimates are approximate and were obtained by Munson et al. (1990) who treated the body as a collection of parts; for instance, the drag of an aircraft was approximated by adding the drag contributions produced by wings, fuselage, tail, etc.

Many different methods for the determination of wall-shear stress have been developed, and the required spatial and temporal resolutions are specific for each application and environment. In laminar flows the sensors must be capable of measuring the time-averaged shear stress, while in turbulent flows both the mean shear and its fluctuating component are of interest. An attempt to classify the techniques available for wall-shear stress measurements was made by Haritonidis (1989), who divided the technologies into direct or indirect methods, depending upon whether the method measures the wall-shear stress directly or infers it from other measured properties. Figure 6.17 shows a bird's eye view of Haritonidis' classification.

Since the number of different methods is enormous, it is out of the scope of the current chapter to review each technique, but a trend common to all measurements and methods can be highlighted. Since the mid-1950s, the evolution of the probes used has been directed towards utilizing smaller and smaller sensors in order to improve accuracy, flexibility, and resolution. There are several review papers describing shear-stress sensors and discussing in detail the merits and drawbacks of the methods used in a vast variety of flow situations. To cite a few, Winter (1977) gives a comprehensive review of available conventional methods and a good discussion of measurements in turbulent flows. Haritonidis (1989) summarizes conventional methods and introduces the first micromachined wall-shear stress sensor. Hakkinen (1991) lists the merits and drawbacks of conventional techniques. More recently, Hanratty and Campbell (1996) discuss the relevant experimental issues associated with the use of various wall-shear stress sensors.

A major problem faced by wall-shear stress measurements is that conventional fabrication of sensors and balances allows only a certain degree of miniaturization since these devices are more or less hand-made.

TABLE 6.1 Operation and Detection Principle of Silicon Micromachined Wall-Shear-Stress Sensors in Chronological Order

Author	Operation	Detection Principle	Active Sensor Area (mm²)
Oudheusden & Huijsing (1988)	Thermal	Thermopile	4×3
Schmidt et al (1988)	Floating element	Capacitive	0.5×0.5
Ng et al. (1991)	Floating element	Piezoresistive	0.12×0.14
Liu et al. (1995)	Thermal	Anemometer	0.2×0.2
Pan et al. (1995)	Floating element	Capacitive	$\approx 0.1 \times 0.1$
Padmanabhan (1997)	Floating element	Optical	0.12×0.12
Kälvesten (1994)	Thermal	Anemometer	$\approx 0.1 \times 0.1$

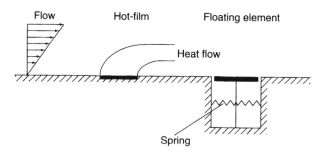

FIGURE 6.18 Wall-shear stress sensors of the hot-film and floating-element type.

Better spatial and temporal resolutions than what conventional probes can provide are needed in typical turbulence studies and reactive flow control. MEMS has the potential of circumventing the conventional sensor limitations because microfabrication offers a high degree of miniaturization and associated increased resolution. This is in addition to the fact that the sensors are not hand-made, implying that each unit is fabricated to extremely low tolerance at reasonable cost. Table 6.1 lists the operation and detection principles as well as the sensor area of some recently developed MEMS-based shear-stress probes.

A glance at the classification scheme of Figure 6.17 shows that the floating-element principle and the heat-transfer method (or more correctly thermal-element method) are both well suited for MEMS fabrication. Figure 6.18 shows a schematic of the principle of operation of the two methods. The thermal-element method relies on measuring the amount of energy necessary for keeping a wall-mounted, electrically heated resistor at constant temperature, despite a time-dependent, convective heat transfer to the flow. A high freestream velocity yields steep velocity and temperature gradients at the heated surface; this results in a good cooling which requires more energy, while the opposite is true for a low freestream velocity. In the floating-element principle, a flush-mounted wall-element — moveable in the plane of the wall-shear stress — is displaced laterally by the tangential viscous force. This movement can be measured using either resistive, capacitive, or optical detection principles.

In the following subsubsections, we describe the principles of the thermal sensor and the floating element, since both strategies are particularly suited for measurements of fluctuating wall-shear stress as well as for MEMS fabrication. In the description of the thermal-sensor method, we briefly recall the classical equations and based on these we highlight possible uncertainties and sources of errors in measuring instantaneous wall-shear stress. For the floating-element method, we discuss the general principle and pitfalls associated with steady and time-dependent measurements. For both principles we review current MEMS-based sensors. Since calibration is a crucial part of wall-shear stress measurements, particularly in the determination of fluctuating quantities, we summarize some common calibration methods and discuss possibilities for dynamic calibration. Finally, we provide an outlook for the use of MEMS-based wall-shear stress sensors for future turbulence studies and flow control.

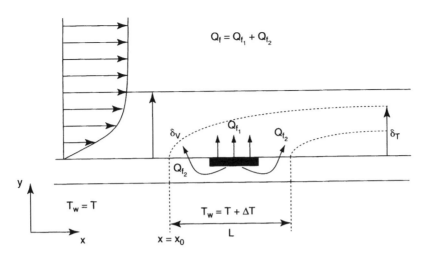

FIGURE 6.19 An illustration of a surface mounted hot-film.

6.4.3.2 Thermal-Sensor Principle

The thermal or hot-film sensor benefits from the fact that the heat transfer from a sufficiently small heated surface depends only on the flow characteristics in the viscous region of the boundary layer. A schematic drawing of a thermal sensor is depicted in Figure 6.19. The hot-film gauge consists of a thin metallic film positioned into a substrate. Usually the gauge forms one part of a Wheatstone bridge, and an electric current is passed through the film in order to maintain it at a constant temperature as heat is continuously being transferred from the film to the moving fluid. The ohmic heating in the device, q_W, is transferred both to the fluid and to the surrounding substrate. This can be expressed as:

$$q_W = q_T + q_C \tag{6.42}$$

where q_T represents the heat transferred to the fluid directly from the heated surface and indirectly through the heated portion of the substrate, and q_C represents the heat lost to the substrate. If the flow is steady and laminar and if the streamwise pressure gradient can be neglected, the resulting heat transfer rate to the fluid q_T has been shown to be related to the wall-shear stress according to the classical relation:

$$\tau_w \propto q_T^3 \tag{6.43}$$

This expression has been derived by many researchers [e.g. Ludwieg, 1950; Liepmann and Skinner, 1954; Bellhouse and Schultz, 1966].[1] However, this simple relation breaks down in unsteady and turbulent flows as has been pointed out by Bellhouse and Schultz (1966). In a turbulent environment, most assumptions made in the derivation are violated and the heat conduction to the substrate changes instantaneously. In spite of this, we will use an extended version of Equation 6.43 to illustrate the assumptions made and to pinpoint possible pitfalls in using hot-film gauges for fluctuating wall-shear-stress measurements. The following argument follows a path outlined by Bellhouse and Schultz (1966) and later by Menendez and Ramaprian (1985), who derived the extended version for boundary-layer flows subjected to a periodic freestream velocity of the form:

$$U_a(t) = U_o(1 + \varepsilon \sin \omega t) \tag{6.44}$$

where ω is the angular frequency of the oscillation, U_o is the mean freestream velocity, and ε is the relative amplitude of the oscillation. The derivation below is strongly abbreviated, and the reader is referred to Menendez and Ramaprian (1985) for a more elaborate version. The problem to be considered is shown in Figure 6.19. A thermal boundary layer develops within a turbulent boundary layer over a heated film

---

[1]See also the asymptotic solution to the conjugate heat transfer problem derived by Stein et al. (2002).

having an effective streamwise length L. The thermal boundary layer is produced by a sudden jump in the surface temperature, from a constant value equal to the ambient temperature T_a to the higher constant value T_w at the location $x = x_o$. It is assumed that the thermal boundary layer is totally embedded in the viscous region, so effects of turbulent diffusion can be neglected. A relation between the local wall-shear stress τ_w and the heat transfer rate from the wall to the fluid q_T is derivable from the first law of thermodynamics. For a two-dimensional flow, neglecting dissipation and compressibility effects and applying the usual boundary-layer approximation, the energy equation reads:

$$\frac{\partial T}{\partial t} + U\frac{\partial T}{\partial x} + V\frac{\partial T}{\partial y} = \frac{\kappa_f}{\rho c_p}\frac{\partial^2 T}{\partial y^2} \tag{6.45}$$

where $T(x, y, t)$ is the desired temperature distribution, t is time, U and V are the velocity components along the coordinates x and y, respectively, and ρ, κ_f, and c_p are respectively the fluid density, thermal conductivity and specific heat. Using the equation of continuity and integrating Equation 6.45 across the thermal boundary layer up to a point $y = y_e$ beyond the edge of the thermal boundary layer, we get:

$$\frac{\partial}{\partial t}\int_0^{y_e}(T - T_0)dy + \frac{\partial}{\partial x}\int_0^{y_e}U(T - T_e)dy = -\frac{1}{\rho c_p}q_w \tag{6.46}$$

where T_e is the temperature at the edge of the thermal boundary layer. It is now assumed that within the viscous region where the thermal boundary layer resides, inertial effects can be neglected. The velocity profile can be approximated by:

$$U = \frac{\tau_w}{\mu}y + \frac{1}{2\mu}\left(\frac{\partial p}{\partial x}\right)y^2 \tag{6.47}$$

where p is the pressure, and μ is the dynamic viscosity of the fluid. It is further assumed that the temperature distribution in the thermal boundary layer is self-similar at any instant of time, and that the thermal boundary layer thickness δ_T can be expressed as follows:

$$\frac{T - T_e}{T_w - T_e} = f(\xi, t) \tag{6.48}$$

$$\frac{\delta_T(x, t)}{\delta_T(x_o + L, t)} = \left[\frac{x - x_o}{L}\right]^n \tag{6.49}$$

where $\xi \equiv y/\delta_T$ is the similarity coordinate, and n is an unknown exponent which can in general depend on time. After some algebra and assuming n is independent of time, an expression for the local wall-shear stress at $x = x_o + L$ is obtained as a function of the heat transfer rate from the wall to the fluid:

$$\tau_w = (A_1 q_w^3) + A_2\frac{\cos\omega t}{q_w} + A_3\frac{\partial q_w}{\partial t} \tag{6.50}$$

where:

$$A_1 \equiv \frac{\mu L}{(1 - n)a\rho c_p\kappa^2(\Delta T_o)^3} \tag{6.51}$$

$$A_2 \equiv \frac{b\kappa(\Delta T_o)a\rho\varepsilon\omega U_o}{2a} \tag{6.52}$$

$$A_3 \equiv \frac{\mu Lc}{(1 + n)a\kappa(\Delta T_o)} \tag{6.53}$$

In these expressions the "shape parameters" a, b, and c are basically functions of the streamwise coordinate x. According to Menendez and Ramaprian (1985), Equation 6.50 can be regarded as the basic relationship between the instantaneous value of the wall-shear stress and the wall heat flux. By taking ensemble average of Equation 6.50 over a large number of cycles and assuming that the turbulent fluctuations are small, a linearized expression for the periodic wall-shear stress is obtained.

To summarize, the classical expression, Equation 6.43, gives a relation between the wall-shear stress and the heat transfer from the wall. This expression assumes steady, laminar, zero-pressure-gradient flow, and is not valid in a turbulent environment. Menendez and Ramaprian (1985) have derived an extended version of Equation 6.43 valid for a periodically fluctuating freestream velocity, Equation 6.50. However, the latter relation contains some assumptions that are questionable for turbulent flows. For instance, in the thermal boundary layer it is assumed that the temperature distribution is self-similar and that the local thickness varies linearly. These assumptions are relevant for a streamwise velocity oscillation and a weak fluctuation, but certainly not in a turbulent flow, which is strongly unstable in all directions.

6.4.3.3 Calibration

Several methods and formulas are in use for calibrating hot-film shear probes operated in the constant-temperature mode and the choice of method depends on flow conditions and sensor used. In this subsubsection, two static calibration methods are discussed. Both are based on the theoretical analysis leading to the relation between rate of heat transfer and wall-shear stress, as discussed in the last subsubsection. The challenge is of course to be able to use the shear-stress sensor in a turbulent environment.

If a laminar flow facility is used to calibrate the wall-shear-stress sensor, then Equation 6.43 can be re-written more conveniently in time-averaged form:

$$\overline{\tau_w}^{1/3} = A\overline{e^2} + B \tag{6.54}$$

where $\overline{\tau_w}$ is the desired mean wall-shear stress, $\overline{e^2}$ is the square of the mean output voltage, and A and B are calibration constants. The term B represents the heat loss to the substrate in a quiescent surrounding, and this procedure is similar to a conventional calibration of a hot-wire [King, 1916]. However, a laminar flow is often difficult to realize in the desired range of turbulence wall-shear stress: it is more practical to calibrate without moving the sensor between calibration site and measurement site. In that case, the calibration is made in a high-turbulence environment and the high-order moments of the voltage must also be considered.

Ramaprian and Tu (1983) proposed an improved calibration method, and the instantaneous version of Equation 6.54, can be re-written and time averaged to give:

$$\overline{\tau_w} = \overline{(Ae^2 + B)^3} = A^3\overline{e^6} + 3A^2B\overline{e^4} + 3AB^2\overline{e^2} + B^3 \tag{6.55}$$

This can be rewritten as:

$$\overline{\tau_w} = C_6\overline{e^6} + C_4\overline{e^4} + C_2\overline{e^2} + C_0 \tag{6.56}$$

where the C's are the new calibration constants. The high-order moments of the voltage can easily be calculated with a computer, but care must be taken that they are fully converged. The relation between e and τ_w can also be represented by a full polynomial function, so if the frequency response of the sensor is sufficiently flat, it is possible to write:

$$\overline{\tau_w} = C_0 + C_1e + C_2\overline{e^2} + \cdots + C_M\overline{e^N} \tag{6.57}$$

where M is the number of calibration points and N is the order of the polynomial above. A system of linear equations is obtained where the calibration coefficients can be computed by a numerical least-square method. For example, the mean wall-shear stress on the left-hand side of Equation 6.57 can be measured with a Preston tube using the method of Patel (1965). The second calibration technique described here is called "stochastic" calibration by Breuer (1995), who also demonstrated that the validity of the calibration polynomial may extend well beyond the original calibration range, although this requires careful determination of the higher-order statistics as well as a thorough understanding of the sensor response function.

The experiments of Bremhorst and Gilmore (1976) showed that the static and dynamic calibration coefficients for hot-wires agree to within a standard error of 3% for the velocity range of 3–32 m/s. Thus, they recommend the continued use of static calibration for dynamic measurements. As pointed out in previous section, this is not true for a wall-mounted hot-film. One solution to this problem may be to calibrate the hot-film in pulsatile laminar flow with a periodic freestream velocity $U_a(t)$, and make use of the Menendez and Ramaprian's formula, Equation 6.50:

$$\overline{\tau_w} = (A\overline{e^2} + B)^3 + \frac{c_1}{A\overline{e^2} + B}\frac{dU_a}{dt} + c_2 A\frac{d\overline{e^2}}{dt}$$

to obtain the additional calibration constants c_1 and c_2. This step is necessary in order to characterize the gauge dynamic response at relatively high frequencies. The constants A and B are obtained from a steady-state calibration.

A difficulty in performing a dynamic calibration is to generate a known sinusoidal wall-shear stress input. Bellhouse and Rasmussen (1968) and Bellhouse and Schultz (1966) achieved this in two different ways. One method is to mount the hot-film on a plate which can be oscillated at various known frequencies and amplitudes. The main drawback to this arrangement is the limited amplitudes and frequencies that can be achieved when attempting to vibrate a relatively heavy structure. An alternative strategy is to generate the shear stress variations by superimposing a monochromatic sound field of different frequencies on a steady, laminar flow field. A hot-wire close to the wall can be used as a reference.

6.4.3.4 Spatial Resolution

Equation 6.50 is an extended version of Equation 6.43, and by assuming a steady flow and zero-pressure-gradient in the streamwise direction it reduces to the classical formula. The latter expression can be used to estimate the effective streamwise length of a hot-film:

$$\tau_w = -A_1 q_W^3 \tag{6.58}$$

By assuming a stepwise temperature variation and introducing the average heat flux $\overline{q_W}$ over the heated area which is assumed to have a streamwise length L, the desired relation reads:

$$\overline{q_W} = 0.807\kappa_f\Delta T_w Pr^{1/3}\left(\frac{\tau_w}{Lv\mu}\right) \tag{6.59}$$

It is convenient to re-write the above expression into a non-dimensional form, and in doing so we keep L explicit:

$$\overline{Nu} \equiv \frac{h_c L}{\kappa_f} = 0.807(PrL^{+2})^{1/3} \tag{6.60}$$

where $\overline{Nu}$ is the Nusselt number averaged over the heated area, L^+ is the streamwise length of the heated area normalized with the viscous length-scale v/u_τ, and h_c is the convective heat transfer coefficient. This equation has been derived for flows with pressure gradient by Brown (1967). The dimensionless sensor length L^+ is a crucial parameter when examining the assumptions made.

The lower limit on L^- is imposed by the boundary-layer approximation since there is an abrupt change of temperature close to the leading and trailing edge zones of the heated strip where the neglected diffusive terms in Equation 6.45 become significant. Tardu et al. (1991) have conducted a numerical simulation of the heat transfer from a hot-film, and found a peak of the local heat transfer at the leading and trailing edges. They conclude that if the hot-film is too narrow, the heat transfer would be completely dominated by these edge effects. Ling (1963) has studied the same problem in a numerical investigation, and concludes that the diffusion in the streamwise direction can be neglected if the Péclet number is larger than 5000. (The Péclet number here is defined as the ratio of heat transported by convection and by molecular diffusion, $Pe = PrL^{+2}$, [Brodkey, 1967]). Pedley (1972) concludes that, provided $0.5\,Pe^{-0.5} < x/L < (1 - 0.7Pe^{-0.5})$, there exists a central part of the hot-film where the boundary-layer solution predicts the heat transfer within 5%. Figure 6.20 shows the relation proposed by Pedley (1972); it can be seen that the heat transfer is correctly described over a large part of the hot-film area, but as the Péclet number decreases, the influence

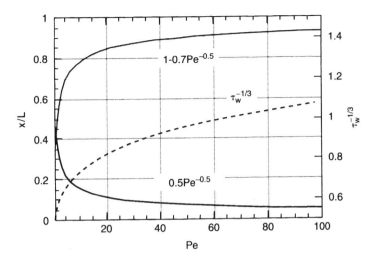

FIGURE 6.20 Pedley's (1972) relation showing the region of validity of the boundary-layer solution as a function of Péclet number. In the area in between the two continuous curves, the boundary-layer approximation predicts the heat transfer accurately to within 5% of the correct value.

from the diffusive terms must be considered. For Péclet numbers larger than 40, the heat transfer is correctly described by more than 80% of the heated area.

The upper limit of L^+ is crucial since there the hot-film thermal boundary layer may not be entirely submerged in the viscous sublayer. Equation 6.58 has been included in Figure 6.20, and it can be seen that Péclet numbers larger than 40 correspond approximately to $\overline{\tau_w}^{1/3} > 0.95$. A relatively simple calculation of the upper limit of L^+ can then be made by assuming that the viscous sublayer is about five viscous units, which yields an upper limit of L^+ which for air can be estimated to be approximately 47.

It can be concluded that the streamwise extent of the hot-film cannot be too small, otherwise the boundary-layer approximations made are not applicable. On the other hand, a sensor that is too large will cause the thermal boundary layer to grow beyond the viscous region. Additionally, the spatial resolution will be adversely affected if the sensor is too large, since the smallest eddies imposed by the flow structures above the wall will then be integrated along the sensor length.

6.4.3.5 Temporal Resolution

The temporal resolution of the thermal probe is affected by the different time-constants of the hot-film and the substrate. The hot-film usually has a much shorter time-constant than the substrate. The higher the percentage of the total heat that leaks into the substrate, the lower is the sensitivity of the device to shear-stress fluctuations; this changes the sensor characteristics sufficiently to invalidate the static calibration. An example of this phenomenon is given by Haritonidis (1989), who showed that a hot-film sensor in a fluctuating wall-shear stress environment will respond quickly to the instantaneous shear stress, while the substrate will react slowly due to its much larger thermal inertia. Haritonidis (1989) also showed that the ratio of the fluctuating sensitivity, S_f, to the average sensitivity, S_a, can be related to the ratio of the effective lengths under dynamic and static conditions:

$$\frac{S_f}{S_a} = \left(\frac{L_f}{L_a}\right)^{2/3} < 1 \tag{6.61}$$

where L_a is the average effective length during static calibration and L_f is the effective length during dynamic calibration. Due to this, the hot-film becomes less sensitive to shear-stress fluctuations at higher frequencies and the static calibration in a laminar flow will not give a correct result. These length-scales can be considerably larger than the probe true extent. For example, Brown (1967) reported that the effective length-scale from a static calibration was about twice the physical length.

Both the substrate material and the amount of heat that is lost to the substrate are crucial when determining the temporal or frequency response of the hot-film sensor. At low frequencies, the thermal waves through the substrate and into the fluid are quasi-static, which means that the fluctuating sensitivity of the hot-film is determined by the first derivative of the static calibration curve. In this range, on the order of a couple of cycles per second, the heat transfer through the substrate responds without time lag to wall-shear stress fluctuations. Basically, this frequency range does not cause any major problems.

For the high-frequency range at the other end of the spectrum, it is possible to estimate the substrate role by considering the propagation of heat waves through a semi-infinite solid slab subjected to periodic temperature fluctuations at one end. This has been reported by Blackwelder (1981), who compared the wavelength of the heat wave to the hot-film length. He showed that the amplitude of the thermal wave would attenuate to a fraction of a percent over a distance equal to its wavelength. A relevant quantity to consider in this context turned out to be the ratio of the wavelength to the length of the substrate since this is an indication of the extent to which the substrate will partly absorb heat from the heated surface and partly return it to the flow. Haritonidis (1989) computed this ratio for a number of fluids and films and concluded that at high frequencies the substrate would not participate in the heat transfer process. However, this conclusion should be viewed with some caution because the frequencies studied were the highest that could be expected in a wall flow.

The most difficult problem occurs for frequencies in the intermediate range, resulting in a clear substrate influence and an associated deviation from the static calibration. Hanratty and Campbell (1996) showed that damping by the thermal boundary layer for pipe flow turbulence is important when:

$$[L^{+2}(5.65 \times 10^{-2})^3 Pr]^{1/2} \leq 1 \tag{6.62}$$

For $Pr = 0.72$, this requires the dimensionless length in the streamwise direction, L^+, to be less than 90. For turbulence applications, this is most disturbing since it is in this frequency range where the most energetic eddies are situated. The primary conclusion from this discussion is that all wall-shear stress measurements in turbulent flows require dynamical calibration of the hot-film sensor.

6.4.3.6 MEMS Thermal Sensors

Kälvesten (1996) and Kälvesten et al. (1996b) have developed a MEMS-based, flush-mounted wall-shear stress sensor that relies on the same principle of operation as the micro-velocity sensor presented by Löfdahl et al. (1992). The shear sensor is based on the cooling of a thermally insulated, electrically heated part of a chip. As depicted in Figure 6.21, the heated portion of the chip is relatively small, $300 \times 60 \times 30\,\mu m^3$, and is thermally insulated by polyimide-filled, KOH-etched trenches. The rectangular top area, with a side-length to side-length ratio of 5:1, yields a directional sensitivity for the measurements of the two perpendicular in-plane components of the fluctuating wall-shear stress. Due to the etch properties of KOH, the $30\,\mu m$-deep, thermally-insulating trenches have sloped walls with a bottom and top width of about $30\,\mu m$. The sensitive part of the chip is electrically heated by a polysilicon piezoresistor and its temperature is measured by an integrated diode. For the ambient temperature, a reference diode is integrated on the substrate chip, far away from the heated portion of the chip. (Note that for backup, two hot diodes and two cold diodes are fabricated on the same chip.)

Kälvesten (1996) performed a static wall-shear stress calibration in the boundary layer of a flat plate. A Pitot tube and a Clauser plot were used to determine the time-averaged wall-shear stress. The power consumed to maintain the hot part of the sensor at a constant temperature was measured and Figure 6.22 shows the data for two different probe orientations. For a step-wise increase of electrical power, the response time was about 6 ms, which is double the calculated value. This response was considerably shortened to $25\,\mu s$ when the sensor was operated in a constant-temperature mode using feedback electronics. Table 6.2 lists some calculated and measured characteristics of the Kälvesten MEMS-based wall-shear stress sensor.

Jiang et al. (1994; 1996) have developed an array of wall-shear stress sensors based on the thermal principle. The primary objective of their experiment was to map and control the low-speed streaks in the wall region of a turbulent channel flow. To properly capture the streaks, each sensor was made smaller than a typical streak width. For a Reynolds number based on the channel half-width and centerline velocity of 10^4, the streaks are estimated to be about 1 mm in width, so each sensor was designed to have a length less than $300\,\mu m$.

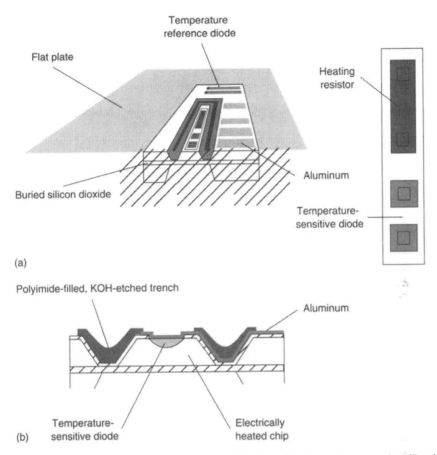

FIGURE 6.21 Flush-mounted wall-shear stress sensor. (a) Top-view; (b) schematic cross-section. (Reprinted with permission from Kälvesten, E. [1996] "Pressure and Wall Shear Stress Sensors for Turbulence Measurements," Royal Institute of Technology, TRITA-ILA-9601, Stockholm Sweden.)

Figure 6.23 shows a schematic of one of Jiang et al.'s sensors. It consists of a diaphragm with a thickness of 1.2 μm and a side-length of 200 μm. The polysilicon resistor wire is located on the diaphragm and is 3 μm wide and 150 μm long. Below the diaphragm there is a 2 μm-deep vacuum cavity so that the device will have a minimal heat conduction loss to the substrate. When the wire is heated electrically, heat is transferred to the flow by heat convection resulting in an electrically measurable power change which is a function of the wall-shear stress. Figure 6.24 shows a photograph of a portion of the 2.85 × 1.00 cm² streak-imaging chip containing just one probe. The sensors were calibrated in a fully-developed channel flow with known average wall-shear stress values. Figure 6.25 depicts the calibration results for 10 sensors in a row. The output of these sensors is sensitive to the fluid temperature, and the measured data must be compensated for this effect. Measurements of the fluctuating wall-shear stress using the sensor of Jiang et al. (1997) have been reported by Österlund (1999) and Lindgren et al. (2000).

6.4.3.7 Floating-Element Sensors

The floating-element technique is a direct method for sensing skin friction, which means a direct measurement of the tangential force exerted by the fluid on a specific portion of the wall. The advantage of this method is that the wall-shear stress is determined without having to make any assumptions about either the flow field above the device or the transfer function between the wall-shear stress and the measured quantity. The sensing wall-element is connected to a balance which determines the magnitude of the applied force. Basically two arrangements are distinguished to accomplish this: displacement balance, which is

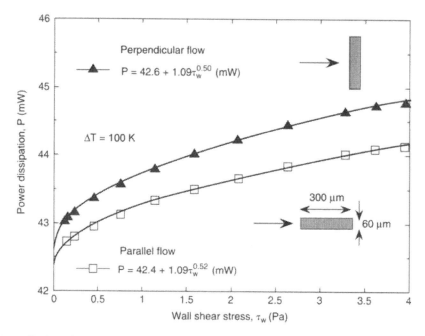

FIGURE 6.22   Total steady-state power dissipation calibration as a function of the wall-shear stress for two different probe orientations. (Reprinted with permission from Kälvesten, E. [1996] "Pressure and Wall Shear Stress Sensors for Turbulence Measurements," Royal Institute of Technology, TRITA-ILA-9601, Stockholm Sweden.)

TABLE 6.2   Some Calculated and Measured Characteristics of the MEMS-based Wall-Shear Stress Sensors Fabricated by Kälvesten et al. (1994)

	Theory		Measurements	
Heated chip top-area, $A = w \times \ell$ (μm^2)	300×60	1200×600	300×60	1200×600
Heated chip thickness, h (μm)	30	30	30	30
Thermal conduction conductance, G_c ($\mu W/K$)	372	510	426	532
Thermal convection conductance, G_f ($\mu W/K$)				
Perpendicular Configuration at 50 m/s	6,7	207	21.0	231
Parallel Configuration at 50 m/s	5.6	192	17.3	160
Thermal capacity, ($\mu J/K$)	1.2	37	—	—
Thermal time-constant at zero flow, (ms)	3	72	6	55
Time response (electronic feedback), (μs)	—	—	25	—

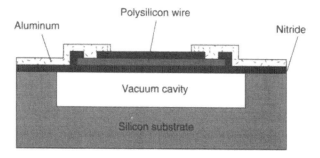

FIGURE 6.23   Cross-sectional structure of a single shear-stress sensor. (Reprinted with permission from Jiang, F., Tai, Y-C, Walsh, K., Tsao, T., Lee, G.B., and Ho, C.-H. [1997] "A Flexible MEMS technology and Its First Application to Shear Stress Skin," in *Proc. IEEE MEMS Workshop (MEMS '97)*, pp. 465–470.)

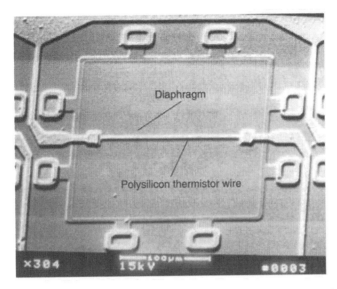

FIGURE 6.24   SEM photo of a single wall-shear stress sensor. (Reprinted with permission from Jiang, F., Tai, Y-C, Walsh, K., Tsao, T., Lee, G.B., and Ho, C.-H. [1997] "A Flexible MEMS technology and Its First Application to Shear Stress Skin," in *Proc. IEEE MEMS Workshop (MEMS '97)*, pp. 465–470.)

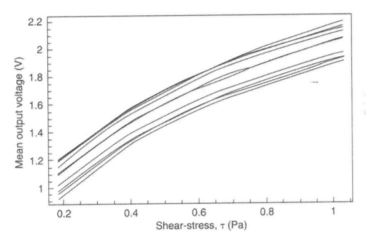

FIGURE 6.25   Calibration curves of 10 different sensors in an array. (Reprinted with permission from Jiang, F., Tai, Y-C, Walsh, K., Tsao, T., Lee, G.B., and Ho, C.-H. [1997] "A Flexible MEMS technology and Its First Application to Shear Stress Skin," in *Proc. IEEE MEMS Workshop (MEMS '97)*, pp. 465–470.)

a direct measurement of the distance the wall-element is moved by the wall-shear stress; or null balance which is the measurement of the force required to maintain the wall-element at its original position when actuated on by the wall-shear stress.

The principle of a floating-element balance is shown in Figure 6.26. In spite of the fact that the force measurement is simple, the floating-element principle is afflicted with some severe drawbacks which strongly limit its use as has been summarized by Winter (1977). It is difficult to choose the relevant size of the wall-element in particular when measuring small forces and in turbulence applications. Misalignments and the gaps around the element, especially when measuring small forces, are constant sources of uncertainty and error. Effects of pressure gradients, heat transfer, and suction or blowing cause large uncertainties in the measurements as well. If the measurements are conducted in a moving frame of reference, effects of

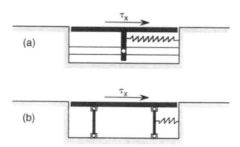

FIGURE 6.26   Schematic drawing of a floating-element device, with (a) pivoted support; and (b) parallel linkage support. (Reprinted with permission from Haritonidis, J.H. [1989] "The Measurements of Wall-Shear Stress," in *Advances in Fluid Mechanics Measurements*, M. Gad-el-Hak, ed., pp. 229–261, Springer-Verlag, Berlin.)

gravity, acceleration, and large transients can also severely influence the results. Haritonidis (1989) discussed the mounting of floating-element balances and errors associated with the gaps and misalignments. In addition, floating-element balances fabricated with conventional techniques have in general poor frequency response and are not suited for measurements of fluctuating wall-shear stress. To summarize, the idea of direct force measurements by a floating-element balance is excellent in principle, but all the drawbacks taken together make them difficult and cumbersome to work with in practice. It was not until the introduction of microfabrication in the late 1980s that floating-element force sensors achieved a revitalized interest in particular for turbulence studies and reactive flow control.

6.4.3.8   MEMS Floating-Element Sensors

Schmidt et al. (1988) were the first to present a MEMS-based floating-element balance for operation in low-speed turbulent boundary layers. A schematic of their sensor is shown in Figure 6.27. A differential capacitive sensing scheme was used to detect the floating element movements. The area of the floating element used was $500 \times 500\,\mu m^2$, and it was suspended by four tethers, which acted both as supports and restoring springs. The floating element had a thickness of $30\,\mu m$, and was suspended $3\,\mu m$ above the silicon substrate on which it was fabricated. The gap on either side of the tethers and between the element and surrounding surface was $10\,\mu m$, while the element top was flush with the surrounding surface within $1\,\mu m$. The element and its tethers were made of polyimide, and the sensor was designed to have a bandwidth of 20 kHz. A static calibration of this force gauge indicated linear characteristics, and the sensor was able to measure a shear stress as low as 1 Pa. However, the sensor showed sensitivity to electromagnetic interference due to the high-impedance capacitance used, and drift problems attributed to water-vapor absorption by the polyimide were observed. No measurements of fluctuating wall-shear stress were made because the signal amplitude available from the device itself was too low in spite of the fact that the first-stage amplification was fabricated directly on the chip.

Since the introduction of Schmidt et al.'s sensor, other floating-element sensors based on transduction, capacitive, and piezoresistive principles have been developed [Ng et al., 1991; Goldberg et al., 1994; Pan et al., 1995]. Ng et al.'s sensor was small and had a floating element with a size of $120 \times 40\,\mu m^2$. It operated on a transduction scheme and was basically designed for polymer-extrusion applications so it operated in the shear stress range of 1–100 kPa. Goldberg et al.'s sensor had a larger floating element size, $500 \times500\,\mu m^2$. It had the same application and the same principle of operation as the Ng et al.'s balance. Neither of these two sensors is of interest in turbulence and flow control applications since their sensitivity is far too low. The capacitive floating-element sensor of Pan et al. (1995) is a force-rebalance device designed for wind-tunnel measurements, and is fabricated using a surface micromachining process. Unfortunately, this particular fabrication technique can lead to non-planar floating-element structures. The sensor has only been tested in laminar flow and no dynamic response of this device has been reported.

Recently, Padmanabhan (1997) has presented a floating-element wall-shear stress sensor based on optical detection of instantaneous element displacement. The probe is designed specifically for turbulent boundary

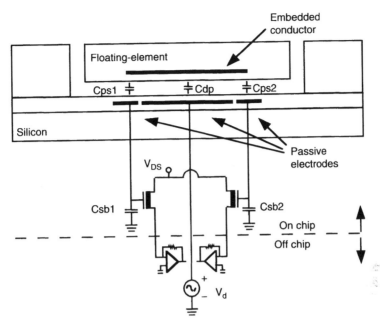

FIGURE 6.27 Schematic of the MEMS floating-element balance. of Schmidt et al. (1988).(Reprinted with permission from Schmidt, M., Howe, R., Senturia, S., and Haritonidis, J. [1988] "Design and Calibration of a Microfabricated Floating-Element Shear-Stress Sensor," *IEEE Trans. on Electron Devices* 35, pp. 750–757.)

layer research and has a measured resolution of 0.003 Pa and a dynamic response of 10 kHz. A schematic illustrating the sensing principle is shown in Figure 6.28. The sensor is comprised of a floating element which is suspended by four support tethers. The element moves in the plane of the chip under the action of wall-shear stress. Two photodiodes are placed symmetrically underneath the floating element at the leading and trailing edges, and a displacement of the element causes a "shuttering" of the photodiodes. Under uniform illumination from above, the differential current from the photodiodes is directly proportional to the magnitude and sign of shear stress. Analytical expressions were used to predict the static and dynamic response of the sensor; based on the analysis, two different floating-element sizes were fabricated, 120 × 120 × 7 μm³ and 500 × 500 × 7 μm³. The device has been calibrated statically in a laminar flow over a stress range of four orders of magnitude, 0.003–10 Pa. The gauge response was linear over the entire range of wall-shear stress. The sensor also showed good repeatability and minimal drift.

A unique feature of the shear sensor just described is that its dynamic response has been experimentally determined to 10 kHz. Padmanabhan (1997) described how oscillating wall-shear stress of a known magnitude and frequency can be generated using an acoustic plane-wave tube. A schematic of the calibration experiment is shown in Figure 6.29. The set-up is comprised of an acrylic tube with a speaker-compression driver at one end and a wedge-shaped termination at the other end. The latter is designed to minimize reflections of sound waves from the tube end and thereby set up a purely travelling wave in the tube. A signal generator and an amplifier drive the speaker to radiate sound at different intensities and frequencies. At some distance downstream, the waves become plane; at this location a condenser microphone (which measures the fluctuating pressure) and the shear-stress sensor are mounted. The flow field inside the plane-wave tube is very similar to a classical fluid dynamics problem — the Stokes second problem. The only difference is that instead of an oscillating wall with a semi-infinite stationary fluid, the plane-wave tube has a stationary wall and oscillating fluid particles far away from the wall. Solutions to the Stokes problem can be found in many textbooks [Brodkey, 1967; Sherman, 1990; and White, 1991]. Padmanabhan (1997) converted the boundary conditions and derived corresponding analytical expression for the plane-wave tube. The analytical solution of the fluctuating wall-shear stress was compared to the measured output of the shear-stress sensor as a function of frequency and a transfer function of

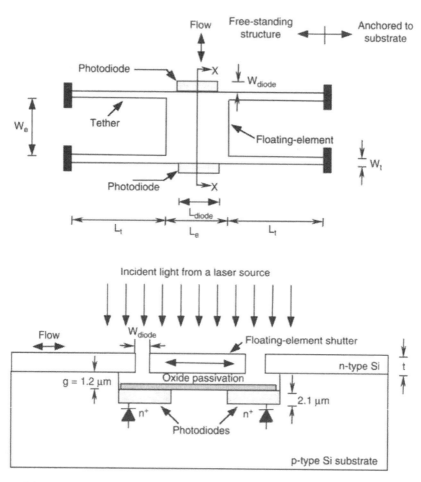

FIGURE 6.28   Schematic of the floating-element balance of Padmanabhan (1997). (Reprinted with permission from Padmanabhan, A. [1997] Silicon Micromachined Sensors and Sensor Arrays for Shear-Stress Measurements in Aerodynamic Flows, Ph.D. Thesis, Department of Aeronautics and Astronautics, Massachusetts Institute of Technology, Cambridge, Massachusetts.)

the sensor was determined. As expected, the measured shear stress showed a square-root dependence on frequency.

6.4.3.9   Outlook for Shear-Stress Sensors

Since coherent structures play a significant role in the dynamics of turbulent shear flows, the ability to control these structures will have important technological benefits such as drag reduction, transition control, mixing enhancement, and separation delay. In particular, the instantaneous wall-shear stress is of interest for reactive control of wall-bounded flows to accomplish any of those goals. An anticipated scenario to realize this vision would be to cover a fairly large portion of a surface, for instance parts of an aircraft wing or fuselage, with sensors and actuators. Spanwise arrays of actuators would be coupled with arrays of wall-shear stress sensors to provide a locally controlled region. The basic idea is that sensors upstream of the actuators detect the passing coherent structures, and sensors downstream of the actuators provide a performance measure of the control. Fast, small, and inexpensive wall-shear stress sensors like the microfabricated thermal or floating-element sensors discussed in this section would be a necessity in accomplishing this kind of futuristic control system. Control theory, control algorithms, and the use of microsensors and microactuators for reactive flow control are among the topics discussed in several chapters within this handbook.

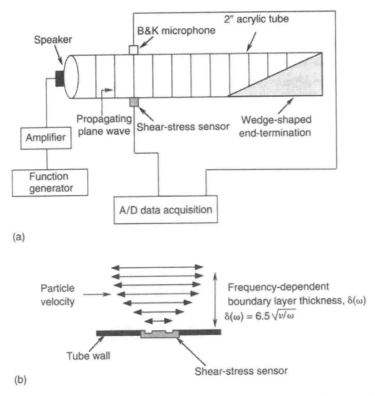

FIGURE 6.29   An acoustic plane-wave tube for the calibration of a floating-element shear sensor. (a) Overall experiment set-up; (b) probe region. (Reprinted with permission from Padmanabhan, A. [1997] Silicon Micromachined Sensors and Sensor Arrays for Shear-Stress Measurements in Aerodynamic Flows, Ph.D. Thesis, Department of Aeronautics and Astronautics, Massachusetts Institute of Technology, Cambridge, Massachusetts.)

As mentioned in the introduction, the fluctuating wall-shear stress is an indicator of the turbulence activity of the flow. In the ongoing development of existing turbulence models for application in high-Reynolds-number flows, absolute values of the fluctuating wall-shear stress are of significant interest. To accomplish this kind of measurements, reliable methods for conducting dynamical calibration of wall-shear stress sensors are needed and must be developed. The recent calibration method of Padmanabhan (1997) is of interest and other strategies are likely to develop in the future. Another intriguing possibility that would be challenging is to design a microsensor where the dynamic effects of the gauge could be controlled in such a way that only a static calibration of the sensor would be adequate.

6.4.4   Pressure Sensors

6.4.4.1   Background

In turbulence modeling, flow control, and aeroacoustics, the fluctuating wall-pressure beneath a wall-bounded flow is a crucial parameter. By measuring this random quantity much information can be gleaned about the boundary layer itself without disturbing the interior of the flow. The fluctuating wall-pressure is coupled via a complex interaction to gradients of both mean-shear and velocity fluctuations as described by the transport equations for Reynolds stresses [Tennekes and Lumley, 1972; Hinze, 1975; Pope, 2000].

The characteristics of the fluctuating wall-pressure field beneath a turbulent boundary layer have been extensively studied in both experimental and theoretical investigations, and reviews of earlier work may be found in Blake (1986), Eckelmann (1990), and Keith et al. (1992). However, from the experimental perspective, knowledge of pressure fluctuations is far from being as comprehensive as that of velocity fluctuations, since there is a lack of a generally applicable pressure-measuring instrument that can be used in the same wide variety of circumstances as the hot-wire anemometer. In spite of this, some general facts have been established for wall-pressure fluctuations. For example, the order of magnitude of the root-mean-square (rms) value, the general shape of the power spectra, and the space-time correlation characteristics [Harrison, 1958; Willmarth and Wooldridge, 1962; Bull, 1967; Bull and Thomas, 1976; Schewe, 1983; Blake, 1986; Lauchle and Daniels, 1984; Farabee and Casarella, 1991].

A clear shortcoming in many of the experiments designed to measure pressure fluctuations is the quality of the data. In the low-frequency range, the data may be contaminated by facility-related noise, while in the high-frequency range the spatial resolution of the transducers limits the accuracy. The former difficulty is usually circumvented by noise cancellation techniques [Lauchle and Daniels, 1984] or by using a free-flight glider as an experimental platform. This problem may not be considered as a major obstacle today. At the other end of the spectrum, the spatial-resolution problem is more difficult to handle. The main criticism raised is that in many experiments the size of the pressure transducer used has been far too large in relation to the thickness of the boundary layer in context, let alone in relation to the characteristic small-scale. The ultimate solution is of course to use small sensors, but testing in tunnels containing fluids with high viscosity — to increase the viscous length-scale in the flow — has also been tried. Using highly viscous fluids creates its own set of problems. Very specialized facilities and instrumentation are needed when oil or glycerin, for example, are used as the working fluid. An oil tunnel is expensive to build and to operate. Moreover, the Reynolds numbers achieved are generally low, and it is not clear how to extrapolate the results to higher-Reynolds-number flows [Gad-el-Hak and Bandyopadhyay, 1994].

Pinhole microphones have also been utilized in an attempt to improve the sensor's spatial resolution. Unfortunately, results from this type of arrangement are questionable, and to this end the use of pinhole microphones must be considered as an open question. Bull and Thomas (1976) concluded that the use of pinhole sensors in air may lead to severe errors in the measured spectra, while Farabee (1986), Gedney and Leehey (1989), and Farabee and Casarella (1991) all claimed that the pinhole sensors are most effective for wall-pressure measurements.

Based on the above arguments, it seems then that the only realistic solution to improve measurements of fluctuating wall-pressure is to use small sensors. For this reason microfabrication offers a unique opportunity for reducing the diaphragm size by at least one order of magnitude. MEMS also provides an opportunity to fabricate inexpensive, dense arrays of pressure sensors for correlation measurements and studies of coherent structures in turbulent boundary layers. In this subsection, we describe the basic principles used for MEMS-based pressure sensors/transducers/microphones. We look specifically at the design of pressure sensors utilizing the piezoresistive principle, which is particularly suited for measurements of wall-pressure fluctuations in turbulent flows. The section contains also a summary of measurements conducted with MEMS pressure sensors and when possible a comparison to conventional data. Finally, we provide an outlook for the use of MEMS-based pressure sensors for turbulence measurements and flow control.

6.4.4.2 Pressure-Sensor Principles

Many different methods have been advanced for the detection of pressure fluctuations [Sessler, 1991]. The principles available are based on detecting the vibrating motion of a diaphragm using piezoelectric, piezoresistive, and capacitive techniques. These principles were already known at the beginning of the twentieth century, but the introduction of photolithographic fabrication methods during the last two decades has provided strong impetus since this technology offers sensors fabricated to extremely low tolerance, increased resolution due to a high degree of miniaturization, and low unit-cost. This method of fabrication is also compatible with other IC techniques so electronic circuitry like pre-amplifiers can be integrated close together with the sensor, an important factor for improving pressure sensor performance. This section will provide a short background of the previously mentioned principles for pressure sensor operations. The word "microphone" will be used for a device whereby sound waves are caused to generate an electric current for the purpose of transmitting or recording sound.

A piezoelectric sensor consists of a thin diaphragm, which is either fabricated in a piezoelectric material or mechanically connected to a cantilever beam consisting of two layers of piezoelectric material with opposite polarization. A vertical movement of the diaphragm causes a stress in the piezoelectric material and generates an electric output voltage. Royer et al. (1983) presented the first MEMS-based piezoelectric sensor shown in Figure 6.30. This sensor consists of a $30\,\mu m$-thick silicon diaphragm with a diameter of 3 mm. On top of the diaphragm a layer of 3–$5\,\mu m$ ZnO was deposited, sandwiched between two SiO_2 layers that contained the upper and lower aluminum electrodes. The sensor can be provided with an integrated preamplifier, and is basically used for microphone applications. A sensitivity of 50–$250\,\mu V/Pa$, and a frequency response in the range of $10\,Hz$–$10\,kHz$ (flat within 5 dB) were recorded. Other researchers have presented similar piezoelectric silicon microphones [Kim et al., 1991; Kuhnel, 1991; Schellin and Hess, 1992; Schellin et al., 1995], and the sensitivities of these microphones were in the range of 0.025–1 mV/Pa. However, applications of the piezoelectric microphone to turbulence measurements are strongly limited because of its high noise level, which has been found to be in the range of 50–72 dB(A)SPL. These noise levels are most commonly measured using an A-weighted filter, in dBs relative to $2 \times 10^{-5}\,Pa$, which is the lowest sound level detectable by the human ear. The A-weighted filter corrects for the frequency characteristics of the human ear and provides a measure of the audibility of the noise.

A piezoresistive sensor consists of a diaphragm which is usually provided with four piezoresistors in a Wheatstone bridge configuration. One common arrangement is to locate two of the gauges in the middle, and two at the edge of the diaphragm. When the diaphragm deflects, the strains at the middle and at the edge of the diaphragm would have opposite signs which cause an opposite effect on the piezoresistive gauges. The most important advantage of this detection principle is its low output-impedance and its high sensitivity. The main drawback is that the piezoresistive material is sensitive to both stress and temperature. Unfortunately, this gives the piezoresistive sensor a strong temperature dependency. Schellin and Hess (1992) presented the first MEMS-fabricated piezoresistive sensor, which is shown in Figure 6.31. This particular sensor was used as a microphone and it had a diaphragm made of $1\,\mu m$-thick, highly boron-doped silicon with an area of $1\,mm^2$. The diaphragm was equipped with 250-nm-thick, p-type polysilicon resistors, which were isolated from the diaphragm by a 60-nm silicon dioxide layer. Using a bridge supply voltage of

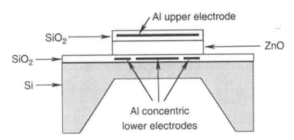

FIGURE 6.30 Cross-sectional view of a piezoelectric silicon microphone. (Reprinted with permission from Royer, M., Holmen, P., Wurm, M., Aadland, P., and Glenn, M. [1983] "ZnO on Si Integrated Acoustic Sensor," *Sensors and Actuators* **4**, pp. 357–362.)

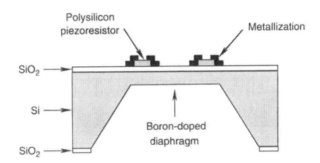

FIGURE 6.31 Cross-sectional view of a piezoresistive silicon microphone. (Reprinted with permission from Schellin, R., and Hess, G. [1992] "A Silicon Microphone Based on Piezoresistive Polysilicon Strain Gauges," *Sensors and Actuators A* **32**, pp. 555–559.)

6 V, this transducer showed a sensitivity of 25 µV/Pa and a frequency response in the range of 100 Hz–5 kHz (+3dB). However, the sensitivity was lower than expected by a factor of 10, which was explained by the initial static stress in the highly boron-doped silicon diaphragm. To improve the sensitivity of the piezo-esistive sensor, different diaphragm materials have been explored such as polysilicon and silicon nitride [Guckel, 1987; Sugiyama et al., 1993].

Most MEMS-fabricated pressure sensors are based on the capacitive detection principle, and a vast major-ity of these sensors are used as microphones. Figure 6.32 shows a cross-sectional view of such condenser microphone together with the associated electrical circuit. The latter must be included in the discussion since the preamplifier constitutes a vital part in determining the sensitivity of the capacitive probe. Basically, a condenser microphone consists of a backchamber (with a pressure equalizing hole), a backplate (with acoustic holes), a spacer, and a diaphragm covering the air gap created by the spacer located on the back-plate. The condenser, C_m, and a DC-voltage source, V_b, constitute the sensing part of the microphone. Fluctuations in the flow pressure field above the diaphragm cause it to deflect which in turn changes the capacitance, C_m. These changes are amplified in the preamplifier, H_o, which acts as an impedance converter with a bias resistor, R_b, and an input capacitance, C_i. In this figure, C_p is a parasitic capacitance which is of interest when determining the microphone attenuation. In discussing the sensitivity of a capacitive sensor, the open-circuit sensitivity is a relevant quantity and is considered to consist of two components, namely the mechanical sensitivity, S_m, and the electrical sensitivity, S_e. The total sensitivity is a weighted value of both. The former sensitivity, S_m, is defined as the increase of the diaphragm deflection dw result-ing from an increase in the pressure, dp, acting on the microphone:

$$S_m = \frac{dw}{dp}$$

(6.63)

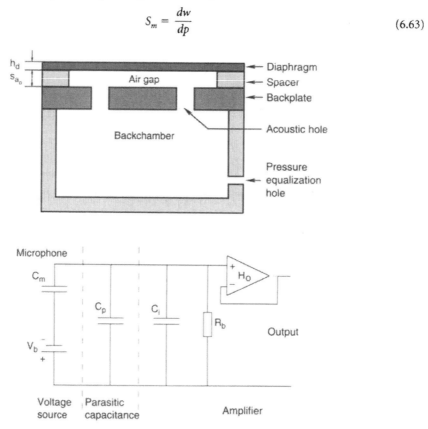

FIGURE 6.32 Cross-sectional view of a condenser microphone. The microphone is connected to an external d.c. bias voltage source, and loaded by a parasitic capacitance, bias resistor, and a preamplifier with an input capacitance. (Reprinted with permission from Löfdahl, L., Kälvesten, E., and Stemme, G. [1996] "Small Silicon Pressure Transducers for Space-Time Correlation Measurements in a Flat Plate Boundary Layer," *J. Fluids Eng.* 118, pp. 457–463.)

From Figure 6.32 we obtain the relation $dw = -ds_{a_o}$, where S_{a_o} is the thickness of the air gap between the diaphragm and the backplate. The electrical sensitivity of the microphone is given by the change in the voltage across the air gap dV resulting from a change in the air-gap thickness ds_{a_o}. Thus:

$$S_e = \frac{dV}{ds_{a_o}} \qquad (6.64)$$

The quasi-static, open-circuit sensitivity S_{open} of a condenser microphone may be defined as $S_{open} = -S_m S_e$. The output signal e_m can then be expressed as:

$$e_m = S_m S_e p \qquad (6.65)$$

where p is the fluctuating pressure. For details of this derivation, see Scheeper et al. (1994).

Hohm (1985) presented the first electret microphone based on MEMS technology. The backplate, (1×1)-cm^2 silicon, was provided with one circular acoustic hole with a diameter of 1 mm. A 2 μm-thick SiO_2 layer was used as electret and was charged to about 350 V. The diaphragm was a metallized 13 μm-thick foil with a diameter of 8 mm. Later polymer foil diaphragms were used in condenser microphones by Sprenkels (1988) and Murphy et al. (1989). In these microphones, the fabrication was made more compatible with standard thin-film technology. Bergqvist and Rudolf (1991) showed that MEMS-fabricated microphones can achieve a high sensitivity. For example, microphones with a (2×2)-cm^2 diaphragm had an open-circuit sensitivity in the range of 1.4–13 mV/Pa. In these microphones, diaphragms with thickness of 5–8 μm were fabricated using an anisotropic etching in a KOH solution and applying an electrochemical etch-stop. More details on the design and performance of the capacitive pressure sensors can be found in Scheeper et al. (1994).

An overview of the most significant dimensions, measured sensitivity, noise level, and high frequency response of first-generation MEMS pressure sensors is provided in Table 6.3.

The data from this table indicate that the piezoelectric sensors seem to have the highest noise levels. Although simple in design, they have fairly low sensitivity and relatively large spatial extension of their diaphragms. The piezoresistive sensors seem to be most flexible since the major advantage of locating

Table 6.3 Summary of Silicon Micromachined Microphones in Chronological Order

Author	Transducer Principle	Diaphragm Area (mm^2)	Upper Frequency (kHz)	Sensitivity (mV/Pa)	Equivalent Noise Level [dB(A)]
Royer et al. (1983)	Piezoelectric	3×3	40	0.25	66
Hohm (1985)	Capacitive	0.8×0.8	>20	4.3	54
Muller (1987)	Piezoelectric	3×3	7.8	0.05	72
Franz (1988)	Piezoelectric	0.8×0.9	45	0.025	68
Sprenkels (1988)	Capacitive	3×3	>10	7.5	25
Sprenkels et al. (1989)	Capacitive	3×3	>10	25	—
Hohm and Hess (1989)	Capacitive	0.8×0.8	2	1	—
Voorthuyzen et al. (1989)	Capacitive	2.45×2.45	15	19	60
Murphy et al. (1989)	Capacitive	3×3	>15	4–8	30
Bergqvist and Rudolf (1991)	Capacitive	2×2	4	13	31.5
Schellin and Hess (1992)	Piezoresistive	1×1	10	0.025 at 6 V	—
Kuhnel (1991)	FET	0.85×1.3	>20	5	62
Kuhnel and Hess (1992)	Capacitive	0.8×0.8	16	0.4–10	<25
Scheeper et al. (1992)	Capacitive	1.5×1.5	14	2	35
Bourouina et al. (1992)	Capacitive	1×1	2.5	3.5	35
Ried et al. (1993)	Piezoelectric	2.5×2.5	18	0.92	57
Kälvesten (1994)	Piezoresistive	0.1×0.1	>25	0.0009 at 10 V	90
Scheeper et al. (1994)	Capacitive	2×2	14	5	30
Bergqvist (1994)	Capacitive	2×2	>17	11	30
Schellin et al. (1995)	Piezoelectric	1×1	10	0.1	60
Kovacs and Stoffel (1995)	Capacitive	0.5×0.5	20	0.065	58
Kronast et al. (1995)	FET	0.5×0.5	25	0.01	69

piezoresistive gauges on a diaphragm is the high sensitivity that is accomplished with this arrangement. The drawback is the high temperature sensitivity, although this problem can be circumvented by a Wheatstone bridge arrangement of the gauges. The capacitive sensors are most common since these sensors have the largest commercial interest and are mainly designed to operate in conventional microphone applications such as music and human communications with pressure amplitudes as low as $20\,\mu$Pa. The capacitive sensors have relatively large spatial extension and are commonly equipped with on-chip electronics for signal amplification. This implies a reduction of the noise levels, but unfortunately it also yields very complex fabrication process. The main advantages of the capacitive sensors are their high sensitivity to pressure and low sensitivity to temperature. Their primary disadvantages are the relatively large diaphragm areas and the decreased sensitivity for high frequencies due to the air-streaming resistance of the narrow air-gap inherent in the principle.

6.4.4.3   Requirements for Turbulent Flows

There is no simple way to calculate the required pressure range of a sensor for turbulence and flow control applications. Tennekes and Lumley (1972) estimate the fluctuating pressure to be a weighted integral of the Reynolds stresses, so its length-scales should in general be larger than those of the velocity fluctuations. Moreover, it is plausible to assume that the order of magnitude of the fluctuating pressure would not be less than the Reynolds stresses, giving a good hint of the intensity of the fluctuating pressure. This intensity depends strongly on the particular flow under consideration, but for a typical flat-plate boundary layer at $Re_\theta = 4000$, this implies that the fluctuating pressure root-mean-square would be of the order of 10 Pa. Higher Reynolds numbers yield higher magnitudes of the fluctuating wall-pressure. As compared to other applications, where the fluctuating pressure has a major interest as in for example, combustion processes, the pressure magnitudes in incompressible turbulent flows are extremely small. There is also a spatial constraint in the sensor design. For the same boundary layer considered here, we have a Kolmogorov length-scale of about $50\,\mu$m, requiring a diaphragm size in the range of 100–$300\,\mu$m. The required temporal resolution of the pressure sensor is probably the easiest to estimate since we have access to good turbulence kinetic energy spectra and these show that the energy content in the flow above 10 kHz is almost negligible. Based on these physical arguments, a frame for the design of a pressure transducer for turbulence applications can be established: the sensor should have a pressure sensitivity of ± 10 Pa, a diaphragm size of $100\,\mu$m, and a flat frequency characteristics in the range of 10 Hz–10 kHz. Of course, the signal-to-noise ratio must be sufficiently high so that an ordinary data acquisition can be made.

A critical scrutiny of the different principles for designing a pressure sensor shows that the most suitable principle for turbulence applications is the piezoresistive, since the required spatial and temporal resolutions can easily be achieved, a simple fabrication in MEMS is possible, and the temperature drift can be controlled.

6.4.4.4   Piezoresistive Pressure Sensors

Micromachined piezoresistive pressure sensors were introduced on the market in the late 1950s by companies such as Kulite, Honeywell, and MicroSystems. These early transducers were all very simple devices and typically the piezoresistive gauges were glued by hand to the diaphragm. As the technology developed, more advanced fabrication methods were introduced, and today, pressure transducers are fabricated using batch-processing technologies with no hand assembly required. A large variety of different pressure transducers is available commercially and the range of applications is very wide, spanning from relatively slow but stable tools for meteorological observations to sensors used to optimize rapid combustion processes. However, an examination of these commercial sensors with turbulence and control applications will show that very few (almost none) can yet be applied for these specific purposes. If the sensors have either a frequency response or sensitivity that is acceptable, then the sensing area is far too large and the necessary spatial resolution is not fulfilled. Or vice versa, if the diaphragm size is acceptable — which is very rare — then either the frequency response or the sensitivity are far too low for the fulfillment of, respectively, the required temporal resolution or an acceptable signal-to-noise ratio. Moreover, a literature review will confirm this statement. The sensors of Table 6.3 are the ones that are fabricated using MEMS technology and fall in the range of having a certain interest for turbulence and flow control applications.

Based on the earlier mentioned criteria, it may be concluded that a piezoresistive sensor for turbulence applications should be fabricated with a diaphragm size of $100 \times 100 \times 0.4\,\mu m^3$. In order to give the pressure sensor the necessary sensitivity, the diaphragm must be very thin, and the air-gap behind the diaphragm (the cavity) must be relatively deep (say, $4\,\mu m$) to reduce the air stiffening effect. For equalization of the static pressure between the cavity and the ambient, a narrow vent channel with a length of 3 mm and a cross-section of $5 \times 0.1\,\mu m^2$ must be included in the design. An isometric view of an acoustic sensor built to those specifications by Kälvesten (1994) is shown in Figure 6.33.

When a diaphragm is deflected by a pressure load, stress is induced in the diaphragm and this should be detected by piezoresistive-strain-gauges located at appropriate positions on the diaphragm. Noting that the geometrical piezoresistive effect in semiconductors can be neglected, the general relation between the relative change of resistance and the strain components in the longitudinal (streamwise), $\langle \varepsilon_L \rangle$, and transverse (spanwise), $\langle \varepsilon_T \rangle$, directions can, according to Guckel (1991), be expressed as:

$$\frac{\Delta R}{R} = G_{par}\langle \varepsilon_L \rangle + G_{per}\langle \varepsilon_T \rangle \tag{6.66}$$

where G_{par} and G_{per} are the longitudinal and transverse gauge-factors. The pressure sensors fabricated by Kälvesten (1994) use two active strain gauges positioned at the points of maximum stress of the diaphragm, i.e., at the middle of the edges of the diaphragm and the other two are integrated beside the diaphragm to serve as passive reference gauges (Figure 6.34). This arrangement makes it possible to connect them in a half-active Wheatstone bridge for high-pressure and low-temperature sensitivity.

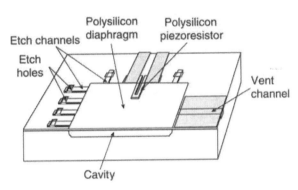

FIGURE 6.33 Isometric view of a pressure sensor/microphone for turbulence applications. The diaphragm side-length is $100\,\mu m$ and its thickness is $0.4\,\mu m$. (Reprinted with permission from Löfdahl, L., Kälvesten, E., and Stemme, G. [1996] "Small Silicon Pressure Transducers for Space-Time Correlation Measurements in a Flat Plate Boundary Layer," *J. Fluids Eng.* 118, pp. 457–463.)

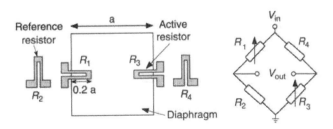

FIGURE 6.34 Top view of the piezoresistor layout with the electrical Wheatstone bridge configuration. (Reprinted with permission from Löfdahl, L., Kälvesten, E., and Stemme, G. [1996] "Small Silicon Pressure Transducers for Space-Time Correlation Measurements in a Flat Plate Boundary Layer," *J. Fluids Eng.* 118, pp. 457–463.)

TABLE 6.4 Some Calculated and Measured Characteristics of the MEMS-Based Microphone Fabricated by Kälvesten (1994)

	Theory		Measurements	
Diaphragm side-length, a (μm)	100	300	100	300
Diaphragm thickness, t (μm)	0.4	0.4	0.4	0.4
Lower frequency limit (Hz)	2×10^{-3}	5×10^{-0}	10^a	10^a
Upper frequency limit (Hz)	1310	894	25^a	25^a
Acoustical sensitivity at 10 V (μV/Pa)	1.0	0.2	0.9	0.3
Static sensitivity at 10 V (μV/Pa)	1.7	2.2	1.2	1.4
Noise (A-weighted), [dB(A)]	—	—	≈90	≈90

[a]Limits set by the calibration measurement setup.

The expected theoretical acoustical pressure sensitivity, S, for small pressure amplitudes has been derived by Kälvesten (1994) and reads:

$$S = \frac{1}{V_{in}} \frac{\partial V_{out}}{\partial p} \approx \frac{0.077 G_{par}(1 - \mu^2)}{\left(\frac{Et^2}{a^2} + \frac{3(1 + \mu)E\varepsilon_o}{4\pi} + \frac{\alpha^{-1}(1 - \mu^2)a^2}{16\pi^2 td} \right)} \qquad (6.67)$$

where V_{in} and V_{ou} are respectively the bridge supply and output voltages, p is the pressure acting on the diaphragm, α is the isothermal compressibility coefficient of air, d is the cavity depth, and G_{par} is the longitudinal gauge-factor of the polysilicon piezoresistor. For the polysilicon diaphragm, μ is the Poisson's ratio, E is the Young's modulus, ε_o is the built-in strain, a is the side-length, and t is the thickness. Assuming a linear diaphragm deflection, this equation can be used to derive the static pressure sensitivity of the sensor. In doing so it is necessary to assume a vacuum-sealed cavity by setting the third term of the denominator (air-gap compressive forces) to zero, see Rossi (1988). Table 6.4 compares the measured and calculated characteristics of the microphone developed by Kälvesten (1994).

A piezoresistive sensor is a combination of mechanical (the small and thin diaphragm); acoustical (the air-gap cavity, the air in the front of the diaphragm, and the narrow vent channel); and electrical (polysilicon piezoresistors) elements. In order to optimize the dimensions of the sensor, an electrical analogy model is very useful for the task [Rossi, 1988]. The important advantage of this analogy — from which the microphone frequency response and the diaphragm center deflection can be derived — is that it allows a good overview of even very complex systems. The analog model relies on the fact that there is a relationship between the geometry of the mechano-acoustical elements and their equivalent electrical impedance. Hence, pressure p [N/m^2] is equivalent to voltage, flow rate q [m^3/s] to current, and acoustic mass M [kg/m^4] to inductance. Damping in the system, which is due to the viscosity of the gas, is described by resistances R [N s/m^5] and capacitors C [m^5/N]. These quantities then describe the diaphragm flexibility and the gas compressibility. However, a limitation of this "equivalent circuit method" is that analytical expressions for the equivalent electrical impedance are available only for simple mechanical and acoustical structures. One possibility for the derivation of the equivalent impedance is demonstrated in Kälvesten (1994) and Kälvesten et al. (1995; 1996a), where an energy method is used for the calculation of the energy and power contributions of the different elements. These expressions are then identified as equivalent electrical impedance. Figure 6.35 shows a simplified circuit diagram of their microphone.

The piezoresistive pressure probe fabricated by Kälvesten (1994) has perhaps the smallest diaphragm built specifically for turbulence diagnosis. Its spatial resolution is adequate to measure the small scales, and the small diaphragm has correspondingly high resonance frequency which leads to a wide frequency bandwidth. The main drawbacks of this small sensor are its relatively low acoustical sensitivity and high equivalent noise level. However, for turbulence applications, the pressure is of the order of 1 Pa, while pressure amplitudes as low as 20 μPa are encountered in human communication and music applications.

The frequency characteristics of Kälvesten's pressure sensor-microphone were determined by using a Bruel & Kjaer, Type 4226 closed coupler with a built-in reference microphone at a sound pressure level of 114 dB. All measurements were performed for a bridge supply voltage of 10 V, and show acoustical sensitivities

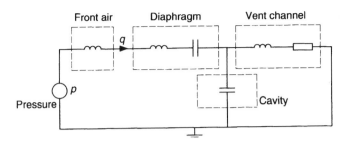

FIGURE 6.35 An electrical analogy describing the equivalent acoustical circuit of the pressure sensor/microphone. (Reprinted with permission from Löfdahl, L., Kälvesten, E., and Stemme, G. [1996] "Small Silicon Pressure Transducers for Space-Time Correlation Measurements in a Flat Plate Boundary Layer," *J. Fluids Eng.* **118**, pp. 457–463.)

of about 0.9 and 0.3 μV/Pa for diaphragm side-lengths of, respectively, 100 and 300 μm. The deviations from the theoretical estimates for the acoustical sensitivities can be explained by the approximations made in the sensitivity calculation and the uncertainty of the level of the built-in stresses in the polysilicon diaphragm. As can be seen from Table 6.4, and as is further elucidated in Kälvesten et al. (1995; 1996a), the required frequency range for turbulence applications of 10 Hz $< f <$ 10 kHz is safely covered by the theoretical range of about 1 Hz $< f <$ 1 MHz. The noise levels of the sensors are nearly the same for the two sensor versions, corresponding to an acoustic level of about 90 dB(A). More details on this sensor can be found in Kälvesten (1994) and Löfdahl et al. (1996).

6.4.4.5 MEMS Pressure Sensors

Genuine MEMS sensors for turbulence measurements were introduced in the 1990s and accordingly, very few reliable measurements for studying wall-pressure fluctuations beneath turbulent boundary layers have been conducted. Otherwise, many MEMS devices have been launched, but the fluid dynamics verification of these devices has been made at a surprisingly low level, typically in some undefined channel or pipe flow.

Schellin et al. (1995) presented a subminiature microphone which was based on the piezoresistive effect in polysilicon using only one chip. This sensor was classified as an acoustical sensor, which was fabricated with a CMOS-process and standard microfabrication technology. The diaphragm area was one square millimeter and the sensitivity was 0.025 mV/Pa at 6 V, so for turbulence applications, this sensor is a bit on the large side as well as too insensitive. The frequency response was determined to be flat from 100 Hz to 5 kHz. Unfortunately, no fluid dynamics measurements have been reported with this sensor.

Kälvesten (1994), Kälvesten et al. (1995; 1996a) and Löfdahl et al. (1996) have designed, fabricated, and used a silicon-based pressure transducer for studies focused on the high-frequency portion of the wall pressure spectrum in a two-dimensional, flat-plate boundary layer. The momentum thickness Reynolds number in their study was $Re_\theta = 5072$. A large value of the ratio between the boundary layer thickness δ and the diaphragm side-length d was used. The side-lengths of the diaphragm were 100 μm ($d^+ = 7.2$) and 300 μm ($d^+ = 21.6$). This gives a ratio of the boundary-layer thickness to the diaphragm side-length of the order of 240 and a resolution of eddies with wavenumbers less than ten viscous units. Power spectra were measured for the frequency range 13 Hz $< f <$ 13 kHz, and these were scaled in outer and inner variables. A clear overlap region between the mid- and high-frequency parts of the spectrum was found, and in this region the slope was estimated to be ω^{-1}. For the high-frequency region, the slope was proportional to ω^{-5}. The normalized rms pressure fluctuations was shown to depend strongly on the dimensionless diaphragm size with an increase connected to the resolution of the high-frequency region as shown in Figure 6.36. Classical data in the field are also plotted in the same figure for comparison.

Correlation measurements in both the longitudinal and transverse directions were performed by Löfdahl et al. (1996). Longitudinal space-time correlations, including the high-frequency range, indicated an advection velocity of the order of half the freestream velocity as can be seen in Figure 6.37. In this figure, U_c is the advection velocity, U_∞ is the freestream velocity, x_1 is the streamwise distance between two probes, and δ^* is the local displacement thickness. The advection velocities computed by Löfdahl et al. (1996) are consistently

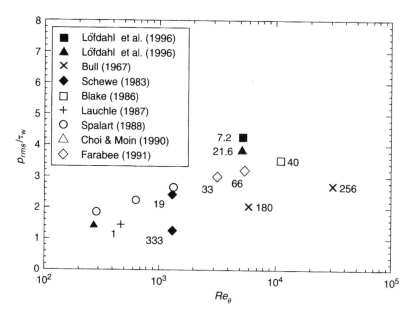

FIGURE 6.36 Dependence of normalized rms pressure fluctuations on Reynolds number. Uncertainty less than 10%. The numbers next to selected data points indicate the diaphragm side-length in wall units. (Reprinted with permission from Löfdahl, L., Kälvesten, E., and Stemme, G. [1996] "Small Silicon Pressure Transducers for Space-Time Correlation Measurements in a Flat Plate Boundary Layer," *J. Fluids Eng.* **118**, pp. 457–463.)

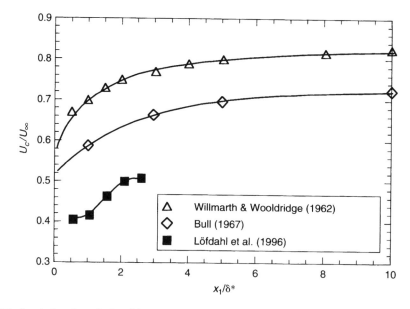

FIGURE 6.37 Local advection velocity of the pressure producing eddies for various frequency ranges. Uncertainty less than 10%. (Reprinted with permission from Löfdahl, L., Kälvesten, E., and Stemme, G. [1996] "Small Silicon Pressure Transducers for Space-Time Correlation Measurements in a Flat Plate Boundary Layer," *J. Fluids Eng.* **118**, pp. 457–463.)

lower than those obtained by Willmarth and Wooldridge (1962) and Bull (1967). A broad-band filtering of the longitudinal correlation confirmed that the high-frequency part of the spectrum is associated with the smaller eddies from the inner part of the boundary layer, resulting in a reduction of the correlation as shown in Figure 6.38.

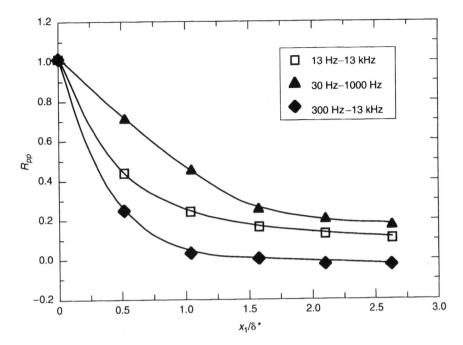

FIGURE 6.38 Longitudinal spatial correlation of the pressure fluctuations in three frequency ranges: 30 Hz < f < 1000 Hz; 300 Hz < f < 13 kHz; and 13 Hz < f < 13 kHz. Momentum thickness Reynolds number is Re = 5072. (Reprinted with permission from Löfdahl, L., Kälvesten, E., and Stemme, G. [1996] "Small Silicon Pressure Transducers for Space-Time Correlation Measurements in a Flat Plate Boundary Layer," *J. Fluids Eng.* **118**, pp. 457–463.)

Recently, Miller et al. (1997) have designed, fabricated, and tested a unique micromachined sensor-array for making optical measurements of surface-pressure distributions. Each sensor element consists of a Fabry-Perot etalon fabricated from single-crystal silicon. Illuminating the sensor array with a near-infrared tunable diode-laser and detecting the reflected light-intensity with an infrared camera, allows remote, simultaneous reading of the reflected signals from the entire sensor array. The results obtained demonstrate the basic feasibility of optically measuring surface pressure using micromachined Fabry-Perot pressure sensor arrays. Future work in this context will concentrate on demonstrating temporally-resolved pressure measurements. This technology is interesting because it offers the possibility of putting the arrays on curved surfaces.

6.4.4.6 Outlook for Pressure Sensors

MEMS pressure transducers have certainly opened many doors in formulating relationships between wall pressure and turbulence. In this subsection, we focus on some issues that should be solved in the near future by further miniaturization and array arrangements of pressure transducers. The Reynolds number effects on the rms values of the wall pressure is one key issue. The data shown in Figure 6.36 seem to suggest a trend that is also depicted schematically in the figure, viz. that they come from a family of d^+. In a review of Reynolds number effects, Gad-el-Hak and Bandyopadhyay (1994) concluded that meaningful trends in the inner layer can be extracted only when the sensors are of the order of the viscous sublayer thickness. At all Reynolds numbers in the figure, the asymptotic values are reached by sensors approaching the size d^+ = 5, and the slope of the family of d^+ lines drops as d^+ values are increased indicating a decreasing sensitivity of sensors to Reynolds number effects. It should be noted that the data shown in Figure 6.36 come from various sources with varying degrees of background noise, differences in data acquisition, corrections applied, and different inherent errors in the sensors and instrumentation. Hence, the uncertainties in each data set are different and probably not known accurately. There are also gaps in the data; thus, a useful contribution would be a systematic variation of the sensor dimension and Reynolds number so the data of Figure 6.36 could be completed.

Using arrays of wall-pressure sensors would allow more extensive studies of space-time correlations of pressure fluctuations. The primary contribution of the small transducers is that the eddies of the flow field can be resolved with a spatial resolution of about five viscous units, i.e., the thickness of the viscous sublayer. Based on the measured correlations, it is possible to examine the advection velocities of the different wavenumbers. It has earlier been shown [e.g., Bull, 1967] that the advection velocities of the turbulent eddies based on broad- and narrow-band frequency analysis approaches 80% and 60% of the freestream velocity, respectively. This can be questioned since in all earlier investigations sensors that were too large have been used. In other words, the ratio between the boundary layer thickness and the characteristic dimension of the sensor has been too low. With MEMS sensors, about ten times larger values of this ratio can be achieved as compared with conventional sensors.

Having access to well-designed pressure-sensor arrays, the advection velocities for the different wavenumbers can be determined. Such data also offer the possibility to examine whether the structures are produced by pressure sources inside the boundary layer with a wide range of advection velocities. The so-called "two family" concept suggested by Bull (1967) can then be scrutinized. It was claimed by Bull (1967) that one family of high-wavenumber components was associated with the turbulent motion in the constant-stress layer. Those components were longitudinally coherent for times proportional to the time needed for the structure to be advected distances equal to their wavelengths, and laterally coherent over distances proportional to their own wavelengths. The other wavenumber family was associated with large-scale eddy motion in the boundary layer, and loses coherence as a group independent of the wavelength. To this end, it has not been possible to question Bull's hypothesis mainly because of the lack of small pressure transducers and associated arrays.

The concepts discussed in relation to MEMS-based pressure sensors are all key issues and significant to aeroacoustics, energy transport in vibrating structures interacting with turbulent flows, turbulence modeling of the PVC terms, and last but certainly not least, to reactive control of the flow.

6.5 Microactuators for Flow Control

6.5.1 Background on Three-Dimensional Structures

It has recently been stated by Busch-Vishniac (1998) that it is generally more difficult to control a system than monitor it. For reactive control, the state of the system at a point or points in space-time must be monitored, then accordingly changed by a suitable actuation process. In the last decade it has been more challenging for scientists and engineers to design and build microsensors than microactuators, and as a consequence the progress in sensors has left behind that of actuators. In this section, the terse coverage of microactuators relative to that of microsensors is a reflection of the relative progress in the two fields.

To reactively control a flow field to reduce the drag for example, it is necessary to develop specific actuators. Most conveniently the actuators should be flush mounted on the body and aimed to influence near-wall phenomena, which in turbulent flows are dominated by small scales. To actuate these scales, it is required that the spatial extent of the actuator be of the same order as the small scales. As mentioned, the small scales decrease in spatial extension as the Reynolds number increases so sensor-actuator miniaturization is crucial for most field applications of the targeted control idea. In addition, the actuators are commonly aimed to operate in a feedback loop with sensors on an extremely short time-scale, implying that the operation of small actuators must additionally be very fast. Other constraints of control actuators are that they should yield minimal pressure drag especially when not in use and that they should be fabricated in large numbers at low-cost so large portions of drag-inducing surfaces may be covered. Additionally, the actuator must be able to apply a significant disturbance to the flow while expending little energy when in operation. Finally, the sensor-actuator system must be able to withstand the typically harsh field environment. Taking all this together in the design of actuators for reactive flow control constitutes a real challenge.

Numerous devices for actuating a variety of flow fields have been advanced during the last few years. This has mostly been motivated by the quest to achieve efficient reactive control of turbulent flows. Wiltse and Glezer (1993) originated an interesting path when they used piezoelectric-driven cantilevers as flow control actuators to beneficially affect free-shear layers. Following the idea of Wiltse and Glezer (1993), Jacobson and Reynolds (1995) designed an actuator that has the piezoelectric cantilever mounted flush with a boundary layer surface. The cantilever was oriented in the streamwise direction, fixed to the wall at its upstream end, and could flex out into the flow and down into a cavity in the wall. The under-cantilever cavity was filled with fluid which was connected to the flow through the gaps on the sides and at the tip of the cantilever. Hence, when the cantilever flexes into the cavity, it forces fluid out of the cavity into the external flow, and when the cantilever flexes out of the cavity, fluid is drawn into the cavity from the external flow. Neither Wiltse and Glezer (1993) nor Jacobson and Reynolds (1995) used MEMS technology to fabricate their actuators. However, the main conclusion drawn in Jacobson and Reynolds (1995) was that future actuators must be MEMS-based in order to match the relevant scales of turbulent flows.

The major challenge in designing MEMS-based actuators is that a motion perpendicular to the plane of the surface is needed to influence selective flow structures. It is implicitly complex to accomplish this, since micromachining and MEMS devices are basically two-dimensional structures. Some kind of quasi-three-dimensional structures have been fabricated using bulk and surface micromachining (see, for example, the passive cantilever structures developed by Bandyopadhyay, 1995). However, these techniques usually have good resolution in the surface plane but in the direction normal to this plane the resolution is less impressive. Genuine three-dimensional structures can be achieved by using the so-called out-of-plane rotation of microstructures. A major advantage of this method is that it utilizes conventional high-resolution surface lithography and thus simultaneously provides an access to the third dimension. A number of techniques based on out-of-plane rotation or folding of silicon microstructures have been advanced, and in this section we present the characteristics of some of these three-dimensional methods which may form a base for future development of MEMS actuators for reactive flow control. Actuating mechanisms based on the bi-layer effect, electrostatic or electromagnetic forces, mechanical folding, and one-layer structures are presented and discussed. The section will conclude with a brief look at some recently developed microturbomachines for which the primary applications may be in fields other than flow control.

6.5.2 The Bi-Layer Effect

Out-of-plane rotation of microstructures can be achieved by controlled stress-engineering in sandwiched structures, the bi-layer effect method. The volume change between the two different materials used is the key to control the rotation, and several methods are available for this. A significant advantage of this technique is that the microstructure is self-assembled, so no external manipulators are needed to raise the structure out-of-plane. However, since the bi-layer effect is based on the volume change of different materials, the rotation is limited to fairly large radius of curvature which for certain applications is certainly a severe drawback.

It has been more than 70 years since Timoshenko (1925) presented the basic theory for bending sandwiched structures which consist of two materials with different coefficients of thermal expansion, the bi-metal effect. By heating a bi-metal cantilever, controlled bending can be achieved. This has been studied, for example, by Riethmuller and Benecke (1988) and Ataka et al. (1993), who used different standard IC materials. Commonly aluminum, chromium, or gold — all having relatively high coefficient of thermal expansion — are used in combination with silicon, polysilicon, silicon dioxide, or silicon nitride, which all have low thermal expansion coefficient. As mentioned, the drawback of the bi-layer method is that the bending radii are always large, and very thin and fragile structures are needed to reduce the radius of curvature. This pitfall reduces the impact of this particular technology.

Figure 6.39 illustrates a surface-micromachined actuator achieved by utilizing the large residual stress between two different polyimide films. The out-of-plane curling can be controlled either with thermal actuation using an integrated heater and the bi-metal effect, or by electrostatic actuation [Elwenspoek et al., 1992; Lin et al., 1995]. Unfortunately, the electrostatic actuation requires relatively high driving voltage. To circumvent this, Yasuda et al. (1997) have proposed using serial connection of several bending actuators as depicted in Figure 6.40.

Smela et al. (1995) have shown that when a conjugated or conducting polymer is changed from insulating to metallic by applying a small voltage, the volume of the polymer is changed. A bi-layer structure, consisting

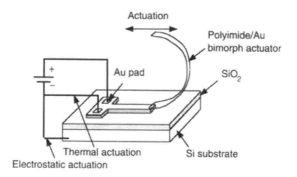

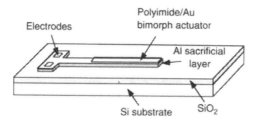

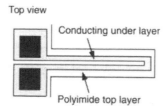

FIGURE 6.39 Surface micromachined three-dimensional structures using a bi-morph structure consisting of two different polyimides with different thermal expansion coefficients. Control of the out-of-plane rotation can be obtained either with integrated gold heater or by electrostatic forces. (Reprinted with permission from Lin, G., Kim, C.J., Konishi, S., and Fujita, H. [1995] "Design, Fabrication and Testing of a C-Shaped Actuator," *Technical Digest, Eighth Inter. Confer. on Solid-State Sensors and Actuators [Transducers '95 and Eurosensors 9]*, pp. 416–419, 25–29 June, Stockholm, Sweden.)

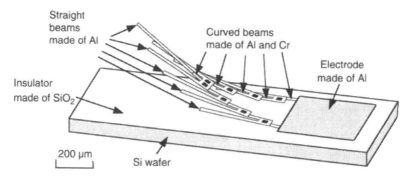

FIGURE 6.40 The difference between residual tensile stresses in the Al–Cr films initially curls the beam upwards. Electrostatic forces obtained through an applied voltage between the serially connected beams and the Si wafer are used to control the out-of-plane rotation. (Reprinted with permission from Yasuda, T., Shimonoyama, I., and Mirura, H. [1997] "CMOS Drivable Electrostatic Microactuator with Large Deflection," in *Proc. IEEE Tenth Inter. Workshop MEMS [MEMS '97]*, 26–30 January, Nagoya, Japan.)

of the polymer and a gold layer will curl and this can be used as a hinge. Large reversible bending angles with small radii of curvature can be obtained with this technique. However, the price to pay is that the tuning requires an electrolyte, which normally is not compatible with sensor applications. The polymer material is also more unstable than conventional integrated circuit materials and has a large potential for degradation.

Volume changes which result in bending can be achieved by using controlled heating to obtain a phase change in a material, so-called "Shape Memory Alloys" (SMA) [Otsuka, 1995]. An integrated heater is used to obtain the phase transformation of the SMA alloy. Additionally, volume changes can also be obtained by applying a voltage over a piezoelectric material deposited on a thin cantilever. When such material expands, the cantilever bends and it is possible to obtain a voltage-controlled, out-of-plane rotation. Unfortunately, both the SMA and piezoelectric techniques give relatively large bending radii.

6.5.3 Electrostatic and Magnetic External Forces

The most straightforward way of bending a structure is simply to apply an active force on the structure. Different forces can be used to repel or attract the free-standing structure so that they bend out of the wafer plane. Both electrostatic and magnetic forces have been utilized. Electrostatic actuators are easy to fabricate, but their operation is limited by small force and displacement outputs. The electromagnetic actuators are, on the other hand, more robust and can be used when larger forces and displacements are required, for example for separation control or drag reduction. MEMS-based electrostatic actuators have been used to control aerodynamic instabilities. Huang et al. (1996) used an electrostatic actuator to control screech in high-speed jets. With the design of a 70 μm peak-to-peak displacement at the resonant frequency of 5 kHz, it has been shown that disturbances in the jet can be suppressed. Figure 6.41 illustrates the external-force principle using a permanent magnet to achieve an out-of-plane rotation [Liu et al., 1995]. Miller et al. (1996) used a microelectromagnetic flap with a 30-turn copper coil and a layer of permalloy on a (4 × 4) mm^2 silicon plate to obtain a 2 mm tip motion of the actuator. These devices have been shown to be effective for changing the location of flow separation [Liu et al., 1995] and to achieve drag reduction [Tsao et al., 1994; Ho et al., 1997].

To summarize, the electrostatic actuators seem to be easy to fabricate. However, their operations are limited by small force and displacement outputs. The electromagnetic actuators are more robust and can be used when larger forces and displacements are required, e.g., for separation control or drag reduction. Recently promising results have been reported by Yang et al. (1997), who used thermopneumatic actuators or microballoons. Typically, these actuators have a heater in a cavity which is covered by a diaphragm; when the fluid inside the cavity is heated, a significant pressure rise is achieved. This causes a deflection of the diaphragm, and for silicon rubber diaphragms more than one mm, displacement has been measured by Yang et al. (1997). Figure 6.42 depicts the fabrication steps and the deflection achieved by Yang et al.'s microballoon actuator.

6.5.4 Mechanical Folding

A simple method to achieve out-of-plane bending of a microstructure is to use an external manipulator and fold a flexible or elastic joint-hinge as shown in Figure 6.43. Unfortunately, this technique relies on the use of an external manipulator and is accordingly not batch compatible. However, to solve this problem different methods with integrated actuators have been presented by Fukuta et al. (1995; 1997) and Garcia (1998). The main drawback of these methods seems to be that they are quite complex and fragile. Elastic joints in polyimide were developed by Suzuki et al. (1992) as part of an attempt to fabricate a microrobot mimicking insects. Figure 6.44 shows their three-dimensional structure with the elastic polyimide-joint and the surface micromachining steps involved in fabricating it. This structure is manually assembled into its final position just like paper folding, and bending angles up to 70° can be achieved.

An approach similar to the elastic polyimide-joint was employed to fabricate three-dimensional piezoresistive sensors using post-process etching of standard CMOS-wafers [Hoffman et al., 1995; Kruglick et al.,

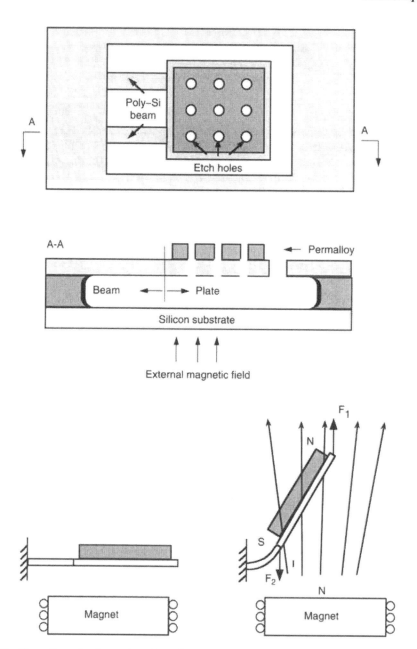

FIGURE 6.41 Three-dimensional actuation achieved by applying external magnetic forces. (Reprinted with permission from Liu, C., Tsao, T., Tai, Y.-C., Liu, W., Will, P., and Ho, C.-M. [1995] "A Micromachined Permalloy Magnetic Actuator Array for Micro Robotic Assembly System," *Technical Digest, Eighth Inter. Confer. on Solid-State Sensors and Actuators [Transducers '95 and Eurosensors 9]*, pp. 328–331, 25–29 June, Stockholm, Sweden.)

1998]. External manipulators were used to rotate the structures out-of-plane with a flexible aluminum hinge, as shown in Figure 6.45. The advantage here is that three-dimensional actuators can readily be integrated on the same chip with the electronics. Moreover, the electrical interconnection to the folded structure is automatically obtained by the aluminum hinge. However, the main drawback is the non-self-assembling fabrication involved which requires interlocking braces. Also, the resulting structure is usually extremely

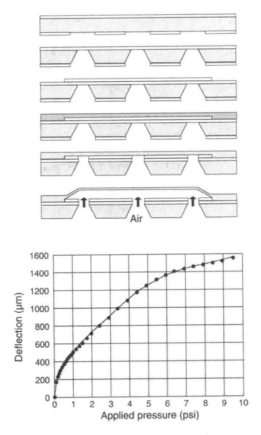

FIGURE 6.42 Steps involved in fabricating a microballoon actuator (top figure), and applied pressure vs. deflection (bottom plot) for a 2.3 × 8.6 mm² balloon actuator. (Reprinted with permission from Yang, X., Grosjean, C., Tai, Y.-C., and Ho, C.-M., [1997] "A MEMS Thermopneumatic Silicone Membrane Valve," in *Proc. IEEE Tenth Inter. Workshop MEMS [MEMS '97]*, 26–30 January, Nagoya, Japan.)

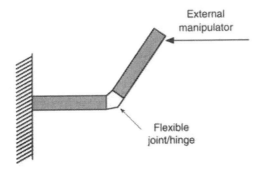

FIGURE 6.43 A three-dimensional microstructure undergoing a mechanical folding caused by external manipulator. (Reprinted with permission from Ebefors, T., Kälvesten, E., and Stemme, G. [1997a] "Dynamic Actuation of Polyimide V-Grooves Joints by Electrical Heating," in *Eurosensors XI*, September 21–24, Warsaw, Poland.)

fragile. A micromachined polysilicon hinge has been presented by Pister et al. (1990). This hinge is fabricated using surface micromachining with two layers of polysilicon and posphosilicate glass as sacrificial layer, as shown in Figure 6.46. A variety of applications using the polysilicon hinge has been presented, for example micro-optical devices such as mirrors and lenses [Lin et al., 1995; Wu et al., 1995; Gunawan

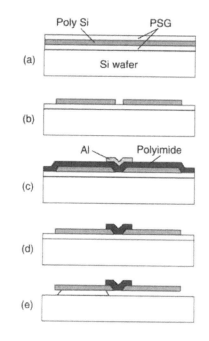

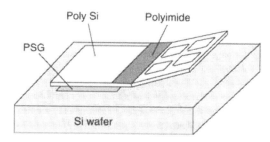

FIGURE 6.44   Steps involved in fabricating a three-dimensional structure that uses mechanical folding of an elastic polyimide joint (top figure), and isometric view of the resulting microactuator (bottom figure). (Reprinted with permission from Suzuki, K., Shimoyama, I., Miuara, H., and Ezura, Y. [1992] "Creation of an Insect-Based Microrobot with an External Skeleton and Elastic Joints," in *Proc. IEEE Fifth Inter. Workshop on MEMS [MEMS '92]*, pp. 190–195, 4–7 February, Travemunde, Germany.)

et al., 1995; Solgaard et al., 1995; Tien et al., 1996a; 1996b; Fan et al., 1997]. However, a drawback of these surface-made microstructures is that they cannot stand by themselves, so they require some kind of inter-locking arrangement to stay bent out of the wafer, which complicates the assembly process. The structure usually consists of many stacked layers resulting in rugged surfaces. To be useful for three-dimensional appli-cations, the difficulties in achieving electrical interconnections to the assembled structure have to be resolved.

Fukuta et al. (1995) have presented a reshaping technique using Joule's heating to effect thermal plastic deformation for permanent three-dimensional polysilicon structures. Figure 6.47 illustrates the experimen-tal procedure for the three-dimensional structure realization. To achieve self-assembling structures without the drawback of using an external micromanipulator, a technique using an integrated scratch-drive actuator for folding was presented by Akiyama et al. (1997). Figure 6.48 shows the principle of the self-assembling process which produces permanent three-dimensional structure by using integrated scratch-drive actuator and reshaping technology. Fukuta et al. (1995) also discuss the self-assembling of three-dimensional structures with reshaping technology.

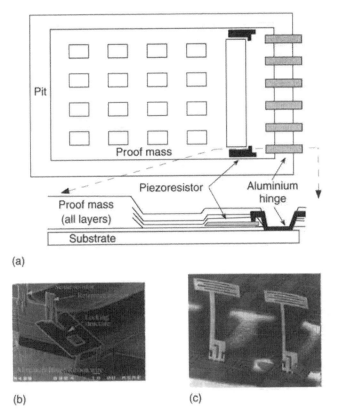

(a)

(b)   (c)

FIGURE 6.45 (a) Overview of a three-dimensional accelerometer design based on aluminum hinges with piezoresistive detection. The aluminum hinges are formed by front-bulk silicon etching of standard CMOS-wafers. (Reprinted with permission from Kruglick, E.J.J., Warneke, B.A., and Pister, K.S.J. [1998] "CMOS 3-Axis Accelerometers with Integrated Amplifier," in *Proceedings of the IEEE Eleventh Inter. Workshop on MEMS [MEMS '98]*, pp. 631–636, 25–29 January, Heidelberg, Germany.) (b) Close-up of the aluminum hinge with an interlocking brace [Kruglick et al., 1998]. (c) SEM-photo of raised test structures. (Reprinted with permission from Hoffman, E., Warneke, B., Kruglick, E., Weigold, J., and Pister, K.S.J. [1995] "3D Structures with Pizoresistive Sensors in Standard CMOS," in *Proc. IEEE Eighth Inter. Workshop on MEMS [MEMS '95]*, pp. 288–293, 29 January–2 February, Amsterdam, The Netherlands.)

6.5.5 One-Layer Structures

All the techniques described in the previous subsection need external manipulators to raise the microstructures. To overcome this drawback, a self-assembling technique for fabrication of three-dimensional structures was presented by Green et al. (1996). The method is based on the surface tension force of a molten solder. The main drawback is that non-standard IC material is used, meaning that very accurate process control is required. In addition, it seems difficult to achieve electrical connections to the out-of-plane rotated structures. To circumvent the lack of IC compatibility, Syms (1998) proposed using glass as the meltable material to achieve a more controlled assembly. However, the glass requires high temperature to melt and this makes it impossible to integrate metal interconnections and electronics together with the three-dimensional structure. Figure 6.49 shows the preparation of a three-dimensional structure based on surface tension forces. The structure after the out-of-plane rotation is shown in the bottom of this figure.

Another self-assembling microjoint technique has recently been presented by Ebefors et al. (1998). In this technique, polyimide is used to create controlled out-of-plane rotations yielding small silicon structures with detailed features. The basic principle of the polyimide joint is shown in Figure 6.50. Polyimide in the

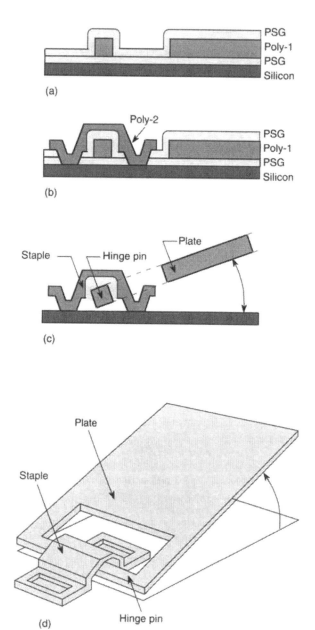

FIGURE 6.46 (a)–(c) Cross-Sections of the hinge during fabrication. (d) A perspective view of a hinged plate after release by external manipulator. (Reprinted with permission from Pister, K.S.J., Judy, S.R., Burgett, S.R., and Fearing, R.S. [1992] "Micro-Fabricated Hinges," *Sensors and Actuators A* 33, pp. 249–256.)

V-grooves shrinks when cured in such a way that the absolute contraction length of the polyimide is larger at the top of the groove than at the bottom. This creates a rotation of the material which implies a bending of the free-standing structure out of the wafer plane. Larger bending angles can be achieved by connecting several V-grooves in a row. Figure 6.51 shows a schematic view of a polyimide joint with three V-grooves. The static, irreversible bending angle is obtained during the thermal curing process when cross-links between the molecules are effected and different solvents in the polyimide are out-gassed yielding a reduction in the polyimide volume. The shrinkage depends on the curing temperature, which

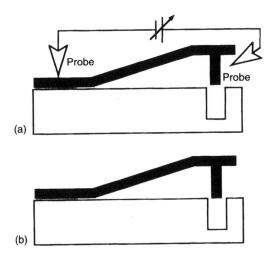

FIGURE 6.47 An external manipulator is used to fold and supply the structure with a reshaping voltage. The Joule's heating raises the temperature of the arm to cause annealing effect which results in stress release and plastic deformation. After removal of the voltage, the arm cools down and the structure retains its three-dimensional shape. (Reprinted with permission from Fukuta, Y., Akiyama, T., and Fujita, H. [1995] "A Reshaping Technology with Joule Heat for Three-Dimensional Silicon Structures," *Technical Digest, Inter. Confer. on Solid-State Sensor and Actuator [Tranducers '95]*, pp. 174–177, 25–29 June, Stockholm, Sweden.)

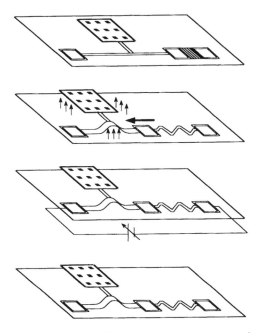

FIGURE 6.48 Illustration of a three-dimensional self-assembled polysilicon structure based on scratch-drive actuators. (Reprinted with permission from Akiyama, T., Collard, D., and Fujita, H. [1997] "Scratch Drive Actuator with Mechanical Links for Self-Assembly of Three-Dimensional MEMS," *J. MEMS* 6, pp. 10–17.)

makes it possible to control the final bending angle rather well. A reversible bending angle can be obtained as a result of the thermal expansion of the cured polyimide. A metal conductor is used as a resistive heater which produces local power dissipation in the joint. A temperature increase results in an expansion of the polyimide and a dynamic change in the bending angle.

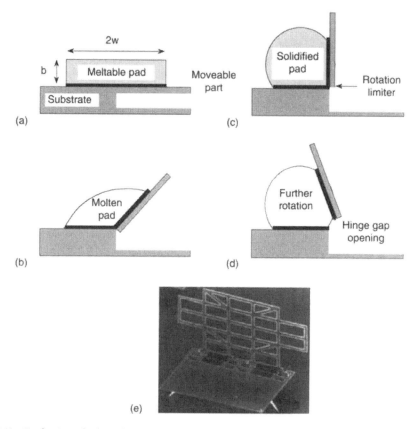

FIGURE 6.49 Production of a three-dimensional structure with Si mechanical parts based on surface tension forces (top figure), and SEM photo of the resulting three-dimensional structure (bottom figure). (Reprinted with permission from Syms, R.R.A. [1998] "Demonstration of Three-Dimensional Microstructure Self-Assembly," *Sensors and Actuators A* 65, pp. 238–243.)

6.6   Microturbomachines

6.6.1   Micropumps

There have been several studies of microfabricated pumps. Some of them use non-mechanical effects. The so-called Knudsen pump uses the thermal-creep effect to move rarefied gases from one chamber to another. Ion-drag is used in electrohydrodynamic pumps [Bart et al., 1990; Richter et al., 1991; Fuhr et al., 1992]. These rely on the electrical properties of the fluid and are thus not suited for many applications. Valveless pumping by ultrasound has also been proposed by Moroney et al. (1991), but produces very little pressure difference. Mechanical pumps based on conventional centrifugal or axial turbomachinery will not work at micromachine scales where the Reynolds numbers are typically small, on the order of 1 or less. Centrifugal forces are negligible and, furthermore, the Kutta condition through which lift is normally generated is invalid when inertial forces are vanishingly small. In general there are three ways in which mechanical micropumps can work: positive-displacement pumps; continuous, parallel-axis rotary pumps; and continuous, transverse-axis rotary pumps.

Positive-displacement pumps are mechanical pumps with a membrane or diaphragm actuated in a reciprocating mode and with unidirectional inlet and outlet valves. They work on the same physical principle as their larger cousins. Micropumps with piezoelectric actuators have been fabricated [Van Lintel et al., 1988; Esashi et al., 1989); Smits, 1990]. Other actuators, such as thermopneumatic, electrostatic,

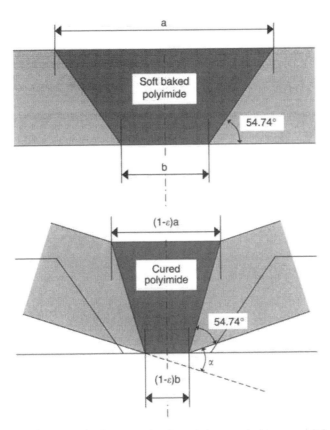

FIGURE 6.50 Principle of the polyimide V-groove joint. The polyimide in the V-groove shrinks when the polyimide is cured. The absolute lateral-contraction length of the polyimide is larger at the top of the V-groove than at the bottom, resulting in a rotation which bends the free-standing structure out-of-the-wafer plane. (Reprinted with permission from Ebefors, T., Kälvesten, E., and Stemme, G. [1997a] "Dynamic Actuation of Polyimide V-Grooves Joints by Electrical Heating," in *Eurosensors XI*, September 21–24, Warsaw, Poland.)

electromagnetic, or bimetallic can be used [Pister et al., 1990; Döring et al., 1992; Gabriel et al., 1992]. These exceedingly minute positive-displacement pumps require even smaller valves, seals, and mechanisms, a not-too-trivial micromanufacturing challenge. In addition there are long-term problems associated with wear or clogging and consequent leaking around valves. The pumping capacity of these pumps is also limited by the small displacement and frequency involved. Gear pumps are a different kind of positive-displacement device.

A continuous, parallel-axis rotary pump is a screw-type, three-dimensional device for low Reynolds numbers and was proposed by Taylor (1972) for propulsion purposes and shown in his seminal film. It has an axis of rotation parallel to the flow direction implying that the powering motor must be submerged in the flow, the flow turned through an angle, or that complicated gearing would be needed.

Continuous, transverse-axis rotary pumps are a machines that have been developed by Sen et al. (1996). They have shown that a rotating body, asymmetrically placed within a duct, will produce a net flow due to viscous action. The axis of rotation can be perpendicular to the flow direction and the cylinder can thus be easily powered from outside a duct. A related viscous-flow pump was designed by Odell and Kovasznay (1971) for a water channel with density stratification. However, their design operates at a much higher Reynolds number and is too complicated for microfabrication.

As evidenced from the third item above, it is possible to generate axial fluid motion in open channels through the rotation of a cylinder in a viscous fluid medium. Odell and Kovasznay (1971) studied a pump based on this principle at high Reynolds numbers. Sen et al. (1996) carried out an experimental study of

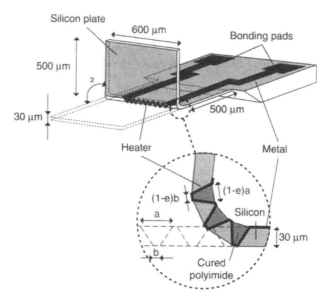

FIGURE 6.51   Isometric view of a three-dimensional structure based on a polyimide joint with three V-grooves. The metal lead-wires through the V-grooves are used to realize electrical connections to the out-of-plane rotated structure. Exploiting the thermal expansion of the polyimide, the metal can also be used as a heater in the V-grooves to obtain reversible movements. (Reprinted with permission from Ebefors, T., Kälvesten, E., Vieider, C., and Stemme, G. [1997b] "New Robust Small Radius Joints Based on Thermal Shrinkage of Polyimide in V-grooves," in *Transducers '97*, June 16–19, Chicago.)

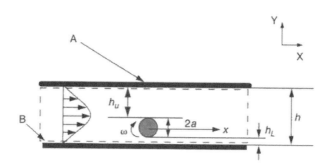

FIGURE 6.52   Schematic of micropump developed by Sen et al. (Reprinted with permission from Sen, M., Wajerski, D., and Gad-el-Hak, M. [1996] "A Novel Pump for MEMS Applications," *J. of Fluids Eng.* 118, pp. 624–627.)

a different version of such a pump. The novel viscous pump, shown schematically in Figure 6.52, consists simply of a transverse-axis cylindrical rotor eccentrically placed in a channel, so that the differential viscous resistance between the small and large gaps causes a net flow along the duct. The Reynolds numbers involved in Sen et al.'s work were low ($0.01 < Re \equiv 2\omega a^2/v < 10$, where ω is the radian velocity of the rotor, and a is its radius), typical of microscale devices, but achieved using a macroscale rotor and a very viscous fluid. The bulk velocities obtained were as high as 10% of the surface speed of the rotating cylinder. Sen et al. (1996) have also tried cylinders with square and rectangular cross-sections, but the circular cylinder delivered the best pumping performance.

A finite-element solution for low-Reynolds-number, uniform flow past a rotating cylinder near an impermeable plane boundary has already been obtained by Liang and Liou (1995). However, a detailed two-dimensional Navier–Stokes simulations of the pump described above have been carried out by Sharatchandra et al. (1997), who extended the operating range of Re beyond 100. The effects of varying the channel height

H and the rotor eccentricity ε have been studied. It was demonstrated that an optimum plate spacing exists and that the induced flow increases monotonically with eccentricity; the maximum flowrate being achieved with the rotor in contact with a channel wall. Both the experimental results of Sen et al. (1996) and the two-dimensional numerical simulations of Sharatchandra et al. (1997) have verified that, at $Re < 10$, the pump characteristics are linear and therefore kinematically reversible. Sharatchandra et al. (1997; 1998a; 1998b) also investigated the effects of slip flow on the pump performance as well as the thermal aspects of the viscous device. Wall slip does reduce the traction at the rotor surface and thus lowers the performance of the pump somewhat. However, the slip effects appear to be significant only for Knudsen numbers greater than 0.1, which is encouraging from the point of view of microscale applications.

In an actual implementation of the micropump, several practical obstacles need to be considered. Among those are the larger stiction and seal design associated with rotational motion of microscale devices. Both the rotor and the channel have a finite, in fact rather small, width. DeCourtye et al. (1998) numerically investigated the viscous micropump performance as the width of the channel, *W*, becomes exceedingly small. The bulk flow generated by the pump decreased as a result of the additional resistance to the flow caused by the side walls. However, effective pumping was still observed with extremely narrow channels. Finally, Shartchandra et al. (1998b) used a genetic algorithm to determine the optimum wall shape to maximize the micropump performance. Their genetic algorithm uncovered shapes that were nonintuitive but yielded vastly superior pump performance.

Though most of the previous micropump discussion is of flow in the steady state, it should be possible to give the eccentric cylinder a finite number of turns or even a portion of a turn to displace a prescribed minute volume of fluid. Numerical computations will easily show the order of magnitude of the volume discharged and the errors induced by acceleration at the beginning of the rotation and deceleration at the end. Such a system can be used for microdosage delivery in medical applications.

6.6.2 Microturbines

DeCourtye et al. (1998) have described the possible utilization of the inverse micropump device as a turbine. The most interesting application of such a microturbine would be as a microsensor for measuring exceedingly small flowrates on the order of nanoliters (i.e., microflow metering for medical and other applications).

The viscous pump described operates best at low Reynolds numbers and should therefore be kinematically reversible in the creeping-flow regime. A microturbine based on the same principle should therefore, lead to a net torque in the presence of a prescribed bulk velocity. The results of three-dimensional numerical simulations of the envisioned microturbine are summarized in this subsection. The Reynolds number for the turbine problem is defined in terms of the bulk velocity, since the rotor surface speed is unknown in this case:

$$Re = \frac{\bar{U}(2a)}{v} \tag{6.69}$$

where $\bar{U}$ is the prescribed bulk velocity in the channel, *a* is the rotor radius, and *v* is the kinematic viscosity of the fluid.

Figure 6.53 shows the dimensionless rotor speed as a function of the bulk velocity for two dimensionless channel widths $W = \infty$ and $W = 0.6$. In these simulations, the dimensionless channel depth is $H = 2.5$ and the rotor eccentricity is $\varepsilon/\varepsilon_{max} = 0.9$. The relation is linear as was the case for the pump problem. The slope of the lines is 0.37 for the two-dimensional turbine and 0.33 for the narrow channel with $W = 0.6$. This means that the induced rotor speed is, respectively, 0.37 and 0.33 of the bulk velocity in the channel. (The rotor speed can never exceed the fluid velocity even if there is no load on the turbine. Without load, the integral of the viscous shear stress over the entire surface area of the rotor is exactly zero, and the turbine achieves its highest albeit finite rpm.) For the pump, the corresponding numbers were 11.11 for the two-dimensional case and 100 for the three-dimensional case. Although it appears that the side walls have bigger influence on the pump performance, in the turbine case, a vastly higher pressure drop is required in the three-dimensional duct to yield the same bulk velocity as that in the two-dimensional duct (dimensionless pressure drop of $\Delta p^{\star} \equiv \Delta p(2a)^2/\rho v^2 = -29$ vs. $\Delta p^{\star} = -1.5$).

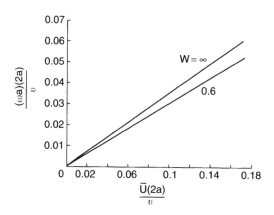

FIGURE 6.53 Turbine rotation as a function of the bulk velocity in the channel. (Reprinted with permission from DeCourtye, D., Sen, M., and Gad-el-Hak, M. [1998] "Analysis of Viscous Micropumps and Microturbines," *Inter. J. Comp. Fluid Dyn.* **10**, pp. 13–25.)

The turbine characteristics are defined by the relation between the shaft speed and the applied load. A turbine load results in a moment on the shaft, which at steady state balances the torque due to viscous stresses. At a fixed bulk velocity, the rotor speed is determined for different loads on the turbine. Again, the turbine characteristics are linear in the Stokes (creeping) flow regime, but the side walls have weaker, though still adverse, effect on the device performance as compared to the pump case. For a given bulk velocity, the rotor speed drops linearly as the external load on the turbine increases. At large enough loads, the rotor will not spin, and maximum rotation is achieved when the turbine is subjected to zero load.

At present it is difficult to measure flowrates on the order of 10^{-12} m³/s (1 nanoliter/s). One possible way is to directly collect the effluent over time. This is useful for calibration but is not practical for on-line flow measurement. Another is to use heat transfer from a wire or film to determine the local flowrate as in a thermal anemometer. Heat transfer from slowly moving fluids is mainly by conduction so that temperature gradients can be large. This is undesirable for biological and other fluids easily damaged by heat. The viscous mechanism that has been proposed and verified for pumping may be turned around and used for measuring. As demonstrated in this subsection, a freely rotating cylinder eccentrically placed in a duct will rotate at a rate proportional to the flowrate due to a turbine effect. In fact other geometries such as a freely rotating sphere in a cylindrical tube should also behave similarly. The calibration constant, which depends on system parameters such as geometry and bearing friction, should be determined computationally to ascertain the practical viability of such a microflow meter. Geometries that are simplest to fabricate should be explored and studied in detail.

6.6.3 Microbearings

Many of the micromachines use rotating shafts and other moving parts that carry a load and need fluid bearings for support, most of them operating with air or water as the lubricating fluid. The fluid mechanics of these bearings are very different compared to that of their larger cousins. Their study falls in the area of microfluid mechanics, an emerging discipline which has been greatly stimulated by its applications to micromachines and which is the subject of this chapter.

Macroscale journal bearings develop their load-bearing capacity from large pressure differences which are a consequence of the presence of a viscous fluid, an eccentricity between the shaft and its housing, a large surface speed of the shaft, and a small clearance to diameter ratio. Several closed-form solutions of the no-slip flow in a macrobearing have been developed. Wannier (1950) used modified Cartesian coordinates to find an exact solution to the biharmonic equation governing two-dimensional journal bearings in the no-slip, creeping flow regime. Kamal (1966) and Ashino and Yoshida (1975) worked in bipolar coordinates; they assumed a general form for the streamfunction with several constants which were determined using the boundary conditions. Although all these methods work if there is no slip, they cannot be readily adapted to

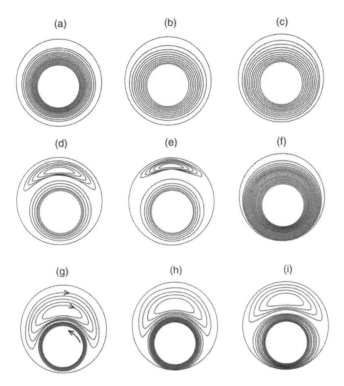

FIGURE 6.54 Effect of slip factor and eccentricity on the microbearing streamlines. From top to bottom, eccentricity changes as $\varepsilon = 0.2, 0.5, 0.8$. From left to right, slip factor changes as $S \equiv [(2 - \sigma_v)/\sigma]$, $Kn = 0, 0.1, 0.5$. (Reprinted with permission from Maureau, J., Sharatchandra, M.C., Sen, M., and Gad-el-Hak, M. [1997] "Flow and Load Characteristics of Microbearings with Slip," *J. Micromech., and Microeng.* 7, pp. 55–64.)

slip flow. The basic reason is that the flow pattern changes if there is slip at the walls and the assumed form of the solution is no longer valid.

Microbearings are different in the following aspects: (1) being so small, it is difficult to manufacture them with a clearance that is much smaller than the diameter of the shaft; (2) because of the small shaft size, their surface speed, at normal rotational speeds, is also small (the microturbomachines being developed presently at MIT operate at shaft rotational speeds on the order of 1 million rpm, and are therefore operating at different flow regime from that considered here); and (3) air bearings in particular may be small enough for non-continuum effects to become important. For these reasons the hydrodynamics of lubrication are very different at microscales. The lubrication approximation that is normally used is no longer directly applicable and other effects come into play. From an analytical point of view there are three consequences of the above: fluid inertia is negligible, slip flow may be important for air and other gases, and relative shaft clearance need not be small.

In a recent study, Maureau et al. (1997) analyzed microbearings represented as an eccentric cylinder rotating in a stationary housing. The flow Reynolds number is assumed small, the clearance between shaft and housing is not small relative to the overall bearing dimensions, and there is slip at the walls due to nonequilibrium effects. The two-dimensional governing equations are written in terms of the streamfunction in bipolar coordinates. Following the method of Jeffery (1920), Maureau et al. (1997) succeeded in obtaining an exact infinite-series solution of the Navier–Stokes equations for the specified geometry and flow conditions. In contrast to macrobearings and due to the large clearance, flow in a microbearing is characterized by the possibility of a recirculation zone which strongly affects the velocity and pressure fields. For high values of the eccentricity and low slip factors, the flow develops a recirculation region, as shown in the streamlines plot in Figure 6.54.

From the infinite-series solution, the frictional torque and the load-bearing capacity can be determined. The results show that both are similarly affected by the eccentricity and the slip factor: they increase with the former and decrease with the latter. For a given load, there is a corresponding eccentricity which generates a force sufficient to separate shaft from housing, i.e., sufficient to prevent solid-to-solid contact. As the load changes, the rotational center of the shaft shifts a distance necessary for the forces to balance.

For a weight that is vertically downwards, the equilibrium displacement of the center of the shaft is in the horizontal direction. This can lead to complicated rotor dynamics governed by mechanical inertia, viscous damping, and pressure forces. A study of these dynamics may be of interest. Real microbearings have finite shaft lengths, and end walls and other three-dimensional effects influence the bearing characteristics. Numerical simulations of the three-dimensional problem can readily be carried out and may also be of interest to the designers of microbearings. Other potential research includes determination of a criterion for onset of cavitation in liquid bearings. From the results of these studies, information related to load, rotational speed, and geometry can be generated that would be useful for the designer.

Finally, Piekos et al. (1997) have used full Navier–Stokes computations to study the stability of ultra-high-speed gas microbearings. They conclude that it is possible, despite significant design constraints, to attain stability for specific bearings to be used with the MIT microturbomachines [Epstein and Senturia, 1997; Epstein et al., 1997], which incidentally operate at much higher Reynolds numbers (and rpm) than the micropumps, microturbines, and microbearings considered thus far in this chapter. According to Piekos et al. (1997), high-speed bearings are more robust than low-speed ones due to their reduced running eccentricities and the large loads required to maintain them.

6.7 Conclusions

In a presentation to the 1959 annual meeting of the American Physical Society, Richard Feynman anticipated the extension of electronic miniaturization to mechanical devices. That vision is now a reality. Microelectromechanical systems, a fledgling field that took off just this decade, are already finding numerous applications in a variety of industrial and medical fields. This chapter focused on MEMS-based sensors and actuators especially as used for the diagnosis and control of turbulent flows. The miniaturization of sensors leads to improved spatial and temporal resolutions for measuring useful turbulence quantities at high Reynolds numbers. The availability of inexpensive, low-energy-usage microsensors and microactuators that can be packed densely on a single chip promises a quantum leap in the performance of reactive flow control systems. Such control is now in the realm of the possible for future vehicles and other industrial devices. In a turbulent flow, an increase in Reynolds number will automatically generate smaller length-scales and shorter time-scales, which both in turn require small and fast sensors for a correct resolution of the flow field. MEMS offer a solution to this problem since sensors with length- and time-scales of the order of the relevant Kolmogorov microscales can now be fabricated. Additionally, these sensors are produced with high accuracy at a relatively low cost per unit. For instance, a MEMS pressure sensor can be used to determine fluctuating pressures beneath a turbulent boundary layer with a spatial resolution that is about one order-of-magnitude finer than what can be achieved with conventional microphones.

In this chapter, we have reviewed the state-of-the-art of microsensors used to measure the instantaneous velocity, wall-shear stress, and pressure, which are quantities of primary importance in turbulence diagnosis. For each group, we provided general background, design criteria, calibration procedure, and examples of measurements conducted with MEMS-based sensors and when possible compared the results to conventional measurements. Microsensors can be fabricated at low unit-cost and can be spaced close together in dense arrays. These traits are particularly useful for studies of coherent structures in wall-bounded turbulent flows.

Reactive flow control is another application where microdevices may play a crucial future role. MEMS sensors and actuators provide opportunities for targeting the small-scale coherent structures in macroscopic turbulent shear flows. Detecting and modulating these structures may be essential for a successful control of wall-bounded turbulent flows to achieve drag reduction, separation delay, and lift enhancement. To cover areas of significant spatial extension, many devices are needed requiring small-scale, low-cost, and low-energy-use components. In this context, the miniaturization, low-cost fabrication, and low-energy consumption of microsensors and microactuators are of utmost interest and promise a quantum leap in control system performance. Combined with modern computer technologies, MEMS yield the essential matching of the length- and time-scales of the phenomena to be controlled.

Numerous actuators have been developed during the past few years. This chapter reviewed the state-of-the-art of microactuators based on the bi-layer effect, electrostatic or electromagnetic forces, mechanical folding, and one-layer structures. We have also briefly described recently advanced ideas for viscous micropumps and microturbines. Future challenges include achieving significant actuation perpendicular to the plane of what is basically a two-dimensional chip, further reducing unit cost and energy expenditure of microactuators, and designing microdevices that are capable of withstanding the harsh field environment of, for example, an aircraft. These are not easy tasks, but the payoff if air, water, or land-vehicle drag for example, could be reduced by a mere few percentage points, would translate into fuel savings in the billions of dollars as well as tremendous benefits to the environment.

Microelectromechanical systems have witnessed phenomenal advances in a mere ten-year period. The 1960s and 1970s were arguably the decades of the transistor and it is likely that the first few years of the third millennium will be the MEMS decades. Medical and industrial breakthroughs are inevitable with every advance in MEMS technology, and the future worldwide market for micromachines is bound to be in the tens of billions of dollars.

Acknowledgments

The authors would like to acknowledge the valuable help of Dr. Andrey Bakchinov and Mr. Peter Johansson for preparing the figures. Our thanks are extended to Professors Haim Bau, Ali Beskok, Kenneth Breuer, Chih-Ming Ho, Stuart Jacobson, and George Karniadakis, who all shared with us several of their reports and papers. Our sincere appreciation to Professor Mihir Sen for sharing his ideas regarding shear-sensor calibration and microturbomachines.

References

Akiyama, T., Collard, D., and Fujita, H. (1997) "Scratch Drive Actuator with Mechanical Links for Self-Assembly of Three-Dimensional MEMS," *J. MEMS*, **6**, pp. 10–17.

Aronson, D., and Löfdahl, L. (1994) "The Plane Wake of a Cylinder: An Estimate of the Pressure Strain Rate Tensor," *Phys. Fluids*, **6**, pp. 2716–2721.

Aronson, D., Johansson, A.V., and Löfdahl, L. (1997) "A Shear-Free Turbulent Boundary Layer-Experiments and Modeling," *J. Fluid Mech.*, **338**, pp. 363–385.

Ashino, I., and Yoshida, K. (1975) "Slow Motion between Eccentric Rotating Cylinders," *Bull. JSME*, **18**, no. 117, pp. 280–285.

Ataka, M., Omodaka, A., and Fujita, H. (1993) "A Biomimetic Micro System — a Ciliary Motion System," in *Proc. Seventh Inter. Confer. on Solid-State Sensors and Actuators (Transducers '93)*, pp. 38–41, Yokohama, Japan.

Bandyopadhyay, P.R. (1995) "Microfabricated Silicon Surfaces for Turbulence Diagnostics and Control," in *Proc. Inter. Symp. on Active Control of Sound and Vibration (Active 95)*, S. Sommerfeldt and H. Hamada, eds., pp. 1327–1338, 6–8 July, Newport Beach, CA.

Bart, S.F., Tavrow, L.S., Mehregany, M., and Lang, J.H. (1990) "Microfabricated Electrohydrodynamic Pumps," *Sensors Actuators A*, **21–23**, pp. 193–197.

Bellhouse, B., and Schultz, L. (1966) "Determination of Mean and Dynamic Skin Friction Separation in Low-Speed Flow with a Thin-Film Heated Element," *J. Fluid Mech.*, **24**, pp. 379–400.

Bellhouse, B.J., and Rasmussen, C.G. (1968) "Low-Frequency Characteristics of Hot-Film Anemometers," *DISA Inf.*, **6**, pp. 3–10.

Bergqvist, J. (1994) Modelling and Micromachining of Capacitive Microphones, Ph.D. Thesis, Uppsala University, Uppsala, Sweden.

Bergqvist, J., and Rudolf, F. (1991) "A New Condenser Microphone in Silicon," *Sensors Actuators A*, 21, pp. 123–125.

Blackwelder, R.F. (1981) "Hot-Wire and Hot-Film Anemometers," in *Methods of Experimental Physics — Fluid Dynamics*, R.J. Emrich, ed., vol. 18A, pp. 259–314, Springer-Verlag, Berlin.

Blake, W.K. (1986) *Mechanics of Flow Induced Sound and Vibrations*, Academic Press, New York.

Bourouina, T., Spirkovitch, S., Baillieu, F., and Vauge, C. (1992) "A New Condenser Microphone with a p^+ Silicon Membrane," *Sensors Actuators A*, 31, pp. 149–152.

Bradshaw, P. (1971) "An Introduction to Turbulence and Its Measurement," Pergamon Press, Oxford.

Bremhorst, K., and Gilmore, D.B. (1976) "Comparison of Dynamic and Static Hot-Wire Anemometer Calibrations for Velocity Perturbation Measurements," *J. Phys. E: Sci. Instrum.*, 9, pp. 1097–1100.

Breuer, K.S. (1995) "Stochastic Calibration of Sensors in Turbulent Flow Fields," *Exp. Fluids*, 19, pp. 138–141.

Brodkey, R.S. (1967) *The Phenomena of Fluid Motions*, Addison-Wesley, New York.

Brown, G.L. (1967) "Theory and Application of Heated Films for Skin Friction Measurements," in *Proc. Heat Transfer and Fluid Mechanics Inst.*, pp. 361–381.

Bryzek, J., Peterson, K., and McCulley, W. (1994) "Micromachines on the March," *IEEE Spectrum*, 31, May, pp. 20–31.

Bull, M.K. (1967) "Wall-Pressure Fluctuations Associated with Subsonic Turbulent Boundary Layer Flow," *J. Fluid Mech.*, 28, pp. 719–754.

Bull, M.K., and Thomas, A.S.W. (1976) "High Frequency Wall-Pressure Fluctuations in Turbulent Boundary Layers," *Phys. Fluids*, 19, pp. 597–599.

Busch-Vishniac, I.J. (1998) "Trends in Electromechanical Transduction," *Phys. Today*, 51, July, pp. 28–34.

Champagne, F.H. (1965) "Turbulence Measurements with Inclined Hot-Wires," Boeing Scientific Research Laboratories, Flight Science Laboratory Report No. 103, Boeing Company, Seattle, Washington.

Collis, D.C., and Williams, M.J. (1959) "Two-Dimensional Convection from Heated Wires at Low Reynolds Number," *J. Fluid Mech.*, 6, pp. 357–384.

Comte-Bellot, G. (1976) "Hot-Wire Anemometry," *Annu. Rev. Fluid Mech.*, 8, pp. 209–231.

Comte-Bellot, G., Strohl, A., and Alcarez, E. (1971) "On Aerodynamic Disturbances Caused by a Single Hot-Wire Probe," *J. Appl. Mech.*, 93, pp. 767–774.

Corrsin, S. (1959) "Outline of Some Topics in Homogenous Turbulent Flow," *J. Geophys. Res.*, 64, pp. 2134–2150.

DeCourtye, D., Sen, M., and Gad-el-Hak, M. (1998) "Analysis of Viscous Micropumps and Microturbines," *Int. J. Comp. Fluid Dyn.*, 10, pp. 13–25.

Diem, B., Rey, P., Renard, S., Viollet, P., Bosson, M., Bono, H., Michel, F., Delaye, M., and Delapierre, G. (1995) "SOI 'SIMOX'; from Bulk to Surface Micromachining, a New Age for Silicon Sensors and Actuators," *Sensors Actuators A*, 46–47, pp. 8–16.

Dowling, A.P., and Ffowcs-Williams, J.E. (1983) *Sound and Sources of Sound*, Wiley, New York.

Drubka, R.E., Tan-Atichat, J., and Nagib, H.M. (1977) "On Temperature and Yaw Dependence of Hot-Wires," IIT Fluid and Heat Transfer Report no. R77-1, Illinois Institute of Technology, Chicago.

Dryden, H.L., and Kuethe, A.M. (1930) "The Measurement of Fluctuations of Air Speed by Hot-Wire Anemometer," NACA Technical Report No. NACA-TR-320, Washington, D.C.

Döring, C., Grauer, T., Marek, J., Mettner, M.S., Trah, H.-P., and Willmann, M. (1992) "Micromachined Thermoelectrically Driven Cantilever Structures for Fluid Jet Deflection," in *Proc. IEEE MEMS '92*, pp. 12–18, 4–7 February, Travemunde, Germany.

Ebefors, T., Kälvesten, E., and Stemme, G. (1997a) "Dynamic Actuation of Polyimide V-Grooves Joints by Electrical Heating," in *Eurosensors XI*, September 21–24, Warsaw, Poland.

Ebefors, T., Kälvesten, E., and Stemme, G. (1998) "New Small Radius Joints Based on Thermal Shrinkage of Polyimide in V-Grooves for Robust Self-Assembly 3-D Microstructures," *J. Micromech. Microeng.*, 8, pp. 188–194.

Ebefors, T., Kälvesten, E., Vieider, C., and Stemme, G. (1997b) "New Robust Small Radius Joints Based on Thermal Shrinkage of Polyimide in V-grooves," in *Transducers '97*, June 16–19, Chicago.

Eckelmann, H. (1990) "A Review of Knowledge on Pressure Fluctuations," in *Near-Wall Turbulence*, S.J. Kline and N.H. Afgan, eds., pp. 328–347, Hemisphere, New York.

Ehrfeld, W., Götz, F., Münchmeyer, D., Schleb, W., and Schmidt, D. (1988) "LIGA Process: Sensor Construction Techniques via X-ray Lithography," *Technical Digest, Solid-State Sensor and Actuator Workshop*, pp. 1–4, 6–9 June, Hilton Head, South Carolina.

Elwenspoek, M., Smith, L., and Hök, B. (1992) "Active Joints for Microrobot Limbs," *J. Micromech. Microeng.*, 2, pp. 221–223.

Epstein, A.H., and Senturia, S.D. (1997) "Macro Power from Micro Machinery," *Science*, 276, pp. 1211.

Epstein, A.H., Senturia, S.D., Al-Midani, O., Anathasuresh, G., Ayon, A., Breuer, K., Chen, K.-S., Ehrich, F.F., Esteve, E., Frechette, L., Gauba, G., Ghodssi, R., Groshenry, C., Jacobson, S.A., Kerrebrock, J.L., Lang, J.H., Lin, C.-C., London, A., Lopata, J., Mehra, A., Mur Miranda, J.O., Nagle, S., Orr, D.J., Piekos, E., Schmidt, M.A., Shirley, G., Spearing, S.M., Tan, C.S., Tzeng, Y.-S., and Waitz, I.A. (1997) "Micro-Heat Engines, Gas Turbines, and Rocket Engines-The MIT Microengine Project," AIAA paper No. 97-1773, AIAA, Reston, Virginia.

Esashi, M., Shoji, S., and Nakano, A. (1989) "Normally Closed Microvalve Fabricated on a Silicon Wafer," *Sensors Actuators*, 20, pp. 163–169.

Fabris, G. (1978) "Probe and Method for Simultaneous Measurement of 'True' Instantaneous Temperature and Three Velocity Components in Turbulent Flow," *Rev. Sci. Instrum.*, 49, pp. 654–664.

Fan, L., Wu, M.C., Choquette, K.D., and Crawford, M.H. (1997) "Self-Assembled Microactuated XYZ Stages for Optical Scanning and Alignment," *Technical Digest, Ninth Inter. Confer. on Solid-State Sensor and Actuator (Tranducers '97)*, pp. 319–322, June 16–19, Chicago.

Farabee, T.M. (1986) "An Experimental Investigations of Wall Pressure Fluctuations Beneath Non-Equilibrium Turbulent Flows," David Taylor Naval Ship Research and Development Center, DTNSRDC Technical Report No. 86/047, Bethesda, Maryland.

Farabee, T.M., and Casarella, M. (1991) "Spectral Features of Wall Pressure Fluctuations Beneath Turbulent Boundary Layers," *Phys. Fluids*, 3, pp. 2410–2419.

Fingerson, L.M., and Freymuth, P. (1996) "Thermal Anemometers," in *Fluid Mechanics Measurements*, 2nd ed., R.J. Goldstein, ed., pp. 115–173, Taylor and Francis, Washington, D.C.

Franz, J. (1988) "Aufbau, Funktionsweise und technische Realisierung eines piezoelektrischen Siliziumsensors fur akustische Grössen," VDI-Berichte no. 677, pp. 299–302, Germany.

Freymuth, P. (1977) "Frequency Response and Electronic Testing for Constant-Temperature Hot-Wire Anemometers," *J. Phys. E: Sci. Instrum.*, 10, pp. 705–710.

Freymuth, P. (1983) "History of Thermal Anemometry," in *Handbook of Fluids in Motion*, M.P. Cheremisinoff, ed., Ann Arbor Publishers, Ann Arbor, Michigan.

Freymuth, P. (1992) "A Bibliography of Thermal Anemometry," TSI Incorporated, St. Paul, Minnesota.

Fuhr, G., Hagedorn, R., Muller, T., Benecke, W., and Wagner, B. (1992) "Microfabricated Electrohydrodynamic (EHD) Pumps for Liquids of Higher Conductivity," *J. MEMS*, 1, pp. 141–145.

Fujita, H., and Kovasznay, L.S.G. (1968) "Measurement of Reynolds Stress by a Single Rotated Hot-Wire Anemometer," *Rev. Scie. Instrum.*, 39, pp. 1351–1355.

Fukuta, Y., Akiyama, T., and Fujita, H. (1995) "A Reshaping Technology with Joule Heat for Three-Dimensional Silicon Structures," *Technical Digest, Inter. Confer. on Solid-State Sensor and Actuator (Tranducers '95)*, pp. 174–177, 25–29 June, Stockholm, Sweden.

Fukuta, Y., Collard, D., Akiyama, T., Yang, E.H., and Fujita, H. (1997) "Microactuated Self-Assembling of 3-D Polysilicon Structures with Reshaping Technology," in *Proc. IEEE Tenth Inter. Workshop on MEMS (MEMS '97)*, pp. 477–481, 26–30 January, Nagoya, Japan.

Gabriel, K.J., Tabata, O., Shimaoka, K., Sugiyama, S., and Fujita, H. (1992) "Surface-Normal Electrostatic/ Pneumatic Actuator," in *Proc. IEEE MEMS '92*, pp. 128–131, 4–7 February, Travemunde, Germany.

Gad-el-Hak, M. (1994) "Interactive Control of Turbulent Boundary Layers: A Futuristic Overview," *AIAA J.*, 32, pp. 1753–1765.

Gad-el-Hak, M. (1998) "Frontiers of Flow Control," in *Flow Control: Fundamentals and Practices*, M. Gad-el-Hak, A. Pollard, and J.-P. Bonnet, eds., Lecture Notes in *Physics*, vol. 53, pp. 109–153, Springer-Verlag, Berlin.

Gad-el-Hak, M. (2000) "Flow Control: Passive, Active, and Reactive Flow Management," Cambridge University Press, London, United Kingdom.

Gad-el-Hak, M., and Bandyopadhyay, P.R. (1994) "Reynolds Number Effect in Wall Bounded Flows," *Appl. Mech. Rev.*, 47, pp. 307–365.

Gad-el-Hak, M., and Tsai, H.M. (2005) *Transition and Turbulence Control*, World Scientific Publishing, Singapore.

Gad-el-Hak, M., Pollard, A., and Bonnet, J.-P. (1998) *Flow Control: Fundamentals and Practices*, Springer-Verlag, Berlin.

Garcia, E.J. (1998) "Micro-Flex Mirror and Instability Actuation Technique," in *Proc. IEEE Eleventh Inter. Workshop on MEMS (MEMS '98)*, pp. 470–475, 25–29 January, Heidelberg, Germany.

Gedney, C.J., and Leehey, P. (1989) "Wall Pressure Fluctuations During Transition on a Flat Plate," in *Proc. Symp. on Flow Induced Noise due to Laminar-Turbulent Transition Process*, ASME NCA, Vol. 5, ASME, New York.

Goldberg, H.D., Breuer, K.S., and Schmidt, M.A. (1994) "A Silicon Wafer-Bonding Technology for Microfabricated Shear-Stress Sensors with Backside Contacts," *Technical Digest, Solid-State Sensor and Actuator Workshop*, pp. 111–115.

Goldstein, M.E. (1976) *Aeroacoustics*, McGraw-Hill, New York.

Green, P.W., Syms, R.R.A., and Yeatman, E.M. (1996) "Demonstration of Three-Dimensional Microstructure Self-Assembly," *J. MEMS*, 4, pp. 170–176.

Guckel, H. (1987) "Fine Grained Films and its Application to Planar Pressure Transducers," in *Proc. Fourth Inter. Confer. on Solid-State Sensors and Actuators (Transducers '87)*, pp. 277–282, 2–5 June, Tokyo, Japan.

Guckel, H. (1991) "Surface Micromachined Pressure Transducers," *Sensors Actuators A*, 28, pp. 133–146.

Gunawan, D.S., Lin, L.Y., Lee, S.S., and Pister, K.S.J. (1995) "Micromachined Free-Space Integrated Micro-Optics," *Sensors Actuators A*, 50, pp. 127–134.

Hakkinen, R.J. (1991) "Survey of Skin Friction Measurements Techniques," AIAA Minisymposium, Dayton, Ohio.

Hallbäck, M. (1993) Development of Reynolds Stress Closures of Homogenous Turbulence through Physical and Numerical Experiments, Ph.D. Thesis, Department of Mechanics, Royal Institute of Technology, Stockholm, Sweden.

Hanratty, T.J., and Campbell, J.A. (1996) "Measurement of Wall-Shear Stress," in *Fluid Mech. Measurements*, 2nd ed., R. Goldstein, ed., pp. 575–640, Taylor and Francis, Washington, D.C.

Haritonidis, J.H. (1989) "The Measurements of Wall-Shear Stress," in *Advances in Fluid Mechanics Measurements*, M. Gad-el-Hak, ed., pp. 229–261, Springer-Verlag, Berlin.

Harrison, M. (1958) "Pressure Fluctuations on the Wall Adjacent to Turbulent Boundary Layers," David Taylor Naval Ship Research and Development Center, DTNSRDC Tech. Rep. No. 1260, Bethesda, Maryland.

Hinze, J.O. (1975) *Turbulence*, 2nd ed., McGraw-Hill, New York.

Ho, C.-M., and Tai, Y.-C. (1996) "Review: MEMS and its Applications for Flow Control," *J. Fluids Eng.*, 118, pp. 437–447.

Ho, C.-M., and Tai, Y.-C. (1998) "Micro-Electro-Mechanical Systems (MEMS) and Fluid Flows," *Annu. Rev. Fluid Mech.*, 30, pp. 579–612.

Ho, C.-M., Tung, S., Lee, G.B., Tai, Y.-C., Jiang, F., and Tsao, T. (1997) "MEMS-A Technology for Advancements in Aerospace Engineering," AIAA Paper No. 97-0545, AIAA, Reston, Virginia.

Hoffman, E., Warneke, B., Kruglick, E., Weigold, J., and Pister, K.S.J. (1995) "3D Structures with Pizoresistive Sensors in Standard CMOS," in *Proc. IEEE Eighth Inter. Workshop on MEMS (MEMS '95)*, pp. 288–293, 29 January–2 February, Amsterdam, The Netherlands.

Hohm, D. (1985) "Subminiatur-Silizium-Kondensatormikrofon, Fortschritte der Akustik," in *DAGA, '85*, pp. 847–850.

Hohm, D., and Hess, G. (1989) "A Subminiature Condenser Microphone with Silicon-Nitride Membrane and Silicon Backplate," *J. Acoust. Soc. Am.*, 85, pp. 476–480.

Holmes, P., Lumley, J., and Berkooz, G. (1996) *Turbulence, Coherent Structures, Dynamical Systems and Symmetry*, Cambridge University Press, London.

Huang, J.-B., Tung, S., Ho, C.-M., Liu, C., and Tai, Y.-C. (1996) "Improved Micro Thermal Shear-Stress Sensor," *IEEE Trans. Instrum. Meas.*, **45**, pp. 570–574.

Israelachvili, J.N. (1991) *Intermolecular and Surface Forces*, 2nd ed., Academic Press, New York.

Jacobson, S.A., and Reynolds, W.C. (1995) "An Experimental Investigation Towards the Active Control of Turbulent Boundary Layers," Department of Mechanical Engineering Report No. TF-64, Stanford University, Stanford, California.

Jeffery, G.B. (1920) "Plane Stress and Plane Strain in Bipolar Co-ordinates," *Philos. Trans. Roy. Soc. A*, **221**, pp. 265–289.

Jiang, F., Tai, Y.-C., Ho, C.-M., and Li, W. (1994) "A Micromachined Polysilicon Hot-Wire Anemometer," *Solid-State Sensor and Actuator Workshop*, Hilton Head, South Carolina.

Jiang, F., Tai, Y.-C., Gupta, B., Goodman, R., Tung, S., Huang, J., and Ho, C.-M. (1996) "A Surface Micromachined Shear-Stress Imager," in *Proc. IEEE Ninth Inter. Workshop on MEMS*, pp. 110–115, February, San Diego, California.

Jiang, F., Tai, Y-C, Walsh, K., Tsao, T., Lee, G.B., and Ho, C.-H. (1997) "A Flexible MEMS technology and Its First Application to Shear Stress Skin," in *Proc. IEEE MEMS Workshop (MEMS '97)*, pp. 465–470.

Jörgensen, F.E. (1971) "Directinal Sensitivity of Wire and Fiber Film Probes," *DISA Inf.*, **11**, pp. 31–37.

Kälvesten, E. (1994) "Piezoresistive Silicon Microphones for Turbulence Measurements," Royal Institute of Technology, TRITA–IL-9402, Stockholm, Sweden.

Kälvesten, E. (1996) "Pressure and Wall Shear Stress Sensors for Turbulence Measurements," Royal Institute of Technology, TRITA-ILA-9601, Stockholm Sweden.

Kälvesten, E., Löfdahl, L., and Stemme, G. (1995) "Small Piezoresistive Silicon Microphones Specially Designed for the Characterization of Turbulent Gas Flows," *Sensors Actuators A*, **46–47**, pp. 151–155.

Kälvesten, E., Löfdahl, L., and Stemme, G. (1996a) "An Analytical Characterization of a Piezoresistive Square Diaphragm Silicon Microphone," *Sensors Mater.*, **8**, pp. 113–136.

Kälvesten, E., Vieider, C., Löfdahl, L., and Stemme, G. (1996b) "An Integrated Pressure-Velocity Sensor for Correlation Measurements in Turbulent Gas Flows," *Sensors Actuators A*, **52**, pp. 51–58.

Kamal, M.M. (1966) "Separation in the Flow Between Eccentric Rotating Cylinders," in *Trans. ASME Ser. D*, **88**, pp. 717–724.

Keith, W., Hurdis, D., and Abraham, B. (1992) "A Comparison of Turbulent Boundary Layer Wall-Pressure Spectra," *J. Fluids Eng.*, **114**, pp. 338–347.

Keller, C.G., and Howe, R.T. (1995) "Nickel-Filled Hexsil Thermally Actuated Tweezers," in *Eighth Inter. Confer. on Solid-State Sensors and Actuators and Eurosensors-9*, Digest of Technical Papers, vol. 2, pp. 376–379, Stockholm, Sweden.

Kim, E.S., Kim, J.R., and Muller, R.S. (1991) "Improved IC-Compatible Piezoelectric Microphones and CMOS Process," in *Proc. Sixth Inter. Confer. on Solid-State Sensors and Actuators (Transducers '91)*, 24–28 June, San Fransisco, Calfornia.

King, L.V. (1914) "On the Convection of Heat from Small Cylinders in a Stream of Fluid: Determination of the Convection of Small Platinum Wires with Applications to Hot-Wire Anemometry," *Philos. Trans. Roy. Soc. London*, **214**, pp. 373–432.

King, L.V. (1915) "On the Precition Measurement of Air Velocity by Means of the Linear Hot-Wire Anemometer," *Philos. Mag., J. Sci. Sixth*, **29**, pp. 556–577.

King, L.V. (1916) "The Linear Hot-Wire Anemometer and its Applications in Technical Physics," *J. Franklin Inst.*, **181**, pp. 1–25.

Kolmogorov, A.N. (1941a) "Local Structure of Turbulence in Incompressible Fluid at very high Reynolds Numbers," *Dokl. Akad. Nauk. SSSR*, **30**, pp. 299–303.

Kolmogorov, A.N. (1941b) "Energy Dissipation in Locally Isotropic Turbulence *Dokl. Akad. Nauk. SSSR*, **32**, pp. 19–21.

Kovacs, G.T.A. (1998) *Micromachined Transducers Sourcebook*, McGraw-Hill, New York.

Kovacs, A., and Stoffel, A. (1995) "Integrated Condenser Microphone with Polysilicon Electrodes," in *MME '95, Micromechanics Europe*, pp. 132–135, 3–5 September, Copenhagen, Denmark.

Kronast, W., Muller, B., and Stoffel, A. (1995) "Miniaturized Single-Chip Silicon Membrane Microphone with Integrated Field-effect Transistor," in *MME '95, Micromechanics Europe*, pp. 136–139, 3–5 September, Copenhagen, Denmark.

Kramers, H. (1946) "Heat Transfer from Spheres to Flowing Media," *Physica*, 12, pp. 61–80.

Kruglick, E.J.J., Warneke, B.A., and Pister, K.S.J. (1998) "CMOS 3-Axis Accelerometers with Integrated Amplifier," in *Proc. IEEE Eleventh Inter. Workshop on MEMS (MEMS '98)*, pp. 631–636, 25–29 January, Heidelberg, Germany.

Kuhnel, W. (1991) "Silicon Condenser Microphone with Integrated Field-Effect Transistor," *Sensors Actuators A*, 25–27, pp. 521–525.

Kuhnel, W., and Hess, G. (1992) "Micromachined Subminiature Condenser Microphones in Silicon," *Sensors Actuators A*, 32, pp. 560–564.

Lauchle, G.C., and Daniels, M.A. (1984) "Wall-Pressure Fluctuations in Turbulent Pipe Flow," *Phys. Fluids*, 30, pp. 3019–3024.

Lesieur, M. (1991) *Turbulence in Fluids*, 2nd ed., Kluwer, Dordrecht, The Netherlands.

Leveque, M.A. (1928) "Transmission de Chaleur par Convection" *Ann. Mines*, 13, pp. 283.

Liang, W.J., and Liou, J.A. (1995) "Flow Around a Rotating Cylinder Near a Plane Boundary," *J. Chinese Inst. Engrs.*, 18, pp. 35–50.

Liepman, H.W., and Skinner, G.T. (1954) "Shearing-Stress Measurements by Use of a Heated Element," NACA Technical Note No. 3268, Washington, D.C.

Ligrani, P.M., and Bradshaw, P. (1987) "Spatial Resolution and Measurements of Turbulence in the Viscous Sublayer using Subminiature Hot-Wire Probes," *Exp. Fluids*, 5, pp. 407–417.

Lin, G., Kim, C.J., Konishi, S., and Fujita, H. (1995) "Design, Fabrication and Testing of a C-Shaped Actuator," *Technical Digest, Eighth Inter. Confer. on Solid-State Sensors and Actuators (Transducers '95 and Eurosensors 9)*, pp. 416–419, 25–29 June, Stockholm, Sweden.

Linder, C., Paratte, L., Gretillat, M., Jaecklin, V., and de Rooij, N. (1992) "Surface Micromachining," *J. Micromech. Microeng.*, 2, pp. 122–132.

Lindgren, B., Österlund, J., and Johansson, A.V. (2000) "Flow Structures in High Reynolds Number Turbulent Boundary Layers," in *Proc. Eighth European Turbulence Confer., Adv. in Turbulence 8*, pp. 399–402, 27–30 June, Barcelona, Spain.

Ling, S.C. (1963) "Heat Transfer from a Small Isothermal Spanwise Strip on an Insulated Boundary," *J. Heat Transfer C*, 85, pp. 230–236.

Liu, C., Tsao, T., Tai, Y.-C., Liu, W., Will, P., and Ho, C.-M. (1995) "A Micromachined Permalloy Magnetic Actuator Array for Micro Robotic Assembly System," *Technical Digest, Eighth Inter. Confer. on Solid-State Sensors and Actuators (Transducers '95 and Eurosensors 9)*, pp. 328–331, 25–29 June, Stockholm, Sweden.

Löfdahl, L., Kälvesten, E., Hadzianagnostakis, T., and Stemme, G. (1994a) "An Integrated Silicon Based Pressure-Shear Stress Sensor for Measurements in Turbulence Flows," *Symp. on Application of Micro-Fabrication to Fluid Mechanics*, ASME Winter Annual Meeting, pp. 245–253, 17–22 November, Atlanta, Georgia.

Löfdahl, L., Kälvesten, E., and Stemme, G. (1994b) "Small Silicon Based Pressure Transducers for Measurements in Turbulent Boundary Layers," *Exp. Fluids*, 17, pp. 24–31.

Löfdahl, L., Kälvesten, E., and Stemme, G. (1996) "Small Silicon Pressure Transducers for Space-Time Correlation Measurements in a Flat Plate Boundary Layer," *J. Fluids Eng.*, 118, pp. 457–463.

Löfdahl, L., Stemme, G., and Johansson, B. (1992) "Silicon Based Flow Sensors for Mean Velocity and Turbulence Measurements," *Exp. Fluids*, 12, pp. 391–393.

Lomas, C.G. (1986) *Fundamentals of Hot-Wire Anemometry*, Cambridge University Press, London.

Ludwieg, H. (1950) "Instrument for Measuring the Wall Shearing Stress of Turbulent Boundary Layers," NACA Technical Memorandum No. 1284, Washington, D.C.

Lueptov, R.M., Breuer, K.S., and Haritonides, J.H. (1988) "Computer-Aided Calibration of X-Probes Using Look-Up Table," *Exp. Fluids*, 6, pp. 115–188.

Madou, M. (2002) *Fundamentals of Microfabrication*, 2nd ed., CRC Press, Boca Raton, Florida.

Maureau, J., Sharatchandra, M.C., Sen, M., and Gad-el-Hak, M. (1997) "Flow and Load Characteristics of Microbearings with Slip," *J. Micromech. Microeng.*, 7, pp. 55–64.

McComb, W.D. (1990) *The Physics of Fluid Turbulence*, Claredon Press, Oxford, United Kingdom.

Menendez, A.N., and Ramaprian, B.R. (1985) "The Use of Flush-Mounted Hot-Film Gauges to Measure Skin Friction in Unsteady Boundary Layers," *J. Fluid Mech.*, 161, pp. 139–159.

Miller, M.F., Allen, M.G., Arkilic, E.B., Breuer, K.S., and Schmidt, M.A. (1997) "Fabry-Perot Pressure Sensor Arrays for Imaging Surface Pressure Distributions," in *Proc. Inter. Confer. on Solid-State Sensors and Actuators (Transducers '97)*, pp. 1469–1472, 16–19 June, Chicago.

Miller, R., Burr, G., Tai, Y.-C., and Psaltis, D. (1996) "Electromagnetic MEMS Scanning Mirrors for Holographic Data Storage," *Sensors and Actuators Workshop*, Catalog No. 96 TRF-0001.

Moin, P., and Bewley, T. (1994) "Feedback Control of Turbulence," *Appl. Mech. Rev.*, 47, no. 6, part 2, pp. S3–S13.

Monin, A.S., and Yaglom, A.M (1975) *Statistical Fluid Mechanics: Mechanics of Turbulence*, vols. 1 and 2, MIT Press, Cambridge, MA.

Moroney, R.M., White, R.M., and Howe, R.T. (1991) "Ultrasonically Induced Microtransport," in *Proc. IEEE MEMS '91*, (Nara, Japan), pp. 277–282, IEEE, New York.

Muller, R.S. (1987) "Strategies for Sensor Research," in *Proc. Fourth Inter. Confer. on Solid-State Sensors and Actuators (Transducers '87)*, pp. 107–111, 2–5 June, Tokyo, Japan.

Munson, B.H., Young, D.F., and Okiishi, T.H. (1990) *Fundamentals of Fluid Mechanics*, Wiley, New York.

Murphy, P., Hubschi, K., DeRooij, N., and Racine, C. (1989) "Subminiature Silicon Integrated Electret Microphone," *IEEE Trans. Electron. Insul.*, 24, pp. 495–498.

Ng, K., Shajii, K., and Schmidt, M. (1991) "A Liquid Shear-Stress Sensor Fabricated Using Wafer Bonding Technology," in *Proc. Sixth Inter. Confer. on Solid-State Sensors and Actuators (Transducers '91)*, pp. 931–934, 24–27 June, San Francisco.

O'Connor, L. (1992) "MEMS: Micromechanical Systems," *Mech. Eng.* 114, February, pp. 40–47.

Odell, G.M., and Kovasznay, L.S.G. (1971) "A New Type of Water Channel with Density Stratification," *J. Fluid Mech.*, 50, pp. 535–543.

Österlund, J. (1999) Experimental Studies of Zero Pressure-Gradient Turbulent Boundary Layer Flow, Ph.D. Thesis, Department of Mechanics, Royal Institute of Technology, Stockholm, Sweden.

Otsuka, K. (1995) "Fundamentals of Shape Memory Alloys-in View of Intelligent Materials," in *Proc. Inter. Symp. on Microsystems, Intelligent Materials and Robots*, pp. 225–230, 27–29 September, Sendai, Japan.

Oudheusden, B., and Huijsing, J. (1988) "Integrated Flow Friction Sensor," *Sensors Actuators A*, 15, pp. 135–144.

Padmanabhan, A. (1997) Silicon Micromachined Sensors and Sensor Arrays for Shear-Stress Measurements in Aerodynamic Flows, Ph.D. Thesis, Department of Aeronautics and Astronautics, Massachusetts Institute of Technology, Cambridge, Massachusetts.

Pan, T., Hyman, D., Mehregany, M., Reshotko, E., and Willis, B. (1995) "Calibration of Microfabricated Shear Stress Sensors," in *Proc. Eighth Inter. Confer. on Solid-State Sensors and Actuators (Transducers '95)*, pp. 443–446, 25–29 June, Stockholm, Sweden.

Patel, V.C. (1965) "Calibration of the Preston Tube and Limitations on its use in Pressure Gradients," *J. Fluid Mech.*, 23, pp.185–208.

Pedley, T.J. (1972) "On the Forced Heat Transfer from a Hot-Film Embedded in the Wall in Two-Dimensional Unsteady Flow," *J. Fluid Mech.*, 55, pp. 329–357.

Perry, A.E. (1982) *Hot-Wire Anemometry*, Claredon Press, Oxford, United Kingdom.

Petersen, K.E. (1982) "Silicon as a Mechanical Material," in *Proc. IEEE 70*, pp. 420–457, IEEE, New York.

Piekos, E.S., Orr, D.J., Jacobson, S.A., Ehrich, F.F., and Breuer, K.S. (1997) "Design and Analysis of Microfabricated High Speed Gas Journal Bearings," AIAA Paper No. 97-1966, AIAA, Reston, Virginia.

Pister, K.S.J., Judy, S.R., Burgett, S.R., and Fearing, R.S. (1992) "Micro-Fabricated Hinges," *Sensors Actuators A*, 33, pp. 249–256.

Pister, K.S.J., Fearing, R.S., and Howe, R.T. (1990) "A Planar Air Levitated Electrostatic Actuator System," IEEE Paper No. CH2832-4/90/0000-0067, IEEE, New York.

Pope, S.B. (2000) *Turbulent Flows*, Cambridge University Press, London, United Kingdom.

Ramaprian, B.R., and Tu, S.W. (1983) "Calibration of a Heat Flux Gage for Skin Friction Measurement," *J. Fluids Eng.*, 105, pp. 455–457.

Richter, A., Plettner, A., Hofmann, K.A., and Sandmaier, H. (1991) "A Micromachined Electrohydrodynamic (EHD) Pump," *Sensors Actuators A*, 29, pp. 159–168.

Ried, R., Kim, E., Hong, D., and Muller, R. (1993) "Pizoelectric Microphone with On-Chip CMOS Circuits," *J. MEMS*, 2, pp. 111–120.

Riethmuller, W., and Benecke, W. (1988) "Thermally Exited Silicon Microactuators," *IEEE Trans. Electron. Devices*, 35, pp. 758–763.

Robinson, S.K. (1991) "Coherent Motions in the Turbulent Boundary Layer," *Annu. Rev. Fluid Mech.*, 23, pp. 601–639.

Rossi, M. (1988) *Acoustics and Electroacoustics*, Artech House, Boston.

Royer, M., Holmen, P., Wurm, M., Aadland, P., and Glenn, M. (1983) "ZnO on Si Integrated Acoustic Sensor," *Sensors Actuators*, 4, pp. 357–362.

Scheeper, P., van der Donk, A., Olthuis, W., and Bergveld, P. (1992) "Fabrication of Silicon Condenser Microphones Using Silicon Wafer Technology," *J. MEMS*, 1, pp. 147–154.

Scheeper, P., van der Donk, A., Olthuis, W., and Bergveld, P. (1994) "A Review of Silicon Microphones," *Sensors Actuators A*, 44, pp. 1–11.

Schellin, R., and Hess, G. (1992) "A Silicon Microphone Based on Piezoresistive Polysilicon Strain Gauges," *Sensors Actuators A*, 32 pp. 555–559.

Schellin, R., Pedersen, M., Olthuis, W., Bergveld, P., and Hess, G. (1995) "A Monolithically Integrated Silicone Microphone with Piezoelectric Polymer Layers," in *MME '95, Micromechanics Europe*, pp. 217–220, 3–5 September, Copenhagen, Denmark.

Schewe, G. (1983) "On the Structure and Resolution of Wall-Pressure Fluctuations Associated with Turbulent Boundary-Layer Flow," *J. Fluid Mech.*, 134, pp. 311–328.

Schmidt, M., Howe, R., Senturia, S., and Haritonidis, J. (1988) "Design and Calibration of a Microfabricated Floating-Element Shear-Stress Sensor," *IEEE Trans. Electron. Devices*, 35, pp. 750–757.

Sen, M., Wajerski, D., and Gad-el-Hak, M. (1996) "A Novel Pump for MEMS Applications," *J. Fluids Eng.*, 118, pp. 624–627.

Sessler, G. (1991) "Acoustic Sensors," *Sensors Actuators A*, 25–27, pp. 323–330.

Sharatchandra, M.C., Sen, M., and Gad-el-Hak, M. (1997) "Navier–Stokes Simulations of a Novel Viscous Pump," *J. Fluids Eng.*, 119, pp. 372–382.

Sharatchandra, M.C., Sen, M., and Gad-el-Hak, M. (1998a) "Thermal Aspects of a Novel Micropumping Device," *J. Heat Transfer*, 120, pp. 99–107.

Sharatchandra, M.C., Sen, M., and Gad-el-Hak, M. (1998b) "A New Approach to Constrained Shape Optimization Using Genetic Algorithms," *AIAA J.*, 36, pp. 51–61.

Sherman, F.S. (1990) *Viscous Flow*, McGraw-Hill, New York.

Smela, E., Inganäs, O., and Lundström, I. (1995) "Controlled Folding of Micrometer-Size Structures," *Science*, 268, pp. 221–223.

Smits, J.G. (1990) "Piezoelectric Micropump with Three Valves Working Peristaltically," *Sensors Actuators A*, 21–23, pp. 203–206.

Solgaard, O., Daneman, M., Tien, N.C., Friedberger, A., Muller, R.S., and Lau, K.Y. (1995) "Optoelectronic Packing Using Silicon Surface-Micromachined Alignment Mirrors," *IEEE Photon. Tech. Lett.*, 7, pp. 41–43.

Speziale, C.G. (1991) "Analytical Methods for the Development of Reynolds Stress Closures in Turbulence," *Annu. Rev. Fluid Mech.*, 23, pp. 107–157.

Speziale, C.G., Sarkar, S., and Gatski, T.B. (1991) "Modelling the Pressure Strain Correlation of Turbulence: An Invariant Dynamical System Approach," *J. Fluid Mech.*, 227, pp. 245–272.

Sprenkels, A.J. (1988) A Silicon Subminiature Electret Capacitive Microphone, Ph.D. Thesis, University of Utrecht, Utrecht, The Netherlands.

Sprenkels, A.J., Groothengel, R.A., Verloop, A.J., and Bergveld, P. (1989) "Development of an Electret Microphone in Silicon," *Sensors Actuators A*, 17, pp. 509–512.

Stanton, T.E., Marshall, D., and Bryant, C.N. (1920) "On the Conditions at the Boundary of a Fluid in Turbulent Motion," *Proc. Roy. Soc. London A*, **97**, pp. 413–434.

Stein, C.F., Johansson, P., Bergh, J., Löfdahl, L., Sen, M., and Gad-el-Hak, M. (2002) "An Analytical Asymptotic Solution to a Conjugate Heat Transfer Problem," *Int. J. Heat Mass Transfer*, **45**, pp. 2485–2500.

Sugiyama, S., Shimaoka, K., and Tabata, O. (1993) "Surface Micromachined Microdiaphragm Pressure Sensor," *Sensors Mater.*, **4**, pp. 265–275.

Suzuki, K., Shimoyama, I., Miuara, H., and Ezura, Y. (1992) "Creation of an Insect-Based Microrobot with an External Skeleton and Elastic Joints," in *Proc. IEEE Fifth Inter. Workshop on MEMS (MEMS '92)*, pp. 190–195, 4–7 February, Travemunde, Germany.

Syms, R.R.A. (1998) "Demonstration of Three-Dimensional Microstructure Self-Assembly," *Sensors Actuators A*, **65**, pp. 238–243.

Tardu, S., Pham, C.T., and Binder, G. (1991) "Effects of Longitudinal Diffusion in the Fluid and Heat Conduction to the Substrate on Response of Wall Hot-Film Gauges," in *Advances in Turbulence 3*, A.V. Johansson and P.H. Alfredsson, eds., Springer-Verlag, Berlin.

Taylor, G.I. (1935) "Statistical Theory of Turbulence," in *Proc. Roy. Soc. London A*, 151, pp. 421–478.

Taylor, G. (1972) "Low-Reynolds-Number Flows," in *Exp. in Fluid Mech.*, pp. 47–54, Nat. Com. Fluid Mech. Films, MIT Press, Cambridge, Massachusetts.

Tennekes, H., and Lumley, J.L. (1972) *A First Course in Turbulence*, MIT Press, Cambridge.

Tien, N.C. (1997) "Silicon Micromachined Thermal Sensors and Actuators," *Microscale Thermophys. Eng.*, 1, pp. 275–292.

Tien, N.C., Kiang, M.H., Daneman, M.J., Solgaard, O., Lau, K.Y., and Muller, R.S. (1996a) "Actuation of Polysilicon Surface-Micromachined Mirrors," *SPIE Proc. Miniaturized System with Micro-Optical and Micromechanics* 2687 pp. 53–59, 30–31 January, San Jose, California.

Tien, N.C., Solgaard, O., Kiang, M.H., Daneman, M.J., Lau, K.Y., and Muller, R.S. (1996b) "Surface Micromachined Mirrors for Laser Beam Positioning," *Sensors Actuators A*, **52**, pp. 76–80.

Timoshenko, S. (1925) "Analysis of Bi-Metal Thermostats," *J. Opt. Soc. Am.*, 11, pp. 233–255.

Townsend, A.A. (1976) *The Structure of Turbulent Shear Flows*, 2nd ed., Cambridge University Press, London.

Tsao, T., Liu, C., Tai, Y.-C., and Ho, C.-M. (1994) "Micromachined Magnetic Actuator for Active Flow Control," in *ASME FED-197*, pp. 31–38, ASME, New York.

Van Lintel, H.T.G., Van de Pol, F.C.M., and Bouwstra, S. (1988) "A Piezoelectric Micropump Based on Micromachining of Silicon," *Sensors Actuators*, **15**, pp. 153–167.

Wannier, G.H. (1950) "A Contribution to the Hydrodynamics of Lubrication," *Q. Appl. Math.*, **8**, pp. 1–32.

White, F.M. (1991) *Viscous Fluid Flow*, 2nd ed., McGraw-Hill, New York.

Wilkinson, S.P. (1990) "Interactive Wall Turbulence Control," in *Viscous Drag Reduction in Boundary Layers*, D.M. Bushnell and J.N. Hefner, eds., pp. 479–509, AIAA, Washington, D.C.

Willmarth, W.W., and Wooldridge, C.E. (1962) "Measurements of the Fluctuating Pressure at the Wall Beneath a Thick Turbulent Boundary Layer," *J. Fluid Mech.*, **14**, pp. 187–210.

Wiltse, J.M., and Glezer, A. (1993) "Manipulation of Free Shear Flows Using Piezoelectric Actuators," *J. Fluid Mech.*, **249**, pp. 261–285.

Winter, K. (1977) "An Outline of the Techniques Available for the Measurement of Skin Friction in Turbulent Boundary Layers," *Prog. Aerosp. Sci.*, **18**, pp. 1–55.

Voorthuysen, J.A., Bergveld, P., and Sprenkels, A.J. (1989) "Semiconductor-Based Electret Sensors for Sound and Pressure," *IEEE Trans. Electr. Insul.*, **24**, pp. 267–276.

Wu, M.C., Lin, L.Y., Lee, S.S., and Pister, K.S.J. (1995) "Micromachined Free-Space Integrated Micro-Optics," *Sensors Actuators A*, **50**, pp.127–134.

Yang, X., Grosjean, C., Tai, Y.-C., and Ho, C.-M., (1997) "A MEMS Thermopneumatic Silicone Membrane Valve," in *Proc. IEEE Tenth Inter. Workshop MEMS (MEMS '97)*, 26–30 January, Nagoya, Japan.

Yasuda, T., Shimonoyama, I., and Mirura, H. (1997) "CMOS Drivable Electrostatic Microactuator with Large Deflection," in *Proc. IEEE Tenth Inter. Workshop MEMS (MEMS '97)*, 26–30 January, Nagoya, Japan.

7

Microrobotics

Thorbjörn Ebefors
SILEX Microsystems AB

Göran Stemme
Royal Institute of Technology

7.1   Introduction

The microelectromechanical systems (MEMS) field has traditionally been dominated by silicon micro-machining. In the early days, efforts were concentrated on fabricating various silicon structures and rel-atively simple components and devices were developed. For describing this kind of microelectromechanical structures the acronym MEMs is used. A growing interest in manufacturing technologies other than the integrated circuit (IC)-inspired silicon wafer and batch MEMs fabrication is evident in the microsystem field today. This interest in alternative technologies has surfaced with the desire to use new MEMs mate-rials that enable a greater degree of geometrical freedom than materials that rely on planar photolitho-graphy as a means to define the structure. One such new MEMs material is plastic, which can be used to produce low-cost, disposable microdevices through microreplication.

The micromachining field has also matured and grown from a technology used to produce simple devices to a technology used for manufacturing complex miniaturized systems which has shifted the acronym from representing structures to microelectromechanical systems. Microsystems encompass microoptical systems (microoptoelectromechanical systems, MOEMS), microfluidics (micro-total analysis systems, μ-TAS), etc. These systems contain micromechanical components including moveable mirrors and lenses, sensors, light sources, pumps and valves, and passive components such as optical and fluidic waveguides, as well as electrical components and power sources of various types.

With the growing focus on systems perspective — for example, the integration of various functions such as sensors, actuators, and processing capabilities into miniaturized systems (which may also be fabricated by different fabrication technologies in the MEMS field) — new problems arise. For example, how does one assemble these very small devices to form larger systems? As early as 1959 Professor Feynman addressed "the problem of manipulating and controlling things on small scale" in his famous APS talk [Feynman, 1960]. Feynman's solution to microassembling problems was the use of micromachines (or microrobotic devices as they will be named in this chapter), which would be used to build or assemble other micromachines (microsystems) consisting of different microdevices. This approach can be regarded as a hybrid technology or serial (one-at-a-time) fabrication and assembly, as opposed to the monolithic approach, where all integration is done on one single silicon chip, preferably using wafer-level assembly. Wafer-level assembly performed on several wafers at the same time (so-called batch fabrication) is a parallel (several-at-a-time) manufacturing process originally developed for the IC industry but now also commonly used for MEMS fabrication. General and good reviews of the different assembly approaches for MEMS have been presented [Cohn et al., 1998; MSTnews, 2000]. The different aspects of MEMS assembly will be further discussed in Section 7.3.2.

All systems, no matter what kind of components they consist of, must be assembled in some way. The assembly of complete microsystems can be done either monolithically (several systems simultaneously assembled at wafer level) or in a hybrid fashion (by serial assembly of several individual microcomponents fabricated by different technologies and on different wafers). In both cases microrobotic devices could simplify the assembly process (i.e., microrobots for making MEMS); however, these microrobots must be highly miniaturized. A common way to obtain miniaturization is by using MEMS technologies (MEMS for making microrobots). The benefits of scaling down all the subsystems of a robot to the same scale as the control systems by integrating motors, sensors, logic, and power supplies onto a single piece of silicon have been discussed by Flynn et al. (1989). Enormous advantages can be obtained in the form of mass produce-ability, lower costs, and fewer connector problems encountered when interconnecting these discrete subsystems in comparison to the macro- (or miniature) robot concepts.

The hybrid microassembling approach, allowing microdevices to be fabricated by various techniques and in various materials to be assembled into more complex systems, is one of the driving motives for robotic miniaturization in general, and MEMS-based microrobotics in particular. Of course, miniaturized robots are also very attractive for assembling all kinds of miniaturized components (e.g., watch components). Besides the use of microrobotic devices for assembling purposes, other application fields such as medical technology may benefit greatly from the use of miniaturized robot devices. One such example is the steerable catheters used for minimal invasive surgery (MIS) shown in Figure 7.1. Such microrobotic catheters have been fabricated using MEMS technologies [Haga et al., 1998; Park and Esashi, 1999].

7.2   What is Microrobotics?

Today we can see a growing worldwide interest in microrobotic devices and their potential, including micromanipulation tools, microconveyers, and microrobots as locomotive mechanisms [Ebefors, 2000]. The term microrobotics covers several different types of small robot devices and systems. The term mainly refers to the complete robot system including what type of work the microrobots perform, how to accomplish those tasks, and other system-oriented aspects. Often, the term microrobotics also includes more fundamental building blocks, such as steerable links, microgrippers, conveyer systems, and locomotive robots, as well as the technologies used to fabricate these devices and the control algorithms used to carry out the robot's task. The term microrobotic can be compared with the MEMS (or MEMs) term in that the former often encompasses not only the system itself, but also fabrication and material issues, as well as simple devices used as building blocks for the system. Although the microrobotics field currently is an area of intensive research, no clear definition of the term microrobot exists. As stated by other authors reviewing the microrobotics field [Dario et al., 1992; Fatikow and Rembold, 1997], many "micro" terms such as micromechatronics, micromechanism, micromachines, and microrobots are used synonymously to indicate a wide range of devices that have a function related to the "fuzzy" concept of operating at a "small" scale; however, "small" scale is a relative term so a more clear definition is needed.

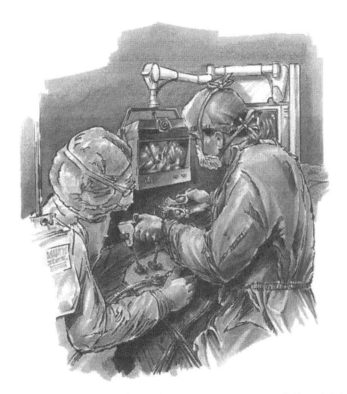

FIGURE 7.1 Example of a microrobotic application. In minimal invasive surgery (MIS), MEMS-based microrobotic devices such as steerable catheters, microgrippers, and other microtools are expected to have a deep impact on endoscopic applications in the near future. (Illustration printed with permission from Surgical Vision, Ltd., U.K. courtesy of Graham Street.)

7.2.1 Task-Specific Definition of Microrobots

The basic definition of a microrobot parallels the features attributed to a robot in the macroworld; a microrobot should be able to move, apply forces, manipulate objects, etc. in the same way as a "macrorobot." The obvious difference between a macrorobot and a microrobot is the size of the robot. Thus, one definition of a microrobot is "a device having dimensions smaller than classical watch-making parts (i.e., μm to mm) and having the ability to move, apply forces, and manipulate objects in a workspace with dimensions in the micrometer or submicrometer range" [Johansson, 1995]. However, in many cases it is important that the robot can move over much larger distances. This task-specific definition is quite wide and includes several types of very small robots as well as stationary micromanipulation systems, which are a few decimeters in size but can carry out very precise manipulation (in the micron or even nanometer range) [Fatikow and Rembold, 1997]. Because the fabrication technologies used to fabricate these devices play an important role, another more precise subdivision is desirable in order to help identify the practical capabilities of the different technologies.

7.2.2 Size and Fabrication Technology Based Definitions of Microrobots

Following the classification scheme of microrobots used by Dario et al. (1992) and Fatikow and Rembold (1997), one can categorize the microrobot into many different groups. In the definition made by Dario et al., the microrobots were separated into three different subcategories, each characterized by the fabrication technology used to obtain the robot and the size of the device, as illustrated in Table 7.1.

Some good examples of "miniature robots" are those that competed in the annual "Micro Robot Maze Contest 1992–99" at the International Symposium on Micromachine and Human Science (MHS) in

TABLE 7.1   Classification of Microrobots According to Size and Fabrication Technology

Robot Class	Size and Fabrication Technology
Miniature robots or minirobots	Having a size on the order of a few cubic centimeters and fabricated by assembling conventional miniature components as well as some micromachines (such as MEMS based microsensors)
MEMS-based microrobots (or microrobots[a])	A sort of "modified chip" fabricated by silicon MEMS-based technologies (such as batch-compatible bulk or surface micromachining or by micromolding and/or replication method) having features in the micrometer range
Nanorobots	Operating at a scale similar to the biological cell (on the order of a few hundred nanometers) and fabricated by nonstandard mechanical methods such as protein engineering

[a]To distinguish a MEMS-based microrobot with micrometer-sized components from the whole class of microrobots (including mini-, micro- and nanorobots), several more or less confusing notations have been proposed. In this publication, the term "MEMS-based microrobot" is introduced and used. The term "MEMS-based microrobot" differs from the notation originally used by Dario et al., but the content is the same.
Source: Adapted from Dario et al., 1992.

Nagoya, Japan [Dario et al., 1998; Ishihara, 1998]. MEMS-based microrobots will be further described in the rest of this chapter.

To distinguish a MEMS-based microrobot with micrometer-sized components from the whole class of microrobots (including mini-, micro-, and nanorobots), several more or less confusing notations have been proposed. In this publication, the term "MEMS-based microrobot" is introduced and used. The term "MEMS-based microrobot" differs from the notation originally used by Dario et al., but the content is the same.

While MEMS-based microrobots consist of billions of atoms and are fabricated using photolithography techniques and various etching methods, nanorobots will most likely be built by assembling individual atoms or molecules one at a time. Instead of using conventional mechanical principles, an approach based on chemical self-assembly will most likely be the technology used to fabricate nanorobots. Examples of these nanorobots (nanomachines or molecular machines) and some proposed fabrication methods used to achieve the different components, such as gears, bearings, and harmonic drives, needed for the realization of a nanorobot are found in Drextler's textbook (1992). In 1999, Kelly et al. presented the first reports of a successful molecular motor. That nanomotor, consisting of a 78-atom molecular paddle wheel, was able to rotate one third of a complete revolution in approximately 3 to 4 hours.

Researchers in the field of theoretical physics pushing for further miniaturization have introduced the concept of quantum machines and quantum computers, which will consist of stretched parts of a single atom [Wilson, 1997]. If we stretch the term microrobotic a bit, devices such as the atomic force microscope (AFM) or scanning tunneling microscope (STM) could also be categorized as microrobotic tools. The sharp probe tips of these microscope devices, which most often are fabricated using MEMS technologies, have been used to manipulate single atoms [Eigler and Schweizer, 1990]. To extend this further, nanorobots and other nanoelectromechanical systems (NEMS) may be fabricated using sharp silicon tips and other microrobotic tools (MEMS for nanorobots and NEMS).

7.2.3 Mobility- and Functional-Based Definition of Microrobots

Besides classification by task or size, microrobots can also be classified by their mobility and functionality [Hayashi, 1991; Dario et al., 1992]. Many robots, including the three different classes of microrobots mentioned in Table 7.1, usually consist of sensors and actuators, a control unit and an energy source. Depending on the arrangement of these components, one can classify microrobots according to the following criteria:

- Locomotive and positioning possibility (yes or no)
- Manipulation possibility (yes or no)
- Control type (wireless or wires)
- Autonomy (energy source onboard or not onboard)

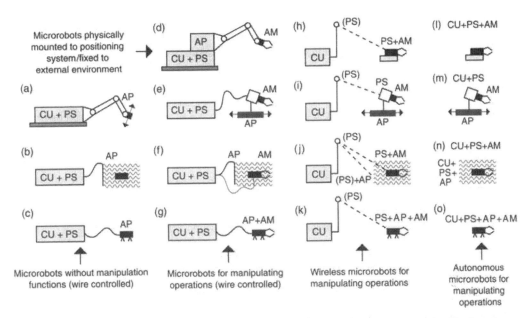

FIGURE 7.2 Classification of microrobots by functionality (modification of earlier presented classification schemes [Hayashi, 1991; Dario et al., 1992; Fatikow and Rembold, 1997]). CU indicates the control unit; PS, the power source or power supply; AP, the actuators for positioning; AM, the actuators for manipulation.

Figure 7.2 illustrates 15 different possible microrobot configurations by combining the four criteria. As depicted in Figure 7.2, the classification is dependent on the following microrobot components: the control (CU), the power source (PS), the actuators necessary for moving the robot platform (robot drive for locomotion and positioning; AP), and the actuators necessary for operation (manipulation using robot arms and hands; AM). Besides the different actuation functions, sensory functions are also needed — for example, tactile sensors for microgrippers or charge-coupled device (CCD) cameras for endoscope applications (compare Figures 7.2d and a).

The ultimate goal is to create a fully autonomous, wireless mobile microrobot equipped with suitable microtools according to Figure 7.2o. Because this is a very difficult task, a good start is to investigate the possibility of making silicon microrobot platforms that are steered and powered through wires, like the one in Figure 7.2c, and to study their locomotion capability.

The majority of MEMS-based microrobotic devices developed so far could be categorized as moveable links: microcatheters [Haga et al., 1998; Park and Esashi, 1999], according to Figure 7.2a; microgrippers [Kim et al., 1992; Keller and Howe, 1995; 1997; Greitmann, 1996; Keller, 1998a; 1998b; Ok et al., 1999] as those in Figure 7.2d; or the microgrippers [Jager, 2000a, 2000b] shown in Figure 7.2e. Among the research publications covering locomotive microrobots, most publications have addressed microconveyance systems (Figure 7.2b) [Kim et al., 1990; Pister et al., 1990; Ataka et al., 1993a; 1993b; Konishi and Fujita, 1993; 1994; Goosen and Wolffenbuttel, 1995; Liu et al., 1995; Böhringer et al., 1996; 1997; Nakazawa et al., 1997; Suh et al., 1997; 1999; Hirata et al., 1998; Kladitis et al., 1999; Nakazawa et al., 1999; Ruffieux and Rooij, 1999; Smela et al., 1999]: see Table 7.3 for more details. Robots using external sources for locomotion could be used (compare Figures 7.2b, f, j, n).

According to Fatiokow and Rembold (1997), several researchers are working on methods to navigate micromechanisms through human blood vessels; however, these microrobots are difficult to control. Examples of partially autonomous systems (compare Figure 7.2j) are the concept for so-called "smart pills." Centimeter-sized pills for sensing temperature or pH inside the body have been presented [Zhou, 1989; Uchiyama, 1995] as well as pills equipped with video cameras [Carts-Powell, 2000]. The pill is swallowed and transported to the part of the body where one wants to measure or record a video sequence. The information of the measured parameter or the signals from the camera is then transmitted (telemetrically) out of the body. More sophisticated approaches involving actuators (AM) for drug delivery of various kinds have also been proposed [Uchiyama, 1995; Fatikow and Rembold, 1997]. The position of the pill inside the body is located by an X-ray monitor or ultrasound. As soon as the pill reaches an infected area, a drug encapsulated in the pill can be released by the actuators onboard. External communication could be realized through radio signals.

Several important results have been presented regarding walking microrobots (Figures 7.2c, g, k) fabricated by MEMS technologies and batch manufacturing. Different approaches for surface-micromachined robots [Yeh et al., 1996; 2000; Kladitis, 1999] and for a piezoelectric dry-reactive-ion-etched microrobot should be mentioned. A suitable low-power Application Specific Integrated Circuit (ASIC) for robot control has been successfully tested and is planned for integration on a walking microrobot [Ruffieux and Rooij, 1999]. The large European Esprit project MINIMAN (1997) is planning to develop moveable microrobotic platforms with integrated tools with six degrees of freedom for applications such as microassembly within a scanning electron microscope (SEM). This project involves different MEMS research groups from several universities and companies across Europe. One of the MINIMAN robots, as well as other MEMS-based microrobots, will be further described in Section 7.5.3.

Further, miniature robot systems with MEMS/MST (MicroSystem Technology) components have been developed [Breguet, 1996; Breguet and Renaud, 1996]. Several research publications on "gnat" minirobots and actuator technologies for MEMS microrobots [Flynn et al., 1989; 1992] were reported by U.S. researchers in the early 1990s. Also, several groups in Japan are currently developing miniaturized robots based on MEMS devices [Takeda, 2001]. In Japan, an extensive 10-year program on "micromachine technology," supported by the Ministry of International Trade and Industry (MITI), started in 1991. One of the goals of this project is to create micro- and miniature robots for microfactory, medical technology, and maintenance applications. Several microrobotic devices including locomotive robots, and microconveyers have been produced within this program. Miniature robot devices or vehicles [Teshigahara et al., 1995] for locomotive tasks, containing several MEMS components, have been presented.

Even though great efforts have been made on robot miniaturization using MEMS technologies, no experimental results on MEMS batch-fabricated microrobots suitable for autonomous walking (i.e., robust enough to be able to carry its own power source or to be powered by telemetric means) have been presented yet (as of year 2000). The first batch-fabricated MEMS-based microrobot platform able to walk was presented in 1999 [Ebefors et al., 1999] (Figures 7.17 and 7.23). However, this robot was powered through wires and not equipped with manipulation actuators. Besides the walking microrobotic devices, several reports on flying robots [Arai et al., 1995; Mainz, 1999; Miki and Shimoyama, 1999; 2000] and swimming robots [Fukuda et al., 1994; 1995] have been published. Micromotors and gear boxes made using LIGA technology (a high-precision, lithographically defined plating technology) are used to build small flying microhelicopters, which are commercially available from the Institute of Microtechnology in Mainz, Germany, as rather expensive demonstration objects. Besides the pure mechanical microrobots, hybrid systems consisting of electromechanical components and living organisms such as cockroaches have also been reported [Shimoyama, 1995a].

7.3   Where To Use Microrobots?

7.3.1   Applications for MEMS-Based Microrobots

Already in the late 1950s, Feynman (1961) foresaw the utility and possibility for miniaturization of machines and robots. According to Feynman, the concepts of microfactory and microsurgery (based on tiny microdevices inserted in a patient's blood vessels) were very interesting possibilities for small machines.

Microrobotic research has gone from being only theoretical ideas in the 1960s and 1970s to actual building blocks when complex systems and "micromachines" on a chip started to emerge in the late 1980s. These building blocks took the form of surface-micromachined micromotors made of polysilicon fabricated on a silicon chip [Muller, 1990]. In the late 1980s and early 1990s, more concrete suggestions on how one should realize MEMS-based microrobotic devices using such micromotors and potential applications were published [Flynn et al., 1989; Trimmer and Jebens, 1989]. Now, 10 years later, one finds research publications on microrobotic devices for practical use in various fields. One of these application fields is medical technology [Dario et al., 1997; Haga et al., 1998; Tendick et al., 1998; Park and Esashi, 1999]. In surgery, the use of steerable catheters and endoscopes, as illustrated in Figure 7.1, is very attractive and they may be further miniaturized by MEMS-based microrobotic devices, allowing advanced computer-assisted surgery (CAS) and even surgery over the Internet. For such applications, smaller and more flexible active endoscopes that can react to instructions in real time are needed in order to assist the surgeon. They may enter into the blood vessels and enter various cavities (angioplasty) by remote control, where they carry out complex measurements and manipulations (gripping, cutting, applying tourniquets, making incisions, suction, and rinsing operations, etc.). To meet these demands, an intelligent endoscope must have a microprocessor, several sensors and actuators, a light source, and possibly an integrated image processing unit. These microrobotic devices will revolutionize classical surgery, but their realization is still a problem because of friction, poor navigability, biocompatibility issues, etc.; they are also not small enough yet [Fatikow and Rembold, 1997].

Other interesting areas for microrobotic devices are production (microassembly [Fatikow and Rembold, 1996; Cohn et al.,1998] and microfactory [Ishikawa, 1997; Kawahara et al., 1997]); metrology (automated testing of microelectronic chips, surface characterization, etc.) [Li et al., 2000]; inspection and maintenance [Suzumori et al., 1999; Takeda, 2001]; biology (capturing, sorting and combining cells [Ok, 1999]); bio-engineering [Takeuchi and Shimoyama, 1999]; and microoptics (positioning of microoptical chips, microlenses, and prisms) [Frank, 1998]. Many of these applications require automated handling and assembly of small parts with accuracy in the submicron range. Different approaches for microassembly, among them microrobotic-assisted assembly, are described in the following section.

7.3.2 Microassembly

Two different aspects of MEMS microrobotics and microassembly should be addressed. The first aspect is the use of MEMs for making microrobots using the silicon monolithic chip fabrication technology (several robots at a time; see Table 7.1). Because many microrobotic devices need three-dimensional (i.e., out-of-plane) arms and legs, the folding of these components is an essential assembly issue and constitutes one side of the assembling technology (i.e., MEMs for making microrobots). The second aspect is the use of microrobots for assembling more complex microsystems (microrobots for making MEMS). The use of microrobots for assembly of microcomponents and devices (fundamental building blocks) into complete microsystems is classified as hybrid manufacturing (one system at a time).

The MEMS fabrication technology, especially surface-micromachining, allows batch-fabricated mechanisms to be preassembled in situ, and all structures on the entire wafer are released simultaneously (parallel assembly) by etching of sacrificial layers. This, together with the ability to integrate electronics on the chip, is a very attractive feature. Micromechanical systems of impressive complexity such as linear motors, rotating gears, and linkages between these components have been made with alignment tolerances in the micrometer range by the use of parallel fabrication methods [Rodgers et al., 1999]. In the MOEMS and microrobotic fields, several devices make use of both parallel and serial assembling. To fold micromirrors out of the wafer plane to achieve, for example, microoptical benches, several serially addressed motors fabricated by standard parallel processing (lithography, thin-film depositions and etching) can be used to achieve the assembling of each individual micromirror. These mirrors are controlled externally by electrical signals connected to the integrated micromotors [Rodgers et al., 1999]. Figure 7.3f shows a complex free-space microoptical bench (FSMOB) that integrates several different optical elements [Lin et al., 1996]. To implement free-space optics monolithically, tall out-of-plane optical elements (moveable mirrors lenses, etc.) with optical axes parallel to the substrate are needed. A common way to realize this kind of system is by using the surface-micromachined microhinge technology [Pister et al., 1992], which enables out-of-plane rotated optical elements to be fabricated using planar processes as shown in Figures 7.3a–e. The planar elements are then assembled into three-dimensional structures after fabrication either by built in-motors for self-assembly or by the use of microrobotic devices and external manipulation. Also integrated

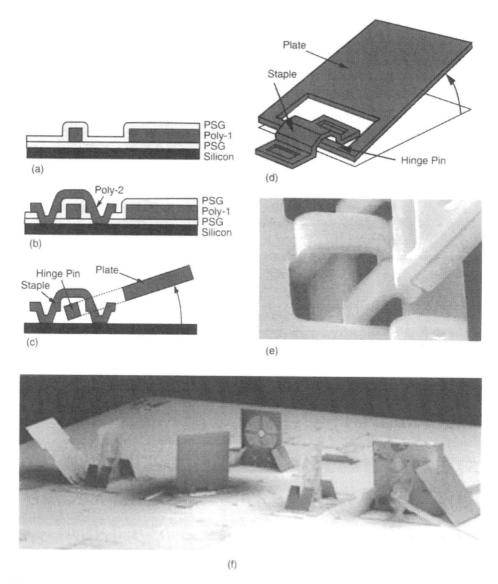

FIGURE 7.3   (a)–(c) Cross sections of a microhinge during fabrication. (d) A perspective view of a hinged plate after release by external manipulators. (e) SEM photograph of a polysilicon hinge. (f) Example of a complex microoptical system based on out-of-plane rotated optical elements using the surface-micromachined microhinge and hybrid mounting of a laser [Lin et al., 1996]. (Photograph of the optical bench taken by David Scharf and printed with permission. Courtesy of M.C. Wu, UCLA and K. Pister, BSAC-Berekley.)

in the FSMOB system (Figure 7.3), is a self-alignment technique that incorporates hybridly mounted active optoelectronic devices such as semiconductor lasers and vertical cavity surface-emitting lasers [Lin et al., 1996]. One key advantage of FSMOB is that the optics can be "prealigned" during the design stage by CAD.

For some applications with very complex out-of-plane folding, the serial self-assembly technology using integrated micromotors is not feasible; therefore, quasi-manual or semi-automated assembly using external manipulators is required. To assemble several folded microstructures into more complex out-of-plane devices, MEMS-based microrobotic tools are used mainly because of their precise and small dimensions. Different MEMS technologies can be used to achieve out-of-plane rotated structures [Ebefors, 2000]. Besides MOEMS applications, out-of-plane rotated microstructures are also essential for the realization of most MEMS-based microrobotic devices. However, for most microrobot applications, the robustness of surface-micromachined legs and arms that are folded out of the wafer are limited. Therefore, bulk micromachining is an attractive technology to fabricate more robust legs and arms for various microrobots.

Besides by use of surface-micromachining, impressive and complex microsystems can also be achieved using bulk micromachining where different kinds of wafers can be stacked on top of each other by bonding [Tong and Gösle, 1999]. However, by using bulk micromachining, the size is generally one order of magnitude larger than for surface-micromachined devices. Several packaging schemes have been used to assemble complete microsystems on the wafer level (parallel batch fabrication) [Corman, 1999]. However, for many applications the use of lithography, etching, and bonding alone cannot fulfill the requirements on the system level. Monolithic integration of electronic and micromechanical functions is not always the most suitable. By combining microelectronics, micromechanics, microfluidics, or microoptics into a single fabrication process, one faces the risk that one may have to compromise the performance of all subsystems.

A disadvantage for all integration of electronic and mechanical components is that the electronics (complementary metal oxide semiconductor [CMOS], bipolar CMOS [BiCMOS], etc.) and MEMS fabrication processes often are not compatible, due to thermal budgets, wafer size, etc. Wafer bonding (flip chip, fusion bonding, etc.) and the monolithic integration approaches of electronics and MEMS on the same chip are then not realistic assembling alternatives. In those cases, the serial assembly ("pick-and-place") can be the only assembly approach that allows integration of electrical and mechanical components. Such micro pick-and-place systems can be achieved by microrobotic devices in the form of microtweezers and microgrippers [Kim et al., 1992; Keller and Howe, 1995; 1997; Greitmann, 1996; Keller, 1998a; 1998b; Ok et al., 1999] and will be further described in Section 7.5.1. To extend this serial assembling approach further, other microrobotic components such as conveyers and miniaturized robots (preferably fabricated using MEMS technologies) are also required; see Section 7.6 regarding the concepts on microfactories and desktop micromanipulation stations.

Besides the monolithic assembling on the wafer level and the serial pick-and-place assembly, parallel approaches for hybrid assembling are also possible. Pister et al. [see Tahhan et al., 1999] presented a concept where folded boxes using three-dimensional microstructures based on aluminum hinges were used to obtain wafer-sized pallets for automated microparts assembly and inspection, as illustrated in Figure 7.4. The microparts (small diced MEMS or electronic chips) are supposed to be randomly transported on the pallet by an external vibration field or by integrated microconveyers (see Section 7.5.2). The boxes with three walls, connected with microlocks, have one opening to capture the conveyed microparts. The microparts are sorted by their size or geometrical dimensions and kept in place by the walls of the boxes for handling, testing, or assembling tasks. In the proposed concept, the fixtures should have integrated sensors and actuators by which the clamps actively close the box when the parts enter the fixture. For this concept, large fixture arrays and electrostatic or optical sensors integrated into the fixture cell should trigger the clamping function. Each cell operates autonomously and no global control should then be necessary.

Parallel sorting and assembling can also be obtained by other MEMS-based microconveyer strategies. Autonomous distributed micromachines (ADMs) are composed of several microcells and are smart enough to decide their behavior by themselves (by the use of integrated sensors) [Konishi and Fujita, 1995]. This concept has been validated for conveyance systems using a computer simulation model. Theories on programmable vector fields for advanced control of microconveyance systems have been presented [Böhringer et al., 1997]. Recently, these algorithms, which do not require sensing or feedback, were experimentally tested using polyimide thermal bimorph ciliary microactuator arrays with integrated CMOS electronics [Suh et al., 1999].

Besides the assembling of complete microsystems, other types of assembling are also required in MEMS. One such example, which already has been mentioned, is the need for three-dimensional MEMS structures for which one wants to create structures or sensor elements with feature sizes so small (micron or submicron) that they require photolithography in all three dimensions. Because normal IC-inspired

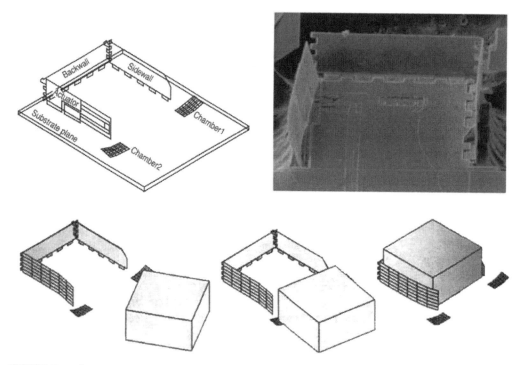

FIGURE 7.4 Illustrations and SEM micrograph of a folded fixture cell for automated microparts assembly and inspection. (Illustration printed with permission and courtesy of K. Böhringer, University of Washington.)

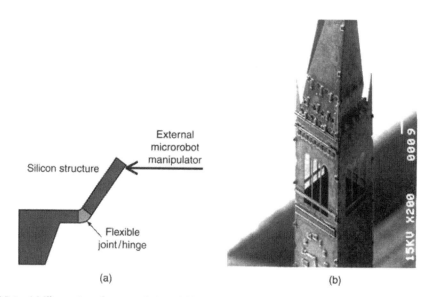

FIGURE 7.5 (a) Illustration of an out-of-plane folded silicon microstructure using an external microrobot tool for the assembly manipulation. The three-dimensional silicon structure which is structured in-plane could be fabricated both with surface- and bulk-micromachining techniques. (b) Example of a complex three-dimensional structure assembled by a MEMS-robotic tool. The three-dimensional model shows the University of California, Berkeley, Companile clock tower having a total height of 1.8 mm above the wafer surface. The microassembly was made manually by the use of the microtool shown in Figure 7.9. (Printed with permission and courtesy of Hui Elliot, BSAC-Berkeley, CA.)

lithography is planar, a commonly used method is to fabricate all structures requiring high-resolution patterns in the plane and then in a final process step-fold these structures out-of-plane using various types of microhinges or microjoints [Ebefors et al., 1998b]. The out-of-plane microstructure assembling can be obtained in the same way as the system level assembling by: (1) serial self-assembling (built-in actuation by micromotors); (2) parallel batch self-assembly (using built-in mechanisms such as surface tension forces [Green et al., 1995; Syms, 1998; 1999] or volume shrinkage effects in polymers [Ebefors, 1998a]) for the assembling; or (3) the use of external manipulators (not self-assembly). Self-assembling of three-dimensional microstructures have been obtained using a variety of techniques [Ebefors, 2000]. All of these techniques are of great interest, as manual assembly is very time, labor, and cost intensive. However, since self-assembly is not feasible in all applications, quasi-manual or semi-automatic assembling using microrobotic tools is one way of reducing cost and time.

Complex three-dimensional structures, such as the clock tower model shown in Figure 7.5, have been assembled using MEMS-based microrobotic tools connected to motorized micromanipulators, as illustrated in Figure 7.9 [Hui et al., 2000]. For most MEMS-based microrobotic devices robust out-of-plane working structures are essential (arms and legs fabricated by bulk micromachining). To be useful as microrobots, these structures also need to produce large strokes and forces when actuated. The next section describes how this can be realized.

7.4 How To Make Microrobots

During the design of a MEMS-based microrobotic device, trade-offs among several parameters such as range of motion, strength, speed (actuation frequency), power consumption, control accuracy, system reliability, robustness, force generation, and load capacity must be taken into consideration. These parameters strongly depend on the actuation principle. In the following section, the physical parameters influencing the performance, as well as different aspects, of actuators are described.

7.4.1 Arrayed Actuator Principles for Microrobotic Applications

For most microactuators, the force that can be generated and the load the actuator withstands are limited. This constitutes a fundamental drawback for all miniaturization in the robotic field. However, by using MEMS microfabrication it is easy to fabricate large arrays of actuators where the actuators are working together and thereby increase the total force and load capability. This approach, which takes advantage of lithographic fabrication, also has the potential for integrating electronics on the robot chip to carry out more sophisticated tasks (e.g., autonomous microrobots). The concept of array configuration of microactuators, where the cooperative work of many coordinated simple actuators generates interaction with the macro world, was introduced by Fujita and Gabriel in 1991. They called their technique distributed micromotion systems (DMMS). The driving schemes for the actuators in the array can be in either a synchronous mode (all actuators are switched on and off at the same time) or an asynchronous mode (different actuators are switched on and off at different times). Most often, the asynchronous mode is favorable as it is more effective and smoother movements can be achieved. Intelligent control of the actuation schemes for the actuators can be achieved by integration of sensors in the system [Suh et al., 1997]. Based on information from these sensors (object weighing or positioning), advanced control of each actuator can be obtained for improved functionality of the micromotion system.

7.4.2 Microrobotic Actuators and Scaling Phenomena

Several theoretical reports on the scale effects and the applicability of various actuation principles for microrobotic devices have been published [Trimmer and Jebens, 1989; Dario et al., 1992; Shimoyama, 1995b; Fearing, 1998; Thornell, 1998]. As stated in the previous section, most microactuators generate relatively small forces and strokes, but these could somewhat be increased by array configurations. However, there are large variations between how different actuation principles scale to microscale; some principles scale more favorably than others. Some actuators (e.g., piezoelectric- and electrostatic-based actuators) have the advantage of low power consumption and can be driven at high speeds (kHz regime and above). However, they generally show relatively low force capacity and small strokes, while others such as magnetic and thermal actuation principles have the potential to exert large forces and displacements when the driving speed is the limiting factor. The thermal and magnetic actuator principles rely on significant currents and power levels which may require forced cooling. During the design of a microrobotic device, the trade-offs among range of motion, strength, speed (actuation frequency), power consumption, control accuracy, system reliability, robustness, load capacity, etc. must be taken into consideration. In an extensive actuator review focusing on microrobotics [Fearing, 1998], the general conclusion was that actuators generating large strokes and high forces are best suited for microrobotics applications. The speed criteria are of less importance as long as the actuation speed is in the range of a couple of hertz and above. Table 7.2 provides some data on stroke and force generation capability as well as power densities and efficiencies for a small selection of actuators suitable for microrobotic applications.

TABLE 7.2 Comparison of a Selection of Microactuators for Microrobotic Applications

Actuator Type	Volume $(10^{-9} m^3)$	Speed $(s^{-1}$ or rad/s*)	Force (N) Torque (Nm)**	Stroke (m)	Power Density (W/m^3)	Power Consumption (W)	Ref.
Linear electrostatic	400	5000	10^{-7}	6×10^{-6}	200	NA	[Kim et al., 1992]
Rotational electrostatic	$\pi/4 \times 0.5^2 \times 3$	40*	2×10^{-7}**		900	NA	[Nakamura et al., 1995]
Rotational piezoelectric	$\pi/4 \times 1.5^2 \times 0.5$	30*	2×10^{-11}**		0.7	NA	[Udayakumar et al., 1991]
Rotational piezoelectric	$\pi/4 \times 4.5^2 \times 4.5$	1.1*	3.75×10^{-3}**		90×10^3	2.5% efficiency	[Bexell and Johansson, 1999; Johansson, 2000]
Linear magnetic	$0.4 \times 0.4 \times 0.5$	1000	2.9×10^{-6}	10^{-4}	3000	NA	[Liu et al., 1994]
Scratch drive actuator (SDA)	$0.07 \times 0.05 \times 0.5$	50	6×10^{-5}	160×10^{-9}	300	NA	[Akiyama and Fujita, 1995]
Rotational magnetic	$2 \times 3.7 \times 0.5$	150*	10^{-6}**		3×10^3	0.002% efficiency	[Teshigahara et al., 1995]
Rotational magnetic	$10 \times 2.5 \times ??$	20*	$350 \times 10^{-}$**		3×10^4	8% efficiency	[Stefanni et al., 1996]
EAP (Ppy–Au bimorph)	$1.91 \times 0.04 \times 0.00008$	0.2		1.25×10^{-3}	1.4×10^4	0.2% efficiency	[Smela et al., 1999]
Thermal polysilicon heatuator	Approx. $0.27 \times 0.02 \times 0.002$	2	$1/96 \times 30 \times 10^{-3}$	3.75×10^{-6}	2×10^4	NA	[Kladitis et al., 1999]
Thermal bimorph polyimide	Approx. $0.4 \times 0.4 \times 0.01$	1–60	69×10^{-6} N/mm^2	$2.6–9 \times 10^{-6}$	$<10^4$	$\approx 16.7 \times 10^{-3}$	[Suh et al., 1997]
PGV-joint actuator	$0.75 \times 0.6 \times 0.03$	3–300	10^{-3}	$10–150 \times 10^{-6}$	10^5	200 mW; 0.001% efficiency	[Ebefors, 2000]

* A rotational actuator that has speed measured in rad/s compared to a linear actuator (s^{-1}).

** It is impossible to separate the torque = fore × stroke into force and stroke for a rotational actuator; therefore, only the torque is given.

7.4.3 Design of Locomotive Microrobot Devices Based on Arrayed Actuators

Living organisms very often offer good models for designing microrobotic systems [Ataka et al., 1993a; Zill and Seyfarth, 1996]. Mimicry of a six-legged insect's gait [Zill, 1996] has been proposed for the design of multilegged robots implemented using microfabrication techniques [Kladitis, 1999]. The first proposed [Benecke, 1988b] and realized [Ataka et al., 1993] MEMS contact conveyance transportation system was based on the ciliary motion principle adopted from nature. The principle for a ciliary motion system (CMS) is illustrated in Figure 7.6. The CMS principle relies on an asynchronous driving technique which requires at least two spatially separated groups of actuators turned on and off at different times, alternately holding and driving the device. Higher speeds and smoother motions can be achieved with such asynchronous driving than with synchronous driving.

For locomotive microrobotic applications such as conveyance systems and walking robots it is essential to have actuators that can both generate forces to lift an object or the robot itself out from the plane (i.e., to avoid surface sticking) and also generate forces that cause in-plane movements. Two fundamental principles exist: contact-free (CF) systems and contact (C) systems. In CF systems, different force fields, such as electrostatic, magnetic, or pneumatic forces, are used to create a cushion to separate the object from the surface (Figure 7.7). To drive the device in the in-plane direction, these force fields need to have a direction dependence to force the device forward (directed air streams for a pneumatic system [Hirata et al., 1998] or the Meissner effect for levitation and orthogonal Lorentz force for driving a magnetic actuator system [Kim et al., 1990]).

The second fundamental principle, C systems, consists of a structure in contact with the moving object (legs for walking), as illustrated in Figure 7.6c. To avoid surface stiction, these structures must create out-of-plane movements. A review of different techniques available to create such three-dimensional actuators

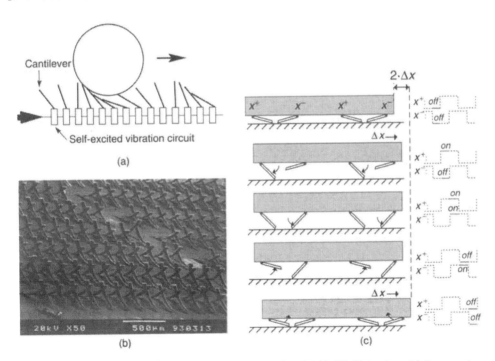

FIGURE 7.6 The ciliary motion principle used in most contact-based microrobotic systems. (a) Conveyance system used in nature to convey object. The cantilevers (ciliary hair) are actuated by a certain phase shift. (b) SEM photograph of the first microconveyer system based on the ciliary motion principle using bimorph polyimide legs. (Illustrations published with permission and courtesy of H. Fujita, University of Tokyo, Japan.) (c) A two-phase CMS for a walking microrobot platform when implemented using MEMS technology for out-of-plane working three-dimensional actuators. By using two rows of legs it is possible to steer in the forward–backward, left–right and up–down directions.

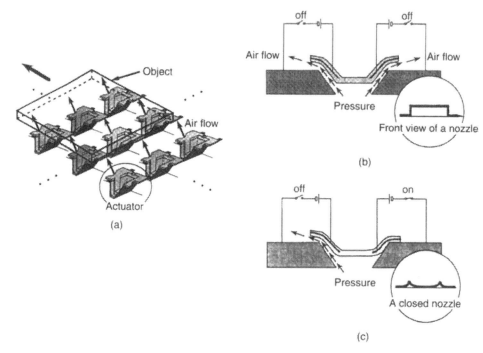

FIGURE 7.7 Electrostatic-controlled pneumatic actuators for a one-dimensional contact-free conveyance system. (a) Concept for the arrayed pneumatic conveyer (i.e., contact-free operation). (b) and (c) Mechanism for flow control by electrostatic actuation of nozzle when (b) the electrostatic nozzle is in the normal situation (off) and (c) when voltage is applied to one electrode (on). (Illustration printed with permission and courtesy of H. Fujita, University of Tokyo.)

has been presented [Ebefors, 2000]. Most of these techniques could be arranged in array configurations for distributed micromotion systems (DMMS).

For out-of-plane actuators using an external force field it is difficult to control each individual actuator in a large array of folded structures. Therefore, a synchronous jumping mode is used to convey the objects or move the device itself. This jumping mode involves quick actuation of all the actuators simultaneously, which forces the object to jump. When the object lands on the actuators (located in their off position), the object has moved a small distance and the actuators can be actuated again to move (walk or convey) further [Liu et al., 1995].

A critical aspect of large distributed micromotion systems based on arrayed actuators and distributed (or collective) actuation is the problem associated with the need for the very high yield of the actuators [Ruffieux and Rooij, 1999]. Just one nonworking actuator could destroy the entire locomotion principle. Therefore, special attention must be paid to achieve high redundancy by parallel designs wherever possible. These aspects will be further addressed in Section 7.5.3.

7.5 Microrobotic Devices

As pointed out in Section 7.2, a microrobotic device can be either a simple catheter with a steerable joint (Figure 7.2a), or a complex autonomous walking robot equipped with various microtools as in Figure 7.2o. Between these two extremes are microgrippers and microtools of various kinds, as well as micro-conveyers and walking microrobot platforms. Each of these three microrobotic devices will be presented more in detail in the following discussion. Section 7.6 describes more complex microrobotic systems where both microtools and actuators for locomotion are integrated to form so-called microfactories or desk-to-manipulation stations. Also, multirobot systems and communication between microrobots in such multirobot systems will be discussed.

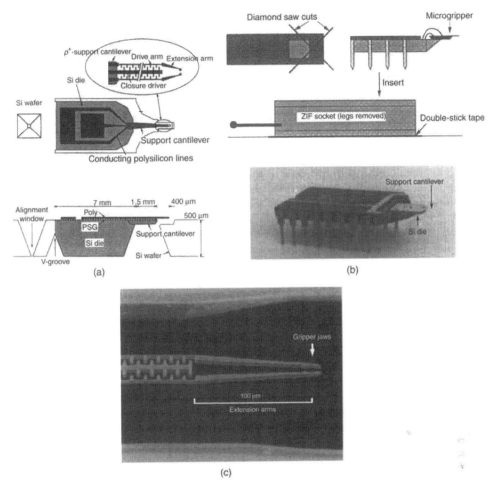

FIGURE 7.8 (a) Schematics of the microgripper unit (top and cross-sectional view). (b) Schematic showing gripper packaging and electrical access and a photograph of the packaged gripper. (c) SEM photograph showing a close-up of the gripper jaws and the comb-drive structures and extension arms. (Illustration printed with permission and courtesy of C.-J. Kim, UCLA.)

7.5.1 Microgrippers and Other Microtools

The first presented microrobotic device was based on in-plane electrostatic actuation [Kim et al., 1992]. This microgripper had two relatively thin gripping arms (thin-film deposited polysilicon) as shown in Figure 7.8. Other microgrippers based on quasi-three-dimensional structures with high-aspect-ratios fabrication techniques (beams perpendicular rather than parallel to the surface) have also been presented [Keller and Howe, 1995; 1997] (Figure 7.9). These kinds of grippers, so-called over-hanging tools, are formed by etching away the substrate under the gripper.

One critical parameter for the in-plane technique is how to achieve actuators with large displacement and force generation capabilities. Thermal actuators are known for their ability to generate high forces. A thermal actuator made from a single material would be easy to fabricate, but the displacement due to thermal expansion of a simple beam, for example, is quite small. This is a general drawback for in-plane actuators that occurs independently of the fabrication technique used. However, by using mechanical leverage, large displacements can be obtained, as was demonstrated by Keller and Howe (1995). They used a replication and micromolding technique, named HEXSIL, to fabricate thermally actuated microtweezers made from nickel and later in polysilicon [Keller and Howe, 1997]. In the HEXSIL process [Keller, 1998a]

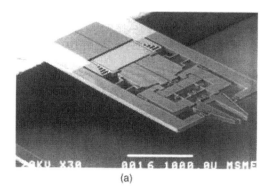
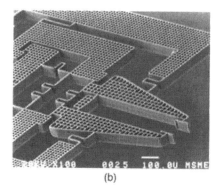

(a)   (b)

FIGURE 7.9   Photograph of fabricated HEXSIL tweezers. (a) Overview of the overhanging microtweezers with a compliant linkage system. (b) Close-up of the 80-μm-tall HEXSIL tweezers. The tip displacement between the closed and open position is typically 40 μm with a time constant <0.5 s for a typical actuation power of 0.5 A at 6 V. (Reprinted with permission from MEMS Precision Instruments, Berkeley, CA and courtesy of C. Keller.)

the mold is formed by deep trench etching in the silicon substrate. A sacrificial layer of oxide is deposited in the silicon mold which is then filled with deposited polysilicon. Then the polysilicon structure is released from the mold by sacrificial etching of the oxide. Afterwards, the mold can be reused by a new oxide and polysilicon deposition process. One advantage of this process is the ability to make thick (100 μm or greater) polysilicon structures (quasi-three-dimensional structures) on which electronics can be integrated. Figure 7.9b shows a close-up of the leverage design for the HEXSIL microtweezer; a large beam is resistively heated by the application of current, and subsequently expansion causes other beams in the link system to rotate and open the tweezer tips. When cooled, the contraction of the thermal element closes the tweezers.

Leverage and linkage systems (sometimes combined with gears for force transfer) are useful techniques for obtaining large displacements or forces that can be used for thermal actuation as well as electrostatic comb-drive actuators [Rodgers et al., 1999]. Several publications on design optimization schemes for various leverage techniques applied to thermal actuators (so-called compliant microstructures) have been presented [Jonsmann et al., 1999]. Another way to achieve a leverage effect is to use clever geometrical designs for single material expansion. One such method is the polyimide-filled V-groove (PVG) joint technology shown in Figure 7.10. The PVG joint technique has also been used for microconveyers and walking microrobots, as will be described in Sections 7.5.2 and 7.5.3. The purpose of the PVG joint microgripper in Figure 7.10 is easy integration with a walking microrobot platform.

Several publications on LIGA-based microgrippers have been presented. The reason for using LIGA is to get quasi-three-dimesnsional structures (thick structures) similar to the HEXIL tweezers in Figure 7.9. The LIGA process has also been used to produce single material (unimorph) in-plane thermal actuators for micropositioning applications. Guckel et al. (1992) presented an asymmetric LIGA structure with one "cold" and one "hot" side to generate large displacements (tenths of a millimeter) with relatively low power consumption, as illustrated in Figure 7.20a. More recently, this approach was used by Comotis and Bright (1996) for surface micromachined polysilicon thermal actuators. With this in-plane actuator they have successfully fabricated over-hanging microgrippers.

As an alternative to single-material expansion actuators, bimorph structures could also be used for out-of-plane acting gripping arms [Greitmann and Buser, 1996]. A bimorph microgripper for automated handling of microparts is shown in Figure 7.11. This device consists of two gripping arm chips assembled together. Each gripper arm has integrated heating resistors for actuation of the bimorph and tactile piezoresistors for force sensing.

Several different approaches to obtain three-dimensional microgrippers working out-of-plane like the ones in Figures 7.10 and 7.11 exist. One commonly used approach is use of the surface-micromachined polysilicon microhinge technology shown in Figures 7.3a–e. Such microhinges have been used both for microgrippers and for articulated microrobot components [Pister et al., 1992; Yeh et al., 1996].

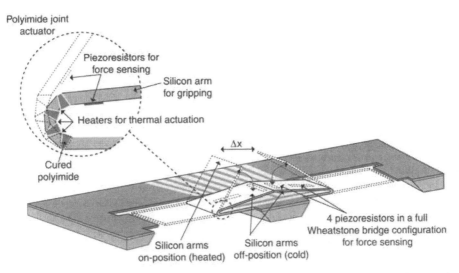

FIGURE 7.10 Concept of a microgripper fabricated by polyimide V-groove joints. (Ebefors, T. et al. [2000] "A Robust Micro Conveyer Realized by Arrayed Polyimide Joint Actuators," *IOP Journal of Micromechanical Microengineering* 10[3], pp. 77–349.) The polyimide in the V-grooves expands due to heating and the gripping arms are opened. Self-assembling out-of-plane rotation of the arms as well as the leverage effect for single material expansion are accomplished by the well-controlled geometrical shape of the V-groove. Polysilicon resistors are used both as resistive heaters and as strain gauges for force sensing.

FIGURE 7.11 Photograph of a microgripper based on bimorph thermal actuation and piezoresistive tactile sensing. (Reprinted with permission from Greitmann, G., and Buser, R. [1996] *Sensors and Actuators A* 53[1–4], pp. 410–415.)

One major drawback of these microgrippers (mainly based on thermal or electrostatic actuation) is found in biological applications. Microgrippers based on thermal, magnetic, or high-voltage electric actuation could easily kill or destroy biological and living samples. The pneumatic microgripper presented by Kim et al. [Ok et al., 1999] avoids such problems. An alternative to heating grippers (such as the ones shown in Figures 7.10 and 7.11, which require relatively high heating temperatures) is the use of shape memory alloy (SMA) actuators. Microgrippers based on SMA often require lower temperatures than thermally actuated bimorph or unimorph grippers. SMA-based three-dimensional microgrippers have been used to grip (clip) an insect nerve for recording the nerve activity of various insects [Takeuchi and Shimoyama, 1999].

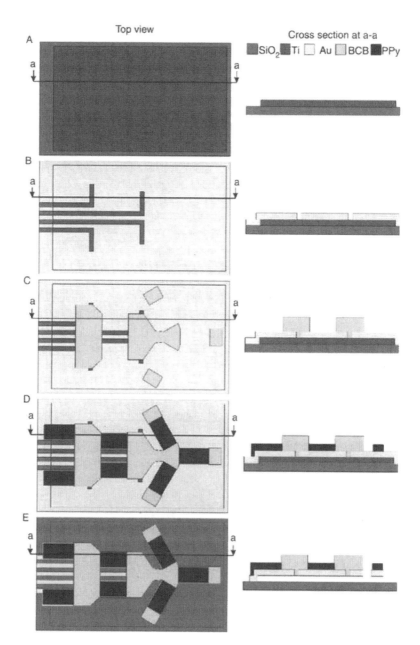

FIGURE 7.12 Schematic drawing of the process steps for fabricating microrobotic arms (in this case, an arm with three fingers arranged in a 120° configuration) based on hinges (micromuscle joints) consisting of PPy(DBS)/Au bimorph structures. (A) Deposition and patterning of a sacrificial Ti layer. (B) Deposition of a structural Au layer and etching of the isolating slits. (C) Patterning of BCB rigid elements. (D) Electrodeposition of PPy (conductive polymer). (E) Etching of the final robot and electrode structure and removal of the sacrificial layer. Each microactuator is 100 μm × 50 μm. The total length of the robot is 670 μm, and the width at the base is either 170 or 240 μm (depending on the wire width). (Reprinted with permission and courtesy of E. Jager, LiTH-IFM, Sweden.)

In the biotechnology field — for example, the growing area of genomics and proteomics — microtools for manipulation of single cells are of major importance. In particular, massive parallel single-cell manipulation and characterization by the use of microrobotic tools are very attractive. In this type of application, the microgrippers usually must operate in aqueous media. Most of the microgrippers presented so far in this

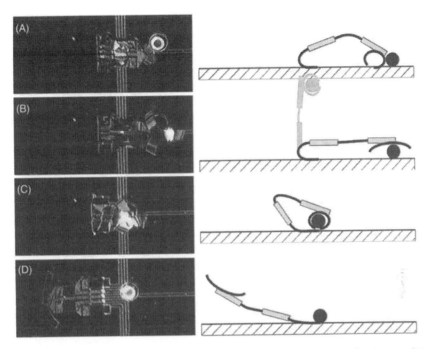

FIGURE 7.13   (A)–(D) Sequence of pictures (left) showing the grabbing and lifting of a 100-μm glass bead and schematic drawings (right) of the motion. In this case, the arm has three fingers, placed 120° from each other (Figure 7.12). The pictures do not illustrate the fact that the bead is actually lifted from the surface before it is placed at the base of the robot arm. This is illustrated in the second sketch on the right where the lifting stage is shown in gray. (Reprinted with permission and courtesy of E. Jager, LiTH-IFM, Sweden.)

review cannot operate in water because of electrical short-circuiting, etc. One possible solution is to use conductive polymers. Such conductive polymers, which undergo volume changes during oxidation and reduction, are often referred to as electroactive polymers (EAPs) or micromuscles. These kinds of micromuscles have been used as joint material for microrobotic arms for single-cell manipulation devices [Smela et al., 1995; Jager, 2000a,b]. Figure 7.12 describes the fabrication of a microrobot arm based on a polypyrole (PPy) conductive polymer. During electrochemical doping of PPy, volume changes take place which can be used to achieve movement of micrometer-size actuators. The actuator joints consist of a PPy and gold bimorph structure, and the rigid parts between the joints consist of benzocyclobutene (BCB). The conjugated polymer is grown electrochemically on the gold electrode, and the electrochemical doping reactions take place in a water solution of a suitable salt. The voltages required to drive the motion are in the range of a few volts.

One of the many experiments conducted with the various robot arms fabricated with the PPy micromuscles is shown in Figure 7.13. The drawback of microrobotic devices based on the conductive polymer hinge (or "micromuscles") is that they cannot operate in dry media.

7.5.2   Microconveyers

Recently, a variety of MEMS concepts for realization of locomotive microrobotic systems in the form of microconveyers have been presented [Riethmüller and Benecke, 1989; Kim et al., 1990; Pister et al., 1990; Ataka et al., 1993a,b; Konishi and Fujita, 1993; 1994; Goosen and Wolffenbuttel, 1995; Liu et al., 1995; Böhringer et al., 1996; 1997; Nakazawa et al., 1997; 1999; Suh et al., 1997; 1999; Hirata et al., 1998; Kladitis et al., 1999; Ruffieux and Rooij, 1999; 2000; Smela et al., 1999; Ebefors et al., 2000]. The characteristics for some of these devices are summarized in Table 7.3, where the microconveyers are classified in two groups: contact-free or contact systems, depending on whether the conveyer is in contact with the moving object or not, and synchronous or asynchronous, depending on how the actuators are driven. Examples of both contact and contact-free microconveyance systems were shown in Figures 7.6 and 7.7, respectively.

TABLE 7.3 Overview of Some Microconveyance Systems

Principle[a]	Maximum Velocity	Moved Object/Load Capacity	Length per Stroke/ Frequency	No. of Actuators/Size	Ref.
CF: pneumatic air bearing (for low-friction levitation) + electrostatic force for driving	Slow	Flat Si pieces/ <1.8 mg	100–500 μm/ max at 1–2 Hz	Not specified	[Pister et al., 1990]
CF: magnetic levitation (Meissner effect) + magnetic Lorentz force for driving	7.1 mm/s[b]	Nd–Fe–B magnet slider/ 8–17 mg	Not specified	Not specified	[Kim et al., 1990]
C: array of thermobimorph polyimide legs[c] (cantilevers); electrical heating (asyn)	0.027– 0.5 mm/s	Flat Si piece/ 2.4 mg	$\Delta x = 80\,\mu m$ $(f < f_c; 33\,mW)/$ $f_c = 10\,Hz$	8 × 2 × 16 legs/ 500 μm/total area: 5 × 5 mm²	[Ataka et al., 1993]
CF: array of pneumatic valves; electrostatically actuated	Not specified	Flat Si piece/ 0.7 mg	Not specified/ $f = 1\,Hz$ (pressure)	9 × 7 valves/ 100 × 200 μm²/ total area: 2 × 3 mm²	[Konishi and Fujita, 1994]
C: array of magnetic inplane flap actuators; external magnet for actuation (syn)	2.6 mm/s[d]	Flat Si pieces/ <222 mg	$\Delta x = 500\,\mu m/$ $f_c = 40\,Hz$	4 × 7 × 8 flaps/ 1400 μm/total area: 10 × 10 mm²	[Liu et al., 1995]
C: array of torsional 5 μm high; Si-tips; electrostatic actuation (asyn)	Slow	Flat glass piece/ ≈1 mg	$\Delta x = 5\,mm/f_c$ high kHz-range	15,000 tips 180 × 240 mm²/total area: 1000 mm²	[Böhringer et al., 1996; 1997]
C: array of thermobimorph polyimide legs;[c] thermal electrostatic actuation (asyn)	0.2 mm/s	Silicon chips/ 250 μN/mm²	$\Delta x = 20\,\mu m/f_c$ not specified	8 × 8 × 4 legs/ 430 μm/total area: 10 × 10 mm²	[Böhringer et al., 1997; Suh et al., 1997]
CF: Pneumatic (air jets)	35 mm/s[e] (for flat objects)	Sliders of Si/ <60 mg	Not specified	2 × 10 slits/ 50 μm/total area: 20 × 30 mm²	[Hirata et al., 1998]
C: array of erected[f] Si-legs; thermal actuation (asyn)	0.00755 mm/s	Piece of plastic film/3.06 mg	$\Delta x = 3.75\,\mu m$ $(f < f_c; 175\,mW)/$ $f_c = 3\,Hz$	96 legs/270 μm/ total area: 10 × 10 mm²	[Kladitis et al., 1999]
CF: array of planar electromagnets	28 mm/s[g] (unloaded)	Flat magnet + external load/ <1200 mg[g]	Not specified	≈40 × 40 coils/ 1 × 1 mm²/total area: 40 × 40 mm²	[Nakazawa et al., 1997; 1999]
C: array of non-erected Si-legs; piezoelectric or thermal actuation (asyn)	Not specified	Not specified	$\Delta x = 10\,\mu m/$ $(f = 1\,Hz, 20\,mW)$ $f_c \approx 30\,Hz$ (thermal)/f_c high kHz-range (piezoelectric)	125 triangular cells (legs) 400 μm (300 μm) long on a hexagonal chip approx. 18 mm²	[Ruffieux and Rooij, 1999]
C: array of erected[c] Si-legs; thermal actuation of polyimide joints (asyn)	12 mm/s[h]	Flat Si pieces + external load/ 3500 mg	$\Delta x = 170\,\mu m$[h] $(f < f_c; 175\,mW)/$ $f_c = 3\,Hz$[i]	2 × 6 legs/500 μm/ total area: 15 × 5 mm²	[Ebefors et al., 2000]

[a] C = contact; CF = contact-free; asyn = asynchronous; syn = synchronous.

[b] The superconductor requires low temperature (77 K).

[c] Self-assembled erection of the legs.

[d] Estimated cycletime ≈25 ms (faster excitation results in uncontrolled jumping motion) and 0.5 mm movements on 8 cycles [Liu, 1995].

[e] For flat sliders. The velocity depends on the surface of the moving slider (critical tolerances of the slider dimensions).

[f] Manual assembly of the erected leg.

[g] Depends on the magnet and surface treatment.

[h] Possible to improve with longer legs and more V-grooves in the joInternational.

[i] Possible to improve. Thinner legs with smaller polyimide mass to heat would increase the cut-off frequency, f_c => larger displacements at higher frequencies.

Contact-free systems have been realized using pneumatic, electrostatic, or electromagnetic forces to create a cushion on which the mover levitates. Magnetic levitation can be achieved by using permanent magnets, electromagnets, or diamagnetic bodies (a superconductor). The main advantage of the contact-free systems is low friction. The drawback of these systems is their high sensitivity to the cushion thickness (load dependent), while the cushion thickness can also be quite difficult to control. Also, this kind of conveyance system often has low load capacity.

Systems where the actuators are in contact with the moving object have been realized based on arrays of moveable legs erected from the silicon wafer surface. The legs have been actuated by using different principles such as thermal, electrostatic, and magnetic actuation. Both synchronous driving [Liu et al., 1995] and the more complex, but also more effective, asynchronous driving modes have been used.

The magnetic [Nakazawa et al., 1997; 1999] and pneumatic [Hirata et al., 1998] actuation principles for contact-free conveyer systems have a disadvantage that they require a specially designed magnet mover or slider which limits the usefulness. With a contact system based on thermal actuators it is possible to move objects of various kinds (nonmagnetic, nonconducting, unpatterned, unstructured, etc.); however, the increased temperature of the leg in contact with the conveyed object may be a limitation in some applications. The contact-free techniques have been developed mainly to meet the necessary criteria for a cleanroom environment, where a contact between the conveyer and the object may generate particles that would then serve to restrict its applicability for conveyance in clean rooms.

A microconveyer structure based on very robust polyimide V-groove actuators has been realized [Ebefors et al., 2000]. This conveyer is shown in Figure 7.14. In contrast to most of the previously presented

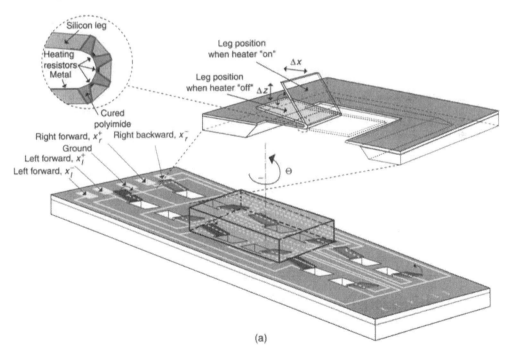

(a)

FIGURE 7.14 (a) Principle for the rotational movements on a test conveyer using robust PVG joints. The left and right side can be actuated separately like a caterpillar. Each leg has a size of $500 \times 600 \times 30\,\mu m$. (b) Photograph showing different (undiced) structures used to demonstrate the function of the polyimide-joint-based microconveyer. One conveyer consists of two rows of legs (12 silicon legs in total). Two sets of legs (six each of x^+ and x^-) are indicated in the photograph (compare Figure 7.6c). For the conveyer with five bonding pads, the right and left rows of legs can be controlled separately for possible rotational conveyance. (c) SEM photographs showing Si legs with a length of $500\,\mu m$. (d) The microconveyer during a load test. The 2-g weight shown in the photograph is equivalent to 350 mg on each leg or 16,000 times the weight of the legs. (Note: Videos of various experiments involving this microconveyer are available at http://www.s3.kth.se/mst/research/gallery/conveyer.html/ or http://www.iop.org/Journals/jm.)

FIGURE 7.14   Continued.

microconveyance systems, the PVG joint approach has the advantage of producing robust actuators with high load capacity. Another attractive feature of this approach is the built-in self-assembly by which one avoids time-consuming manual erection of the conveyer legs out of plane. Some of the conveyers listed in Table 7.3 require special movers (e.g., magnets or sliders with accurate dimensions). The PVG joint conveyer solution is more flexible because one can move flat objects of almost any material and shape. The large actuator displacement results in a fast system that is less sensitive to the surface roughness of the moving object. By using individually controlled heaters in each actuator, an efficient asynchronous driving mode has been realized, which also allows a parallel design giving relatively high redundancy for actuator failure. The first experiments with the conveyer showed good performance, and one of the highest reported load capacities for MEMS-based microconveyers was obtained. The maximum load successfully conveyed on the structure had a weight of 3500 mg and was placed on a 115-mg silicon mover, as shown in Figure 7.14d. Conveyance velocities up to 12 mm/s have been measured. Both forward–backward and simple rotational conveyance movements have been demonstrated. The principle for rotating an object by a two-row conveyer is shown in Figure 7.14a. The lifetime of the PVG joints actuator exceeds 2×108 load cycles and so far no device has broken due to fatigue.

The most sophisticated microrobotic device fabricated to date is the two-dimensional microconveyer system with integrated CMOS electronics for control which has been fabricated by Suh et al. (1999). The theories on programmable vector fields for advanced control of microconveyance systems presented by Böhringer et al. (1997) were tested on this conveyer. Several different versions of these conveyers have been fabricated throughout the years [Suh et al., 1997; 1999]. All versions are based on polyimide thermal bimorph ciliary microactuator arrays, as shown in Figure 7.15.

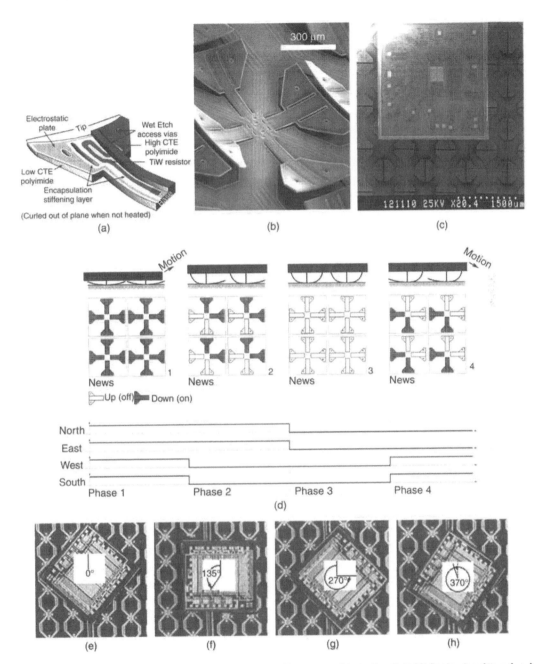

FIGURE 7.15 (a) Principal of operation of an organic microactuator (bimorph polyimide legs) using thermal and electrostatic forces for actuation. Half of the upper polyimide and silicon nitride encapsulation/stiffening layer have been removed along the cilium's axis of symmetry to show the feature details. (b) SEM photograph of a polyimide cilia motion pixel. Four actuators in a common center configuration make up a motion pixel. Each cilium is 430 μm long and bends up to 120 μm out of the plane. (c) Micro cilia device moving an ADXL50 accelerometer chip. (d) The CMS-principle for two-dimensional conveyance (compare Figure 7.6). The state of the four actuators (north, east, west, south) is encoded with small letters (e.g., n) for down, and capital letters (e.g., S) for up. (e)–(h) Images (video frames) of a 3 × 3-mm² IC chip rotating from a rotation demonstration. (Printed with permission and courtesy of Suh, Böhringer and Kovacs, Stanford University.)

7.5.3 Walking MEMS Microrobots

In principle, most of the microconveyer structures described in the previous section could be turned upside down to realize locomotive microrobot platforms. For contact systems, that means that the device will have legs for walking or jumping. The contact-free systems relying on levitation forces will "float" over the surface rather than walk. Such systems seem more difficult to realize than the contact-operating robots. The focus for the rest of this section is therefore on contact microrobot systems for walking.

Although it seems straightforward to turn a microconveyer upside down, most of the existing conveyers do not have enough load capacity to carry their own weight. Further, there are problems on how to supply the robot with the required power. As illustrated in Figure 7.16, power supply through wires may influence the robot operation range, and the stiffness of the wires may degrade the controllability too much. On the other hand, telemetric or other means of wireless power transmission require complex electronics on the robot. Because many actuators proposed for microrobotics require high power consumption, the limited amount of power that can be transmitted through wireless transmission is a big limitation for potential autonomous robot applications. To avoid the need for interconnecting wires, designs based on solar cells have been proposed, as have low-power-consuming piezoelectric actuators [Ruffieux and Rooij, 1999; Ruffieux, 2000], electrostatic comb-drives [Yeh et al., 1996], or inch-worm actuators [Yeh and Pister, 2000] suited for such wireless robots. For wire-powered robots, a limited amount of wires is preferable, which implies that simple leg actuation schemes (on–off actuation as used for the walking robot platform in Figure 7.6c) are required if complex onboard steering electronics are to be avoided. Several proposals for making totally MEMS-based microrobots include the possibility of locomotion (e.g., walking) powered with or without wires.

7.5.3.1 Examples of Walking Microrobots Designs

Several different principles used to actuate the different legs on a walking microrobot have been proposed, most of them mimicking principles used in nature. Some of the most feasible principles for walking MEMS microrobot platforms are listed below.

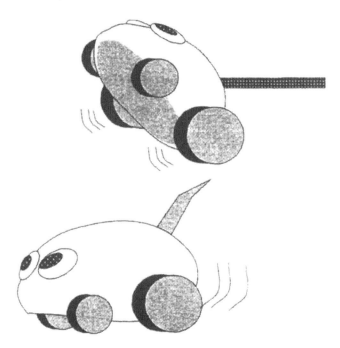

FIGURE 7.16 Influence of energy cable on microrobots. (Illustration printed with permission and courtesy of T. Fukuda, Nagoya University, Japan.)

Ciliary motion was used by Ebefors et al. (1999) for eight-legged robots, as illustrated in Figures 7.6 and 7.17.

Elliptical leg movements adopted from the animal kingdom have been proposed by Ruffieux and Rooij (1999) and Ruffieux (2000) and are illustrated in Figure 7.18. Similar to this concept is the principle used by Simu and Johansson (2001). Their microrobot concept consists of both a walking and a microtool unit for highly flexible micromanipulation. Instead of using thin-film piezolayers and silicon legs as did Ruffieux, Johansson and Simu have made robot legs in a solid, multilayer, piezoceramic material mounted on a glass body [Simu and Johansson, 1999] (see Figure 7.22).

FIGURE 7.17   Photograph of the microrobot platform, used for walking, during a load test. The load of 2500 mg is equivalent to maximum 625 mg/leg (or more than 30 times the weight of the robot itself). The power supply is maintained through three 30-μm-thin and 5- to 10-cm-long bonding wires of gold. The robot walks using the asynchronous ciliary motion principle described in Figure 7.6(c). The legs are actuated using the polyimide V-groove joint technology described in Figure 7.10. (Photograph by P. Westergård and published with permission.) (Note: Videos of various experiments performed using the microrobot shown here are available at www.s3.kth.se/mst/research/gallery/microrobot_video.html.)

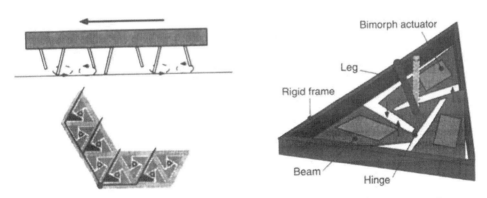

FIGURE 7.18   The rotational leg walking principle. A few hundred cells (shown in the right picture) are arranged into a hexagonal array, inside a triangular grid frame that provides stiffness and room for interconnection. The simplest form of gait requires two phases so that half the actuators are in contact with the ground, where friction transmits their motion to the device, while the other half is preparing for the next step (compare to the CMS technique depicted in Figure 7.6). The bottom left figure illustrates the serial interconnections of the actuator resulting in six independent groups of beams (From Ruffieux, D. et al. [2000] in *Proceedings IEEE 13th International Conference on Micro Electromechanical Systems [MEMS 2000]*, pp. 662–667. With permission.)

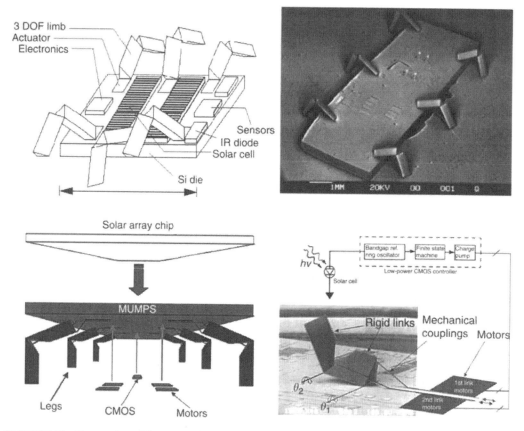

FIGURE 7.19   First version of the microrobot prototype based on surface micromachined microhinges for joints; each leg has three degrees of freedom and is comprised of two 1.2-mm-long, rigid polysilicon links and electrostatic step motors for movement (not included in the SEM-photo). (Bottom) Concept for a new design based on a solar-powered silicon microrobot. Various components can be made separately and then assembled. Surface micromachined hinges are used for folding the legs (see Figures 7.3d and e). Each leg has two links, and each link will be actuated by an inchworm motor [Yeh and Pister, 2000]. (Printed with permission and courtesy of K. Pister, BSAC-Berkeley, CA.)

Gait mimicry of six-legged insects (similar to that of a crab [Zill and Seyfarth, 1996]) has been proposed by several research groups — for example, the six-legged microrobot by Yeh et al. (1996) and Yeh and Pister (2000) and the multilegged microrobot prototype by Kladitis et al. (1999). These concepts are illustrated in Figures 7.19 and 7.20.

Inch-worm robots [Thornell, 1998] or slip-and-stick robots [Breguet and Renaud, 1996] that mimick an inch-worm or caterpillar are attractive microrobot walking principles, as these techniques take advantage of the frictional forces rather than trying to avoid them. The scratch-drive actuator principle [Mita et al., 1999] works according to this principle by firmly attaching to the surface only half of the robot or actuator body, then extending the spine (or middle part) of the body before anchoring the other half, releasing the first grip, shortening the spine, swapping the grip, and so on in a repeatable forward motion.

Vibration fields and resonating thin-film silicon legs with polyimide joints have been used by Shimoyama et al. [Yasuda et al., 1993], as illustrated in Figure 7.21.

Because friction in many cases poses a restriction on microrobots due to their small size, solutions that take advantage of this effect (e.g., inch-worm robots) rather than trying to avoid it are the best suited for microgait. This trend can also be seen in the evolution of micromotors for microrobotic and other applications. Currently, many researchers try to avoid bearings or sliding contacts in their motors and instead

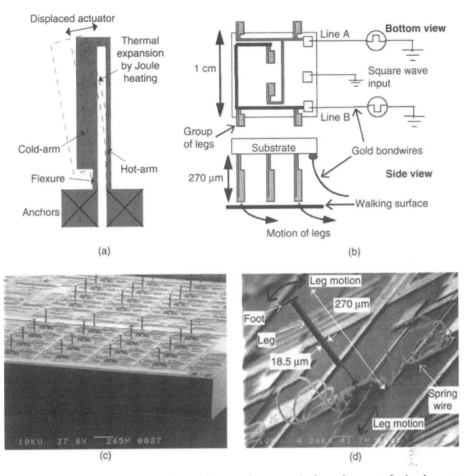

FIGURE 7.20 (a) Schematic illustration of a single-material asymmetric thermal actuator for in-plane actuation (also called "heatuator"). The actuator structure was originally made of nickel in a LIGA process, although surface-micromachined polysilicon structures have also been used. (b) An out-of-plane folded heatuator for microrobotic applications. (c) Photograph of the microrobotic device in a conveyance mode (belly up). (d) Close-up of one of the heatuator legs erected out of plane. (Part (a) printed with permission from Guckel, H. et al. [1992] in *Technical Digest, Solid-State Sensor and Actuator Workshop*, pp. 73–75.) (Part (b) printed with permission from Kladitis, P. et al. [1999] in *Proceedings IEEE 12th International Conference on Micro Electro Mechanical Systems (MEMS '99)*, pp. 570–575.) (Photographs courtesy of P. Kladitis and V. Bright, MEMS Group at the University of Colorado at Boulder.)

make use of friction in wobble and inch-worm motors [Yeh and Pister, 2000] or actuators with flexible joints [Suzuki et al., 1992; Ebefors et al., 1999] without the drawback of wearing out.

The main problem associated with the fabrication of silicon robots is to achieve enough strength in the moveable legs and in the rotating joints. Most efforts to realize micromachined robots utilize surface micromachining techniques which results in relatively thin and fragile legs. [Yeh et al., 1996] have proposed the use of surface micromachined microhinges for joints, poly-Si beams for linkage to the triangular polysilicon legs, and linear electrostatic stepper motor actuation for the realization of a microrobot, as illustrated in Figure 7.19.

A similar approach for walking microrobots was used by Kladitis et al. (1999). They also used the microhinge technique to fold the leg out of plane, but instead of area-consuming comb drives for actuation, they used the thermal "heatuator" principle integrated in the leg, as illustrated in Figure 7.20. To further improve the robustness and load capacity of their robot, they used 96 legs arranged in six groups instead of using a

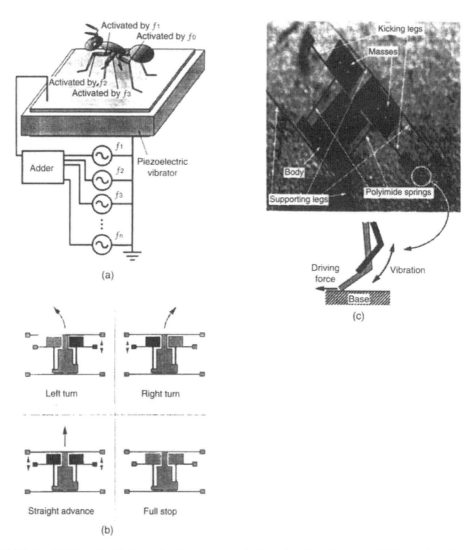

FIGURE 7.21 (a) The principle for a selective power supply through a vibrating energy field [Yasuda et al., 1993]. The microrobot has several resonant actuators with mutually exclusive resonance frequencies. The power and control signals to the robot are obtained via the vibrating table. (b) The four possible kinds of actions. (c) Photograph of the 1.5×0.7-mm^2 surface micromachined microrobot. The legs have different spring constants and masses, resulting in different resonance frequencies. Polyimide is used for the soft springs and the polyimide joints are used for the erected leg. (Printed with permission from Yasuda et al. [1993] *Technical Digest, 7th International Conference on Solid-State Sensors and Actuators (Transducers '93)*, pp. 42–45, June 7–10, Yokohama, Japan.)

six-legged robot. However, that robot structure could only withstand a load four times the dead weight of the robot. That was not enough to obtain locomotive walking but was high enough for conveyance applications.

For both microrobots described above, which were based on surface micromachined polysilicon microhinges, the polysilicon legs were manually erected out of plane. The use of the microhinge technique may cause problems because of wear after long-term actuation. Shimoyama et al. [Suzuki et al., 1992; Miura et al., 1995] introduced the concept of creating insect-like microrobots with exoskeletons made from surface-micromachined polysilicon plates and rigid polyimide joints that have low friction and therefore lower the risk for wearing out. These microrobots were powered externally by a vibrating field (no cables were needed) as illustrated in Figure 7.21. By cleverly designed robot legs having different masses and spring constants and thus different mechanical resonance frequencies, the leg to be actuated can be selected by applying a certain

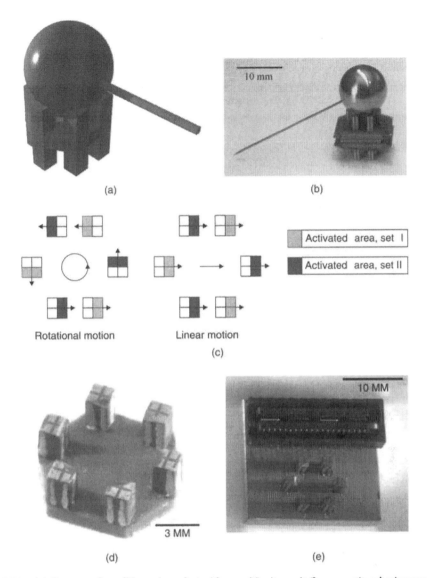

FIGURE 7.22 (a) Concept of a walking microrobot with a positioning unit (lower part) and micromanipulating unit (upper part) for turning the ball equipped with a tool. (b) A first prototype of the robot; this version is unable to walk. (c) Driving mechanism for the stator unit. (d) A photograph showing a six-legged piezoelement stator. (e) Experimental setup for the micropositioning unit. Note the amount of interconnection wires for this prototype. (Courtesy of U. Simu, Uppsala University, Sweden.)

external vibrating (resonance) frequency. The robot can then be "shaken" in a controlled way to walk forward and left or right, as illustrated in Figure 7.21b. The need for a vibrating table to achieve the locomotion strongly limits the applications for this robot. Also, the thin legs cause the robot to stick easily to the vibrating table by surface forces, which means that the surface on which the robot walks must be insulating.

In a large European microrobot project named MINIMAN, different miniaturized microrobots have been developed and more are under development. By the year 2000, a few demonstrators of minirobots and also a microrobot, having five degrees of freedom micromanipulation capability, had been fabricated. This prototype microrobot is illustrated in Figure 7.22. Milling with a high-precision CNC (computer numerical controlled) machine tool in a green ceramic body has been used to make a stator unit where the drive elements are monolithically integrated with the base plate. The robot is made out of two almost identical piezoelectric stator units (Figure 7.22d), which are put together back to back. One stator unit consists of six individually controlled legs. The lower stator, the micropositioning unit, should move laterally on a flat surface while the upper stator, the micromanipulating unit, will turn a ball equipped with a tool arbitrarily. Figure 7.22c illustrates the driving mechanism of the six-legged piezoceramic positioning and manipulating units to gain rotation and linear motion. Each three-axial element (leg) in the stator unit has four electrically separated parts able to produce the different motions needed for control of the robot. The elements are divided into two sets (different shading for sets I and II) phase-shifted 180°. Therefore, three elements are always in friction contact with the counter surface, producing a conveying (or walking) motion of the ball (or the robot itself).

The mechanical walking mechanism with solid multilayer piezoactuators is designed to give high handling speed, force, and precision. A future version will include integration of the power drivers with the control circuitry and soldering the unit to a flexible printed circuit connecting the actuator elements in the stator unit. The electronic chip would then be placed between the two stator units.

Figure 7.22e shows the experimental setup for testing the first prototype of the micromanipulating stator unit. These stators in combination with prototypes of ICs intended for integration have been used to evaluate the performance. This setup will also be combined with a larger positioning platform, MINI-MAN III (a walking minirobot platform), to obtain flexible microrobotic devices for micromanipulation.

7.5.3.2   Redundancy Criteria for Walking Microrobots

As was already pointed out in this chapter, one critical aspect when designing large distributed micromotion systems based on arrayed actuators and distributed (or collective) actuation is the problem associated with the need for a very high yield of the actuators [Ruffieux and Rooij, 1999]. One nonworking actuator can undo the entire walking principle. Therefore, special attention has to be paid to achieve high redundancy by parallel designs wherever possible. The conveyer legs of the robot prototype design presented in Figure 7.22 are fabricated by stacking several layers of thick-film piezoelectric layers on top of each other with metal electrodes in between. Because one layer is approximately 50 μm thick, a piezostructure for a 1-mm-long leg consists of several layers and electrodes, which require very high yield in the fabrication. The metal electrodes are connected in parallel by the metal on the edges of the piezolegs to obtain higher redundancy for failure. Besides the need for parallel designs, the interconnecting wires must be kept at a minimum for walking microrobots (Figure 7.16). For the piezorobot prototype shown in Figure 7.22e, the stator unit is soldered to a flexible printed circuit board as it will be in the complete microrobot. The only difference is that the conductors do not lead to a surface-mounted connector, but instead to surface mounted integrated circuits. By using such a configuration, the number of cables required to steer the robot will be minimized. In the current prototype version, the amount of interconnecting wires is quite large, as can be seen in Figure 7.22e.

For the microrobots shown in Figures 7.17 and 7.23, all the legs are connected in parallel so all of the actuators or legs are not needed to have perfect functionality. If one leg is broken, all the others are still functional, as the power supply line is connected in parallel (Figure 7.23) instead of the serial connection, which is the most often used design for distributed microsystems. However, it is important that the nonfunctional (or broken) legs do not limit the walking function by blocking the action of the other legs. To further increase redundancy, each heating resistor in the PVG joint (Figures 7.10 and 7.14) is divided into separate parts, allowing functional heating even if some part of the resistors (or the via contacts between the resistor and the metal) is not working.

7.5.3.3   Steering of Microrobots

As shown in Figure 7.23, the simplest way to steer a CMS-based walking microrobot is by controlling the phase of the two rows of actuators on the robot. By driving one side forward and the other side backwards (or by not actuating it at all) the robot makes turns imitating a caterpillar. Other approaches to steer CMS microrobots to the left and right include:

- Power: Longer stroke lengths (changes in the x direction, or Δx) on one side can be produced by increasing the power, but this would also produce higher steps (changes in the y direction, or Δy), so the robot would walk with a stoop.

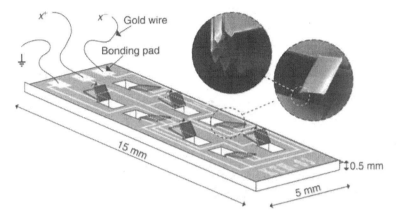

FIGURE 7.23 An upside-down view of the PVG joint-based microrobot platform used to study walking microrobots. The robot consists of two sets of legs (four of each x^+ and x^-). The metal interconnection is done in a parallel way to obtain the highest possible redundancy against failure. With three bonding pads the robot can walk forward and backward as well as lift up and down. By driving the legs on the left and right side at different speeds or stroke lengths, like a caterpillar (requires five wires), the robot can make left–right turns. The SEM photographs show the silicon leg and a close-up of a five-V-groove polyimide joint. (Printed with permission from Ebefors, T. et al. [1999] "A Walking Silicon Micro-Robot," *10th International Conference on Solid-State Sensors and Actuators (Transducers '99)*, June 7–10, pp. 1202–1205, Sendai, Japan.)

- Frequency: Increasing the number of steps on one side would produce smaller steps (both Δx and Δy), so again the robot would walk with a stoop.
- Combination of frequency and power: Equal stroke lengths, but faster on one side, would allow "smooth" robot movements.

Another approach to obtain steerable walking robots is the use of large two-dimensional arrays with several rows of legs oriented in both the x and y directions. For these systems, based on the principle described in Figure 7.15, onboard electronics are probably required for minimizing the amount of wires required for the steering signals. Monolithic integration of CMOS electronics has been shown to work for conveyer systems [Suh et al., 1999], but the integration is not an easy task. Hybrid integration of electronics on a separate chip, as is planned for the microrobot shown in Figure 7.22, is not an easy task either, when many interconnect wires are needed.

7.6 Multirobot System (Microfactories and Desktop Factories)

In the famous 1959 American Physical Society presentation by Feynman (1961), describing a new and exciting field of physics (miniaturization of systems or "there's plenty of room at the bottom"), the use of small machines to build smaller machines or systems was anticipated as a way to obtain miniaturization. However, this has not yet been realized in the MEMS field. Instead, we use larger and larger machines to achieve miniaturization and study physics at a very small scale (large vacuum chambers to evaporate the very thin films used in MEMS or the huge accelerator facilities used to study the smallest particles in the atom). However, at the research level, interest in such microfactories, where small machines build even smaller machines, has been growing since the mid-1990s.

Often MEMS is seen as the next step in the silicon system revolution (miniaturization) in that silicon integration breaks the confines of the electrical world and allows the ability to interact with the environment by means of sensors and actuators. Most of the commercially successful processes employed in

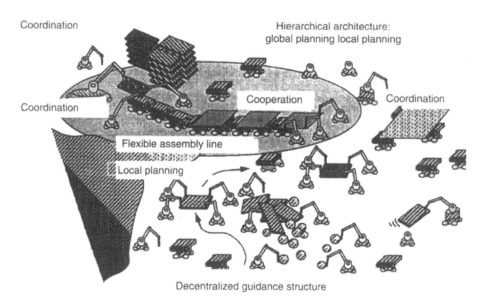

Coordination   Hierarchical architecture:
                                                global planning local planning

Coordination   Cooperation   Coordination

Flexible assembly line

Local planning

Decentralized guidance structure

FIGURE 7.24 Concept for flexible decentralized multi-microrobot system. (Translated with permission from Fatikow, S., *Mikroroboter and Mikromontage*, B.G. Tenbner, Stuttgart-Leipzig in German.)

MEMS technology have been adopted from the fabrication scheme of integrated circuits. This kind of processing equipment is usually very large and expensive but has the very attractive feature of batch fabrication, which allows production of low-cost devices in large quantities. However, as already described in this chapter, manufacturing concepts other than the IC-inspired batch silicon microfabrication are being developed today. All IC-based micromachining techniques require large volumes to be cost effective, as the equipment is large and expensive. Emerging alternative fabrication technologies allow for more flexible geometry (true three-dimensional structures) not feasible by planar IC process and integration of materials with properties other than those of semiconductors and thin metal films [Thornell and Johansson, 1998; Thornell, 1999]. These technologies utilize miniaturized versions of traditional tools typical for local processing (such as drilling, milling, etc. [Masuzawa, 1998]) rather than the parallel (batch) approach used in the IC industry. The main challenge in adopting these processes for commercial application is the slow processing time associated with high-precision machining. One potential solution to this problem might be to use microfactories, where a great number of microrobotic devices (conveyers and robots) are working in parallel to support the microtools with material [Hirai et al., 1993; Aoyama et al., 1995; Ishihara et al., 1995]. This kind of multirobot system, illustrated in Figure 7.24, introduces special requirements for the microrobots. These requirements include a wireless power supply (onboard power sources or telemetrically supplied power) and interrobot communication.

During the last decade, efforts have been made to develop various multi-microrobot system concepts, especially in Japan [Ishikawa and Kitahara, 1997; Kawahara et al., 1997], where the term microfactories is used, but also in Europe and U.S., where the most frequently used term is desktop station fabrication. The idea behind the microfactory approach is to allow production of small devices. Often, other factors such as saving energy, space, and resources are also considered important, so the manufacturing process should also be cost effective for low-volume microsystems. It is still impossible to predict if these techniques will ever be competitive with standard parallel-batch MEMS processing, but two important advantages have already been demonstrated: flexibility and geometrical freedom.

Similar to the microfactory concept is the "automated desktop station using micromanipulation robots" proposed by Fatikow et al. [Fatikow and Rembold, 1996; 1997; Fatikow, 2000], as illustrated in Figure 7.25. Here, the focus is on assembly of different types of microdevices, fabricated by batch MEMS techniques, to build various microsystems rather than fabricate subcomponents in the systems.

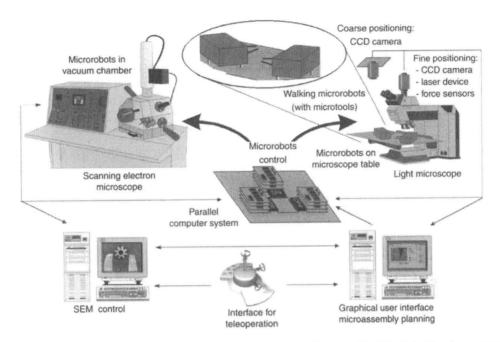

FIGURE 7.25 Concept of the flexible microrobot-based microassembly station (FMMS). (Printed with permission and courtesy of S. Fatikow, University of Kassel, Germany.)

7.6.1 Microrobot Powering

The ultimate goal for many microrobotics researchers is to achieve true autonomous microrobots like the ones shown in Figure 7.2o for work in multirobot systems such as microfactories. Some of the locomotive microrobot platforms presented in this chapter can be used as first steps toward that goal, but much research remains. Currently, there are no energy sources available to power a fully autonomous microrobot. Some robot actuators, such as the polyimide V-groove joint actuator shown in Figure 7.17 and other thermal actuators, have the drawback of high power consumption. Further research concerning the reduction of power is needed. This is a general problem: robust and efficient microactuators usually have such high energy consumption that an onboard power supply cannot provide enough electric power over an extended period of time. Using cables to supply the energy to a microrobot obstructs the action (especially in multirobot systems) of the microrobot due to the stiffness of the cables; therefore, the robot has to be supplied with energy without using cables. Several research groups are working on developing microenergy sources — for example, various microgenerators, pyroelectric elements, solar batteries, fuel cells, and other power-MEMS concepts [Lin et al., 1999]. An overview of thin-film batteries that may be suitable for microrobot applications is given by Kovacs (1998). Low-weight batteries have been used for a flying microrobot [Pornsin-Sirirak et al., 2000]. Even though the lifetime of such batteries and dc–dc converter systems is short, it demonstrates the possibilities for battery-powered microrobots. Solar cells made of a-Si–H junction for microrobot applications have also been proposed [Ruffieux and Rooij, 1999; Ruffieux, 2000]. By connecting several a-Si–H junctions a voltage of about 10 V is expected to be generated.

7.6.2 Microrobot Communication

Besides the problems associated with the power sources, more research is required on the wireless transmission of steering signals to the microrobot and other means of communication between microrobots. Wireless microrobots and other microdevices can be achieved by using acoustic [Suzumori et al., 1999], optical [Ishihara and Fukuda, 1998], electromagnetic [Hierold et al., 1998], or thermal transmission of the control signal to the robot.

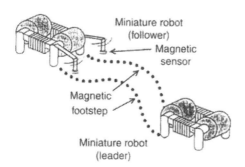 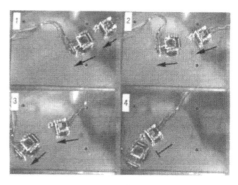

FIGURE 7.26 Communication between miniature robots. The following robot follows the leader by a magnetic trace. (Printed with permission and courtesy of H. Aoyama, University of Electro-Communications, Tokyo.)

Common to all multirobot system concepts is the big challenge of how to obtain autonomous communication between different microrobotic devices. Normally, various telemetric solutions similar to those described in the previous section for wireless power transfer are also proposed for communication, but other approaches seem to be feasible as well. In nature, the ant creates an odor trail to guide fellow ants. Shimoyama (1995a,b) has been working on hybrid systems consisting of mechanical parts and living organisms. By using "living sensors" in the form of insect antennae, Shimoyama and co-workers were able to steer and control a mobile robot by pheromone stimulation [Kuwana et al., 1995]. Aoyama et al. (1997) presented inch-sized robots that leave behind a magnetic footprint which can be detected by other robots for communication and motion guidance, as illustrated in Figure 7.26. These footprints fade out with time, allowing continuous and advanced motion involving several miniature robots.

7.7 Conclusion and Discussion

Even though the microrobotic field and especially MEMS-based microrobotics are areas of constant expansion and we see a growing worldwide interest in the concept and possible applications, it still remains to be seen if these devices will come to practical and commercial use in applications such as hybrid assembling and manufacturing of microsystems. In this review, two fundamentally different aspects of microrobotics have been addressed: (1) MEMs for making microrobots and (2) microrobots for making MEMS. The MEMs fabrication technology is an attractive approach for obtaining miniaturized robots. By scaling down the remaining robotic subsystems to the same scale as the control systems by integrating actuators, sensors, logic, and power supplies onto a single piece of silicon, enormous advantages can be obtained in terms of mass produce-ability, lower costs, and fewer connector problems when interconnecting these discrete subsystems. The other aspect of MEMS and microrobotics is the application of these microrobots. Among several applications for microrobots, their use in medical applications (minimal invasive surgery) and biotechnology (cell-sorting) should be mentioned as the most promising for future commercial success of MEMS-based microrobotic devices. Further, the use of various microrobotic devices and robot systems for assembling complex microsystems and perhaps also for more effective fabrication by hybrid technologies is an attractive approach for making more cost-effective devices and systems when low-volume or prototyping devices and systems are needed (microrobots for making MEMS). The micromanipulation tools, microconveyers, and microrobots for locomotive mechanisms described in this chapter should be seen as the first steps in the long chain of developments needed to achieve autonomous microrobots for work in multirobotic systems such as microfactories for fabrication of other microsystems. However, much research remains in this field and it could be worth remembering the words of the great physicist Feynman when he "invented" the field of MEMS back in 1959:

> What are the possibilities of small but movable machines? They may or they may not be useful, but they surely would be fun to make.

The devices described in this chapter illustrate the basic building blocks and some of the progress made in the young research field of MEMS-based microrobotics during the 1990s. Therefore, let us have fun, feel the wonder of microrobotics when "small movable machines" are fabricated, and when new frontiers for science and research are explored. Beside the element of fun, there are tremendous opportunities for microrobotic applications, thus a large potential for industrial innovation and competition.

Defining Terms

Ciliary motion system (CMS): A transportation system based on the ciliary motion principle adopted from nature. The CMS principle relies on an asynchronous driving technique that requires at least two spatially separated groups of actuators that are turned on and off at different times, alternately holding and driving the device. Higher speeds and smoother motions can be achieved with such asynchronous driving than with synchronous driving.

Distributed micromotion systems (DMMS): The concept of an array configuration of microactuators, where the cooperative work of many coordinated simple actuators generates interaction with the macroworld.

Hybrid assembly: Assembly of several individual microcomponents fabricated by different technologies and on different wafers. If microrobotic devices are used for the assembly (microrobots for making MEMS), extremely small components can be assembled.

MEMS-based microrobotics: A sort of "modified chip" fabricated by silicon MEMS-based technologies (such as batch-compatible bulk- or surface-micromachining or by micromolding and/or replication method) having features in the micrometer range.

Microfactory: A complete automatic factory containing conveyers, robot arms and other machine tools at a very small level. A synonym for microfactory is desktop factory.

Microrobotics: A general term for devices having dimensions smaller than classical watch-making parts (i.e., μm to mm) with the ability to move, apply forces and manipulate objects in a workspace with dimensions in the micrometer or submicrometer range. Included in the term are both system aspects and more fundamental building-block aspects of a robot system. Examples of these fundamental building blocks in the microrobotic field are microgrippers, microconveyers and locomotive robots (arms for pick-and-place and/or legs for walking).

Monolithic assembly: All integration is done on one single (silicon) chip preferably using wafer-level assembly. This technology is preferable when fabricating microrobotic devices that rely on actuator arrays (MEMS for making microrobots).

Parallel (batch) manufacturing: Fabrication performed on several wafers at the same time (several-devices-at-a-time process) originally developed for the IC industry but now also commonly used for MEMS fabrication.

Serial manufacturing: One-device-at-a-time fabrication/assembly contrary to the monolithic approach where all integration is done on one single silicon chip using wafer-level assembly.

References

Akiyama, T., and Fujita, H. (1995) "A Quantiative Analysis of Scratch Drive Actuator Using Buckling Motion," *Proc. of IEEE 8th Int. Workshop on Micro Electro Mechanical Systems (MEMS '95)*, Jan. 29–Feb. 2, pp. 310–315, Amsterdam, the Netherlands.

Aoyama, H., Iwata, F., and Sasaki, A. (1995) "Desktop Flexible Manufacturing System by Movable Miniature Robots — Miniature Robots with Micro Tool and Sensor," *IEEE Int. Conference on Robotics and Automation*, May 21–27, pp. 660–665, Nagoya, Aichi, Japan.

Aoyama, H., Santo, T., Iwata, F., and Sasaki, A. (1997) "Pursuit Control of Micro-Robot Based Magentic Footstep," *Proc. of International Conference on Micromechatronics for Information and Precision Equipment (MIPE '97)*, July 20–23, pp. 256–260, Tokyo, Japan.

Arai, K.I., Sugawara, W., and Honda, T. (1995) "Magnetic Small Flying Machines," *Tech. Digest Transducers '95 and Eurosensors IX*, June 25–29, pp. 316–319, Stockholm, Sweden.

Ataka, M., Omodaka, A., and Fujita, H. (1993) "A Biometric Micro System — A Ciliary Motion System," *Techn. Digest, 7th Int. Conf. Solid-State Sensors and Actuators (Transducers '93)*, June 7–10, pp. 38–41, Yokohama, Japan.

Ataka, M., Omodaka, A., Takeshima, N., and Fujita, H. (1993) "Fabrication and Operation of Polyimide Bimorph Actuators for a Ciliary Motion System," *IEEE J. MEMS*, 2(4 (December 1993)), pp. 146–150.

Benecke, W.R. a. W. (1988) "Thermally Excited Silicon Microactuators," *IEEE Trans. Electron. Devices*, (vol. **35**, 6), pp. 758–763.

Bexell, M., and Johansson, S. (1999) "Fabrication and Evaluation of a Piezoelectric Miniature Motor," *Sensors Actuators A*, **A** 75, pp. 8–16.

Böhringer, K.F., Donald, B.R., and MacDonald, N.C. (1996) "Single-Crystal Silicon Actuator Arrays for Micro Manipulation Tasks," *Proc. of IEEE 9th Int. Workshop on Micro Electro Mechanical Systems (MEMS '96)*, Feb. 11–15, pp. 7–12, San Diego, USA.

Böhringer, K.F., Donald, B.R., MacDonald, N.C., Kovacs, G.T.A., and Suh, J.W. (1997) "Computational Methods for Design and Control of MEMS Micromanipulator Arrays," *IEEE Computational Science & Engineering*, Jan.–March 1997, pp. 17–29.

Breguet, J.-M. (1996) "Stick and Slip Actuators and Parallel Architectures Dedicated to Microrobotics," *SPIE Conference on Microrobotics: Components and Applications, part of SPIE's Photonics East*, November, pp. 13–24, Boston.

Breguet, J.-M., and Renaud, P. (1996) "A 4 Degrees-of-Freedoms Microrobot with Nanometer Resolution," *Robotics*, **14**, pp. 199–203.

Carts-Powell, Y. (2000) "Tiny Camera in a Pill Extends Limits of Endoscopy," *SPIE: OE-Reports* August 2000 (No. 200): available on Internet at: http://www.spie.org.

Cohn, M.B., Böhringer, K.F., Noworolski, J.M., Singh, A., Keller, C.G., Goldberg, K.Y., and Howe, R.T. (1998) "Microassembly Technologies for MEMS," *SPIE Conference on Microfluidic Devices and Systems*, Sept. 21–22, pp. 2–16, Santa Clara, California, USA.

Comotis, J., and Bright, V. (1996) "Surface Micromachined Polysilicon Thermal Actuator Arrays and Applications," *Solid State Sensor and Actuator Workshop*, June 2–6, pp. 174–177, (Hilton Head), Hilton Head, South Carolina, USA.

Corman, T. (1999) Vacuum-Sealed and Gas-Filled Micromachined Devices, Instrumentation Laboratory, Department of Signals, Sensors and Systems (S3), Royal Institute of Technology (KTH), Stockholm, Sweden.

Dario, P., Valleggi, R., Carrozza, M.C., Montesi, M.C., and Cocco, M. (1992) "Review 'Microactuators for Microrobots: A Critical Survey'," *J. Micromech. Microeng.*, 2(3 (September)), pp. 141–157.

Dario, P., Carrozza, M.C., Lencioni, L., Magnani, B., and D'Attanasio, S. (1997) "A Micro Robotic System for Colonoscopy," *Int. Conf. on Robotics and Automation (ICRA '97)*, April 20–25, pp. 1567–1572, Albuquerque, New Mexico, USA, IEEE.

Dario, P., Carrozza, M.C., Stefanini, S., and D'Attanasio, S. (1998) "Contest Winner A Mobile Microrobot Actuated by a New Electromagnetic Wobbler Micromotor," *IEEE/ASME Transaction on Mechatronics*, 3(1, March) ?–?

Drextler, E. (1992) "Nanosystems — Molecular Machinery, Manufacturing, and Computation," John Wiley & Sons, Inc, New York.

Ebefors, T. (2000) "Polyimide V-Groove Joints for Three-Dimensional Silicon Transducers — Exemplified through a 3D Turbulent Gas Flow Sensor and Micro-Robotic Devices," Instrumentation Laboratory, Department of Signals, Sensors and Systems (S3), Stockholm, Sweden, Royal Institute of Technology (KTH).

Ebefors, T., Kälvesten, E., and Stemme, G. (1998) "New Small Radius Joints Based on Thermal Shrinkage of Polyimide in V-grooves for Robust Self-Assembly 3-D Microstructures," *J. Micromech. Microeng.*, 8(3 (September)), pp. 188–194.

Ebefors, T., Kälvesten, E., and Stemme, G. (1998) "Three Dimensional Silicon Triple-Hot-Wire Anemometer Based on Polyimide Joints," *IEEE 11th International Workshop on Micro Electro Mechanical Systems (MEMS '98)*, Jan. 25–29, pp. 93–98, Heidelberg, Germany, IEEE.

Ebefors, T., Mattsson, J., Kälvesten, E., and Stemme, G. (1999) "A Walking Silicon Micro-Robot," *The 10th International Conference on Solid-State Sensors and Actuators (Transducers '99)*, June 7–10, pp. 1202–1205, Sendai, Japan.

Ebefors, T., Mattsson, J., Kälvesten, E., and Stemme, G. (2000) "A Robust Micro Conveyer Realized by Arrayed Polyimide Joint Actuators," *IOP — J. Micromech. Microeng.*, vol. 10(No. 3 (September 2000)), pp. 337–349.

Eigler, D.M., and Schweizer, E.K. (1990) "Positioning Single Atoms with a Scanning Tunneling Microscope," *Nature*, **344**, pp. 524–526.

Fatikow, S. (2000) Mikroroboter und Mikromontage, B.G. Teubner, Stuttgart, Germany.

Fatikow, S., and Rembold, U. (1996) "An Automated Microrobot-Based Desktop Station for Micro Assembly and Handling of Micro-Objects," *IEEE Conference on Emerging Technologies and Factory Automation (EFTA '96)*, Nov. 18–21, pp. 586–592, Kauai, Hawai, USA.

Fatikow, S., and Rembold, U. (1997) *Microsystem Technology and Microrobotics*, Springer-Verlag, Berlin, Germany.

Fatikow, S., and Rembold, U. (1997 (Chapter 8 "Microrobotics")). *Microsystem Technology and Microrobotics*, Springer-Verlag, Berlin, Germany.

Fearing, R.S. (1998) "Powering 3-Dimensional Microrobots: Power Density Limitations," *IEEE Int. Conf. Robotics and Automation*.

Feynman, R.P. (1960) "There's Plenty of Room at the Bottom," in Miniaturization. "There's plenty of room at the bottom" at the Annual meeting of the American Physical Society (APS) at the California Institute of Technology, Dcember 26, 1959. Published as a chapter in the Reinhold Publishing Corporation book. Miniaturization. Horance D. Gilbert, ed. Also reprinted in IEEE Journal of MEMS, vol. 1, no. 1, March 1992, pp. 60–66. e.b. H.D. Gilbert, Reinhold Publishing Corp., New York: (Reprinted by S. Senturia in *IEEE/ASMA Journal Microelectromechanical Systems*, vol. 1, No. 1, March 1992, pp. 60–66).

Flynn, A.M., Brooks, R.A., III, W.M.W., and Barrett, D.S. (1989) "The World's Largest One Cubic Inch Robot," *Proc. of IEEE 2nd Int. Workshop on Micro Electro Mechanical Systems (MEMS '89)*, Feb. 20–22, pp. 98–101, Salt Lake City, USA, IEEE.

Flynn, A., Tavrow, L., Bart, S., Brooks, R., Ehrlich, D., Udayakumar, K.R., and Cross, E. (1992) "Piezoelectric Micromotors for Microrobots," *IEEE J. MEMS*, 1(1 (March)), pp. 44–51.

Frank, T. (1998) "Two-Axis Electrodynamic Micropositioning Devices," *J. Micromech. Microeng.*, 8, pp. 114–118.

Fujita, H., and Gabriel, K.J. (1991) "New Opportunities for Micro Actuators," *Tech. Digest, 6th Int. Conf. Solid-State Sensors and Actuators (Transducers '91)*, June 24–27, pp. 14–20, San Francisco, CA, USA.

Fukuda, T., Kawamoto, A., Arai, F., and Matsuura, H. (1994) "Mechanism and Swimming Experiment of Micro Mobile Robot in Water," *The 7th IEEE International Micro Electro Mechanical Systems Workshop (MEMS '94)*, Jan. 25–28, 1994, pp. 273–278, Oiso, Japan.

Fukuda, T., Kawamoto, A., Arai, F., and Matsuura, H. (1995) "Steering Mechanism and Swimming Experiment of Micro Mobile Robot in Water," *Proc. of IEEE 8th Int. Workshop on Micro Electro Mechanical Systems (MEMS '95)*, Jan. 29–Feb. 2, pp. 300–305, Amsterdam, the Netherlands.

Goosen, J., and Wolffenbuttel, R. (1995) "Object Positioning Using a Surface Micromachined Distributed System," *Tech. Digest, 8th Int. Conf. on Solid-State Sensors and Actuators (Transducers '95 and Eurosensors IX)*, June 25–29, pp. 396–399, Stockholm, Sweden.

Green, P.W., Syms, R.R.A., and Yeatman, E.M. (1995) "Demonstration of Three-Dimensional Microstructure Self-Assembly," *J. MEMS*, 4, No. 4, pp. 170–176.

Greitmann, G., and Buser, R. (1996) "Tactile Microgripper for Automated Handling of Microparts," *Sensors Actuators A*, 53(Nos. 1–4), pp. 410–415.

Greitmann, G., and Buser, R.A. (1996). "Tactile microgripper for Automated Handling of Microparts" *Sensors Actuators A*, 53 pp. 410–415.

Guckel, H., Klein, J., Christenson, T., Skrobis, K., Laudon, M., and Lovell, E. (1992) "Termo-Magnetic Metal Flexure Actuators," *Tech. Digest, Solid-State Sensors and Actuator Workshop*, June 13–16, pp. 73–75, (Hilton Head '92), Hilton Head, SC, USA.

Haga, Y., Tanahashi, Y., and Esashi, M. (1998) "Small Diameter Active Catheter Using Shape Memory Alloy," *IEEE 11th International Workshop on Micro Electro Mechanical Systems (MEMS '98)*, Jan. 25–29, pp. 419–424, Heidelberg, Germany, IEEE.

Hayashi, T. (1991) "Micro Mechanism," *J. Robotics Mechatronics*, 3, pp. 2–7.

Hierold, C., Clasbrummel, B., Behrend, D., Scheiter, T., Steger, M., Opperman, K., Kapels, H., Landgraf, E., Wenzel, D., and Etzrodt, D. (1998) "Implantable Low Power Integrated Pressure Sensor System for Minimal Invasive Telemetric Patient Monitoring," *IEEE 11th International Workshop on Micro Electro Mechanical Systems (MEMS '98)*, Jan. 25–29, pp. 568–573, Heidelberg, Germany, IEEE.

Hirai, S., Sakane, S., and Takase, K. (1993) "Cooperative Task Execution Technology for Multiple Micro Robot Systems," *Proc. of the IATP Workshop on Micromachine Technologies and Systems*, pp. 32–37, Tokyo, Japan.

Hirata, T., Akashi, T., Bertholds, A., Gruber, H.P., Schmid, A., Grétillat, M.-A., Guenat, O.T., and Rooij, N.F.d. (1998) "A Novel Pneumatic Actuator System Realized by Microelectrodischarge Machining," *Proc. of IEEE 11th Int. Workshop on Micro Electro Mechanical Systems (MEMS '98)*, Jan. 25–29, pp. 160–165, Heidelberg, Germany.

Hui, E., Howe, R.T., and Rodgers, M.S. (2000) "Single-Step Assembly of Complex 3-D Microstructures," *The 13th IEEE International Conference on Micro Electro Mechanical Systems (MEMS'2000)*, Jan. 23–27, pp. 602–607, Miyazaki, Japan.

Ishihara, H. (1998, June 1998) "International Micro Robot Maze Contest Information," Retrieved March 10th, 2000, from http://www.mein.nagoya-u.ac.jp/maze/index.html.

Ishihara, H., and Fukuda, T. (1998) "Optical Power Supply System for Micromechanical System," *SPIE Conference on Microrobotics and Micromanipulation*, Nov. 4–5, pp. 140–145, Boston, Massachusetts, USA.

Ishihara, H., Fukuda, T., Kosuge, K., Arai, F., and Hamagishi, K. (1995). "Approach to Distributed Micro Robatic System — Development of Micro Line Trace Robot and Autonomous," *IEEE Int. Conference on Robotics and Automation*, May 21–27, pp. 375–380, Nagoya, Aichi, Japan.

Ishikawa, Y., and Kitahara, T. (1997) "Present and Future of Micromechatronics," *International Symposium on Micromechanics and Human Science (MHS '97)*, Oct. 5–8, pp. 13–20, Nagoya, Japan, EEE, Piscataway, NJ, USA.

Jager, E., Inganäs, O., and Lundström, I. (2000). "Microrobots for Micrometer-Size Objects in Aqueous Media: Potential Tools for Single Cell Manipulation," *Science*, vol. 288(30 June 2000), pp. 2335–2338.

Jager, E.W.H., Semla, E., and Inganäs, O. (2000). "Microfabricating Conjugated Polymer Actuators," *Science*, 290(November 2000), pp. 1540–1545.

Johansson, S. (1995) "Micromanipulation for Micro- and Nanomanufacturing," *INRIA/IEEE Symposium on Emerging Technologies and Factory Automation (ETFA '95)*, Oct. 10–13, pp. 3–8, Paris, France, IEEE.

Johansson, S. (2000). Private Communication, April 2000.

Jonsmann, J., Sigmund, O., and Bouwstra, S. (1999) "Multi Degrees of Freedom Electro-Thermal Microactuators," *The 10th International Conference on Solid-State Sensors and Actuators (Transducers '99)*, June, 7–10, pp. 1373–1375, Sendai, Japan.

Kawahara, N., Suto, T., Hirano, T., Ishikawa, Y., Kitahara, T., Ooyama, N., and Ataka, T. (1997) "Microfactories: New Applications of Micromachine Technology to the Manufacture of Small Products," *Research J. Microsyst. Technol.*, 3(2 (Feb. 1997)), pp. 37–41. Concepts of microfactories consisting of micro-devices made by micromachine technologies are discussed. A microfactory is a small production system whose size is very small with respect to the dimensions of the small products. Typical examples of microfactories are classified according to production types. Microfactory systems have a great possibility to innovate the production systems of small products by making the best use of the inherent properties of the systems such as miniaturized facilities, mobility and flexibility. The microfactory saves energy, manufacturing space and mineral resources with decreasing size of the factory facilities.

Keller, C. (1998) "Microfabricated High Aspect Ratio Silicon Flexures: Hexsil, RIE, and KOH Etched Design & Fabrication," *MEMS PI*, http://www.memspi.com/book.html.

Keller, C. (1998) "Microfabricated Silicon High Aspect Ratio Flexures for In-Plane Motion," Dept. of Materials Science and Mineral Engineering, Berkeley, CA, USA, Univ. of California at Berkeley.

Keller, C.G., and Howe, R.T. (1995) "Nickel-Filled Hexsil Thermally Actuated Tweezers," *Tech. Digest Transducers '95 and Eurosensors IX*, pp. 376–379, June 25–29, Stockholm, Sweden.

Keller, C.G., and Howe, R.T. (1997) "Hexsil Tweezers for Teleoperated Micro-assembly," *Proc. of IEEE 10th Int. Workshop on Micro Electro Mechanical Systems (MEMS '97)*, pp. 72–77, Jan. 26–30, Nagoya, Japan.

Kelly, T.R., Silva, H.D., and Silva, R.A. (1999 (Sept. 9)) "Unidirectional Rotary Motion in a Molecular System," *Nature*, 401(6749), pp. 150–152.

Kim, C.-J., Pisano, A., and Muller, R. (1992) "Silicon-Processed Overhanging Microgripper," *IEEE/ASME J. MEMS*, 1(1), pp. 31–36.

Kim, Y.-K., Katsurai, M., and Fujita, H. (1990) "Fabrication and Testing of a Micro Superconducting Actuator Using the Meissner Effect," *Proc. of IEEE 3rd Int. Workshop on Micro Electro Mechanical Systems (MEMS '90)*, pp. 61–66, Feb. 11–14, Napa Valley, USA.

Kladitis, P., Bright, V., Harsh, K., and Lee, Y.C. (1999) "Prototype Microrobots for Micro Positioning in a Manufacturing Process and Micro Unmanned Vehicles," *Proc. of IEEE 12th Int. Conference on Micro Electro Mechanical Systems (MEMS '99)*, pp. 570–575, Jan. 17–21, Orlando, USA.

Konishi, S., and Fujita, H. (1993) "A Conveyance System Using Air Flow Based on the Concept of Distributed Micro Motion Systems," *Tech. Digest, 7th Int. Conf. Solid-State Sensors and Actuators (Transducers '93)*, pp. 28–31, June 7–10, Yokohama, Japan.

Konishi, S., and Fujita, H. (1994) "A Conveyance System Using Air Flow Based on the Concept of Distributed Micro Motion Systems," *IEEE J. MEMS*, 3(2 (June 1994)), pp. 54–58.

Konishi, S., and Fujita, H. (1995) "System Design for Cooperative Control of Arrayed Microactuators," *Proc. of IEEE 8th Int. Workshop on Micro Electro Mechanical Systems (MEMS '95)*, pp. 322–327, Jan. 29–Feb. 2, Amsterdam, The Netherlands.

Kovacs, G.T.A. (1998) Chapter 8.3.3 "Thin-Film Batteries," *Micromachined Transducers Sourcebook*, pp. 745–750, McGraw-Hill, New York.

Kuwana, Y., Shimoyama, I., and Miura, H. (1995) "Steering Control of a Mobile Robot Using Insect Antennae," *IEEE/RSJ International Conference on Intelligent Robots and Systems 95: "Human Robot Interaction and Cooperative Robots,"* pp. 530–535.

Li, M.-H., Wu, J.J., and Gianchandani, Y.B. (2000) "High Performance Scanning Thermal Probe Using a Low Temperature Polyimide-Based Micromachining Process," *The 13th IEEE International Conference on Micro Electro Mechanical Systems (MEMS '2000)*, pp. 763–768, Jan. 23–27, Miyazaki, Japan.

Lin, C.-C., Ghodssi, R., Ayon, A., Chen, D.-Z., Jacobsen, S., Breuer, K., Epstein, A., and Schmidt, M. (1999) "Fabrication and Characterization of a Micro Turbine/Bearing Rig," *The 12th IEEE International Micro Electro Mechanical System Conference (MEMS '99)*, pp. 529–533, Jan. 17–21, 1999, Orlando, Florida, USA.

Lin, L.Y., Shen, J.L., Lee, S.S., and Wu, M.C. (1996) "Realization of Novel Monolithic Free-Space Optical Disk Pickup Heads by Surface Micromachining," *Opt. Lett.*, vol. 21(No. 2 (January 15)), pp. 155–157.

Liu, C., Tsao, T., Tai, Y.-C., and Ho, C.-H. (1994) "Surface Micromachined Magnetic Actuators," *The 7th IEEE International Micro Electro Mechanical Systems Workshop (MEMS '94)*, pp. 57–62, Jan. 25–28, 1994, Oiso, Japan.

Liu, C., Tsao, T., Tai, Y.-C., Liu, W., Will, P., and Ho, C.-H. (1995) "A Micromachined Permalloy Magnetic Actuator Array for Micro Robotics Assembly System," *Tech. Digest, 8th Int. Conf. Solid-State Sensors and Actuators (Transducers '95 and Eurosensors IX)*, pp. 328–331, June 25–29, Stockholm, Sweden.

Mainz, I.f.M., and GmbH. (1999, Sept 10, 1999). "Micro-Motors: The World's Tiniest Helicopter," Retrieved March 16, 2000, 2000, from http://www.imm-mainz.de/english/developm/products/hubi.html.

Masuzawa, T. (1998) Chapter 3. "Micromachining by Machine Tools," *Micromechanical Systems — Principles and Technology*, 6, pp. 63–82, T. Fukuda and W. Menz, Elsevier Science, Amsterdam, The Netherlands.

Miki, N., and Shimoyama, I. (1999) "Flight Performance of Micro-Wings Rotating in an Alternating Magnetic Field," *Proc. of IEEE 12th Int. Conference on Micro Electro Mechanical Systems (MEMS '99)*, pp. 153–158, Jan. 17–21, Orlando, USA.

Miki, N., and Shimoyama, I. (2000) "A Micro-Flight Mechanism with Rotational Wings," *The 13th IEEE International Conference on Micro Electro Mechanical Systems (MEMS '2000)*, pp. 158–163, Jan. 23–27, Miyazaki, Japan.

Miniman (1997 (Dec)) "MINIMAN — **Mini**aturised Robot for Micro **MAN**ipulation.

Mita, Y., Oba, T., Hashiguchi, G., Mita, M., Minotti, P., and Fujita, H. (1999) "An Inverted Scratch-Drive-Actuators Array for Large Area Actuation of External Objects," *Tech. Digest of Transducers '99*, pp. 1196–1197, June 7–10, Sendai, Japan.

Miura, H., Yasuda, T., Kubo, Y., and Shimoyama, I. (1995) "Insect-Model-Based Microrobot," *Tech. Digest, 8th Int. Conf. Solid-State Sensors and Actuators (Transducers '95)*, pp. 392–395, June 16–19, Stockholm, Sweden.

MSTnews (2000) MATAS: A modular assembly technology for hybrid μTAS: Special volume dedicated to Packaging and Modular Microsystems.

Muller, R.S. (1990) "Microdynamics," *Sensors Actuators A*, A21–A23, pp. 1–8.

Nakamura, K., Ogara, H., Maeda, S., Sangawa, U., Aoki, S., and Sato, T. (1995) "Evaluation of the Micro Wobbler Motor Fabricated by Concentric Buildup Process," *Proc. of IEEE 8th Int. Workshop on Micro Electro Mechanical Systems (MEMS '95)*, pp. 374–379, Jan. 29–Feb. 2, Amsterdam, the Netherlands.

Nakazawa, H., Wantanabe, Y., Morita, O., Edo, M., and Yonezawa, E. (1997) "The Two-Dimensional Micro Conveyer," *Tech. Digest, 9th Int. Conf. Solid-State Sensors and Actuators (Transducers '97)*, pp. 33–36, June 16–19, Chicago, USA.

Nakazawa, H., Wantanabe, Y., Morita, O., Edo, M., Yushina, M., and Yonezawa, E. (1999) "Electromagnetic Micro-Parts Conveyer with Coil-Diod Modules," *Tech. Digest, 10th Int. Conf. Solid-State Sensors and Actuators (Transducers '99)*, pp. 1192–1195, June 7–10, Sendai, Japan.

Ok, J., Chu, M., and Kim, C.-J. (1999) "Pneumatically Driven Microcage for Micro-Objects in Biological Liquid," *The 12th IEEE International Micro Electro Mechanical System Conference (MEMS '99)*, pp. 459–463, Jan. 17–21, 1999, Orlando, Florida, USA, IEEE.

Park, K.-T., and Esashi, M. (1999) "A Multilink Active Catheter with Polyimide-Based Integrated CMOS Interface Circuits," *IEEE J. MEMS*, **8**(Dec)(4), pp. 349–357.

Pister, K., Fearing, R., and Howe, R. (1990). "A Planar Air Levitated Electrostatic Actuator System," *Proc. of IEEE 3rd Int. Workshop on Micro Electro Mechanical Systems (MEMS '90)*, pp. 67–71, Feb. 11–14, Napa Valley, USA.

Pister, K.S.J., Judy, M.W., Burgett, S.R., and Fearing, R.S. (1992) "Micro-Fabricated Hinges," *Sensors Actuators A*, A 33, pp. 249–256.

Pornsin-Sirirak, N., Lee, S., Nassef, H., Grasmeyer, J., Tai, Y.C., Ho, C.M., and Keennon, M. (2000) "MEMS Wing Technology for a Battery-Powered Ornithopter," *The 13th IEEE International Conference on Micro Electro Mechanical Systems (MEMS '2000)*, pp. 799–804, Jan. 23–27, Miyazaki, Japan.

Riethmüller, W., and Benecke, W. (1989) "Application of Silicon-Microactuators Based on Bimorph Structures," *Proc. of IEEE 2nd Int. Workshop on Micro Electro Mechanical Systems (MEMS '89)*, pp. 116–120, Feb. 20–22, Salt Lake City, USA, IEEE.

Rodgers, S., Sniegowski, J., Allen, J., Miller, S., Smith, J., and McWhorter, P. (1999) "Intricate Mechanisms-on-a-Chip Enabled by 5-Level Surface Micromachining," *The 10th International Conference on Solid-State Sensors and Actuators (Transducers '99)*, pp. 990–993, June 7–10, Sendai, Japan.

Ruffieux, D., and Rooij, N.F.d. (1999) "A 3-DOF Bimorph Actuator Array Capable of Locomotion," *The 13th European Conference on Solid-State Transducers (Eurosensors XIII)*, pp. 725–728, Sept. 12–15, The Hauge, The Netherlands.

Ruffieux, D., Dubois, M.A., and Rooij, N.F.d. (2000) "An AlN Piezoelectric microacctuator array," *The 13th IEEE International Conference on Micro Electro Mechanical Systems (MEMS '2000)*, pp. 662–667, Jan. 23–27, Miyazaki, Japan.

Shimoyama, I. (1995) "Hybrid System of Mechanical Parts and Living Organisms for Microrobots," *Proc. of IEEE 6th International Symposium on Micro Machine and Human Science (MHS '95)*, p. 55, Oct. 4–6, Nagoya, Japan.

Shimoyama, I. (1995) "Scaling in Microrobots," *IEEE 95*, pp. 208–211, IEEE.

Simu, U., and Johansson, S. (1999) "Multilayered Piezoceramic Microactuators Formed by Milling in the Green State," *SPIE Conference on Devices and Process Technologies for MEMS and Microelectronics*, part of SPIE's Int. Symp. on Microelectronics and MEMS (MICRO/MEMS '99), Oct. 27–29, 1999, Royal Pines Resort, Queensland, Australia.

Simu, U., and Johansson, S. (2001) "A Monolithic Piezoelectric Miniature Robot with 5 DOF," Submitted for conference presentation.

Smela, E., Inganäs, O., and Lundström, I. (1995) "Controlled Folding of Micrometer-Size Structures," *Science*, **268**(June), pp. 1735–1738.

Smela, E., Kallenbach, M., and Holdenried, J. (1999) "Electrochemically Driven Polypyrrole Bilayers for Moving and Positioning Bulk Micromachined Silicon," *J. MEMS*, **8**(4 (December)), pp. 373–383.

Stefanni, C., Carroza, M.C., and Dario, P. (1996) "A Mobile Micro-Robot Driven by a New Type of Electromagnetic Micromotor," *Proc. of IEEE 7th International Symposium on Micro Machine and Human Science (MHS '96)*, Nagoya, Japan.

Suh, J., Glader, S., Darling, R., Storment, C., and Kovacs, G. (1997) "Organic Thermal and Electrostatic Ciliary Microactuator Array for Object Manipulation," *Sensors Actuators A*, **58**, pp. 51–60.

Suh, J., Darling, B., Böhringer, K., Donald, B., Baltes, H., and Kovacs, G. (1999) "CMOS Integrated Ciliary Actuator Array as a General-Purpose Micromanipulation Tool for Small Objects," *IEEE/ASME J. MEMS*, **8**(4 (December 1999)), pp. 483–496.

Suzuki, K., Shimoyama, I., Miura, H., and Ezura, Y. (1992) "Creation of an Insect-based Microrobot with an External Skeleton and Elastic Joints," *Proc. of IEEE 5th Int. Workshop on Micro Electro Mechanical Systems (MEMS '92)*, pp. 190–195, Feb. 4–7, Travemünde, Germany.

Suzumori, K., Miyagawa, T., Kimura, M., and Hasegawa, Y. (1999) "Micro Inspection Robot for 1-in Pipes," *IEEE/ASME Trans. Mechatronics*, **4**(3 (September)), pp. 286–292.

Syms, R.R.A. (1998) "Rotational Self-Assembly of Complex Microstructures by the Surface Tension of Glass," *Sensors Actuators A*, **65**(Nos. 2,3 (15 March 1998)), pp. 238–243.

Syms, R.R.A. (1999) "Surface Tension Powered Self-Assembly of 3-D Micro-Optomechanical Structures," *J. MEMS*, **8**(4 (December)), pp. 448–455.

Tahhan, I.N., Zhuang, Y., Böhringer, K.F., Pister, K.S.J., and Goldberg, K. (1999). "MEMS Fixtures for Handling and Assembly of Microparts," *Micromachined Devices and Components V*, pp. 129–139 Santa Clara, CA, USA, SPIE.

Takeuchi, S., and Shimoyama, I. (1999) "Three Dimensional SMA Microelectrodes with Clipping Structure for Insect Neural Recording," *The 12th IEEE International Micro Electro Mechanical System Conference (MEMS '99)*, pp. 464–469, Jan. 17–21, 1999, Orlando, Florida, USA, IEEE.

Tendick, F., Sastry, S.S., Fearing, R.S., and Cohn, M. (1998) "MIS Application of Micromechatronics in Minimally Invasive Surgery," *IEEE/ASME Trans. Mechatronics*, **3**(1, March) ?–?.

Teshigahara, A., Watanable, M., Kawahara, N., Ohtsuka, Y., and Hattori, T. (1995) "Performance of a 7-mm Microfabricated Car," *IEEE/ASME J. MEMS*, **4**(2 (June 1995)), pp. 76–80.

Thornell, G. (1998) "Lilliputian Reflections," *Micro Structure Workshop (MSW '98)*, pp. 24.1–24.6, March 24–25, Uppsala, Sweden.

Thornell, G. (1999) Minuscular Sculpturing, Acta Univertsitatis Upsaliensis, Faculty of Science and Technology, Uppsala University, Uppsala, Sweden.

Thornell, G., and Johansson, S. (1998) "Microprocessing at the Fingertips," *J. Micromech. Microeng.*, **8**, pp. 251–262.

Tong, Q.-Y., and Gösle, U. (1999) *Semiconductor Wafer Bonding — Science and Technology*, John Wiley & Sons, Inc, New York.

Trimmer, W., and Jebens, R. (1989) "Actuators for Micro Robots," *IEEE 2nd International Workshop on Micro Electro Mechanical Systems (MEMS '89)? IEEE*.

Uchiyama, A. (1995) "Endoradiosonde Needs Micro Machine Technology," *IEEE Proceedings of the Sixth International Symp. on Micro Machine and Human Science (MHS '95)*, pp. 31–37.

Udayakumar, K.R., Bart, S.F., Flynn, A.M., Chen, J., Tavrow, L.S., Cross, L.E., Brooks, R.A., and Ehrlich, D.J. (1991) "Ferroelectric Thin-Film Ultrasonic Micromotors," *Proc. of IEEE 4th Int. Workshop on Micro Electro Mechanical Systems (MEMS '91)*, pp. 109–113, Jan. 30–Feb. 2, Nara, Japan.

Wilson, J. (1997) "Shrinking Micromachines," *Popular Mechanics*, Nov 1997 (available on Internet at: http://popularmechanics.com/popmech/sci/9711STROP.html).

Yasuda, T., Shimoyama, I., and Miura, H. (1993) "Microrobot Actuated by a Vibration Field," *Tech. Digest, 7th Int. Conf. Solid-State Sensors and Actuators (Transducers '93)*, pp. 42–45, June 7–10, Yokohama, Japan.

Yeh, R. (2000, Last updated: 3/17/2000) "Photo Gallery," Retrieved April 2000.

Yeh, R., Kruglick, E.J.J., and Pister, K.S.J. (1996) "Surface-Micromachined Components for Articulated Microrobots," *J. MEMS*, 5(1 (March)), pp. 10–17.

Zhou, G.-X. (1989) "Swallowable or Implantable Body Temperature Telemeter-Body Temperature Radio Pill," *IEEE Proceedings of the Fifteenth Annual Northeast Bioengineering Conference*, pp. 165–166.

Zill, S.N., and Seyfarth, E.-A. (1996) "Exoskeltal Sensors for Walking," *Scientific American*, (July 1996), pp. 70–74.

Further Information

A good introduction to MEMS-based microrobotics is presented in Microsystem Technology and Microrobotics (Fatikows and Rembold, Springer-Verlag, Berlin, 1997) and Mikroroboter und Mikromontage (in German); Fatikow, B.G. Teubner, Stuttgart-Leipzig, 2000). A good review on actuators suitable for microrobots as well as different aspects of the definition of microrobots is found in the paper, "Review — Microactuators for Microrobots: A Critical Survey," by Dario et al., and published in *IOP Journal of Micromechanics and Microengineering (JMM)*, Vol. 2, pp. 141–157, 1992. Proceedings from the "Microrobotics and Micromanipulation" conference have been published annually by SPIE (the International Society for Optical Engineering) since 1995. These proceedings document the latest development in the field of microrobotics each year.

8

Microscale Vacuum Pumps

E. Phillip Muntz and
Marcus Young
University of Southern California

Stephen E. Vargo
Siimpel Corporation

8.1 Introduction

There are numerous potential applications for meso- and microscale sampling instruments based on mass spectrometry [Nathanson et al., 1995; Ferran and Boumsellek, 1996; Orient et al., 1997; Freidhoff et al., 1999; Wiberg et al., 2000; White et al., 1998; Short et al., 1999; Piltingsrud, 1997] and gas chromatography [Terry et al., 1979]. Other miniaturized instruments utilizing electron optics [Chang, 1989; Chang et al., 1990; Park et al., 1997; Callas, 1999] will require both high vacuum and repeated solid sample transfers from higher pressure environments. The mushrooming interest in chemical laboratories on chips will likely result in manifestations requiring vacuum capabilities. At present, there are no microscale vacuum pumps to pair with the embryonic instruments and laboratories being developed. However, there is now a wider variety of available mesoscale vacuum pumps. Certainly, small vacuum pumps will not always be necessary. Some of the new devices are attractive because of low quantities of waste and rapidity of analysis; not directly because they are small, energy efficient, or need to be portable. However, for other applications involving portability or autonomous operations, small vacuum pumps with suitably low power requirements will be necessary.

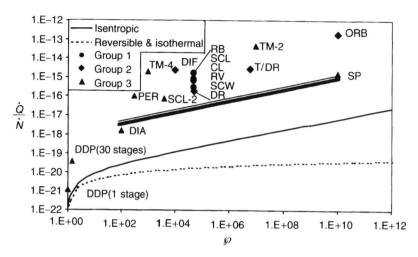

FIGURE 8.1 Representative performance ($\dot{Q}/\dot{N}$) of selected macro- and mesoscale vacuum pumps (from Table 8.1) as a function of pressure ratio ($\wp$).

This chapter will address how to approach identifying microscale and mesoscale vacuum pumping capabilities, consistent with the volume and energy requirements of meso- and microscale instruments and processes. The mesoscale pumps now available are discussed. Existing microscale pumping devices are not reviewed because none are available with attractive performance characteristics (a review of the attempts has recently been presented by Vargo, 2000; Young et al., 2001; Young, 2004; see also NASA/JPL, 1999).

In the macroscale world, vacuum pumps are not very efficient machines, ranging in thermal efficiencies from very small fractions of one percent to a few percent. They generally do not scale advantageously to small sizes, as is discussed in the section on Pump Scaling. Because there is a continuing effort to miniaturize instruments and chemical processes, there is not much desire to use oversized, power intensive vacuum pumps to permit them to operate. This is true even for situations where the pump size and power are not critical issues. Because at present microscale pump generated vacuums are unavailable, serious limits are currently imposed on the potential microscale applications of many high performance analytical instruments and chemical processes where portability or autonomous operations are necessary.

To illustrate this point, the performance characteristics of several types of macroscale and mesoscale vacuum pumps are presented in Figure 8.1 and Table 8.1. Pumping performance is measured by $\dot{Q}/\dot{N}$, which is derived from the power required, $(\dot{Q},W)$, and the pump's upflow in molecules per unit time ($\dot{N}$, #/s). Representative vacuum pumping tasks are indicated by the inlet pressure, p_i, and the pressure ratio, $\wp$, through which $\dot{N}$ is being pumped. The reversible, constant temperature compression power required per molecule of upflow ($\dot{Q}/\dot{N}$) as a function of $\wp$ is depicted in Figure 8.1. The adiabatic reversible (isentropic) compression power that would be required is also shown in Figure 8.1. The constant temperature comparison is most appropriate for vacuum systems. Note the 3- to -5 decade gap in the ideal $\dot{Q}/\dot{N}$ and the actual values for the macroscale vacuum pumps. For low pump inlet pressures the gases are very dilute and high volumetric flows are required to pump a given $\dot{N}$. The result is relatively large machines with significant size and frictional overheads. Unfortunately, scaling the pumps to smaller sizes generally increases the overhead relative to the upflow $\dot{N}$.

The current state of the art in mesoscale vacuum pump technology has been achieved primarily by shrinking macroscale pump technologies. A closer look at the two technologies that have recently been successfully shrunk to the mesoscale — diaphragm pumps and scroll pumps — will demonstrate the scaling possibilities of macroscale pump technologies. A KNF Neuberger diaphragm pump (DIA) is an example of a mesoscale diaphragm gas roughing pump. The diaphragm pump occupies a volume of 973 cm³, consumes 35 Watts of power, has a maximum pumping speed of 4.8 L/s, and reaches an ultimate

TABLE 8.1 Data with Sources for Conventional Vacuum Pumps that Might be Considered for Miniaturization

Type	p_P, p_E (mbar)	S_P (l/s)	$\dot{Q}$ (W)	$\dot{Q}/\dot{N}$ (W/#/s)	$\wp$ (—)	Comments and Sources
	Group 1: Macroscale Positive Displacement					
Roots Blower (RB)	2E-2, 1E3	2.8	2100	1.6E-15	5E4	5 stage, Lafferty p. 161
Claw (CL)	2E-2, 1E3	6.1	2100	7.5E-16	5E4	4 stage, Lafferty p. 165
Screw (SCW)	2E-2, 1E3	3.6	500	3E-16	5E4	Lafferty, p. 167
Scroll (SCL)	2E-2, 1E3	1.4	500	7.7E-16	5E4	Lafferty, p. 168
Rotary Vane (RV)	2E-2, 1E3	0.7	250	6.2E-16	5E4	Catalog
	Group 2: Macroscale Kinetic and Ion					
Drag (DR)	2E-2, 1E3	36	3300	1.9E-16	5E4	Molecular/Regenerative Lafferty p. 253
Diffusion (DIF)	1E-5, 1E-1	2.5E4	1.4E3	2.5E-15	1E4	Zyrianka, catalog
Turbo/drag (T/DR)	1E-5, 4.5E1	30	20	2.5E-15	4.5E6	Alcatel, catalog (30H+30)
Orbitron (ORB)	1E-7, 1E3	1700	750	2E-13	1E10	Denison, 1967
	Group 3: Mesoscale Pumps					
Turbomolecular (TM-4)	1E-5, 1E-2	4	2	2E-15	1E3	Experimental, $f_P \approx$ 1.7E3, 4 cm dia., Creare Website
Peristaltic (PER)	1.6, 1E3	3.3E-3	20	1E-16	6E2	Piltingsrud, 1996
Sputter Ion (SP)	1E-7, 1E3	2.2	1.1E-2	1.7E-15	1E10	Based on Suetsugu, 1993, 1.5 cm dia., 3.1 cm length
Turbomolecular (TM-2)	1E-6, 1E1	4	7	5E-14	1E7	Kenton, 2003
Scroll (SCL-2)	1E-1, 1E3	.12	25	7.8E-17	1E4	Air Squared Website
Diaphragm (DIA)	1E1, 1E3	.08	34.8	1.6E-18	1E2	KNF Neuberger Website
Dual Diaphragm (DDP)	9.9E2, 1E3	5.E-4	8.0E-3	6.0E-22	1.01	Cabuz et al., 2001

pressure of 1.5 Torr. The energy efficiency of the diaphragm pump, 1.6×10^{-18} W/#/s, is consistent with the energy efficiency of macroscale pumps operating with the limited pressure ratio, $\wp = 100$. Similar diaphragm pumps are available at smaller sizes, but with ever increasing ultimate pressures.

Honeywell is currently developing the smallest published mesoscale diaphragm pump, the Dual Diaphragm Pump (DDP) [Cabuz et al., 2001]. A single stage of the DDP measures 1.5 cm × 1.5 cm × 0.1 cm, and is manufactured using an injection molding process. The pump is driven by the controlled electrostatic actuation of two thin diaphragms with non-overlapping apertures in a sequence that first fills a pumping chamber with gas, and then expels the gas from the chamber. The DDP has a pumping speed of 30 sccm at a power consumption of 8 mW. It is, however, only able to maintain a maximum pressure difference of 14.7 Torr per stage, making it strictly a low pressure ratio pump. Cascades of thirty stages have been manufactured to increase the total pressure ratio; the corresponding energy efficiency is given in Figure 8.1. This again illustrates the capability of making mesoscale diaphragm pumps, but with ever increasing ultimate pressures as the volume is decreased below roughly one liter.

Scroll pumps have been successfully shrunk to the same length scale as diaphragm pumps. Air Squared has a variety of commercially available mesoscale scroll pumps. The smallest Air Squared scroll pump (SCL-2) occupies a volume of 1580 cm³, consumes 25 W of power, has a pumping speed of 7 L/s, and can reach an ultimate pressure of 10 mTorr. The physical dimensions, power consumption, and throughput are all similar to the mesoscale diaphragm pumps, but the achievable ultimate pressure is lower by several orders of magnitude. The energy efficiency of the scroll pump, 7.8×10^{-17} W/#/s, appears to be better than macroscale scroll pumps operating over similar pressure ratios.

The limit in scalability of scroll pumps is illustrated by a mesoscale scroll pump that recently has been proposed by JPL and USC [Moore et al., 2002, 2003]. The diameter of the scroll section is 1.2 cm. The main concerns for this mesoscale scroll pump are the coupled issues of the manufacturing tolerances required to provide sufficient sealing and the anticipated lifetime of effectively sealing scrolls. Initial performance estimates made using an analytical performance model, experimentally validated with macroscale scroll pumps, indicate that the gap spacing (including both manufacturing tolerances and the effects of the rotary motion of the stages) must be held under $2\,\mu m$ for the pump to be viable. These manufacturing tolerances cannot be met with current technologies and is the main focus of the development work with the pump. Because of the required micrometer sized clearances at even the mesoscale, it appears unlikely that scroll pumps will be scaled to the microscale.

In addition to the power requirement, vacuum pumps tend to have large volumes, so that an additional indicator of relevance to miniaturized pumps is a pump's volume (V_p) per unit upflow ($V_p/\dot{N}$). The two measures $\dot{Q}/\dot{N}$ and $V_p/\dot{N}$ will be used throughout the following discussions for evaluating different approaches to the production of microscale vacuums.

This chapter will address only the production of appropriate vacuums where throughput or continuous gas sampling, or alternatively multiple sample insertions, are required. In some cases so-called capture pumps (sputter ion pumps, getter pumps) may provide a convenient high- and ultra-high vacuum pumping capacity. However, because of the finite capacity of these pumps before regeneration, they may or may not be suitable for long duration studies. An example of a miniature cryosorption pump has been discussed by Piltingsrud (1994). Because such trade-offs are very situation-dependent, only the sputter ion and orbitron ion capture pumps are considered in the present study. Both of these pump "active" (N_2, O_2, etc.) and inert (noble and hydrocarbon) gases, whereas other non-evaporable and evaporable getters only pump the active gases efficiently [Lafferty, 1998].

8.2 Fundamentals

8.2.1 Basic Principles

There are several basic relationships derived from the kinetic theory of gases [Bird, 1998; Cercignani, 2000; Lafferty, 1998] that are important to the discussion of both macroscale and microscale vacuum pumps. The *conductance* or volume flow in a channel under free molecule or collisionless flow conditions can be written as:

$$C_L = C_A \alpha \tag{8.1}$$

where C_L is the channel volume flow in one direction for a channel of length L. The conductance of the upstream aperture is C_A and α is the probability that a molecule, having crossed the aperture into the channel, will travel through the channel to its end (this includes those that pass through without hitting a channel wall and those that have one or more wall collisions). Employing the kinetic theory expression for the number of molecules striking a surface per unit time per unit area ($n_g\overline{C}'/4$), the aperture conductance is:

$$C_A = (\overline{C}'/4)A_A = \{(8kT_g/\pi m)^{1/2}/4\}A_A \tag{8.2}$$

where A_A is the aperture's area and $\overline{C}' = \{8kT_g/\pi m\}^{1/2}$ is the mean thermal speed of the gas molecules of mass m, k is Boltzmann's constant and T_g and n_g are the gas temperature and number density. The probability α can be determined from the length and shape of the channel and the rules governing the reflection at the channel's walls [Lafferty, 1998; Cercignani, 2000].

Several terms associated with wall reflection will be used. Diffuse reflection of molecules is when the angle of reflection from the wall is independent of the angle of incidence, with any reflected direction in the gas space equally probable per unit of projected surface area in that direction. The reflection is said to be *specular* if the angle of incidence equals the angle of reflection and both the incident and reflected velocities lie in the same plane and have equal magnitude.

The condition for effectively collisionless flow (no significant influence of intermolecular collisions) is reached when the mean free path (λ) of the molecules between collisions in the gas is significantly larger than a representative lateral dimension (l) of the flow channel. Usually this is expressed by the Knudsen number (Kn), such that:

$$Kn_l = \lambda/l \geqslant 10 \tag{8.3}$$

The mean free path λ can have, for present purposes, the elementary kinetic theory form:

$$\lambda = 1/(\sqrt{2}\Omega n_g) \tag{8.4}$$

with Ω being the temperature dependent hard sphere total collision cross-section of a gas molecule (for a hard sphere gas of diameter d, $\Omega = \pi d^2$). As an example the mean free path for air at 1 atm and 300 K is $\lambda \approx 0.06\,\mu\text{m}$.

Expressions for conductance analogous to Equation (8.1) can be obtained for transitional ($10 > Kn_l > 0.001$) flows and continuum ($Kn_l \leqslant 10^{-3}$) viscous flows (see Cercignani, 2000, and the references therein). For the present discussion the major interest is in collisionless and early transitional flow ($Kn \geqslant 0.1$).

The performance of a vacuum pump is conventionally expressed as its pumping speed or volume of upflow (S_P) measured in terms of the volume flow of low pressure gas from the chamber that is being pumped (in detail there are specifications about the size and shape of the chamber [Lafferty, 1998]). Following the recipe of Equation (8.1), the pumping speed can be written as:

$$S_P = C_{AP}\alpha_P \tag{8.5}$$

Once a molecule has entered the pump's aperture of conductance C_{AP}, it will be "pumped" with probability α_P. Clearly, $(1 - \alpha_P)$ is the probability that the molecule will return or be backscattered to the low pressure chamber. A pump's upflow in this chapter will generally be described in terms of the molecular upflow in molecules per unit time ($\dot{N}$, #/s). For a chamber pressure of p_I and temperature T_I the number density n_I is given by the ideal gas equation of state, $p_I = n_I k T_I$, and the molecular upflow is:

$$\dot{N} = S_P n_I \tag{8.6}$$

8.2.2 Conventional Types of Vacuum Pumps

The several types of available vacuum pumps have been classified [c.f. Lafferty, 1998] into convenient groupings, from which potential candidates for microelectromechanical systems (MEMS) vacuum pumps can be culled. The groupings include:

- Positive displacement (vane, piston, scroll, Roots, claw, screw, diaphragm)
- Kinetic (vapor jet or diffusion, turbomolecular, molecular drag, regenerative drag)
- Capture (getter, sputter ion, orbitron ion, cryopump)

In systems requiring pressures $<10^{-3}$ mbar the positive displacement, molecular, and regenerative drag pumps are used as "backing" or "fore" pumps for turbomolecular or diffusion pumps. The capture pumps in their operating pressure range ($<10^{-4}$ mbar) require no backing pump but have a more or less limited storage capacity before needing "regeneration." Also, some means to pump initially to about 10^{-4} mbar from the local atmospheric pressure is necessary. For further discussion of these pump types, refer to Lafferty's excellent book [Lafferty, 1998]. The historical roots of this reference are also of interest [Dushman, 1949; Dushman and Lafferty, 1962]. For this discussion, it should be noted that the positive displacement pumps mechanically trap gas in a volume at a low pressure, the volume decreases, and the trapped gas is rejected at a higher pressure. The kinetic pumps continuously add momentum to the pumped gas so that it can overcome adverse pressure gradients and be "pumped." The storage pumps trap gas or ions on and in a nanoscale lattice, or in the case of cryocondensation pumps, simply condense the gas. In either case the storage pumps have a finite capacity before the stored gas has to be removed (pump regeneration) or fresh adsorption material supplied.

8.2.3 Pumping Speed and Pressure Ratio

For all pumps, except ion pumps, there is a trade between upflow, S_P, and the pressure ratio, $\wp$, that is being maintained by the pump. One can identify a pump's performance by two limiting characteristics

[Bernhardt, 1983]: first, the maximum upflow, $S_{P,MAX}$, which is achieved when the pressure ratio $\wp = 1$; and second, the maximum pressure ratio, $\wp_{MAX}$, which is obtained for $S_P = 0$. In many cases a simple expression relating pumping speed (S_P) and pressure ratio ($\wp$) to $S_{P,MAX}$ and $\wp_{MAX}$ describes the trade between speed and pressure ratio:

$$S_P/S_{P,MAX} = \frac{(1 - \wp/\wp_{MAX})}{(1 - 1/\wp_{MAX})} \tag{8.7}$$

This relationship is not strictly correct because in many pumps the conductances that result in backflow losses relative to the upflow change dramatically as pressure increases. For the critical lower pressure ranges (10^{-1} mbar in macroscale pumps but significantly higher in microscale pumps), Equation (8.7) is a reasonable expression for the trade between speed and pressure ratio. Equation (8.7) is convenient because $\wp_{MAX}$ and $S_{P,MAX}$ are identifiable and measurable quantities which can then be generalized by Equation (8.7).

8.2.4　Definitions for Vacuum and Scale

The terms vacuum and MEMS have both flexible and strict definitions: for the present discussion, the following categories of vacuum in reduced scale devices will be used. The pressure range from 10^{-2} to 10^3 mbar will be defined as low or roughing vacuum. For the range from 10^{-2} mbar to 10^{-7} mbar the terminology will be high vacuum and for pressure below 10^{-7} mbar, ultra high vacuum.

For the foreseeable future most small-scale vacuum systems are unlikely to fall within the strict definition of MEMS devices (maximum component dimension $< 100 \,\mu m$). Typically device dimensions somewhat larger than 1 cm are anticipated. They will be fabricated using MEMS techniques but the total construct will be better termed mesoscale. Device scale lengths 10 cm and greater indicate macroscale devices.

8.3　Pump Scaling

In this section the sensitivities of performance to size reduction of several generic, conventional vacuum pump configurations are discussed. Positive displacement pumps, turbomolecular (also molecular drag kinetic pumps), sputter ion, and orbitron ion capture pumps are the major focus. Other possibilities, such as diffusion kinetic pumps, diaphragm positive displacement pumps, getter capture pumps, cryocondensation and cryosorption pumps do not appear to be attractive for MEMS applications. This is due to vaporization and condensation of a separate working fluid (diffusion pumps); large backflow due to valve leaks relative to upflow (diaphragm pumps); low saturation gas loadings (getter capture pumps); an inability to pump the noble gases (getter capture pumps); and the difficulty in providing energy efficient cryogenic temperatures for MEMS scale cryocondensation or cryosorption pumps.

8.3.1　Positive Displacement Pumps

Consider a generic positive displacement pump that traps a volume, V_T, of low pressure gas with a frequency, f_T, trappings per unit time. In order to derive a phenomenological expression for pumping speed there are several inefficiencies that need to be taken into account. These include backflow due to clearances, which is particularly important for dry pumps; and the volumetric efficiency of the pump's cycle, including both time dependent inlet conductance effects and dead volume fractions. The generic positive displacement pump is illustrated in Figure 8.2. In general, the pumping speed, S_P, for an intake number density, n_I, and an exhaust number density, n_E (or $\wp$ corresponding to the pressures, p_I, and p_E, since the process gas temperature is assumed to be constant in the important low pressure pumping range) can be derived:

$$S_P = (1 - \wp \wp_G^{-1})(1 - e^{-(C_{II}/V_{T,I}f_T)\beta_1})V_{T,I}f_T - (\wp - 1)C_{LB}\beta_2 \tag{8.8}$$

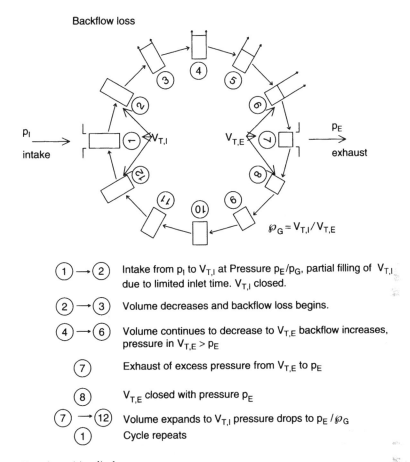

Backflow loss

$\wp_G = V_{T,I}/V_{T,E}$

(1) → (2)   Intake from p_I to $V_{T,I}$ at Pressure p_E/p_G, partial filling of $V_{T,I}$ due to limited inlet time. $V_{T,I}$ closed.

(2) → (3)   Volume decreases and backflow loss begins.

(4) → (6)   Volume continues to decrease to $V_{T,E}$ backflow increases, pressure in $V_{T,E}$ > p_E

(7)   Exhaust of excess pressure from $V_{T,E}$ to p_E

(8) $V_{T,E}$ closed with pressure p_E

(7) → (12)   Volume expands to $V_{T,I}$ pressure drops to $p_E / \wp_G$

(1)   Cycle repeats

FIGURE 8.2 Generic positive displacement vacuum pump.

In Equation (8.8), $\wp$ is the pressure ratio $p_E/p_I \equiv n_E/n_I$ (assuming $T_I = T_E$); C_{LI} is the pump's inlet conductance; $V_{T,I}$ is the trapping volume at the inlet; C_{LB} is the conductance of the backflow channels between exhaust and inlet pressures; $\wp_G = V_{T,I}/V_{T,E}$ is the geometric trapped volume ratio between inlet and exit; β_1 is the fraction of the trapping cycle during which the inlet aperture is exposed; and β_2 is the fraction of the cycle during which the backflow channels are exposed to the pressure ratio $\wp$.

The pumping speed expression of Equation (8.8) applies most directly to a single compression stage. The backflow conductance is assumed to be constant because the flow is in the "collisionless" flow regime, which exists in the first few stages of a typical dry pumping system. The inlet conductance for a dry microscale system will be in the collisionless flow regime at low pressure (say 10^{-2} mbar). The inlet conductance per unit area can increase significantly (amount depends on geometry) for transitional inlet pressures [Lafferty, 1998; Sone and Itakura, 1990; Sharipov and Seleznev, 1998]. The performance of macroscale (inlet apertures of several cm and larger) positive displacement pumps at a given inlet pressure will thus benefit from the increased inlet conductance per unit area compared to their reduced scale counterparts in the important low pressure range of 10^{-3} to 10^{-1} mbar.

The term $(1 - \wp \wp_G^{-1})$ in Equation (8.8) represents an inefficiency due to a finite dead volume in the exhaust portion of the cycle. The effect of incomplete trapped volume filling during the open time of the inlet aperture is represented by $(1 - e^{-(C_{LI}/V_{T,I}f_T)\beta_1})$. The ideal (no inefficiencies) pumping speed is $V_{T,I}f_T$. The backflow inefficiency is $(\wp - 1)C_{LB}\beta_2$. The dimensions of all groupings are volume per unit time. As in all pumps a maximum upflow $(S_{P,MAX})$ can be found by assuming $\wp = 1$ in Equation (8.8). Similarly the maximum pressure ratio ($\wp_{MAX}$) can be obtained from Equation (8.8) by setting $S_P = 0$. With some manipulations Equation (8.8) can be rewritten as:

$$S_P = V_{T,I} f_T \wp_G^{-1} \{1 + (C_{LB} \beta_2 / V_{T,I} f_T) \wp_G - \exp(-C_{LI} \beta_1 / V_{T,I} f_T)\} (\wp_{MAX} - \wp) \qquad (8.9)$$

The relationship between $\wp$, $\wp_{MAX}$, S_P, and $S_{P,MAX}$ can be found by setting $\wp = 1$ in Equation (8.9). Substituting back into Equation (8.9) gives the same expression as in Equation (8.7).

For the purposes of this chapter, the upflow of molecules per unit time ($\dot{N} = S_P n_I$) is a useful measure of pumping speed. The form of Equation (8.9) has been checked by fitting it successfully to observed pumping curves (S_P vs. p_I) for several positive displacement pumps (using data in Lafferty, 1998) between inlet pressures of 10^{-2} and 10^{0} mbar. This is done by following Equation (8.9) and re-plotting the experimental results using S_P and $(\wp_{MAX} - \wp)$ as the two variables. The variation of inlet conductance with Kn, $C_{LI}(Kn)$, is important in matching Equation (8.9) to the observed pumping performances.

At low inlet pressures, the energy use of positive displacement pumps is dominated by friction losses due to the relative motion of their mechanical components. Taking the two possibilities of sliding and viscous friction as limiting cases, the frictional energy losses can be represented as:

$$\dot{Q}_{sf} = \bar{u} \mu_{sf} A_s \tilde{F}_N \qquad (8.10a)$$

for sliding friction; and for viscous friction:

$$\dot{Q}_\mu = \bar{u} \mu A_s (\bar{u}/h) \qquad (8.10b)$$

Here $\dot{Q}_{sf}$ and $\dot{Q}_\mu$ are the powers required to overcome sliding friction and viscous friction respectively. The coefficients are respectively μ_{sf} and μ, A_s is the effective area involved, $\tilde{F}_N$ is the normal force per unit area, and $\bar{u}$ is a representative relative speed of the two surfaces that are in contact for sliding friction or separated by a distance, h, for the viscous case. The contribution to viscous friction in clearance channels exposed to the process gas at low pumping pressures is usually not important, but there may be significant viscous contributions from bearings or lubricated sleeves. An estimate of the power use per unit of upflow can be obtained by combining Equations (8.9) and (8.10) to give ($\dot{Q}/\dot{N}$), with units of power per molecule per second or energy per molecule.

Consider the geometric scaling to smaller sizes of a "reference system" macroscopic positive displacement vacuum pump. The scaling is described by a scale factor, s_i, applied to all linear dimensions. Inevitable manufacturing difficulties when s_i is very small are put aside for the moment. As a result of the geometric scaling the operating frequency, $f_{T,i}$, needs to be specified. It is convenient to set $f_{T,i} = s_i^{-1} f_{T,R}$, which keeps the component speeds constant between the reference and scaled versions. This may not be possible in many cases, another more or less arbitrary condition would be to have $f_{T,i} = f_{T,R}$. Using the scaling factor and Equation (8.9), an expression for the scaled pump upflow, $\dot{N}_i$, can be written as a function of s_i and values of the pump's important characteristics at the reference or $s_i = 1$ scale (e.g. $V_{T,I,i} = s_i^3 V_{T,I,R} s_i^{-1} f_{T,R}$). For the case of $f_{T,i} = s_i^{-1} f_{T,R}$:

$$\dot{N}_i = n_I s_i^3 V_{T,I,R} s_i^{-1} f_{T,R} \wp_G^{-1} [1 + (s_i^2 C_{LB,R} \beta_2 / s_i^3 V_{T,I,R} s_i^{-1} f_{T,R}) \wp_G$$
$$- \exp\{-(s_i^2 C_{LI,R} \beta_1 / s_i^3 V_{T,I,R} s_i^{-1} f_{T,R})\}] (\wp_{MAX} - \wp) \qquad (8.11)$$

Similarly the scaled version of Equation (8.10b) becomes:

$$\dot{Q}_{\mu,i} = \bar{u}^2 \mu s_i^2 A_{S,R} / s_i h_R \qquad (8.12)$$

Forming the ratios $(\dot{Q}_{\mu,i}/\dot{Q}_{\mu,R})/(\dot{N}_i/\dot{N}_R) \equiv (\dot{Q}_{\mu,i}/\dot{N}_i)/(\dot{Q}_{\mu,R}/\dot{N}_R)$ eliminates many of the reference system characteristics and highlights the scaling. In this case, for the viscous energy dissipation per molecule of upflow in the scaled system compared to the same quantity in the reference system, a scaling relationship is obtained:

$$(\dot{Q}_{\mu,i}/\dot{N}_i)/(\dot{Q}_{\mu,R}/\dot{N}_R) = s_i^{-1} \qquad (8.13)$$

TABLE 8.2a   Effect of scaling on performance

Type	$\dfrac{\wp_{MAX,i}}{\wp_{MAX,R}}$	$\dfrac{S_{P,MAX,i}}{S_{P,MAX,R}}$	$\dfrac{(\dot{Q}_{sf}/\dot{N}_{MAX})_i}{(\dot{Q}_{sf}/\dot{N}_{MAX})_R}$	$\dfrac{(\dot{Q}_{\mu}/\dot{N}_{MAX})_i}{(\dot{Q}_{\mu}/\dot{N}_{MAX})_R}$	$\dfrac{(V_P/\dot{N}_{MAX})_i}{(V_P/\dot{N}_{MAX})_R}$
		$f_{T,i} = s_i^{-1} f_{T,R}$	$(\bar{u}_i = \bar{u}_R)$		
Positive Displacement	1	s_i^2	1	s_i^{-1}	s_i
Turbomolecular	1	s_i^2	1	s_i^{-1}	s_i
		$f_{T,i} = f_{T,R}$	$(\bar{u}_i = s_i \bar{u}_R)$		
Positive Displacement	$\mathcal{O}[1]$ to $\mathcal{O}[s_i]$	$\mathcal{O}[s_i^3]$	$>\mathcal{O}[1]$	$>\mathcal{O}[1]$	$>\mathcal{O}[1]$
Turbomolecular	$(\wp_{MAX,R})^{(s_i-1)}$	$\mathcal{O}[s_i^4]$	$>\mathcal{O}[s_i^{-1}]$	$>\mathcal{O}[s_i^{-1}]$	$>\mathcal{O}[s_i^{-1}]$
	$\dfrac{\wp_{MAX,i}}{\wp_{MAX,R}}$	$\dfrac{S_{P,MAX,i}}{S_{P,MAX,R}}$	$(\dot{Q}/\dot{N}_{MAX})_i/(\dot{Q}/\dot{N}_{MAX})_R$		$\dfrac{(V_P/\dot{N}_{MAX})_i}{(V_P/\dot{N}_{MAX})_R}$
Sputter Ion (inactive and active gases)	1	s_i^2	$V_{D,i}/V_{D,R} \approx 1$		s_i
Orbitron Ion Active gases	1	s_i^2	1		s_i
Obitron Ion Inactive gases (ions)	1	s_i	1		s_i^2

Notes: For the case of $f_{T,i} = f_{T,R}$ the expressions that result are sensitive to the particular importance of backflow in each case. Estimates have been made, using typical values for the losses, of the order of magnitude of these expressions in order to simplify the presentation. The detailed expressions appear below in Table 8.2b. The scaling of $V_{P,i}/V_{P,R}$ is assumed to be as s_i^3 when obtaining the $(V_P/\dot{N}_{MAX})_i/(V_P/\dot{N}_{MAX})_R$ scaling.

A summary of the results of this type of scaling analysis applied to positive displacements pumps is presented in Table 8.2a, along with all the other types of pumps that are considered below. In Table 8.2a, the performance can be summarized using $\wp_{MAX}$ ($S_P = 0$) and $S_{P,MAX}$ or $\dot{N}_{MAX}$ ($\wp = 1$) in order to eliminate $\wp$ appearing explicitly as a variable in the scaling expressions through the dependency of $\dot{N}$ on pressure ratio (refer to Equation 8.11).

8.3.2 Kinetic Pumps

Because of their sensitivity to orientation and their potential for contamination, diffusion pumps are not suitable for MEMS scale vacuum pumps, except possibly for situations permitting fixed installations. The other major kinetic pumps, turbomolecular and molecular drag, require high rotational speed components but are dry; in macroscale versions they can be independent of orientation, at least in time independent situations. Only the turbomolecular and molecular drag pumps will be discussed in this chapter.

Bernhardt (1983) developed a simplified model of turbomolecular pumping. Following this description the maximum pumping speed ($\wp = 1$) of a turbomolecular pumping stage (rotating blade row and a stator row) can be written as:

$$S_{P,MAX} = A_I (\overline{C}'/4)(v_c/\overline{C}')/[(1/qd_f) + (v_c/\overline{C}')] \tag{8.14}$$

In Equation (8.14), A_I is the inlet area to the rotating blade row, v_c is an average tangential speed of the blades, q is the trapping probability of the rotating blade row for incoming molecules, besides blade geometry it is a function of ($v_c/\overline{C}'$). The term, d_f, accounts for a reduction in transparency due to blade thickness. It is assumed in Equation (8.14) that the blades are at an angle of 45° to the rotational plane of the blades. As a convenience, writing $v_r = (v_c/\overline{C}')$, v_r is similar to but not identical with the Mach number.

The maximum pressure ratio ($S_P = 0$) can be written as:

$$\wp_{MAX} = e^{\xi v_r} \tag{8.15}$$

where ξ is a constant that depends on blade geometry [Bernhardt, 1983]. Dividing Equation (8.14) by $A_I(\overline{C}'/4)$ gives the pumping probability:

$$\alpha_{P,MAX} = v_r/[(1/qd_f) + v_r] \tag{8.16}$$

TABLE 8.2b Detailed Expressions for Size Scaling with $f_{T,i} = f_{T,R}$

Type	$\dfrac{\wp_{MAX,i}}{\wp_{MAX,R}}$	$\dfrac{S_{P,MAX,i}}{S_{P,MAX,R}}$	$\dfrac{(\dot{Q}_{sl}/\dot{N}_{MAX})_i}{(\dot{Q}_{sl}/\dot{N}_{MAX})_R}$	$\dfrac{(\dot{Q}_{sl}/\dot{N}_{MAX})_i}{(\dot{Q}_{sl}/\dot{N}_{MAX})_R}$
Positive Displacement	$\left[\dfrac{1-\exp(-s_i^{-1}K_{I,R})+s_i^{-1}K_{B,R}}{1-\exp(-K_{I,R})+K_{B,R}}\right] \times$ $\left[\dfrac{1-\exp(-K_{I,R})+\wp_G K_{B,R}}{1-\exp(-s_i^{-1}K_{I,R})+s_i^{-1}\wp_G K_{B,R}}\right]$	$s_i^3\left[\dfrac{1-\exp(-s_i^{-1}K_{I,R})}{1-\exp(-K_{I,R})}\right]$	$\left[\dfrac{1-\exp(-K_{I,R})}{1-\exp(-s_i^{-1}K_{I,R})}\right]$	$\left[\dfrac{1-\exp(-K_{I,R})}{1-\exp(-s_i^{-1}K_{I,R})}\right]$
Turbomolecular	$(\wp_{MAX,R})^{(s_i-1)}$	$s_i^3\left[\dfrac{(q_R d_{I,R})^{-1}+(v_{c,R}/\bar{C}')}{(q_R d_{I,R})^{-1}+s_i(v_{c,R}/\bar{C}')}\right]$	$\left[\dfrac{(q_R d_{I,R})^{-1}+(s_i v_{c,R}/\bar{C}')}{(q_R d_{I,R})^{-1}+(v_{c,R}/\bar{C}')}\right]$	$\left[\dfrac{(q_R d_{I,R})^{-1}+(s_i v_{c,R}/\bar{C}')}{(q_R d_{I,R})^{-1}+(v_{c,R}/\bar{C}')}\right]$

Notes: The $K_{I,R}$ and $K_{B,R}$ are associated with the inlet losses and backflow losses respectively (larger K_I leads inlet losses but the larger $K_{B,R}$ the more serious the backflow loss). $K_{I,R} = C_{LI,R}\beta_1/V_{TI,R}, f_{T,R}K_{B,R} = C_{LB,R}\beta_2/V_{TI,R}f_{T,R}$. The symbols are defined in the section on positive displacement pumps.

The $\wp_{MAX}$ for turbomolecular stages is generally large ($> \mathcal{O}[10^5]$) for gases other than He and H_2 due to the exponential expression of Equation (8.15). During operation in a multi-stage pump the stages can be employed at pressure ratios $\wp \ll \wp_{MAX}$. Thus, from Equation (8.7) (which also applies to turbomolecular drag stages) the pumping speed approaches $S_{P,MAX}$.

The scaling characteristics of turbomolecular pumps can be derived using Equations (8.14) and (8.15). It is assumed that the pump blades remain similar during the scaling. The tangential speed (v_c) will be written as $2\pi R f_p$, where R is a characteristic radius of the blade row and f_p is the rotational frequency in rps. From Equations (8.14) and (8.15):

$$(S_{P,MAX})_i = s_i^2 A_{I,R}\,(\overline{C}'/4)\{[2\pi s_i R_R f_{P,i}/\overline{C}']/[(1/q_i d_{f,i}) + (2\pi s_i R_R f_{P,i}/\overline{C}')]\}$$

$$\wp_{MAX,i} = \exp(\xi 2\pi s_i R_R f_{P,i}/\overline{C}') \tag{8.17}$$

For example, the case of $f_{P,i} = f_{P,R}$, $d_{f,i} = d_{f,R}$, gives:

$$\frac{(S_{P,MAX})_i}{(S_{P,MAX})_R} = s_i^3[(1/q_R d_{f,R}) + (v_{c,R}/\overline{C}')]/[(1/q_i d_{f,R}) + (s_i v_{c,R}/\overline{C}')] \tag{8.18}$$

$$\wp_{MAX,i} = (\wp_{MAX,R})^{s_i-1} \tag{8.19}$$

For the case of $f_{P,i} = s_i^{-1} f_{P,R}$ (constant v_c):

$$\frac{(S_{P,MAX})_i}{(S_{P,MAX})_R} = s_i^2, \qquad \wp_{MAX,i} = \wp_{MAX,R} \tag{8.20}$$

A complete set of scaling results is presented in Table 8.2a and 8.2b.

8.3.2 Capture Pumps

8.3.2.1 Sputter Ion Pumps

The sputter ion pump (SIP) is an option for high vacuum MEMS pumping. The application of simple scaling approaches to these pumps is difficult; however centimeter scale pumps are already available. The SIP has a basic configuration illustrated in Figure 8.3. A cold cathode discharge (Penning discharge), self

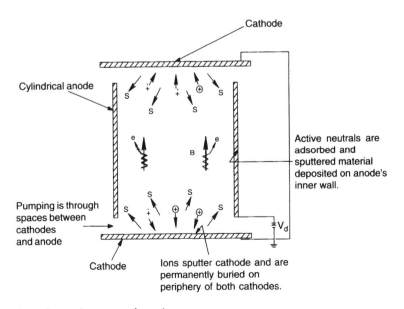

FIGURE 8.3   Figure Sputter ion pump schematic.

maintained by a several thousand volt potential difference and an externally imposed magnetic field that restricts the loss of discharge electrons, causes ions created in the discharge by electron-neutral collisions to bombard the cathodes. The energetic ions (with energy some fraction of the driving potential) both sputter cathode material (usually Ti) and imbed themselves in the cathodes. Sputtered material deposits on the anode and portions of the opposing cathodes. The freshly deposited material acts as a continuously refreshed adsorption pumping surface for "active" gases (most things other than the noble gases and hydrocarbons). Small (down to an 8 mm diameter anode cylinder and a cathode separation of 3.6 cm) pumps have been studied theoretically by Suetsugu (1993). His results compare reasonably well to experimental results for the particular case of a 1.5 cm diameter anode. For a discharge voltage of 3000 V, a magnetic field strength of 0.3 T and a 0.8 cm diameter anode a pumping speed of slightly greater than 0.5 l/s is predicted at 10^{-8} Torr. For these conditions the discharge is operating in the high magnetic field (HMF) mode, which results in a maximum pumping speed. The pumping speed increases slowly as pressure increases.

SIP's have a finite but relatively long life; they may be useful when ultra high vacuums are required in small scale systems. Their scaling to true MEMS sizes is uncertain because they require several thousand volts to operate reasonably effectively (ion impact energies approaching 1000 V are required for efficient sputtering and rare gas ion burial). The description of SIP operation developed by Suetsugu (1993) can be used to provide a scaling expression (that needs to be employed cautiously). The power used by the discharge can be obtained knowing the applied potential difference, V_D, and the ion current, I_{ion}. Suetsugu (1993) gives for the pumping speed:

$$S_P = (K_G q \eta I_{ion})/(3.3 \times 10^{19})ep_l \tag{8.21}$$

where p_l is the pressure in Torr, η is the sputtering coefficient of cathode material due to the impact of energetic ions, and q is the sticking coefficient for active gases on the sputtered material. The charge on an electron is e, K_G is a non-dimensional geometric parameter derived from the electrode configuration that remains constant with geometric scaling. An expression for the power required per pumped molecule becomes:

$$(\dot{Q}/\dot{N})_i = (V_D I_{ion})_i/[\{(K_G q \eta I_{ion})/((3.3 \times 10^{19})ep_l)\}_i\{10^{-3}n_l\}] \tag{8.22}$$

and

$$(\dot{Q}/\dot{N})_i/(\dot{Q}/\dot{N})_R = V_{D,i}/V_{D,R} \tag{8.23}$$

This assumes $q_i = q_R$, $\eta_i = \eta_R$.

The scaling expression for pumping speed becomes:

$$S_{P,i}/S_{P,R} \equiv S_{P,MAX,i}/S_{P,MAX,R} = (I_{ion})_i/(I_{ion})_R \tag{8.24}$$

where the $S_{P,MAX}$ is employed to be consistent with the previous useage for other pumps, although the Equation (8.7) relationship does not really apply in this case.

The ion current is obtained by an iterative numerical solution involving the number density of trapped electrons [Suetsugu, 1993]. The scaling expression for $\dot{Q}/\dot{N}$ in Equation (8.23) is particularly simple because the ion currents cancel. The scaling represented by Equation (8.23) appears reasonably valid providing η remains relatively constant, which implies that the ion energy should be relatively constant. Ion burial in the cathodes, which is the mechanism by which SIPs pump rare gases, is not discussed in detail in this chapter but can be considered within the framework of Suetsugu's analysis. Scaling results are summarized in Table 8.2a.

8.3.2.2   Orbitron Ion Pump

The orbitron ion pump [Douglas, 1965; Denison, 1967; Bills, 1967] was developed based on an electrostatic electron trap best known for application in ion pressure gauges. A sketch is presented in Figure 8.4.

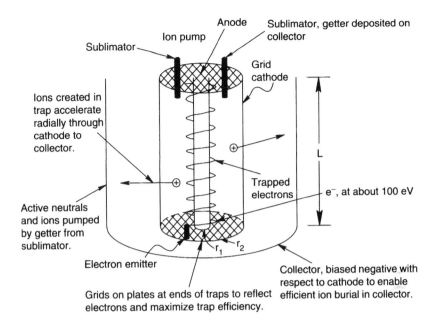

FIGURE 8.4 Orbitron pump schematic.

Injected electrons orbit an anode, the triode version illustrated in Figure 8.4 [Denison, 1967; Bills, 1967] has an independent sublimator that provides a continuous active getter (Ti) coating of the ion collector. The getter permits active gas pumping as well as permanent burial of rare gas and other ions that are accelerated out of the trap through the cathode mesh by the radial electric fields. Initial work on reducing the size of an orbitron to MEMS scales has recently been reported by Wilcox et al. (1999).

For a given geometry and potential difference between the anode and cathode mesh the cylindrical capacitor represented by the trap geometry has a limiting maximum net negative charge of orbiting electrons. The corresponding ionization rate in the trap can be written as:

$$\dot{N}_{MAXions} = \frac{2X\pi\varepsilon_0 V_D^{3/2}\Omega_I L n_g}{(em_e)^{1/2}\ln(r_2/r_1)^{3/2}} \tag{8.25}$$

where X is the fraction of the maximum charge that permits stable electron orbits (less than 0.5), V_D is the applied potential difference, m_e is the electron mass, ε_0 is the permittivity of free space, Ω_I is the electron-neutral ionization cross-section, L is the trap's length, r_2 and r_1 are the radii of the trap's cathode and anode respectively (Figure 8.4).

The orbitron's noble gas pumping speed and the trap's volume can be scaled based on the expression for ionization rate in Equation (8.25):

$$\frac{(S_{P,MAXions})_i}{(S_{P,MAXions})_R} = s_i\frac{\Omega_i(V_D^{3/2})_i}{\Omega_R(V_D^{3/2})_R}; \qquad \frac{(V_P/\dot{N}_{MAXions})_i}{(V_P/\dot{N}_{MAXions})_R} = s_i^2\frac{\Omega_{I,i}(V_D)_i^{3/2}}{\Omega_{I,R}(V_D)_R^{3/2}} \tag{8.26}$$

Note that this is favorable scaling.

The sublimator's scaling, assuming the temperature of the sublimating getter is constant, can be written for neutrals and ions as:

$$\frac{(\dot{Q}_s/\dot{N}_{MAXneut})_i}{(\dot{Q}_s/\dot{N}_{MAXneut})_R} = 1; \qquad \frac{(\dot{Q}_{ion}/\dot{N}_{MAXions})_i}{(\dot{Q}_{ion}/\dot{N}_{MAXions})_R} = 1 \tag{8.27}$$

8.4   Pump-Down and Ultimate Pressures for MEMS Vacuum Systems

What is the consequence of size scaling a vacuum system? Consider an elementary system made up of a pump and a vacuum chamber of volume V_c and surface area A_{sc}. The pump is modeled by writing the pumping speed as:

$$S_{P,i} = \{A_{I,P}(\overline{C}'/4)\alpha_P\}_i \tag{8.28}$$

where $A_{I,P}$ is the area of the pump's inlet aperture from the chamber and α_P is the probability that once through the aperture a molecule will be pumped. For a geometrically similar size change $S_{P,i} = s_i^2(\alpha_{P,i}/\alpha_{P,R})S_{P,R}$, and the pumping speed per unit surface area of the vacuum chamber is:

$$(S_P/A_{sc})_i/(S_P/A_{sc})_R = \alpha_{P,i}/\alpha_{P,R} \tag{8.29}$$

Assuming equal outgassing rates per unit area for the reference and scaled systems, the ultimate system pressure will only depend on the pumping probability ratio $\alpha_{P,i}/\alpha_{P,R}$. The pump-down time for the system can be measured using the ratio $S_P/V_c = \tau_P$:

$$\tau_{P,i}/\tau_{P,R} = s_i(\alpha_{P,R}/\alpha_{P,i}) \tag{8.30}$$

For geometric scaling and the same outgassing rates a MEMS system will have a significantly shorter pump-down time assuming $\alpha_{P,i}/\alpha_{P,R}$ can be kept near one.

In practice MEMS scale vacuum systems are likely to have pump apertures relatively much larger than their macroscopic counterparts. The economic deterrent to having large aperture pumps that exists at macroscales does not apply at MEMS scale. At MEMS scales it appears that technical issues associated with pump construction will favor making the pumps as large as possible. Consequently, relatively large pump apertures with areas about the same as the cross section of the pumped volume are anticipated.

8.5   Operating Pressures and $\dot{N}$ Requirements in MEMS Instruments

The selection of vacuum pumps for MEMS instruments and processes will depend on operating pressure and $\dot{N}$ requirements. Since this can be determined reliably only when the task, instrument, and detector or a particular process have been specified, it is virtually impossible to discuss significant general size scaling tendencies. For example, there has been speculation [R.M. Young, 1999] that a MEMS mass spectrometer sampling instrument might operate at upper pressures specified by keeping the Knudsen number based on the quadrupole length constant compared to similar macroscale instruments. This can typically lead to tolerable upper operating pressures for microscale instruments of 10^{-3} to 10^{-2} mbar, depending on the scaling factor (see also Ferran and Boumsellek, 1996). On the other hand the default response of many mass spectroscopists is 10^{-5} mTorr, independent of scaling. Vargo (2000) based $\dot{N}$ requirements for a miniaturized sampling mass spectrometer on the goal of replacing the entire volume of gas in the instrument (30 cm^3 in Vargo's case) every second at an operating pressure of 10^{-4} mTorr, giving $\dot{N} = 1.4 \times 10^{14}$ molecules/s. A point to remember is that for a constant Kn system the equilibrium quantity of adsorbed gas in the system increases compared to unadsorbed gas as the s_i decreases [Muntz, 1999].

A careful consideration of the operating pressure and $\dot{N}$ requirements for a particular situation is important, but impossible within the confines of this chapter. Because of the difficulty in supplying volume and power compatible microscale vacuum systems, it will be important for overall system design to define operating conditions that are based on real needs.

8.6 Summary of Scaling Results

The scaling analyses previously outlined have been applied to several pump types, with the results appearing in Table 8.2a and b. The operating frequencies were selected to give two extremes: maintaining a constant average speed $\bar{u}_i = \bar{u}_R$ (tangential for rotating devices or linear for reciprocating), by using the frequency scaling, $f_{T,i} = s_i^{-1} f_{T,R}$; or maintaining a constant frequency, $f_{T,i} = f_{T,R}$, resulting in $\bar{u}_i = s_i \bar{u}_R$. Two alternative types of frictional drag — sliding and viscous — have been included, again as extremes of the likely possibilities.

The scaling expressions for the case $\bar{u}_i = \bar{u}_R$ are simple when normalized by their respective reference scale values. For the second case where $\bar{u}_i = s_i \bar{u}_R$, the expressions are more complex and the results depend on the relative magnitude of the quantities $K_{I,R}$, $K_{B,R}$, $q_R d_{f,R}$, etc. (Table 8.2(b)). For the cases involving more complex expressions order of magnitude estimates of the scaling based on typical pump characteristics have been included in Table 8.2(a).

The sputter ion pumps have been included assuming that permanent magnets provide the required field strengths. Note that the mesoscale SIP performance presented in Figure 8.1 and Table 8.1 is for HMF operation.

To put the scaling results in Table 8.2 in perspective, remember that they are for geometrically accurate scale reductions. It is assumed that the relative dimensional accuracy of the components is the same in the reduced scale realization as in the reference macroscopic pumps. This is a very idealized assumption. The dimensional accuracy that can currently be attained in micromechanical parts as a function of size is illustrated in Figure 8.5 (derived in part from Madou, 1997). It is clear from Figure 8.5 that the scaling results of Table 8.2 may be very optimistic if true MEMS scale pumps (component sizes 100 µm or less) are required. On the other hand the scalings do represent the best scaled performances that could be expected and are a useful guide. Note from Figure 8.5 that the smallest fractional tolerances can be achieved by precision machining techniques for approximately 1 cm size components. As a result, mesoscale pumps may be possible from a tolerance (although perhaps not economic) perspective using precision machining techniques.

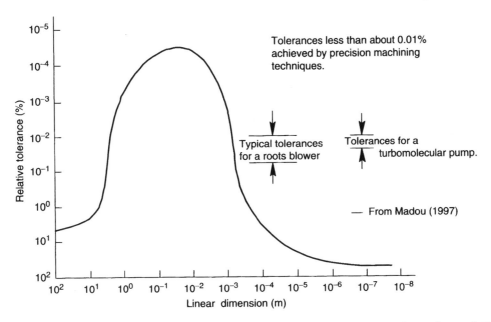

FIGURE 8.5 Dimensional accuracy of manufactured components as a function of size. (After Madou, M. [1997] *Fundamentals of Microfabrication*, CRC Press, Boca Raton, Florida.)

Several points should be noted from Table 8.2. For the case of $\overline{u}_i = \overline{u}_R$, the ideal scalings to small sizes are reasonable (remember s_i will range between 10^{-1} and 10^{-4}). In the case of viscous friction losses, the energy use per molecule of upflow becomes large at small scales (increases as s_i^{-1}) while the pump volume per unit upflow decreases as s_i decreases. For the case of positive displacement pumps and $\overline{u}_i = s_i\overline{u}_R$, the energy use per molecule of upflow scales satisfactorily, but upflow scales as s_i^3 so that the volume scaling is of $\mathcal{O}[1]$ rather than the s_i^{-1} for the $\overline{u}_i = \overline{u}_R$ case. The $\wp_{MAX}$ scaling to small scales for $\overline{u}_i = s_i\overline{u}_R$ is a disaster for turbomolecular pumps. Also the upflow, energy, and pump volume all scale badly for the turbomolecular pump with $\overline{u}_i = s_i\overline{u}_R$. These scalings are all a result of the trapping coefficient $q \sim s_i$ for low peripheral speeds. Although not explicitly included, molecular drag pumps can be expected to scale similarly to the turbomolecular pump. For positive displacement pumps the pressure ratio scales well if there are no losses but can scale as badly as s_i depending on specific pump characteristics. For the cases where the pressure ratio scales badly, more pump stages would be required for a given task, leading to larger pumps as indicated by the $>$ symbol in the energy use and pump volume per unit upflow columns.

The sputter ion pump scales well to smaller sizes. The major concern for microscales will be the fundamental requirement for relatively high voltages. Also, thermal control will be difficult as it is complicated by the need for high field strengths (0.5–1 T) using co-located permanent magnets.

The orbitron ion pump scales well to smaller sizes but unfortunately as seen in Figure 8.1 and Table 8.1 begins with a very poor performance as measured by $\dot{Q}/\dot{N}$.

Generally, for the positive displacement and turbomolecular pumps, the idealized geometric scaling results in Table 8.2 demonstrate that there is a mixed bag of possibilities, ranging from decreased performance to maintaining performance, with a few cases showing improvement on macroscale performance by going to smaller scales. From a vacuum pump perspective with ideal scaling there is little to no advantage based on performance to go to small scales, except for the ion pumps.

The actual performance of small-scale pumps is likely to be significantly poorer than the idealized scaling results shown in Table 8.2. For instance it is very difficult to attain the high rotational speeds necessary to satisfy the $\overline{u}_i = \overline{u}_R$ requirements in MEMS scale devices; on the other hand, recent progress in air bearing technology has been reported for mesoscale gas turbine wheels [Fréchette et al., 2000]. Mesoscale sputter ion pumps have been operated and the investigation of orbitron scaling is just beginning. Whether either can be scaled to true MEMS sizes is unclear, but they may be the only alternative for achieving high vacuum with MEMS pumps.

Keeping the preceding comments in mind it is useful to re-visit the macroscale vacuum pump performances reviewed in Figure 8.1. Consider a typical energy requirement from Figure 8.1 of 3×10^{-15} W/molecule of upflow for macroscale systems; assume that this can be maintained at mesoscales to pump through a pressure ratio of 10^6 (10^{-3} mbar to 1 bar). A typical upflow, assuming a 3 cm^3 volume at 10^{-3} mbar is changed every second, is 1.1×10^{14} molecules/s and the required energy is 0.33 W. This is somewhat high but tolerable for a mesoscale system. However, with the expected degradation of the performance of complex macroscale pumps at meso- and microscales, it is clear that it is important to search for alternative, unconventional pumping technologies that will be both buildable and operate efficiently at small scales.

8.7 Alternative Pump Technologies

The previous section on scaling indicates that searching for appropriate alternative technologies as a basis for MEMS vacuum pumps is necessary. There has been some effort in this regard during the past decade. In 1993, Muntz, Pham-Van-Diep, and Shiflett hypothesized that the rarefied gas dynamic phenomenon of thermal transpiration might be particularly well suited for MEMS scale vacuum pumps. Thermal transpiration is the application of a more general phenomenon — thermal creep — that can be used to provide a pumping action in flow channels for Knudsen numbers ranging from very large to about 0.05. The observation resulted in a publication [Pham-Van-Diep, 1995], which led to the construction of a prototype micromechanical pump stage by Vargo (2000) and Vargo et al. (1999). A 15-stage radiantly driven Knudsen Compressor along with a complete cascade performance model has been developed recently by M. Young

(2004), Young et al., 2004. A MEMS thermal transpiration pump has also been proposed by R.M. Young (1999). There is one fundamental problem with thermal transpiration or thermal creep pumps: they are staged devices that require part of each stage to have a minimum size corresponding to a dimension greater than about 0.2 molecular mean free paths (λ) in the pumped gas. At 1 mTorr (1.32×10^{-3} mbar) $\lambda \approx 0.05$ m in air, resulting in required passages no smaller than about 1 cm. Thus, at low inlet pressures the pumps can be unacceptably large for MEMS applications. This issue has been discussed by Han et al., 2004.

An interestingly different version of a thermal creep pump has been suggested by Sone and his co-workers [Sone et al., 1996; Aoki et al., 2000; Sone and Sugimoto, 2003], although it also has a low pressure use limit similar to the one mentioned previously.

Another alternative, the accommodation pump, which is superficially similar to thermal transpiration pumps but based on a different physical phenomenon, has been investigated [Hobson, 1970]. It can in principle be used to provide pumping at arbitrarily low pressures without minimum size restrictions.

8.7.1 Outline of Thermal Transpiration Pumping

Two containers, one with a gas at temperature T_L and one at T_H are separated by a thin diaphragm of area A_i in which there are single or multiple apertures that each have an area A_a and a size $\sqrt{A_a} \ll \lambda_L$ or λ_H (Figure 8.6). The number of molecules hitting a surface per unit time per unit area in a gas is $n\bar{C}'/4$. In Figure 8.6 this means that there are $(n_L \bar{C}'_L/4)$ molecules passing from cold to hot through an aperture. Similarly, there are $(n_H \bar{C}'_H/4)A_a$ molecules per unit time passing from hot to cold into the cold chamber.

Assume m is the same for the molecules in both chambers and assume that the inlet and outlet are adjusted so that $p_L = p_H$. Under these circumstances the net number flow of molecules from cold to hot is, with the help of the equation of state:

$$\dot{N}_{MAX} = A_a(2\pi mk)^{-1/2} p_L[T_H^{1/2} - T_L^{1/2}]/(T_L T_H)^{1/2} \tag{8.31}$$

If on the other hand there is no net flow:

$$(p_H/p_L)_{\dot{N}=0} = \wp_{MAX} = (T_H/T_L)^{1/2} \tag{8.32}$$

For p_H between p_L and $p_L(T_H/T_L)^{1/2}$ there will be both a pressure increase and a net flow, which is the necessary condition for a pump! This effect is known as thermal transpiration. If there are Γ apertures the total upflow of molecules is:

$$\dot{N} = \Gamma A_a(2\pi mk)^{-1/2}[p_L/T_L^{1/2} - p_H/T_H^{1/2}] \tag{8.33}$$

If the hot gas at T_H is allowed to cool and sent to another stage as indicated in Figure 8.7, for the condition $\lambda \ll \sqrt{A_j}$, where A_j is the stage area (but also $\lambda \gg \sqrt{A_{aj}}$) the pressure $p_{L,j+1} = p_{H,j}$. Thus a cascade of stages with net upflow is a pump with no net temperature increase over the pump cascade. This cascade

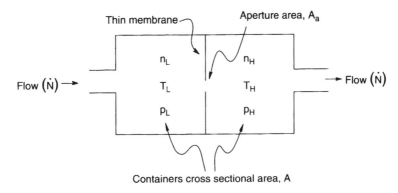

Thin membrane

Aperture area, A_a

n_L n_H

Flow ($\dot{N}$) → T_L T_H → Flow ($\dot{N}$)

p_L p_H

Containers cross sectional area, A

FIGURE 8.6 Elementary single stage of a thermal transpiration compressor.

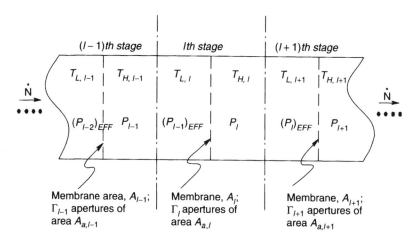

FIGURE 8.7 Cascade of thermal transpiration stages to form a Knudsen Compressor (the pressure difference, $(p_\ell)_{EFF} - p_\ell$, drives the flow through the connector as illustrated in Figure 8.8).

of thermal transpiration pump stages has no moving parts and no close tolerances between shrouds and impellars, so it is an ideal candidate for a MEMS pump. However, the compressor stages sketched in Figure 8.7 have a serious problem. The thin films containing the apertures are not practical in most applications because of heat transfer considerations. A more practical stage is shown in Figure 8.8 where the thin membrane has been replaced by a bundle of capillary tubes. The temperature and pressure profiles in the stage are also illustrated in Figure 8.8 along with nomenclature that will be used below.

The stage configuration in Figure 8.8 can be put in a cascade (Figure 8.7) to form a pump as proposed by Pham-Van-Diep et al. (1995). It was called a Knudsen Compressor after original work by Knudsen, (1910a and b). Several investigations implementing thermal transpiration in macroscale pumps have been made over the years [Baum, 1957; Turner, 1966; Hopfinger, 1969; Orner, 1970], but with no result beyond initial analysis and laboratory experimental studies. For the Knudsen Compressor a performance analysis was presented for the case of collisionless flow in the capillaries and continuum flow in the connectors [Pham-van-Diep, 1995]. Energy requirements per molecule of upflow ($\dot{Q}/\dot{N}$) were estimated. Recently Muntz and several collaborators [Muntz et al., 1998; Vargo, 2000; Muntz et al., 2002; Han et al., 2004] have extended the analysis to situations where the flow in both the capillaries and connector section can be in the transitional flow regime ($10 > Kn > 0.05$). An initial look at the minimization of cascade energy consumption and volume has been conducted by searching for minimums in $\dot{Q}/\dot{N}$ and $V_p/\dot{N}$ [Muntz, 1998; Vargo, 2000; Muntz et al., 2002]. Based on these studies a preprototype micromechanical Knudsen Compressor stage has been constructed and tested [Vargo, 2000; Vargo and Muntz, 1996; Vargo and Muntz, 1998]. Additional optimization analysis has been recently completed to provide designs for Knudsen Compressors operating at different pressure ratio-gas upflow conditions [Young et al., 2001; Young et al., 2003; Young et al., 2004; Young, 2004].

For transitional flow in a capillary tube with a longitudinal temperature gradient imposed on the tube's wall, flow is driven from the cold to hot ends as illustrated in Figure 8.9. The increased pressure at the hot end drives the return flow. For small temperature differences and thus small pressure increases, the Boltzmann equation can be linearized and results obtained for the flow through a cylindrical capillary driven by small wall temperature gradients [Sone and Itakura, 1990; Loyalka and Hamoodi, 1990; Loyalka and Hickey, 1991]. The definition of the flow coefficients (Q_T and Q_P) are implied in the following expression [Sone, 1968]:

$$\dot{M} = p_{AVG}(2(k/m)T_{AVG})^{-1/2} A \left[\frac{L_r}{T_{AVG}} \frac{dT}{dx} Q_T - \frac{L_r}{p_{AVG}} \frac{dp}{dx} Q_P \right] \tag{8.34}$$

Here A is the cross-sectional area of the capillary, L_r its radius, Q_P is the backflow due to the pressure increase, and Q_T is the thermally driven upflow near the walls. The sum determines the net mass flow

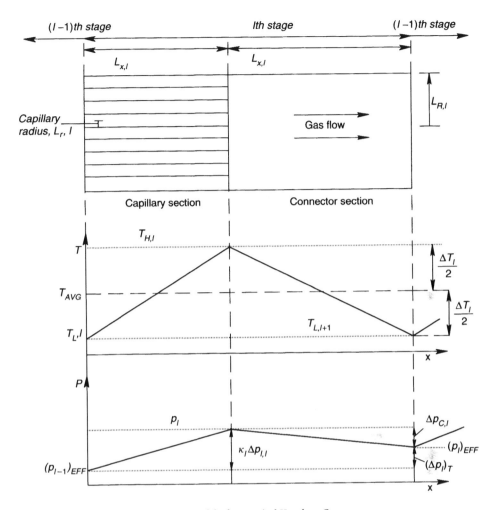

FIGURE 8.8   Capillary tube assembly as a model of a practical Knudsen Compressor.

through the tube from cold to hot. For large $Kn(= \lambda/L_r)$ the $\dot{M} = 0$ pressure increase provided by a tube is $p_H/p_C = Q_T/Q_P = (T_H/T_C)^{1/2}$, identical to the aperture case discussed earlier. In the large Kn case the backflow and upflow completely mingle over the entire tube cross-section. For transitional Kn's the upflow is confined near the wall in a layer on the order of λ thick, as illustrated in Figure 8.9.

The values of Q_T and Q_P vary markedly throughout the transitional flow regime as shown in Figure 8.10. The details of their functional variation as well as the ratio Q_T/Q_P are important to any pump using thermal transpiration. Their roles are best illustrated by the expression for pumping speed and pressure ratio of a Knudsen Compressor's j'th stage [Muntz et al., 1998; Muntz et al., 2002]:

$$S_{P,j} = \pi^{1/2}(1 - \kappa_j)\left|\frac{\Delta T}{T_{AVG}}\right|\left[\frac{Q_T}{Q_P} - \frac{Q_{T,C}}{Q_{P,C}}\right]_j\left[\frac{L_x/L_r}{FQ_P} + \frac{L_x/L_R}{F_C(A_c/A)Q_P}\right]_j^{-1}(C_A)_j \qquad (8.35)$$

In Equation (8.35) the κ_j is a parameter that sets the fraction of the $\dot{M} = 0$ pressure rise that is realized in a finite upflow situation. The pressure ratio for the stage, assuming $|\Delta T/T_{AVG}| \gg 1$ is:

$$\wp_j = 1 + \kappa_j\left\{\left|\frac{\Delta T}{T_{AVG}}\right|\left[\frac{Q_T}{Q_P} - \frac{Q_{T,C}}{Q_{P,C}}\right]_j\right\} \qquad (8.36)$$

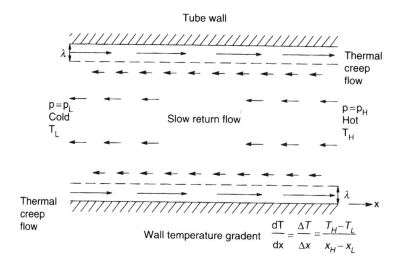

FIGURE 8.9   Transitional flow configuration in a capillary tube.

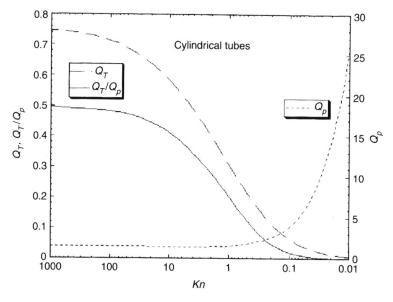

FIGURE 8.10   Transitional flow coefficients as a function of Knudsen number ($Kn = \lambda/L_r$) for cylindrical capillaries. (After Vargo, S.E. [2000] The Development of the MEMS Knudsen Compressor as a Low Power Vacuum Pump for Portable and In Situ Instruments, Ph.D. Thesis, University of Southern California, Los Angeles.]

Other symbols in Equations (8.35) and (8.36) are: $|\Delta T|$, the temperature change across both the capillary section and the connector section; F, the fraction of the capillary section's area; A_j, which corresponds to the open area of the capillary tubes; F_C, the fraction of the connector section's area; and $A_{C,j}$, which is open (usually $F_{C,j} = 1$). The dimensions $L_{x,j}$, $L_{X,j}$, $L_{r,j}$, and $L_{R,j}$ are defined in Figure 8.8.

From Equation (8.36), if $\kappa_j = 0$, then $\wp_j = 1$. The corresponding maximum upflow is obtained by substituting $\kappa_j = 0$ into Equation (8.35) to give $(S_{P,MAX})_j$. The result is:

$$S_{P,j}/(S_{P,MAX})_j = (1 - \kappa_j) \tag{8.37}$$

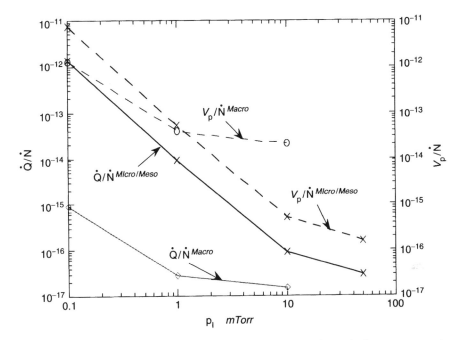

FIGURE 8.11 Simulation results for several macro- and microscale pumping tasks from p_I to $p_E = 1$ atm using Knudsen compressors. (After results from Table 3 of Vargo, S.E. [2000] The Development of the MEMS Knudsen Compressor as a Low Power Vacuum Pump for Portable and In Situ Instruments, Ph.D. Thesis, University of Southern California, Los Angeles.]

which can be combined with the expression for $\wp_{j,MAX,j}$ (using $\kappa_j = 1$ in Equation [(8.35]) to give the familiar relationship for a vacuum pumping system (Equation [8.7]) between $S_{P,j}$, $S_{P,MAX,j}$, $\wp_j$ and $\wp_{MAX,j}$.

The expression for pumping speed in Equation (8.35) can also be used to compute the pumping probability, $\alpha_{P,j}$, where from Equation (8.5) $\alpha_{P,j} = S_{P,j}/C_{A,j}$. Using Equations (8.35) and (8.36) and remembering that $\dot{N}_j = S_{P,j}n_{I,j}$, where $n_{I,j}$ is the number density at the inlet to the stage, cascade performance calculations can be accomplished. The approach is presented in detail by Muntz et al. (2002). The results of several cascade simulations are given by Vargo (2000). As shown by Muntz, the majority of the energy required is the thermal conduction loss across the capillary tube bundle. The physical realization of the capillary section of a Knudsen Compressor is frequently not a bundle of capillary tubes, but rather a porous membrane. Nevertheless the theoretical behavior based on "equivalent" capillary tubes appears to be consistent with the results obtained experimentally [Vargo's Ph.D. Thesis, Vargo 2000, and the discussions presented therein].

Selected results of micro-, meso-, and macroscale pump cascade simulations reported by Vargo (2000) are plotted in Figure 8.11 and presented in Table 8.3. Note that all the Knudsen Compressor cascade results are from simulations; a micromechanical cascade has yet to be constructed although several mesoscale stages have been operated in series [Vargo, 1996]. More recently, a 15-stage conventionally machined radiantly driven Knudsen Compressor has been demonstrated by Young (2004). In his thesis, Vargo (2000) describes the construction of a laboratory prototype micromechanical Knudsen Compressor stage and the experimental results. A scheme for constructing a cascade is also presented, along with predicted energy requirements, but this has yet to be built and tested. The Knudsen Compressor performance presented in Figure 8.11 is based on a significant body of preliminary work, but has been only partially demonstrated experimentally [Young, 2004; Young et al., 2004]. The Figure 8.11 results are for varying initial pressures p_I and pumping to 1 atm, the detailed conditions are presented in Table 8.3. In MEMS devices pressures below 10 mTorr extract a significant energy penalty, tending to preclude

TABLE 8.3 Results of Simulations of Micro/meso and Macro Scale Knudsen Compressors for Several Pumping Tasks (from Vargo, 2000)

p_I (mTorr)	p_E (mTorr)	$\dot{Q}/\dot{N}$ W/(#/s)	$V_P/\dot{N}$ cm³/(#/s)	$\wp$
Micro/Meso scale (from Vargo Tables 3, 4 and 9)				
0.1	7.6E5	1.4E-12	7.8E-12	7.6E6
1.0	7.6E5	9E-15	5.2E-14	7.6E5
10	7.6E5	9E-17	5.2E-16	7.6E4
10	5.25E3	1.4E-17	1.6E-16	5.25E2
50	7.6E5	2.9E-17	1.6E-16	1.5E4
Macroscale (from Vargo Table 3)				
0.1	7.6E5	9.1E-16	1.3E-12	7.6E6
1.0	7.6E5	2.8E-17	4.1E-14	7.6E5
10	7.6E5	1.5E-17	2.1E-14	7.6E4

Notes: Micro/Meso scale cascade: $L_r = 500\,\mu m$, $L_R/L_r = 5$ from p_I to 50 mTorr, then constant Kn; Macroscale cascade: $L_r = 5\,mm$, $L_R/L_r = 20$ from p_I to 10 mTorr, then constant Kn_i.

Knudsen Compressor applications to lower pressures at MEMS scale. This restriction, however, depends on the $\dot{N}$ that is required. The energy penalty originates from the difficulty in reaching effectively continuum flow in the connector section for a reasonable size. Note that the macroscale simulations presented in Figure 8.11 give efficient pumping to significantly lower p_I (higher $\wp$).

The size scaling of Knudsen Compressors at low inlet pressures is dominated by the ratio L_R/L_r, that can be achieved in the early stages of a particular design. The ratio L_R/L_r determines the Knudsen number ratio which in turn determines the Q_T/Q_P ratio (Figure 8.10). By referring to Equation (8.36) the effects on the stage $\wp_j$ are immediately clear. In the limit of $L_R/L_r \to 1$, $\wp = 1$. For $L_R/L_r \gg 1$, $Q_{T,C}/Q_{P,C} \ll 1$ and $\wp_j$ reaches a maximum value that depends only on κ_j and $|\Delta T/T_{AVG}|$.

The minimization of a Knudsen Compressor cascade's energy requirement depends on a number of important parameters describing individual stage configurations. The results reported by Vargo (2000) are only a beginning attempt at optimizing Knudsen Compressor performance (there are extended discussions of optimization issues in Muntz et al.'s report [1998, 2002]). The predicted Knudsen Compressor Performance presented in Figure 8.11 is quite promising. Generally, as illustrated in Figure 8.12, the energy requirement for the micro-mesoscale version is at least an order of magnitude better than macroscale, positive displacement mechanical backing pumps. It also appears very competitive with the macroscale molecular drag kinetic backing pumps used in conjunction with turbomolecular pumps.

Another MEMS thermal transpiration compressor has been proposed by R.M. Young (1999). Its configuration is different from the Knudsen Compressor, relying on temperature gradients established in the gas rather than along a wall. No experimental or theoretical analysis of the flow or pressure rise characteristics of this configuration has been reported. There are theoretical reasons to believe that the flow effects induced by thermal gradients in a gas are significantly smaller than the effects supported by wall thermal gradients [Cercignani, 2000].

8.7.2 Accommodation Pumping

In 1970 Hobson introduced the idea of high vacuum pumping by employing a characteristic of surface scattering that had been observed in molecular beam experiments (c.f. the review by Smith and Saltzburg, 1964). For some surfaces that give quasi specular reflection under controlled conditions, when the beam temperature differs from the surface temperature, the quasi specular scattering lobe moves towards the surface normal for cold beams striking hot surfaces. For hot beams striking cold surfaces, the lobe moves away from the normal. Hobson's initial study was followed in short order by several investigations adopting the same approach [Hobson and Pye, 1972; Hobson and Salzman, 2000; Doetsch and Ryce, 1972; Baker et al., 1973]. Other investigators, relying on the same phenomenon, have used different

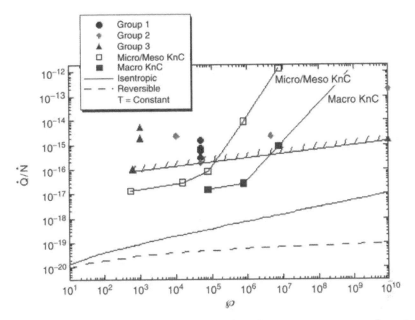

FIGURE 8.12 Comparison of macro- and microscale Knudsen Compressor performances from simulations to representative available macro- and mesoscale vacuum pumps (Vargo, S.E. [2000] The Development of the MEMS Knudsen Compressor as a Low Power Vacuum Pump for Portable and In Situ Instruments, Ph.D. Thesis, University of Southern California, Los Angeles.]

geometrical arrangements [Tracy, 1974; Hemmerich, 1988]. Recently Hudson and Bartel (1999) have analyzed a sampling of the previous experiments with the Direct Simulation Monte Carlo technique. To date the most successful pumping arrangement, which is sketched in Figure 8.13, has been that of Hobson.

Following Hobson's lead it is easy to write expressions for $\wp_{MAX,j} = p_3/p_1$ and $S_{P,MAX,j}$ for the stage in Figure 8.13. The accommodation pump works best under collisionless flow conditions so assume $\lambda \geq 10d$, where d is the tube diameter. Assuming steady state conditions, conservation of molecule flow equations for each volume can be solved simultaneously for zero net upflow to give:

$$\wp_{MAX,j} = \frac{\alpha_{23}}{\alpha_{32}} \cdot \frac{\alpha_{12}}{\alpha_{21}} = \frac{\alpha_{12}}{\alpha_{21}} \tag{8.38}$$

Here α_{kl} is the probability for a molecule having entered a tube from the k'th chamber reaching the l'th chamber. In this case, since the tube joining chambers 2 and 3 is assumed always to reflect diffusely, $\alpha_{23} = \alpha_{32}$.

The same molecule flow equations can be solved along with Equation (8.38) to provide maximum upflow or pumping speed ($\wp_j = 1$):

$$S_{P,MAX,j} = \frac{\overline{C}_H' A_j}{4} \{(\wp_{MAX,j} - 1)\alpha_{2,1}/[1 + (\alpha_{2,1})/(8L_r/3L_x)]\} \tag{8.39}$$

From the earlier discussions it is reasonable to anticipate that $\alpha_{2,1}$ will have a value that approximates the result for diffuse reflection, which for a long tube is $(8L_r/3L_x)$. A simple approximate expression for maximum pumping speed becomes:

$$S_{P,MAX,j} = (\overline{C}_H'/4)A_j(4L_r/3L_x)(\wp_{MAX,j} - 1) \tag{8.40}$$

If $\wp_{MAX,j}$ is known from experiments, $S_{P,MAX,j}$ can be found from Equation (8.40).

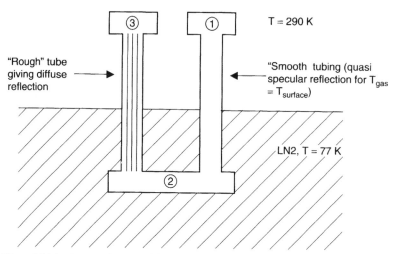

"Smooth" tube is normal pyrex tubing.
"Rough" tube is leached pyrex to provide suppression of specular reflection for the
temperature difference of the experiment.

FIGURE 8.13 Schematic representation of Hobson's accommodation pump configuration. (After Hobson, J.P., and Salzman, D.B. [2000] "Review of Pumping by Thermomolecular Pressure," Journal of Vacuum Science and Technology A **18**(4), pp. 1758–1765.)

Scaling accommodation pumping is not difficult because, unlike the Knudsen Compressor, the flow is collisionless everywhere. If the minimum radius of the two channels in a stage is such that $L_r \geq 100\,\mathrm{nm}$, there is little reason to believe that $\wp_{MAX,j}$ or the transmission probabilities will vary with scale. For geometric scaling:

$$\frac{S_{P,MAX,j}}{S_{P,MAX,R}} = s_i^2 \qquad (8.41)$$

No analysis or measurements of the energy requirements of accommodation pumping are available but the temperature differences required to obtain an effect are quite large (the temperature ratio in various experiments ranges from 3 to 4). Careful thermal management would be required at MEMS scales. Additionally, the stage $\wp_{MAX,j}$ values are quite low (Hobson's are typically around 1.15). For an upflow of $S_{P,MAX,j}/2$, to pump through a pressure ratio of 10^2 would require about 125 stages. For comparison, a Knudsen Compressor operating between the same temperatures would require 10 stages. Of course, as outlined earlier, the Knudsen Compressor could not pump effectively at very low pressures whereas accommodation pumps can.

Using the miniature Creare turbomolecular pump scaled down with constant $f_{P,i}$ to a 2 cm diameter, the authors have estimated an air pumping probability of about 0.05. A Hobson-type accommodation pump with 2 cm diameter, $L_r/L_x = 0.05$ pyrex tubes, and operating between room, and LN_2 temperatures would have, from Equation (8.40), a pumping probability of about 0.01.

Scaling Hobson's pump to MEMS scales would permit operation with molecular flow up to pressures much greater than $10^{-2}\,\mathrm{mbar}$, providing the surface reflection characteristics were not negatively affected by adsorbed gases. This is a serious uncertainty since all of the available experiments have been reported for pressures between 1 and several orders of magnitude less than $10^{-2}\,\mathrm{mbar}$.

8.8 Conclusions

The performances of a representative sampling of macroscale vacuum pumps have been reviewed based on parameters relevant to micro- and mesoscale vacuum requirements. If the macroscale performances of these generally complex devices could be maintained at meso- and microscales they might be barely compatible with the energy and volume requirements of meso- and microsampling instruments. A general scaling analysis of macroscale pumps considered as possible candidates for scaling (although going to microscales with these scaled pumps is currently beyond the state of the manufacturing art) is presented and provides little to no incentive for simply trying to scale existing pumps, with the possible exception of ion pumps.

When the difficulty in developing MEMS manufacturing techniques with adequate tolerances is considered, the possibility of maintaining the macroscale performance at microscales is remote. As a consequence two alternative pump technologies are examined. The first, the Knudsen Compressor, is based on thermal creep flow or thermal transpiration, and appears attractive as a result of both analysis and initial experiments. The second, also a thermal creep pump, is under study but requires further analysis before its microscale suitability can be established. Both have however, a characteristic that prevents them from operating efficiently when pumping high vacuums ($<10^{-3}$ to 10^{-2} mbar).

The second possibility, accommodation pumping, although superficially similar to the Knudsen Compressor, is based on surface molecular reflection characteristics and will pump to ultra high vacuums. Further analysis and experiments are required to be able to provide performance estimates of accommodation pumping in meso- and microscale contexts.

The alternative technology pumps have no moving parts; no liquids are required for seals, pumping action, or lubrication. These are characteristics shared with the ion pumps. They all have a capability of providing mesoscale pumping and the potential for truly MEMS scale pumping. To realize these alternative technology pumps significant, focused research and development efforts will be required.

For Further Information

A concise up-to-date review of the mathematical background of internal rarefied gas flow is presented in the recent book, *Rarefied Gas Dynamics*, by Carlo Cercignani (2000).

Vacuum terminology and important considerations for the description of vacuum system performance and pump types are presented in the excellent publication of J.M. Lafferty, *Foundations of Vacuum Science and Technology* [Lafferty, 1998].

A detailed presentation of work on rarefied transitional flows is documented by the proceedings of the biannual International Symposia on Rarefied Gas Dynamics (various publishers over the last 45 years, all listed in Cercignani, 2000). This reference source is generally only useful for those with the time and inclination for academic study of undigested research, although there are several review papers in each publication.

Defining Terms

Microscale, Mesoscale and Macroscale: In reference to a device these terms as used in this paper correspond to the following typical dimensions:
> **Microscale:** smallest components $< 100\,\mu m$, device < 1 cm;
> **Mesoscale:** smallest component $100\,\mu m$ to 1 mm, device 1 cm to 10 cm
> **Macroscale:** smallest components > 1 mm, device > 10 cm.

Capture pumps: Vacuum pumps where pumping action relies on sequestering pumped gas in a solid matrix until it can be removed in a separate, off-line operation.

Regeneration: Off-line removal of sequestered gas from a capture pump.

Conductance: Measure of the capability of a vacuum system component (tube, channel) for handling gas flows, has units of gas volume per unit time.

Pumping speed: Measure of the pumping ability of a vacuum pump in terms of volume per unit time of inlet pressure gas that can be pumped.

Knudsen number: Indication of the degree of rarefaction of flow in a vacuum system component, for $Kn > 10$ molecules collide predominantly with the walls, for $Kn < 10^{-3}$ intermolecular collisions dominate.

Vacuum: For the purposes of this chapter the pressure ranges corresponding to various descriptions attached to vacuums are:

low or roughing 10^3 to 10^{-2} mbar;
high 10^{-2} to 10^{-7} mbar;
ultra high $<10^{-7}$ mbar

Acknowledgments

The authors thank Andrew Jamison for preparing the figures and Tariq El-Atrache and Tricia Harte for patiently working on the manuscript through many revisions. Prof. Andrew Ketsdever, Prof. Geoff Shiflett, and Dean Wiberg have provided helpful comments. Discussions with Dr. David Salzman over several years have been both entertaining and informative.

References

Aoki, K., Sone, Y., Takata, S., Takahashi, K., and Bird, G.A. (2000) "One-way Flow of a Rarefied Gas Induced in a Circular Pipe with a Periodic Temperature Distribution," *Proceedings of the 22nd Rarefied Gas Dynamics Symposium*, G.A. Bird, T.J. Bartel, M.A. Gallis, eds., AIP July 9–14, Sydney, Australia, to be published.

Baker, B.G., Hobson, J.P., and Pye, A.W. (1973) "Further Measurements of Physical Factors Influencing Accommodation Pumps," *J. Vac. Sci. Technol.*, 10(1), pp. 241–245.

Baum, H. (1957) *Vakuum-Technick* 7, pp. 154–159.

Bernhardt, K.H. (1983) "Calculation of the Pumping Speed of Turbomolecular Vacuum Pumps by Means of Simple Mechanical Data," *J. Vac. Sci. Technol. A*, 1(2), pp. 136–139.

Bird, G.A. (1998) *Molecular Gas Dynamics and the Direct Simulation of Gas Flows*, Clarendon Press, Oxford, United Kingdom.

Bills, D.G. (1967) "Electrostatic Getter-Ion-Pump design," *J. Vac. Sci. Technol.*, 4(4), pp. 149–155.

Cabuz, C., Herb, W.R., Cabuz, E.I., and Son Thai Lu (2001) "The Dual Diaphragm Pump," Presented at the 14th IEEE International Conference on Micro Electro Mechanical Systems, Interlaken, Switzerland.

Callas, J.L. (1999) "Vacuum System Requirements for a Miniature Scanning Electron Microscope," in *NASA/JPL Miniature Vacuum Pumps Workshop*, Glendale, California, JPL website http://cissr.jpl. nasa.gov/pumpsworkshop/.

Cercignani, C. (2000) *Rarefied Gas Dynamics*, Cambridge University Press, Cambridge, United Kingdom.

Chang, T.H.P., Kern, D.P., and McCord, M.A. (1989) "Electron Optical Performance of a Scanning Tunneling Microscope Controlled Field Emission Microlens System," *J. Vac. Sci. Technol. B*, 7, pp. 1855–1861.

Chang, T.H.P., Kern, D.P., and Muray L.P. (1990) "Microminiaturization of Electron Optical Systems," *J. Vac. Sci. Technol. B*, 8, pp. 1698–1705.

Denison, D.R. (1967) "Performance of a New Electrostatic Getter-Ion Pump," *J. Vac. Sci. Technol.*, 4(4), pp. 156–162.

Doetsch, I.H., and Ryce, S.A. (1972) "Separation of Gas Mixtures by Accommodation Pumping," *Can. J. Chem.*, 50(7), pp. 957–960.

Douglas, R.A. (1965) "Orbitron Vacuum Pump," *Rev. Sci. Instrum.*, 36(1), pp. 1–6.

Dushman, S. (1949) *Scientific Foundations of Vacuum Technique*, John Wiley & Sons, New York.

Dushman, S., and Lafferty, J.M. (1962) *Scientific Foundations of Vacuum Technique*, John Wiley & Sons, New York.

Ferran, R.J., and Boumsellek, S. (1996) "High-Pressure Effects in Miniature Arrays of Quadrupole Analyzers for Residual Gas Analysis from 10^{-9} to 10^{-2} Torr," *J. Vac. Sci. Technol.*, 14, pp. 1258–1265.

Fréchette, L.G., Stuart, A.J., Breuer, K.S., Ehrich, F.F., Ghodssi, R., Kanna, R., Wang, C.W., Zhong, X., Schmidt, M.A., and Epstein, A.H., (2000) "Demonstration of a Microfabricated High-Speed Turbine Supported on Gas Bearings," in *Solid-State Sensor and Actuator Workshop*, pp. 43–47, Hilton Head Island.

Freidhoff, C.B., Young, R.M., Sriram, S.S., Braggins, T.T., O'Keefe, T.W., Adam, J.D., Nathanson, H.C., Symms, R.R.A., Tate, T.J., Ahmad, M.M., Taylor, S., and Tunstall, J. (1999) "Chemical Sensing Using Nonoptical Microelectromechanical Systems," *J. Vac. Sci. Technol. A*, 17, pp. 2300–2307.

Han, Y-L., Young, M., Muntz, E.P., and Shiflett, G. (2004) "Knudsen Compressor Performance at Low Pressures," Rarefied Gas Dynamics, Capitelli, M., ed., AIP CP 762, pp. 162–167.

Hemmerich, J.L. (1988) "Primary Vacuum Pumps for the Fusion Reactor Fuel Cycle," *J. Vac. Sci. Technol. A*, 6(1), pp. 144–153.

Hobson, J.P. (1970) "Accommodation Pumping — A New Principle for Low Pressure," *J. Vac. Sci. Technol.*, 7(2), pp. 301–357.

Hobson, J.P., and Pye, A.W. (1972) "Physical Factors Influencing Accommodation Pumps," *J. Vac. Sci. Technol.*, 9(1), pp. 252–256.

Hobson, J.P., and Salzman, D.B. (2000) "Review of Pumping by Thermomolecular Pressure," *J. Vac. Sci. Technol. A*, 18(4), pp. 1758–1765.

Hopfinger, E.J., and Altman, M. (1969) "A Study of Thermal Transpiration for the Development of a New Type of Gas Pump," *J. Eng. Power, Trans. ASME*, 91, pp. 207–215.

Hudsen, M.L., and Bartel T.J. (1999) "DSMC Simulation of Thermal Transpiration Pumps," in *Rarefied Gas Dynamics* 1, R. Brun, R. Campargue, and J.C. Lengrand, eds., pp. 719–726, Cépaduès Éditions, Toulouse, France.

Knudsen, M. (1910a) "Eine Revision der Gleichgewichtsbedingung der Gase. Thermische Molekularströmung," *Ann. Phys.*, 31, pp. 205–229.

Knudsen, M. (1910b) "Thermischer Molekulardruck der Gase in Röhre," *Ann. Phys.*, 33, pp. 1435–1448.

Lafferty, J.M. (1998) *Foundations of Vacuum Science and Technology*, John Wiley & Sons, New York.

Loyalka, S.K., and Hamoodi, S.A. (1990) "Poiseuille Flow of a Rarefied Gas in a Cylindrical Tube: Solution of Linearized Boltzmann Equation," *Phys. Fluids A*, 2(1)1, pp. 2061–2065.

Loyalka, S.K., and Hickey, K.A. (1991) "Kinetic Theory of Thermal Transpiration and the Mechanocaloric Effect: Planar Flow of a Rigid Sphere Gas with Arbitrary Accommodation at the Surface," *J. Vac. Sci. Technol. A*, 9(1), pp. 158–163.

Madou, M. (1997) *Fundamentals of Microfabrication*, CRC Press, Boca Raton, Florida.

Moore, E., Muntz, E.P., Eyre, B.F., Myung, N., Orient, O., Shcheglov, K., and Wiberg, D. (2002) "Performance Analysis for Meso-Scale Scroll Pumps," Presented at the 2nd NASA/JPL Miniature Pumps Workshop, Pasadena, CA.

Moore, E.J., Muntz, E.P., Eyre, B., Myung, N., Orient, O., Shcheglov, K., and Wiberg, D. (2003) "Analysis of a Two Wrap Meso Scale Scroll Pump," in Rarefied Gas Dynamcis, Ketsdever, A., and Muntz, E.P., eds., AIP Conference Proceedings 663, Melville, N.Y., pp. 1033–1040.

Muntz, E.P., Sone, Y., Aoki, K., and Vargo, S. (1998) "Performance Analysis and Optimization Considerations for a Knudsen Compressor in Transitional Flow," *USC Aerospace and Mechanical Engineering Report No. 98001*, University of Southern California, Los Angeles.

Muntz, E.P. (1999) "Surface Dominated Rarefied Flows and the Potential of Surface Nanomanipulations," in *Rarefied Gas Dynamics* 1, R. Brun, R. Campargue and J.C. Lengrand, eds., pp. 3–15, Cépaduès Éditions, Toulouse, France.

Muntz, E.P., Sone, Y., Aoki, K., Vargo, S., and Young, M. (2002) "Performance Analysis and Optimization for the Knudsen Compressor in Transitional Flow," *J. Vac. Sci. Tech. A*, 20(1), Jan/Feb, pp. 214–224.

NASA/JPL (1999) *NASA/JPL Miniature Vacuum Pumps Workshop*, http://cicsr.jpl.gov.

Nathanson, H.C., Liberman, I., and Freidhoff, C. (1995) "Novel Functionality Using Micro Gaseous Devices," in *The 8th Annual International Workshop on Micro Electro Mechanical Systems — MEMS '95*, pp. 72–76, IEEE, Amsterdam, Netherlands.

Orient, O.J., Chutjian, A., and Garkanian, V. (1997) "Miniature, High-Resolution, Quadrupole Mass-Spectrometer Array," *Rev. Sci. Instrum.*, 68, pp. 1393–1397.

Orner, P.A., and Lammers, G.B. (1970) "The Application of Thermal Transpiration to a Gaseous Pump," *J. Basic Eng. Trans. ASME*, 92, pp. 294–302.

Park, J.-Y., Choi, H.-J., Lee, Y., Kang, S., Chunk, K., Park, S.W., and Kuk, Y. (1997) "Fabrication of Electron-Beam Microcolumn Aligned by Scanning Tunneling Microscope," *J. Vac. Sci. Technol. A*, **15**, pp. 1499–1502.

Pham-Van-Diep, G., Keeley, P., Muntz, E.P., and Weaver, D.P. (1995) "A Micromechanical Knudsen Compressor," in *Proceedings 19th International Symposium on Rarefied Gas Dynamics*, J. Harvey and G. Lord, eds., pp. 715–721, Oxford University Press, Oxford, United Kingdom.

Piltingsrud, H.V. (1994) "Miniature Cryosporption Vacuum Pump for Portable Instruments," *J. Vac. Sci. Technol. A*, **12**, pp. 235–240.

Piltingsrud, H.V. (1996) "Miniature Peristaltic Vacuum Pump for Use in Portable Instruments," *J. Vac. Sci. Technol. A*, **14**, pp. 2610–2617.

Piltingsrud, H.V. (1997) "A Field Deployable Gas Chromatograph/Mass Spectrometer for Industrial Hygiene Applications," *Am. Ind. Hyg. Assoc. J.*, **58**, pp. 564–577.

Sharipov, F., and Seleznev, V. (1998) "Data on Internal Rarefied Gas Flows," *J. Phys. Chem. Ref. Data*, **27**, pp. 657–706.

Short, R.T., Fries, D.P., Toler, S.K., Lembke, C.E., and Byrne, R.H. (1999) "Development of an Underwater Mass-Spectrometry System for In Situ Chemical Analysis," *Meas. Sci. Technol.*, **10**, pp. 1195–1201.

Smith, J.N., and Saltzburg, H. (1964) "Recent Studies of Molecular Beam Scattering from Continuously Deposited Gold Films," in *Proceedings 4th International Symposium on Rarefied Gas Dynamics* 2, J.H. de Leeuw, ed., pp. 491–504, Academic Press, New York.

Sone, Y., and Itakura, E. (1990) "Analysis of Poiseuille and Thermal Transpiration Flows for Arbitrary Knudsen Numbers by a Modified Knudsen Number Expansion Method and Their Database," *J. Vac. Soc. Jpn.*, **33**, pp. 92–94.

Sone, Y., and Sugimoto, H. (2003) "Vacuum Pump without Moving Parts and its Performance," in *Rarefied Gas Dynamics*, Ketsdever, A., and Muntz, E.P., eds., AIP Conference Proceedings 663, Melville, N.Y., pp. 1041–1048.

Sone, Y., Waniguchi, Y., and Aoki, K. (1996) "One-Way Flow of a Rarefied Gas Induced in a Channel with a Periodic Temperature Distribution," *Phys. Fluids*, **8**, pp. 2227–2235.

Sone, Y., and Yamamoto, K. (1968) "Flow of a Rarefied Gas Through a Circular Pipe," *Phys. Fluids*, **11**, pp. 1672–1678.

Suetsugu, Y. (1993) "Numerical Calculation of an Ion Pump's Pumping Speed," *Vacuum*, **46**(2), pp. 105–111.

Terry, S.C., Jerman, J.H., and Angell, J.B. (1979) "A Gas Chromatographic Air Analyzer Fabricated on a Silicon Wafer," *IEEE Transactions on Electron. Devices*, **ED-26**, pp. 1880–1886.

Tracy, D.H. (1974) "Thermomolecular Pumping Effect," *J. Phys. E: Sci. Instrum.*, 7, pp. 533–563.

Turner, D.J. (1966) "A Mathematical Analysis of a Thermal Transpiration Vacuum Pump," *Vacuum*, **16**, pp. 413–419.

Vargo, S.E. (2000) The Development of the MEMS Knudsen Compressor as a Low Power Vacuum Pump for Portable and In Situ Instruments, Ph.D. Thesis, University of Southern California, Los Angeles, California.

Vargo, S.E., Muntz, E.P., Shiflett, G.R., and Tang, W.C. (1999) "The Knudsen Compressor as a Micro and Macroscale Vacuum Pump Without Moving Parts or Fluids," *J. Vac. Sci. Technol. A*, 17, pp. 2308–2313.

Vargo, S.E., and Muntz, E.P. (1996) "An Evaluation of a Multi-Stage Micromechanical Knudsen Compressor and Vacuum Pump," in *Proceedings 20th International Symposium on Rarefied Gas Dynamics*, C. Shen, ed., pp. 995–1000, Peking University Press, Beijing, China.

Vargo, S.E., and Muntz, E.P. (1998) "Comparison of Experiment and Prediction for Transitional Flow in a Single Stage Micromechanical Knudsen Compressor," in *Proceedings 21st International Symposium on Rarefied Gas Dynamics*, R. Brun, R. Campargue and J.C. Lengrand, eds., pp. 711–718, Cépaduès Éditions, Marseille, France.

White, A.J., Blamire, M.G., Corlette, C.A., Griffiths, B.W., Martin, D.M., Spencer, S.B., and Mullock, S.J. (1998) "Development of a Portable Time-of-Flight Membrane Inlet Mass Spectrometer for Environmental Analysis," *Rev. Sci. Instrum.*, **69**, pp. 565–571.

Wiberg, D., Scheglov, K., White, V., Orient, O., and Chutjian, A. (2000) "Toward a Micro Gas Chromatograph/Mass Spectrometer (GC/MS) System," presented at the 3rd Annual International Conference on Integrated Nano/Microtechnology for Space Applications, Houston, TX.

Wilcox, J.Z., Feldman, J., George, T., Wilcox, M., and Sherer, A. (1999) "Miniaturization of High Vacuum Pumps: Ring-Anode Orbitron," presented at *AVS 46th International Symposium*, Seattle, Washington, Oct. 25–29.

Young, M., Han, Y-L., Muntz, E.P., and Shiflett, G. (2003) "Characteristics of a Radiantly Driven Multi-Stage Knudsen Compressor," IMECE2003-41486.

Young, M., Han, Y-L., Muntz, E.P., and Shiflett, G. (2004) "Characterization and Optimization of a Radiantly Driven Multi-Stage Knudsen Compressor," Rarefied Gas Dynamics, Capitelli, M., ed., AIP CP 762, pp. 174–179.

Young, M., Vargo, S.E., Shiflett, G., Muntz, E.P., and Green, A. (2001) "The Knudsen Compressor as an Energy Efficient Micro-Scale Vacuum Pump," the 2001 ASME International Mechanical Engineering Congress and Exposition (IMECE2001), New York.

Young, M. (2004) Investigation of Several Important Phenomena Associated with the Development of Knudsen Compressors, Ph.D. Thesis, University of Southern California, Los Angeles.

Young, R.M. (1999) "Analysis of a Micromachine Based Vacuum Pump on a Chip Actuated by the Thermal Transpiration Effect," *J. Vac. Sci. Technol. B*, 17, pp. 280–287.

9

Nonlinear Electrokinetic Devices

Yuxing Ben
Massachusetts Institute of Technology

Hsueh-Chia Chang
University of Notre Dame

9.1 Introduction

There is considerable interest in exploiting the attractive features of electrokinetics to develop micro-devices such as pumps, mixers, particle sorters/detectors, and others for miniature applications in the life-science and homeland-security industries (specific applications include drug design, drug delivery and detection, medical diagnostic devices, high-performance liquid and capillary chromatographs, and explosives detection) and for combinatorial synthesis (such as rapid chemical analyses and high throughput screening) [Koch et al., 2000; Cheng and Kricka, 2001; Stone and Kim, 2001; Trau and Battersby, 2001; Delgado, 2002; Nguyen and Wereley, 2002, Reyes et al., 2002; Stone et al., 2004]. Since electrokinetic liquid or particle motion is imparted by implantable mircoelectrodes, it is far easier to control, direct, and meter than the other microfluidic transportation mechanisms like syringe-displaced, peristaltic, centrifugally driven, and thermal Marangoni-driven flows.

However, electrokinetic devices also suffer from several shortcomings. The efficiency of electrokinetic pumps is generally lower than that of mechanical pumps because the hydrodynamic viscous length scale is now the double layer thickness rather than the transverse channel dimension, leading to extremely large wall shear and viscous dissipation. The direct-current (DC) electrokinetics necessary involve a steady current sustained by electrode reactions. Such reactions can often produce bubbles and ion contaminants that must be bled from the device or neutralized with proper buffer solutions. DC electro-osmotic flow has little hydrodynamic shear outside the double layer. This produces low dispersion in capillary chromatographs but is responsible for low mixing efficiencies in electrokinetic devices. Analyte dispersion and particle aggregation are also issues for electrokinetic flows. The objective of this review is to assess the performance of some electrokinetic devices and to summarize some proposed design strategies to circumvent them. The focus will be on the new field of nonlinear electrokinetics. We will address two specific nonlinear electrokinetic phenomena and the new devices they have inspired.

Generally speaking, electrokinetics is the motion of liquid or particles under an electric field. By linear electrokinetics, we mean that the velocity is linearly dependent on the applied field. Theory for linear electrokinetics has been well established [Probstein, 1994]. Most dielectric substances possess surface charges that will attract screening counterions when brought into contact with an aqueous (polar) medium. Some of the charging mechanisms include ionization, ion adsorption, and dissolution [Russel et al., 1989]. The effect of any charged surface in an electrolyte solution will be to influence the distribution of nearby ions in the solution. Ions of opposite charge to that of the surface (counterions) are attracted toward the surface while ions of like charge (coions) are repelled from the surface. This attractive and repulsive electrostatic force, when balanced by the usual diffusive chemical potential gradient, leads to the formation of an electric double layer known as the Debye layer. If an electric field is applied tangentially along a charged surface, then the electric field will exert a force on the charge in the diffuse layer. This layer is a part of the electrolyte solution, and the migration of the mobile ions will drag the solvent with them. For a sufficiently thin Debye layer, the velocity profile approaches a constant away from the Debye layer. This asymptotic U may be regarded as a "slip velocity" of the outer fluid velocity relative to the surface, which can be described as

$$U = -\frac{\varepsilon}{\mu}\varsigma E_t \qquad (9.1)$$

where ε is the permittivity of the electrolyte; μ is the viscosity; ς is the zeta potential, which is determined by the surface charge of the material (it has the same sign) and the electrolyte concentration; and E_t is the electric field tangential to the charged surface. This is the *Smoluchowski slip velocity*. This slip velocity is linear with respect to the applied tangential field and involves an equilibrium Boltzmann distribution of charges in the Debye layer. The movement of liquid relative to a stationary charged surface by an applied electric field is termed *electroosmosis*. The movement of a charged surface plus attached material relative to stationary liquid by an applied electric field is termed *electrophoresis*. The electrophoretic velocity of a dielectric sphere also has the same form but with a different coefficient and sign if the material remains the same. As a point of reference, for a typical field of 100 V/cm (higher values cause protein denaturing and Joule heating), the linear electrokinetic velocity is typically several hundred microns per second for a typical buffer solution and a typical zeta potential of 100 mV, corresponding to a 10 nm thick double layer.

It is important to note that the slip velocity and the tangential Maxwell stress that drives it are possible only for electrolytes with mobile ions in the double layers. For dielectric liquids without space charge, the double layer is absent unless the dielectric liquid is extremely polar and can ionize the surface groups and dissolve the resulting ions. In general, however, interfacial charges in dielectric liquids result from field-induced internal molecular or atomic dipole formation. The ponderomotive force that such dielectric polarization produces would only be in the direction of the permittivity gradient or normal to the surface [Johns, 1995], and there would be no tangential force to drive the flow. Consequently, ponderomotive forces can only drive dielectric liquids in electrowetting when the front interface produces a permittivity gradient in the tangential direction of the channel.

These linear electrokinetic phenomena for electrolytes, due to equilibrium ion distributions established by a large surface field, have many interesting features. With uniform surface charge and zeta potential and without applied pressure gradient, Equation (9.1) implies that the applied field lines are identical to the streamlines [Morrison, 1970]. As the electric field is irrotational in the electroneutral bulk (the potential is a harmonic function satisfying the Laplace equation), the electrokinetic flow is actually a potential flow despite a miniscule Reynolds number, and vortices are not possible under these conditions. For unidirectional flow in straight channels, the bulk flow outside the double layer is flat and shear-free as in potential flow. Because hydrodynamic Taylor dispersion [Taylor, 1954; Aris, 1956] of a finite volume of solute results from transverse gradient of longitudinal velocity, namely the shear rate, electroosmotic flows with flat velocity profiles and small shear are preferred in capillary chromatography and have been used successfully in separating DNAs and other molecules [Rhodes and Snyder, 1994].

Other than bubble and ion generation at the DC electrodes, however, linear electrokinetics still have several shortcomings when applied over a chip that is several centimeters long. For example, the channels usually have turns to cover more areas. Taylor dispersion can result at the turns because of the difference in transit time (or streamline velocity over the same length) at the two sides of the turn. Taylor dispersion due to this transverse difference in velocity around turns can reduce the separation efficiency dramatically [Dutta and Leighton, 2001]. With low dispersion, electrokinetic flow suffers from mixing deficiencies that reduce reaction yield and promote colloid/protein aggregation/precipitation [Chang, 2001; Thamida and Chang, 2002; Takhistov et al., 2003]. Such undesirable colloidal segregation and aggregation phenomena in microdevices can be minimized if microvortices can be generated within the flow channels to mix the suspension. However, generation of microvortices is difficult in electrokinetic flow due to its irrotational features. One strategy is to introduce surfaces with nonuniform zeta potential [Ajdari, 1995; Herr et al., 2000; Stroock et al., 2002]. However, such nonuniformities are difficult to impose at junctions and membrane surfaces where vortex mixing is most needed.

Several micropumps based on linear electrokinetics have been reported in recent years [Laser and Santiago, 2004; Zeng et al., 2001]. Because of the large current of such linear electrokinetic pump, bubble and ion generation by electrochemical reaction at the electrodes are major issues. A pH gradient often develops as a result of the ion production and will often produce nonuniform zeta potential due to its sensitivity to pH. Pressure-driven backflow and long circulations can then result due to such pH gradients [Minerick et al., 2002]. Buffer solutions are usually used to neutralize the pH gradient, but depending on the load and sample, utilizing such buffer solutions is not always possible. A high voltage exceeding several hundred volts is usually required for linear electrokinetic pumps, rendering them impractical or unsafe for portable devices. Dense packing using silica particles to maximize the double layer–to–pore-diameter ratio, to reduce the current, and to increase back pressure has been suggested as a partial remedy to these issues [Yao et al., 2003].

Although linear electrophoresis operates best on large-scale channels, such as in DNA gel electrophoresis, it cannot manipulate the nanoscale particles precisely [Hughes, 2003] as it has no sensitivity with respect to the particle size. Separating target cells from complex fluid samples such as human blood remains a significant challenge [Cheng and Kricka, 2001]. Other electrokinetic phenomena such as dielectrophoresis pioneered by Pohl (1978), electrorotation [Wang et al., 1992; Wang et al., 1997; Gimsa, 2001], and traveling wave electrophoresis [Cui and Morgan, 2000; Morgan et al., 2001] have been suggested to achieve better control and sensitivity. These new mechanisms are all nonlinear electrokinetic phenomena and some of them will be described below.

9.2 Nonlinear Electrokinetics

A large family of nonlinear and nonequilibrium electrokinetic phenomena have been found or rediscovered recently. All of them work under the same basic principle: the induction of nonuniform polarization within the double layer with the external field. As a result, potential drop across the polarization layer, or zeta potential, is dependent on the normal external field, as well as the surface field due to surface charges. A direct generalization of Equation (9.1) suggests the resulting electroosmotic and electrophoretic velocities should depend nonlinearly on the external field. We hence expect a much larger velocity than linear electrokinetics at large fields. Boltzmann equilibrium ion distributions [Probstein, 1994] can no longer exist. These nonlinear electrokinetic phenomena are hence often nonequilibrium in nature.

The first example of nonlinear electrokinetic phenomena is dielectrophoresis (DEP) induced by an AC field [Pohl, 1978; Jones, 1995; Zimmermann and Neil, 1996; Gascoyne and Vykoukal, 2002; Morgan and Green, 2003; Hughes, 2003]. Charging and discharging of the double layer by the external fields leads to external field-induced dipoles at the dielectric particle or cell surface. If the dynamic polarization is due entirely to the normal field and is fast compared to the period of the AC field, the Maxwell stress is always in the same direction and has a nonzero time average. As shown in Figure 9.1, if the field is nonuniform, the greater electric field strength across one side of the particle means that the force generated on that side is greater than the force induced on the opposite side of the particle and a net force is exerted toward the region of greatest electric field. This motion of the particle is termed positive dielectrophoresis. However, this double layer polarization mechanism for DEP has not been scrutinized in the literature.

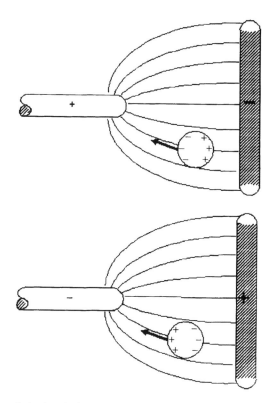

FIGURE 9.1 Double layer polarization during positive dielectrophoresis.

Instead, a dielectric polarization mechanism based on internal atomic and molecular dipoles is typically used. To account for inadequacies of this polarization mechanism, conductivity is introduced to produce a Maxwell-Wagner factor that basically models a capacitor and a resistor in parallel on both sides of the surface. Such equivalent circuit models, as in similar models we will develop in section 9.3, should model the charging dynamics of the double-layer capacitor. However, a direct link has not been established. In any case, using the semiempirical Maxwell-Wagner factor, the dielectrophoretic force F_{DEP} [Johns, 1995] upon a dielectric sphere of permittivity ε_2 and radius R suspended in a medium of permittivity ε_1 and subjected to an electric field E, is

$$F_{DEP} = 2\pi\varepsilon_1 R^3 K \nabla E^2 \tag{9.2}$$

where K is $Re\left(\frac{\varepsilon_2 - \varepsilon_1}{\varepsilon_2 + \varepsilon_1}\right)$, the Clausius–Mossotti function, and generalizes to the Maxwell–Wagner factor if the permittivity ε is complex to include conductivity $\varepsilon - \sigma/i\omega$, where ω is the frequency of the AC field. The dielectrophoretic velocity is then proportional to the divergence of the square of the electric field intensity, quite distinct from linear electrokinetic velocity of Equation (9.1). Dielectrophoretic velocity is also sensitive to the frequency of the applied field via the Maxwell-Wagner factor, cell or particle size R, and electrical properties. A recent review by Gascoyne and Vykoukal (2002) shows that dielectrophoresis has tremendous applications in cell sorting and separations.

Electrorotation occurs when a dipole is induced by a rotating electric field [Wang et al. 1992; Zimmermann and Neil, 1996; Wang et al., 1997; Gimsa and Wachner, 1998; Gimsa, 2001; Hughes, 2003]. A lag between the orientation of the electric field and that of the dipole moment develops due to the charge relaxation times of the double layer, and thus a torque is induced as the dipole moves to reorient itself with the electric field. These relaxation times for particles in an electrolyte probably correspond to electromigration and diffusion times across the double layer and across the surface of the particles. As in DEP, a definitive

FIGURE 9.2 Experimental snapshot of the microchannel junction. The silica channel and the latex colloids are oppositely charged, and hence both electrophoretic and electroosmotic motions are in same direction. The picture shows a spiral colloidal aggregation at the inner corner.

analysis of how the double-layer dynamics affect rotation is still lacking. Owing to the continuous rotation of the electric field, the torque is induced continually and the cell rotates. Electrorotation has been used to study the dielectric properties of matter, such as the interior properties of biological cells and biofilms.

Nonlinear electrokinetic phenomena exist under a DC field as well. One example of nonlinear DC electrokinetics is polarization at nearly insulate wedges [Thamida and Chang, 2002; Takhistov et al., 2003]. A large field penetration exists near sharp channel corners even for channels made with low-permittivity dielectrics. Hence, the external field can penetrate the double layers on both sides of the corner and also through the corner dielectric in between. This normal field penetration is inward at one side and outward on the other. As such, its field-induced polarization is of opposite charge on the two sides. This produces a converging nonlinear electroosmotic flow that yields an observable microjet and vortex at the corner — both are impossible with linear electrokinetics. Significant particle aggregation occurs at this corner, as seen in Figure 9.2, due to the converging stagnation flow of the nonlinear electrokinetics. The long-range hydrodynamic convection transports the particles to the corner, where a DC dielectrophoresis force traps them. This long-range trapping effect of converging stagnation flow will be employed in several other designs to be discussed later. Because it is this flow that convects the particles toward the point, the trapping mechanism is far longer range and stronger than the AC dielectrophoresis (DEP) force in Equation (9.2), whose range is determined by the field gradient length scale and whose amplitude scales as the third power of the particle size R. Aggregation is absent away from the corner, as is consistent with linear electrokinetics, but it occurs at the corner due to the localized nonlinear electrokinetics.

Another example is the "electrokinetic phenomenon of the second kind" first envisioned by Dukhin (see review by Dukhin in 1991). It involves a highly conductive and ion-selective granule that permits the external field and diffusion to drive a flux of counterions (a current) into half of a granule. (The coions

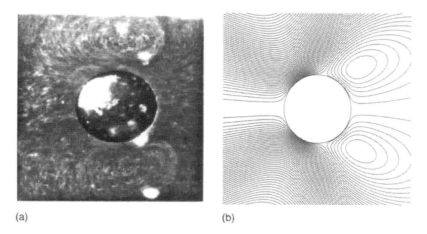

(a)   (b)

FIGURE 9.3 (a) Illuminated titanium oxide powder brings out the streamlines of electrokinetic flow around 1 mm spherical granule housed in a slot in a field of about 100 V/cm. (b) Computed streamlines.

cannot be driven into the other half due to the ion specificity.) This steady flux of ions immediately renders the potential and concentration distributions within the double layer different from the Boltzmann equilibrium distributions that cannot sustain a flux. Since this flux is provided by the electromigration of ions driven by the external field, the external field necessarily penetrates the double layer, and the latter's polarization is dependent on the normal external field. Dukhin's theory yields a prediction that the electrophoretic velocity of a spherical granule of radius a scales as

$$U \sim \frac{\varepsilon}{\mu} E^2 a \qquad (9.3)$$

Many of the features expected of nonlinear electrokinetics have been observed for this DC electrokinetic phenomenon of the second kind. Large vortices on the side were observed by Mishchuk and Takhistov (1995), and the electrophoretic velocity, which is not linearly dependent on the applied field, was measured by Barany et al. (1998). Ben and Chang studied this problem theoretically in detail (2002) (Figure 9.3). A mixing device based on the strong mixing action of the vortices in Figure 9.3 is developed [Wang et al., 2004] for microfluidic applications.

Nonlinear flows near polarized particles have been analyzed for two decades (see review by Murtsovkin, 1996). Recently, Bazant and Squires [Bazant and Squires, 2004; Squires and Bazant, 2004] studied flow around metal objects (termed IECO), with possible microfluidic applications in mind. The surface of a metal has a constant potential; that is, every ion that is driven into the polarized layer by the external field will be compensated by an opposite charge that moves even more rapidly to the surface on the solid side. This compensation would ensure there is no net charge on two sides of the surface and the potential remains the same. However, the number of ions that can be driven into the polarized layer can, in principle, be increased arbitrarily by raising the applied field. This field-dependent polarization accounts for the nonlinear dependence of the electrokinetic velocity. They predict that the velocity will scale as E^2, where E is the external field. Vortices will occur around a spherical metal because of the geometry. Their work shows pumping of fluid in a particular direction to be possible with an asymmetric design.

Linear DC electroosmotic flow around particles of the same Zeta potential toward an electrode surface with a different polarization can produce vortices when the particles are close to the electrode surface [Solomentsev et al., 1997; Solomentsev et al., 2000]. These vortices are on the side of the particles away from the surface. They can hence induce parallel motion of the particles due to hydrodynamic interaction between two adjacent particles. This hydrodynamic interaction is attractive and leads to much larger lateral particle velocities, but the external-field induced nonuniform polarization produces parallel dipoles on two adjacent particles. Electrostatic interaction between these induced dipoles is attractive for

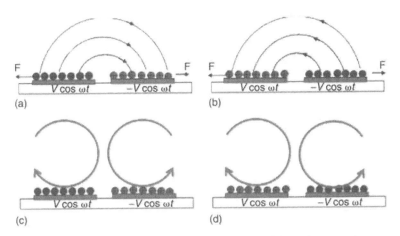

FIGURE 9.4 (**See color insert following page 2-12.**) Schematic illustration of the capacitive charging: (a) and (b) demonstrate the electric field, and F represents time averaged Maxwell force; (c) and (d) demonstrate the flow profile.

two particles along the same field line. This interaction is responsible for linear self-assembly along field lines [Minerick et al., 2003]. The electrostatic interaction is repulsive for parallel self assembly of particles on different field lines. The electrostatic repulsion between these dipoles then competes with the attractive hydrodynamic forces in the parallel self-assembly dynamics. However, as observed and analyzed by Trau et al. (1997), Yeh et al. (1997), and Nadal et al. (2002), spontaneous self-assembly of colloids on electrode surfaces and even in the bulk [Hu et al., 1994] occurs when a converging stagnation flow appears for the nonlinear AC electroosmotic flow field. Hence, self-assembly seems to occur more readily due to the induced electrostatic dipoles and the hydrodynamic vortices generated by AC nonlinear electrokinetics.

Nonlinear electrokinetic phenomena also occur at the electrodes supplying the AC field, and this has spurred considerable interest in designing such AC electrokinetic devices. In contrast to DC electrokinetics, at a sufficiently high frequency electrochemical reactions do not occur at the electrode, and the key bubble/ion generation problem is removed. This is by far the most attractive feature of AC electrokinetic phenomena. The field still penetrates the electrolyte to charge and discharge ions onto the electrodes. Such capacitive charging produces a very polarized double layer over the electrode, which can again drive a tangential flow as in linear electroosmosis, as shown in Figure 9.4. As the polarization and the effective zeta potential are now field-dependent, Equation (9.1) then stipulates that all AC electrokinetic phenomena are nonlinear. The induced polarization is much stronger than the usual polarization due to surface charges on typical dielectric surfaces. Velocity can be several hundred microns per second at the moderate applied electrode potential of several volts. This AC polarization leads to strong electroosmotic vortices on the micron-sized electrodes that have been observed and analyzed [Ramos et al., 1999; Ajdari, 2000; Brown et al., 2000; Gonzalez et al., 2000; Green et al., 2000; Green et al., 2002; Studer et al., 2002; Mpholo et al., 2003; Ramos et al., 2003]. The vortices have the same size as the electrodes. The dynamic charging and discharging of ions into the double layer by the external AC field yields an interesting dynamic screening phenomenon that develops over a time scale of $\lambda L/D$ [Gonzalez et al., 2000], where D is the diffusivity of the ions.

With potential microfluidic applications in mind, Ajdari (2000) predicted that asymmetric AC electroosmotic vortices on asymmetric planar electrodes can lead to a net flow instead of the closed circulation within vortices. This AC electroosmotic pump was constructed and experimentally verified by Brown et al. (2000).

Other forms of nonlinear electrokinetics such as electrospray [Yeo et al., 2004], electroporation [Chang, 1989], and electrowetting [Jones et al., 2004] have also been scrutinized recently. All of them have potential applications in microfluidic lab-on-a-chip devices.

It is clear from the above review that nonlinear electrokinetics is a rich new field with many potential applications and many clear advantages over linear electrokinetics (lack of bubble and ion generation at electrodes). In the next section, we will focus on a specific AC electroosmotic flow on electrodes due to Faradaic charging by electrochemical reactions. Although reactions are now present at the electrodes because high voltages and low frequencies are used, there is little net production of bubbles and ions. The product ions of one half-cycle are consumed in the next, and the number of gas molecules generated in each half-cycle is not sufficient to nucleate gas bubbles. Yet, significant transient polarization occurs at each half-cycle due to the reactions. As a result of this strong Faradaic polarization, very high fields can be employed to drive high flow without bubble and ion generation; this combines the advantages of DC electrokinetics and nonreactive AC electrokintetics. Two kinds of microfluidic devices will be discussed: one is for assembling and dispersing particles on the electrode surface or concentrating bacteria; the other is for pumping fluid.

9.3   AC Electrokinetics on Electrodes: Effects of Faradaic Reaction

Consider two parallel, infinitely long electrodes with an AC field $V \cos \omega t$ applied as shown in Figure 9.4. In the first half-cycle after turning on the AC field, the electric field line is from the left electrode to the right, as shown in Figure 9.4a. Cations due to electromigration move to the right electrode along the field lines, and coions move in the opposite direction. As a result, cations accumulate on the double layer of the right electrode, and anions accumulate on the other electrode. The electrode geometry then introduces a tangential field that will move the accumulated ions. These ions in turn drive the liquid motion due to viscous effects. In the next half cycle, as shown in Figure 9.4b.The electric field changes direction; however, the polarity of the accumulated charges also changes accordingly. As the electric force is equal to the electric field times the charge, the instantaneous and time-averaged electric force remains in the same direction at the same electrode. This produces an inward slip velocity on the electrode. One would expect the field and the slip velocity to weaken toward the outer edges of the electrode pair. Continuity dictates that a large flow into a region with a weak driving force (slip velocity) must produce a large pressure-driven backflow in the opposite direction. This opposing flow produces converging stagnation lines and vortices, one example of which already has been seen in Figure 9.3, for nonuniform polarization and slip velocity on an ion-exchange granule. A similar vortex motion on the parallel electrodes is sketched in Figure 9.4c and d.

Usually, the thickness of the double layer is much less than the electrode width such that the tangential current can be neglected. One can model this charging mechanism with a distributed system of a capacitor and resistor in series: the double/diffuse layer behaves as a capacitor, and the electroneutral region behaves as a resistor. A charge balance in the normal direction across the double layer results in

$$\sigma \frac{\partial \phi}{\partial n} = \frac{\partial q}{\partial t} \tag{9.4}$$

where σ is the conductivity, ϕ is the electric potential, q is the charge per unit area in the double layer, and n represents the normal direction to the electrode surface. If the voltage drop across the diffuse double layer is sufficiently small ($\Delta \phi < RT/F = 0.025V$), there is a linear relationship between the charge and the voltage from the Debye–Huckel approximation of the Boltzmann charge distribution, i.e. $q = C_{DL}(\phi - V)$. Equation (9.4) can then be written in the complex form of a Fourier series as

$$\sigma \frac{\partial \phi}{\partial y} = i\omega q = i\omega C_{DL}(\phi - V) \tag{9.5}$$

where $C_{DL} = \frac{\varepsilon}{\lambda}$ is the capacitance per unit of area of the total double layer, ε is the dielectric permittivity of the solvent, and λ is the double layer thickness. To make it more accurate, sometimes C_{DL} is given by a combination of two capacitors in series — the Stern or compact layer capacitance C_s, and the diffuse double layer capacitance C_d, which is $\frac{\varepsilon}{\lambda}$ [Gonzalez et al., 2000],

$$C_{DL} = \frac{C_s C_d}{C_s + C_d} \tag{9.6}$$

Although experimentally the potential drop across the diffuse double layer can exceed 0.025 V, the linear analysis still gives useful information on the flow. The nonlinearity will become more important at higher applied potential as suggested in [Bazant et al., 2004].

From Gonzalez et al. (2000), the velocity scales as E^2, where E is the external field. At lower frequencies, the potential drop will mostly occur across the double layer, so the tangential field is small. At higher frequencies, there is not enough time for the charges to migrate into and accumulate within the double layer, so the potential drop will mostly occur in the bulk. There hence exists an optimum frequency at which the velocity has a maximum. This optimum frequency is determined by the diffusivity D, the double layer thickness λ, and the macroscopic length scale L and can be described by $\frac{D}{\lambda L}$.

However, if the applied voltage is larger than the ionization potential of the electrodes or the ion species in the electrolyte, reaction at the electrolyte–electrode interface will produce or consume ions. In the first half-cycle, the left electrode has a positive potential; that is, the electric field is from the left to the right electrode. During the anodic reaction cycle for this electrode, metal can lose electrons and eject metal ions. A possible anodic reaction is

$$M \rightarrow M^{+n} + ne^-. \tag{9.7}$$

Water electrolysis may also happen at the acidic conditions

$$3H_2O \rightarrow 2H_3O^+ + 1/2O_2 + 2e^-, \tag{9.8}$$

or for the basic condition,

$$2OH^- \rightarrow H_2O + 1/2O_2 + 2e^-. \tag{9.9}$$

The coions in the solution still move to the left electrode. However, if the reactions in Equations (9.7) and (9.8) dominate at higher potentials, as shown in Figure 9.5a, the net ions or net charges accumulated on the electrode will be positive, which is the opposite of capacitive charging in Figure 9.4a. On the right electrode, cathodic reactions occur. The reaction could be metal deposition:

$$M^{+n} + ne^- \rightarrow M. \tag{9.10}$$

Water electrolysis reaction for acidic condition could be

$$2H_3O^+ + 2e^- \rightarrow 2H_2O + H_2, \tag{9.11}$$

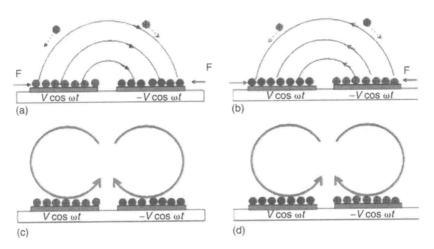

FIGURE 9.5 (See color insert following page 2-12.) Schematic illustration of the Faradaic charging: (a) and (b) on the left, anions are driven to the same electrode surface where cations are produced by a Faradaic anodic reaction during the half-cycle when the electrode potential is positive; (c) and (d) the flow directions are opposite to those in Figure 9.4.

or, for the basic conditions,

$$2H_2O + 2e^- \rightarrow 2OH^- + H_2. \tag{9.12}$$

Like the left electrode, cations still move to this electrode such that positive charges accumulate on the surface. Only reaction (9.12) produces negative ions. If this reaction dominates at higher potentials, the net accumulated charges on this electrode will be negative, as shown in Figure 9.5b. In Figure 9.4a and b and Figure 9.5a and b, the directions of the electric field are the same, so they will produce opposite flow fields, as shown in Figure 9.4c and Figure 9.5c and d. Moreover, positive ions produced in the anodic reaction cycle will increase the local potential, and negative ions produced in the cathodic reaction cycle will reduce it. The resultant electric field outside the double layer is increased by this effect because the charge in the double layer has the same polarization as the electrode, thus amplifying the effective electrode field. In contrast, at lower frequencies, capacitive charging of ions with opposite polarization to the electrode tends to screen the electrode field. Hence, flow due to such Faradaic reaction charging exists at any frequency that is less than the inverse reaction time.

The flow does not necessarily reverse because of the reaction. For example, if the reaction in Equation (9.10) dominates at the left electrode during its anodic reaction cycle, the flow will be in the same direction as that for capacitive charging. In the same sense, if the reaction in Equations (9.10) or (9.11) dominates in its cathodic cycle, the flow will not reverse either. For a detailed description, see [Ben et al., 2004, Ben, 2004].

A simple zeroth-order reaction model is discussed below. This model assumes that ion production dominates ion consumption at the same electrodes, which produces a specific polarization and flow direction that is not shared by a general Faradaic charging mechanism. Nevertheless, it is a basic model that captures a large class of Faradaic reactions. A charge balance without considering the detail structure of the double layer can be written as follows,

$$\frac{\varepsilon}{\lambda} \frac{\partial}{\partial t} (\phi - V \cos \omega t) \mp zRF \cos \omega t = \sigma \frac{\partial \phi}{\partial n} \tag{9.13}$$

where z is the valence and R is the reaction constant for a constant reaction that is independent of concentration and electrode potential, except that it has an opposite sign on the two electrodes. The $-$ sign applies to the right electrode at $a < x < L + a$, and the $+$ sign to the left electrode at $-L - a < x < -a$. To simplify the problem, we have assumed that the reaction constants are equal for anodic and cathodic reactions. The first term in Equation (9.13) represents the charge accumulation as that in Equation (9.5); the third term represents the Ohmic current; and the second term, the reaction. With Faradaic reaction, the double layer behaves as a capacitor and an ion source/sink in parallel with a resistor in series. Upon Fourier transform in time $\left(\frac{\partial}{\partial t} \rightarrow i\omega \right)$, Equation (9.13) becomes an effective boundary condition involving complex coefficients,

$$i\omega(\phi - V) \mp RF = \sigma \frac{\partial \phi}{\partial n}. \tag{9.14}$$

We shall use the simple model Equation (9.14) in some of our global flow field calculations. It replaces the purely capacitive charging model of Equations (9.4) and (9.5). As such, the potential ϕ becomes a complex potential.

We return to the two parallel, symmetric, and infinitely long electrodes on a nonconducting substrate, as shown schematically in Figure 9.4. We designate the separation between the two electrodes as $2a$, the width of each electrode as L.

In the electroneutral Ohmic bulk region outside the double layer, the electric potential ϕ satisfies the Laplace equation

$$\nabla^2 \phi = 0 \tag{9.15}$$

With a thin polarized layer approximation, the Maxwell stress in the polarized layer has been shown [Gonzalez et al., 2000] to produce a time-average slip velocity on the wall,

$$\langle u \rangle = -\frac{\varepsilon}{4\mu} \nabla_s (|\Delta V|)^2 = -\frac{\varepsilon}{4\mu} \nabla_s |\phi \mp V|^2 \tag{9.16}$$

where ε is the dielectric constant, μ is the viscosity, ∇V is the potential drop across the polarization layer of each electrode (corresponding to the different signs), and ∇_s denotes a surface gradient and the absolute sign is taken of the complex variables within. As the electrode potential amplitude V is constant, the stagnation points correspond to regions where the tangential gradient of the outer ohmic potential (i.e., the outer tangential field) $\nabla_s \phi$ vanishes.

Equation (9.16) is coupled to the bulk potential in Equation (9.15) and represents the effective slip condition for the bulk creeping flow equation $\mu \nabla^2 u = \nabla p$. Combined with Equation (9.14), the effective boundary condition for electric potential ϕ, the bulk problem is closed.

If Faradaic charging dominates there is no charge accumulation, and the Faradaic charge generation is balanced by electromigration away from the electrode to produce a constant current density boundary condition

$$\frac{\partial \phi}{\partial y} = \pm \frac{RF}{\sigma} \tag{9.17}$$

The Faradaic polarization and current are hence in phase and both persist at low frequencies.

If the Faradaic reaction is weak at low voltages, the current $\sigma \frac{\partial \phi}{\partial y}$ changes direction such that it now charges the electrode. A low-frequency expansion of electric potential $\phi = \phi_0 + \omega \phi_1 + \omega^2 \phi_2 \ldots$, after a Fourier transform then yields

$$\sigma \frac{\partial \phi_1}{\partial y} = \mp iCV, \quad \sigma \frac{\partial \phi_{n+1}}{\partial y} = \mp iC\phi_n \quad (n > 0) \tag{9.18}$$

where the first nontrivial capacitive current $\sigma \frac{\partial \phi_1}{\partial y}$ is out-of-phase with the voltage and all the other ϕ_{n+1} have a 90° lag to the previous ϕ_n. Potential on the nonconducting surfaces without a metal cover for $|x| < a$ and $|x| > L + a$ satisfies the insulation condition

$$\frac{\partial \phi}{\partial y} = 0. \tag{9.19}$$

Hence, at every order, the normal field $\frac{\partial \phi_n}{\partial y} = 0$ is specified at the boundary, and a Neumann problem results.

The solution to the bulk Laplace Equation (9.15) with Neumann conditions Equations (9.17)–(9.19) can be conveniently represented by the double layer Green's integral formula:

$$\phi = \int_{-L-a}^{L+a} \frac{\partial \phi}{\partial y}(x_0) G(x_0, x, y) dx_0 \tag{9.20}$$

where the Green's function due to a unit line source is

$$G = \frac{1}{2\pi} \ln((x - x_0)^2 + y^2). \tag{9.21}$$

For Faradaic charging, the solution is

$$\phi = \frac{RF}{2\pi\sigma} \left\{ (L + a - x)\log\left[(L + a - x)^2 + y^2\right] + 2y\arctan\frac{L + a - x}{y} - (a - x)\log\left[(a - x)^2 + y^2\right] \right.$$

$$- 2y\arctan\frac{a - x}{y} + (a + x)\log\left[(a + x)^2 + y^2\right] + 2y\arctan\frac{a + x}{y}$$

$$\left. - (L + a + x)\log\left[(L + a + x)^2 + y^2\right] - 2y\arctan\frac{L + a + x}{y} \right\} \tag{9.22}$$

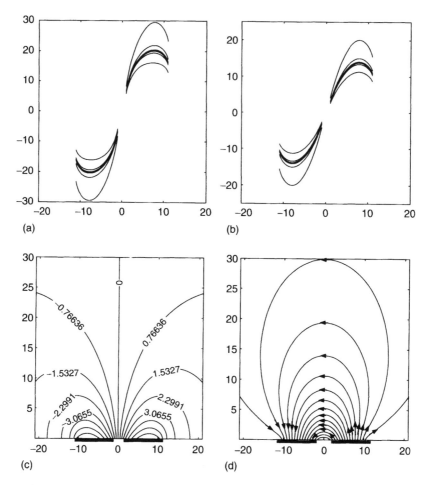

FIGURE 9.6 (a) and (b): Oscillatory convergence of imaginary part and real part of ϕ for the low-frequency expansion ansatz of capacitive charging at the frequency $\omega = 0.12$ and at an applied voltage of $V = 40$. Here, ω and V are dimensionless variables. The characteristic frequency is $D/\lambda L$ and the characteristic potential is $RT/F = 25\,mV$. The thicker lines are the converged ϕ_i and ϕ_r respectively. The potential and field lines of Faradaic charging are shown in (c) and (d) where the stripes represent the electrodes. The dimensionless value of a is 1, and that of L is 10. The electrode tangential field vanishes at ± 8.

The resulting equal potential lines are plotted in Figure 9.6c. As expected from two isolated collinear line sources with uniform intensity, the electric potential is antisymmetric, and there are two potential extrema on the electrode surface with the same value but with different signs. This means that the surface electric field is zero and changes directions at the extrema as shown in Figure 9.6d.

By letting the tangential field $\frac{d\phi}{dx}$ at $y \to 0$ equal zero, we obtain the locations of the extrema and, from (9.22), the stagnation points:

$$x_{stag} = \pm \frac{\sqrt{2}}{2} \sqrt{(L + a)^2 + a^2}. \tag{9.23}$$

When $L \gg a$, $x_{stag} = \pm L/\sqrt{2}$. Equation (9.16) shows that slip velocity is proportional to $\frac{d\phi}{dx}$ and the surface flow on the electrode is inward between the two stagnation points and outward outside the stagnation points. Two pairs of large vortices, each with size $\frac{L}{\sqrt{2}}$ and $\left(1 - \frac{1}{\sqrt{2}}\right)L$, will then be driven by this surface flow above the electrodes.

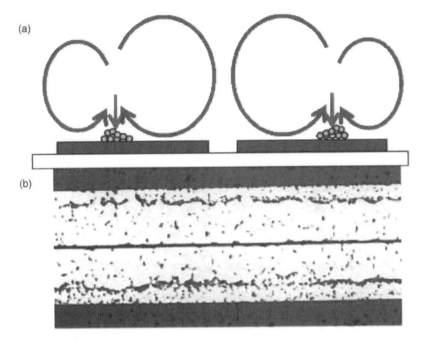

FIGURE 9.7 (See color insert following page 2-12.) Particle focusing lines along the stagnation points for capacitive charging. The vertical force toward the electrode is a weak DEP or gravitational force. The circulation is opposite for Faradaic charging. An actual image of the assembled particles is shown below.

For capacitive charging, a series of Neumann problems have to be solved for ϕ_n in Equation (9.18). The leading-order solution ϕ_1 corresponds to the low-frequency limit when the double layer is fully polarized and the external field is screened. Therefore, the total potential drop $2V$ is entirely across the two symmetric double layers and the resulting field drives the current that charges the double layer, that is $\frac{\partial \phi_1}{\partial y} = \mp iCV/\sigma$. As in the constant current density condition Equation (9.17) of Faradaic charging, this current density condition for capacitive charging produces the same extrema in ϕ_1 as in Equation (9.23). However, a recursive condition Equation (9.18) for subsequent orders ϕ_n that captures finite accumulation effects shows that the normal field $\frac{\partial \phi_{n+1}}{\partial y}$ in the next order is proportional to the potential (charge) of the previous order. As a result, nonuniform charging occurs at all higher order terms, but the extrema remain invariant at Equation (9.23) for all orders. Since the stagnation line is independent of the magnitude of the specified flux in the Neumann condition, our small ω expansion shows that it exists at any frequency lower than the characteristic frequency $D/\lambda L$ for capacitive charging [Ajdari, 2002; Morgan, 2003]. For higher frequencies, the stagnation lines also could exist, although this is beyond our current theory.

The circulation intensity for capacitive charging is different, however, from Faradaic charging. Furthermore, the circulation direction is exactly opposite. Capacitive flow at the stagnation point x_{stag}, converges toward a vertical plane. Particles convected by the capacitive flow would need to move vertically upward in this plane. However, the vertical flow is weak at the stagnation point and an opposite vertical force can prevent this vertical motion away from the electrodes. The particles would then be trapped at the stagnation line, as seen in Figure 9.7. Such an opposite vertical force on the particle can be a DEP force [Pohl, 1978], which is a short-range force as described in Equation (9.3). It also can be gravitational force if the particle is not negatively buoyant and the electrode is at the bottom of the container. These weak forces on the particle are overwhelmed by AC eletroosmotic convection due to capacitive charging everywhere except at the stagnation lines.

In contrast, Faradaic flow at the stagnation point x_{stag} is a diverging flow toward the electrode surface. On the electrode surface, the tangential electroosmotic convection is perpendicular to DEP or any other

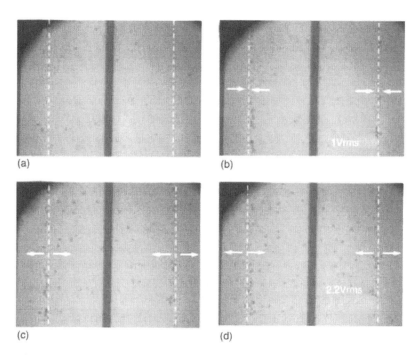

FIGURE 9.8 (See color insert following page 2-12.) The writing and erasure processes for Au electrodes at $\omega = 100\,Hz$. The frames are taken at 0 s, 5 s, 10 s, and 15 s after the field is turned on. The initial voltage is 1.0 Vrms and is increased to 2.2 Vrms at 7.0 s. Particles on the electrode in the first two frames (a) and (b) move in directions consistent with electro-osmotic flow due to capacitive charging and assemble into lines. They are erased by Faradaic charging in the next two frames (c) and (d). The arrows demonstrate the direction of particle motion. The dashed lines are located at the theoretical $L/\sqrt{2}$.

weak vertical force. Hence, particle convection cannot be eliminated by these other forces. Therefore, one does not expect any trapping with Faradaic flow. In fact, any particle assembly at x_{stag} is expected to be swept away by the Faradaic flow in a reverse erasure mechanism.

To verify this prediction, 100 nm Al and Ti/Au electrodes are fabricated using photolithography and liftoff. The separation between the two electrodes is 20 µm and the electrode width L is 300 µm. An AC electric potential at 100 Hz was applied between two electrodes. Latex particles 5 microns in diameter and of concentration of about 10^6 particles per cc are suspended in DI water and are allowed to settle by gravity onto the electrodes. For all electrodes, linear self-assembly is observed at low voltages and, although the demarcation voltage varies with electrode material. Erasure is not observed until $V = 3.6$ Vrms for Al but occurs at $V = 2.2$ Vrms for Au. In Figure 9.8, the formation and stage-wise erasing of the lines are imaged for Au electrodes. Before their collection onto the electrodes, the particles are also observed to be convected by the vortices with circulations consistent with capacitive and Faradaic charging during the writing and erasing conditions respectively.

Concentrating and segregating microsized particles in electrolytes are extremely slow and difficult. Natural or imposed attractive forces (electrostatic, gravitational, electrophoretic, dielectrophoretic, etc.) are very weak and short range due to Debye screening and high-power (quadratic or cubic) scaling with respect to particle size. Yet the vortices generated by our AC electrokinetic mechanisms have normal dimensions as large as the electrode width, which can be increased almost indefinitely. Their linear velocity also can reach as high as 1 mm/s. The use of such a long-range and strong hydrodynamic force to focus particles first to a plane and then to a line (3D to 1D projection) offers a very attractive mechanism to rapidly trap and concentrate particles in dilute suspensions. With AC capacitive charging, Wu et al. (2004) can concentrate bacteria of 10^4 cells per cc. Two concentrated bacteria lines are shown in Figure 9.9.

FIGURE 9.9 (See color insert following page 2-12.) Bacteria trapping by AC electroosmotic flow.

FIGURE 9.10 Fabricated planar T electrode arrays made of Aluminum in 1 μm thickness for AC micropump. The electrodes are 100 microns wide; the length of each element within the array is 1 mm; the gap between T elements is 300 μm; and the entire array is 2 cm in length. Every other T element has the same polarity.

We have shown that AC electroosmotic flow tends to generate closed vortices. These closed vortices are good for micromixers but do not produce the net flow required for pumps. Ajdari (2000) and Ramos et al. (1999) have proposed a solution by using nonsymmetric electrodes and lowering the upper channel wall to suppress the back flow of the larger vortex on the larger electrode. However, AC electroosmotic due to capacitive charging only exist in a narrow frequency window. As shown before, reactions will not screen the electric field at the lower frequency and reverse the flow directions. With the reaction considered, several designs can be used to break the symmetry to obtain a net flow:

1. An asymmetric electrode geometry can be exploited.
2. The applied potential can be programmed asymmetrically such as by imposing a traveling wave electric field.
3. A combination of breaking both the geometric symmetry and the electric field symmetry can be used.

An optimum design has yet to be proposed. Lastochkin et al. (2004) reported a new AC pump design based on AC Faradaic reactions on the electrodes by breaking the geometry symmetry with series of orthogonal electrode elements.

The T electrode can be regarded as the most asymmetric design and produces net flow down the tip without a top wall, as shown in Figure 9.10. The transverse flows along the horizontal arm cancel each other while the longitudinal flow down the tip is unabated above the planar electrode. In the experiments, a voltage larger than 10 volt is used. The reaction constant should be an exponential function of $\pm V \cos \omega t - \phi$, the potential drop across the polarization layer. To the leading order, ϕ is assumed to be constant along the electrode (Ben, 2004). Hence, the reaction constant is an exponential function of the applied voltage. Although a stagnation point exists on the orthogonal electrode, it is expected that little back flow would develop because the flow is three dimensional.

The pump with 16 T-shaped aluminum electrodes, each measuring 100 microns in width, has been fabricated on glass with standard lithography techniques in the clean room as shown in Figure 9.10 and successfully tested. Every other T electrode has the same polarity. Frequencies between 200 KHz and

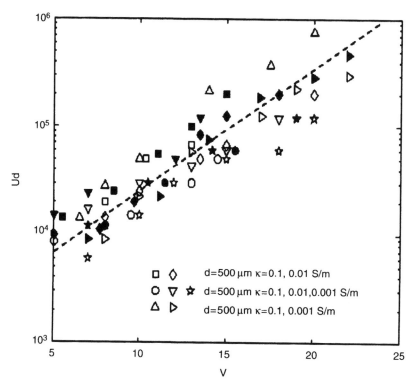

FIGURE 9.11 Experimental measured velocities at different frequencies and different voltages before bubbles occur.

1 MHz were used in experiments, and the electrolyte was NaCl. The fluid was flowing continually from right to the left. The velocities are plotted in a semi-log scale as a function of velocity at the deferent frequencies in Figure 9.11. The results indicate an exponential dependence on the voltage because the reaction constant is an exponential function of the applied potential as expected. This is larger than the usual quadratic scaling expected from non-Faradaic AC electroosmosis like AC capacitive charging. It is, in fact, consistent with the Arhenius exponential dependence of redox or dissolution reactions on voltage. Due to this exponential scaling, velocities exceeding 1 mm/s can be achieved with the T-electrode micropump. The operating conditions are such that bubbles are not observed and conductivity measurement after operating the pump for hours shows little net ion generation even though there is obviously ion production at the electrode during each half-cycle. Electrolytes with different conductivities are chosen. The velocity reaches a constant asymptote or blows up as the frequency decreases — in contrast to the capacitive charging mechanism, which would produce a negligible velocity at low frequency due to external field screening. This is consistent with the nonscreening Faradaic charging mechanism, as a longer period increases the Faradaic polarization without screening the external field. Lastochkin et al. (2004) observe the AC electroosmotic velocity to increase as conductivity increases from 0.001 S/m to 0.01 S/m. This feature is again different from linear electrokinetics and AC capacitive charging mechanisms where a higher conductivity usually means higher electrolyte concentration and a thinner double layer. One explanation is that the reaction kinetic constant depends on the potential drop across the Stern layer. The higher the conductivity of the solution, the lower the potential drop in the electroneutral region. As a result, the potential drop across the Stern layer increases and the reaction constant increases such that more charges are produced by Faradaic charging. Finally, Lastochin et al. are able to increase their rms voltage to 20 V, corresponding to a large field in excess of 300 V/cm for their electrode separation of less than 100 microns, without producing bubbles. In contrast, bubbles would appear at a DC electrode beyond 1 or 2 volts. Linear electrokinetic pumps even with their high voltages rarely employ a field strength that exceeds

100 V/cm. Fields higher than 100 V/cm often denaturalize proteins and produce significant Joule heating. Nevertheless, the AC Faradaic micropumps of Lastochkin et al. show that nonlinear electrokinetics by the AC Faradaic mechanism can allow high fields to achieve high flow and velocity for high-conductivity biofluids without generating bubbles or ions — all clear advantages over linear electrokinetic pumps and AC pumps without Faradaic reaction.

9.4 Summary

We have reviewed the general features, advantages, and shortcomings of electrokinetic devices for microfluidic applications. Two specific designs based on AC Faradaic charging are discussed in detail. One captures, concentrates, and disperses bioparticles, and the other pumps fluids at high speed without bubble generation and ion contamination. It is expected that many more such nonlinear electrokinetic devices will be proposed in the near future.

Acknowledgments

Our electrokinetics research at the Center for Micro-Fluidics and Medical Diagnostics at the University of Notre Dame has benefited from an excellent group of Ph.D. and postdoctoral students. We would like to acknowledge those who have contributed to the work reported here (with their current affiliation if they have left Notre Dame) in chronological order: P. Takhistov (Rutgers), A. Indeikina, A. Minerick (Missippi State), S. Thamida (Unilever), J. Wu (Tennessee), S.-C. Wang (Chong Cheng), D. Lastochkin, R. Zhou, and P. Wang.

References

Ajdari, A. (1995) "Electro-Osmosis on Inhomogeneously Charged Surfaces," *Phys. Rev. Lett.*, 75, p. 755.

Ajdari, A. (2000) "Pumping Liquids Using Asymmetric Electrode Arrays," *Phys. Rev. E*, 61, p. R45.

Ajdari, A. (2002) "Electrokinetic 'Ratchet' Pumps for Microfluidics," *Appl. Phys. A*, 75, p. 271.

Aris, R. (1956) "On the Dispersion of a Solute in a Fluid Flowing Through a Tube," *Proc. Roy. Soc. London A*, 235, p. 67.

Barany, S. Mishchuk, N.A., and Prieve, D.C. (1998) "Superface Electrophoresis of Conducting Dispersed Particles," *J. Colloid. Interf. Sci.*, 207, p. 240.

Bazant, M.Z., and Squires, T.M. (2004) "Induced-Charge Electrokinetic Phenomena: Theory and Microfluidic Applications," *Phys. Rev. Lett.*, 92, p. 066101.

Bazant, M.Z., Thornton, K., and Ajdari, A. (2004) "Diffuse-Charge Dynamics in Electrochemical Systems," preprint.

Ben, Y. (2004) Nonlinear Electrokinetic Phenomena in Microfluidic Devices, Ph.D. dissertation, University of Notre Dame.

Ben, Y., and Chang, H.-C. (2002) "Nonlinear Smoluchowski Slip Velocity and Micro-Vortex Generation," *J. Fluid Mech.*, 461, p. 229.

Ben, Y., Wu, J., and Chang, H.-C. (2004) "Linear Particle Assembly and Erasure by AC Electroosmotic Flow," in preparation.

Brown, A.B.D., Smith, C.G., and Rennie, A.R. (2000) "Pumping of Water with AC Electric Fields Applied to Asymmetric Pairs of Microelectrodes," *Phys. Rev. E*, 63, p. 016305.

Chang, D.C. (1989) "Cell Fusion and Cell Poration by Pulsed Radio-Frequency Electric Fields," in *Electroporation and Electrofusion in Cell Biology*, Neumann, E., Sowers, A.E., and Jordan, C.A., eds., Plenum Press, New York.

Chang, H.-C. (2001) "Bubble/Drop Transport in Microchannels," in *The MEMS Handbook*, 1st ed., p. 11–1, CRC Press, Boca Raton.

Cheng, J., and Kricka, L.J. (2001) *Biochip Technology*, Harwood Academic Publishers.

Cui, L., and Morgan, H. (2000) "Design and Fabrication of Travelling Wave Dielectrophoresis Structure," *J. Micromech. Microeng.*, 10, p. 72.

Delgado, A.V. (2002) *Interfacial Electrokinetics and Electrophoresis*, Marcel Dekker, Inc, New York.

Dukhin, S.S. (1991) "Electrokinetic Phenomena of the Second Kind and Their Application," *Adv. Colloid. Interf. Sci.*, 35, p. 173.

Dutta, D., and Leighton, D.T. (2001) "Dispersion Reduction in Pressure Driven Flow through Microetched Channels," *Anal. Chem.*, 73, p. 504.

Gascoyne, P.R.C., and Vykoukal, J. (2002) "Particle Separation by Dielectrophoresis," *Eletrophoresis*, 23, p. 1973.

Gimsa, J., and Wachner, D. (1998) "A Unified Resistor-Capacitor Model for Impedance, Dielectrophoresis, Electrorotation, and Induced Transmembrane Potential," *Biophys. J.*, 75, p. 1107.

Gimsa, J. (2001) "A Comprehensive Approach to Electro-Orientation, Electrodeformation, Dielectrophoresis, and Electrorotation of Ellipsoidal Particles and Biological Cells," *Bioelectrochemistry*, 54, p. 23.

Gonzalez, A., Ramos, A., Green, N.G., Castellanos, A., and Morgan, H. (2000), "Fluid Flow Induced by Nonuniform AC Electric Fields in Electrolytes on Microelectrodes: 2. A Linear Double-Layer Analysis," *Phys. Rev. E*, 61, p. 4019.

Green, N.G., Ramos, A., Gonzalez, A., Morgan, H., and Castellanos, A. (2000) "Fluid Flow Induced by Nonuniform AC Electric Fields in Electrolytes on Microelectrodes: 1. Experimental Measurements," *Phys. Rev. E*, 61, p. 4011.

Green, N.G., Ramos, A., Gonzalez, A., Morgan, H., and Castellanos, A. (2002) "Fluid Flow Induced by Nonuniform AC Electric Fields in Electrolytes on Microelectrodes: 3. Observation of Streamlines and Numerical Simulation," 66, p. 026305-1.

Herr, A.E., Molho, J.I., Santiago, J.G., Mungal, M.G., Kenny, T.W., and Garguilo, M.G. (2000) "Electroosmotic Capillary Flow with Nonuniform Zeta Potential," *Anal. Chem.*, 72, p. 1052.

Hu, Y., Glass, J.L., and Griffith, A.E. (1994) "Observation and Simulation of Electrohydrodynamic Instabilities in Aqueous Colloidal Suspensions," *J. Chem. Phys.*, 100, p. 4674.

Hughes, M.P. (2002) "Strategies for Dielectrophoretic Separation in Laboratory-on-a-chip Systems," *Electrophoresis*, 23, p. 2569.

Hughes, M.P. (2003) *Nanoelectromechanics in Engineering and Biology*, CRC Press, Boca Raton.

Jones, T.B. (1995) *Electromechanics of Particles*, Cambridge University Press, New York.

Jones, T.B., Wang, K.-L., and Yao, D.-J. (2004) "Frequency-Dependent Electromechanics of Aqueous Liquids: Electrowetting and Dielectrophoresis," *Langmuir*, 20, p. 2813.

Koch, M., Evans A., and Brunnschweiler, A. (2000) *Microfluidic Technology and Applications*, Research Studies Press LTD, Philadelphia.

Laser, D.J., and Santiago, J.G. (2004) "A Review of Micropumps," *J. Micromech. Microeng.*, 14, p. R35.

Lastochkin, D., Zhou, R., Wang, P., Ben, Y., and Chang, H.-C. (2004) "Electrokinetic Micropump and Micro-Mixer Design Based on AC Faradaic Polarization," accepted, *J. Appl.Phys. D*.

Minerick, A.R., Ostafin, A.E, and Chang, H.-C. (2002) "Electrokinetic Transport of Red Blood Cells in Microcapillaries," *Electrophoresis*, 23, p. 2165.

Minerick, A.R., Zhou, R., Takhistov, P., and Chang, H.-C. (2003) "Manipulation and Characterization of Red Blood Cells with AC Fields in Micro-Devices," *Electrophoresis*, 24, p. 3703.

Mishchuk, N.A., and Takhistov, P.V. (1995) "Electroosmosis of the Second Kind," *Colloid. Surf. A*, 95, p. 119.

Morgan, H., and Green, N.G. (2003) *AC Electrokinetics: Colloids and Nanoparticles*, Research Studies Press, Ltd, Philadelphia.

Morgan, H., Izquierdo, A.G., Bakewell, D., Green, N.G., and Ramos, A. (2001) "The Dielectrophoretic and Travelling Wave Forces Generated by Interdigitated Electrode Arrays: Analytical Solution Using Fourier Series," *J. Phys. D: Appl. Phys.*, 34, p. 1553

Morrison, F.A., Jr. (1970) "Electrophoresis of a Particle of Arbitrary Shape," *J. Colloid. Interf. Sci.*, 34, 210.

Mpholo, M., Smith, C.G., and Brown, A.B.D. (2003) "Low Voltage Plug Flow Pumping Using Anisotropic Electrode Arrays," *Sensors Actuators B*, 92, p. 262.

Murtsovkin, V.A. (1996) "Nonlinear Flows Near Polarized Disperse Particles," *Colloid J.*, 58, p. 341.

Nadal, F., Argoul, F., Hanusse, P., Pouligny, B., and Ajdari, A. (2002) "Electrically Induced Interactions between Colloidal Particles in the Vicinity of a Conducting Plane," *Phys. Rev. E*, 64, p. 061409.

Nguyen, N.T., and Wereley, S.T. (2002) *Fundamentals and Applications of Microfluidics*, Artech House Publisher, Boston.

Pethig, R. (1996) "Dielectrophoresis: Using Inhomogeneous AC Electrical Fields to Separate and Manipulate Cells," *Crit. Rev. Biotechnol.*, 16, p. 331.

Pohl, H.A. (1978) *Dielectrophoresis*, Cambridge University Press, Cambridge.

Probstein, R.F. (1994) *Physicochemical Hydrodynamics*, Wiley-Interscience, New York.

Ramos, A., Gonzalez, A., Castellanos, A., Green, N.G., and Morgan, H. (2003) "Pumping of Liquids with AC Voltages Applied to Asymmetric Pairs of Microelectrodes," *Phys. Rev. E*, **67**, p. 056302-1.

Ramos, A., Morgan, H., Green, N.G., and Castellanos, A. (1999) "AC Electric-Field-Induced Fluid Flow in Microelectrodes, *J. Colloid. Interf. Sci.*, **217**, p. 420.

Reyes, D.R., Iossifidis, D., Auroux, P.-A., and Manz, A. (2002) "Micro Total Analysis Systems. 1. Introduction, Theory, and Technology," *Anal. Chem.*, **74**, 2623.

Rhodes, P.H., and Snyder, R.S. (1994) "Theoretical and Experimental Studies on the Stabilization of Hydrodynamic Flow in Free Fluid Electrophoresis," in *Cell Electrophoresis*, by J. Bauer, p. 75, Boca Raton.

Russel, W.B., Saville, D.A., and Schowalter, W.R. (1989) *Colloidal Dispersion*, Cambridge University Press, Cambridge, New York.

Sides, P.J. (2001) "Electrohydrodynamic Particle Aggregation on an Electrode Driven by an Alternating Electric Field Normal to It," *Langmuir*, **17**, p. 5791.

Sides, P.J. (2003) "Calculation of Electrohydrodynamic Flow around a Single Particle on an Electrode," *Langmuir*, **19**, p. 2745.

Solomentsev, Y., Bohmer, M., and Anderson, J.L. (1997) "Particle Clustering and Pattern Formation during Electrophoretic Deposition: A Hydrodynamic Model," *Langmuir*, **13**, p. 6058.

Solomentsev, Y., Guelcher, S.A., Bevan, M., and Anderson, J.L. (2000) "Aggregation Dynamics for Two Particles during Electrophoretic Deposition Under Steady Fields," *Langmuir*, **16**, p. 9208.

Squires, T., and Bazant, M.Z. (2004) "Induced Charge Electro-Osmosis", *J. Fluid Mech.*, **509**, p. 217.

Studer, V., Pepin, A., Chen Y., and Ajdari, A. (2002) "Fabrication of Microfluidic Devices for AC Electrokinetic Fluid Pumping," *Microelectron. Eng.*, **61**, p. 915.

Stone, H.A., and Kim, S. (2001) "Microfluidics: Basic Issues, Applications, and Challenges," *AICHE J.*, **47**, p. 1250.

Stone, H.A., Strook, A.D., and Ajdari, A. (2004) "Engineering Flow in Small Devices: Microfluidics towards Lab-on-a-Chip," *Annu. Rev. Fluid Mech.*, **36**, p. 381.

Stroock, A.D. , Dertinger, S.K.W, Ajdari, A, Mezic, I., and Stone, H.A., Whitesides, G.M. (2002) "Chaotic Mixer for Microchannels," *Science*, **295**, p. 647.

Takhistov, P., Duginova, K., and Chang, H.-C. (2003) "Electrokinetic Mixing Vortices Due to Electrolyte Depletion at Microchannel Junctions," *J. Colloid. Interf. Sci.*, **263**, pp. 133–43.

Taylor, G.I. (1954) "The Dispersion of Matter in Turbulent Flow through a Pipe," *Proc. Roy. Soc. London A*, **223**, p. 446.

Thamida, S.K., and Chang, H.-C. (2002) "Nonlinear Electrokinetic Ejection and Entrainment Due to Polarization at Nearly Insulated Wedges," *Phys.Fluids*, **14**, p. 4315.

Trau, M., and Battersby, B.J. (2001) "Novel Colloid Materials for High-Throughput Screening Applications in Drug Discovery and Geomics," *Adv. Mater.*, **13**, p. 975.

Trau, M., Saville, D. A., and Aksay, I.A. (1996) "Field-Induced Layering of Colloidal Crystals," *Science*, **272**, p. 706.

Trau, M., Saville, D.A., and Aksay, I.A. (1997) "Assembly of Colloidal Crystals at Electrode Interfaces," *Langmuir*, **13**, p. 637.

Wang, S.-C., Lai, Y.-W., Ben, Y., and Chang, H.-C. (2004) "Microfluidic Mixing by AC and DC Nonlinear Vortex Flows," *Ind. Eng. Chem. Res.*, accepted, **43**, p. 2902.

Wang, X., Wang, X.B., and Gascoyne, P.R. (1997) "General Expressions for Dielectrophoretic Force and Electrorotational Torque Derived Using the Maxwell Stress Tensor Method," *J. Electrostat.*, **39**, p. 277.

Wang, X.B., Pethig, R., and Jones, T.B. (1992) "Relationship of Dielectrophoretic and Electrorotational Behavior Exhibited by Polarized Particles," *J. Phys. D: Appl. Phys.*, **25**, p. 905.

Wu, J., Ben, Y., Battigelli, D., and Chang, H.-C. (2005) "Long-Range AC Electro-Osmotic Trapping and Detection of Bioparticles," submitted to *Ind. Eng. Chem. Res.*, **44**, p. 2815.

Yao, S.H., Hertzog, D.E., Zeng, S.L., Mikkelsen, J.C., and Santiago, J.G. (2003) "Porous Glass Electroosmotic Pumps: Design and Experiments," *J. Colloid. Interf. Sci.*, **268**, pp. 143–53.

Yeh, S.R., Seul, M., and Shraiman, B.I. (1997) "Assembly of Ordered Colloidal Aggregates by Electric-Field-Induced Fluid Flow," *Nature*, **386**, p. 57.

Yeo, L.Y., Lastochkin, D., Wang, S.-C., and Chang, H.-C. (2004) "A New AC Electrospray Mechanism by Maxwell-Wagner Polarization and Capillary Resonance," *Phys. Rev. Lett.*, **92**, p. 133902.

Zeng, S., Chen, C.-H., Mikkelsen, J.C., and Santiago, J.G. (2001) "Fabrication and Characterization of Electroosmotic Micropumps," *Sensors Actuators B*, **79**, p. 107.

Zimmermann, U., and Neil, G.A. (1996) *Electromanipulation of Cells*, CRC Press, Boca Raton.

10

Microdroplet Generators

Fan-Gang Tseng
National Tsing Hua University

10.1 Introduction

Microdroplet generators are becoming an important research area in MEMS (microelectromechanical-systems), not only because of their historically valuable marketing device, the ink-jet printhead, but also because of many other emerging applications in precise- or micro-amount fluidic control. The history of microdroplet generators began with the idea's inception by Sweet (1964, 1971) using piezo actuation. In the late 1970s HP and the Cannon Corporation [Nielsen et al., 1985] began using thermal bubble actuation. Since then, numerous research activities regarding ink-jet applications have been conducted. Recently, with the emerging applications in biomedicine, fuel injection, chemistry, pharmaceuticals, electronic fabrication, microoptical devices, IC cooling, and solid free form, much research has focused on microdroplet generators. Thus, many new operation principles, designs, fabrication processes, and materials related to microdroplet generation were explored and developed during the last decade.

In this chapter, microdroplet generators are defined as devices that generate microsized droplets in a controllable manner; that is, droplet size and number can be controlled and counted accurately. Thus, atomizers, traditional fuel injectors, or similar droplet generation devices that do not meet those criteria are not discussed.

Microdroplet generators usually employ mechanical actuation to generate high pressure that overcomes liquid surface tension and viscous force and permits droplet ejection. Depending on the droplet size, the applied pressure is usually higher than several atmospheres. The operation principles, structure/process designs, and materials often play key roles in the performance of droplet generators.

In addition to the well-known ink-jet printing, applications for microdroplet generators cover a wide spectrum in various fields such as direct writing, fuel injection, solid free form, solar cell fabrication, LEPD fabrication, packaging, microoptical components, particle sorting, microdosage, plasma spraying, drug screening/delivery/dosage, micropropulsion, integrated circuit cooling, and chemical deposition. Many of these applications may become key technologies for integrated microsystems in the near future.

This chapter provides an overview of the operation principles, physical properties, design issues, fabrication processes and issues, characterization methods, and applications of microdroplet generators.

10.2 Operation Principles of Microdroplet Generators

There have been many attempts to generate controllable micro droplets [Myers et al., 1984; Buehner et al., 1977; Twardeck et al., 1977; Carmichael et al., 1977; Ashley et al., 1977; Bugdayci et al., 1983; Darling et al., 1984; Lee et al., 1984; Nielsen et al., 1985; Bhaskar et al., 1985; Allen et al., 1985; Krause et al., 1995; Chen et al., 1995; Tseng et al., 1996; Hirata et al., 1996; Zhu et al., 1996]. Most of these have followed the principle of creating pressure differences by either lowering the outer pressure of a nozzle or increasing the inner pressure to push or pull liquid out of the nozzle to form droplets. Typical techniques are pneumatic, piezoelectric, thermal bubble, thermal buckling, focused acoustic wave, and electrostatic actuations. The basic principles of those droplet generators are introduced in the following sections. Also included in the last section is an acceleration-based ejection method that employs inertial force for droplet generation.

10.2.1 Pneumatic Actuation

Spray nozzles are presently one of the most commonly used devices for generating droplets, as in airbrushes or sprayers. Two types of spray nozzles are shown in Figure 10.1. Figure 10.1a shows an airbrush that lowers

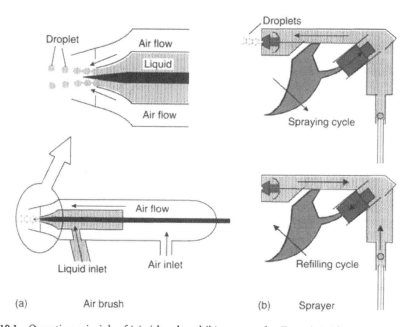

FIGURE 10.1   Operation principle of (a) airbrush and (b) sprayer, after Tseng (1998d).

pressure at the outer area of the capillary tube by blowing air across the tube's end; this forces the liquid to move out of the tube and form droplets. The second device, a sprayer shown in Figure 10.1b, employs high pressure to push liquid through a small nozzle to form droplets. Typical sizes of the droplets generated by spray nozzles range from around tens to hundreds of μm in diameter. These devices can be fabricated in micro sizes by micromachining technology. However, controlling individual nozzles in an array format is difficult.

10.2.2 Piezoelectric Actuation

Droplet ejection by piezoelectric actuation was invented by Sweet in 1964. Two types of droplet ejection devices are based on piezoelectric technology. One is the continuous ink-jet [Buehner et al., 1977; Twardeck et al., 1977; Carmichael et al., 1977; Ashley et al., 1977]. Figure 10.2a schematically shows the operational principle of this type of ink-jet. Conductive ink is forced out of the nozzles by pressure. The jet would break up continuously into droplets with random sizes and spacing, but uniform size and spacing of the droplets is controlled by applying an ultrasonic wave at a fixed frequency to the ink through a piezoelectric transducer. The continuously generated droplets pass through a charge plate, and only the desired droplets are charged by the electric field and deflected to print out, while the undesired droplets are collected by a gutter and recycled. One piezoelectric transducer can support multiple nozzles, so the

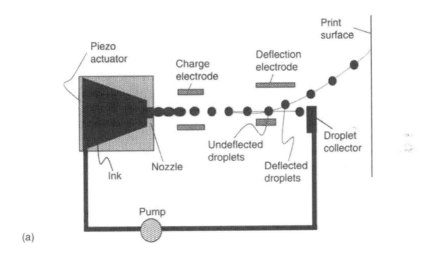

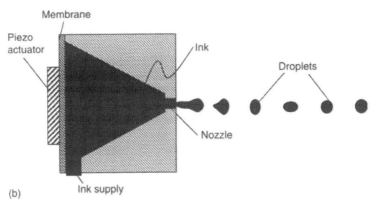

FIGURE 10.2 (a) Operation principle of a piezoelectric-actuated continuous droplet generator, after Buehner (1977). (b) Operation principle of a piezoelectric-actuated droplet-on-demand droplet generator, after Lee (1984).

nozzle spacing can be as small as desired for high-resolution arrays. The complexity of this device's droplet charging and collecting system is the major obstacle to its practical application.

The other device, the droplet-on-demand ink-jet, utilizes a piezoelectric tube or disc for droplet ejection only when printing a spot is desired [Bugdayci et al., 1983; Darling et al., 1984; Lee et al., 1984]. Figure 10.2(b) shows a typical drop-on-demand drop generator. The operational principle is based on the generation of an acoustic wave in a fluid-filled chamber by a piezoelectric transducer through the application of a voltage pulse. The acoustic wave interacts with the free meniscus surface at the nozzle to eject a single drop. The major advantage of the drop-on-demand method is that it does not require a complex system for droplet deflection. Its main drawback is that the size of piezoelectric transducer tube or disc, in the order of sub mm to several mm, is too large for high-resolution applications. It was reported that the typical frequency for a stable operation of piezoelectric ink-jet would be tens of kHz [Chen et al., 1999].

10.2.3 Thermal-Bubble Actuation

Thermal bubble jet technology has been studied by HP in the United States and by CANON in Japan since the early 1980s [Nielsen et al., 1985; Bhaskar et al., 1985; Allen et al., 1985]. Many other designs also are reported in the literature [for example, Krause et al., 1995; Chen et al., 1995; Tseng et al., 1996, 1998]. Figure 10.3 shows the cross-section of a thermal bubble jet device. Liquid in the chamber is heated by applying a pulse current to the heater under the chamber. The temperature of the liquid covering the surface of the heater rises to around the liquid critical point in microseconds, and then a bubble grows on the surface of the heater, which serves as a pump. The bubble pump pushes liquid out of the nozzle to form a droplet. After the droplet is ejected, the heating pulse is turned off and the bubble starts to collapse. Liquid refills the chamber when surface tension on the free surface of the meniscus returns it to the original position. A second pulse generates another droplet. The energy consumption for ejecting each

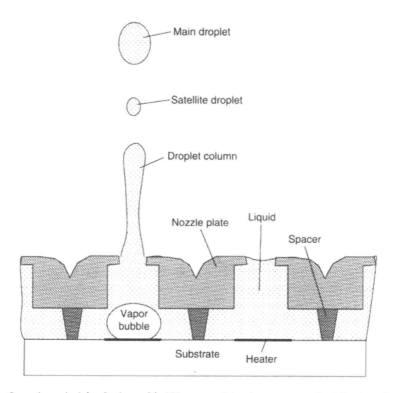

FIGURE 10.3   Operation principle of a thermal-bubble-actuated droplet generator, after Allen (1985).

droplet is around 0.04 mJ for an HP ThinkJet printhead. Because bubbles can deform freely, the chamber size of the thermal bubble jet is smaller than that of other actuation means, which is important for high-resolution applications. The resolution reported in the literature ranges from 150 to 600 dpi [Krause et al., 1995] and 1016 dpi [Chen et al., 1995]. The typical operational frequency for contemporary thermal bubble jets is around several to tens of kHz.

10.2.4 Thermal-Buckling Actuation

Hirata et al. (1996) employed a buckling diaphragm for droplet generation. Figure 10.4 schematically shows the technique's basic operational principle. A composite circular membrane consisting of a silicon dioxide and nickel layer is fixed on the border and remains separated from the substrate by a small gap. A heater is placed at the center of the composite membrane and electrically isolated from it. Pulsed current is sent to the heater, and then the membrane is heated for several μs. When the thermally induced stress is greater than the critical stress, the diaphragm buckles abruptly and ejects a droplet out of the nozzle. The power required to generate a droplet at a speed of 10 m/s is around 0.1 mJ using a 300 μm diameter diaphragm. The power consumption and the size of the buckling membrane ink-jet device are much larger than those of the thermal bubble jet. The reported frequency response of a membrane buckling jet ranges from 1.8 to 5 kHz depending on the desired droplet velocity.

10.2.5 Acoustic-Wave Actuation

Figure 10.5 shows a lensless liquid ejector using constructive interference from acoustic waves to generate droplets [Zhu et al., 1996]. A PZT thin-film actuator with the help of an on-chip Fresnel lens was employed to generate and focus acoustic waves on the air–liquid interface for droplet formation. The actuation results from the excitation of a piezoelectric film under a burst of RF signal. The device does not need a nozzle to define droplets, reducing the clogging problems that trouble most droplet generators employing nozzles. Droplet size also can be controlled by using acoustic waves with specific frequencies. However, due to the acoustic waves' vigorous agitation of the liquid, it is difficult to maintain a quiet interface for reliable and repeatable droplet generation. As a result, a "nozzle area" is still needed to maintain a stable interface.

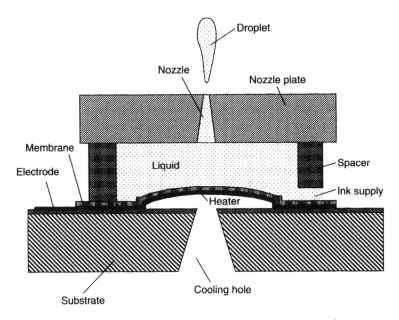

FIGURE 10.4   Operation principle of a thermal-buckling-actuated droplet generator, after Hirata (1996).

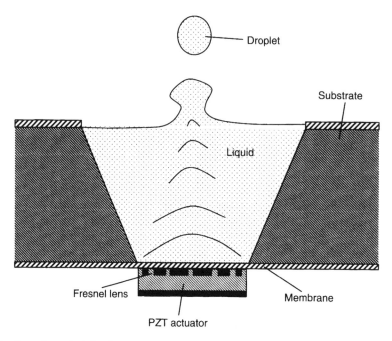

FIGURE 10.5   Operation principle of an acoustic-wave-actuated droplet generator, after Zhu (1996).

The applied RF frequency ranges from 100 to 400 MHz, and the burst period is 100 μs. The power consumption for one droplet is around 1 mJ, which is high compared to other techniques. The droplet size ranges from 20 to 100 μm depending on the RF frequency. The reported size of the device is 1×1 mm², which is much larger than the droplet generators mentioned previously.

10.2.6   Electrostatic Actuation

The electrostatically driven ink-jet printhead was first introduced by the Seiko Epson Corporation [Kamisuki et al., 1998, 2000] for commercial printing. As shown in Figure 10.6, the actuation is initiated by applying a DC voltage between an electrode plate and a pressure plate; this deflects the pressure plate for ink filling. When the voltage stops, the pressure plate reflects back and pushes the droplet out of the nozzle. This device has been developed for use in electric calculators due to its low power consumption, less than 0.525 mW/ nozzle. The driving voltage for a SEAJet is 26.5 V, and the driving frequency can be up to 18 kHz with uniform ink ejection. A 128-nozzle chip with 360 dpi pitch resolution also has been demonstrated. This device was claimed to offer high printing quality (for bar code), high speed printing, low power consumption, and long lifetime (more than 4 billion ejections) under heavy-duty usage as well as low acoustic noise. However, the fabrication comprises a complex bonding process among three different micromachined pieces. Further, making the pressure plate requires a very precise etching process to control the accuracy and uniformity of its thickness. Due to the deformation limitation of solid materials and alignment accuracy in the bonding process, the nozzle pitch may not be easily reduced any further for higher resolution applications.

10.2.7   Inertial Actuation

Inertial droplet actuators apply a high acceleration to the nozzle chip to cause droplet ejection. This type of apparatus is shown in Figure 10.7. The print module consists of large reservoirs on the top plate, which is connected to nozzles on the bottom plate. The print module is mounted on a long cantilever beam with a piezo-bimorph-actuator for acceleration generation. Generating 1 nl droplets from 100 μm diameter nozzles requires 500 μs. Twenty-four liquid droplets of different solution types were demonstrated to be ejected simultaneously from the nozzles in a 500 μm pitch. The smallest droplet claimed to be generated is

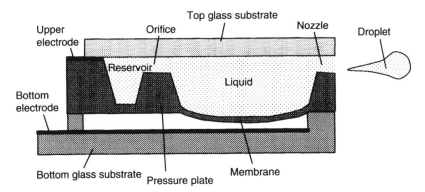

FIGURE 10.6 Operation principle of an electrostatic-actuated droplet generator, after Kamisuki (1998).

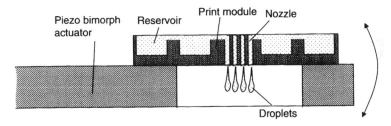

FIGURE 10.7 Operation principle of an inertia-actuated droplet generator, after Gruhler (1999).

100 pl from 50 μm diameter nozzles. This principle provides a gentle ejection process for bioreagent applications. However, the ejection of smaller droplets may encounter strong surface tension and flow-drag forces in micro scale that are much larger than the droplet inertial. That the droplets cannot be selectively and individually ejected from the desired nozzles also limits this technique's applications.

10.3   Physical and Design Issues

Droplet generation involves wide varieties of physical issues and design concerns including micro fluid flow, heat transfer, wave propagation, surface properties, material properties, and structure strength. The following sections discuss frequency response, thermal cross talk, hydraulic cross talk, overfill, satellite droplets, puddle formation, and material issues commonly seen in microdroplet generators.

10.3.1   Frequency Response

The frequency response of droplet generators is one of the important measures of their performance. The reported typical frequency response for thermal bubble jets, piezo jets, thermal buckling jets, acoustic wave jets, electrostatic, and inertial jets ranges from kHz to tens of kHz. Piezo and acoustic wave jets typically have higher frequency response than the others. Recently, a novel design by Tseng et al. (1998) improved the frequency response of thermal bubble jets by about three times (35 kHz), which is compatible with the speed of piezo jets. Higher speed devices (in the range of hundreds of kHz) are currently under development by major ink-jet printer makers and will see further breakthroughs in the next few years.

Three important time constants that affect the frequency response of microdroplet generators are actuation, droplet ejection, and liquid refilling, as shown in Figure 10.8.

The typical time constant from heating to bubble formation in a thermal bubble jet is around 2–10 μs for chamber sizes ranging from 20 to 100 μm [Tseng et al., 1998d]. Thermal buckling and electrostatic type microdroplet generators have actuation times of around tens to hundreds of μs [estimated from Hirata

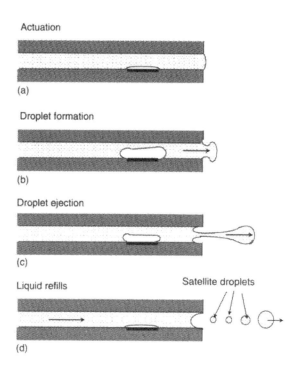

FIGURE 10.8 The sequence of droplet generation.

et al., 1996], owing to the large actuation plate required for sufficient displacement to form droplets. Inertial type generators required even longer actuation times [estimated from Gruhler et al., 1999], typically hundreds to thousands of μs, due to the large cantilever structure for generating large droplets with sufficient inertial force to overcome liquid surface tension and viscosity. The actuation time for piezo or acoustic wave type may be shorter [Darling et al., 1984; Zhu et al., 1996], around μs to tens of μs.

After the application of actuation pressure to the liquid, the droplet starts to eject. The ejection sequence usually takes from a couple to hundreds of μs for droplet volumes from 1 pl to 1 nl [Tseng et al., 1998d], which does not vary much among the different operational principles.

The liquid refills automatically due to surface tension force after the droplet ejects. Liquid refilling time can vary by 3 orders of magnitude (e.g., from less than 10 μs to over 1 s), depending on the length and geometry of the refilling path. In most the commercial ink-jet printhead designs, a chamber neck [Nielsen et al., 1985], an elongated chamber channel [Nielsen et al., 1985], or a physical valve [Karz et al., 1994] is used to prevent hydraulic cross talk and maintain a high pressure in the firing chamber. However, if not arranged properly, those designs may greatly increase the refilling time causing a reduction in the frequency response. As a result, preventing hydraulic cross talk without sacrificing device speed becomes an important issue in the design of super-high-speed and high-resolution droplet generators. Tseng et al. (1998a, 1998b, and 1998c) introduced the concept of a virtual chamber neck to speed the refilling process of thermal bubble jets and suppress cross talk. The virtual neck consists of vapor bubbles, forms for pressurization while the droplet is ejecting, and opens to reduce flow resistance while the liquid is refilling, thus increasing the frequency response. This concept is shown in Figure 10.9.

In simulating the droplet actuation and formation sequence, much work [Curry et al., 1977; Lee et al., 1977; Levanoni et al., 1977; Pimbley et al., 1977; Fromm et al., 1984; Bogy et al., 1984; Asai et al., 1987–92; Mirfakhraee et al., 1989; Chen et al., 1997–99; Rembe et al., 2000] has been conducted on bubble formation, droplet generation, and the aerodynamics of droplets traveling in the atmosphere both for thermal-bubble- and piezo-type jets. Interested readers can refer to those works for details.

Typical bubble formation, growth, and collapse sequences in a thermal bubble jet are shown in Figure 10.10 (Yang et al., 2004); three different heaters with length-width ratios of 1, 2, and 3 respectively

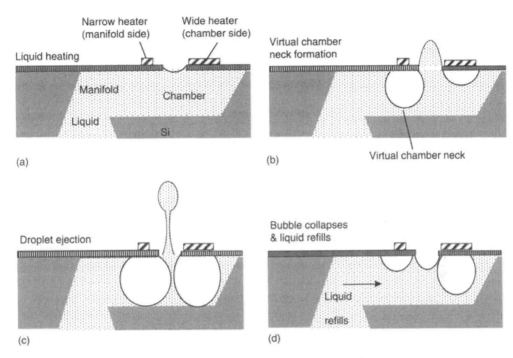

FIGURE 10.9   Operation principle of virtual chamber neck, after Tseng (1998c).

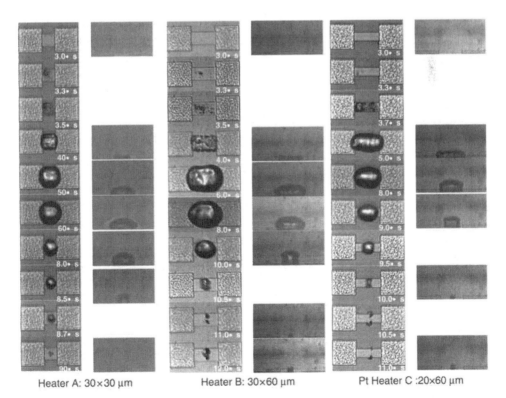

FIGURE 10.10   Bubble formation sequences for three different heater designs under same heat flux of 1.2 GW/m^2, courtesy Yang, et al. (2004).

are illustrated. Under the same heat flux condition, maximum bubble volumes are different among these three cases. As shown in Figure 10.11, heater type B with a length/width aspect ratio of 2 has the largest volume variation under the same heat flux, implying that it is the most efficient of the three designs (Yang, et al., 2004).

10.3.2 Thermal/Hydraulic Cross Talk and Overfill

When nozzle pitch becomes small, two types of cross talk, hydraulic and (for thermal bubble jet devices) thermal, become significant in multiple-nozzle droplet generators. Hydraulic cross talk relates to the transportation of pressure waves from the firing chamber to the neighboring chambers, as shown in Figure 10.12. The vibration of the meniscus of the neighboring chambers may result in poor droplet volume control or, even worse, unexpected droplet ejection. Thermal cross talk, which appears only in thermal bubble jet devices, is the transportation of thermal energy from the firing chamber to neighboring chambers, resulting in poor droplet volume control. After droplet ejection, the liquid refilling process sometimes causes meniscus oscillation, posing another problem — overfill. Overfill, similar to cross talk, increases the waiting time for the next droplet ejection and even causes undesired droplet ejection. The phenomenon of overfill is schematically illustrated in Figure 10.13.

These issues stem from inadequate flow compliance among nozzles. One solution by IBM, Inc., [Neilsen et al., 1985] is to lengthen the channels between the reservoirs and the chambers. However, the longer channels also increase the flow resistance and inertia significantly, adding to the liquid refilling time.

HP, Inc., tried to solve this problem by using either a parallel compliant (reservoir) or a chamber neck. Figure 10.14 shows a slot beside the nozzles; the slot is used as a reservoir to store energy while the bubble

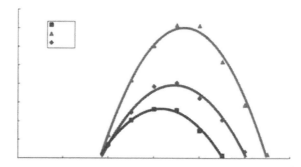

FIGURE 10.11 (See color insert following page 2-12.) Bubble volume variation versus time for three different heater designs under same heat flux of $1.2\,GW/m^2$, courtesy Yang, et al. (2004).

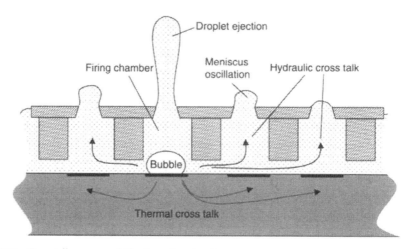

FIGURE 10.12 Cross talk among neighboring microchambers.

is exploding and to release energy while the bubble is collapsing [Neilsen et al., 1985]. Figure 10.15 shows the second approach, which involves narrowing the inlet of the chamber to form a chamber neck.

The effects of these methods were simulated by Buskirk et al. (1988). Figure 10.16 shows the simulation results for the meniscus-position of the firing chamber and the neighboring chamber in different chamber designs. In this figure, since the chambers are connected to one another through the liquid supply, the meniscuses on the firing chamber and the neighboring chambers encounter large fluctuation owning to pressure transmission during the droplet ejection from the firing chamber. Doubling the channel length (series compliance) damped down the fluctuation some, but not completely. Only the chamber neck design almost completely damped down the meniscus fluctuation; however, it has the slowest response of the three designs.

In contrast to the fixed chamber neck design, Xerox Corp., [Karz et al., 1994] used a flexible plate as a valve to address the cross talk issue. However, this design may suffer from low frequency response and material reliability problems. Tseng et al. (2000) used a novel virtual chamber neck employing bubble as a virtual valve to reduce the cross talk problem while maintaining the high frequency response of the droplet generator as shown in Figure 10.9. Their work demonstrated that the bubble responds faster and operates more reliably than solid valves in micro scale.

In addition to hydraulic cross talk, thermal cross talk in Tseng's work was also reduced by placing the heater on the chamber on top of a thin film with low thermal conduction instead of leaving it on the thick

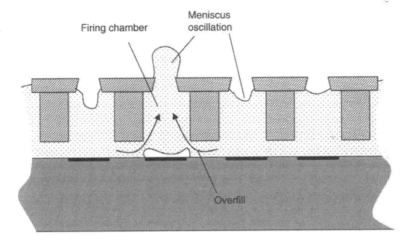

FIGURE 10.13 Overfill phenomenon.

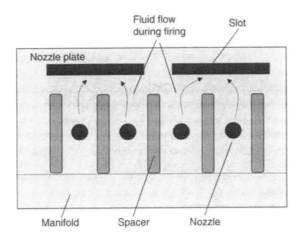

FIGURE 10.14 Method of applying parallel compliant (reservoir) to overcome cross talk.

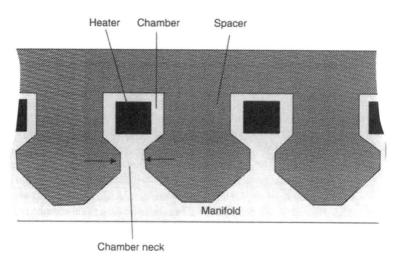

FIGURE 10.15   The cross talk overcome by employing chamber neck.

substrate, through which heat can be conducted to the neighbors [Tseng et al., 1998c, 2001a]. As the nozzle pitch becomes smaller, the cross talk problem clearly becomes more severe in the operation of very high-resolution and high-speed droplet generators.

10.3.3   Satellite Droplets

Satellite droplets result from the breakdown of the long ejected liquid column. This breakdown is due to the interaction among surface tension, air drag force, and inertial force. The velocity mismatch along the liquid column, resulting from the variation of actuation velocity, promotes the breakdown. The droplet ejection sequence captured from a commercial ink-jet reveals the steps of satellite droplet formation, as shown in Figure 10.17 [Tseng et al., 1998c]. When applied for printing, the quality is degraded by the occurrence of satellite drops as revealed in Figure 10.18. Satellite droplets also reduce the precision of liquid dispensing control.

There have been many attempts to predict droplet formation. Asai et al. (1987–1992) conducted both numerical simulation and experimental measurements to obtain the temporal variation of droplet length at a thermal bubble jet. However, their work does not include the formation process of satellite droplets. For drop-on-demand ink-jet devices, Fromm et al. (1984) and Chen et al. (1997–1999) solved the Navier–Stokes equation to predict the droplet formation and demonstrated the process of satellite droplet formation. Pimbley et al. (1977) and Chen et al. (1997b) demonstrated the evolution of droplets as well as satellite droplet formation by flow visualization. However, they did not discuss a detailed method for eliminating the separation of satellite droplets from the main droplet.

Different methods have been tried to eliminate satellite droplets in commercial products. In piezo droplet generators, triangular waves were used to eliminate satellite drops [Chen et al., 1999]. In thermal bubble jets, Tseng et al. (1998a, 1998c, 2000) proposed a novel method employing bubble as a trimmer to cut off long droplet tails and eliminate satellite droplets as shown in Figure 10.19. The shortened tail of the liquid column in Tseng's work is drawn back by surface tension into the main droplet, thus eliminating satellite drops.

10.3.4   Puddle Formation

Liquid puddle forms when liquid is pushed to flow outward and accumulates on the nozzle's outside surface. Puddle exerts great blocking force on droplet ejection, causing the distortion or even the cessation of droplet ejection. One of the major reasons for puddle formation is the hydrophilic nozzle surface.

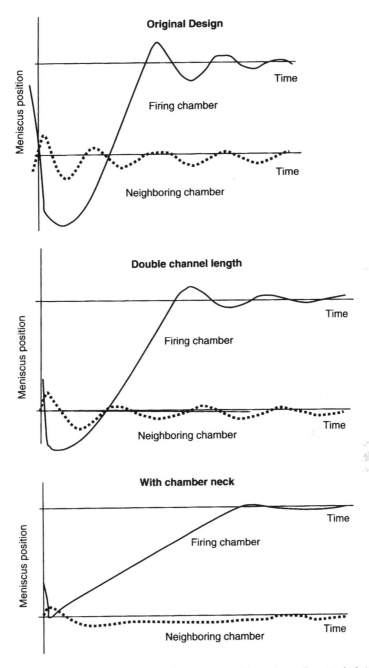

FIGURE 10.16 Meniscus oscillation in different chamber structures, data redrawn from Buskirk (1988).

When the nozzle's outer surface contacts working fluid from the chamber, it accumulates liquid puddle. It was observed by Tseng et al. (1998d) that the puddle appears after several continuous operations; that is, the chamber surface does not accumulate puddle until it gets wet after several runs. If the operation of droplet-ejection operation stops, puddle is drawn back into the chamber by surface tension. However, once the chamber surface becomes wet, puddle always forms when the operation starts again. Figure 10.20 shows puddle formation during the running of the microinjector [Tseng et al., 1998d]. Notice that the droplet ejection position is away from the nozzle due to the distortion by liquid puddle.

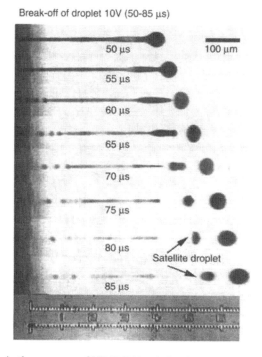

FIGURE 10.17   Droplet ejection sequence of HP 51626A printhead, courtesy Tseng (1998c).

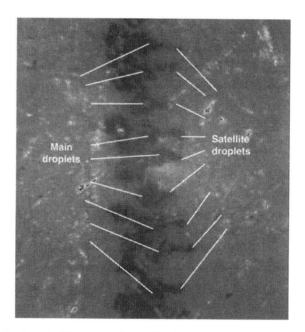

FIGURE 10.18   A printed vertical line smeared by satellite droplets, courtesy Tseng (1998c).

One way to eliminate puddle formation is to coat the chamber's outer surface with a nonwetting material. (The inner surface of the chamber needs to remain hydrophilic for liquid refill.) However, even with this coating, there is still no guarantee that puddle will not form. More research is underway to fully understand the mechanism in the puddle formation process.

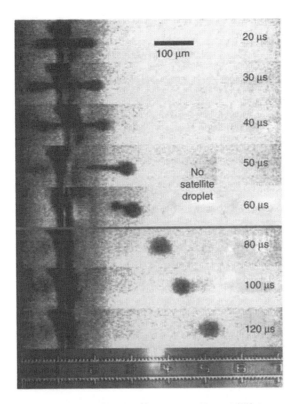

FIGURE 10.19 Droplet ejection without satellite droplets, courtesy Tseng (1998a).

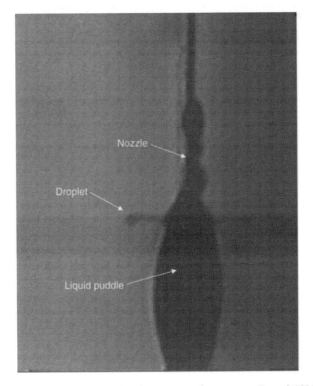

FIGURE 10.20 Liquid puddle formation outside the micronozzles, courtesy Tseng (1998d).

10.3.5   Material Issues

Material issues including stress, erosion, durability, and compatibility are very complex problems in the design of microdroplet generators.

In the area of processing, material compatibility, stress, and durability problems are commonly discussed. Material compatibility issues result from the processing temperature, processing environment (oxidation, reactive gas, etc.), etching method used, and adhesion ability; stress issues usually concern the processing temperature as well as the doping condition; material durability issues are due either to the material's intrinsic properties or the mechanical forces induced during the process (i.e., fluid flow force, surface tension force, vacuum forces, or handling force). To eliminate the material issues, much care needs to be taken in designing the process flow, such as compensating for material stress during or after the fabrication process; performing high-temperature processes before introducing the low temperature material; finishing aggressive wet etching before metal film deposition; or using low temperature bonding material and processes to protect IC and microdevices.

In selecting chamber materials, metals such as nickel and stainless steel have been widely employed for different microdroplet generators due to their ease of fabrication (such as by wet etching or electroplating), high mechanical strength, resistance to erosion by certain bases or acids, imperviousness to solvents, and high durability in cycling operations, etc. However, owing to cost and fabrication-precision considerations, the chamber materials of many commercial ink-jets have been changed to polymers, such as polyimide or PMMA in recent years. As a result, chamber formation by bonding a polymer thin plate on the actuator substrate and nozzle-array formation by laser drilling on those polymer thin plates are now very common in the ink-jet industry. Nevertheless, the bonding and laser drilling processes may encounter issues such as bonding nonuniformity, time consuming serial laser drilling processes, and alignment limitations. Therefore, various polymer-MEMS processes for fabricating multiple embedded chambers integrated with nozzles have been developed; these not only simplify the fabrication process but also improve the accuracy of alignment and nozzle fabrication on the chamber structures. Employing double exposures (one full power, and the other partial power) on a single SU-8 resist layer with antireflection coating in the resist–substrate interface, Chuang, et al. (2003) demonstrated that embedded microchannel structures can be fabricated easily, as shown in Figure 10.21. The thickness of chamber roof can be controlled from 14 to 60 μm in a 2 μm resolution.

Durability, stress, and erosion issues are the major problems concerning operation. Due to the cycling nature of the droplet generation process, the materials chosen for actuation face challenges not only from stress but also from fatigue. The HP Corporation reported that possible reasons for failure of the heater passivation material are cavitation and thermal stress [Bhaskar et al., 1985]. Silicon, low-stress silicon nitride, silicon carbide, silicon dioxide, and some metals, are usually used to overcome these problems. In addition to selecting proper materials, reducing sharp corners in the design is an important key to eliminating stress concentration points and thus preventing material from cracking. The working fluid's erosion of structural materials is another serious issue. Lee et al. (1999) reported the erosion of the spacer material in a commercial ink-jet head while using diesel fuel as working fluid. In contrast, materials including silicon and silicon nitride used by Tseng et al. (1998c, 1998d) and Lee et al. (1999) in the microinjector are free of this problem and can also be used with a wide variety of fluids including solvents and chemicals. Selecting materials wisely, ordering them correctly in the process, and properly designing the materials in the microstructures are the three primary measures for reducing material issues.

10.4   Fabrication of Micro-Droplet Generators

The structures in microdroplet generators commonly include a manifold for storing liquid, microchannels for transporting liquid, microchambers for holding liquid, nozzles for defining droplet size and direction, and actuation mechanisms for generating droplets. Occasionally, droplet generators may not have nozzles but generate droplets locally by energy focusing means, such as acoustic wave droplet generators [Zhu et al., 1996]. Before micromachining processes became widely used, most processes for fabricating microdroplet generators

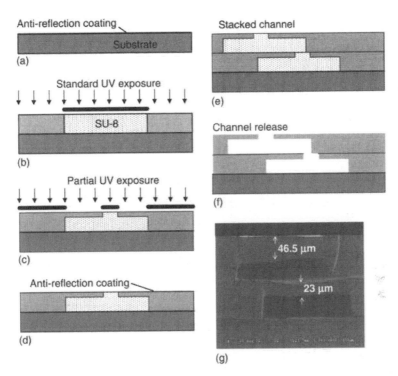

FIGURE 10.21 (a)–(f) Fabrication process of polymer multiple embedded microchambers and (g) fabricated chambers, courtesy Chuang et al. (2003).

followed the same general method: nozzle plates, fluid handling plates, and actuation plates are manufactured separately and then integrated into a single device. However, as the nozzle resolution becomes finer, bonding processes pose severe alignment, yield, and material problems as well as IC compatibility issues. On the other hand, the interconnection lines may not have enough space to fan out from each chamber when nozzle resolution is higher than 600 dpi. As a result, monolithic methods for fabricating high resolution IC integrated droplet generators have become very important. The following sections introduce examples of different fabrication techniques.

10.4.1 Multiple Pieces

Figure 10.22 schematically shows the traditional method of fabricating microdroplet generators by bonding separately fabricated pieces [Tseng, 1998d]. In this process, actuation plates are fabricated separately from the nozzle plates. In the thermal bubble jet, heaters are usually sputtered or evaporated and then patterned with an IC circuit on the bottom plate; piezo, thermal buckling, electrostatic, and inertial actuators consist of more complex structures, such as piezo disks, thin plate structures, or cantilever beams. Nozzles are fabricated by electroforming [Ta et al., 1988], molding, or laser drilling [Keefe et al., 1997]. These separately processed pieces are assembled either by using intermediate layers of polymer spacer material [Siewell et al., 1985; Askeland et al., 1988; Hirata et al., 1996; Keefe et al., 1997] or directly adhering several pieces through anodic bonding [Kamisuki et al., 1998, 2000], fusion bonding [Gruhler et al., 1999], eutectic bonding, or low temperature chemical bonding. However, most of the bonding methods are chip-level rather than wafer-level processes and face challenges of alignment, bonding quality, and material–process compatibility. As the nozzle resolution becomes higher than 600 dpi, alignment accuracy approaching 4 μm (10% of the nozzle pitch) becomes difficult to attain. Higher alignment accuracy significantly increases the fabrication cost, especially for the chip-level process. Bonding quality is another important issue affecting the fabrication yield of large array and high-resolution devices. Additionally, the bonding materials (mostly polymers)

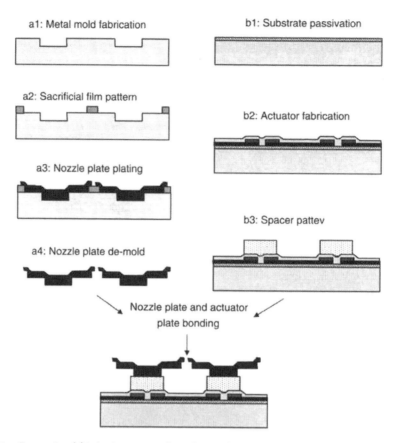

FIGURE 10.22   Conventional fabrication process flow of microdroplet generators.

chosen must be suitable for the application environments and working fluids. Finally, bonding processes involving heat, pressure, high voltage, or chemical situations restrict IC integration with the droplet generators, and the IC integration is essential for large-array and high resolution applications.

10.4.2 Monolithic Fabrication

To address the problems inherent in using multiple pieces, monolithic processes utilizing micromachining technology have been widely employed since the early 1990s. Two primary monolithic methods have been introduced: one combines bulk and surface micromachining and the other uses bulk micromachining and the deep UV lithography associated with electroforming (or UV lithography only).

For example, Tseng et al. (1998d) combined surface and bulk micromachining to fabricate a microdroplet generator array with potential nozzle resolution up to 1200 dpi (printing resolution can be 2400 dpi or higher). This design used double bulk micromachining processes to fabricate the fluid handling system, including the manifold, microchannels, and microchambers. Surface micromachining, on the other hand, was used for fabricating heaters, interconnection lines, and nozzles. The whole process was finished on (100) crystal orientation silicon wafers. Figures 10.23 and 10.24 show the three-dimensional structure of the microinjectors and the monolithic fabrication process respectively. The ejection of 0.9 pl droplets has also been demonstrated by Tseng et al. (2001b) using the high-resolution microinjectors. The structural materials used in the microinjector are silicon, silicon nitride, and silicon oxide, which are durable in high temperature and suitable for various liquids (even some harsh chemicals). Using this device, ICs can be easily integrated on the same silicon substrate.

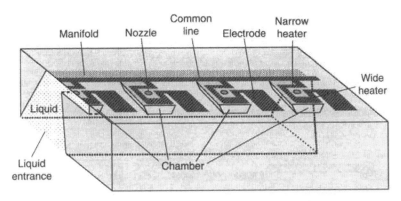

FIGURE 10.23   Schematic three-dimensional structure view of microinjectors, after Tseng (1998c).

The second primary method can be found in Lee's (1999) work. This process used multiexposure and signal development (MESD) lithography to define microchannel and microchamber structures (photoresist as sacrificial layer) and constructed the physical structures with electroformed metal. The manifold was manufactured from the wafer's backside by electrochemical methods [Lee et al., 1995]. This device also demonstrated a capacity for very high-resolution arrays and compatibility with the IC process. Another method, using photoresist as sacrificial layer and polyimide as structure layer, was introduced by Chen et al. (1998) for high resolution and IC compatible applications.

10.5   Characterization of Droplet Generation

Droplet trajectory, volume, ejection direction, and ejection sequence/velocity are four important quantitative measures of the ejection quality of microdroplet generators. The following sections briefly introduce the basic methods for testing droplet generation.

10.5.1   Droplet Trajectory

Droplet trajectory can be visualized by directing a flashing light on the ejection stream, as shown in Figure 10.25 [Tseng et al., 1998a]. The white dots in Figure 10.26 show the visualized droplet stream. The visualized droplet trajectory follows an exponential curve that is very different from the parabolic curve expected for normal sized objects with a similar initial horizontal velocity. Tseng et al. (1998a) also estimated droplet trajectory by solving a set of ordinary differential equations from the balance of horizontal and vertical forces on a single droplet flying through air.

From this analysis, the vertical position Y and horizontal position X of the droplet can be expressed by the following equations:

$$Y = U_{v\infty}\left[t - \frac{U_{v\infty}}{g}\left(1 - e^{\frac{-6\pi\mu r_0}{m}t} \right) \right] \tag{10.1}$$

$$X = \frac{U_{H0}m}{6\pi\mu r_0}\left(1 - e^{\frac{-6\pi\mu r_0}{m}t} \right) \tag{10.2}$$

where g is the acceleration due to gravity, t is the time, m is the mass and r_0 is the radius of the droplet, μ is the viscosity of air, $U_{v\infty} = \frac{mg}{6\pi\mu r_0}$ is the droplet terminal velocity, and U_{H0} is the initial horizontal velocity. The trajectory determined by the experiment is drawn in Figure 10.26 and fits the visualized trajectory well except at the end, suggesting the interaction among droplets. From this simple analysis, the maximum flying distance of a droplet with a known diameter can be estimated as:

$$X_{max} = \frac{U_{H0}m}{6\pi\mu r_0} = \frac{2\rho_{liquid}}{9\mu_{air}}(U_{H0}r_0^2), \tag{10.3}$$

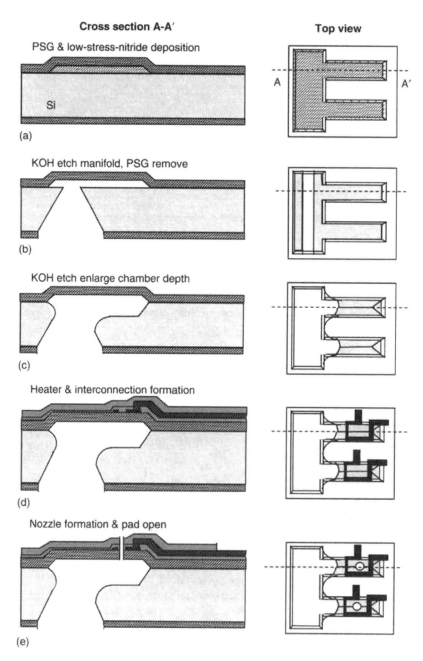

FIGURE 10.24   Fabrication process flow of monolithic microinjectors, after Tseng (1998c).

when $t \sim \infty$. Here the maximum distance is proportional to the droplet velocity and droplet radius to the second power. For different droplet sizes with the same initial velocity, the maximum flying distance of smaller droplets decreases very fast. To obtain 1 mm flying distance, the droplet with 10 m/s initial velocity needs a minimum radius of 2.7 μm. From the above estimation, droplet size should be maintained above a certain value to ensure enough flying distance for printing. Printing with very fine droplets (diameter smaller than a couple of micrometers) requires either increasing the droplets' initial velocity or printing in a special vacuum environment to overcome air drag.

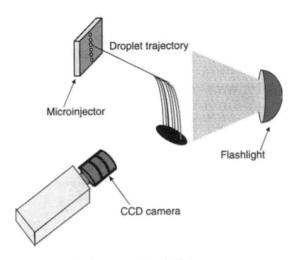

FIGURE 10.25 Experimental setup for droplet stream visualization.

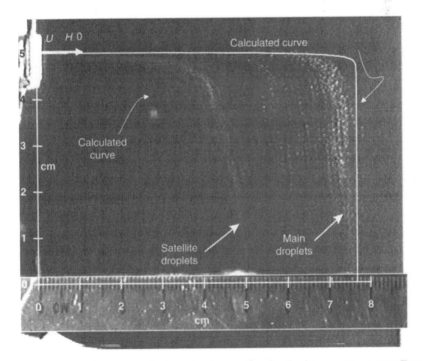

FIGURE 10.26 Visualized droplet stream and estimated trajectory of a microdroplet generator, courtesy Tseng (1998a).

10.5.2 Ejection Direction

Droplet direction can be determined by the visualized trajectory. Many parameters, including nozzle shape, roughness, aspect ratio, and wetting property as well as actuation direction and chamber design, affect droplet direction. In general, symmetric structure design and accurate alignment can help control droplet direction.

10.5.3 Ejection Sequence/Velocity and Droplet Volume

To characterize the detailed droplet ejection sequence, a visualization system [P.-H. Chen et al., 1997b, Tseng et al., 1998c] as shown in Figure 10.27 has been widely used. In this system, an LED was placed under the

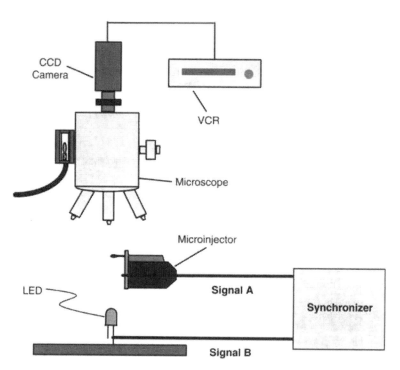

FIGURE 10.27   Experimental setup for droplet ejection sequence visualization.

droplet generator to back-illuminate the droplet stream. Two signals synchronized with adjustable time delay were sent to a microinjector and an LED respectively. Droplets were ejected continuously from a droplet generator, and the droplet images were frozen by the LED's flashing light at specified time delays, as shown in Figure 10.17. Droplet volume can be determined from the images by assuming the droplet is axi-symmetric, or from weighing certain numbers of droplets. Droplet velocity can be estimated by measuring the difference in flying distance of the droplet fronts in two succeeding images.

10.5.4   Flow Field Visualization

Flow field visualization is one of the most direct and effective tools to better understand flow properties such as cross talk, actuation sequence, liquid refill, and droplet formation inside microdroplet generators. Flow visualization in small scale presents some difficulties that do not occur in large scale, such as limited viewing angles, impossible to generate light sheet, reflection from the particles trapped on the wall, short response time, and small spatial scale. Meinhart et al. (2000) adopted a micrometer resolution particle image velocimetry system to measure instantaneous velocity fields in an electrostatically actuated ink-jet head. The system introduced 700 nm diameter fluorescent particles for flow tracing; the spatial as well as temporal resolutions of the image velocimetry are 5–10 μm and 2–5 μs respectively. The four primary phases of ink-jet operation — infusion, inversion, ejection, and relaxation — were clearly captured and quantitatively analyzed.

10.6   Applications

More than a hundred applications for microdroplet generators have been explored. This section summarizes some of them.

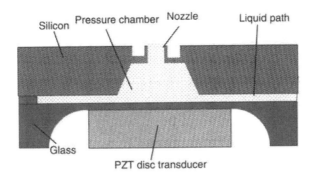

FIGURE 10.28 The injector design for mass spectrometry, after Luginbuhl (1999).

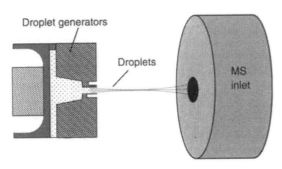

FIGURE 10.29 Operation principle of mass spectrometry using microdroplet generators.

10.6.1 Ink-jet Printing

Ink-jet printing, which involves arranging small droplets on a printing medium to form texts, figures, or images, is the most well-known microdroplet application. The smaller and cleaner the droplets are, the sharper the printing is. However, smaller droplets cover a smaller printing area and thus require more printing time. Therefore in printing, high-speed microdroplet generation with stable and clean micro sized droplets is desired for fast and high quality printing. The printing media can be paper, textile, skin, cans or other surfaces that can adsorb or absorb printing solutions. Ink-jet printing generated revenues of more than $10 billion worldwide in 2000 and will keep growing in the future.

10.6.2 Biomedical and Chemical Sample Handling

The application of microdroplet generators in biomedical sample handling is an emerging field that has drawn much attention in the past few years. Many research efforts have focused on droplet volume control, droplet size miniaturization, compatibility issues, the variety of samples, and high-throughput parallel methods.

Luginbuhl et al. (1999), Miliotis et al. (2000), and Wang et al. (1999) developed piezo- and pneumatic-type droplet injectors respectively for mass spectrometry. Figure 10.28 schematically shows the design of the injectors, which generate submicron to micron sized bioreagent droplets for sample separation and analysis in a mass spectrometer, as shown in Figure 10.29. Luginbuhl et al. (1999) employed silicon bulk micromachining to fabricate silicon nozzle plates and Pyrex glass actuation plates, while Wang et al. (1999) employed a combination of surface and bulk micromachining to fabricate the droplet generator. These injectors are part of the lab-on-the-chip system for incorporating microchips with macroinstruments.

Microdroplet generators were also used by Koide et al. (2000), Nilsson et al. (2000), Goldmann et al. (2000), and Szita et al. (2000) for the accurate dispensing of biological solutions. Piezo- and thermal-type injectors were used in those investigations for protein, peptide, enzyme, or DNA dispensing. With an operation principle similar to ink-jet printing, the devices provided for precisely dispensing and depositing a

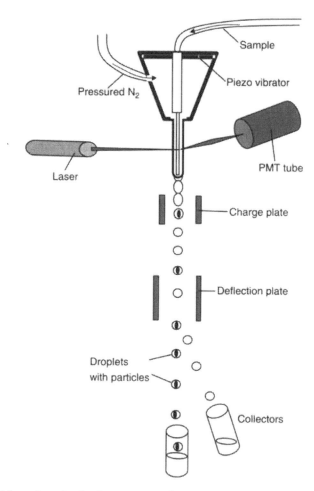

FIGURE 10.30   Particle sorting using droplet generators, after Asano (1995).

single biological droplet onto a desired medium, and they also could dispense droplet arrays. The arrayed bioreagents can be bioprocessed further for high-throughput analysis.

Continuous jet-type droplet generators were reported by Asano et al. (1995) to effectively focus and sort particles by using the electrostatic force. The experimental setup is shown in Figure 10.30. A syringe pump pressurizes the sample fluid to pass through a nitrogen sheath flow for focusing, and then the sample is ejected from a piezoelectric transducer disturbed nozzle to form droplets. Droplets containing the desired particles were charged at the breakup point and deflected into collectors. The reported separation probability for 5, 10, and 15 μm particles can be as high as 99%. However, the inner jet diameter limits the particle size for separation. Other than solid particle separation, this method potentially can be applied to cell sorting for biomedical applications.

In addition to biomedical reagent handling, microdroplet generators were widely used in chemical handling. For example, Shah et al. (1999) used an ink-jet system to print catalyst patterns for electroless metal deposition. This system used a commercial ink-jet printhead to ejecte a Pt solution as a seed layer for Cu electroless plating. The lines produced by this method were reported to be 100 μm wide and 0.2–2 μm high.

10.6.3   Fuel Injection and Mixing Control

Microdroplet generators used for fuel injection can dispense controllable and uniform droplets, which are important for mixing and combustion applications.

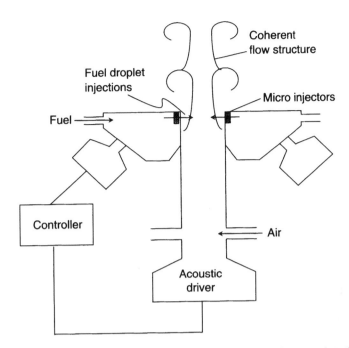

FIGURE 10.31 Control of mixing and fuel injection by microdroplet generators, after Tseng (1996).

Combustion efficiency depends on the mixing rate of the reactants. The reactants in a shear flow are first entrained by large vortical structures (Brown and Roshko, 1974) and then mixed by fine scale eddies. The entrainment can be greatly enhanced by controlling the evolution of large-scale vortices either actively (Ho et al., 1982) or passively (Ho et al., 1987). The effectiveness of controlling large-scale vortical structures for increasing combustion efficiency has also been experimentally demonstrated (Shadow et al., 1987). Although much work has been done on improving the mixing efficiency in combustion chambers, improving the small scale mixing and reducing the evaporation time of liquid fuel remain great challenges in combustion research.

Traditional injectors with nozzle diameters of around tens to hundreds of μms can neither supply uniform microdroplets for reducing evaporation time and fine scale mixing nor eject droplets that can be controlled individually to modulate vortex structure [Lee et al., 1999]. To overcome those limitations, Tseng et al. (1996) proposed a microdroplet injector array fabricated by the micromachining technologies used for fuel injection. The droplets ejected from microinjectors are uniform and the diameter can be from μm to tens of μms, which is close to the micro scale of small turbulence eddies. The fine scale mixing can be carried out by the reaction of the small turbulence eddies directly with the microdroplets. The smaller, more uniform droplets increase the overall size of the evaporation surface and thus greatly reduce evaporation time. In addition, an appropriate arrangement of the microinjectors around the nozzle of a dump combustor (Figure 10.31) provides spatially coherent perturbations to control the large vortices. Two types of coherent structures, spanwise and streamwise vortices, can be influenced by imposing subharmonics of the air jet's most unstable instability frequency. Control of the spanwise vortices can be accomplished by applying temporal amplitude modulation on injection. If the ejecting phases of the microdroplets along the azimuthal direction are the same, the mode zero instability (Brown et al., 1974) is enhanced. Imposing a certain defined phase lag on these microinjectors generates higher mode instability (Brown et al., 1974) waves, which are usually beneficial for mass transfer enhancement. Since about 1000 injectors are placed around the nozzle, the spatial modulation in the azimuthal direction can perturb the streamwise vortices. The interaction of streamwise and spanwise vortices by microinjectors creates fine scale mixing.

10.6.4   Direct Writing and Packaging

Micro droplet generators offer an alternative to lithography for electronics and opto-electronics manufacturing. This approach has the advantages of precise volume control of dispensed materials, data driven flexibility, low cost, high speed, and low environmental impact, as described by Hayes and Cox (1998). The materials have been demonstrated for the application to manufacturing processes including adhesive for component bonding, filled polymer systems for direct resistor writing and oxide deposition, and solder for solder bumping of flip-chips, BGAs (ball grid arrays), PCBs (printed circuit boards), and CSPs (chip-scale-packages) [Hays et al., 1999; Teng et al., 1988]. In those printings, the desired temperature is 100~200°C, and the viscosity of the fluids needs to be around 40 cps; in some cases, an inert process environment, such as nitrogen flow, is required to protect the materials from oxidation.

Through direct writing by ink-jet printing, the difficulty of fabrication from photolithography or screen-printing processes for solar cell metallization and LEP (light emitting polymer) deposition of LEPD (light emitting polymer displays) can be easily eliminated. In solar cell metallization, metalo-organic decomposition (MOD) silver ink was used to ink-jet print directly onto solar cell surfaces and thus avoid p-n junction degradation in the traditional screen printing method requiring firing temperatures of 600–800°C. Ink-jet printing also provides for the formation of a uniform line film on rough solar cell surfaces [Tang et al., 1988a, 1988b; Somberg et al., 1990], which is not easy to achieve using traditional photolithography.

On the other hand, organic light emitting devices requiring the deposition of multiple organic layers to perform full color operation present similar problems. Due to the organic layers' solubility in many solvents and aqueous solutions, conventional methods, such as photolithography, screen printing and evaporation, that require a wet patterning process are not compatible with them on the same substrate [Hebner et al., 1998; Shimoda et al., 1999; Kobayashi et al., 2000]. Thus direct writing of organic materials by ink-jet printing becomes one of the promising solutions for providing a safe patternable process without wet etching. However, owing to the pinholes that appear on the patterned materials, high quality polymer devices may not be easily ink-jet printed. Yang's group proposed combining an ink-jet printed layer with a uniform spin coated polymer layer to overcome this problem [Bharathan et al., 1998]. In such a system, the uniform layer serves as a buffer layer to seal the pinholes, and the ink-jet-printing layer contains the desired patterns [Bharathan et al., 1998].

10.6.5   Optical Component Fabrication and Integration

Integrated microoptics has become a revolutionary concept in the optics field because it provides the advantages of low cost, miniaturization, improved spatial resolution and time response, and a reduced assembly process that is not possible by traditional means. As a result, fabricating and integrating miniaturized optical components with performance similar to or better than traditional components are critical issues in integrated microoptics systems. Standard bulk or surface micromachining provides various ways to fabricate active/passive micromirrors, wave-guides, and Fresnel lenses, but making a refractive lens with curved surfaces is not easy. Compared to photolithography, which utilizes a patterned and melted photoresist column as the lens, the ink-jet printing method allows more flexibility in process design, material choices, and system integration. Cox et al. employed ink-jet printing technology to eject heated polymer material in fabricating a micro lenslet array [Cox et al., 1994; Hays et al., 1998]. The shape of the lens was controlled by the viscosity of the droplets at the impact point, the substrate wetting condition, and the cooling/curing rate of the droplets [Hays et al., 1998]. A 70–150 μm diameter lens has been successfully fabricated with a density greater than 15,000/cm^2 and focal lengths between 50 and 150 μm. Wave-guides using ink-jet technology also have been demonstrated by Cox et al. (1994).

Since optical components with varying properties can be selectively deposited onto the desired region, integration of those components with fabricated ICs or other devices is possible and efficient.

10.6.6   Solid Free Forming

Not only two-dimensional patterns but also three-dimensional solid structures can be generated by microdroplet generators. Orme et al. (1993) and Marusak et al. (1993) reported the application of molten

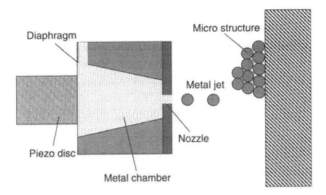

FIGURE 10.32 Operation principle of three-dimensional structure fabrication by microdroplet generator, after Yamaguchi (2000).

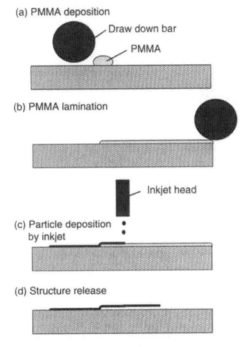

FIGURE 10.33 Fabrication of three-dimensional structures by droplet ejection and polymer lamination, after Fuller (2000).

metal drops for solid free form fabrication. Evans' group demonstrated the application of continuous and drop-on-demand ink-jets for ceramic printing to fabricate 3-D structures as well as functionally graded materials [Blasdell et al., 2000; Mott et al., 1999]. Yamaguchi et al. (2000) used metal jets to print functional three-dimensional microstructures; Figure 10.32 shows the operation principle. Yamaguchi et al. (2000) proposed employing multijets for structure and sacrificial material deposition to print an overhanging structure, while Fuller et al. (2000) used laminated PMMA film as the supporting material for the ejection of metal cantilever beams. The fabrication principle is shown in Figure 10.33.

10.6.7 Manufacturing Process

Droplet generators also provide novel ways of material processing. For example, submicron ceramic particles can be plasma sprayed for surface coating, as introduced by Blazdell and Juroda [Blazdell et al., 2000].

The operation principle is shown in Figure 10.34. A continuous jet printer was used for droplet formation from ceramic solution. The ceramic stream was delivered into the hottest part of the plasma jet and then sprayed onto the working piece. Splats from the plasma spray are claimed to be similar in morphology to those produced using conventional plasma spraying of a coarse powers, but they are significantly smaller, which may impart unique characteristics such as extension of solid state solubility, refinement of grain size, formation of metastable phases, and high concentration of point defects [Blazdell et al., 2000].

10.6.8 IC Cooling

Conventionally, blowing fans and fins are widely used for cooling IC chips, especially in CPUs. Recently, as heating power has increased greatly with increasing CPU size, more advanced methods, such as heat pipes, CPL, and impinging air jets, have been introduced for quick heat removal. However, no matter how the designs improve, the upper limit of of those devices' heat removal ability is in the order of tens of W/cm^2. In addition, as chips become larger, detecting hot spots and selectively removing the heat only from hot regions

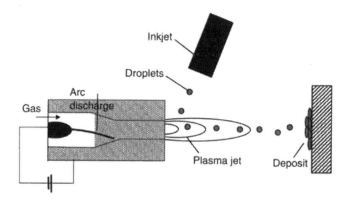

FIGURE 10.34   Operation principle of plasma spraying by microdroplet generators, after Blazdell (2000).

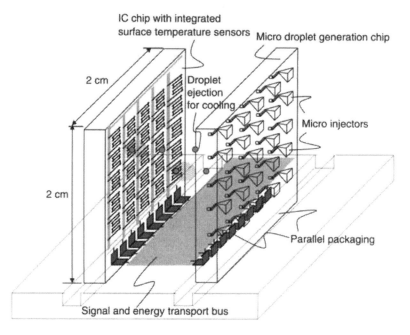

FIGURE 10.35   Conceptual design of IC cooling by microdroplet injections.

to preserve energy is highly desired but difficult to perform using traditional means. Transportating latent heat through the droplet evaporation process is a promising new way of doing this. This method can in principle remove three to four orders more heat than conventional means. Further, the cooling spot can be selected and monitored by an integrated micro temperature sensor and IC array. The conceptual design by Tseng (2001c), as shown in Figure 10.35, used a two-dimensional array of microinjectors to selectively deposit liquid droplets onto the chip surface. The frequency and number of droplets applied can be adjusted to maintain a dry chip surface with a constant temperature. The estimated maximum heat removed by this device is around 300,000 W/cm^2, more than 1000 times more than by conventional means. The temperature sensors and control circuits can be fabricated on the same chip to form a self-contained smart system.

10.7   Concluding Remarks

Droplet generators are important fluid handling devices for precise liquid dosing control. MEMS technology makes micro sized droplet generators possible and popular for many applications. Various techniques for droplet generation, including piezoelectric, thermal bubble, thermal buckling, focused acoustic wave, electrostatic, and inertial actuation, have been employed. Compared to the other methods, thermal bubble generation has larger actuation deformation, simpler design/fabrication, and fewer limitations on chamber volume, but it has the drawbacks of temperature sensitivity and of being influenced by the liquid's properties. The piezo-type jet has the advantages of high frequency response, controllable droplet size, and free of satellite drops, but its limitations on finite actuation deformation thus restrict the chamber volume's miniaturization. Electrostatic and thermal bulking jets have size limitations similar to piezo jets. The electrostatic jet has the benefit of low power consumption, but both it and the thermal bulking jet have limited frequency response due to their size restrictions. The acoustic wave droplet generator, on the other hand, is not fully developed and stable enough for commercial applications, while the inertial actuation method's operation principle limits its miniaturization. More types of microdroplet generators are being developed, and they may one day replace the ones we have been using for decades.

Physical properties, design issues, and manufacturing problems are important concerns in the design and fabrication of microdroplet generators. The associated issues including frequency response, cross talk, satellite droplets, puddle formation, material selection, and integration require great care at all levels of design and fabrication. MEMS technology provides some key solutions to those practical issues.

Many applications and potential applications have made microdroplet generators important and exciting since their inception. Ink-jet printing, the traditional application, has generated huge amounts of revenue. Moreover, hundreds of promising applications have been emerging, including bioreagent handling, fine chemical handling, drug delivery, direct writing, solid free form, IC cooling, and fuel injection. Many exciting applications of microdroplet generators are yet to be discovered.

References

Allen, R.R., Meyer, J.D., and Knight, W.R. (1985) "Thermodynamics and Hydrodynamics of Thermal Ink Jets," *Hewlett-Packard J.*, May, pp. 21–27.

Asai, A. (1989) "Application of the Nucleation Theory to the Design of Bubble Jet Printers," *Jpn. J. Appl. Phys.*, 28, pp. 909–15.

Asai, A. (1991) "Bubble Dynamics in Boiling under High Heat Flux Pulse Heating," *ASME J. Heat Transfer*, 113, pp. 973–79.

Asai, A. (1992) "Three-Dimensional Calculation of Bubble Growth and Drop Ejection in a Bubble Jet Printer," *Trans. ASME*, 114, pp. 638–41.

Asai, A., Hara, T., and Endo, I. (1987) "One Dimensional Model of Bubble Growth and Liquid Flow in Bubble Jet Printers," *Jpn. J. Appl. Phys.*, 26, pp. 1794–1801.

Asai, A., Hirasawa, S., and Endo, I. (1988) "Bubble Generation Mechanism in the Bubble Jet Recording Process," *J. Imag. Tech.*, 14, pp. 120–24.

Asano, K., Funayama, Y., Yatsuzuka, K., and Higashiyama, Y. (1995) "Spherical Particle Sorting by Using Droplet Deflection Technology," *J. Electrostat.*, 35, pp. 3–12.

Ashley, C.T., Edds, K.E., and Elbert, D.L. (1977) "Development and Characterization of Ink for an Electrostatic Ink Jet Printer," *IBM J. Res. Dev.*, 21, pp. 69–74.

Askeland, R.-A., Childers W.-D., and Sperry, W.-R. (1988) "The Second-Generation Thermal InkJet Structure," *Hewlett-Packard J.*, August, pp. 28–31.

Bharathan, J., and Yang, Y. (1998) "Polymer Electroluminescent Devices Processed by Inkjet Printing: 1. Polymer Light-Emitting Logo," *Appl. Phys. Lett.*, 72, pp. 2660–62.

Bhaskar, E.V., and Aden, J.S. (1985) "Development of the Thin-film Structure for the Thinkjet Printhead," *Hewlett-Packard J.*, May, pp. 27–33.

Blazdell, P.F., and Evans, J.R.G. (2000) "Application of a Continuous Ink Jet Printer to Solid Freefroming of Ceramics," *J. Mater. Process. Technol.*, 99, pp. 94–102.

Blazdell, P., and Kuroda, S. (2000) "Plasma Spraying of Submicron Ceramic Suspensions Using a Continuous Ink Jet Printer," *Surf. Coat. Technol.*, 123, pp. 239–46.

Bogy, D.B., and Talke, F.E. (1984) "Experimental and Theoretical Study of Wave Propagation Phenomena in Drop-On Demand Ink Jet Devices," *IBM J. Res. Dev.*, 28, pp. 314–21.

Brown, G.L., and Roshko, A. (1974) "On Density Effects and Large Structure in Turbulent Mixing Layers," *J. Fluid Mech.*, 64, pp. 775–816.

Buehner, W.L., Hill, J.D., Williams, T.H., and Woods, J.W. (1977) "Application of Ink Jet Technology to a Word Processing Output Printer," *IBM J. Res. Dev.*, 21, pp. 2–9.

Bugdayci, N., Bogy. D.B., and Talke, F.E. (1983) "Axisymmetric Motion of Radially Polarized Piezoelectric Cylinders Used in Ink jet Printing," *IBM J. Res. Dev.*, 27, pp. 171–80.

Buskirk, W.-A., Hackleman, D.-E., Hall, S.-T., Hanarek, P.-H., Low, R.-N., Trueba, K.-E., and Van de Poll, R.-R. (1988) "Development of a High-Resolution Thermal Inkjet Printhead," *Hewlett-Packard J.*, Oct., pp. 55–61.

Carmichael, J.M. (1977) "Controlling Print Height in an Ink Jet Printer," *IBM J. Res. Dev.*, 21, pp. 52–55.

Chen, J.-K., and Wise, K.D. (1995) "A High-Resolution Silicon Monolithic Nozzle Array for Inkjet Printing," *The 8th International Conference on Solid-State Sensors and Actuators, and Eurosensors 9*, June 25–29, Stockholm, pp. 321–24.

Chen, J.-K., Juan, W., Kubby, J., and Hseih, B.-C. (1998) "A Monolithic Polyimide Nozzle Aray for Inkjet Printing," *Tech. Dig. IEEE Solid State Sensor and Actuator Workshop*, pp. 308–11, June 8–11, Hilton Head Island, SC.

Chen, P.-H., Chen, W.-C., and Chang, S.-H. (1997a) "Bubble Growth and Ink Ejection Process of a Thermal Ink Jet Printhead," *Int. J. Mech. Sci.*, 39, pp. 687–95.

Chen, P.-H., Chen, W.-C., and Chang, S.-H. (1997b) "Visualization of Drop Ejection Process of a Thermal Bubble Ink Jet Printhead," *Proc. The 1st Pacific Symp. Flow Visualization and Image Processing, Honolulu*, pp. 132–37, February 22–26.

Chen, P.-H., Chen, W.-C., Ding, P.-P., and Chang, S.-H. (1998) "Droplet Formation of a Thermal Sideshooter Inkjet Printhead," *Int. J. Heat Fluid Flow*, 19, pp. 382–90.

Chen, P.-H., Peng, H.-Y., Liu, H.-Y., Chang, S.-L., Wu, T.-I., and Cheng C.-H. (1999) "Pressure Response and Droplet Ejection of a Piezoelectric Inkjet Printhead," *Int. J. Mec. Sci.*, 41, pp. 235–48.

Chuang, Y.-J., Tseng, F.-G., Cheng, J.-H., and Lin, W.-K. (2003) "A Novel Fabrication Method of SU-8 Stacked Micro Channels By UV Dosage Control," *Sensors Actuators A*, 103, pp. 64–69.

Cox, W.R., Chen, T., Hayes, D.J., MacFarlane, D.L., Narayan, V., and Jatum, J.A. (1994) "Microjet Fabrication of Microlens Arrays," *IEEE Photon. Technol. Lett.*, 6, pp. 1112–14.

Curry, S.A., and Portig, H. (1977) "Scale Model of an Ink Jet," *IBM J. Res. Dev.*, 21, pp. 10–20.

Darling, R.H., Lee, C.-H., and Kuhn, L. (1984) " Multiple-Nozzle Ink Jet Printing Experiment," *IBM J. Res. Dev.*, 28, pp. 300–6.

Fromm, J.E. (1984) "Numerical Calculation of the Fluid Dynamics of Drop-On Demand Jets," *IBM J. Res. Dev.*, 28, pp. 322–33.

Fuller, S., and Jacobson, J. (2000) "Ink Jet Fabricated Nanoparticle MEMS," *Proc. IEEE MEMS Conference*, pp. 138–41, January 23–27, Kyoto, Japan.

Goldmann, T., and Gonzalez, J.S. (2000) "DNA-Printing: Utilization of a Standard Inkjet Printer for the Transfer of Nucleic Acids to Solid Supports," *J. Biochem. Biophys. Methods*, **42**, pp. 105–10.

Gruhler, H., Hey, N., Muller, M., Bekesi, S., Freygang, M., Sandmaier, H., and Zengerle, R. (1999) "Topspot: A New Method for the Fabrication of Biochips," *Proc. IEEE MEMS '99*, pp. 7413–17, January 17–21, Orlando, Florida.

Hays, D.J., and Cox, W.R. (1998) "Micro Jet Printing of Polymers for Electronics Manufacturing," *Proceedings of 3rd international Conference on Adhesive Joining and Coating Technology in Electronics Manufacturing*, pp. 168–73, September 21–30, Binghamton, NY.

Hayes, D.J., Grove, M.E., and Cox, W.R. (1999) "Development and Application by Ink-Jet Printing of Advanced Packaging Materials," *1999 International Symposium on Advanced Packaging Materials*, pp. 88–93, March 14–17, Chateau Elan, Braelton, Georgia.

Hebner, T.R., Wu, C.C., Marcy, D., Lu, M.H., and Sturm, J.C. (1998) "Ink-Jet Printing of Doped Polymer for Organic Light Emitting Devices," *Appl. Phys. Lett.*, **72**, pp. 519–21.

Hirata, S., Ishii, Y., Matoba, H., and Inui, T. (1996) "An Ink-Jet Head Using Diaphragm Microactuator," *Proc. of the 9th IEEE Micro Electro Mechanical Systems Workshop*, pp. 418–23, February 11–15, San Diego.

Ho, C.M., and Gutmark, E. (1987) "Vortex Induction and Mass Entrainment in a Small Aspect Ratio Elliptic Jet," *J. Fluid Mech.*, **179**, pp. 383–405.

Ho, C.M., and Huang, L.S. (1982) "Subharmonics and Vortex Merging in Mixing Layers," *J. Fluid Mech.*, **119**, pp. 443–73.

Kamisuki, S., Fuji, M, Takekoshi, T., Tezuka, C., and Atobe, M. (2000) "A High Resolution, Electrostatically-Driven Commercial Inkjet Head," *Proc. IEEE MEMS '00*, pp. 793–98, January 23–27, Miyazaki, Japan.

Kamisuki, S., Hagata, T., Tezuka, C., Nose, Y., Fuji, M., and Atobe, M. (1998) "A Low Power, Small, Electrostatically-Driven Commercial Inkjet Head," *Proc. IEEE MEMS '98*, pp. 63–68, January 25–29, Heidelberg, Germany.

Karz, R.S., O'Neill, J.F., and Daneile, J.J. (1994) "Ink Jet Printhead with Ink Flow Direction Valves," U. S. Patent No. 5,278,585.

Keefe, B.-J., Ho, M.-F., Courian, K.-J., Steinfield, S.-W., Childers, W.-D., Tappon, E.-R., Trueba, K.-E., Chapman, T.-I., Knight, W.-R., and Moritz, J.-G. (1997) "Inkjet Printhead Architecture for High Speed and High Resolution Printing," U.S. Patent No. 5,648,805.

Kobayashi, H., et al. (2000) "A Novel RGB Multicolor Light-Emitting Polymer Display," *Synth. Met.*, **111–112**, pp. 125–28.

Koide, A., Sasaki, Y., Yoshimura, Y., Miyake R., and Terayama, T. (2000) "Micromachined Dispenser with High Flow Rate and High Resolution," *Proc. IEEE MEMS '00*, pp. 424–28, January 23–27, Miyazaki, Japan.

Krause, P., Obermeier, E., and Wehl, W. (1995) "Backshooter: A New Smart Micromachined Single-Chip Inkjet Printhead," *The 8th International Conference on Solid-State Sensors and Actuators, and Eurosensors 9*, pp. 325–28, June 25–29, Stockholm.

Lee, F.C., Mills, R.N., and Talke, F.E. (1984) "The Application of Drop-On Demand Ink Jet Technology to Color Printing," *IBM J. Res. Dev.*, **28**, pp. 307–13.

Lee, H.C. (1977) "Boundary Layer Around a Liquid Jet," *IBM J. Res. Dev.*, **21**, pp. 48–51.

Lee, J.-D., Lee H.-D., Lee H.-J., Yoon, J.-B., Han, K.-H., Kim, J.-K., Kim, C.-K., and Han, C.-H. (1995) "A Monolithic Thermal Inkjet Printhead Utilizing Electrochemical Etching and Two-Step Electroplating Techniques," *J. MEMS*, **8**, pp. 601–04.

Lee, J.-D., Yoon, J.-B., Kim, J.-K., Chung, H.-J., Lee, C.-S., Lee, H.-D., Lee, H.-J., Kim, C.-K., and Han, C.-H. (1999) "A Thermal Inkjet Printhead with a Monolithically Fabricated Nozzle Plate and Self-Aligned Ink Feed Hole," *J. MEMS*, **8**, pp. 229–36.

Lee, Y.-K., Yi, U., Tseng, F.-G., Kim, C.J., and Ho, C.-M. (1999) "Diesel Fuel Injection by a Thermal Microinjector," *Proc. MEMS, ASME IMECE '99*, pp. 419–25, November 14–19, Nashville, Tennessee.

Levanoni, M. (1977) "Study of Fluid Flow through Scaled-Up Ink Jet Nozzles," *IBM J. Res. Dev.*, **21**, pp. 56–68.

Luginbuhl, P. et al. (1999) "Micromachined Injector for DNA Spectrometry," *IEEE Transducers '99*, pp. 1130–33, June 7–10, Sendai, Japan.

Marusak, R.E. (1993) "Picoliter Solder Droplet Dispensing," *Proceedings of The Solid Freeform Fabrication Symposium*, pp. 81–87, August 9–11, Austin.

Meinhart, C.-D., and Zhang H. (2000) "The Flow Structure Inside a Microfabricated Inkjet Printhead," *J. MEMS*, 9, pp. 67–75.

Miliotis, T., Eficsson, D., Marko-Varga, G., Ekstrom, S., Nilsson, J., and Laurell, T. (2000) "Interfacing Protein and Peptide Separation to Maldi-tof MS Using Microdispensing and On-Target Enrichment Strategies," *Proc. μTAS '00*, pp. 387–91, May 14–18, Enschede, The Netherlands.

Mirfakhraee, A. (1989) Growth and Collapse of Vapor Bubbles in Ink-Jet Printers, Ph.D. dissertation, Univeristy of California, Berkeley.

Mott, M., and Evans, J.R.G. (1999) "Zirconia/Alumina Functionally Graded Material Made by Ceramic Ink Jet Printing," *Mater. Sci. Eng. A*, 271, pp. 344–52.

Myers, R.A., and Tamulis, J.C. (1984) "Introduction to Topical Issue on Non-Impact Printing Technologies," *IBM J. Res. Dev.*, 28, pp. 234–40.

Nielsen, N.J. (1985) "History of ThinkJet Printhead Development," *Hewlett-Packard J.*, May, pp. 4–10.

Nilsson, J., Bergkvist, J., and Laurell, T. (2000) "Optimization of the Droplet Formation in a Piezo-Electric Flow-Through Microdispenser," *Proc. μTAS '00*, pp. 75–78, May 14–18.

Orme, M., and Huang, C. (1993) "Thermal Design Parameters Critical to the Development of Solid Freefrom Fabrication of Structural Materials with Controlled Nano-Liter Droplets," *Proceedings of The Solid Freeform Fabrication Symposium*, pp. 88–94, August 9–11, Austin.

Pimbley, W.T., and Lee, H.C. (1977) "Satellite Droplet Formation in a Liquid Jet," *IBM J. Res. Dev.*, 21, pp. 21–30.

Rembe, C., Siesche, S.A.D., and Hofer, E.P. (2000) "Thermal Ink Jet Dynamics: Modeling, Simulation, and Testing," *Microelectron. Reliab.*, 40, pp. 525–32.

Shadow, K.C., Gutmark, E., Wilson, K.J., Parr, D.M., Mahan, V.A., and Ferrell, G.B. (1987) "Effect of Shear-Flow Dynamics in Combustion Processes," *Combust. Sci. Technol.*, 54, pp. 103–16.

Shah, P., Kevrekides, Y., and Benziger J. (1999) "Ink-Jet Printing of Catalyst Patterns for Electroless Metal Deposition," *Langmuir*, 15, pp. 1584–87.

Shimoda, T., Kimura M., Seki S., Kobayashi, H., Kanbe, S., and Miyahita, S. (1999) "Technology for Active Matrix Light Emitting Polymer Displays," *IEEE IEDM '99*, pp. 107–10, December 5–8, Washington, DC.

Siewell, G.-L., Boucher, W.R., and McClelland, P.H. (1985) "The ThinkJet Orifice Plate: A Part with Many Functions," *Hewlett-Packard J.*, May, pp. 33–37.

Somberg, H. (1990) "Inkjet Printing for Metallization on Very Thin Solar Cells," Photovoltaic Specialists Conference, *Conference Record of the Twenty First IEEE*, pp. 666–67, May 21–25, Kissimmee, Florida.

Sweet, R.G. (1964) "High Frequency Recording with Electrostatically Deflected Ink Jets," *Stanford Electronics Laboratories Technical Repost No. 1722-1*, Stanford University.

Sweet, R.G. (1971) "Fluid Droplet Recorder," U. S. Patent No. 3,576,275.

Szita, N., Sutter, R., Dual, J., and Buser, R. (2000) "A Fast and Low-Volume Pipettor with Integrated Sensors for High Precision," *Proc. IEEE MEMS '00*, pp. 409–13, January 23–27, Miyazaki, Japan.

Ta, C.-C., Chan, L.-W., Wield, P.-J., and Nevarez R. (1988) "Mechanical Design of Color Graphics Printer," *Hewlett-Packard J.*, August, pp. 21–27.

Teng, K.F., Azadpour, M.A., and Yang, H.Y. (1988) "Rapid Prototyping of Multichip Packages Using Computer-Controlled, Ink Jet Direct-Write," *Proceedings of the 38th Electronics Components Conference*, pp. 168–73, May 9–11, LA, CA.

Teng, K.F., and Vest, R.W. (1988a) "Metallization of Solar Cells with Ink Jet Printing and Silver Metallo-Organic Inks," *IEEE Trans. Components, Hybrids, Manuf. Technol.*, 11, pp. 291–97.

Teng, K.F., and Vest, R.W. (1988b) "Application of Ink Jet Technology on Photovoltaic Metallization," *IEEE Electron. Device Lett.*, 9, pp. 591–93.

Tseng, F.-G., Linder, C., Kim, C.-J., and Ho, C.-M (1996) "Control of Mixing with Micro Injectors for Combustion Application," *Micro-Electro-Mechanical Systems (MEMS) DSC-ASME IMECE*, pp. 183–87, November 17–22, Atlanta.

Tseng, F.-G., Kim, C.-J., and Ho, C.-M (1998a) "A Microinjector Free of Satellite Drops and Characterization of the Ejected Droplets," *Proc. MEMS, ASME IMECE '98*, pp. 89–95, November 15–20, Anaheim, CA.

Tseng, F.-G., Shih, C., Kim, C.-J., and Ho, C.-M (1998b) "Characterization of Droplet Injection Process of a Microinjector," in *Abstract Book of 13th U.S. Congress of Applied Mechanics*, pp. TB3, University of Florida.

Tseng, F.-G., Kim, C.-J., and Ho, C.-M (1998c) "A Novel Microinjector with Virtual Chamber Neck," *Technical Digest of The 11th IEEE International Workshop on Micro Electro Mechanical Systems*, pp. 57–62, January 25–29, Heidelberg, Germany.

Tseng, F.-G. (1998d) A Micro Droplet Generation System, Ph.D. dissertation, University of California, Los Angeles.

Tseng, F.-G., Kim, C.-J., and Ho, C.-M (2000) "Apparatus and Method for Using Bubble as Virtual Valve in Micro Injector to Eject Fluid," U.S. and international patents pending.

Tseng, F.-G., Kim, C.-J., and Ho, C.-M (2001a) "A Monolithic, High Frequency Response, High-Resolution Microinjector Array Ejecting Sub Pico-Liter Droplets Without Satellite Drops: Part 1. Concepts, Designs and Molding," *J. Microelectromech. Syst.*, 11, pp. 427–36.

Tseng, F.-G., Kim, C.-J., and Ho, C.-M (2001b) "A Monolithic, High Frequency Response, High-Resolution Microinjector Array Ejecting Sub Pico-Liter Droplets Without Satellite Drops: Part 2. Fabrication, Characterization, and Performance Comparison," *J. Microelectromech. Syst.*, 11, pp. 437–47.

Tseng, F.-G. (2001c) "Droplet Impinging Micro Cooling Arrays," proposal for NSC project.

Twardeck, T.G (1977) "Effect of Parameter Variations on Drop Placement in an Electrostatic Ink Jet Printer," *IBM J. Res. Dev.*, 21, pp. 31–36.

Yamaguchi, K., Sakai, K., Ymanaka, T., and Hirayama, T. (2000) "Generation of Three-Dimensional Micro Structure Using Metal Jet," *Precision Eng.*, 24, pp. 2–8.

Yang, I.-D., Tseng, F.-G., and Chieng, C.-C. (2004) "Micro Bubble Formation Dynamic under High Heat Flux on Heaters with Different Aspect Ratios," submitted to *Microscale Thermophysical Engineering*.

Wang, X.Q., Desai, A., Tai, Y.C., Licklider, L., and Lee, T.D. (1999) "Polymer-Based Electrospray Chips for Mass Spectrometry," *Proc. IEEE MEMS '99*, pp. 523–28, January 17–21, Orlando, Florida.

Zhu, X., Tran, E., Wang, W., Kim, E.S., and Lee, S.Y. (1996) "Micromachined Acoustic-Wave Liquid Ejector," Solid-State Sensor and Actuator Workshop, pp. 280–82, June 2–6, Hilton Head, South Carolina.

11

Micro Heat Pipes and Micro Heat Spreaders

G.P. Peterson
Rensselaer Polytechnic Institute

Choondal B. Sobhan
National Institute of Technology Calicut

11.1 Introduction

A heat pipe is a device with very high effective thermal conductivity that is capable of transferring large quantities of heat over considerable distances without an appreciable temperature gradient. The high effective thermal conductance of the heat pipe maintains the vapor core temperature at an almost uniform temperature while transferring heat, making it capable of being used also as a "heat spreader." As described by Peterson (1994), a heat pipe operates in a closed two-phase cycle in which heat added to the evaporator region causes the working fluid to vaporize and move to the cooler condenser region, where the vapor condenses, giving up its latent heat of vaporization. In traditional heat pipes, the capillary forces existing in a wicking structure pump the liquid back to the evaporator. While the concept of utilizing a wicking structure as part of a device capable of transferring large quantities of heat with a minimal temperature drop was first introduced by Gaugler (1944), it was not until much more recently that the concept of combining phase change heat transfer and microscale fabrication techniques (i.e., MEMS devices for the dissipation and removal of heat), was first proposed by Cotter (1984). This initial introduction envisioned a series of very small "micro" heat pipes incorporated as an integral part of semiconductor devices. While no experimental results or prototype designs were presented, the term *micro heat pipe* was first defined as a heat pipe "so small that the mean curvature of the liquid–vapor interface is necessarily comparable in magnitude to the reciprocal of the hydraulic radius of the total flow channel" [Babin et al., 1990]. Early proposed applications of these devices included the removal of heat from laser diodes [Mrácek, 1988] and other small localized heat generating devices [Peterson, 1988a, 1988b]; the thermal control of photovoltaic cells [Peterson, 1987a, 1987b]; the removal or dissipation of heat from the leading edge of hypersonic aircraft [Camarda et al., 1997]; applications involving the nonsurgical treatment of cancerous tissue through either hyper- or hypothermia [Anon., 1989; Fletcher and Peterson, 1994]; and space applications in which heat pipes are embedded in silicon radiator panels to dissipate the large amounts of waste heat generated [Badran et al., 1993].

While not all of these applications have been implemented, micro heat pipes ranging in size from 30 μm to 1 mm in characteristic cross-sectional dimensions and from 10 mm to 60 mm in length have been analyzed, modeled, and fabricated; the larger of these are currently commonplace in commercially available products, such as laptop computers or high precision equipment where precise temperature control is essential. Reported studies include those on individual micro heat pipes and micro heat pipe arrays made as an integral part of silicon substrates. Theoretical and experimental analysis has led to the characterization of the influence of geometrical and operational parameters on the performance of these devices. Determination of the operating limitations of micro heat pipes also has been an objective of the research on micro heat pipes. More recently, this work has been expanded to include micro heat spreaders fabricated in silicon or in new metallized polymeric materials, which can be used to produce highly conductive, flexible heat spreaders capable of dissipating extremely high heat fluxes over large areas, thereby reducing the source heat flux by several orders of magnitude.

Since the initial introduction of the micro heat pipe concept, the study of micro scale heat transfer has grown enormously and has encompassed not only phase change heat transfer but the entire field of heat transfer, fluid flow, and in particular, a large number of fundamental studies in thin film behavior, as described elsewhere in this book. Microscale fluid behavior and heat transfer at the microscale, along with the variations between the behavior of bulk thermophysical properties and those that exist at the micro- or nano-scale levels are all areas of considerable interest. While the division between micro- and macroscale phase-change behavior is virtually indistinguishable, in applications involving phase change heat transfer devices, such as micro heat pipes and micro heat spreaders, it can best be described by applying the dimensionless expression developed by Babin and Peterson (1990) and described later in this chapter. This expression relates the capillary radius of the interface and the hydraulic radius of the passage and provides a good indicator of when the forces particular to the microscale begin to dominate.

A number of previous reviews have summarized the literature published prior to 2000 [Peterson and Ortega, 1990; Peterson, 1992; Cao et al., 1993; Peterson et al. 1996; Faghri, 2001; Garimella and Sobhan, 2001]; however, significant advances have been made over the past few years, particularly in developing a better understanding of the thin film behavior that governs the operation of these devices. The following review begins with a very brief overview of the early work in this area and then looks at advances made in individual micro heat pipes and arrays of micro heat pipes and more recent investigations of flat plate microscale heat spreaders.

For heat pipes operating in steady state, a number of fundamental mechanisms limit the maximum heat transfer. These have been summarized and described by Marto and Peterson (1988) in a concise format that will be summarized here; they include the capillary wicking limit, viscous limit, sonic limit, entrainment limit, and boiling limit. The first two of these deal with the pressure drops occurring in the liquid and vapor phases. The sonic limit results from pressure gradient induced vapor velocities that may result in choked vapor flow, while the entrainment limit focuses on the entrainment of liquid droplets in the vapor stream, which inhibits the return of the liquid to the evaporator and ultimately leads to dry-out. Unlike these limits, which depend upon the axial transport, the   limit is reached when the heat flux applied in the evaporator portion is high enough that nucleate boiling occurs in the evaporator wick, creating vapor bubbles that partially block the return of fluid.

While a description of the transient operation and start-up dynamics of these devices is beyond the scope of this work, it is appropriate to include a brief description of the fundamentals of heat pipe operation and the methods for determining the steady-state limitations. For additional information on the theory and fundamental phenomena that cause each of these limitations, refer to Tien (1975), Chi (1976), Dunn and Reay (1982), Peterson (1994), and Faghri (1995).

11.1.1 Fundamentals of Heat Pipe Operation

The heat pipe is a passive, two-phase heat transfer device that utilizes the liquid–vapor phase change processes occurring in working fluid to transfer heat and then to pump the working fluid through the capillary action of the wick. In a conventional heat pipe, the required capillary action, or *wicking*, is obtained by the use of a capillary wick, which could be, for example, a metallic screen wick or a porous wick. Many kinds of wick structures have been tested and utilized in the conventional heat pipe. The heat added externally at the evaporator section of the heat pipe vaporizes the liquid inside the heat pipe, which is at the saturation temperature. The vapor moves to the condenser section as more and more liquid evaporates along the evaporator section. At the condenser section, due to external heat transfer from the heat pipe body, the vapor condenses and the liquid is circulated back to the evaporator section through the capillary wick. Thus the working fluid undergoes a thermodynamic cycle, and the physical processes involved in the working of a heat pipe are heat transfer, phase change, and capillary induced fluid flow. The effective thermal conductivity of the heat pipe is normally many times that of the wall material, as most of the heat is transferred by the thermodynamic cycle that the working fluid undergoes.

Though the basic working principles of both micro heat pipes and larger, conventional heat pipes are very similar, there is an essential difference in that the micro heat pipe typically does not employ a wicking structure for the circulation of the working fluid but depends upon small liquid arteries. The micro heat pipe is essentially a channel of a polygonal cross-section that contains a small, predetermined quantity of saturated working fluid. Heat added to the evaporator section of the micro heat pipe results in the vaporization of a portion of the working fluid. The vapor then flows through the central portion of the channel cross-section. The return flow of the liquid formed in the condenser is accomplished by utilizing the capillary action at the narrow corner regions of the passage. Thus, in the micro heat pipe, wicking is provided by the corners of the passage, thus avoiding the need for a wick structure for liquid recirculation. The vapor and liquid flow in the micro heat pipe are also characterized by the varying cross-sectional areas of the two fluid paths, unlike the flow of the vapor and liquid confined to the core and the wick regions of the conventional heat pipe.

The required condition for micro heat pipe operation is that the average radius of the liquid–vapor meniscus formed at the corners of the channel is comparable in magnitude with the reciprocal of the hydraulic radius (i.e., the characteristic dimension) of the total flow channel [Cotter, 1984]. The operation of the micro heat pipe will be described in detail under Section 11.2.

11.1.1.1 Operating Limits

The major limitations on the operation of micro heat pipes as well as conventional heat pipes are the capillary, viscous, sonic, entrainment, and boiling limitations. These limitations and the methods to determine them are explained in the sections that follow.

11.1.1.1.1 Capillary Limitation

The operation and performance of heat pipes depend on many factors including the shape, working fluid, and wick structure. The primary mechanism by which these devices operate results from the difference in the capillary pressure across the liquid–vapor interfaces in the evaporator and condenser. The pressure distributions that prevail in the heat pipe are shown in Figure 11.1. To operate properly, this pressure difference must exceed the sum of all the pressure losses throughout the liquid and vapor flow paths. This relationship can be expressed as

$$\Delta P_c \geqslant \Delta P_+ + \Delta P_- + \Delta P_l + \Delta P_v \tag{11.1}$$

With the liquid and vapor pressure drops in differential form, Equation (11.1) can be written as:

$$\Delta P_c \geqslant \Delta P_+ + \Delta P_- + \int_{L_{eff}} \frac{\partial P_l}{\partial x} dx + \int_{L_{eff}} \frac{\partial P_v}{\partial x} dx + \Delta P_v \tag{11.1.1}$$

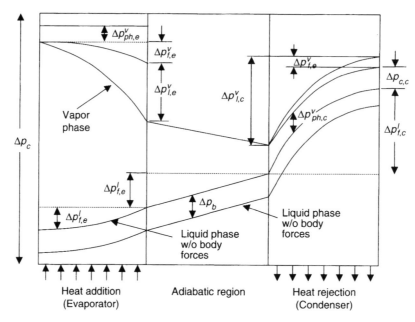

FIGURE 11.1   Pressure distribution in operating heat pipes. (Reprinted with permission from Peterson, G.P. [1994] *An Introduction to Heat Pipes: Modeling, Testing and Applications*, John Wiley & Sons, New York.)

where

ΔP_c = net capillary pressure difference,

ΔP_+ = normal hydrostatic pressure drop,

ΔP_- = axial hydrostatic pressure drop,

$\Delta P_l = \int_{L_{eff}} \dfrac{\partial P_l}{\partial x} dx$ = viscous pressure drop occurring in the liquid phase, and

$\Delta P_v = \int_{L_{eff}} \dfrac{\partial P_v}{\partial x} dx$ = viscous pressure drop occurring in the vapor phase.

As long as this condition is met, liquid is returned to the evaporator. For situations where the summation of the viscous pressure losses, ΔP_l and ΔP_v, and the hydrostatic pressure losses, ΔP_+ and ΔP_-, is greater than the capillary pressure difference between the evaporator and condenser, the wicking structure becomes starved of liquid and dries out. This condition, referred to as the capillary wicking limitation, varies according to the wicking structure, working fluid, evaporator heat flux, and operating temperature.

11.1.1.1.1.1 Capillary Pressure   The capillary pressure difference at a liquid–vapor interface ΔP_c is defined by the Laplace–Young equation, which for most heat pipe applications reduces to

$$\Delta P_{c,m} = \left(\frac{2\sigma}{r_{c,e}} \right) - \left(\frac{2\sigma}{r_{c,c}} \right) \tag{11.2}$$

where $r_{c,e}$ and $r_{c,c}$ represent the radii of curvature in the evaporator and condenser regions respectively.

During normal heat pipe operation, the vaporization occurring in the evaporator causes the liquid meniscus to recede into the wick, reducing the local capillary radius, $r_{c,e}$, while condensation in the condenser results in increases in the local capillary radius $r_{c,c}$. It is this difference in the two radii of curvature that "pumps" liquid from the condenser to the evaporator. During steady-state operation, it is generally assumed that the capillary radius in the condenser $r_{c,c}$ approaches infinity, so that the maximum capillary pressure for a heat pipe operating at steady-state can be expressed as a function of only the capillary radius of the evaporator wick,

$$\Delta P_{c,m} = \left(\frac{2\sigma}{r_{c,e}} \right) \tag{11.3}$$

Values for the effective capillary radius $r_{c,e}$ can be found theoretically for simple geometries [Chi, 1976] or experimentally for other more complicated structures.

11.1.1.1.1.2 Hydrostatic Pressure Drops The normal and axial hydrostatic pressure drops ΔP_+ and ΔP_- are the result of the local gravitational body force. The normal and axial hydrostatic pressure drop can be expressed as

$$\Delta P_+ + \rho_l g d_v \cos \psi \tag{11.4}$$

and

$$\Delta P_- = \rho_l g L \sin \psi \tag{11.5}$$

where ρ_l is the density of the liquid, g is the gravitational acceleration, d_v is the diameter of the vapor portion of the pipe, ψ is the angle the heat pipe makes with respect to the horizontal, and L is the length of the heat pipe.

In a gravitational environment, the axial hydrostatic pressure term may either assist or hinder the capillary pumping process depending upon whether the tilt of the heat pipe promotes or hinders the flow of liquid back to the evaporator (i.e., whether the evaporator lies either below or above the condenser). In a zero-g environment or in cases where the surface tension forces dominate, such as micro heat pipes, both of these terms can be neglected.

11.1.1.1.1.3 Liquid Pressure Drop As the liquid returns from the condenser to the evaporator, it experiences a viscous pressure drop ΔP_l, which can be written in terms of the frictional drag,

$$\frac{dP_l}{dx} = - \frac{2\tau_l}{(r_{h,l})} \tag{11.6}$$

where τ_l is the frictional shear stress at the liquid–solid interface and $r_{h,l}$ is the hydraulic radius, defined as twice the cross-sectional area divided by the wetted perimeter.

This pressure gradient is a function of the Reynolds number Re_l and drag coefficient f_l defined as

$$Re_l = \frac{2(r_{h,l})\rho_l V_l}{\mu_l} \tag{11.7}$$

and

$$f_l = \frac{2\tau_l}{\rho_l V_l^2} \tag{11.8}$$

respectively, where V_l is the local liquid velocity, which is related to the local heat flow

$$V_l = \frac{q}{\varepsilon A_w \rho_l \lambda} \tag{11.9}$$

A_w is the wick cross-sectional area; ε is the wick porosity; and λ is the latent heat of vaporization.

Combining these expressions yields an expression for the pressure gradient in terms of the Reynolds number, drag coefficient, and thermophysical properties

$$\frac{dP_l}{dx} = \left(\frac{(f_l\, Re_l)\mu_l}{2\varepsilon\, A_w (r_{h,l})^2 \lambda \rho_l} \right) q \tag{11.10}$$

which can in turn be written as a function of the permeability K as

$$\frac{dP_l}{dx} = \left(\frac{\mu_l}{K A_w \lambda \rho_l} \right) q \tag{11.11}$$

where the permeability is expressed as

$$K = \frac{2\varepsilon (r_{h,l})^2}{f_l Re_l} \tag{11.12}$$

For steady-state operation with constant heat addition and removal, Equation (11.11) can be integrated over the length of the heat pipe to yield

$$\Delta P_l = \left(\frac{\mu_l}{K A_w \lambda \rho_l} \right) L_{eff} q \tag{11.13}$$

where L_{eff} is the effective heat pipe length defined as

$$L_{eff} = 0.5 L_e + L_a + 0.5 L_c \tag{11.14}$$

11.1.1.1.1.4 Vapor Pressure Drop The method for calculating the vapor pressure drop in heat pipes is similar to that used for the liquid pressure drop described above, but it is complicated by the mass addition and removal in the evaporator and condenser respectively and by the compressibility of the vapor phase. As a result, accurate computation of the total pressure drop requires that the dynamic pressure be included. In-depth discussions of the methodologies for determining the overall vapor pressure drop have been presented previously by Chi (1976), Dunn and Reay (1983), and Peterson (1994). The resulting expression is similar to that developed for the liquid

$$\Delta P_v = \left(\frac{C(f_v Re_v) \mu_v}{2(r_{h,v})^2 A_v \rho_v \lambda} \right) L_{eff} q \tag{11.15}$$

where $(r_{h,v})$ is the hydraulic radius of the vapor space and C is a constant that depends on the Mach number.

Unlike the liquid flow, which is driven by the capillary pressure difference and hence is always laminar, the vapor flow is driven by the temperature gradient, and for high heat flux applications may result in turbulent flow conditions. As a result, it is necessary to determine the vapor flow regime as a function of the heat flux by evaluating the local axial Reynolds number, defined as

$$Re_v = \frac{2(r_{h,v}) q}{A_v \mu_v \lambda} \tag{11.16}$$

Due to compressibility effects, it is also necessary to determine if the flow is compressible. This is accomplished by evaluating the local Mach number, defined as

$$Ma_v = \left(\frac{q}{A_v \rho_v \lambda (R_v T_v \gamma_v)} \right)^{1/2} \tag{11.17}$$

where R_v is the gas constant, T_v is the vapor temperature, and γ_v is the ratio of specific heats, which is equal to 1.67, 1.4, or 1.33 for monatomic, diatomic, and polyatomic vapor respectively [Chi, 1976]. A Mach number value greater than 0.2 is normally used as the criterion for considering the compressibility effects.

Previous investigations summarized by Kraus and Bar-Cohen (1983) have demonstrated that the following combinations of these conditions can be used with reasonable accuracy.

Laminar-incompressible

$$Re_v < 2300, \quad Ma_v < 0.2$$
$$(f_v Re_v) = 16 \tag{11.18}$$
$$C = 1.00$$

Laminar-compressible

$$Re_v < 2300, \quad Ma_v > 0.2$$
$$(f_v Re_v) = 16 \tag{11.19}$$
$$C = \left[1 + \left(\frac{\gamma_v - 1}{2} \right) Ma_v^2 \right]^{-1/2}$$

Turbulent-incompressible

$$Re_v > 2300, \quad M_v < 0.2$$
$$(f_v Re_v) = 0.038 \left(\frac{2(r_{h,v})q}{A_v \mu_v \lambda} \right)^{3/4} \tag{11.20}$$
$$C = 1.00$$

Turbulent-compressible

$$Re_v > 2300, \quad Ma_v > 0.2$$
$$(f_v Re_v) = 0.038 \left(\frac{2(r_{h,v})q}{A_v \mu_v \lambda} \right)^{3/4}$$
$$C = \left[1 + \left(\frac{\gamma_v - 1}{2} \right) Ma_v^2 \right]^{-1/2} \tag{11.21}$$

The solution procedure is to first assume laminar, incompressible flow and then to compute the Reynolds and Mach numbers. Once these values have been found, the initial assumptions of laminar, incompressible flow can be evaluated and the appropriate modifications made.

11.1.1.1.2 Viscous Limitation

At very low operating temperatures, the vapor pressure difference between the closed end of the evaporator (the high pressure region) and the closed end of the condenser (the low pressure region) may be extremely small. Because of this small pressure difference, the viscous forces within the vapor region may prove to be dominant and hence limit the heat pipe operation. Dunn and Reay (1983) discuss this limit in more detail and suggest the criterion

$$\frac{\Delta P_v}{P_v} < 0.1 \tag{11.22}$$

for determining when this limit might be of a concern. For steady-state operation or applications in the moderate operating temperature range, the viscous limitation normally will not be important.

11.1.1.1.3 Sonic Limitation

The sonic limitation in heat pipes is the result of vapor velocity variations along the length of the heat pipe due to the axial variation of the vaporization and condensation. Much like the effect of decreased outlet pressure in a converging-diverging nozzle, decreased condenser temperature results in a decrease in the evaporator temperature up to, but not beyond, that point where choked flow occurs in the evaporator causing the sonic limit to be reached. Any further decreases in the condenser temperature do not reduce either the evaporator temperature or the maximum heat transfer capability, due to the existence of choked flow.

The sonic limitation in heat pipes can be determined as

$$q_{s,m} = A_v \rho_v \lambda \left(\frac{\gamma_v R_v T_v}{2(\gamma_v + 1)} \right)^{1/2} \tag{11.23}$$

where T_v is the mean vapor temperature within the heat pipe [Chi, 1976].

11.1.1.1.4 Entrainment Limitation

In an operating heat pipe, the liquid and vapor typically flow in opposite directions resulting in a shear stress at the interface. At very high heat fluxes, liquid droplets may be picked up or entrained in the vapor flow. This entrainment results in dry-out of the evaporator wick due to excess liquid accumulation in the condenser. The Weber number, *We*, which represents the ratio of the viscous shear force to the force resulting from the liquid surface tension, can be used to determine at what point this entrainment is likely to occur.

$$We = \frac{2(r_{h,w})\rho_v V_v^2}{\sigma} \tag{11.24}$$

To prevent the entrainment of liquid droplets in the vapor flow, the Weber number must therefore be less than one, which implies that the maximum heat transport capacity based on the entrainment limitation may be determined as

$$q_{e,m} = A_v \lambda \left(\frac{\sigma \rho_v}{2(r_{h,w})} \right)^{1/2} \tag{11.25}$$

where $(r_{h,w})$ is the hydraulic radius of the wick structure [Dunn and Reay, 1983].

11.1.1.1.5 Boiling Limitation

As mentioned previously, all of the limits discussed so far depend upon the axial heat transfer. The boiling limit, however, depends upon the evaporator heat flux and occurs when the nucleate boiling in the evaporator wick creates vapor bubbles that partially block the return of fluid. The presence of vapor bubbles in the wick requires both the formation of bubbles and also the subsequent growth of these bubbles. Chi (1976) has developed an expression for the boiling limit, which can be written as

$$q_{b,m} = \left(\frac{2\pi L_{eff} k_{eff} T_v}{\lambda \rho_v \ln(r_i/r_v)} \right) \left(\frac{2\sigma}{r_n} - \Delta P_{c,m} \right) \tag{11.26}$$

where k_{eff} is the effective thermal conductivity of the liquid–wick combination, r_i is the inner radius of the heat pipe wall, and r_n is the nucleation site radius [Dunn and Reay 1983].

11.1.1.2 Heat Pipe Thermal Resistance

Once the maximum heat transport capacity is known, it is often useful to determine the temperature drop between the evaporator and condenser. The overall thermal resistance for a heat pipe is comprised of nine resistances of significantly different orders of magnitude arranged in a series/parallel combination. These resistances are shown in Figure 11.2 and can be summarized as follows:

R_{pe} — The radial resistance of the pipe wall at the evaporator

R_{we} — The resistance of the liquid–wick combination at the evaporator

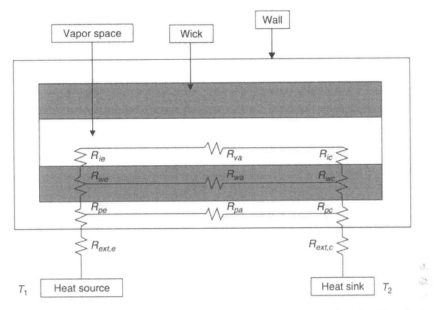

FIGURE 11.2   Thermal resistances. (Reprinted with permission from Peterson, G.P. [1994] *An Introduction to Heat Pipes: Modeling, Testing and Applications*, John Wiley & Sons, New York.)

R_{ie} — The resistance of the liquid–vapor interface at the evaporator
R_{va} — The resistance of the adiabatic vapor section
R_{pa} — The axial resistance of the pipe wall
R_{wa} — The axial resistance of the liquid–wick combination
R_{ic} — The resistance of the liquid–vapor interface at the condenser
R_{wc} — The resistance of the liquid–wick combination at the condenser
R_{pc} — The radial resistance of the pipe wall at the condenser

Previous investigations have indicated that typically the resistance of the vapor space, the axial resistances of the pipe wall, and liquid–wick combinations can all be neglected. In addition, the liquid–vapor interface resistances and the axial vapor resistance can, in most situations, be assumed to be negligible. This leaves only the pipe wall radial resistances and the liquid–wick resistances at both the evaporator and condenser.

As presented by Peterson (1994), the radial resistances at the pipe wall can be computed from Fourier's law as

$$R_{pe} = \frac{\delta}{k_p A_e} \tag{11.27}$$

for flat plates, where δ is the plate thickness and A_e is the evaporator area, or

$$R_{pe} = \frac{\ln(d_o/d_i)}{2\pi L_e k_p} \tag{11.28}$$

for cylindrical pipes, where L_e is the evaporator length. An expression for the equivalent thermal resistance of the liquid–wick combination in circular pipes is

$$R_{we} = \frac{\ln(d_o/d_i)}{2\pi L_e k_{eff}} \tag{11.29}$$

where k_{eff} is the effective conductivity of the liquid wick combination. Various models for the effective thermal conductivity of standard heat pipe wicks have been presented by Chi (1976), Dunn and Reay (1983), and Peterson (1994).

Combining these individual resistances allows the overall thermal resistance to be determined, which when combined with the maximum heat transport found previously, will yield an estimation of the overall temperature drop.

11.2 Individual Micro Heat Pipes

The earliest embodiments of micro heat pipes typically consisted of a long thin tube with one or more small noncircular channels that utilized the sharp angled corner regions as liquid arteries. While they were initially quite novel in size (see Figure 11.3), it soon became apparent that devices with characteristic diameters of approximately 1 mm functioned in nearly the same manner as larger, more conventional liquid artery heat pipes. Heat applied to one end of the heat pipe vaporizes the liquid in that region and forces it to move to the cooler end where it condenses and gives up the latent heat of vaporization. This vaporization and condensation process causes the liquid–vapor interface in the liquid arteries to change continually along the pipe, as illustrated in Figure 11.4, resulting in a capillary pressure difference between the evaporator and condenser regions. This capillary pressure difference promotes the flow of the working fluid from the condenser back to the evaporator through the triangular corner regions. These corner regions serve as liquid arteries; thus no wicking structure is required [Peterson, 1990, 1994]. The following sections summarize the analytical and experimental investigations conducted on individual micro heat pipes, arrays of micro heat pipes, flat plate microscale heat spreaders, and the latest advances in the development of highly conductive, flexible phase-change heat spreaders.

11.2.1 Modeling Micro Heat Pipe Performance

The first steady-state analytical models of individual micro heat pipes utilized the traditional pressure balance approach developed for use in more conventional heat pipes and described earlier in this chapter. These models provided a mechanism by which the steady-state and transient performance characteristics of micro heat pipes could be determined, and they indicated that while the operation was similar to that observed in larger, more conventional heat pipes, the relative importance of many of the parameters is quite different. Perhaps the most significant difference was the relative sensitivity of the micro heat pipes to the amount of working fluid present. These early steady-state models later led to the development of both transient numerical models and 3-D numerical models of the velocity, temperature, and pressure distribution within individual micro heat pipes [Peterson, 1992, 1994; Peterson et al., 1996; Longtin et al., 1994; Sobhan et al. 2000; Sobhan and Peterson, 2004].

11.2.1.1 Steady-State Modeling

The first steady-state model specifically designed for use in modeling of micro heat pipes was developed by Cotter (1984). Starting with the momentum equation and assuming a uniform cross-sectional area and no slip conditions at the boundaries, this expression was solved for both the liquid and vapor pressure

FIGURE 11.3   Micro heat pipe cooled ceramic chip carrier. (Reprinted with permission from Peterson, G.P. [1994] *An Introduction to Heat Pipes: Modeling, Testing and Applications*, John Wiley & Sons, New York.)

differential and then combined with the continuity expression. The result was a first order ordinary differential equation that related the radius of curvature of the liquid–vapor interface to the axial position along the pipe. Building upon this model, Peterson (1988a) and Babin et al. (1990) developed a steady-state model for a trapezoidal micro heat pipe using the conventional steady-state modeling techniques outlined by Chi (1976) and described earlier in this chapter. The resulting model demonstrated that the capillary pumping pressure governed the maximum heat transport capacity of these devices.

The performance limitations resulting from the models presented by Cotter (1984) and by Babin et al. (1990) were compared and indicated significant differences in the capillary limit predicted by the two models. These differences have been analyzed and found to be the result of specific assumptions made in the initial formulation of the models [Peterson, 1992].

A comparative analysis of these two early models was performed by Gerner et al. (1992), who indicated that the most important contributions of Babin et al. (1990) were the inclusion of the gravitational body force and the recognition of the significance of the vapor pressure losses. In addition, the assumption that the pressure gradient in the liquid flow passages was similar to that occurring in Hagen–Poiseuille flow was questioned, and a new scaling argument for the liquid pressure drop was presented. In this development, it was assumed that the average film thickness was approximately one-fourth the hydraulic radius, resulting in a modified expression for the capillary limitation.

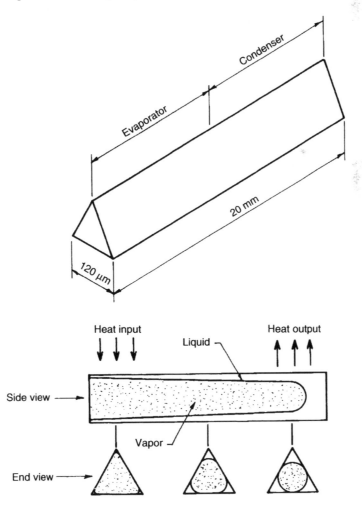

FIGURE 11.4 Micro heat pipe operation. (Reprinted with permission from Peterson, G.P., Swanson, L.W., and Gerner, F.M. (1996) "Micro Heat Pipes," in *Microscale Energy Transport*, C.L. Tien, A. Majumdar, and F.M. Gerner eds., Taylor-Francis Publishing Co., Washington D.C.)

A significant contribution made by Gerner et al. (1992) was the recognition that the capillary limit may never actually be reached due to Kelvin–Helmholtz type instabilities occurring at the liquid–vapor interface. Using stability analysis criteria for countercurrent flow in tubes developed by Tien et al. (1979) and minimizing the resulting equations, the wavelength was found to be approximately 1 cm for atmospheric water and methanol. Since this length was long with respect to the characteristic wavelength, it was assumed that gravity was the dominant stabilizing mechanism. The decision as to whether to use the traditional capillary limit proposed by Babin et al. (1990) or the interfacial instability limit proposed by Gerner (1992) should be governed by evaluating the shape and physical dimensions of the specific micro heat pipe being considered.

Khrustalev and Faghri (1994) presented a detailed mathematical model of the heat and mass transfer processes in micro heat pipes that described the distribution of the liquid and the thermal characteristics as a function of the liquid charge. The liquid flow in the triangular corners of a micro heat pipe with polygonal cross-section was considered by accounting for the variation of the curvature of the free liquid surface and the interfacial shear stresses due to the liquid vapor interaction. The predicted results were compared with the experimental data obtained by Wu and Peterson (1991) and Wu et al. (1991) and indicated the importance of the liquid charge, the contact angle, and the shear stresses at the liquid vapor interface in predicting the maximum heat transfer capacity and thermal resistance of these devices.

Longtin et al. (1994) developed a one-dimensional steady-state model for the evaporator section of the micro heat pipe. The governing equations were solved, assuming a uniform temperature along the heat pipe. The solution indicated that the maximum heat transport capacity varied with respect to the cube of the hydraulic diameter of the channel.

An analytical model for the etched triangular micro heat pipe developed by Duncan and Peterson (1995) is capable of calculating the curvature of the liquid–vapor meniscus in the evaporator. This model was used to predict the capillary limit of operation of the heat pipe and to arrive at the optimal value of the liquid charge. In a subsequent work, a hydraulic diameter was defined, incorporating the frictional effects of the liquid and the vapor, and was used in a model for predicting the minimum meniscus radius and maximum heat transport in triangular grooves [Peterson and Ma, 1996b]. The major parameters influencing the heat transport capacity of the micro heat pipe were found to be the apex angle of the liquid arteries, the contact angle, heat pipe length, vapor velocity and the tilt angle. Ma and Peterson (1998) presented analytical expressions for the minimum meniscus radius and the maximum capillary heat transport limit in micro heat pipes that were validated with experimental data.

A detailed steady-state mathematical model for predicting the heat transport capability of a micro heat pipe and the temperature gradients that contribute to the overall axial temperature drop as a function of the heat transfer was developed by Peterson and Ma (1999). The unique nature of this model was that it considered the governing equation for fluid flow and heat transfer in the evaporating thin film region. The model also consisted of an analytical solution of the two dimensional heat conduction in the macro evaporating regions in the triangular corners. The effects of the vapor and liquid flows in the passage, the flow and condensation of the thin film caused by the surface tension in the condenser, and the capillary flow along the axial direction of the micro heat pipe were considered in this model. The predicted axial temperature distribution was compared with experimental data, with very good agreement. The model was capable of calculating both the heat transfer distribution through the thin film region and the heat transfer-operating temperature dependence of the micro heat pipe. It was concluded from the study that the evaporator temperature drop was considerably larger than that at the condenser and that the temperature drops increased with an increase in input power when the condenser is kept at a constant temperature.

The maximum heat transfer capacity of copper–water micro heat pipes was also explored by Hopkins et al. (1999) using a one-dimensional model for predicting the capillary limitation. In this analysis, the liquid–vapor meniscus was divided into two regions depending on whether the contact angle can be treated as a constant at the evaporator or as a variable along the adiabatic and condenser sections.

11.2.1.2 Transient Modeling

As heat pipes diminish in size, the transient nature becomes of increasing interest. The ability to respond to rapid changes in heat flux coupled with the need to maintain constant evaporator temperature in modern high-powered electronics necessitates a complete understanding of the temporal behavior of these devices. The first reported transient investigation of micro heat pipes was conducted by Wu and Peterson (1991). This initial analysis utilized the relationship developed by Collier (1981) and used later by Colwell and Chang (1984) to determine the free molecular flow mass flux of evaporation. The most interesting result from this model was the observation that reverse liquid flow occurred during the start-up of micro heat pipes. As explained in the original reference [Wu et al., 1990], this reverse liquid flow is the result of an imbalance in the total pressure drop and occurs because the evaporation rate does not provide an adequate change in the liquid–vapor interfacial curvature to compensate for the pressure drop. As a result, the increased pressure in the evaporator causes the meniscus to recede into the corner regions forcing liquid out of the evaporator and into the condenser. During start-up, the pressure of both the liquid and vapor are higher in the evaporator and gradually decrease with position, promoting flow away from the evaporator. Once the heat input reaches full load, the reverse liquid flow disappears and the liquid mass flow rate into the evaporator gradually increases until a steady-state condition is reached. At this time the change in the liquid mass flow rate is equal to the change in the vapor mass flow rate for any given section [Wu and Peterson, 1991]. The flow reversal in the early transient period of operation of a micro heat pipe has also been captured by Sobhan et al. (2000) using their numerical model.

Several more-detailed transient models have been developed. Badran et al. (1993) developed a conjugate model to account for the transport of heat within the heat pipe and conduction within the heat pipe case. This model indicated that the specific thermal conductivity of micro heat pipes (effective thermal conductivity divided by the density) could be as high as 200 times that of copper and 100 times that of Gr/Cu composites.

Ma et al. (1996) developed a closed mathematical model of the liquid friction factor for flow occurring in triangular grooves. This model, which built upon the earlier work of Ma et al. (1994), considered the interfacial shear stresses due to liquid–vapor frictional interactions for countercurrent flow. Using a coordinate transformation and the Nachtsheim–Swigert iteration scheme, the importance of the liquid vapor interactions on the operational characteristics of micro heat pipes and other small phase change devices was demonstrated. The solution resulted in a method by which the velocity distribution for countercurrent liquid–vapor flow could be determined, and it allowed the governing liquid flow equations to be solved for cases where the liquid surface is strongly influenced by the vapor flow direction and velocity. The results of the analysis were verified using an experimental test facility constructed with channel angles of 20, 40, and 60 degrees. The experimental and predicted results were compared and found to be in good agreement [Ma and Peterson 1996a, 1996b; Peterson and Ma 1996a].

A transient model for a triangular micro heat pipe with an evaporator and condenser section was presented by Sobhan et al. (2000). The energy equation as well as the fluid flow equations were solved numerically, incorporating the longitudinal variation of the cross-sectional areas of the vapor and liquid flows, to yield the velocity, pressure, and temperature distributions. The effective thermal conductivity was computed and characterized with respect to the heat input and the cooling rate under steady and transient operation of the heat pipe. The reversal in the liquid flow direction as discussed by Wu and Peterson (1991) was also obvious from the computational results.

11.2.1.3 Transient One-Dimensional Modeling of Micro Heat Pipes

A flat micro heat pipe heat sink consisting of an array of micro heat pipe channels was used to form a compact heat dissipation device to remove heat from electronic chips. Each channel in the array served as an independent heat transport device. The analysis presented here examined an individual channel in such an array. The individual micro heat pipe channel analyzed had a triangular cross-section. The channel was fabricated on a copper substrate and the working fluid used was ultrapure water.

11.2.1.3.1 The Mathematical Model

The micro heat pipe consisted of an externally heated evaporator section and a condenser section subjected to forced convective cooling. A one-dimensional model was sufficient for the analysis, as the variations in the field variables were significant only in the axial direction, due to the geometry of the channels. A transient model that proceeded until steady-state was utilized to analyze the problem completely. In this problem, the flow and heat transfer processes are governed by the continuity, momentum, and energy equations for the liquid and vapor phases. A nonconservative formulation can be utilized because the problem deals with low velocity flows. As phase change occurs, the local mass rates of the individual liquid and vapor phases are coupled through a mass balance at the liquid–vapor interface. The cross-sectional areas of the vapor and liquid regions and the interfacial area vary along the axial length due to the progressive phase change occurring as the fluid flows along the channel. These variations in the area can be incorporated into the model through the use of suitable geometric area coefficients, as described in Longtin et al. (1994). The local meniscus radii at the liquid–vapor interface are calculated using the Laplace–Young equation. The friction factor, which appears in the momentum and energy equations is incorporated through appropriate models for fluid friction in varying area channels, as described in the literature.

The governing differential equations can be described as follows:

11.2.1.3.2   Laplace–Young Equation

$$P_v - P_l = \frac{\sigma}{r} \tag{11.30}$$

11.2.1.3.3   Vapor phase equations
Vapor continuity equation: evaporator section

$$\left(\frac{\sqrt{3}}{4}d^2 - \beta_l r^2\right)\frac{\partial u_v}{\partial x} - 2\beta_l u_v r \frac{\partial r}{\partial x} + \beta_i \frac{\rho_l}{\rho_v} r V_{il} = 0 \tag{11.31}$$

Vapor continuity equation: condenser section

$$\left(\frac{\sqrt{3}}{4}d^2 - \beta_l r^2\right)\frac{\partial u_v}{\partial x} - 2\beta_l u_v r \frac{\partial r}{\partial x} - \beta_i \frac{\rho_l}{\rho_v} r V_{il} = 0 \tag{11.32}$$

It should be noted that the vapor continuity equation incorporates the interfacial mass balance equation.

$$\rho_l V_{il} = \rho_v V_{iv} \tag{11.33}$$

Vapor momentum equation:

$$\rho_v \left(\frac{\sqrt{3}}{4}d^2 - \beta_l r^2\right)\frac{\partial u_v}{\partial t} = 2\rho_v \left(\frac{\sqrt{3}}{4}d^2 - \beta_l r^2\right)u_v \frac{\partial u_v}{\partial x} - 2\rho_v \beta_l r u_v^2 \frac{\partial r}{\partial x}$$

$$+ \left(\frac{\sqrt{3}}{4}d^2 - \beta_l r^2\right)\frac{\partial P_v}{\partial x} - \frac{1}{2}\rho_v u_v^2 f_{vw}(3d - \beta_{lw}r)$$

$$- \frac{1}{2}\rho_v^2 u_v^2 f_{vi}\beta_l r \tag{11.34}$$

Vapor energy equation: evaporator section

$$\left(\frac{\sqrt{3}}{4}d^2 - \beta_l r^2\right)\frac{\partial E_v}{\partial t} + \frac{\partial}{\partial x}\left[u_v\left(\frac{\sqrt{3}}{4}d^2 - \beta_l r^2\right)(E_v + P_v)\right]$$

$$= \frac{\partial}{\partial x}\left\{\frac{4}{3}\mu_v\left(\frac{\sqrt{3}}{4}d^2 - \beta_l r^2\right)u_v\frac{\partial u_v}{\partial x} + k_v\left(\frac{\sqrt{3}}{4}d^2 - \beta_l r^2\right)\frac{\partial T_v}{\partial x}\right\}$$

$$+ q(3d - \beta_{lw}r) + h_{fg}V_{il}\rho_l\beta_l r + \frac{1}{2}\rho_v u_v^2 f_{vw}u_v(3d - \beta_{lw}r) + \frac{1}{2}\rho_v u_v^2 f_{vi}u_v\beta_l r \tag{11.35}$$

Vapor Energy equation: condenser section

$$\left(\frac{\sqrt{3}}{4}d^2 - \beta_l r^2\right)\frac{\partial E_v}{\partial t} + \frac{\partial}{\partial x}\left[u_v\left(\frac{\sqrt{3}}{4}d^2 - \beta_l r^2\right)(E_v + P_v)\right]$$

$$= \frac{\partial}{\partial x}\left\{\frac{4}{3}\mu_v\left(\frac{\sqrt{3}}{4}d^2 - \beta_l r^2\right)u_v\frac{\partial u_v}{\partial x} + k_v\left(\frac{\sqrt{3}}{4}d^2 - \beta_l r^2\right)\frac{\partial T_v}{\partial x}\right\}$$

$$+ h_{fg}V_{il}\rho_l\beta_i r - h_o(3d - \beta_{lw}r)\Delta T + \frac{1}{2}\rho_v u_v^2 f_{vw}u_v(3d - \beta_{lw}r) + \frac{1}{2}\rho_v u_v^2 f_{vi}u_v\beta_i r \quad (11.36)$$

11.2.1.3.4 *Liquid Phase Equations*
Liquid continuity equation: evaporator section

$$r\frac{\partial u_l}{\partial x} + 2u_l\frac{\partial r}{\partial x} - \frac{\beta_i}{\beta_l}V_{il} = 0 \quad (11.37)$$

Liquid continuity equation: condenser section

$$r\frac{\partial u_l}{\partial x} + 2u_l\frac{\partial r}{\partial x} + \frac{\beta_i}{\beta_l}V_{il} = 0 \quad (11.38)$$

Liquid momentum equation:

$$\rho_l r\frac{\partial u_l}{\partial t} = -2\rho_l\left(ru_l\frac{\partial u_l}{\partial x} + u_l^2\frac{\partial r}{\partial x}\right) - r\frac{\partial P_l}{\partial x} - \frac{1}{2}\rho_l u_l^2 f_{lw}\frac{\beta_{lw}}{\beta_l} - \frac{1}{2}\rho_l u_l^2 f_{li}\frac{\beta_i}{\beta_l} \quad (11.39)$$

Liquid energy equation: evaporator section

$$\beta_l r^2\frac{\partial E_l}{\partial t} + \frac{\partial}{\partial x}[u_l\beta_l r^2(E_l + P_l)] = \frac{\partial}{\partial x}\left[\frac{4}{3}\mu_l u_l\beta_l r^2\frac{\partial u_l}{\partial x} + k\beta_l r^2\frac{\partial T_l}{\partial x}\right]$$

$$+ q\beta_{lw}r - h_{fg}V_{il}\rho_l\beta_i r + \frac{1}{2}\rho_l u_l^2 f_{lw}u_l\beta_{lw}r$$

$$+ \frac{1}{2}\rho_l u_l^2 f_{li}u_l\beta_i r \quad (11.40)$$

Liquid energy equation: condenser section

$$\beta_l r^2\frac{\partial E_l}{\partial t} + \frac{\partial}{\partial x}[u_l\beta_l r^2(E_l + P_l)] = \frac{\partial}{\partial x}\left[\frac{4}{3}\mu_l u_l\beta_l r^2\frac{\partial u_l}{\partial x} + k\beta_l r^2\frac{\partial T_l}{\partial x}\right]$$

$$+ h_{fg}V_{il}\rho_l\beta_i r - h_o\beta_{lw}r\Delta T + \frac{1}{2}\rho_l u_l^2 f_{lw}u_l\beta_{lw}r$$

$$+ \frac{1}{2}\rho_l u_l^2 f_{li}u_l\beta_i r \quad (11.41)$$

The vapor and liquid pressures can be computed as follows:

1. The ideal gas equation of state is utilized for computing the pressure in the vapor. Because the vapor is either saturated or super heated, the ideal gas state equation is reasonably correct and is used extensively in the analysis.

2. For the liquid phase the Hagen–Poiseuille equation is used as a first approximation, with the local hydraulic diameter for the wetted portion of the liquid-filled region adjacent to the corners. The values of pressure obtained from this first approximation are substituted into the momentum equations and iterated for spatial convergence.

11.2.1.3.5 State Equations
Equation of state for the vapor:

$$P_v = \rho_v R_v T_v \tag{11.42}$$

Hagen–Poiseuielle Equation as first approximation for the liquid flow

$$\frac{\partial P_l}{\partial x} = -\frac{8\mu_l u_l}{\left(\dfrac{D_H^2}{4}\right)} \tag{11.43}$$

The boundary conditions are

$$\text{at } x = 0 \quad \text{and} \quad x = L$$

$$u_l = 0; \quad u_v = 0; \quad \partial T/\partial x = 0$$

The initial conditions are

$$\text{at } t = 0 \quad \text{and} \quad \text{for all } x$$

$$P_l = P_v = P_{sat}; \quad T_l = T_v = T_{amb.}$$

At $x = 0$

$$P_v - P_l = \frac{\sigma}{r_o} \tag{11.44}$$

The value of r_o, the initial radius of curvature of the interface meniscus for the copper–water system, was adopted from the literature.

A numerical procedure based on the finite difference method was used to solve the above system of equations to obtain the transient behavior and field distributions in the micro heat pipe, and the results of computation have been discussed in the literature [Sobhan and Peterson, 2004]. Parametric studies were also presented in this paper.

11.2.1.3.6 Area Coefficients in the Computational Model
Figure 11.5 illustrates the geometric configuration of the vapor and liquid flow in the cross-section of the micro heat pipe, along with the meniscus idealized as an arc of a circle at any longitudinal location. The definitions of the area coefficients, as derived for this configuration, are given below:

Referring to Figure 11.5, the cross section is an equilateral triangle with $\phi = \pi/3 - \alpha$, and $\eta = r \sin \phi$. The total area of the liquid in the cross section is

$$A_l = \beta_l r^2 \tag{11.45}$$

Where, $\beta_l = 3\left[\sqrt{3} \sin^2\left(\frac{\pi}{3} - \alpha\right) + 0.5 \sin 2\left(\frac{\pi}{3} - \alpha\right) - \left(\frac{\pi}{3} - \alpha\right)\right]$

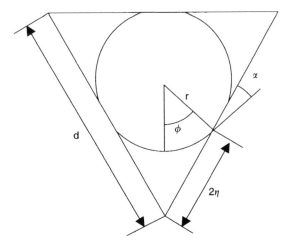

FIGURE 11.5   Cross-sectional geometry of the triangular micro heat pipe for determining the area coefficients. (Reprinted with permission from Longtin, J.P., Badran, B., and Gerner, F.M. [1994] "A One-Dimensional Model of a Micro Heat Pipe During Steady-State Operation," *ASME J. Heat Transfer* 116, pp. 709–15.)

The total area of interface in length dx is

$$A_i = \beta_i r dx, \quad \text{where } \beta_i = 6\left(\frac{\pi}{3} - \alpha\right) \tag{11.46}$$

The total wetted perimeter for the three corners $= \beta_{lw} r$,

$$\text{where } \beta_{lw} = \pi \sin\left(\frac{\pi}{3} - \alpha\right) \tag{11.47}$$

11.2.2   Testing of Individual Micro Heat Pipes

As fabrication capabilities have developed, experimental investigations on individual micro heat pipes have been conducted on progressively smaller and smaller devices, beginning with early investigations on what now appear to be relatively large micro heat pipes, approximately 3 mm in diameter and progressing to micro heat pipes in the 30 μm diameter range. These investigations have included both steady-state and transient investigations.

11.2.2.1   Steady-State Experimental Investigations

The earliest experimental tests of this type reported in the open literature were conducted by Babin et al. (1990), who evaluated several micro heat pipes approximately 1 mm in external diameter. The primary purposes of this investigation were to determine the accuracy of the previously described steady-state modeling techniques, to verify the micro heat pipe concept, and to determine the maximum heat transport capacity. The fabrication techniques used to produce these test articles were developed by Itoh Research and Development Company, Osaka, Japan (Itoh, 1988). As reported previously, a total of four test articles were evaluated, two each from silver and copper. Two of these test pipes were charged with distilled deionized water, and the other two were used in an uncharged condition to determine the effect of the vaporization–condensation process on these devices' overall thermal conductivity. Steady-state tests were conducted over a range of tilt angles to determine the effect of the gravitational body force on the operational characteristics. An electrical resistance heater supplied the heat into the evaporator. Heat rejection was achieved through the use of a constant temperature ethyl–glycol solution, which flowed over the condenser portion of the heat pipe. The axial temperature profile was continuously monitored by five thermocouples bonded to the outer surface of the heat pipe using a thermally conductive epoxy. Three thermocouples were located on the evaporator, one on the condenser, and one on the outer surface of the adiabatic section. Throughout the tests, the heat input was systematically increased and the temperature of the coolant bath adjusted to maintain a constant adiabatic wall temperature (Babin et al., 1990).

The results of this experiment have been utilized as a basis for comparison with a large number of heat pipe models. As previously reported [Peterson et al., 1996], the steady-state model of Babin et al. (1990) over-predicted the experimentally determined heat transport capacity at operating temperatures below 40°C and under-predicted it at operating temperatures above 60°C. These experimental results represented the first successful operation of a micro heat pipe that utilized the principles outlined in the original concept of Cotter (1984) and, as such, paved the way for numerous other investigations and applications.

There has been a large amount of experimental research work on micro heat pipes under steady-state operation following the early experimental studies reported by Babin et al. (1990). Fabricating micro heat pipes as an integral part of silicon wafers provided a means of overcoming the problems imposed by the thermal contact resistance between the heat pipe heat sink and the substrate material, and Peterson et al. (1991) in their first attempt to study the performance of an integral micro heat pipe, compared the temperature distributions in a silicon wafer with and without a charged micro heat pipe channel. The wafer with the integral micro heat pipe showed as much as an 11% reduction of the maximum chip temperature, which worked out to a 25% increase in effective thermal conductivity, at a heat flux rate of 4 W/cm². More extensive experimental work and detailed discussions on triangular and rectangular micro heat pipes and on micro heat pipe arrays fabricated on silicon wafers can be found in Mallik et al. (1992) and Peterson et al. (1993). Peterson (1994) further discussed the fabrication, operation, modeling, and testing aspects of integral micro heat pipes in silicon. Steady-state experiments on a micro heat pipe array fabricated in silicon using the vapor deposition technique were also reported by Mallik and Peterson (1995).

In an attempt to conduct visualization experiments on micro heat pipes, Chen et al. (1992) fabricated a heat transport device by attaching a wire insert to the inner wall of a glass capillary tube, so that capillary action is obtained at the corners formed by the two surfaces. This was also modeled as a porous medium. The device was highly influenced by gravity, as revealed by comparisons of the experimental and predicted results for maximum heat flow with horizontal and vertical orientations. It appears that this device functioned more like a thermosyphon than a capillary driven heat pipe.

Experimental studies were performed on triangular grooves fabricated in a copper substrate with methanol as the working fluid in order to determine the capillary heat transport limit [Ma and Peterson, 1996a]. A parameter, "the unit effective area heat transport," was defined for the grooves ($q_{eff} = q/A_{eff}$, where $A_{eff} = D_H/Le$) to be used as a performance index. An optimum geometry was found that gave the maximum unit effective area heat transport. Further, this maximum depended on the geometrical parameters, namely the tilt angle and the effective length of the heat pipe.

Modifying the analytical model for the maximum heat transport capacity of a micro heat pipe developed by Cotter (1984), a semiempirical correlation was proposed by Ha and Peterson (1998b). The method used was to compare the results predicted by Cotter's model with experimental data and then modify the model to incorporate the effects of the intrusion of the evaporator section into the adiabatic section of the heat pipe under near dry-out conditions. With the proposed semiempirical model, a better agreement between the predicted and experimental results was obtained.

Hopkins et al. (1999) experimentally determined the maximum heat load for various operating temperatures of copper–water micro heat pipes. These micro heat pipes consisted of trapezoidal or rectangular micro grooves and were positioned in vertical or horizontal orientations. The dry-out condition also was studied experimentally. The effective thermal resistance was found to decrease with an increase in the heat load.

11.2.2.2 Transient Experimental Investigations

While the model developed by Babin et al. (1990) was shown to predict the steady-state performance limitations and operational characteristics of the trapezoidal heat pipe reasonably well for operating temperatures between 40 and 60°C, little was known about the transient behavior of these devices. As a result, Wu et al. (1991) undertook an experimental investigation of the devices' transient characteristics. This experimental investigation again utilized micro heat pipe test articles developed by Itoh (1988); however, this particular test pipe was designed to fit securely under a ceramic chip carrier and had small fins at the condenser end of the heat pipe for removal of heat by free or forced convection, as shown in Figure 11.3. Start-up and transient tests were conducted in which the transient response characteristics of the heat pipe as a function of incremental power increases, tilt angle, and mean operating temperature were measured.

Itoh and Polásek (1990a, 1990b), presented the results of an extensive experimental investigation on a series of micro heat pipes ranging in size and shape from 1 to 3 mm in diameter and 30 to 150 mm in length. The investigation utilized both cross-sectional configurations, similar to those presented previously or a conventional internal wicking structure (Polásek, 1990; Fejfar et al., 1990). The unique aspect of this particular investigation was the use of neutron radiography to determine the distribution of the working fluid within the heat pipes [Itoh and Polásek, 1990a; Itoh and Polásek, 1990b; Ikeda, 1990]. Using this technique, the amount and distribution of the working fluid and noncondensale gases were observed during real time operation along with the boiling and/or reflux flow behavior. The results of these tests indicated several important results [Peterson, 1992];

- As is the case for conventional heat pipes, the maximum heat transport capacity is principally dependent upon the mean adiabatic vapor temperature.
- Micro heat pipes with smooth inner surfaces were found to be more sensitive to overheating than those with grooved capillary systems.
- The wall thickness of the individual micro heat pipes had greater effect on the thermal performance than did the casing material.
- The maximum transport capacity of heat pipes utilizing axial channels for return of the liquid to the evaporator were found to be superior to those utilizing a formal wicking structure.

The experimental work on the micro heat pipe array fabricated in silicon using vapor deposition technique [Mallik and Peterson, 1995] was extended to also include the performance under transient conditions. The results of this study were presented in Peterson and Mallik (1995).

11.3 Arrays of Micro Heat Pipes

Apart from theoretical and experimental research on individual micro heat pipes, modeling, fabrication, and testing of micro heat pipe arrays of various designs also have been undertaken. Significant work on these subjects is presented in the following sections.

11.3.1 Modeling of Heat Pipe Arrays

The initial conceptualization of micro heat pipes by Cotter (1984) envisioned fabricating micro heat pipes directly into the semiconductor devices as shown schematically in Figure 11.6. While many of the previously discussed models can be used to predict the performance limitations and operational characteristics of individual micro heat pipes, it is not clear from the models or analyses how the incorporation of an array of these devices might affect the temperature distribution or the resulting thermal performance. Mallik et al. (1991)

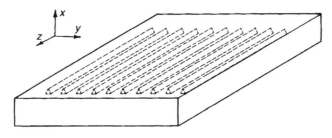

FIGURE 11.6   Array of micro heat pipes fabricated as an integral part of a silicon wafer.

FIGURE 11.7   (See color insert following page 2-12.) Silicon wafer into which an array of micro heat pipes has been fabricated.

developed a three-dimensional numerical model capable of predicting the thermal performance of an array of parallel micro heat pipes constructed as an integral part of semiconductor chips similar to that illustrated in Figure 11.7. In order to determine the potential advantages of this concept, several different thermal loading configurations were modeled and the reductions in the maximum surface temperature, the mean chip temperature, and the maximum temperature gradient across the chip were determined [Peterson, 1994].

Although the previous investigations of Babin et al. (1990), Wu and Peterson (1991), and Wu et al. (1991) indicated that an effective thermal conductivity greater than ten times that of silicon could be achieved, additional analyses were conducted to determine the effect of variations in this value. Steady-state analyses were performed using a heat pipe array comprised of nineteen parallel heat pipes. Using an effective thermal conductivity ratio of five, the maximum and mean surface temperatures were 37.69°C and 4.91°C respectively. With an effective thermal conductivity ratio of ten, the maximum and mean surface temperatures were 35.20°C and 4.21°C respectively. Using an effective thermal conductivity ratio of fifteen, the maximum and mean surface temperatures were 32.67°C and 3.64°C respectively [Peterson, 1994]. These results illustrate how the incorporation of an array of micro heat pipes can reduce the maximum wafer temperature, reduce the temperature gradient across the wafers, and eliminate localized hot spots. In addition, this work highlighted the significance of incorporating these devices into semiconductor chips, particularly those constructed in materials with thermal conductivities significantly less than that of silicon, such as gallium arsenide.

This work was further extended to determine transient response characteristics of an array of micro heat pipes fabricated into silicon wafers as a substitute for polycrystalline diamond or other highly thermally conductive heat spreader materials [Mallik and Peterson 1991; Mallik et al. 1992]. The resulting transient three-dimensional numerical model was capable of predicting the time dependent temperature distribution occurring within the wafer when given the physical parameters of the wafer and the locations of the heat sources and sinks. The model also indicated that significant reductions in the maximum localized wafer temperatures and thermal gradients across the wafer could be obtained through the incorporation of an array of micro heat pipes. Utilizing heat sinks located on the edges of the chip perpendicular to the axis of the heat pipes and a cross-sectional area porosity of 1.85%, reductions in the maximum chip temperature of up to 40% were predicted.

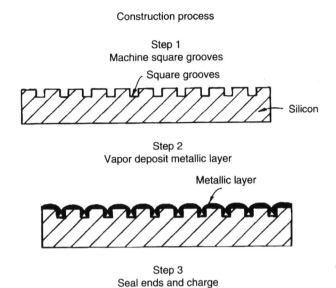

Construction process

Step 1
Machine square grooves

Square grooves

Silicon

Step 2
Vapor deposit metallic layer

Metallic layer

Step 3
Seal ends and charge

FIGURE 11.8 Vapor deposition process for fabricating micro heat pipes.

11.3.2 Testing of Arrays of Micro Heat Pipes

Peterson et al. (1991) fabricated, charged, and tested micro heat pipe arrays incorporated as an integral part of semiconductor wafers. These tests represented the first successful operation of these devices reported in the open literature. In this investigation, several silicon wafers were fabricated with distributed heat sources on one side and an array of micro heat pipes on the other as illustrated in Figure 11.7. Since that time, a number of experimental investigations have been conducted to verify the micro heat pipe array concept and determine the potential advantages of constructing an array of micro heat pipes as an integral part of semiconductor devices [Peterson et al. 1993; Peterson 1994]. The arrays tested have typically been fabricated in silicon and have ranged in size from parallel rectangular channels 30 μm wide, 80 μm deep, and 19.75 mm long, machined into a silicon wafer 20 mm square and 0.378 mm thick with an interchannel spacing of 500 μm to etched arrays of triangular channels 120 μm wide and 80 μm deep machined into 20 mm square silicon wafers 0.5 mm thick [Peterson et al. 1993]. In addition, arrays of micro heat pipes fabricated using a vapor deposition process first proposed by Peterson (1990) and illustrated in Figure 11.8 were tested by Mallik et al. (1995).

In this work, wafers with arrays of 34 and 66 micro heat pipes were evaluated using an IR thermal imaging system in conjunction with a VHS video recorder. These arrays occupied 0.75% and 1.45% of the wafer cross-sectional area respectively. The wafers with micro heat pipe arrays demonstrated a 30% to 45% reduction in the thermal time constant when compared to that obtained for plain silicon wafers, which led to a significant reduction in the maximum wafer temperature. The experimental results were then used to validate the transient numerical model described previously [Peterson and Mallik, 1995].

11.3.3 Fabrication of Arrays of Micro Heat Pipes

Considerable information is available on the methods used to fabricate micro heat pipes with hydraulic diameters on the order of 20 to 150 μm in diameter into silicon or gallium arsenide wafers. These early investigations included the use of conventional techniques such as the machining of small channels [Peterson, 1988b; Peterson et al., 1991]; the use of directionally dependent etching processes to create rectangular or triangular shaped channels [Peterson, 1988b; Gerner, 1990; Mallik et al., 1991; Gerner et al., 1992]; or other more elaborate techniques that utilize a multisource vapor deposition process illustrated in Figure 11.8 [Mallik et al., 1991; Weichold et al., 1992] to create an array of long narrow channels of triangular cross-section lined with a thin layer of copper. Peterson (1994) has summarized these. The earliest fabricated arrays were machined into a silicon wafer 2 cm square and 0.378 mm thick, with an interchannel spacing of 500 μm. Somewhat later, Adkins et al. (1994) reported on a different fabrication process used for an array of heat pipes with a segmented vapor space. Peterson (1988b), Gerner (1990), Peterson et al. (1993), Ramadas et al. (1993), and Gerner et al. (1994) have described other processes. All of these techniques were similar in nature and typically utilized conventional photolithography masking techniques coupled with an orientation dependent etching technique.

Perhaps the most important aspects of these devices are the shape and relative areas of the liquid and vapor passages. A number of investigations have been directed at the optimization of these grooves. These include investigations by Ha and Peterson (1994), which analytically evaluated the axial dry-out of the evaporating thin liquid film; one by Ha and Peterson (1996), which evaluated the interline heat transfer; and others that examined other important aspects of the problem [Ha and Peterson 1998a, 1998b; Peterson and Ha, 1998; Ma and Peterson 1998]. These studies and others have shown both individual micro heat pipes and arrays of micro heat pipes to be extremely sensitive to flooding [Peterson, 1992]. For this reason, several different charging methods have been developed and described in detail [Duncan and Peterson, 1995]. These vary from those that are similar to the methods utilized on larger more conventional heat pipes to one in which the working fluid is added and then the wafer is heated to above the critical temperature of the working fluid so that the working fluid is in the supercritical state and exists entirely as a vapor. The array is then sealed and allowed to cool to below the critical temperature, allowing the vapor to cool and condense. When in the critical state, the working fluid is uniformly distributed throughout the individual micro heat pipes, so the exact charge can be carefully controlled and calculated.

11.3.4   Wire Bonded Micro Heat Pipe Arrays

One of the designs that has been developed and evaluated for use in both conventional electronic applications and for advanced spacecraft applications consists of a flexible micro heat pipe array fabricated by sintering an array of aluminum wires between two thin aluminum sheets as shown in Figure 11.9. In this design, the sharp corner regions formed by the junction of the plate and the wires act as the liquid arteries. When made of aluminum with ammonia or acetone as the working fluid, these devices become excellent candidates for use as flexible radiator panels for long-term spacecraft missions, and they can have a thermal conductivity that greatly exceeds the conductivity of an equivalent thickness of any known material.

A numerical model combining both conduction and radiation effects to predict the heat transfer performance and temperature distribution of these types of radiator fins in a simulated space environment has been developed [Wang et al., 2001]. Three different configurations were analyzed, experimentally evaluated, and the results compared. Each of the three configurations were modeled both with and without a working fluid charge in order to determine the reduction in the maximum temperature, mean temperature, and temperature gradient on the radiator surface. Table 11.1 lists the physical specifications of the three micro heat pipe arrays fabricated. Acetone was used as the working fluid in both the modeling effort and also in the actual experimental tests. The flexible radiator with the array of micro heat pipes was found to have an effective thermal conductivity of more than 20 times that of the uncharged version and 10 times that of a solid material.

The results of the preliminary tests conducted on these configurations are shown in Figure 11.10. As indicated, the heat transport was proportional to the temperature difference between the evaporator and condenser; that is, the effective thermal conductivity of the micro heat pipe array was constant with respect to the temperature. From the temperature difference and heat transport obtained as shown in Figure 11.10, the effective conductivity was calculated. As illustrated in Figure 11.11, the effective thermal conductivities of micro heat pipe arrays No. 1, No. 2, and No. 3 were 1446.2 W/Km, 521.3 W/Km, and 3023.1 W/Km, respectively. For the micro heat pipe arrays without any working fluid, the effective conductivities in the x-direction were 126.3 W/Km, 113.0 W/Km, and 136.2 W/Km respectively. Comparison of the predicted and experimental results indicated these flexible radiators with the arrays of micro heat pipes have an effective thermal conductivity of between fifteen and twenty times that of the uncharged version. This results in a more uniform temperature distribution that could significantly improve the overall radiation effectiveness, reduce the overall size, and meet or exceed the baseline design requirements for long-term manned missions to Mars.

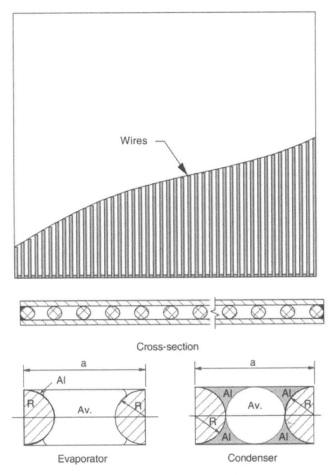

FIGURE 11.9 Flexible wire bonded heat pipe. (Reprinted with permission from Wang, Y., Ma, H.B., and Peterson, G.P. (2001) "Investigation of the Temperature Distributions on Radiator Fins with Micro Heat Pipes," *AIAA J. Thermophysics and Heat Transfer* 15(1), pp. 42–49.)

TABLE 11.1   Configuration of Micro Heat Pipe. Reprinted with Permission from [Wang, Y., Ma, H.B., and Peterson, G.P. (2001) "Investigation of the Temperature Distributions on Radiator Fins with Micro Heat Pipes," *AIAA J. Thermophysics and Heat Transfer* 15(1), pp. 42–49.]

	Prototype		
	No. 1	No. 2	No. 3
Material	Aluminum	Aluminum	Aluminum
Working fluid	Acetone	Acetone	Acetone
Total dimension (mm)	152 × 152.4	152 × 152.4	152 × 152.4
Thickness of sheet (mm)	0.40	0.40	0.40
Diameter of wire (mm)	0.50	0.80	0.50
Number of wires	43	43	95

Wang and Peterson (2002a) presented an analysis of wire-bonded micro heat pipe arrays using a one-dimensional steady state analytical model that incorporated the effects of the liquid–vapor phase interactions and the variation in the cross-section area. The model was used to predict the heat transfer performance and optimum design parameters. An experimental facility was fabricated, and tests were conducted to

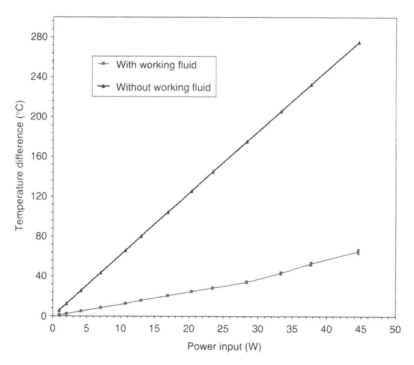

FIGURE 11.10   (See color insert following page 2-12.) Temperature difference of micro heat pipe arrays with or without working fluid. (Reprinted with permission from Wang, Y., Ma, H.B., and Peterson, G.P. (2001) "Investigation of the Temperature Distributions on Radiator Fins with Micro Heat Pipes," *AIAA J. Thermophysics and Heat Transfer* 15(1), pp. 42–49.)

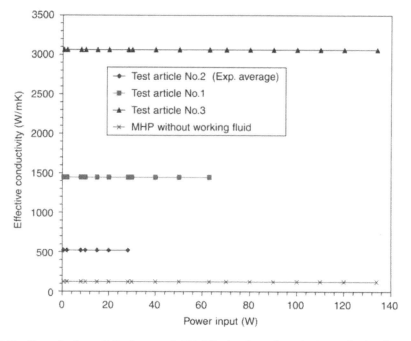

FIGURE 11.11   (See color insert following page 2-12.) Effective thermal conductivity of micro heat pipe arrays. (Reprinted with permission from Wang, Y., Ma, H.B., and Peterson, G.P. (2001) "Investigation of the Temperature Distributions on Radiator Fins with Micro Heat Pipes," *AIAA J. Thermophysics and Heat Transfer* 15(1), pp. 42–49.)

verify the concept as well as to validate the proposed model. The results indicated that the maximum heat transport capacity increased with increases in wire diameter and that the overall value was proportional to the square of the wire diameter. The numerical model indicated that the maximum heat transport capacity increased with increases in the wire spacing and predicted the existence of an optimal configuration for the maximum heat transfer capacity. Further optimization studies on a wire-bonded micro heat pipe radiator in a radiation environment were reported in Wang and Peterson (2002b). A combined numerical and experimental investigation was performed in order to optimize the heat transfer performance of the radiator. The optimal charge volume was found to decrease with increasing heat flux. The overall maximum heat transport capacity of the radiator was found to be strongly governed by the spacing of the wires, the length of the radiator, and the radiation capacity of the radiator surface. The numerical results were consistent with experimental results, which indicated that the uniformity of the temperature distribution and the radiation efficiency both increased with increasing wire diameter. Among the specimens tested, the maximum heat transport capacity of 15.2 W was found to exist for radiators utilizing a wire diameter of 0.635 mm. Comparison of the proposed micro heat pipe radiators with solid conductors and uncharged versions indicated significant improvements in the temperature uniformity and overall radiation efficiency. Aluminum–acetone systems of wire-bonded micro heat pipes were tested in this study.

A flat heat pipe thermal module for use as a cooling device for mobile computers was analyzed by Peterson and Wang (2003). It consisted of a wire-bonded heat pipe and a fin structure to dissipate heat. The temperature and heat flux distributions were calculated, and a performance analysis was done using a resistance model. Effects of the wire diameter, mesh number of the wire configuration, and the tilt angle of the heat pipe on the maximum heat transport capacity were investigated. The effect of the air flow rate on the thermal resistance and the influence of the operating temperature and air flow velocities on the heat dissipation capacity were also studied. Larger wire diameters were found to lead to a significant increase in the maximum heat transport capacity.

11.4 Flat Plate Micro Heat Spreaders

While arrays of micro heat pipes can significantly improve the effective thermal conductivity of silicon wafers and other conventional heat spreaders, they are of limited value in that they provide heat transfer only along the axial direction of the individual heat pipes. To overcome this problem, flat plate heat spreaders capable of distributing heat over a large two-dimensional surface have been proposed by Peterson (1992, 1994). In this application, a wicking structure is fabricated in silicon multichip module substrates to promote the distribution of the fluid and the vaporization of the working fluid (Figure 11.12). This wick structure is the key element in these devices, and several methods for wick manufacture have been considered [Peterson et al. 1996].

In the most comprehensive investigation of these devices to date, a flat plate micro heat pipe similar to that described by Peterson et al. (1996) was fabricated in silicon multichip module (MCS) substrates 5 mm × 5 mm square [Benson et al. 1996a; Benson et al. 1996b]. These devices, which are illustrated in Figure 11.12, utilized two separate silicon wafers. On one of the two wafers, the wick pattern was fabricated leaving a small region around the perimeter of the wafer unpatterned to allow the package to be hermetically sealed. The other silicon wafer was etched in such a manner that a shallow well was formed corresponding to the wick area. The two pieces were then wafer bonded together along the seal ring. Upon completion of the fabrication, the flat plate micro heat pipe was filled through a small laser drilled port located in one corner of the wafer. Because the entire wicking area was interconnected, the volume of the liquid required to charge was of sufficient volume that conventional charging techniques could be utilized [Benson et al. 1996].

11.4.1 Modeling of Micro Heat Spreaders

Analytical investigations of the performance of these micro heat spreaders or flat plate heat pipes have been underway for some time; Benson et al. (1996a), Benson et al. (1996b), and Peterson (1996) have summarized the results. These investigations have demonstrated that these devices can provide an effective mechanism for distributing the thermal load in semiconductor devices and reducing the localized hot spots resulting

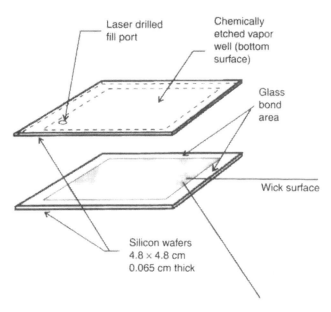

FIGURE 11.12   Flat plate micro heat spreader. (Reprinted with permission from Benson, D.A., Mitchell, R.T., Tuck, M.R., Adkins, D.R., and Palmer, D.W. (1996a) "Micro-machined Heat Pipes in Silicon MCM Substrates," *Proc. IEEE Multichip Module Conference*, 6–7 February, Santa Cruz, CA.)

FIGURE 11.13   Wick pattern prepared with bidirectional saw cuts on a silicon wafer. (Reprinted with permission from Benson, D.A., Mitchell, R.T., Tuck, M.R., Adkins, D.R., and Palmer, D.W. (1996a) "Micro-machined Heat Pipes in Silicon MCM Substrates," *Proc. IEEE Multichip Module Conference*, 6–7 February, Santa Cruz, CA.)

from active chip sites [Peterson, 1996]. The models indicate that the performance of these devices is excellent. In addition, because these devices can be made from silicon, Kovar, or a wide variety of other materials, an excellent match between the coefficient of thermal expansion (CTE) can be achieved while keeping the material and fabrication costs very low. A number of different wicking structures have been considered. Among these are wicks fabricated using a silicon dicing saw (Figure 11.13), wicks fabricated using conventional anisotropic etching techniques (Figure 11.14), and wicks fabricated using a deep plasma etching technique (Figure 11.15). Recent modeling has focused on the development of optimized wicking structures that could be fabricated directly into the wafer and provide maximum capillary pumping while optimizing the thin film region of the meniscus in order to maximize the heat flux [Wayner et al. 1976; Peterson and Ma, 1996b, 1999; Peterson and Ma 1999].

FIGURE 11.14 Chemically etched orthogonal, triangular groove wick. (Reprinted with permission from Mallik, A.K., and Peterson, G.P. [1991] "On the Use of Micro Heat Pipes as an Integral Part of Semiconductors," *3rd ASME-JSME Thermal Engineering Joint Conference Proc.*, vol. 2, pp. 394–401, March 17–22, Reno, Nevada.)

FIGURE 11.15 Wick pattern on silicon prepared by a photomask and deep plasma etch technique. Wick features are 25 microns wide and 50 microns deep wafer. (Reprinted with permission from Benson, D.A., Mitchell, R.T., Tuck, M.R., Adkins, D.R., and Palmer, D.W. (1996a) "Micro-machined Heat Pipes in Silicon MCM Substrates," *Proc. IEEE Multichip Module Conference*, 6–7 February, Santa Cruz, CA.)

The results of these optimization efforts have demonstrated that these micro scale flat plate heat spreaders allow the heat to be dissipated in any direction across the wafer surface, thereby vastly improving performance. The resulting effective thermal conductivities can approach and perhaps exceed that of diamond coatings of equivalent thicknesses. Table 11.2 [Benson et al. 1998] illustrates the relative comparison of these flat plate heat pipes and other types of materials traditionally utilized in the electronics industry for heat spreading. In this comparison, it is important to note that the ideal heat spreader would have the thermal conductivity of diamond, a coefficient of thermal expansion of silicon, and a cost comparable to aluminum. As shown, flat plate heat pipes fabricated in either silicon or Kovar compare very favorably with diamond in terms of thermal conductivity, have a close coefficient of thermal expansion of silicon relatively (or exactly in the case of silicon), and a projected cost that is quite low. Based upon this comparison, it would appear that these flat plate heat pipes have tremendous commercial potential.

11.4.2 Testing of Micro Heat Spreaders

As described by Benson et al. (1998) a number of different flat plate micro heat pipe test articles have been evaluated using an IR camera to determine the spatially resolved temperature distribution. Using this information and a technique initially described by Peterson (1993) for arrays of micro heat pipes, the effective thermal conductivity of charged and uncharged flat plate micro heat pipes, a series of micro heat spreaders were evaluated experimentally. The results indicated that an effective thermal conductivity

TABLE 11.2 Thermal Conductivity, Coefficient of Thermal Expansion, Cost Estimates, and Scaling Trends of Current and Potential Substrate Materials. Reprinted with Permission from Benson, D.A., Adkins, D.R., Mitchell, R.T., Tuck, M.R., Palmer, D.W., and Peterson, G.P. (1998) "Ultra High Capacity Micro Machined Heat Spreaders," *Microscale Thermophys. Eng.*, 2(1), pp. 21–29

Materials	Therm. Conduct. (W/cm-K)	CTE (10^{-6}/K)	Cost Substrate ($/Square Inch)	Scaling with Area Cost Trend
Alumina	0.25	6.7	$0.09	6″ limit
FR-4	Depends on copper	13.0	$0.07	Constant to 36″
A1N	1.00–2.00	4.1	$0.35	6″ limit
Silicon	1.48	4.7	$1.00	6″–10″ limit
Heat pipe in silicon	$8.00 \rightarrow 20.00(?)$	4.7	$3.00	6″–10″ limit
A1	2.37	41.8	$0.0009	Scales as area
Cu	3.98	28.7	$0.0015	Scales as area
Diamond	10.00–20.00	1.0–1.5	$1000.00	Scales as area^2
Kovar	0.13	5.0	$0.027	Scales as area
Heat pipe in Kovar	>8.00	5.0	$0.10	Scales as area
A1SiC	2.00 (at 70%)	7.0 (?)	$1.00	Casting size limited

between 10 and 20 W/cm-K was possible over a fairly broad temperature range. These values of thermal conductivity approach that of polycrystalline diamond substrates, or more than five times that of a solid silicon substrate, even at elevated temperatures (50°C) and power levels (15 W/cm²). The cost of such advanced silicon substrates is estimated at $.60/cm² (see Table 11.2). Any other inexpensive material with a CTE close to that of the chip may also be a potential option for the heat pipe case material. For example, many alloys in the Fe/Ni/Co family have CTEs closely matching those of semiconductor materials [Benson et al. 1996].

As noted by Peterson (1992) several aspects of the technology remain to be examined before flat plate micro heat spreaders can come into widespread use, but it is clear from the results of these early experimental tests that spreaders such as those discussed here, fabricated as an integral part of silicon chips, present a feasible alternative cooling scheme that merits serious consideration for a number of heat transfer applications.

11.4.3 Fabrication of Micro Heat Spreaders

The fabrication of these micro heat spreaders is basically just an extension of the methods used by several early investigations to fabricate individual micro heat pipes with hydraulic diameters on the order of 20 to 150 μm. As discussed previously, a number of different wicking structures have been utilized. These wicking structures included Kovar, silicon, or gallium arsenide and employed conventional techniques such as the machining, directionally dependent etching, plasma etching or multisource vapor deposition processes.

Charging of these devices is somewhat easier than for the individual arrays of micro heat pipes, and while these devices are still sensitive to undercharge, they can accommodate an overcharge much more readily.

11.5 New Designs

In addition to the designs described above, several new designs are currently being developed and evaluated for use in conventional electronic applications, advanced spacecraft applications, and biomedical applications. In electronic applications, the function of the heat pipe design may entail collecting heat from a microprocessor and transporting it to a conventional heat spreader or to a more readily available heat sink, such as the screen of a laptop computer. In the advanced spacecraft applications, these devices may be used to fabricate highly flexible radiator fin structures for use on long-term spacecraft missions.

A design currently being investigated consists of an array of flexible micro heat pipes fabricated in a polymer material; as illustrated in Figure 11.16a. This material is extruded in such a fashion that it has a series of large rectangular grooves that serve as the actual heat pipes, each approximately 200 microns

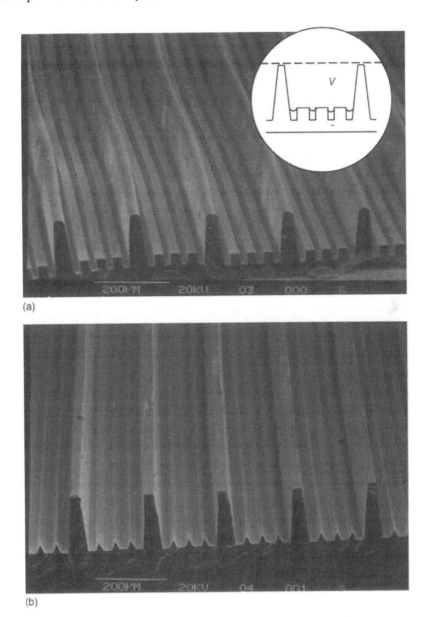

FIGURE 11.16 Flexible polymer micro heat pipe: (a) rectangular grooves, (b) trapezoidal grooves.

wide. Within each of these micro heat pipes is a series of smaller grooves that serve as the liquid arteries (see inset). These grooves can be either rectangular or trapezoidal, as shown in Figure 11.16a or Figure 11.16b. In both cases, the material is polypropylene, and the internal dimension of the individual heat pipes is approximately 200 microns. The smaller grooves within each of the individual heat pipes are designed to transport the fluid from the evaporator to the condenser.

While only preliminary experimental test data are available, this design appears to hold great promise for both spacecraft radiator applications and flexible heat spreaders used in earth-based electronic applications.

In order to understand the heat transfer and fluid flow mechanisms in the microwick structures of flexible micro membrane/thin film heat pipes, experimental and theoretical studies were performed [Wang and Peterson, 2002c; Wang and Peterson, 2003]. Experimental tests were conducted to evaluate the evaporation heat transfer limit in the polymer microfilm with 26 μm capillary grooves. The experiments indicated that the maximum heat transport capacity decreased significantly as the effective length of the

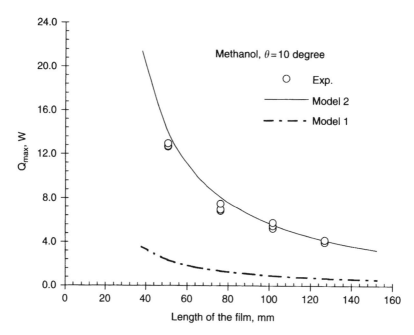

FIGURE 11.17 Comparison of the modeling and experimental results on microchanneled polymer films. (Reprinted with permission from Wang, Y.X., and Peterson, G.P. (2002c) "Capillary Evaporation in Microchanneled Polymer Films," paper no. AIAA-2002-2767, 8th AMSE/AIAA Joint Thermophysics and Heat Transfer Conference, 24–27 June, St. Louis, MO.)

polymer film increased. The experimental observations also indicated that the maximum liquid meniscus radius occurred in the microgrooves just prior to dry-out.

An analytical model based on the Darcy law was used to obtain the pressure gradients, and the experimental results were validated. Two models for predicting the maximum heat transport capacity were developed — one assuming that the liquid fills only the micro grooves, and the other considering flooding of the space above the micro grooves — and the calculated results were compared with experimental values. It was found that the experimentally determined maximum capillary evaporation heat transfer agreed better with the second model, which took into account the flooding effect. Figure 11.17 shows the comparison of the experimental and analytical results.

The analytical model, based on parametric studies, indicated that decreasing the bottom width of trapezoidal grooves very slightly can improve the evaporation heat transfer performance significantly. The analytical models were also used to determine the optimal half-angle of the groove for the best heat transfer performance.

Investigations of polymer-based flexible micro heat pipes for applications in spacecraft radiators have also been undertaken [McDaniels and Peterson, 2001]. Building upon the demonstrated effectiveness of micro heat pipe arrays as heat spreaders in electronics applications, the possibility of use of regions of micro heat pipe arrays in flexible radiators was tested. Analytical modeling suggested that a lightweight polymeric material with imbedded micro heat pipe arrays can meet heat dissipation requirements while contributing less mass than other flexible materials. The capillary pumping limit was estimated as a function of the operating temperature, using the analytical model, with water and methanol as the working fluid. For water, the maximum heat transport was found to be 18 mW per channel, at around 160°C, while for methanol it was 2.2 mW per channel at 120°C. It was shown that the obtained radiator capacity in the range 6.0 kW to 12.2 kW, at source temperatures of 40°C or higher, met or exceeded the dissipation requirements of a reference spacecraft design.

The focus of this investigation consisted of micro heat pipe arrays that were made from a composite of two layers: an ungrooved metal foil and a grooved polymer film. A low heat bonding between a polymer coating of the foil and the raised points of the grooved film formed the micro heat pipe channels. The analysis was used to compute the capillary pumping pressure and the dynamic and frictional pressure drops in the

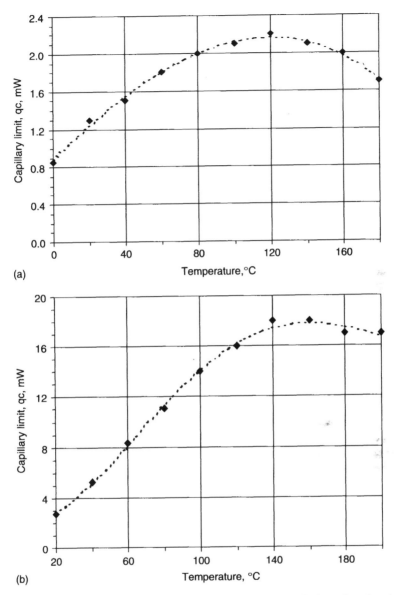

(a)

(b)

FIGURE 11.18 (a) Capillary limitation for the polymer micro heat pipe for a single channel as a function of temperature for the contact angle in the range 0–20 deg. Methanol is the working fluid. (b) Capillary limitation for the polymer micro heat pipe for a single channel as a function of temperature for the contact angle in the range 0–20 deg. With water as the working fluid.

liquid and the vapor. The results showing the variation of the capillary limit with respect to the temperature for the two working fluids are shown in Figure 11.18. Selected results were used, with Reynolds number as the criterion, to assess the validity of the model simplifications regarding the liquid and vapor flow regimes, which assumed laminar flow for liquid and vapor.

Diverse uses for the micro heat pipe and micro heat spreader can be found in biomedical applications. One such application is in catheters that provide a hyperthermia or hypothermia source for use in the treatment of tumors and cancers. Two U.S. patents have been granted for micro heat pipe catheters [Fletcher and Peterson 1990, 1993]. In the first of these, the micro heat pipe catheter enables the hypo- or hyperthermic treatment of cancerous tumors or other diseased tissue. The heat pipe is about the size of a hypodermic needle, and is thermally insulated along a substantial portion of its length. The heat pipe includes a channel that is partially charged with an appropriate working fluid. The device provides the delivery or removal of thermal energy directly to or from the tumor or diseased tissue site. In a second design, the catheter uses a variety of passive heat pipe structures alone or in combination with feedback devices. This catheter is particularly useful in treating diseased tissue that cannot be removed by surgery, such as a brain tumor. Another biomedical application under development is the polymer-based micro heat pipe heat spreader, which is being proposed for the treatment of neocortical seizures by implanting a device that can provide localized cooling.

11.6   Summary and Conclusions

It is clear from the preceding review that the concept of using micro scale heat pipes and/or flat plate micro heat pipe heat spreaders is feasible, practical, and cost effective. A number of different concepts and sizes have been shown to be acceptable from both an experimental and theoretical perspective, and a number of these devices are already in widespread use. Steady-state and transient models have been developed and verified experimentally and are capable of predicting the operational limits and performance characteristics of micro heat pipes with diameters less than 100 microns with a high degree of reliability. These models are currently being expanded for use in both individual heat pipes and arrays of heat pipes constructed as an integral part of semiconductor devices.

In addition to the analytical work, experimental evaluation has indicated that these devices can be effective in dissipating and transporting heat from localized heat sources and are presently being used in a number of commercial applications. Arrays of micro heat pipes on the order of 35 microns have been successfully fabricated, charged, tested, and incorporated as an integral part of semiconductor devices. Extensive testing has indicated that these heat pipes can provide an effective method for dissipating localized heat fluxes, eliminating localized hot spots, reducing the maximum wafer temperatures, and thereby improving the wafer reliability.

Finally, several new designs have been and continue to be developed with uses ranging from spacecraft radiator applications to land-based electronics applications. These new designs incorporate optimized wicking structures and clever new fabrication schemes along with materials not previously utilized for heat pipe applications.

Nomenclature

A = area, m²
C = constant, defined in text, specific heat, J/kg K
d = diameter, m, side of the triangular channel, m
D_H = hydraulic mean diameter of the channel, m
E = total energy per unit volume, J/m³   [$E = \rho(CT + \frac{1}{2}u^2)$]
f = drag coefficient, dimensionless
h_{fg} = latent heat of vaporization, J/kg
h_o = heat transfer coefficient at the condenser, W/m² K
k = thermal conductivity, W/m K
k^* = thermal conductivity ratio with respect to copper
K = wick permeability, m²
L = length, length of the heat pipe, m
Ma = Mach number, dimensionless
P = pressure, Pa
q = heat flow rate, W
q_{in} = heat flux into the heat sink, W/m²
R = thermal resistance, K/W; or universal gas constant, J/kg-K
Re = Reynolds number, dimensionless

r = radius, radius of curvature of the meniscus, m
r_o = initial radius of curvature of the meniscus, m
t = time, s
T = temperature, K
ΔT = temperature difference, $T - T_{amb}$, K
u = axial velocity, m/s
V = velocity, m/s
w = groove width, m; or wire spacing, m
We = Weber number, dimensionless
x = axial co-ordinate

Greek Symbols

β = geometric area coefficient in the micro heat pipe model
λ = latent heat of vaporization, J/kg
μ = dynamic viscosity, kg/m-s
ρ = density, kg/m³
σ = surface tension, N/m
τ = shear stress, N/m²
ψ = angle of inclination, degrees or radians

Subscripts

amb = ambient
b = boiling
c = capillary, capillary limitation, condenser, cross section
e = entrainment, evaporator section
eff = effective
f = fin
h = hydraulic
i = inner, interface
l = liquid
li = liquid-interface
lw = liquid-wall
m = maximum
o = outer
p = pipe
s = sonic
sat = saturation
v = vapor
vi = vapor-interface
vw = vapor-wall
w = wire spacing, wick
$-$ = axial hydrostatic pressure
$+$ = normal hydrostatic pressure

References

Adkins, D.R., Shen, D.S., Palmer, D.W., and Tuck, M.R., (1994) "Silicon Heat Pipes for Cooling Electronics," *Proc. 1st Annual Spacecraft Thermal Control Symp.*, November 16–18, Albuquerque, NM.

Anon. (1989) "Application of Micro Heat Pipes in Hyperthermia," Annual Report of the Itoh Research and Development Laboratory, Osaka.

Babin, B.R., and Peterson, G.P. (1990) "Experimental Investigation of a Flexible Bellows Heat Pipe for Cooling Discrete Heat Sources," *ASME J. Heat Transfer*, 112, pp. 602–7.

Babin, B.R., Peterson, G.P., and Wu, D. (1990) "Steady-State Modeling and Testing of a Micro Heat Pipe," *ASME J. Heat Transfer*, 112, pp. 595–601.

Badran, B., Albayyari, J.M., Gerner, F.M., Ramadas, P., Henderson, H.T., and Baker, K.W. (1993) "Liquid Metal Micro Heat Pipes," 29th National Heat Transfer Conference, *HTD*, Vol. 236, pp. 71–85, August, 8–11, Atlanta, Georgia.

Benson, D.A., Mitchell, R.T., Tuck, M.R., Adkins, D.R., and Palmer, D.W. (1996a) "Micro-Machined Heat Pipes in Silicon MCM Substrates," *Proc. IEEE Multichip Module Conference*, pp. 127–29, February, 6–7, Santa Cruz, CA.

Benson, D.A., Adkins, D.R., Peterson, G.P., Mitchell, R.T., Tuck, M.R., and Palmer, D.W. (1996b) "Turning Silicon Substrates into Diamond: Micromachining Heat Pipes," Advances in Design, Materials, and Processes for Thermal Spreaders and Heat Sinks Workshop, April 19–21, Vail, CO.

Benson, D.A., Adkins, D.R., Mitchell, R.T., Tuck, M.R., Palmer, D.W., and Peterson, G.P. (1998) "Ultra High Capacity Micro Machined Heat Spreaders," *Microscale Thermophys. Eng.*, 2, pp. 21–29.

Camarda, C.J., Rummler, D.R., and Peterson, G.P. (1997) "Multi Heat Pipe Panels," *NASA Tech Briefs*, LAR-14150.

Cao, Y, Faghri, A., and Mahefkey, E.T. (1993) "Micro/Miniature Heat Pipes and Operating Limitations," Heat Pipes and Capillary Pumped Loops, ASME HTD, 236, pp. 55–62.

Chen, H., Groll, M., and Rosler, H. (1992) "Micro Heat Pipe: Experimental Investigation and Theoretical Modeling," *Proc. 8th Int'l. Heat Pipe Conf.*, pp. 396–400, 14–18 September, Beijing.

Chi, S.W. (1976) *Heat Pipe Theory and Practice*, McGraw-Hill, New York.

Collier, J.G. (1981) *Convective Boiling and Condensation*, McGraw-Hill, New York.

Colwell, G.T., and Chang, W.S. (1984) "Measurements of the Transient Behavior of a Capillary Structure Under Heavy Thermal Loading," *Int. J. Heat Mass Transfer*, 27, pp. 541–51.

Cotter, T.P. (1984) "Principles and Prospects of Micro Heat Pipes," *Proc. 5th Int. Heat Pipe Conf.*, pp. 328–35, 14–17 May, Tsukuba, Japan.

Duncan, A.B., and Peterson, G.P. (1995) "Charge Optimization of Triangular Shaped Micro Heat Pipes," *AIAA J. Thermophys. Heat Transfer*, 9, pp. 365–67.

Dunn, P.D., and Reay, D.A. (1982) *Heat Pipes*, 3rd ed., Pergamon Press, New York.

Faghri, A. (1995) *Heat Pipe Science and Technology*, Taylor and Francis, Washington, DC.

Faghri, A. (2001) "Advances and Challenges in Micro/Miniature Heat Pipes," *Ann. Rev. Heat Transfer*, 12, pp. 1–26.

Fejfar, K., Polásek, F., and Stulc, P. (1990) "Tests of Micro Heat Pipes," *Ann. Report of the SVÚSS*, Prague.

Fletcher, L.S., and Peterson, G.P. (1993) "A Micro Heat Pipe Catheter for Local Tumor Hyperthermia," U.S. Patent No. 5,190,539.

Garimella, S.V., and Sobhan, C.B. (2001) "Recent Advances in the Modeling and Applications of Nonconventional Heat Pipes," *Adv. Heat Transfer*, 35, pp. 249–308.

Gerner, F.M. (1990) "Micro Heat Pipes," AFSOR Final Report No. S-210-10MG-066, Wright-Patterson AFB, Dayton, Ohio.

Gerner, F.M., Badran, B., Henderson, H.T., and Ramadas, P. (1994) "Silicon-Water Micro Heat Pipes," *Therm. Sci. Eng.*, 2, pp. 90–97.

Gerner, F.M., Longtin, J.P., Ramadas, P., Henderson, T.H., and Chang, W.S. (1992) "Flow and Heat Transfer Limitations in Micro Heat Pipes," 28th National Heat Transfer Conference, San Diego, Ca., pp. 99–104, August 9–12.

Ha, J.M., and Peterson, G.P. (1994) "Analytical Prediction of the Axial Dryout of an Evaporating Liquid Film in Triangular Micro Channels," *ASME J. Heat Transfer*, 116, pp. 498–503.

Ha, J.M., and Peterson, G.P. (1996) "The Interline Heat Transfer of Evaporating Thin Films along a Micro Grooved Surface," *ASME J. Heat Transfer*, 118, pp. 747–55.

Ha, J.M., and Peterson, G.P. (1998a) "Capillary Performance of Evaporating Flow in Micro Grooves: An Analytical Approach for Very Small Tilt Angles," *ASME J. Heat Transfer*, 120, pp. 452–57.

Ha, J.M., and Peterson, G.P. (1998b) "The Heat Transport Capacity of Micro Heat Pipes," *ASME J. Heat Transfer*, 120, pp. 1064–71.

Hopkins, R., Faghri, A., and Khrustalev, D. (1999) "Flat Miniature Heat Pipes with Micro Capillary Grooves," *ASME J. Heat Transfer*, 121, pp. 102–9.

Ikeda, Y. (1990) "Neutron Radiography Tests of Itoh's Micro Heat Pipes," private communications of the Nagoya University to F. Polásek.

Itoh, A. (1988) "Micro Heat Pipes," *Prospectus of the Itoh R and D Laboratory*, Osaka.

Itoh, A., and Polásek, F. (1990a) "Development and Application of Micro Heat Pipes," *Proc. 7th International Heat Pipe Conf.*, pp. 93–110, May 21–25, Minsk, USSR.

Itoh, A., and Polásek, F. (1990b) "Micro Heat Pipes and Their Application in Industry," *Proc. of the Czechoslovak-Japanese Symposium on Heat Pipes*, Rícany, Czechoslovakia.

Kendall, D.L. (1979) "Vertical Etching of Silicon at Very High Aspect Ratios," *Ann. Rev. Mater. Sci.*, pp. 373–403.

Khrustalev, D., and Faghri, A. (1994) "Thermal Analysis of a Micro Heat Pipe," *ASME J. Heat Transfer*, 116, vol. 9, pp. 189–98.

Longtin, J.P., Badran, B., and Gerner, F.M. (1994) "A One-Dimensional Model of a Micro Heat Pipe During Steady-State Operation," *ASME J. Heat Transfer*, 116, pp. 709–15.

Ma, H.B., Peterson, G.P., and Lu, X.J. (1994) "The Influence of the Vapor–Liquid Interactions on the Liquid Pressure Drop in Triangular Microgrooves," *Int. J. Heat Mass Transfer*, 37, pp. 2211–19.

Ma, H.B., Peterson, G.P., and Peng, X.F. (1996) "Experimental Investigation of Countercurrent Liquid–Vapor Interactions and its Effect on the Friction Factor," *Exp. Therm. Fluid Sci.*, 12, pp. 25–32.

Ma, H.B., and Peterson, G.P. (1996a) "Experimental Investigation of the Maximum Heat Transport in Triangular Grooves," *ASME J. Heat Transfer*, 118, pp. 740–46.

Ma, H.B., and Peterson, G.P. (1996b) "Temperature Variation and Heat Transfer in Triangular Grooves with an Evaporating Film," *AIAA J. Thermophys. Heat Transfer*, 11, pp. 90–98.

Ma, H.B., and Peterson, G.P. (1998) "Disjoining Pressure Effect on the Wetting Characteristics in a Capillary Tube," *Microscale Thermophys. Eng.*, 2, pp. 283–97

Mallik, A.K., and Peterson, G.P. (1991) "On the Use of Micro Heat Pipes as an Integral Part of Semiconductors," *3rd ASME-JSME Thermal Engineering Joint Conference Proc.*, vol. 2, pp. 394–401, March 17–22, Reno, Nevada.

Mallik, A.K., and Peterson, G.P. (1995) "Steady-State Investigation of Vapor Deposited Micro Heat Pipe Arrays," *ASME J. Electron. Packag.*, 117, pp. 75–81.

Mallik, A.K., Peterson, G.P., and Weichold, M.H. (1991) "Construction Processes for Vapor Deposited Micro Heat Pipes," *10th Symp. on Electronic Materials Processing and Characteristics*, June 3–4, 1991, Richardson, TX.

Mallik, A.K., Peterson, G.P., and Weichold, M.H. (1992) "On the Use of Micro Heat Pipes as an Integral Part of Semiconductor Devices," *ASME J. Electron. Packaging*, 114, pp. 436–42.

Mallik, A.K., Peterson, G.P., and Weichold, M.H. (1995) "Fabrication of Vapor Deposited Micro Heat Pipes Arrays as an Integral Part of Semiconductor Devices," *ASME J. Micromech. Syst.*, 4, pp. 119–31.

Marto, P.J., and Peterson, G.P. (1988) "Application of Heat Pipes to Electronics Cooling," in *Advances in Thermal Modeling of Electronic Components and Systems*, chapter 4, pp. 283–336, Bar-Cohen, A., and Kraus, A. D., eds., Hemisphere Publishing Corporation, New York.

McDaniels, D., and Peterson, G.P. (2001) "Investigation Polymer Based Micro Heat Pipes for Flexible Spacecraft Radiators," *Proc. ASME Heat Transfer Div.*, HTD-Vol. 142, New York.

Mrácek, P. (1988) "Application of Micro Heat Pipes to Laser Diode Cooling," Annual Report of the VÚMS, Prague, Czechoslovakia.

Peterson, G.P. (1987a) "Analysis of a Heat Pipe Thermal Switch," *Proc. 6th International Heat Pipe Conference*, vol. 1, pp. 177–83, May 25–28, Grenoble, France.

Peterson, G.P. (1987b) "Heat Removal Key to Shrinking Avionics," *Aerosp. Am.*, 8, no. 10, pp. 20–22.

Peterson, G.P. (1988a) "Investigation of Miniature Heat Pipes," Final Report, Wright Patterson AFB, Contract No. F33615-86-C-2733, Task 9.

Peterson, G.P. (1988b) "Heat Pipes in the Thermal Control of Electronic Components," *Proc. 3rd International Heat Pipe Symposium*, pp. 2–12, September 12–14, Tsukuba, Japan.

Peterson, G.P. (1990) "Analytical and Experimental Investigation of Micro Heat Pipes," *Proc. 7th Int. Heat Pipe Conf.*, Paper No. A-4, May 21–25, Minsk, USSR.

Peterson, G.P. (1992) "An Overview of Micro Heat Pipe Research," *Appl. Mech. Rev.*, 45, no. 5, pp. 175–89.

Peterson, G.P. (1993) "Operation and Applications of Microscopic Scale Heat Pipes," *Encyclopedia of Science and Technology*, vol. 20, pp. 197–200, McGraw-Hill, New York.

Peterson, G.P. (1994) *An Introduction to Heat Pipes: Modeling, Testing and Applications*, John Wiley & Sons, New York.

Peterson, G.P. (1996) "Modeling, Fabrication and Testing of Micro Heat Pipes: An Update," *Appl. Mech. Rev.*, 49, no. 10, pp. 175–83.

Peterson, G.P., Duncan, A.B., Ahmed, A.K., Mallik, A.K., and Weichold, M.H. (1991) "Experimental Investigation of Micro Heat Pipes in Silicon Devices," 1991 ASME Winter Annual Meeting, ASME Vol. DSC-32, pp. 341–348, December 1–6, Atlanta, GA.

Peterson, G.P., Duncan, A.B., and Weichold, M.H. (1993) "Experimental Investigation of Micro Heat Pipes Fabricated in Silicon Wafers," *ASME J. Heat Transfer*, 115, pp. 751–56.

Peterson, G.P., and Ha, J.M. (1998) "Capillary Performance of Evaporating Flow in Micro Grooves: Approximate Analytical Approach and Experimental Investigation," *ASME J. Heat Transfer*, 120, pp. 743–51.

Peterson, G.P., and Ma, H.B. (1999) "Temperature Response and Heat Transfer in a Micro Heat Pipe," *ASME J. Heat Transfer*, 121, pp. 438–45.

Peterson, G.P., and Ma, H.B. (1996a) "Analysis of Countercurrent Liquid-Vapor Interactions and the Effect on the Liquid Friction Factor," *Exp. Therm. Fluid Sci.*, 12, pp. 13–24.

Peterson, G.P., and Ma, H.B. (1996b) "Theoretical Analysis of the Maximum Heat Transport in Triangular Grooves: A Study of Idealized Micro Heat Pipes," *ASME J. Heat Transfer*, 118, pp. 734–39.

Peterson, G.P., and Mallik, A.K. (1995) "Transient Response Characteristics of Vapor Deposited Micro Heat Pipe Arrays," *ASME J. Electron. Packag.*, 117, pp. 82–87.

Peterson, G.P., and Ortega, A. (1990) "Thermal Control of Electronic Equipment and Devices," *Advances in Heat Transfer*, vol. 20, pp. 181–314, Hartnett, J.P., and Irvine, T.F., eds., Pergamon Press, New York.

Peterson, G.P., Swanson, L.W., and Gerner, F.M. (1996) "Micro Heat Pipes," in *Microscale Energy Transport*, Tien, C.L., Majumdar, A., and Gerner, F.M., eds., pp. 295–338, Taylor-Francis, Washington DC.

Peterson, G.P., and Wang, Y.X. (2003) "Flat Heat Pipe Cooling Devices for Mobile Computers," IMECE 2003, Washington, DC.

Polásek, F. (1990) "Testing and Application of Itoh's Micro Heat Pipes," *Ann. Report of the SVÚSS*, Prague, Czechoslovakia.

Ramadas, P, Badran, B., Gerner, F.M., Henderson, T.H., and Baker, K.W. (1993) "Liquid Metal Micro Heat Pipes Incorporated in Waste-Heat Radiator Panels," *Tenth Symposium on Space Power and Propulsion*, Jan. 10–14, Albuquerque, New Mexico.

Sobhan, C.B., Xiaoyang, H., and Liu C.Y (2000) "Investigations on Transient and Steady-State Performance of a Micro Heat Pipe," *AIAA J. Thermophys. Heat Transfer*, 14, pp. 161–69.

Sobhan, C.B., and Peterson, G.P. (2004) "Modeling of the Flow and Heat Transfer in Micro Heat Pipes," *2nd International Conference on Microchannels and Mini Channels*, 15–17 June, Rochester, NY.

Tien, C.L., Chung, K.S., and Lui, C.P. (1979) "Flooding in Two-phase Countercurrent Flows," EPRI NP-1283.

Wang, Y., Ma, H.B., and Peterson, G.P. (2001) "Investigation of the Temperature Distributions on Radiator Fins with Micro Heat Pipes," *AIAA J. Thermophys. Heat Transfer*, 15, pp. 42–49.

Wang, Y.X., and Peterson, G.P. (2002a) "Analysis of Wire-Bonded Micro Heat Pipe Arrays," *AIAA J. Thermophys. Heat Transfer*, 16, pp. 346–55.

Wang, Y.X., and Peterson, G.P. (2002b) "Optimization of Micro Heat Pipe Radiators in a Radiation Environment," *AIAA J. Thermophys. Heat Transfer*, 16, pp. 537–46.

Wang, Y.X., and Peterson, G.P. (2002c) "Capillary Evaporation in Microchanneled Polymer Films," paper no. AIAA-2002-2767, 8th AMSE/AIAA Joint Thermophysics and Heat Transfer Conference, 24–27 June, St. Louis.

Wang, Y.X., and Peterson, G.P. (2003) "Capillary Evaporation in Microchanneled Polymer Films," *AIAA J. Thermophys. Heat Transfer*, 17, pp. 354–59.

Wayner, Jr., P.C., Kao, Y.K., and LaCroix, L.V. (1976) "The Interline Heat-Transfer Coefficient of an Evaporating Wetting Film," *Int. J. Heat Mass Transfer*, 19, pp. 487–92.

Weichold, M.H., Peterson, G.P., and Mallik, A. (1993) *Vapor Deposited Micro Heat Pipes*, U.S. Patent No. 5,179,043.

Wu, D., Peterson, G.P., and Chang, W.S. (1991) "Transient Experimental Investigation of Micro Heat Pipes," *AIAA J. Thermophys. Heat Transfer*, 5, pp. 539–45.

Wu, D., and Peterson, G.P. (1991) "Investigation of the Transient Characteristics of a Micro Heat Pipe," *AIAA J. Thermophys. Heat Transfer*, 5, pp. 129–34.

12

Microchannel Heat Sinks

Yitshak Zohar
University of Arizona

12.1 Introduction

The last decade has witnessed impressive progress in micromachining technology enabling the fabrication of micron-sized mechanical devices, which have become more prevalent in both commercial applications and scientific research. These micromachines have had a major impact on many disciplines, including biology, chemistry, medicine, optics, and aerospace, mechanical, and electrical engineering. This emerging field not only provides miniature transducers for sensing and actuation in a domain that we could not examine in the past but also allows us to venture into research areas where the surface effects dominate most of the physical phenomena [Ho and Tai, 1998]. Fundamental heat-transfer problems posed by the development and processing of advanced integrated circuits (ICs) and microelectromechanical systems (MEMS) are becoming a major consideration in the design and application of such systems. The demands on heat-removal and temperature-control functions in modern devices that have highly transient thermal loads require an approach providing high cooling rates and uniform temperature distributions.

As the field of microfluidics and micro-heat-transfer continues to grow, it becomes increasingly important to understand the mechanisms and fundamental differences involved with heat transfer in single- and two-phase flow in microducts. The idea of fabricating microchannel heat sinks is not new. As early as two decades ago, Tuckerman and Pease (1981) pioneered the use of microchannels for cooling planar integrated circuits. They demonstrated that by flowing water through small cooling channels etched in a silicon substrate, heat-transfer rates of about $10^5 \, \text{W/m}^2\text{K}$ could be achieved. This rate is about two orders of magnitude higher than that in the state-of-the-art commercial technologies for cooling arrays of ICs. Subsequently, similar experiments were conducted by, among others, Goldberg (1984), Wu and Little (1984), Mahalingam (1985), and Nayak et al. (1987), who passed either liquids or gases in channels ranging in cross-sectional size from $100 \, \mu\text{m}$ to $1000 \, \mu\text{m}$. The heat-transfer performance of flows in microchannels has been theoretically analyzed by a number of researchers [e.g., Keyes, 1984; Samalam, 1989; Weisberg et al., 1992]. These works include a number of simplifications and approximations, the major ones being that the heat-transfer coefficient along the channel walls is uniform, the heat transport occurs through the vertical fins, the fluid temperature at each cross-section is uniform, and the gas flow is incompressible. In reality, however, the heat-transfer coefficient is a function of the local velocity profile and varies appreciably along the walls. A significant amount of heat is likely to be transported through the channel top or bottom. The fluid temperature field is not one dimensional but is at least two dimensional. Most of these early studies dealt with single-phase flow, either liquid or gas, through microchannels, as that proved complicated enough. However, it is clear that utilizing the latent heat associated with phase change can dramatically enhance the performance of microchannel heat sinks.

The area of two-phase forced convection heat transfer in microchannels is relatively young, and most of the work has been carried out within the last decade [Stanley et al., 1995]. By far, the majority of the reported research work in this area has been empirical. Peng and Wang (1993) investigated the flow boiling through microchannels with a cross-section of $0.6 \, \text{mm} \times 0.7 \, \text{mm}$. They reported that no partial nucleate boiling existed and that the velocity and liquid subcooling had no obvious effect on the flow nucleate boiling. Moreover, no bubbles were observed throughout the investigation. Peng et al. (1995) argued that the flow boiling was initiated at once and that immediately fully developed nucleate boiling took place. The liquid species and concentration were all found to affect the boiling heat-transfer coefficients [Peng et al., 1996]. Bowers and Mudawar (1994) investigated the pressure drop and critical heat flux in mini- and microchannels. They found that the thermal resistance in a microchannel was lower; however, in the nucleation regime, the pressure drop along the microchannel rose drastically with the increased heat flux. They modeled the pressure drop using a homogeneous equilibrium model to account for acceleration and friction effects. Ravigururajan (1998) studied the impact of channel geometry on two-phase flow in microchannels. The diamond cross-section resulted in a lower heat-transfer coefficient but higher critical heat flux compared to the triangular cross-section. Only recently have efforts to derive analytical models from basic principles rather than empirical correlations been reported. Peles and Haber (2000) calculated the steady-state, one-dimensional, evaporating, two-phase flow in a triangular microchannel. Although the physical model has been simplified to render the mathematical model tractable, it is the first serious attempt to calculate such a complex flow field.

In this chapter, the discussion is limited to microchannel heat sinks under steady-state operation. The supplied heat is removed by either single- or two-phase flow forced through microducts, while the single phase can be either water or vapor. Unsteady phenomena in the operation of microchannel heat sinks are naturally of great interest and should be covered separately, as the aim here is to highlight size effects on heat and mass transport in microchannels.

12.2 Fundamentals of Convective Heat Transfer in Microducts

In the science of thermodynamics, which deals with energy in its various forms and with its transformation from one form to another, two particularly important transient forms are defined: work and heat. These energies are termed transient because by definition they exist only when there is an exchange of energy between two systems or between a system and its surroundings. When such an exchange takes place without the transfer of mass from the system and not by means of a temperature difference, the energy is said to have been transferred through the performance of work. If the exchange of energy between the systems is the result of a temperature difference, the exchange is said to have been accomplished via the transfer

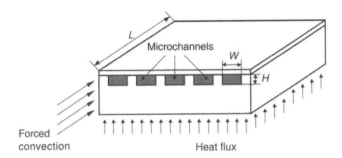

FIGURE 12.1   Typical microchannel heat sink.

of heat. The existence of a temperature difference is the distinguishing feature of the energy exchange form known as heat transfer. Microchannel heat sinks, of which a typical schematic is shown in Figure 12.1, are a class of devices that can be applied for the removal of thermal energy from very small areas.

12.2.1   Modes of Heat Transfer

The mechanism by which heat is transferred in an energy conversion system is complex; however, three basic and distinct modes of heat transfer have been classified: conduction, convection, and radiation [Sadik and Yaman, 1995]. Convection is the heat-transfer mechanism that occurs in a fluid by the mixing of one portion of the fluid with another portion due to gross movements of the mass of fluid. Although the actual process of energy transfer from one fluid particle or molecule to another is heat conduction, the energy may be transported from one point in space to another by the displacement of the fluid itself. An analysis of convective heat transfer is, therefore, more involved than that of heat transfer by conduction alone because the motion of the fluid must be studied simultaneously with the energy transfer process. The fluid motion may be caused by external mechanical means (e.g., pumps), in which case the process is called forced convection. If the fluid motion is caused by density differences created by the temperature differences existing in the fluid mass, the process is termed free, or natural, convection. The important heat transfers in liquid–vapor phase-change processes (i.e., boiling and condensing) are also classified as convective mechanisms because fluid motion is still involved, with the additional complication of a latent heat exchange. Hence, heat transfer in microchannel heat sinks belongs to this class of forced convection heat transfer with or without phase change [Stephan, 1992].

12.2.2   The Continuum Hypothesis

In an analysis of convective heat transfer in a fluid, the motion of the fluid must be studied simultaneously with the heat transport process. In its most fundamental form, the description of the motion of a fluid involves a study of the behavior of all the discrete particles (e.g., molecules) that make up the fluid. The most fundamental approach to analyzing convective heat transfer, therefore, would be to apply the laws of mechanics and thermodynamics to each individual particle or to a statistical group of particles, subsequent to some initial conditions. Such an approach, kinetic theory or statistical mechanics, would give an insight into the details of the energy transfer processes; however, it is not practical for most scientific problems and engineering applications.

In most applications, the primary interest lies not in the molecular behavior of the fluid but rather in the average or macroscopic effects of many molecules. It is these macroscopic effects that we ordinarily perceive and measure. In the study of convective heat transfer, therefore, the fluid is treated as an infinitely divisible substance, a continuum, while the molecular structure is neglected. The continuum model is valid as long as the size and the mean free path of the molecules are small enough compared with other dimensions existing in the medium such that a statistical average is meaningful.

The continuum assumption breaks down, however, whenever the mean free path of the molecules becomes the same order of magnitude as the smallest significant dimension of the problem. In gas flows, the deviation of the state of the fluid from continuum is represented by the Knudsen number, defined as $Kn \equiv \lambda/L$. The mean free path λ is the average distance traveled by the molecules between successive collisions, and L is the characteristic length scale of the flow. The appropriate flow and heat-transfer models depend on the range of the Knudsen number, and a classification of the different gas flow regimes is as follows [Schaaf and Chambre, 1961]:

$Kn < 10^{-3}$ continuum flow
$10^{-3} < Kn < 10^{-1}$ slip flow
$10^{-1} < Kn < 10^{+1}$ transition flow
$10^{+1} < Kn$ free molecular flow

In the slip-flow regime, the continuum flow model is still valid for the calculation of the flow properties away from solid boundaries. However, the boundary conditions have to be modified to account for the incomplete interaction between the gas molecules and the solid boundaries. Under normal conditions, Kn is less than 0.1 for most gas flows in microchannel heat sinks with a characteristic length scale on the order of $1 \mu m$. Therefore, only the slip-flow regime will be discussed, not the transition- or the free-molecular-flow regime. The continuum assumption is of course valid for liquid flows in microchannel heat sinks.

12.2.3 Thermodynamic Concepts

The most convenient framework within which heat-transfer problems can be studied is the system, which is a quantity of matter, not necessarily constant, contained within a boundary. The boundary can be physical, partly physical and partly imaginary, or wholly imaginary. The physical laws to be discussed are always stated in terms of a system. A control volume is any specific region in space across the boundaries of which mass, momentum, and energy may flow and within which mass, momentum, and energy storage may take place and on which external forces may act. The complete definition of a system or a control volume must include at least implicitly the definition of a coordinate system, as the system may be moving or stationary. The characteristic of interest of a system is its state, which is a condition of the system described by its properties. A property of a system can be defined as any quantity that depends on the state of the system and is independent of the path (i.e., previous history) by which the system arrived at the given state. If all the properties of a system remain unchanged, the system is said to be in an equilibrium state.

A change in one or more properties of a system necessarily means that a change in the state of the system has occurred. The path of the succession of states through which the system passes is called the process. When a system in a given initial state goes through a number of different changes of state or processes and finally returns to its initial state, the system has undergone a cycle. The properties describe the state of a system only when it is in equilibrium. If no heat transfer takes place between any two systems when they are placed in contact with each other, they are said to be in thermal equilibrium. Any two systems are said to have the same temperature if they are in thermal equilibrium with each other. Two systems that are not in thermal equilibrium have different temperatures, and heat transfer may take place from one system to the other. Therefore, temperature is a property that measures the thermal level of a system.

When a substance exists as part liquid and part vapor at a saturation state, its quality is defined as the ratio of the mass of vapor to the total mass. The quality χ may be considered a property ranging between 0 and 1. Quality has meaning only when the substance is in a saturated state (i.e., at saturated pressure and temperature). The amount of energy that must be transferred in the form of heat to a substance held at constant pressure so that a phase change occurs is called the latent heat. It is the change in enthalpy, which is a property of the substance at the saturated conditions, of the two phases. The heat of vaporization, boiling, is the heat required to completely vaporize a unit mass of saturated liquid.

12.2.4 General Laws

The general laws when referring to an open system (e.g., microchannel heat sink) can be written in either an integral or a differential form. The law of conservation of mass simply states that in the absence of any mass–energy conversion the mass of the system remains constant. Thus, in the absence of a source or sink, $Q = 0$, the rate of change of mass in the control volume (CV) is equal to the mass flux through the control surface (CS). Newton's second law of motion states that the net force F acting on a system in an inertial coordinate system is equal to the time rate of change of the total linear momentum of the system. Similarly, the law of conservation of energy for a control volume states that the rate of change of the total energy E of the system is equal to the sum of the time rate of change of the energy within the control volume and the energy flux through the control surface.

The first law of thermodynamics, which is a particular statement of conservation of energy, states that the rate of change in the total energy of a system undergoing a process is equal to the difference between the rate of heat transfer to the system and the rate of work done by the system. The second law of thermodynamics leads to the introduction of entropy S as a property of the system. It states that the rate of change in the entropy of the system is either equal to or larger than the rate of heat transfer to the system divided by the system temperature during the heat-transfer process. Even in cases where entropy calculations are not of interest, the second law of thermodynamics is still important because it is equivalent to stating that heat cannot pass spontaneously from a lower to a higher temperature system.

12.2.5 Particular Laws

Fourier's law of heat conduction, based on the continuum concept, states that the heat flux due to conduction in a given direction (i.e., the heat-transfer rate per unit area) within a medium (solid, liquid, or gas) is proportional to temperature gradient in the same direction, namely:

$$q'' = -k\nabla T \tag{12.1}$$

where q'' is the heat flux vector, k is the thermal conductivity, and T is the temperature.

Newton's law of cooling states that the heat flux from a solid surface to the ambient fluid by convection q'' is proportional to the temperature difference between the solid surface temperature T_w and the fluid free-stream temperature T_∞ as follows:

$$q'' = h(T_w - T_\infty) \tag{12.2}$$

where h is the heat transfer coefficient.

12.2.6 Governing Equations

The integral form of the conservation laws is useful for the analysis of the gross behavior of the flow field. However, detailed point-by-point knowledge of the flow field can be obtained only from the equations of fluid motion in differential form. Microchannel heat sinks typically incorporate arrays of elongated microchannels varying in cross-sectional shape; therefore, it is most convenient to use the governing equations derived either in a rectangular or cylindrical coordinate system. The governing equations for forced convection heat transfer in differential form include conservation of mass, momentum, and energy as follows:

$$\frac{D\rho}{Dt} + \rho(\nabla \cdot \mathbf{U}) = 0 \tag{12.3}$$

$$\rho\frac{D\mathbf{U}}{Dt} = -\nabla P + \rho\mathbf{B} + \mu\nabla^2\mathbf{U} + (\mu + \eta)\,\nabla\,(\nabla \cdot \mathbf{U}) \tag{12.4}$$

$$\rho c_p\frac{DT}{Dt} = k\nabla^2 T + \frac{DP}{Dt} + \phi + \theta \tag{12.5}$$

In this set of equations, ρ is the density; P is the thermodynamic pressure; **B** is the body force (e.g., gravity); μ and η are the shear and the bulk viscosity coefficients respectively; c_p is the specific heat; θ is the heat source or sink; and ϕ is the viscous dissipation given by:

$$\phi = 2\mu\left[\left(\frac{\partial u}{\partial x}\right)^2 + \left(\frac{\partial v}{\partial y}\right)^2 + \left(\frac{\partial w}{\partial z}\right)^2 + \frac{1}{2}\left(\frac{\partial u}{\partial y} + \frac{\partial v}{\partial x}\right)^2 + \frac{1}{2}\left(\frac{\partial v}{\partial z} + \frac{\partial w}{\partial y}\right)^2 + \frac{1}{2}\left(\frac{\partial u}{\partial z} + \frac{\partial w}{\partial x}\right)^2\right] \quad (12.6)$$

where u, v and w are the three components of the velocity vector U in a rectangular coordinate system (x, y, z). The state of a simple compressible pure substance or of a mixture of gases is defined by two independent properties. From experimental observations, it has been established that the behavior of gases at low density is closely given by the ideal-gas equation of state:

$$P = \rho RT \quad (12.7)$$

where R is the specific gas constant. At very low density, all gases and vapors approach ideal-gas behavior; however, the behavior may deviate substantially from that at higher densities. Nevertheless, due to its simplicity, the ideal gas equation of state has been widely used in thermodynamic calculations.

12.2.7 Size Effects

Length scale is a fundamental quantity that dictates the type of forces or mechanisms governing physical phenomena. Body forces are scaled to the third power of the length scale. Surface forces depend on the first or the second power of the characteristic length. This difference in slopes means that a body force must intersect a surface force as a function of the length scale. Empirical observations in biological studies and MEMS show that 1 mm is approximately the order of the demarcation scale [Ho and Tai, 1998]. The characteristic scale of microsystems is smaller than 1 mm; therefore, body forces such as gravity can be neglected in most cases, even in liquid flows, in comparison with surface forces. The large surface-to-volume ratio is another inherent characteristic of microsystems. This ratio is typically inversely proportional to the smaller length scale of the device cross-section and is about 1 μm in surface-micromachined devices. The large surface-to-volume ratio in microdevices accentuates the role of surface effects.

12.2.7.1 Noncontinuum Mechanics

The characteristic length scale of a microchannel (i.e., the hydraulic diameter) is typically on the order of a few micrometers. When gas is the working fluid, the mean free path is about 10 to 100 nm, resulting in a Knudsen number of about 0.05. Thus, the flow is considered to be in the slip regime, $0.001 < Kn < 0.1$, where deviations from the state of continuum are relatively small. Consequently, the flow is still governed by Equations (12.3) to (12.5), derived and based on the continuum assumption. The rarefaction effect is modeled through Maxwell's velocity-slip and Smoluchowski's temperature-jump boundary conditions [Beskok and Karniadakis, 1994]:

$$U_s - U_w = \frac{2 - \sigma_U}{\sigma_U}\lambda\frac{\partial U}{\partial n}\bigg|_w \quad (12.8a)$$

$$T_j - T_w = \frac{2 - \sigma_T}{\sigma_T}\frac{2\gamma}{\gamma + 1}\frac{k}{\mu c_p}\lambda\frac{\partial T}{\partial n}\bigg|_w \quad (12.8b)$$

U_w and T_w are the wall velocity and temperature respectively; U_s and T_j are the gas flow velocity and temperature at the boundary; n is the direction normal to the solid boundary; $\gamma = c_p/c_v$ is the ratio of specific heats; and σ_U and σ_T are the momentum and energy accommodation coefficients respectively, which model the momentum and energy exchange of the gas molecules impinging on the solid boundary. Experiments with gases over various surfaces show that both coefficients are approximately 1.0. This essentially means a diffuse reflection boundary condition, where the impinging molecules are reflected at a random angle uncorrelated with the incident angle.

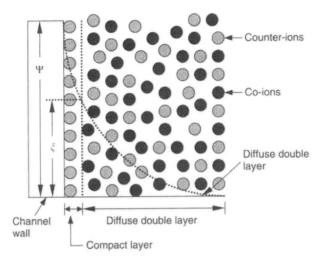

Counter-ions

Co-ions

Diffuse double
layer

Channel
wall

Diffuse double layer

Compact layer

FIGURE 12.2   Electric double layer (EDL) at the channel wall.

12.2.7.2 Electric Double Layer

Most solid surfaces are likely to carry electrostatic charge (i.e., an electric surface potential) due to broken
bonds and surface charge traps. When a liquid containing a small amount of ions is forced through a
microchannel under hydrostatic pressure, the solid-surface charge will attract the counterions in the liquid to establish an electric field. The arrangement of the electrostatic charges on the solid surface and the
balancing charges in the liquid is called the electric double layer (EDL), as illustrated in Figure 12.2.
Counterions are strongly attracted to the surface and form a compact layer, about 0.5 nm thick, of immo-
bile counterions at the solid–liquid interface due to the surface electric potential. Outside this layer, the ions
are affected less by the electric field and are mobile. The distribution of the counterions away from the
interface decays exponentially within the diffuse double layer, with a characteristic length inversely proportional to the square root of the ion concentration in the liquid. The thickness of the diffuse EDL ranges
from a few up to several hundreds of nanometers depending on the electric potential of the solid surface,
the bulk ionic concentration, and other properties of the liquid. Consequently, EDL effects can be neglected
in macrochannel flow. In microchannels, however, the EDL thickness is often comparable to the characteristic size of the channel, and its effect on the fluid flow and heat transfer may not be negligible.

Consider a liquid between two parallel plates, separated by a distance H, containing positive and negative
ions in contact with a planar, positively charged surface. The surface bears a uniform electrostatic potential
ψ_0, which decreases with the distance from the surface. The electrostatic potential ψ at any point near
the surface is approximately governed by the Debye–Huckle linear approximation [Mohiuddin Mala
et al., 1997]:

$$\frac{d^2\psi}{dy^2} = \frac{2n_0\zeta^2 e^2}{\varepsilon\varepsilon_0 k_b T}\,\psi \tag{12.9}$$

where ε is the dielectric constant of the medium, and ε_0 is the permittivity of vacuum; ζ is the valence of
negative and positive ions; e is the electron charge; k_b is the Boltzmann constant; and n_0 is the ionic concentration. The characteristic thickness of the EDL is the Debye length given by $k_d^{-1} = (\varepsilon\varepsilon_0 k_b\,T/2\,n_0\,\zeta^2\,e^2)^{1/2}$.
For the boundary conditions when $\psi = 0$ at the midpoint, $y = 0$, and $\psi = \xi$ on both walls, $y = \pm H/2$,
the solution is

$$\psi = \frac{\xi}{\sin h(k_d H/2)}|\sin h(k_d y)| \tag{12.10}$$

where ξ is the electric potential at the boundary between the diffuse double layer and the compact layer.

12.2.7.3   Polar Mechanics

In classical nonpolar mechanics, the mechanical action of one part of a body on another is assumed to be equivalent to a force distribution only. However, in polar mechanics, the mechanical action is assumed to be equivalent to not only a force but also a moment distribution. Thus, the state of stress at a point in nonpolar mechanics is defined by a symmetric second-order tensor, which has six independent components. On the other hand, in polar mechanics, the state of stress is determined by a stress tensor and a couple-stress tensor. The most important effect of couple stresses is to introduce a size-dependent effect that is not predicted by the classical nonpolar theories [Stokes, 1984].

In micropolar fluids, rigid particles contained in a small volume can rotate about the center of the volume element described by the microrotation vector. This local rotation of the particles is in addition to the usual rigid body motion of the entire volume element. In micropolar fluid theory, the laws of classical continuum mechanics are augmented with additional equations that account for conservation of microinertia moments. Physically, micropolar fluids represent fluids consisting of rigid, randomly oriented particles suspended in a viscous medium, where the deformation of the particles is ignored. The modified momentum, angular momentum, and energy equations are

$$\rho\frac{DU}{Dt} = \nabla \cdot \tau + \rho\mathbf{f} \tag{12.11}$$

$$\rho I\frac{D\mathbf{\Omega}}{Dt} = \nabla \cdot \sigma + \rho\mathbf{g} + \tau_x \tag{12.12}$$

$$\rho c_p \frac{DT}{Dt} = k\nabla^2 T + \tau : (\nabla U) + \sigma : (\nabla\mathbf{\Omega}) - \tau_x \cdot \mathbf{\Omega} \tag{12.13}$$

where $\mathbf{\Omega}$ is the microrotation vector and I is the associated microinertia coefficient; $\mathbf{f}$ and $\mathbf{g}$ are the body and couple force vectors, respectively, per unit mass; τ and σ are the stress and couple-stress tensors; $\tau : (\nabla U)$ is the dyadic notation for $\tau_{ji}U_{i,j}$, the scalar product of τ and ∇U. If $\sigma = 0$ and $\mathbf{g} = \mathbf{\Omega} = 0$, then the stress tensor t reduces to the classical symmetric stress tensor, and the governing equations reduce to the classical model [Lukaszewicz, 1999].

12.3   Single-Phase Convective Heat Transfer in Microducts

Flows completely bounded by solid surfaces are called internal flows and include flows through ducts, pipes, nozzles, diffusers, etc. External flows are flows over bodies in an unbounded fluid. Flows over a plate, a cylinder, or a sphere are examples of external flows, and they are not within the scope of this article. Only internal flows, in either liquid or gas phase, within microducts will be discussed, with an emphasis on size effects, which may potentially lead to behavior that is different than similar flows in macroducts.

12.3.1   Flow Structure

Viscous flow regimes are classified as laminar or turbulent on the basis of flow structure. In the laminar regime, flow structure is characterized by smooth motion in laminae, or layers. The flow in the turbulent regime is characterized by random three-dimensional motions of fluid particles superimposed on the mean motion. These turbulent fluctuations enhance the convective heat transfer dramatically. However, turbulent flow occurs in practice only as long as the Reynolds number, $Re = \rho U_m D_h/\mu$, is greater than a critical value, Re_{cr}. The critical Reynolds number depends on the duct inlet conditions, surface roughness, vibrations imposed on the duct walls, and the geometry of the duct cross-section. Values of Re_{cr} for various duct cross-section shapes have been tabulated elsewhere [Bhatti and Shah, 1987]. In practical applications, though, the critical Reynolds number is estimated to be

$$Re_{cr} = \frac{\rho U_m D_h}{\mu} \cong 2300 \tag{12.14}$$

where U_m is the mean flow velocity and $D_h = 4A/S$ is the hydraulic diameter, with A and S being the cross-section area and the wetted perimeter respectively. Microchannels are typically larger than 1000 μm in length with a hydraulic diameter of about 10 μm. The mean velocity for gas flow under a pressure drop of about 0.5 MPa is less than 100 m/s, and the corresponding Reynolds number is less than 100. The Reynolds number for liquid flow will be even smaller due to the much higher viscous forces. Thus, in most applications, the flow in microchannels is expected to be laminar. Turbulent flow may develop in short channels with large hydraulic diameter under high-pressure drop and therefore will not be discussed here.

12.3.2 Entrance Length

When a viscous fluid flows in a duct, a velocity boundary layer develops along the inside surfaces of the duct. The boundary layer fills the entire duct gradually, as sketched in Figure 12.3. The region where the velocity profile is developing is called the hydrodynamics entrance region, and its extent is the hydrodynamic entrance length. An estimate of the magnitude of the hydrodynamic entrance length L_h in laminar flow in a duct is given by Shah and Bhatti (1987):

$$\frac{L_h}{D_h} = 0.056\, Re \tag{12.15}$$

The region beyond the entrance region is referred to as the hydrodynamically fully developed region. In this region, the boundary layer completely fills the duct and the velocity profile becomes invariant with the axial coordinate.

If the walls of the duct are heated (or cooled), a thermal boundary layer will also develop along the inner surfaces of the duct, shown in Figure 12.3. At a certain location downstream from the inlet, the flow becomes fully developed thermally. The thermal entrance length L_t is then the duct length required for the developing flow to reach fully developed condition. The thermal entrance length for laminar flow in ducts varies with the Reynolds number, Prandtl number ($Pr = \mu c_p / k$) and the type of the boundary condition imposed on the duct wall. It is approximately given by:

$$\frac{L_t}{D_h} \cong 0.05\, Re\, Pr \tag{12.16}$$

More accurate discussion on thermal entrance length in ducts under various laminar flow conditions can be found elsewhere [e.g., Shah and Bhatti, 1987].

In most practical applications of microchannels, the Reynolds number is less than 100 while the Prandtl number is on the order of 1. Thus, both the hydrodynamic and thermal entrance lengths are less than 5 times the hydraulic diameter. Because the length of microchannels is typically two orders of magnitude larger than the hydraulic diameter, both entrance lengths are less than 5% of the microchannel length and can be neglected.

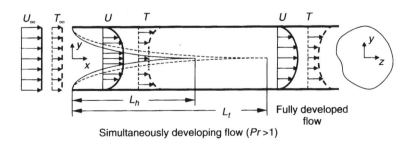

Simultaneously developing flow ($Pr > 1$)

FIGURE 12.3 Hydrodynamically and thermally developing flow, followed by hydrodynamically and thermally fully developed flow.

12.3.3   Governing Equations

Representing the flow in rectangular ducts as flow between two parallel plates, the two-dimensional governing equations can be simplified as follows (Sadik and Yaman, 1995):

Continuity:

$$\frac{\partial(\rho u)}{\partial x} + \frac{\partial(\rho v)}{\partial y} = 0 \tag{12.17}$$

x-momentum:

$$\frac{\partial(\rho u u)}{\partial x} + \frac{\partial(\rho v u)}{\partial y} = -\frac{\partial P}{\partial x} + \mu\left(\frac{\partial^2 u}{\partial x^2} + \frac{\partial^2 u}{\partial y^2}\right) + \frac{\mu}{3}\frac{\partial}{\partial x}\left(\frac{\partial u}{\partial x} + \frac{\partial v}{\partial y}\right) \tag{12.18}$$

y-momentum:

$$\frac{\partial(\rho u v)}{\partial x} + \frac{\partial(\rho v v)}{\partial y} = -\frac{\partial P}{\partial y} + \mu\left(\frac{\partial^2 v}{\partial x^2} + \frac{\partial^2 v}{\partial y^2}\right) + \frac{\mu}{3}\frac{\partial}{\partial y}\left(\frac{\partial u}{\partial x} + \frac{\partial v}{\partial y}\right) \tag{12.19}$$

Energy:

$$u\frac{\partial T}{\partial x} + v\frac{\partial T}{\partial y} = \frac{k}{\rho c_p}\left(\frac{\partial^2 T}{\partial x^2} + \frac{\partial^2 T}{\partial y^2}\right) + \frac{2\mu}{\rho c_p}\left[\left(\frac{\partial u}{\partial x}\right)^2 + \left(\frac{\partial v}{\partial y}\right)^2 + \frac{1}{2}\left(\frac{\partial u}{\partial y} + \frac{\partial v}{\partial x}\right)^2\right] \tag{12.20}$$

12.3.4   Fully Developed Gas Flow Forced Convection

Analytical solution of Equations (12.17) to (12.20) is not available. Some solutions can be obtained upon further simplification of the mathematical model. Indeed, incompressible gas flows in macroducts with different cross-sections subjected to a variety of boundary conditions are available [Shah and Bhatti, 1987]. However, the important features of gas flow in microducts are mainly due to rarefaction and compressibility effects. Two more effects due to acceleration and nonparabolic velocity profile were found to be of second order compared to the compressibility effect (van den Berg et al., 1993). The simplest system for demonstration of the rarefaction and compressibility effects is the two-dimensional flow between parallel plates separated by a distance H, with L being the channel length ($L/H \gg 1$). If $MaKn \ll 1$, all streamwise derivatives can be ignored except the pressure gradient, which is the driving force. The Mach number, $Ma = U/a$, is the ratio between the fluid speed and the speed of sound a. In such a case, the momentum equation reduces to:

$$-\frac{dP}{dx} + \mu\frac{d^2 u}{dy^2} = 0 \tag{12.21}$$

with the symmetry condition at the channel centerline, $y = 0$, and the slip boundary conditions at the walls, $y = \pm H/2$, as follows:

$$\frac{du}{dy} = 0 \quad @ \quad y = 0 \tag{12.22}$$

$$u = -\lambda\frac{du}{dy}\bigg|_{y=H/2} \quad @ \quad y = \pm H/2 \tag{12.23}$$

Integration of Equation (12.21) twice with respect to y, assuming $P = P(x)$, yields the following velocity profile [Arkilic et al., 1997]:

$$u(y) = -\frac{H^2}{8\mu}\frac{dP}{dx}\left[1 - \left(\frac{y}{H/2}\right)^2 + 4Kn(x)\right] \tag{12.24}$$

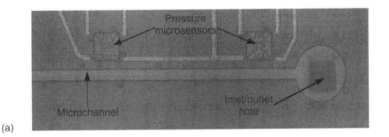

(a)

(b)

FIGURE 12.4 Slip flow effect on a microchannel flow: (a) microchannel, 40 μm wide, integrated with pressure microsensors; (b) a comparison between calculated (dash lines) and measured (symbols) streamwise pressure distributions. (Reprinted by permission of Elsevier Science from Li, X. et al. [2000] "Gas Flow in Constriction Microdevices," *Sensors and Actuators A*, 83, pp. 277–83.)

where $Kn(x) = \lambda(x)/H$. The streamwise pressure distribution $P(x)$ calculated based on the same model is given by:

$$\frac{P(x)}{P_o} = -6Kn_o + \sqrt{\left(6Kn_o + \frac{P_i}{P_o}\right)^2 - \left[\left(\frac{P_i^2}{P_o^2} - 1\right) + 12Kn_o\left(\frac{P_i}{P_o} - 1\right)\right]\left(\frac{x}{L}\right)} \qquad (12.25)$$

where P_i is the inlet pressure, P_o the outlet pressure, and Kn_o is the outlet Knudsen number. It is difficult to verify experimentally the cross-stream velocity distribution $u(y)$ within a microchannel. However, detailed pressure measurements have been reported [Liu et al., 1993; Pong et al., 1994]. A picture of a microchannel integrated with pressure sensors for such experiments is shown in Figure 12.4a. Indeed, the calculated pressure distributions based on Equation (12.25) were found to be in a close agreement with the measured values as shown in Figure 12.4b [Li et al., 2000]. Furthermore, the mass flow rate Q_m as a function of the inlet and outlet conditions is obtained by integrating the velocity profile with respect to x and y as follows:

$$Q_m = \frac{H^3 W P_o^2}{24\mu RTL}\left[\left(\frac{P_i}{P_o}\right)^2 - 1 + 12Kn_o\left(\frac{P_i}{P_o} - 1\right)\right] \qquad (12.26)$$

where W is the width of the channel. This simple equation was found to yield accurate results for three different working gasses: nitrogen, helium, and argon, with ambient temperatures ranging from 20 to 60°C, as demonstrated in Figure 12.5 [Jiang et al., 1999a].

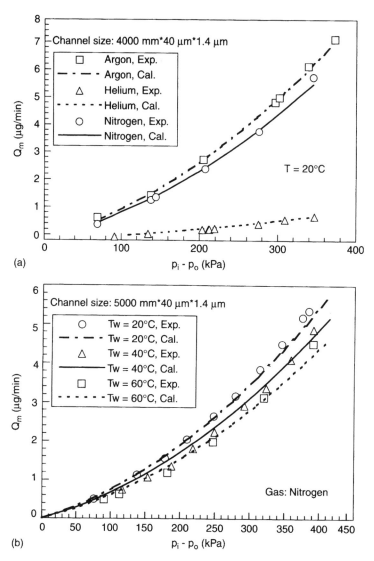

FIGURE 12.5   Slip flow effect on microchannel mass flow rate as a function of the total pressure drop for various working gases (a) and wall temperatures (b). (Reprinted with permission from Jiang, L. et al. [1999] "Fabrication and Characterization of a Microsystem for Microscale Heat Transfer Study" *J. Micromech. Microeng.*, **9**, pp. 422–28.)

The microchannel flow temperature distribution and heat flux depend on the boundary conditions, and extensive analytical work has been conducted (Harley et al., 1995; Beskok et al., 1996). However, closed-formed analytical solutions in general are still not available. Numerical simulations of Equations (12.17) to (12.20) were carried out for constant wall temperature and constant heat flux boundary conditions by Kavehpour et al. (1997), and the results are summarized in Figure 12.6. The heat transfer rate from the wall to the gas flow decreases while the entrance length increases due to the rarefaction effect (i.e., increasing Knudsen number). This may not be a universal result, however, as the slip flow conditions include two competing effects [Zohar et al., 1994]. The velocity slip at the wall increases the flow rate, thus enhancing the cooling efficiency. On the other hand, the temperature jump at the boundary acts as a barrier to the flow of heat to the gas, thus reducing the cooling efficiency. The net result of these effects depends on the specific material properties and specific geometry of the system.

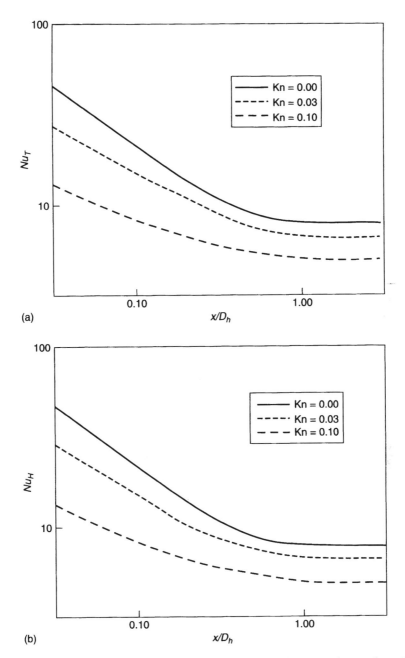

FIGURE 12.6 Numerical simulations of the effect of the inlet Knudsen number Kn_i on the Nusselt number Nu along a microchannel for uniform wall temperature (Nu_T): (a) and heat flux (Nu_H), (b) boundary conditions. (Reprinted by permission of Taylor & Francis, Inc., from Kavehpour, H.P. et al. [1997] "Effects of Compressibility and Rarefaction on Gaseous Flows in Microchannels," *Numerical Heat Transfer A*, **32**, pp. 677–96.)

A microchannel integrated with suspended temperature sensors was constructed (Figure 12.7a) for an initial attempt to experimentally assess the slip-flow effects on heat transfer in microchannels [Jiang et al., 1999a]. The resulting temperature distributions along the microchannel are shown in Figure 12.7b for different wall temperatures and pressure drops. In all cases, the temperature along the channel is almost uniform and equal to the wall temperature, and no cooling effect has been observed. Indeed, on the one

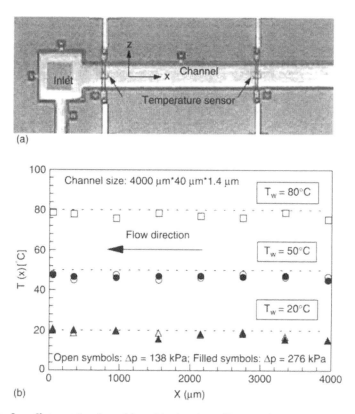

(a)

(b)

FIGURE 12.7 Slip-flow effect on microchannel flow: (a) microchannel integrated with suspended temperature sensors; (b) measured streamwise temperature distributions for different ambient temperature and pressure drop. (Reprinted with permission from Jiang, L. et al. [1999] "Fabrication and Characterization of a Microsystem for Microscale Heat Transfer Study," *J. Micromech. Microeng.* **9**, pp. 422–28.)

hand, the slip flow effects are small, but on the other hand, the sensitivity of the experimental system is not sufficient. Thus, experiments with higher resolution and greater sensitivity are required to accurately verify the weak slip flow effects on the temperature and the heat-transfer coefficient predicted by theoretical analyses and numerical simulations.

12.3.5 Fully Developed Liquid Flow Forced Convection

Liquid flow is considered to be incompressible even in microducts because the distance between the molecules is much smaller than the characteristic scale of the flow. Hence, no rarefaction effect is encountered, and the classical model in Equation (12.21) should be valid. Again, in such a case, extensive data are readily available [Shah and Bhatti, 1987]. However, two unique features of liquid flow in microducts, polarity and EDL, could affect the flow behavior.

The characteristic length scale of the electric double layer is inversely proportional to the square root of the ion concentration in the liquid. For example, in pure water the scale is about 1 μm, while in 1 mole of NaCl solution the EDL length scale is only 0.3 nm. Thus, in microducts, liquid flow with low ionic concentration and the associated heat transfer can be affected by the presence of the EDL. The x-momentum and energy equations for a two-dimensional duct flow can be reduced to [Mohiuddin Mala et al., 1997]:

$$\mu \frac{d^2 u}{dy^2} - \frac{dP}{dx} - \varepsilon \varepsilon_0 \frac{E_s}{L} \frac{d^2 \psi}{dy^2} = 0 \qquad (12.27)$$

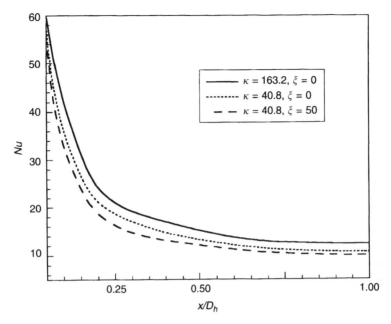

FIGURE 12.8 Electric double layer effect on the variation of the local Nusselt number Nu along the channel length. (Reprinted by permission of Elsevier Science from Mohiuddin Mala, G. et al. [1997] "Heat Transfer and Fluid Flow in Microchannels," *Int. J. Heat Mass Transfer* **40**, pp. 3079–88.)

$$\rho c_p\left(u\frac{\partial T}{\partial x}\right) = k\left(\frac{\partial^2 T}{\partial y^2} + \frac{\partial^2 T}{\partial x^2}\right) + \mu\left(\frac{\partial u}{\partial y}\right)^2 \tag{12.28}$$

where E_s is the steaming potential and L is the duct length. Equation (12.27) was solved analytically, and Equation (12.28) was solved numerically for constant wall temperature boundary condition for a given inlet liquid temperature. The results showed that both the temperature gradient at the wall and the difference between the wall and the bulk temperature decrease with downstream distance. The value of the temperature gradient decreases much faster, resulting in a decreasing Nusselt number, $Nu = hD_h/k$, along the channel, as plotted in Figure 12.8. However, with no double layer effects (i.e., $\xi = 0$) a higher heat-transfer rate (higher Nu) is obtained. The EDL results in a reduced flow velocity (higher apparent viscosity), thus decreasing the heat-transfer rate.

In order to evaluate micropolar effects on microchannel heat transfer, Jacobi (1989) considered the steady fully developed laminar flow in a cylindrical microtube with uniform heat flux, for which the energy equation is given by:

$$\rho c_p\left(u\frac{\partial T}{\partial x}\right) = \frac{k}{r}\left(\frac{\partial T}{\partial r} + r\frac{\partial^2 T}{\partial r^2}\right) \tag{12.29}$$

where r is the radial coordinate. Both the velocity and temperature radial distributions were analytically estimated. Based on the temperature field, the heat-transfer rate was calculated and the results are shown in Figure 12.9 for different values of Γ, a length scale that depends on the viscosity coefficients of the micropolar fluid. The Nusselt number is smaller than the classical value of $Nu = 4.3636$ by as much as 7% for this micropolar flow. Although the micropolar fluid theory has been applied to many situations, however, the drawback to these analyses is still the unknown viscosity coefficients.

Clearly, the EDL and micropolar fluid effects on liquid forced convection in microducts are indirect; namely, the velocity is modified due to these effects and, as a consequence, the heat-transfer rate is affected. Thus, it is important first to verify the hydrodynamic effects. Indeed, it has been suggested in a few reports that theoretical calculations based on the classical model did not agree with experimental

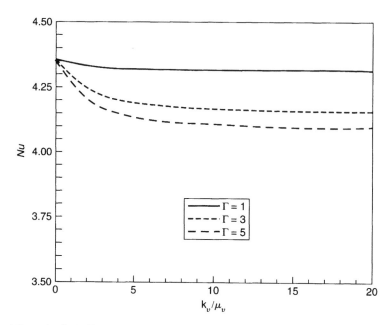

FIGURE 12.9   Micropolar fluid effect on the Nusselt number *Nu* as a function of the viscosity ratio, k_v/μ_v, for different values of Γ. (Reprinted with permission from Jacobi, A.M. [1989] "Flow and Heat Transfer in Microchannels Using a Microcontinuum Approach," *J. Heat Transfer* 111, 1083–85.)

measurements of liquid flow properties in microchannels [Pfahler et al., 1990; Peng et al., 1994; Peng and Peterson, 1996]. An experimental study of water flow in a microchannel with a cross-section area of 600 μm × 30 μm was carried out specifically to evaluate micropolar effects by Papautsky et al. (1998). They concluded that micropolar fluid theory provides a better approximation of the experimental data than the classical theory. However, a close examination of the results shows that the difference between the results of the two theories is smaller than the difference between the experimental data and the predictions of either theory.

In carefully conducted experiments of water flow through a suspended microchannel (Figure 12.10a) with a cross-section area of 20 μm × 2 μm and under a pressure drop of up to 500 psi, none of these effects has been observed [Wu et al., 1998]. The slight mismatch between theory and experiment was found to be a result of the bulging effect of the channel roof under the high pressure. The deformation of the channel roof can be measured accurately. Once the corrected cross-section area has been accounted for adequately in the calculations, the classical theory results agree well with the experimental measurements as evident in Figure 12.10b. However, more research work is required to verify these observations because these discrepancies may have to do more with experimental errors rather than true size effects.

12.4   Two-Phase Convective Heat Transfer in Microducts

Micro heat sinks have been constructed as micro heat exchangers for cooling of thermal microsystems developed and investigated either experimentally or theoretically. It is a common finding that the cooling rates in such microchannel heat exchangers should increase significantly due to a decrease in the convective resistance to heat transport caused by a drastic reduction in the thickness of the thermal boundary layers. The potentially high heat-dissipation capacity of such a micro heat sink is based on the large heat-transfer-surface-to-volume ratio of the microchannel heat exchanger. In order to increase the heat flux from a microchannel with single-phase flow while maintaining practical limits on surface temperature, it is necessary to increase the heat-transfer coefficient by either increasing the flow rate or decreasing the hydraulic diameter. Both are accompanied by a large increase in the pressure drop. However,

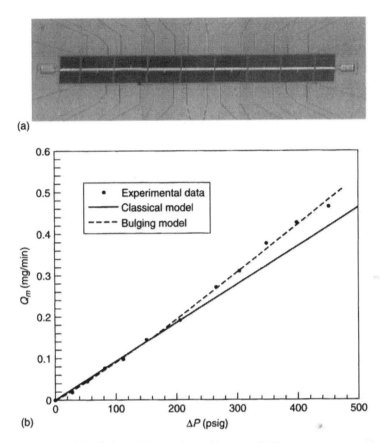

FIGURE 12.10 Microchannel liquid flow: (a) microchannel integrated with temperature sensors on the channel roof; (b) a comparison between liquid flow rate measurements as a function of the pressure drop and theoretical calculations based on classical and bulging models. (Reprinted with permission from Wu, P. et al. [1998] "A Suspended Microchannel with Integrated Temperature Sensors for High-Pressure Flow Studies," in *Proc. 11th Int. Workshop on Micro Electro Mechanical Systems (MEMS '98)*, pp. 87–92. © 1998/2000 IEEE.)

forced-convection flow with phase change can achieve a very high heat-removal rate for a constant flow rate while maintaining a relatively constant surface temperature determined by the saturation properties of the cooling fluid. The advantage of using two-phase over single-phase micro heat sinks is clear. Single-phase heat sinks compensate for high heat flux by a large streamwise increase in both coolant and heat sink temperature. Two-phase heat sinks, in contrast, utilize latent heat exchange, which maintains streamwise uniformity both in the coolant and the heat sink temperature at a level set by the coolant saturation temperature. Therefore, it is expected that two-phase heat transfer may lead to significantly more efficient heat transfer, and a two-phase micro heat exchanger would be the most promising approach for cooling in microsystems [Stanley et al., 1995].

Heat transfer during boiling of a liquid in free convection is essentially determined by the difference between the heating-surface and boiling temperatures, the properties of the liquid, and the properties of the heating surface. Thus, the heat-transfer coefficient can be represented by a simple empirical correlation of the form $h \propto q^m$. During boiling in forced convection, however, the flow velocities of the vapor and liquid phases and the phase distribution play additional roles. Consequently, the mass flow rate and the quality are additional limiting factors, giving rise to a correlation of the form $h \propto q^m Q_m^n f(\chi)$. Forced convection boiling is complex not only due to the coexistence of two separate phases having different properties but also especially to the existence of a highly convoluted vapor–liquid interface resulting in a variety of flow patterns. Typical patterns that have experimentally been observed in macroducts, such as bubble, slug, churn, annular, and drop flow, are sketched in Figure 12.11. Accordingly, flow pattern maps have been suggested in which the duct orientation on heat-transfer boiling is significant due to gravity effects [Stephan, 1992].

12.4.1 Boiling Curves

Forced convection boiling is attractive because it ensures low device temperature for high power dissipation or, alternatively, it allows higher power dissipation for a given device temperature. Measurements of either the inner wall or the fluid bulk temperature distributions along a microduct under forced convection boiling are not available yet, due to the difficulty in integrating sensors at the desired locations. However, measurements of the surface temperature of a microchannel heat sink device have been reported [Jiang et al., 1999b]. A picture of the integrated microsystem consisting of an array of microducts, a local microheater, and an array of temperature microsensors is shown in Figure 12.12. The 35 diamond-shaped microducts, each with a hydraulic diameter of about 40 μm, are buried between two bonded silicon

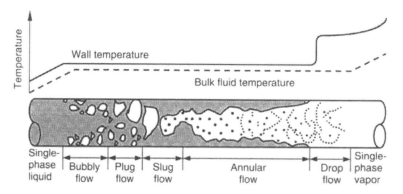

FIGURE 12.11 Wall and mean fluid temperature, flow patterns, and the accompanying heat-transfer ranges in a typical heated duct. (Reprinted with permission from Stephan, K. [1992] *Heat Transfer in Condensation and Boiling*, Springer-Verlag, Berlin.)

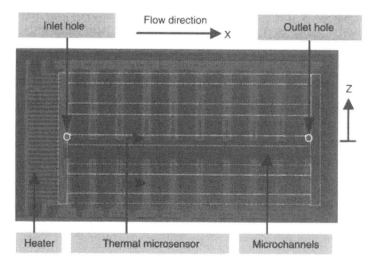

FIGURE 12.12 Photograph of a microchannel heat sink showing the localized heater, the buried microchannel array, and the temperature microsensor array. (Reprinted with permission from Jiang, L. et al. [1999] "Phase Change in Micro-Channel Heat Sinks with Integrated Temperature Sensors," *J. MEMS* **8**, 358–65. © 1999/2000 IEEE.)

wafers. A significant reduction of the device temperature is demonstrated in Figure 12.13. Initially, the device temperature and its temperature gradient for a given power dissipation (3.6 W) is high. The maximum temperature of about 230°C is measured close to the heater. The device temperature drops sharply to about 115°C even for the low flow rate of 0.25 mL/min (average liquid velocity of about 6.7 cm/s within each duct). Increasing the water flow rate leads to further reduction of the device temperature to a level below the saturation temperature of about 100°C. This is expected, as a higher flow rate results in a higher heat-transfer rate. Consequently, the device internal energy (i.e., the device temperature) decreases. Furthermore, the temperature distribution becomes more uniform as well, which suggests that the local heat-transfer rate is highly nonuniform. It should be emphasized, though, that the flow is in single-liquid phase for the high-flow-rate case and in two-phase for the low-flow-rate case, as indicated by the exit fluid quality. Hence, the heat-transfer mechanism changes character as the flow rate varies.

The measured spanwise temperature distributions were found to be uniform, similar to the streamwise temperature distributions plotted in Figure 12.13. Thus, the average temperature along the device centerline can characterize the device temperature. In order to obtain a complete boiling curve, the device temperature was recorded as the input power increased by small increments while maintaining the inlet water flow rate constant at room temperature (22°C). This experiment was repeated several times for different devices with varying flow rates [Jiang et al., 1999b]; the results are summarized in Figure 12.14a. In all curves, the device average temperature increases monotonically, almost linearly, with the power level. At a certain input power known as critical heat flux (CHF), the temperature increases sharply. The exit flow changes from single-liquid phase, quality zero, through two-phase flow of liquid–vapor, to a single vapor phase, quality one, under CHF conditions. These boiling curves are in contrast to the previously reported data of Bowers and Mudawar (1994) plotted in Figure 12.14b for a microchannel 510 μm in diameter. The typical boiling plateau illustrated at the inset of Figure 12.14a has not been observed under all tested conditions. The plateau in the boiling curve is due to the saturated nucleate boiling, where bubbles continuously form, grow and detach such that the temperature is kept uniform and constant although the heat dissipation is increasing until the CHF condition is approached. The curves in Figure 12.14a suggest that the saturated nucleate boiling does not develop in such microducts due to size effect, which could be verified by flow visualization of the boiling pattern.

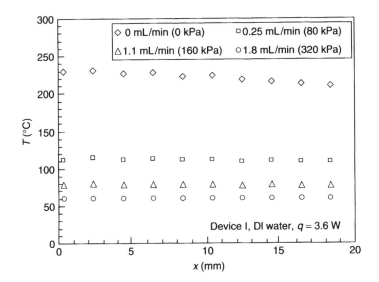

FIGURE 12.13 Flow rate effect on the temperature distribution along the microchannel heat sink centerline. (Reprinted with permission from Jiang, L. et al. [1999] "Phase Change in Micro-Channel Heat Sinks with Integrated Temperature Sensors," *J. MEMS* **8**, 358–65. © 1999/2000 IEEE.)

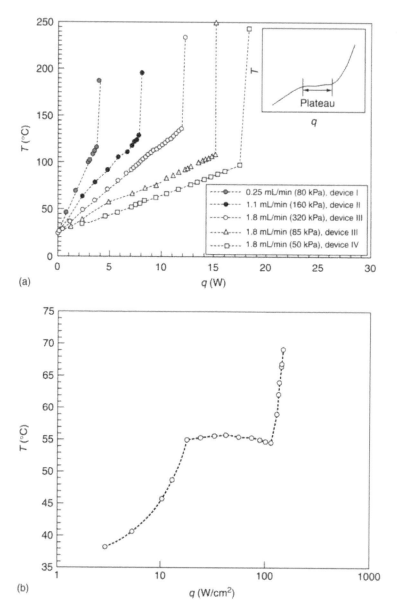

FIGURE 12.14   Boiling curves of device temperature as a function of the input power for microchannels with: (a) water and $D_h = 40\,\mu m$ or $80\,\mu m$ [Jiang et al., 1999b], and (b) R-113 and $D_h = 510\,\mu m$ [Bowers and Mudawar, 1994]. (Reprinted with permission from Elsevier Science.)

12.4.2   Critical Heat Flux

The critical heat flux is the most important factor used to determine the upper limit of the heat sink cooling ability. When the CHF condition is approached, a sudden dry-out takes place at the heat-transfer surface. This is accompanied by a drastic reduction of the heat-transfer coefficient and a sharp rise in surface temperature. The exit flow quality is one, as the entire liquid passing through the heat sink changes phase into vapor. Therefore, it is reasonable that the critical heat flux q_{CHF} increases linearly with the flow rate (Figure 12.15a) because most of the input power is converted into latent heat at about the saturation temperature [Jiang et al., 1999b]. An important parameter associated with the CHF condition is the corresponding

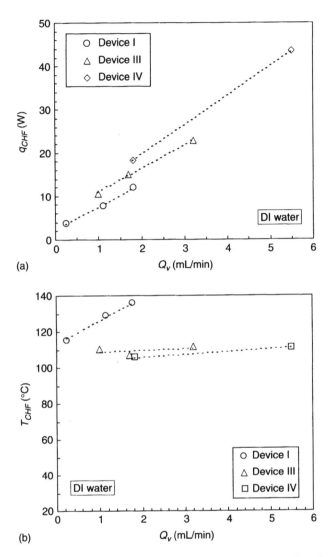

FIGURE 12.15 Flow rate effect on the critical heat flux (a) and the corresponding device temperature (b). (Reprinted with permission from Jiang, L. et al. [1999] "Phase Change in Micro-Channel Heat Sinks with Integrated Temperature Sensors," *J. MEMS* **8**, 358–65. © 1999/2000 IEEE.)

device temperature. The dependence of the average temperature under the CHF condition T_{CHF} on the water flow rate Q_v is shown in Figure 12.15b for three different devices. For the large heat sinks, $D_h = 80\,\mu m$, the CHF temperature depends neither on the flow rate nor on the number of channels. Furthermore, T_{CHF} is slightly higher than the saturation temperature of water under atmospheric pressure, 100°C. The higher CHF temperature may be due to the higher pressure, larger than 1 atm, throughout the microducts. However, for the small heat sink, $D_h = 40\,\mu m$, the CHF temperature increases almost linearly with the water flow rate, which cannot be attributed to the higher pressure. The difference between the two heat sinks is puzzling, and more experiments with wider flow rate ranges are required to confirm this observation.

Bowers and Mudawar (1994) reported similar dependence of the critical heat flux on liquid flow rate in a study comparing the performance of mini- and microchannel heat sinks. The data exhibited a lack of subcooling effect on the CHF for both heat sinks and under all operating conditions. This was attributed to fluid reaching the saturation temperature within a short distance into the heated section of the channel. However, they did notice a distinct separation between mini- and microchannel curves, which was explained as a result of the large difference in L/D ratio (L and D being the channel length and diameter respectively): 3.94 for the minichannel and 19.6 for the microchannel. Consequently, they proposed the following CHF correlation:

$$\frac{q_{mp}}{Gh_{fg}} = 0.16 \; We^{-0.19} \left(\frac{L}{D}\right)^{-0.54} \tag{12.30}$$

where q_{mp} is the CHF based upon the heated channel inside area, G is the mass velocity, and h_{fg} is the latent heat of evaporation. $We = G^2 L / \beta \rho$ is the Weber number, where β and ρ are the liquid surface tension and density respectively. The authors argued that the small diameter of the channels resulted in an increased frequency and effectiveness of droplet impact on the channel wall. This could have increased the heat-transfer coefficient and enhanced the CHF compared to droplet flow regions in larger tubes. The small overall size of the heat sinks seemed to contribute to delaying CHF by conducting heat away from the downstream region undergoing partial or total dry-out to the boiling region of the channel. Thus, a higher heat-transfer rate is required to trigger CHF conditions along the entire microchannel rather than just at the downstream region.

12.4.3 Flow Patterns

Two-phase flow patterns in ducts are the result of the detailed heat transfer between the solid boundary and the working fluid. The flow patterns are important because they directly determine the temperature distributions in both the solid boundary and the fluid flow. Mudawar and Bowers (1999) suggested that low- and high-velocity flows are characterized by drastically different flow patterns as well as unique CHF trigger mechanisms. Whereas the low flow exhibits a succession of bubbly, slug, and annular flow, the high flow is characterized by a bubbly flow near the wall with a liquid core. Unfortunately, limited results of flow patterns have been reported thus far, so it is not clear whether this distinction is valid for microchannel heat sinks.

An integrated microsystem similar to the one shown in Figure 12.12 has been fabricated to study the forced convection boiling flow patterns [Jiang et al., 2000]. The triangular grooves etched in the silicon wafer were covered by a bonding glass wafer rather than a silicon wafer in order to facilitate flow visualizations. In microducts, body forces such as gravity are negligible with respect to surface forces (i.e., surface tension or capillary forces). Consequently, the microduct orientation has little effect on forced convection boiling, and no difference between the flow patterns in horizontal and vertical microducts could be detected experimentally. Furthermore, the boiling modes identified in these microducts are different from the classical patterns sketched in Figure 12.11. At moderate power levels, an annular flow mode with liquid droplets within the vapor core could be observed, as shown in Figure 12.16a, while the vapor–liquid interface in the channel appears to be wavy. This mode should be regarded as an unstable transition stage because it was not always detected. Moreover, when it did appear, it was short lived. An annular flow mode, shown in Figure 12.16b, was observed to be a stable pattern for a wide range of input power levels, $0.6 < q/q_{CHF} < 0.9$. A thin liquid film coated each channel wall, and an interface between the liquid film and the vapor core was clearly distinguishable. No liquid droplets existed within the vapor core, indicating that the vapor-core temperature was higher than the liquid saturation temperature.

Evaporation at the liquid film–vapor core interface dominated the heat transfer from the channel wall to the fluid in the annular flow mode. Because the heat is conducted through the liquid film to the interface, the temperature at the wall has to increase to allow a higher heat-transfer rate enforced by the increased input power. The temperature would increase linearly with the input power if the film thickness stayed constant. However, the film thickness decreased with increased power due to the evaporation process, resulting in higher quality of the two-phase exit flow. Thus, the input power is converted into: (1) latent heat required for evaporation at the liquid–vapor interface due to the phase change, and (2) internal energy of the liquid film manifested by the increased liquid and wall temperature. The combination of the two mechanisms resulted in a monotonic temperature increase with decreasing slope as the input power increased. It is not clear whether the annular flow is a general pattern in microchannels due to size effect or if it is unique only to triangular channel cross-sections due to the strong capillary forces at the sharp corners (similar to micro heat pipes).

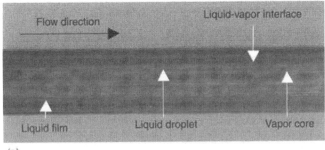

(a)

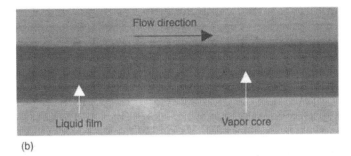

(b)

FIGURE 12.16   Flow patterns during forced convection boiling: (a) unstable annular flow with liquid droplets in the vapor core (q/q_{CHF} = 0.6), and (b) stable annular flow (q/q_{CHF} = 0.8) (channel width, 50 μm; 35 channels in the microdevice). (Reprinted with permission from Jiang, L. et al. [2001] "Forced Convection Boiling in a Microchannel Heat Sink," *J. MEMS* 10, pp. 80–87. © 2000 IEEE.)

12.4.4   Bubble Dynamics

Boiling is a phase-change process in which vapor bubbles are formed either on a heated surface or in a superheated liquid layer adjacent to the heated surface. It differs from evaporation at predetermined vapor–liquid interfaces because it also involves creation of these interfaces at discrete sites on the heated surface. Nucleate boiling is a very efficient mode of heat transfer, and it is used in various energy-conversion systems. The number density of sites that become active increases as wall heat flux or wall superheat increases. Clearly, the addition of new nucleation sites influences the heat-transfer rate from the solid surface to the working fluid. Knowledge of the nucleation site density as a function of wall superheat is, therefore, needed to develop a credible model for predicting the heat flux. Several other parameters also affect the site density, \including the surface finish, surface wettability, and material thermophysical properties. After inception, a bubble continues to grow until the forces causing it to detach from the surface exceed those pushing the bubble against the wall. Bubble dynamics, which plays an important role in determining the heat-transfer rate, includes the processes of bubble growth, bubble departure, and bubble release frequency [Dhir, 1998].

Jiang et al. (2000) reported that the first experimentally observed mode of phase change, local nucleation boiling, was detected in the microchannel heat sink at an input power level as low as $q/q_{CHF} \cong 0.5$. The working fluid was water, and the corresponding device temperature was about 70°C. Bubbles could be seen forming at specific locations along the channel walls at a few active nucleation sites. Bubble generation, growth, and explosion at a fairly high frequency inside the microducts were recorded; a mature bubble is shown in Figure 12.17. However, there were very few, if any, active nucleation sites along the channel walls. Furthermore, most of the nucleation sites became inactive after one or two runs, suggesting that they may have been residues of the fabrication process. Therefore, no attempt was made to characterize the bubble release frequency. At a slightly higher input power level, $0.5 < q/q_{CHF} < 0.6$, large bubbles were generated at the inlet/outlet common passages that connect the microchannel array to the device common

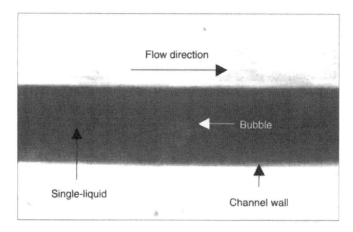

FIGURE 12.17   Active nucleation site within the microchannel exhibiting bubble formation, growth, and explosion. (Reprinted with permission from Jiang, L. et al. [2001] "Forced Convection Boiling in a Microchannel Heat Sink," *J. MEMS* 10, pp. 80–87. © 2000 IEEE.)

inlet/outlet. The boiling activity at these larger passages, shown in Figure 12.18, became more intense with increasing input power. Furthermore, the upstream bubbles were forced through the microducts as shown in Figure 12.19. The bubbles typically grew to a size larger than the microduct cross-section. Therefore, upon departure from their nucleation sites, these bubbles blocked the duct entrances, as pictured in Figure 12.19a, until the upstream pressure was high enough to force them into the microduct. In some cases, the bubbles traveled slowly along the channel as slug flow, as shown in Figure 12.19b. In most instances, however, the bubbles were ejected at high speed through the microduct and could not be detected until they reappeared at the channel exit, as shown in Figure 12.19c. A further increase of the input power level, $q/q_{CHF} > 0.7$, resulted in the annular flow pattern, and the nucleation sites on the duct walls could no longer be observed. The corresponding device temperature was about 90°C. It seems very likely that suppression of the nucleation sites within the microduct was the result of the activity of the upstream bubbles as they passed through the ducts rather than a genuine size effect. Similar bubble activity was reported by Peles et al. (1999), who conducted experiments with an almost identical microchannel heat sink.

It is reasonable to expect the bubble dynamics after inception to be affected by the channel size, unlike the nucleation site density. However, it is clear that in channels with a hydraulic diameter as small as 25 μm, bubble growth and departure have been observed. Thus, the lack of partial nucleate boiling of subcooled liquid flowing through microchannels cannot be attributed to a direct fundamental size effect suppressing bubble dynamics (i.e., bubbles cannot grow and detach due to the small size of the channel). However, it is very plausible that the absence of partial nucleate boiling is an indirect size effect. Namely, another boiling mode such as annular flow becomes dominant due to small channel size (i.e., strong capillary forces), and as a result the bubble dynamics mechanisms are suppressed.

12.4.5   Modeling of Forced Convection Boiling

Phase change from liquid to gas within a microchannel presents a formidable challenge for physical and mathematical modeling. It is not surprising, therefore, that very little work has been reported on this subject. One of the first attempts to address this problem is the derivation of Peles et al. (2000), which was based on fundamental principles rather than empirical formulations. The idealized pattern of the flow in a heated microduct is depicted in Figure 12.20. In this model, the microchannel entrance flow is in single-liquid phase and the exit flow in single-vapor phase. The two phases are separated by a meniscus at a location determined by the heat flux. Such a flow is characterized by a number of specific properties due to the existence of the interfacial surface, which is infinitely thin with a jump in pressure and velocity across the interface while the temperature is continuous. Within the single-liquid or vapor phase, heat transfer from

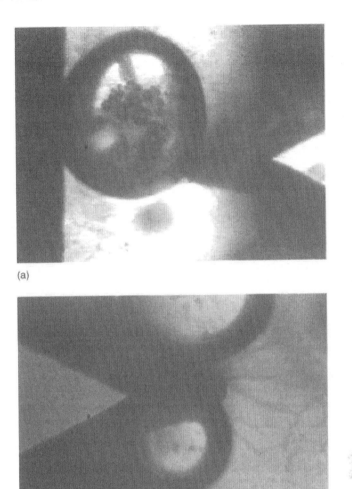

(a)

(b)

FIGURE 12.18 Bubble formation and growth at (a) inlet and (b) outlet common passages of a microchannel array (q/q_{CHF} = 0.5; 35 channels, each 50 μm in width). (Reprinted with permission from Jiang, L. et al. [2001] "Forced Convection Boiling in a Microchannel Heat Sink," *J. MEMS* 10, pp. 80–87. © 2000 IEEE.)

the wall to the fluid is accompanied by a streamwise increase of the liquid or vapor temperature and velocity. At the liquid–vapor interface, heat flux causes the liquid to move downstream and evaporate.

In addition to the standard equations of conservation and state for each phase, the mathematical model includes conditions corresponding to the interface surface. For stationary capillary flow, these conditions can be expressed by the equations of continuity of mass, thermal flux across the interface, and the balance of all forces acting on the interface. For a capillary with evaporative meniscus, the governing equations take the following form [Peles et al., 2000]:

$$\sum_{b=1}^{2} \rho^{(b)} U^{(b)} n_i^{(b)} = 0 \tag{12.31}$$

$$\sum_{b=1}^{2} \left(c_p^{(b)} \rho^{(b)} U^{(b)} T^{(b)} + k^{(b)} \frac{\partial T^{(b)}}{\partial x_i} \right) n_i^{(b)} = 0 \tag{12.32}$$

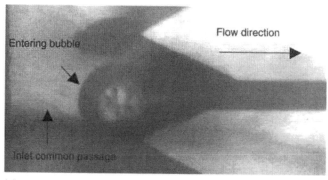

(a)

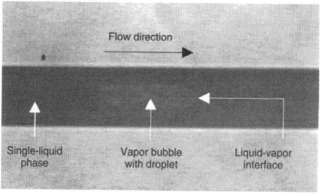

(b)

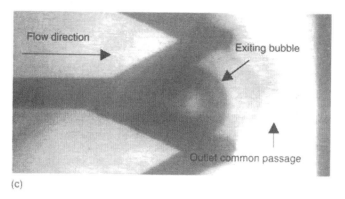

(c)

FIGURE 12.19 Sequence of pictures of a bubble (a) entering, (b) traveling through, and (c) exiting a microchannel 50 μm in width (q/q_{CHF} = 0.5). (Reprinted with permission from Jiang, L. et al. [2001] "Forced Convection Boiling in a Microchannel Heat Sink," *J. MEMS* **10**, pp. 80–87. © 2000 IEEE.)

$$\sum_{b=1}^{2} \left(P^{(b)} + \rho^{(b)} u_i^{(b)} u_j^{(b)} \right) n_i^{(b)} = \left(\sigma_{ij}^{(2)} - \sigma_{ij}^{(1)} \right) n_j + \beta \left(\frac{1}{r_1} + \frac{1}{r_2} \right) n_i^{(2)} + \frac{\partial \beta}{\partial x_i} \quad (12.33)$$

where n_i and n_j correspond to the normal and tangent directions respectively, and σ_{ij} is the tensor of viscous tension. The superscript b represents either vapor ($b = 1$) or liquid ($b = 2$). When the interface surface is expressed by a function, $x = f(y, z)$, the general radii of curvature r_i are found from the equation:

$$A_2 r_i^2 + A_1 r_i + A_0 = 0 \quad (12.34)$$

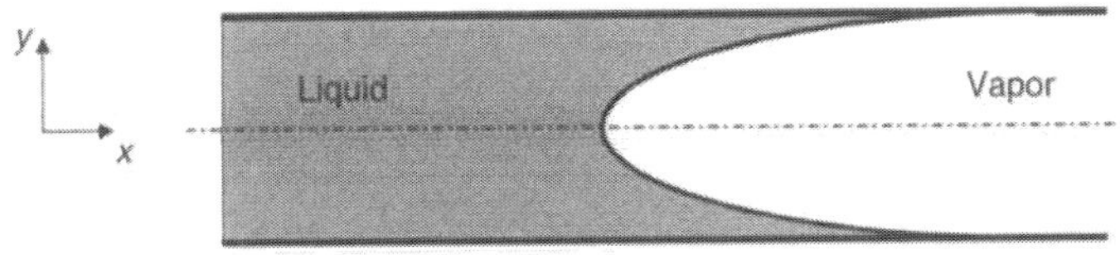

FIGURE 12.20 Physical model of forced convection boiling in a microchannel showing the single-liquid and single-vapor phases separated by an interface. (Reprinted by permission of Elsevier Science from Peles, Y.P. et al. [2000] "Thermodynamic Characteristics of Two-Phase Flow in a Heated Capillary," *Int. J. Multiphase Flow* 26, pp. 1063–93.)

The coefficients A_i depend on the shape of the interface. This model, though simplified a great deal, clearly demonstrates the complexity of the theoretical models that will have to be developed in order to obtain meaningful results. Peles et al. further assumed a quasi-one-dimensional flow and derived a set of equations for the average parameters in order to solve the system of equations. A comparison between the calculations and measurements will not be very useful at this stage, as the physical model on which the mathematical model is based has not been observed experimentally in microducts. However, the reported flow patterns could be the result of the specific triangular cross-section used in the experiment [Peles et al., 1999; Jiang et al., 2000]. It may well be that with a different cross-section shape (e.g., square or triangular) the flow pattern depicted in Figure 12.20 could develop as a stable phase-change mode.

12.5 Summary

Single-phase forced convection heat transfer in macrosystems has been investigated and well documented. The theoretical calculations and the empirical correlations should be applied to microsystems as well, unless a clear size effect has been identified that would require modification of these results. Most known size effects — velocity slip, electric double layer, and micropolar fluid — affect the thermal performance indirectly by modifying the velocity field. The only size effect directly related to the heat-transfer mechanism is the temperature jump boundary condition for gas flow, which is the result of incomplete energy exchange between the impinging molecules and the solid boundary due to the size of the channel. Therefore, research in this area should first concentrate on the size effect on the velocity field. Theoretical calculations of size effects on the velocity and temperature distributions are available to some extent. With the advent of micromachining technology, it is possible to fabricate microchannels integrated with microsensors to collect precise experimental data that can lead to sharp conclusions. Very few studies have been reported to date, and more work is required. Although solid experimental verification is still lacking, initial results show that most size effects are quite small and oftentimes are within the measurement experimental errors.

Two-phase convective heat transfer in microchannels appears to be a technology that can satisfy the demand for the dissipation of high fluxes associated with electronic and laser devices. However, in this area, both experimental and theoretical research related to phase change in microchannels is still very limited. Bubble dynamics during boiling is a complex phenomenon, and size effects can be very significant, much more so than in single-phase forced convection. It is therefore vital to establish a credible set of experimental data to provide guidance for theoretical modeling. Initial results show that classical bubble dynamics could still be observed in microchannels, but classical flow patterns related to bubble activity could not. Again, this might be an indirect effect, as other phase-change modes become more dominant in microchannels. Clearly, local as well as global measurements are required to supply adequate information in order to understand the heat-transfer mechanisms in microchannels. The integration of microsensors (temperature, pressure, capacitance, etc.) in microchannel systems and flow field visualizations (flow patterns, bubble dynamics, phase-change evolution) are becoming key components in the march to understand microscale forced convection heat transfer.

Acknowledgments

The author would like to thank Dr. Linan Jiang for her help with assembling the material and the critical review of the manuscript.

References

Arkilic, E.B., Schmidt, M.A., and Breuer, K.S. (1997) "Gaseous Slip Flow in Long Microchannels," *J. MEMS* **6**, pp. 167–78.

Beskok, A., and Karniadakis, G.E. (1994) "Simulation of Heat and Momentum Transfer in Complex Microgeometries," *J. Thermophys. Heat Transfer* **8**, pp. 647–55.

Beskok, A., Karniadakis, G.E., and Trimmer, W. (1996) "Rarefaction and Compressibility Effects in Gas Microflows," *J. Fluids Eng.* **118**, pp. 448–56.

Bhatti, M.S., and Shah, R.K. (1987) "Turbulent and Transition Flow Convective Heat Transfer in Ducts," in *Handbook of Single Phase Convective Heat Transfer*, Kakac, S., Shah, R.K., and Aung, W., eds., chapter 4, John Wiley & Sons, New York, pp. 4-1–4-166.

Bowers, M.B., and Mudawar, I. (1994) "High Flux Boiling in Low Flow Rate, Low Pressure Drop Mini-Channel and Micro-Channel Heat Sinks," *Int. J. Heat Mass Transfer* **37**, pp. 321–32.

Dhir, V.K. (1998) "Boiling Heat Transfer," *Ann. Rev. Fluid Mech.* **30**, pp. 365–401.

Goldberg, N. (1984) "Narrow Channel Forced Air Heat Sink," IEEE Trans. Components, Hybrids, Manu. Technol. CHMT-7, pp. 154–159.

Harley, J.C., Huang, Y., Bau, H.H., and Zemel, J.N. (1995) "Gas Flow in Micro-Channels," *J. Fluid Mech.* **284**, pp. 257–74.

Ho, C.M., and Tai, Y.C. (1998) "Micro-Electro-Mechanical-Systems (MEMS) and Fluid Flows," *Ann. Rev. Fluid Mech.* **30**, pp. 570–612.

Jacobi, A.M. (1989) "Flow and Heat Transfer in Microchannels Using a Microcontinuum Approach," *J. Heat Transfer* **111**, pp. 1083–85.

Jiang, L., Wang, Y., Wong, M., and Zohar, Y. (1999a) "Fabrication and Characterization of a Microsystem for Microscale Heat Transfer Study," *J. Micromech. Microeng.* **9**, pp. 422–28.

Jiang, L., Wong, M., and Zohar, Y. (1999b) "Phase Change in Micro-Channel Heat Sinks with Integrated Temperature Sensors," *J. MEMS* **8**, pp. 358–65.

Jiang, L., Wong, M., and Zohar, Y. (2001) "Forced Convection Boiling in a Microchannel Heat Sink," *J. MEMS* **10**, pp. 80–87.

Kavehpour, H.P., Faghri, M., and Asako, Y. (1997) "Effects of Compressibility and Rarefaction on Gaseous Flows in Microchannels," *Numerical Heat Transfer A* **32**, pp. 677–96.

Keyes, R.W. (1984) "Heat Transfer in Forced Convection Through Fins," *IEEE Trans. Electron Devices*, ED-31, pp. 1218–21.

Li, X., Lee, W.Y., Wong, M., and Zohar, Y. (2000) "Gas Flow in Constriction Microdevices," *Sensors and Actuators A* **83**, pp. 277–83.

Liu, J.Q., Tai, Y.C., Pong, K.C., and Ho, C.M. (1993) "Micromachined Channel/Pressure Sensor Systems for Micro Flow Studies," in *Proc. 7th Int. Conf. on Solid-State Sensors and Actuators (Transducers '93)*, pp. 995–997, IEEE, Piscataway, NJ.

Lukaszewicz, G. (1999) *Micropolar Fluids: Theory and Applications*, Birkhauser, Boston.

Mahalingam, M. (1985) "Thermal Management in Semiconductor Device Packaging," *Proc. IEEE* **73**, pp. 1396–1404.

Mohiuddin Mala, G., Li, D., and Dale, J.D. (1997) "Heat Transfer and Fluid Flow in Microchannels," *Int. J. Heat Mass Transfer* **40**, pp. 3079–88.

Mudawar, I., and Bowers, M.B. (1999) "Ultra-High Critical Heat Flux (CHF) for Subcooled Water Flow Boiling: 1. CHF Data and Parametric Effects for Small Diameter Tubes," *Int. J. Heat Mass Transfer* **42**, pp. 1405–28.

Nayak, D., Hwang, L., Turlik, I., and Reisman, A. (1987) "A High-Performance Thermal Module for Computer Packaging," *J. Electronic Mater.* **16**, pp. 357–64.

Papautsky, I., Brazzle, J., Ameel, T.A., and Frazier, A.B. (1998) "Microchannel Fluid Behavior Using Micropolar Fluid Theory," *Proc. 11th Int. Workshop on Micro Electro Mechanical Systems (MEMS '98)*, pp. 544–549, IEEE, Piscataway, NJ.

Peles, Y.P., and Haber, S. (2000) "A Steady State, One-Dimensional Model for Boiling Two-Phase Flow in Triangular Micro-Channel," *Int. J. Multiphase Flow* **26**, pp. 1095–1115.

Peles, Y.P., Yarin, L.P., and Hetsroni, G. (1999) "Evaporating Two-Phase Flow Mechanism in Microchannels," in *Proc. SPIE Int. Soc. Opt. Eng.* **3680**, pp. 226–36.

Peles, Y.P., Yarin, L.P., and Hetsroni, G. (2000) "Thermodynamic Characteristics of Two-Phase Flow in a Heated Capillary," *Int. J. Multiphase Flow* **26**, pp. 1063–93.

Peng, X.F., and Peterson, G.P. (1996) "Convective Heat Transfer and Flow Friction for Water Flow in Microchannel Structures," *Int. J. Heat Mass Transfer* **39**, pp. 2599–2608.

Peng, X.F., Peterson, G.P., and Wang, B.X. (1994) "Heat Transfer Characteristics of Water Flowing Through Microchannels," *Exp. Heat Transfer* **7**, pp. 265–83.

Peng, X.F., Peterson, G.P., and Wang, B.X. (1996) "Flow Boiling of Binary Mixtures in Microchanneled Plates," *Int. J. Heat Mass Transfer* **39**, pp. 1257–64.

Peng, X.F., and Wang, B.X. (1993) "Forced Convection and Flow Boiling Heat Transfer for Liquid Flowing Through Microchannels," *Int. J. Heat Mass Transfer* **36**, pp. 3421–27.

Peng, X.F., Wang, B.X., Peterson, G.P., and Ma, H.B. (1995) "Experimental Investigation of Heat Transfer in Flat Plates with Rectangular Microchannels," *Int. J. Heat Mass Transfer* **38**, pp. 127–37.

Pfahler, J., Harley, J.C., Bau, H.H., and Jemel, J.N. (1990) "Liquid Transport in Micron and Submicron Channels," *Sensors and Actuators A* **21–23**, pp. 431–34.

Pong, K.C., Ho, C.M., Liu, J., and Tai, Y.C. (1994) "Non-Linear Pressure Distribution in Uniform Micro-Channels," *ASME FED* **197**, pp. 51–56.

Ravigururajan, T.S. (1998) "Impact of Channel Geometry on Two-Phase Flow Heat Transfer Characteristics of Refrigerants in Microchannel Heat Exchangers," *J. Heat Transfer*, **120**, pp. 485–91.

Sadik, K., and Yaman, Y. (1995) *Convective Heat Transfer*, 2nd ed., CRC Press, Boca Raton.

Samalam, V.K. (1989) "Convective Heat Transfer in Microchannels," *J. Electronic Mater.* **18**, pp. 611–17.

Schaaf, S.A., and Chambre, P.L. (1961) *Flow of Rarefied Gases*, Princeton University Press, Princeton, NJ.

Shah, R.K., and Bhatti, M.S. (1987) "Laminar Convective Heat Transfer in Ducts," in *Handbook of Single Phase Convective Heat Transfer*, Kakac, S., Shah, R.K., and Aung, W., eds., chapter 3, John Wiley & Sons, New York.

Stanley, R.S., Ameel, T.A., and Warrington, R.O. (1995) "Convective Flow Boiling in Microgeometries: A Review and Application," in *Proc. Convective Flow Boiling*, an international conference held April 30–May 5 at Banff, Alberta, Canada, pp. 305–10.

Stephan, K. (1992) *Heat Transfer in Condensation and Boiling*, Springer-Verlag, Berlin.

Stokes, V.K. (1984) *Theories of Fluids with Microstructure*, Springer-Verlag, Berlin.

Tuckerman, D.B., and Pease, R.F.W. (1981) "High-Performance Heat Sinking for VLSI," *IEEE Electron Device Lett.* **2**, pp. 126–29.

van den Berg, H.R., ten Seldam, C.A., and van der Gulik, P.S. (1993) "Compressible Laminar Flow in a Capillary," *J. Fluid Mech.* **246**, pp. 1–20.

Weisberg, A., Bau, H.H., and Zemel, J.N. (1992) "Analysis of Microchannels for Integrated Cooling," *Int. J. Heat Mass Transfer* **35**, pp. 2465–74.

Wu, P., and Little, W.A. (1984) "Measurement of the Heat Transfer Characteristics of Gas Flow in Fine Channel Heat Exchangers Used for Microminiature Refrigerators," *Cryogenics* **24**, pp. 415–20.

Wu, S., Mai, J., Zohar, Y., Tai, Y.C., and Ho, C.M. (1998) "A Suspended Microchannel with Integrated Temperature Sensoaqrs for High-Pressure Flow Studies," in *Proc. 11th Int. Workshop on Micro Electro Mechanical Systems (MEMS '98)*, pp. 87–92, IEEE, Piscataway, NJ.

Zohar, Y., Chu, W.K.H., Hsu, C.T., and Wong, M. (1994) "Slip Flow Effects on Heat Transfer in MicroChannels," *Bull. Am. Phys. Soc.* **39**, p. 1908.

13

Flow Control

Mohamed Gad-el-Hak
Virginia Commonwealth University

"I would like to shock you by stating that I believe that with today's technology we can easily — I say easily — construct motors one fortieth of this size on each dimension. That's sixty-four thousand times smaller than the size of McLellan's motor (400 microns on a side). And in fact, with our present technology, we can make thousands of these motors at a time, all separately controllable. Why do you want to make them? I told you there's going to be lots of laughter, but just for fun, I'll suggest how to do it — it's very easy."

(From the talk "Infinitesmal Machinery," delivered by Richard P. Feynman at the Jet Propulsion Laboratory, Pasadena, California, 23 February 1983.)

13.1   Introduction

The subject of flow control — particularly reactive flow control — is broadly introduced in this chapter, leaving some of the details to other chapters in this as well as in the other two volumes, which will deal with the issues of control theory, distributed and optimal control, soft computing tools such as neural networks, genetic algorithms and fuzzy logic, diagnostic and control of turbulent flows, and futuristic control specifically targeting the coherent structures in turbulent flows. Distributed flow control can greatly benefit from the availability of inexpensive microsensors and microactuators, hence the relevance of this topic to the *Handbook of MEMS*. In reactive control of turbulent flows, large arrays of minute sensors and actuators form feedback or feedforward control loops, which have a performance that can be enhanced by employing optimized control algorithms and soft computing tools, hence the inclusion in this book of chapters on these specialized topics.

The ability to manipulate a flowfield actively or passively to effect a desired change is of immense technological importance, and this undoubtedly accounts for the fact that the subject is more hotly pursued by scientists and engineers than any other topic in fluid mechanics. The potential benefits of realizing efficient flow-control systems range from saving billions of dollars in annual fuel costs for land, air, and sea vehicles to achieving economically and environmentally more competitive industrial processes involving fluid flows. Methods of control to effect transition delay, separation postponement, lift enhancement, drag reduction, turbulence augmentation, and noise suppression are considered. Prandtl (1904) pioneered the modern use of flow control in his epoch-making presentation to the *Third International Congress of Mathematicians* held at Heidelberg, Germany. In just eight pages, Prandtl introduced the boundary-layer theory, explained the mechanics of steady separation, opened the way for understanding the motion of real fluids, and described several experiments in which the boundary layer was controlled. He used active control of the boundary layer to show the great influence such a control exerted on the flow pattern. Specifically, Prandtl used suction to delay boundary-layer separation from the surface of a cylinder.

Notwithstanding Prandtl's success, aircraft designers in the three decades following his convincing demonstration were accepting lift and drag of airfoils as predestined characteristics with which no man could or should tamper [Lachmann, 1961]. This predicament changed mostly due to the German research in boundary-layer control pursued vigorously shortly before and during World War II. In the two decades following the war, extensive research on laminar flow control, where the boundary layer formed along the external surfaces of an aircraft is kept in the low-drag laminar state, was conducted in Europe and the United States, culminating in the successful flight test program of the X-21 where suction was used to delay transition on a swept wing up to a chord Reynolds number of 4.7×10^7. The oil crisis of the early 1970s brought renewed interest in novel methods of flow control to reduce skin-friction drag even in turbulent boundary layers. In the 1990s, the need to reduce the emissions of greenhouse gases and to construct supermaneuverable fighter planes, faster and quieter underwater vehicles, and hypersonic transport aircraft (e.g., the U.S. National Aerospace Plane) provides new challenges for researchers in the field of flow control.

The three books by Gad-el-Hak et al. (1998), Gad-el-Hak (2000) and Gad-el-Hak and Tsai (2005) provide an up-to-date overview of the subject of flow control. In this chapter, following a description of the unifying principles of flow control, we focus on the concept of targeted control in which distributed arrays of microsensors and microactuators, connected in open- or closed-control loops, are used to target the coherent structures in turbulent flows in order to effect beneficial flow changes such as drag reduction, lift enhancement, noise suppression, etc.

13.2 Unifying Principles

A particular control strategy is chosen based on the kind of flow and the control goal to be achieved. Flow control goals are strongly, often adversely, interrelated; there lies the challenge of making the tough compromises. There are several different ways for classifying control strategies to achieve a desired effect. Presence or lack of walls, Reynolds and Mach numbers, and the character of the flow instabilities are all important considerations for the type of control to be applied. All these seemingly disparate issues are what places the field of flow control in a unified framework. They will be discussed in turn in the following five subsections.

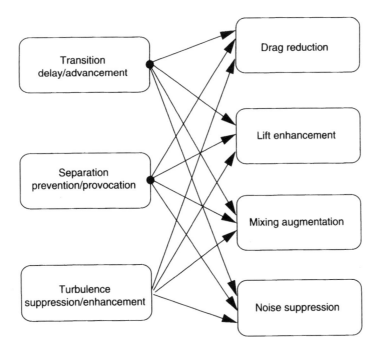

FIGURE 13.1   Engineering goals and corresponding flow changes.

13.2.1   Control Goals and Their Interrelation

What do engineers want to achieve when attempting to manipulate a particular flowfield? Typically they aim at reducing the drag; enhancing the lift; augmenting the mixing of mass, momentum, or energy; suppressing the flow-induced noise; or a combination thereof. To achieve any of these useful end results, for either free-shear or wall-bounded flows, transition from laminar to turbulent flow may have to be either delayed or advanced, flow separation may have to be either prevented or provoked, and finally turbulence levels may have to be either suppressed or enhanced. All those engineering goals and the corresponding flow changes intended to effect them are schematically depicted in Figure 13.1. None of that is particularly difficult if taken in isolation, but the challenge is in achieving a goal using a simple device, inexpensive to build as well as to operate, and, most importantly, has minimum side effects. For this latter hurdle, the interrelation between control goals must be elaborated, and this is what is attempted below.

Consider the technologically very important boundary layers. An external wall-bounded flow, such as that developing on the exterior surfaces of an aircraft or a submarine, can be manipulated to achieve transition delay, separation postponement, lift increase, skin-friction, and pressure[1] drag reduction, turbulence augmentation, heat transfer enhancement, or noise suppression. These objectives are not necessarily mutually exclusive. The schematic in Figure 13.2 is a partial representation of the interrelation between one control goal and another. To focus the discussion further, consider the flow developing on a lifting surface such as an aircraft wing. If the boundary layer becomes turbulent, its resistance to separation is enhanced and more lift could be obtained at increased incidence. On the other hand, the skin-friction drag for a laminar boundary layer can be as much as an order of magnitude less than that for a turbulent one. If transition is delayed, lower skin friction as well as lower flow-induced noise are achieved. However,

---

[1]Pressure drag includes contributions from flow separation, displacement effects, induced drag, wave drag, and, for time-dependent motion of a body through a fluid, virtual mass.

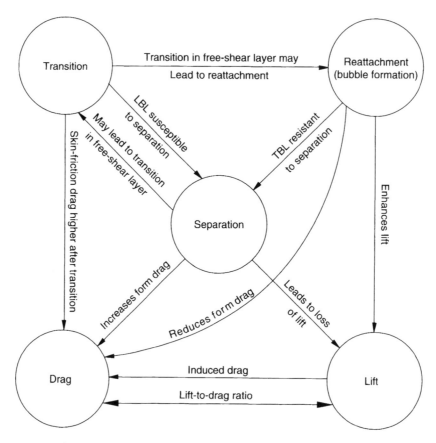

FIGURE 13.2   Interrelation between flow control goals.

the laminar boundary layer can only support very small adverse pressure gradient without separation and subsequent loss of lift and increase in form drag occur.

Once the laminar boundary layer separates, a free-shear layer forms and for moderate Reynolds numbers transition to turbulence takes place. Increased entrainment of high-speed fluid due to the turbulent mixing may result in reattachment of the separated region and formation of a laminar separation bubble. At higher incidence, the bubble breaks down either separating completely or forming a longer bubble. In either case, the form drag increases and the lift-curve's slope decreases. The ultimate goal of all this is to improve the airfoil's performance by increasing the lift-to-drag ratio. However, induced drag is caused by the lift generated on a lifting surface with a finite span. Moreover, more lift is generated at higher incidence but form drag also increases at these angles.

All of the above point to potential conflicts as one tries to achieve a particular control goal only to adversely affect another goal. An ideal method of control that is simple, inexpensive to build and operate, and does not have any tradeoffs does not exist, and the skilled engineer has to make continuous compromises to achieve a particular design goal.

13.2.2   Classification Schemes

There are different classification schemes for flow control methods. One is to consider whether the technique is applied at the wall or away from it. Surface parameters that can influence the flow include roughness, shape, curvature, rigid-wall motion, compliance, temperature, and porosity. Heating and cooling of the surface can influence the flow via the resulting viscosity and density gradients. Mass transfer can take place through a porous wall or a wall with slots. Suction and injection of primary fluid can have significant

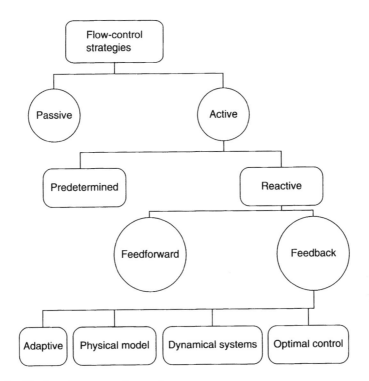

FIGURE 13.3   Classification of flow control strategies.

effects on the flowfield, influencing particularly the shape of the velocity profile near the wall and thus the boundary layer susceptibility to transition and separation. Different additives, such as polymers, sur-factants, micro-bubbles, droplets, particles, dust, or fibers, can also be injected through the surface in water or air wall-bounded flows. Control devices located away from the surface can also be beneficial. Large-eddy breakup devices (also called outer-layer devices, or OLDs), acoustic waves bombarding a shear layer from outside, additives introduced in the middle of a shear layer, manipulation of freestream turbulence levels and spectra, gust, and magneto- and electro-hydrodynamic body forces are examples of flow control strategies applied away from the wall.

A second scheme for classifying flow control methods considers energy expenditure and the control loop involved. As shown in the schematic in Figure 13.3, a control device can be passive, requiring no auxiliary power and no control loop, or active, requiring energy expenditure. For the action of passive devices, some prefer to use the term flow management rather than flow control [Fiedler and Fernholz, 1990], reserving the latter terminology for dynamic processes. Active control requires a control loop and is further divided into predetermined or reactive. Predetermined control includes the application of steady or unsteady energy input without regard to the particular state of the flow. The control loop in this case is open as shown in Figure 13.4a, and no sensors are required. Because no sensed information is being fed forward, this open control loop is not a feedforward one. This subtle point is often confused in the litera-ture, blurring predetermined control with reactive, feedforward control. Reactive control is a special class of active control where the control input is continuously adjusted based on measurements of some kind. The control loop in this case can either be an open, feedforward one (Figure 13.4b) or a closed, feedback loop (Figure 13.4c). Classical control theory deals, for the most part, with reactive control.

The distinction between feedforward and feedback is particularly important when dealing with the control of flow structures that convect over stationary sensors and actuators. In feedforward control, the measured variable and the controlled variable differ. For example, the pressure or velocity can be sensed at an upstream location, and the resulting signal is used together with an appropriate control law to trigger an actuator, which in turn influences the velocity at a downstream position. Feedback control,

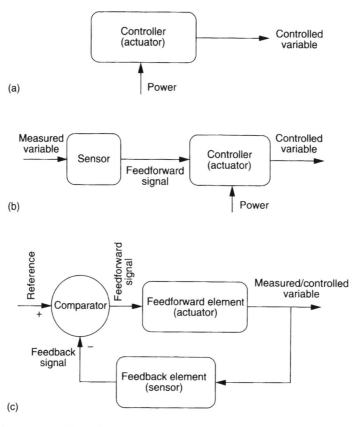

FIGURE 13.4 Different control loops for active flow control. (a) Predetermined, open-loop control; (b) Reactive, feedforward, open-loop control; (c) Reactive, feedback, closed-loop control.

on the other hand, necessitates that the controlled variable be measured, fed back, and compared with a reference input. Reactive feedback control is further classified into four categories: adaptive, physical model based, dynamical systems based, and optimal control [Moin and Bewley, 1994].

A yet another classification scheme is to consider whether the control technique directly modifies the shape of the instantaneous or mean velocity profile or selectively influence the small dissipative eddies. An inspection of the Navier–Stokes equations written at the surface [Gad-el-Hak, 2000], indicates that the spanwise and streamwise vorticity fluxes at the wall can be changed, either instantaneously or in the mean, via wall motion/compliance, suction/injection, streamwise or spanwise pressure-gradient (respectively), normal viscosity-gradient, or a suitable streamwise or spanwise body force. These vorticity fluxes determine the fullness of the corresponding velocity profiles. For example, suction (or downward wall motion), favorable pressure-gradient or lower wall-viscosity results in vorticity flux away from the wall, making the surface a source of spanwise and streamwise vorticity. The corresponding fuller velocity profiles have negative curvature at the wall and are more resistant to transition and to separation but are associated with higher skin-friction drag. Conversely, an inflectional velocity profile can be produced by injection (or upward wall motion), adverse pressure-gradient, or higher wall-viscosity. Such profile is more susceptible to transition and to separation and is associated with lower, even negative, skin friction. Note that many techniques are available to effect a wall viscosity-gradient; for example surface heating or cooling, film boiling, cavitation, sublimation, chemical reaction, wall injection of lower or higher viscosity fluid, and the presence of shear thinning/thickening additive.

Flow control devices can alternatively target certain scales of motion rather than globally changing the velocity profile. Polymers, riblets, and LEBUs, for example, appear to selectively damp only the small dissipative eddies in turbulent wall-bounded flows. These eddies are responsible for the (instantaneous) inflectional profile and the secondary instability in the buffer zone, and their suppression leads to increased scales, a delay in the reduction of the (mean) velocity-profile slope, and consequent thickening of the wall region. In the buffer zone, the scales of the dissipative and energy containing eddies are roughly the same and, hence, the energy containing eddies will also be suppressed resulting in reduced Reynolds stress production, momentum transport and skin friction.

13.2.3 Free-Shear and Wall-Bounded Flows

Free-shear flows, such as jets, wakes, and mixing layers, are characterized by inflectional mean-velocity profiles and are therefore susceptible to inviscid instabilities. Viscosity is only a damping influence in this case: the prime instability mechanism is vortical induction. Control goals for such flows include transition delay or advancement, mixing enhancement, and noise suppression. External and internal wall-bounded flows, such as boundary layers and channel flows, can also have inflectional velocity profiles, but in the absence of adverse pressure-gradient and similar effects, are characterized by non-inflectional profiles; thus, viscous instabilities should then be considered. These kinds of viscosity-dominated wall-bounded flows are intrinsically stable and therefore are generally more difficult to control. Free-shear flows and separated boundary layers, on the other hand, are intrinsically unstable and lend themselves more readily to manipulation.

Free-shear flows originate from some kind of surface upstream be it a nozzle, a moving body, or a splitter plate, and flow control devices can therefore be placed on the corresponding walls albeit far from the fully-developed regions. Examples of such control include changing of the geometry of a jet exit from circular to elliptic [Gutmark and Ho, 1986]; using periodic suction/injection in the lee side of a blunt body to affect its wake [Williams and Amato, 1989]; and vibrating the splitter plate of a mixing layer [Fiedler et al., 1988]. These and other techniques are extensively reviewed by Fiedler and Fernholz (1990), who offer a comprehensive list of appropriate references, and more recently by Gutmark et al. (1995), Viswanath (1995), and Gutmark and Grinstein (1999).

13.2.4 Regimes of Reynolds and Mach Numbers

Reynolds number is the ratio of inertial to viscous forces and — absent centrifugal, gravitational, electromagnetic, and other unusual effects — Re determines whether the flow is laminar or turbulent. It is defined as $Re \equiv v_o L / v$, where v_o and L are respectively suitable velocity and length scales, and v is the kinematic viscosity. For low-Reynolds-number flows, instabilities are suppressed by viscous effects and the flow is laminar, as can be found in systems with large fluid viscosity, small length-scale, or small velocity. The large-scale motion of the highly viscous volcanic molten rock and air or water flow in capillaries and microdevices are examples of laminar flows. Turbulent flows seem to be the rule rather than the exception, occurring in or around most important fluid systems such as airborne and waterborne vessels, gas and oil pipelines, material processing plants, and the human cardiovascular and pulmonary systems.

Because of the nature of their instabilities, free-shear flows undergo transition at extremely low Reynolds numbers as compared to wall-bounded flows. Many techniques are available to delay laminar-to-turbulence transition for both kinds of flows, but none would do that to indefinitely high Reynolds numbers. Therefore, for Reynolds numbers beyond a reasonable limit, one should not attempt to prevent transition but rather deal with the ensuing turbulence. Of course early transition to turbulence can be advantageous in some circumstances, for example to achieve separation delay, enhanced mixing, or augmented heat transfer. The task of advancing transition is generally simpler than trying to delay it. Numerous books and articles specifically address the control of laminar-to-turbulence transition [e.g., Gad-el-Hak, 2000, and references therein]. For now, we briefly discuss transition control for various regimes of Reynolds and Mach numbers.

Three Reynolds number regimes can be identified for the purpose of reducing skin friction in wall-bounded flows. First, if the flow is laminar, typically at Reynolds numbers based on distance from leading edge $<10^6$, then methods of reducing the laminar shear stress are sought. These are usually velocity-profile modifiers, for example adverse-pressure gradient, injection, cooling (in water), and heating (in air), that reduce the fullness of the profile at the increased risk of premature transition and separation. Secondly, in the range of Reynolds numbers from 1×10^6 to 4×10^7, active and passive methods to delay transition as far back as possible are sought. These techniques can result in substantial savings and are broadly classified into two categories: stability modifiers and wave cancellation. The skin-friction coefficient in the laminar flat-plate can be as much as an order of magnitude less than that in the turbulent case. Note, however, that all the stability modifiers, such as favorable pressure-gradient, suction or heating (in liquids), result in an increase in the skin friction over the unmodified Blasius layer. The object is, of course, to keep this penalty below the potential saving; i.e., the net drag will be above that of the flat-plate laminar boundary layer but presumably well below the viscous drag in the flat-plate turbulent flow. Thirdly, for $Re > 4 \times 10^7$, transition to turbulence cannot be delayed with any known practical method without incurring a penalty that exceeds the saving. The task is then to reduce the skin-friction coefficient in a turbulent boundary layer. Relaminarization [Narasimha and Sreenivasan, 1979] is an option, although achieving a net saving here is problematic at present.

The Mach number is the ratio of a characteristic flow velocity to local speed of sound, $Ma \equiv v_o/a_o$. It determines whether the flow is incompressible ($Ma < 0.3$) or compressible ($Ma > 0.3$). The latter regime is further divided into subsonic ($Ma < 1$), transonic ($0.8 < Ma < 1.2$), supersonic ($Ma > 1$), and hypersonic ($Ma > 5$). Each of those flow regimes lends itself to different optimum methods of control to achieve a given goal. Take laminar-to-turbulence transition control as an illustration [Bushnell, 1994]. During transition, the field of initial disturbances is internalized via a process termed receptivity and the disturbances are subsequently amplified by various linear and nonlinear mechanisms. Assuming that by-pass mechanisms, such as roughness or high levels of freestream turbulence, are identified and circumvented, delaying transition then is reduced to controlling the variety of possible linear modes: Tollmien-Schlichting modes, Mack modes, crossflow instabilities, and Görtler instabilities. Tollmien-Schlichting instabilities dominate the transition process for two-dimensional boundary layers having $Ma < 4$, and are damped by increasing the Mach number, by wall cooling (in gases), and by the presence of favorable pressure-gradient. Contrast this to the Mack modes, which dominate for two-dimensional hypersonic flows. Mack instabilities are also damped by increasing the Mach number and by the presence of favorable pressure-gradient, but are destabilized by wall cooling. Crossflow and Görtler instabilities are caused by, respectively, the development of inflectional crossflow velocity profile and the presence of concave streamline curvature. Both of these instabilities are potentially harmful across the speed range, but are largely unaffected by Mach number and wall cooling. The crossflow modes are enhanced by favorable pressure-gradient, while the Görtler instabilities are insensitive. Suction suppresses, to different degrees, all the linear modes discussed in here.

13.2.5 Convective and Absolute Instabilities

In addition to grouping the different kinds of hydrodynamic instabilities as inviscid or viscous, one could also classify them as convective or absolute based on the linear response of the system to an initial localized impulse [Huerre and Monkewitz, 1990]. A flow is convectively unstable if, at any fixed location, this response eventually decays in time; in other words, if all growing disturbances convect downstream from their source. Convective instabilities occur when there is no mechanism for upstream disturbance propagation, as for example in the case of rigid-wall boundary layers. If the disturbance is removed, then perturbation propagates downstream and the flow relaxes to an undisturbed state. Suppression of convective instabilities is particularly effective when applied near the point where the perturbations originate.

If any of the growing disturbances has zero group velocity, the flow is absolutely unstable. This means that the local system response to an initial impulse grows in time. Absolute instabilities occur when a mechanism exists for upstream disturbance propagation, as for example in the separated flow over a backward-facing step where the flow recirculation provides such mechanism. In this case, some of the growing disturbances can travel back upstream and continually disrupt the flow even after the initial disturbance is neutralized. Therefore, absolute instabilities are generally more dangerous and more difficult to control; nothing short of complete suppression will work. In some flows, for example two-dimensional blunt-body wakes, certain regions are absolutely unstable while others are convectively unstable. The upstream addition of acoustic or electric feedback can change a convectively unstable flow to an absolutely unstable one and self-excited flow oscillations can thus be generated. In any case, identifying the character of flow instability facilitates its effective control (i.e., suppressing or amplifying the perturbation as needed).

13.3 The Taming of the Shrew

For the rest of this chapter, we focus on reactive flow control specifically targeting the coherent structures in turbulent flows. By comparison with laminar flow control or separation prevention, the control of turbulent flow remains a very challenging problem. Flow instabilities magnify quickly near critical flow regimes, and therefore delaying transition or separation are relatively easier tasks. In contrast, classical control strategies are often ineffective for fully turbulent flows. Newer ideas for turbulent flow control to achieve, for example, skin-friction drag reduction, focus on the direct onslaught on coherent structures. Spurred by the recent developments in chaos control, microfabrication, and soft computing tools, reactive control of turbulent flows, where sensors detect oncoming coherent structures and actuators attempt to favorably modulate those quasi-periodic events, is now in the realm of the possible for future practical devices.

Considering the extreme complexity of the turbulence problem in general and the unattainability of first-principles analytical solutions in particular, it is not surprising that controlling a turbulent flow remains a challenging task, mired in empiricism and unfulfilled promises and aspirations. Brute force suppression, or *taming*, of turbulence via active, energy-consuming control strategies is always possible, but the penalty for doing so often exceeds any potential benefits. The artifice is to achieve a desired effect with minimum energy expenditure. This is of course easier said than done. Indeed, suppressing turbulence is as arduous as *The Taming of the Shrew*.

13.4 Control of Turbulence

Numerous methods of flow control have already been successfully implemented in practical engineering devices. Delaying laminar-to-turbulence transition to reasonable Reynolds numbers and preventing separation can readily be accomplished using a myriad of passive and predetermined active control strategies. Such classical techniques have been reviewed by, among others, Bushnell (1983; 1994); Wilkinson et al. (1988); Bushnell and McGinley (1989); Gad-el-Hak (1989; 2000); Bushnell and Hefner (1990); Fiedler and Fernholz (1990); Gad-el-Hak and Bushnell (1991a; 1991b); Barnwell and Hussaini (1992); Viswanath (1995); and Joslin et al. (1996). Yet, very few of the classical strategies are effective in controlling free-shear or wall-bounded turbulent flows. Serious limitations exist for some familiar control techniques when applied to certain turbulent flow situations. For example, in attempting to reduce the skin-friction drag of a body having a turbulent boundary layer using global suction, the *penalty* associated with the control device often exceeds the *saving* derived from its use. What is needed is a way to reduce this penalty to achieve a more efficient control.

Flow control is most effective when applied near the transition or separation points; in other words, near the critical flow regimes where flow instabilities magnify quickly. Therefore, delaying or advancing laminar-to-turbulence transition and preventing or provoking separation are relatively easier tasks to accomplish. Reducing the skin-friction drag in a non-separating turbulent boundary layer, where the mean flow is quite stable, is a more challenging problem. Yet, even a modest reduction in the fluid resistance to the motion of, for example, the worldwide commercial airplane fleet is translated into fuel savings estimated to be in the billions of dollars. Newer ideas for turbulent flow control focus on the direct onslaught on coherent structures via reactive control strategies that utilize large arrays of microsensors and microactuators.

The primary objective of this chapter is to advance possible scenarios by which viable control strategies of turbulent flows could be realized. As will be argued in the following sections, future systems for control of turbulent flows in general and turbulent boundary layers in particular could greatly benefit from the merging of the science of chaos control, the technology of microfabrication, and the newest computational tools collectively termed soft computing. Control of chaotic, nonlinear dynamical systems has been demonstrated theoretically as well as experimentally, even for multi-degree-of-freedom systems. Microfabrication is an emerging technology that has the potential for producing inexpensive, programmable sensor/actuator chips that have dimensions of the order of a few microns. Soft computing tools include neural networks, fuzzy logic, and genetic algorithms and are now more advanced as well as more widely used as compared to just few years ago. These tools could be very useful in constructing effective adaptive controllers.

Such futuristic systems are envisaged as consisting of a large number of intelligent, interactive, microfabricated wall sensors and actuators arranged in a checkerboard pattern and targeted toward specific organized structures that occur quasi-randomly within a turbulent flow. Sensors detect oncoming coherent structures, and adaptive controllers process the sensors' information providing control signals to the actuators that in turn attempt to favorably modulate the quasi-periodic events. A finite number of wall sensors perceives only partial information about the entire flowfield above. However, a low-dimensional dynamical model of the near-wall region used in a Kalman filter can make the most of the partial information from the sensors. Conceptually this is not too difficult, but in practice the complexity of such a control system is daunting and much research and development work still remain.

The following discussion is organized into ten sections. A particular example of a classical control system — suction — is described in the following section. This will serve as a prelude to introducing the selective suction concept. The different hierarchies of coherent structures that dominate a turbulent boundary layer and that constitute the primary target for direct onslaught are then briefly recalled. The characteristic lengths and sensor requirements of turbulent flows are then discussed in the two subsequent sections. This is followed by a description of reactive flow control and the selective suction concept. The number, size, frequency, and energy consumption of the sensor/actuator units required to tame the turbulence on a full-scale air or water vehicle are estimated in that same section. This is followed by an introduction to the topic of magnetohydrodynamics and a reactive flow control scheme using electromagnetic body forces. The emerging areas of chaos control and soft computing, particularly as they relate to reactive control strategies, are then briefly discussed in the two subsequent sections. This is followed by a discussion of the specific use of MEMS devices for reactive flow control. Finally, brief concluding remarks are given in the last section.

13.5 Suction

To set the stage for introducing the concept of targeted or selective control, this section will first address global control as applied to wall-bounded flows. A viscous fluid that is initially irrotational will acquire vorticity when an obstacle is passed through the fluid. This vorticity controls the nature and structure of the boundary-layer flow in the vicinity of the obstacle. For an incompressible, wall-bounded flow, the flux of spanwise or streamwise vorticity at the wall, and hence whether the surface is a sink or a source of vorticity, is affected by the wall motion (e.g. in the case of a compliant coating); transpiration (suction or injection); streamwise or spanwise pressure gradient; wall curvature; normal viscosity gradient near the wall (caused by, for example, heating or cooling of the wall or introduction of a shear-thinning/shear thickening additive into the boundary layer); and body forces (such as electromagnetic ones in a conducting fluid). These alterations separately or collectively control the shape of the instantaneous as well as the mean velocity profiles that in turn determines the skin friction at the wall, the boundary layer ability to resist transition and separation, and the intensity of turbulence and its structure.

To illustrate, this section will focus on global wall suction as a generic control tool. The arguments presented here and in subsequent sections are equally valid for other global control techniques, such as geometry modification (body shaping), surface heating or cooling, electromagnetic control, etc. Transpiration provides a good example of a single control technique that is used to achieve a variety of goals. Suction leads to a fuller velocity profile (vorticity flux away from the wall) and can, therefore, be employed to delay laminar-to-turbulence transition, postpone separation, achieve an asymptotic turbulent boundary layer (i.e., one having constant momentum thickness), or relaminarize an already turbulent flow. Unfortunately, global suction cannot be used to reduce the skin-friction drag in a turbulent boundary layer. The amount of suction required to inhibit boundary-layer growth is too large to effect a net drag reduction. This is a good illustration of a situation where the penalty associated with a control device might exceed the saving derived from its use.

Small amounts of fluid withdrawn from the near-wall region of a boundary layer change the curvature of the velocity profile at the wall and can dramatically alter the stability characteristics of the flow. Concurrently, suction inhibits the growth of the boundary layer, so that the critical Reynolds number based on thickness may never be reached. Although laminar flow can be maintained to extremely high Reynolds numbers provided that enough fluid is sucked away, the goal is to accomplish transition delay with the minimum suction flow rate. This will reduce not only the power necessary to drive the suction pump, but also the momentum loss due to the additional freestream fluid entrained into the boundary layer as a result of withdrawing fluid from the wall. That momentum loss is, of course, manifested as an increase in the skin-friction drag.

The case of uniform suction from a flat plate at zero incidence is an exact solution of the Navier–Stokes equation. The asymptotic velocity profile in the viscous region is exponential and has a negative curvature at the wall. The displacement thickness has the constant value $\delta^* = v/|v_w|$, where v is the kinematic viscosity, and $|v_w|$ is the absolute value of the normal velocity at the wall. In this case, the familiar von Kármán integral equation reads: $C_f = 2C_q$. Bussmann and Münz (1942) computed the critical Reynolds number for the asymptotic suction profile to be: $Re_{\delta^*} \equiv U_\infty \delta^*/v = 70{,}000$. From the value of δ^* given above, the flow is stable to all small disturbances if $C_q \equiv |v_w|/U_\infty > 1.4 \times 10^{-5}$. The amplification rate of unstable disturbances for the asymptotic profile is an order of magnitude less than that for the Blasius boundary layer [Pretsch, 1942]. This treatment ignores the development distance from the leading edge needed to reach the asymptotic state. When this is included into the computation, a higher $C_q = 1.18 \times 10^{-4}$ is required to ensure stability [Iglisch, 1944; Ulrich, 1944].

In a turbulent wall-bounded flow, the results of Eléna (1975; 1984) and Antonia et al. (1988) indicate that suction causes an appreciable stabilization of the low-speed streaks in the near-wall region. The maximum turbulence level at $y^+ \approx 13$ drops from 15% to 12% as C_q varies from 0 to 0.003. More dramatically, the tangential Reynolds stress near the wall drops by a factor of 2 for the same variation of C_q. The dissipation length-scale near the wall increases by 40% and the integral length-scale by 25% with the suction.

The suction rate necessary for establishing an asymptotic turbulent boundary layer independent of streamwise coordinate (i.e., $d\delta_\theta/dx = 0$) is much lower than the rate required for relaminarization ($C_q \approx 0.01$), but is still not low enough to yield net drag reduction. For Reynolds number based on distance from leading edge $Re_x = \mathcal{O}[10^6]$, Favre et al. (1966), Rotta (1970), and Verollet et al. (1972), among others, report an asymptotic suction coefficient of $C_q \approx 0.003$. For a zero-pressure-gradient boundary layer on a flat plate, the corresponding skin-friction coefficient is $C_f = 2C_q = 0.006$, indicating higher skin friction than if no suction was applied. To achieve a net skin-friction reduction with suction, the process must be further optimized. One way to accomplish that is to target the suction toward particular organized structures within the boundary layer and not to use it globally as in classical control schemes. This point will be revisited later, but the coherent structures to be targeted and the length-scales to be expected are first detailed in the following two sections.

13.6 Coherent Structures

The previous discussion indicates that achieving a particular control goal is always possible. The challenge is reaching that goal with a penalty that can be tolerated. Suction, for example, would lead to a net drag reduction, if only we could reduce the suction coefficient necessary for establishing an asymptotic turbulent boundary layer to below one-half of the unperturbed skin-friction coefficient. A more efficient way of using suction, or any other global control method, is to target particular coherent structures within the turbulent boundary layer. Before discussing this *selective control* idea, this section and the following shall briefly describe the different hierarchy of organized structures in a wall-bounded flow and the expected scales of motion.

The classical view that turbulence is essentially a stochastic phenomenon having a randomly fluctuating velocity field superimposed on a well-defined mean has been changed in the last few decades by the realization that the transport properties of all turbulent shear flows are dominated by quasi-periodic, large-scale vortex motions [Laufer, 1975; Cantwell, 1981; Fiedler, 1988; Robinson, 1991]. Despite the extensive research work in this area, no generally accepted definition of what is meant by coherent motion has emerged. In physics, coherence stands for well-defined phase relationship. For the present purpose we adopt the rather restrictive definition given by Hussain (1986): "*A coherent structure is a connected turbulent fluid mass with instantaneously phase-correlated vorticity over its spatial extent.*" In other words, underlying the random, three-dimensional vorticity that characterizes turbulence, there is a component of large-scale vorticity, which is instantaneously coherent over the spatial extent of an organized structure. The apparent randomness of the flowfield is, for the most part, due to the random size and strength of the different types of organized structures comprising that field.

In a wall-bounded flow, a multiplicity of coherent structures have been identified mostly through flow visualization experiments, although some important early discoveries have been made using correlation measurements [Townsend, 1961; 1970; Bakewell and Lumley, 1967]. Although the literature on this topic is vast, no research-community-wide consensus has been reached particularly on the issues of the origin of and interaction between the different structures, regeneration mechanisms, and Reynolds number effects. What follows are somewhat biased remarks addressing those issues, gathered mostly via low-Reynolds-number experiments. The interested reader is referred to the book edited by Panton (1997) and the large number of review articles available [e.g., Kovasznay, 1970; Laufer, 1975; Willmarth, 1975a; 1975b; Saffman, 1978; Cantwell, 1981; Fiedler, 1986; 1988; Blackwelder, 1988; 1998; Robinson, 1991; Delville et al., 1998]. The paper by Robinson (1991) in particular summarizes many of the different, sometimes contradictory, conceptual models offered thus far by different research groups. Those models are aimed ultimately at explaining how the turbulence maintains itself, and range from the speculative to the rigorous but none, unfortunately, is self-contained and complete. Furthermore, the structure research dwells largely on the kinematics of organized motion and little attention is given to the dynamics of the regeneration process.

In a boundary layer, the turbulence production process is dominated by three kinds of quasi-periodic — or, depending on one's viewpoint, quasi-random — eddies: (1) the large outer structures; (2) the intermediate Falco eddies; and (3) the near-wall events. The large, three-dimensional structures scale with the boundary-layer thickness, δ, and extend across the entire layer [Kovasznay et al., 1970; Blackwelder and Kovasznay, 1972]. These eddies control the dynamics of the boundary layer in the outer region, such as entrainment, turbulence production, etc. They appear randomly in space and time, and seem to be, at least for moderate Reynolds numbers, the residue of the transitional Emmons spots [Zilberman et al., 1977; Gad-el-Hak et al., 1981; Riley and Gad-el-Hak, 1985].

The Falco eddies are also highly coherent and three dimensional. Falco (1974; 1977) named them typical eddies because they appear in wakes, jets, Emmons spots, grid-generated turbulence, and boundary layers in zero, favorable and adverse pressure gradients. They have an intermediate scale of about 100 v/u_τ (100 wall units; u_τ is the friction velocity, and v/u_τ is the viscous length-scale). The Falco eddies appear to be an important link between the large structures and the near-wall events.

The third kind of eddies exists in the near-wall region ($0 < y < 100\ v/u_\tau$) where the Reynolds stress is produced in a very intermittent fashion. Half of the total production of turbulence kinetic energy ($-\overline{uv}\ \partial \bar{U}/\partial y$) takes place near the wall in the first 5% of the boundary layer at typical laboratory Reynolds numbers (smaller fraction at higher Reynolds numbers), and the dominant sequence of eddy motions there are collectively termed the bursting phenomenon. This dynamically significant process, identified during the 1960s by researchers at Stanford University [Kline and Runstadler, 1959; Runstadler et al., 1963; Kline et al., 1967; Kim et al., 1971; Offen and Kline, 1974; 1975], was reviewed by Willmarth (1975a), Blackwelder (1978), Robinson (1991), and more recently Panton (1997), and Blackwelder (1998).

To focus the discussion on the bursting process and its possible relationships to other organized motions within the boundary layer, refer to the schematic in Figure 13.5 adapted from Blackwelder (1998). Qualitatively, the process, according to at least one school of thought, begins with elongated, counter-rotating, streamwise vortices having diameters of approximately 40 wall units or 40 v/u_τ. The estimate for the diameter of the vortex is obtained from the conditionally averaged spanwise velocity profiles reported by Blackwelder and Eckelmann (1979). There is a distinction, however, between vorticity distribution and a vortex [Saffman and Baker, 1979; Robinson et al., 1989; Robinson, 1991], and the visualization results of Smith and Schwartz (1983) may indicate a much smaller diameter. In any case, referring to Figure 13.6, the counter-rotating vortices exist in a strong shear and induce low- and high-speed regions between them. Those low-speed streaks were first visualized by Francis Hama at the University of Maryland [see Corrsin, 1957], although Hama's contribution is frequently overlooked in favor of the subsequent and more thorough studies conducted at Stanford University and cited above. The vortices and the accompanying eddy structures occur randomly in space and time. However, their appearance is sufficiently regular that an average spanwise wavelength of approximately 80 to 100 v/u_τ has been identified by Kline et al. (1967) and others.

It might be instructive at this point to emphasize that the distribution of streak spacing is very broad. The standard of deviation is 30–40% of the more commonly quoted mean spacing between low-speed streaks of 100 wall units. Both the mean and standard deviation are roughly independent of Reynolds number in the rather limited range of reported measurements (Re_θ = 300–6500, see Smith and Metzler, 1983; Kim et al., 1987). Butler and Farrell (1993) have shown that the mean streak spacing of 100 v/u_τ is consistent with the notion that this is an optimal configuration for extracting "the most energy over an

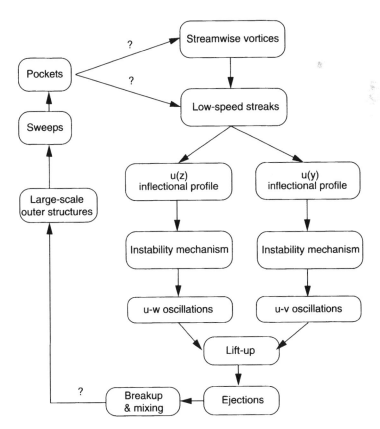

FIGURE 13.5 Proposed sequence of the bursting process. The arrows indicate the sequential events, and the "?" indicates relationships with less supporting evidence. (After Blackwelder, 1998.)

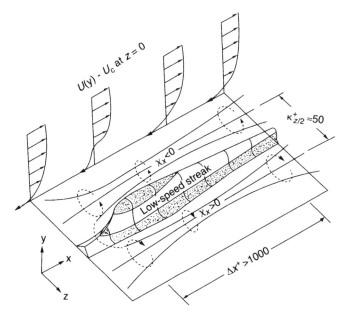

FIGURE 13.6   Model of near-wall structure in turbulent wall-bounded flows. (After Blackwelder, 1978.)

appropriate eddy turnover time." In their work, the streak spacing remains 100 wall units at Reynolds numbers, based on friction velocity and channel half-width, of $a^+ = 180$–360.

Kim et al. (1971) observed that the low-speed regions grow downstream, lift up, and develop (instantaneous) inflectional $U(y)$ profiles.[2] At approximately the same time, the interface between the low- and high-speed fluid begins to oscillate, apparently signaling the onset of a secondary instability. The low-speed region lifts up away from the wall as the oscillation amplitude increases, and then the flow rapidly breaks up into a completely chaotic motion. The streak oscillations commence at $y^+ \approx 10$, and the abrupt breakup takes place in the buffer layer although the ejected fluid reaches all the way to the logarithmic region. Since the breakup process occurs on a very short time-scale, Kline et al. (1967) called it a burst.

Virtually all of the net production of turbulence kinetic energy in the near-wall region occurs during these bursts. Corino and Brodkey (1969) showed that the low-speed regions are quite narrow, i.e., $z = 20 \, v/u_\tau$, and may also have significant shear in the spanwise direction. They also indicated that the ejection phase of the bursting process is followed by a large-scale motion of upstream fluid that emanates from the outer region and cleanses (sweeps) the wall region of the previously ejected fluid. The sweep phase is, of course, required by the continuity equation and appears to scale with the outer-flow variables. The sweep event seems to stabilize the bursting site, in effect preparing it for a new cycle.

Considerably more has been learned about the bursting process during the last two decades. For example, Falco (1980; 1983; 1991) has shown that when a typical eddy, which may be formed in part by ejected wall-layer fluid, moves over the wall it induces a high uv sweep (positive u and negative v). The wall region is continuously bombarded by *pockets* of high-speed fluid originating in the logarithmic and possibly the outer layers of the flow. These pockets appear to scale — at least in the limited Reynolds number range where they have been observed $Re_\theta = \mathcal{O}[1000]$ — with wall variables and tend to promote or enhance the inflectional velocity profiles by increasing the instantaneous shear leading to a more rapidly growing instability. The relation between the pockets and the sweep events is not clear, but it seems that the former forms the highly irregular interface between the latter and the wall-region fluid. More recently, Klewicki et al. (1994) conducted a four-element hot-wire probe measurements in a low-Reynolds-number

---

[2]According to Swearingen and Blackwelder (1984), inflectional $U(z)$ profiles are just as likely to be found in the near-wall region and can also be the cause of the subsequent bursting events (see Figure 13.5).

canonical boundary layer to clarify the roles of velocity–spanwise vorticity field interactions regarding the near-wall turbulent stress production and transport.

Other significant experiments were conducted by Tiederman and his students [Donohue et al., 1972; Reischman and Tiederman, 1975; Oldaker and Tiederman, 1977; Tiederman et al., 1985], Smith and his colleagues [Smith and Metzler, 1982; 1983; Smith and Schwartz, 1983], and the present author and his collaborators. The first group conducted extensive studies of the near-wall region, particularly the viscous sublayer, of channels with Newtonian as well as drag-reducing non-Newtonian fluids. Smith's group, using a unique, two-camera, high-speed video system, was the first to indicate a symbiotic relationship between the occurrence of low-speed streaks and the formation of vortex loops in the near-wall region. Gad-el-Hak and Hussain (1986) and Gad-el-Hak and Blackwelder (1987a) have introduced methods by which the bursting events and large-eddy structures are artificially generated in a boundary layer. Their experiments greatly facilitate the study of the uniquely controlled simulated coherent structures via phase-locked measurements.

Blackwelder and Haritonidis (1983) have shown convincingly that the frequency of occurrence of the bursting events scales with the viscous parameters consistent with the usual boundary-layer scaling arguments. An excellent review of the dynamics of turbulent boundary layers has been provided by Sreenivasan (1989). More information about coherent structures in high-Reynolds number boundary layers is given by Gad-el-Hak and Bandyopadhyay (1994). The book edited by Panton (1997) emphasizes the self-sustaining mechanisms of wall turbulence.

13.7 Scales

In this and the following section, we develop the equations needed to compute the characteristic lengths and sensor requirements of turbulent flows in general and wall-bounded flows in particular. Turbulence is a high-Reynolds-number phenomena that is characterized by the existence of numerous length- and time-scales [Tennekes and Lumley, 1972]. The spatial extension of the length-scales is bounded from above by the dimensions of the flowfield and from below by the diffusive and dissipative action of the molecular viscosity.

If we limit our interest to shear flows, which are basically characterized by two large length-scales, one in the streamwise direction (the convective or longitudinal length-scale) and the other perpendicular to the flow direction (the diffusive or lateral length-scale), we obtain a more well-defined problem. Moreover, at sufficiently high Reynolds numbers the boundary-layer approximation applies and a wide separation between the lateral and the longitudinal length-scales can be assumed. This leads to some attractive simplifications in the equations of motion, for instance that the elliptical Navier–Stokes equations are transferred to the parabolic boundary-layer equations [see e.g. Hinze, 1975]. So in this approximation, the lateral scale is approximately equal to the extension of the flow perpendicular to the flow direction (e.g., the boundary-layer thickness), and the largest eddies have typically this spatial extension. These eddies are most energetic and play a crucial role both in the transport of momentum and contaminants. A constant energy supply is needed to maintain the turbulence, and this energy is extracted from the mean flow into the largest most energetic eddies. The lateral length-scale is also the relevant scale for analyzing this energy transfer. However, there is an energy destruction in the flow due to the action of the viscous forces (the dissipation), and for the analysis of this process other smaller length-scales are needed.

As the eddy size decreases, viscosity becomes a more significant parameter since one property of viscosity is its effectiveness in smoothing out velocity gradients. The viscous and the nonlinear terms in the momentum equation counteract each other in the generation of small-scale fluctuations. While the inertial terms try to produce smaller and smaller eddies, the viscous terms check this process and prevent the generation of infinitely small scales by dissipating the small-scale energy into heat. In the early 1940s, the universal equilibrium theory was developed by Kolmogorov (1941a; 1941b). One cornerstone of this theory is that the small-scale motions are statistically independent of the relatively slower large-scale turbulence. An implication of this is that the turbulence at the small scales depends only on two parameters, namely the rate at which energy is supplied by the large-scale motion and the kinematic viscosity. In addition, the equilibrium theory assumes that the rate of energy supply to the turbulence should be equal to the rate of dissipation. Hence, in the analysis of turbulence at small scales, the dissipation rate per unit mass, ε, is a relevant parameter together with the kinematic viscosity, v. Kolmogorov (1941a) used simple dimensional arguments to derive a length-, a time-, and a velocity-scale relevant for the small-scale motion, respectively given by:

$$\eta = \left(\frac{v^3}{\varepsilon} \right)^{\frac{1}{4}}$$

(13.1)

$$\tau = \left(\frac{v}{\varepsilon} \right)^{\frac{1}{2}}$$

(13.2)

$$\upsilon = (v\varepsilon)^{\frac{1}{4}}$$

(13.3)

These scales are accordingly called the Kolmogorov microscales, or sometimes the inner scales of the flow. As they are obtained through a physical argument, these scales are the smallest scales that can exist in a turbulent flow and they are relevant for both free-shear and wall-bounded flows.

In boundary layers, the shear-layer thickness provides a measure of the largest eddies in the flow. The smallest scale in wall-bounded flows is the viscous wall unit, which will be shown below to be of the same order as the Kolmogorov length-scale. Viscous forces dominate over inertia in the near-wall region, and the characteristic scales there are obtained from the magnitude of the mean vorticity in the region and its viscous diffusion away from the wall. Thus, the viscous time-scale, t_v, is given by the inverse of the mean wall vorticity:

$$t_v = \left[\frac{\partial \bar{U}}{\partial y} \bigg|_w \right]^{-1}$$

(13.4)

where $\bar{U}$ is the mean streamwise velocity. The viscous length-scale, ℓ_v, is determined by the characteristic distance by which the (spanwise) vorticity is diffused from the wall, and is thus given by:

$$\ell_v = \sqrt{v t_v} = \sqrt{\frac{v}{\frac{\partial \bar{U}}{\partial y} \bigg|_w}}$$

(13.5)

where v is the kinematic viscosity. The wall velocity-scale (so-called friction velocity, u_τ) follows directly from the above time- and length-scales:

$$u_\tau = \frac{\ell_v}{t_v} = \sqrt{v \frac{\partial \bar{U}}{\partial y} \bigg|_w} = \sqrt{\frac{\tau_w}{\rho}}$$

(13.6)

where τ_w is the mean shear stress at the wall, and ρ is the fluid density. A wall unit implies scaling with the viscous scales, and the usual $(\cdots)^+$ notation is used; for example, $y^+ = y/\ell_v = yu_\tau/v$. In the wall region, the characteristic length for the large eddies is y itself, while the Kolmogorov scale is related to the distance from the wall y as follows:

$$\eta^+ \equiv \frac{\eta u_\tau}{v} \approx (\kappa y^+)^{\frac{1}{4}}$$

(13.7)

where κ is the von Kármán constant (≈ 0.41). As y^+ changes in the range of 1–5 (the extent of the viscous sublayer), η changes from 0.8 to 1.2 wall units.

We now have access to scales for the largest and smallest eddies of a turbulent flow. To continue the analysis of the cascade energy process, it is necessary to find a connection between these diverse scales. One way of obtaining such a relation is to use the fact that at equilibrium the amount of energy dissipating at high wavenumbers must equal the amount of energy drained from the mean flow into the energetic large-scale eddies at low wavenumbers. In the inertial region of the turbulence kinetic energy spectrum, the flow is almost independent of viscosity and since the same amount of energy dissipated at the high wavenumbers must pass this "inviscid" region, an inviscid relation for the total dissipation may be obtained by the following argument. The amount of kinetic energy per unit mass of an eddy with a wavenumber in the inertial sublayer is proportional to the square of a characteristic velocity for such an eddy, u^2. The rate of transfer of energy is assumed to be proportional to the reciprocal of one eddy turnover time, u/ℓ, where ℓ is a characteristic length of the inertial sublayer. Hence, the rate of energy that is supplied to the small-scale eddies via this particular wavenumber is of order of u^3/ℓ, and this amount of energy must be equal to the energy dissipated at the highest wavenumber, expressed as:

$$\varepsilon \approx \frac{u^3}{\ell} \tag{13.8}$$

Note that this is an inviscid estimate of the dissipation since it is based on large-scale dynamics and does not either involve or contain viscosity. More comprehensive discussion of this issue can be found in Taylor (1935) and Tennekes and Lumley (1972). From an experimental perspective, this is a very important expression since it offers one way of estimating the Kolmogorov microscales from quantities measured in a much lower wavenumber range.

Since the Kolmogorov length- and time-scales are the smallest scales occurring in turbulent motion, a central question will be how small these scales can be without violating the continuum hypothesis. By looking at the governing equations, it can be concluded that high dissipation rates are usually associated with large velocities; this situation is more likely to occur in gases than in liquids so it would be sufficient to show that for gas flows the smallest turbulence scales are normally much large than the molecular scales of motion. The relevant molecular length-scale is the mean free path, L, and the ratio between this length and the Kolmogorov length-scale, η, is the microstructure Knudsen number and can be expressed as (see Corrsin, 1959):

$$Kn = \frac{L}{\eta} \approx \frac{Ma^{\frac{1}{4}}}{Re} \tag{13.9}$$

where the turbulence Reynolds number, Re, and the turbulence Mach number, Ma, are used as independent variables. It is obvious that a turbulent flow will interfere with the molecular motion only at high Mach number and low Reynolds number, and this is a very unusual situation occurring only in certain gaseous nebulae.[3] Thus, under normal conditions the turbulence Knudsen number falls in the group of continuum flows. However, measurements using extremely thin hot-wires, small MEMS sensors, or flows within narrow MEMS channels can generate values in the slip-flow regime and even beyond, and this implies that for instance the no-slip condition may be questioned, as thoroughly discussed in Part I of this book.

13.8 Sensor Requirements

It is the ultimate goal of all measurements in turbulent flows to resolve both the largest and smallest eddies that occur in the flow. At the lower wavenumbers, the largest and most energetic eddies occur, and normally there are no problems associated with resolving these eddies. Basically, this is a question of having access to computers with sufficiently large memory for storing the amount of data that may be

---

[3] Note that in microduct flows and the like, the Re is usually too small for turbulence to even exist. So the issue of turbulence Knudsen number is mute in those circumstances even if rarefaction effects become strong.

necessary to acquire from a large number of distributed probes, each collecting data for a time period long enough to reduce the statistical error to a prescribed level. However, at the other end of the spectrum, both the spatial and the temporal resolutions are crucial, and this puts severe limitations on the sensors to be used. It is possible to obtain a relation between the small and large scales of the flow by substituting the inviscid estimate of the total dissipation rate, Equation 13.8, into the expressions for the Kolmogorov microscales, Equations 13.1–13.3. Thus:

$$\frac{\eta}{\ell} \approx \left(\frac{u\ell}{v}\right)^{-\frac{3}{4}} = Re^{-\frac{3}{4}} \tag{13.10}$$

$$\frac{\tau u}{\ell} \approx \left(\frac{u\ell}{v}\right)^{-\frac{1}{2}} = Re^{-\frac{1}{2}} \tag{13.11}$$

$$\frac{v}{u} \approx \left(\frac{u\ell}{v}\right)^{-\frac{1}{4}} = Re^{-\frac{1}{4}} \tag{13.12}$$

where Re is the Reynolds number based on the speed of the energy containing eddies, u, and their characteristic length, ℓ. Since turbulence is a high-Reynolds-number phenomenon, these relations show that the small length-, time-, and velocity-scales are much less than those of the larger eddies, and that the separation in scales widens considerably as the Reynolds number increases. Moreover, this also implies that the assumptions made on the statistical independence and the dynamical equilibrium state of the small structures will be most relevant at high Reynolds numbers. Another conclusion drawn from the above relations is that if two turbulent flowfields have the same spatial extension (i.e., same large-scale) but different Reynolds numbers, there would be an obvious difference in the small-scale structure in the two flows. The low-Reynolds-number flow would have a relatively coarse small-scale structure, while the high-Re flow would have much finer small eddies.

To spatially resolve the smallest eddies, sensors that are of approximately the same size as the Kolmogorov length-scale for the particular flow under consideration are needed. This implies that as the Reynolds number increases smaller sensors are required. For instance, in the self-preserving region of a plane-cylinder wake at a modest Reynolds number, based on the cylinder diameter, of 1840, the value of η varies in the range of 0.5–0.8 mm [Aronson and Löfdahl, 1994]. For this case, conventional hot-wires can be used for turbulence measurements. However, an increase in the Reynolds number by a factor of ten will yield Kolmogorov scales in the micrometer range and call for either extremely small conventional hot-wires or MEMS-based sensors. Another illustrating example of the Reynolds number effect on the requirement of small sensors is a simple two-dimensional, flat-plate boundary layer. At a momentum thickness Reynolds number of $Re_\theta = 4000$, the Kolmogorov length-scale is typically of the order of $50\,\mu m$, and in order to resolve these scales it is necessary to have access to sensors that have a characteristic active measuring length of the same spatial extension.

Severe errors will be introduced in the measurements by using sensors that are too large, since such sensors will integrate the fluctuations due to the small eddies over their spatial extensions, and the energy content of these eddies will be interpreted by the sensors as an average "cooling." When measuring fluctuating quantities, this implies that these eddies are counted as part of the mean flow and their energy is "lost." The result will be a lower value of the turbulence parameter, and this will wrongly be interpreted as a measured attenuation of the turbulence [see for example Ligrani and Bradshaw, 1987]. However, since turbulence measurements deal with statistical values of fluctuating quantities, it may be possible to loosen the spatial constraint of having a sensor of the same size as η, to allow a sensor dimensions that are slightly larger than the Kolmogorov scale, say on the order of η.

For boundary layers, the wall unit has been used to estimate the smallest necessary size of a sensor for accurately resolving the smallest eddies. For instance Keith et al. (1992) state that ten wall units or less is a relevant sensor dimension for resolving small-scale pressure fluctuations. Measurements of fluctuating velocity gradients, essential for estimating the total dissipation rate in turbulent flows, are another challenging task. Gad-el-Hak and Bandyopadhyay (1994) argue that turbulence measurements with probe lengths greater than the viscous sublayer thickness (about 5 wall units) are unreliable particularly near the surface. Many studies have been conducted on the spacing between sensors necessary to optimize the formed velocity gradients [see Aronson et al., 1997, and references therein]. A general conclusion from both experiments and direct numerical simulations is that a sensor spacing of 3–5 Kolmogorov lengths is recommended. When designing arrays for correlation measurements or for targeted control, the spacing between the coherent structures will be the determining factor. For example, when targeting the low-speed streaks in a turbulent boundary layer, several sensors must be situated along a lateral distance of 100 wall units, the average spanwise spacing between streaks. This requires quite small sensors, and many attempts have been made to meet these conditions with conventional sensor designs. However, in spite of the fact that conventional sensors like hot-wires have been fabricated in the micrometer size-range (for their diameter but not their length), they are usually hand-made, difficult to handle, and are too fragile, and here the MEMS technology has really opened a door for new applications.

It is clear from the above that the spatial and temporal resolutions for any probe to be used to resolve high-Reynolds-number turbulent flows are extremely tight. For example, both the Kolmogorov scale and the viscous length-scale change from few microns at the typical field Reynolds number — based on the momentum thickness — of 10^6, to a couple of hundred microns at the typical laboratory Reynolds number of 10^3. MEMS sensors for pressure, velocity, temperature, and shear stress are at least one order of magnitude smaller than conventional sensors [Ho and Tai, 1996; 1998; Löfdahl et al., 1996; Löfdahl and Gad-el-Hak, 1999]. Their small size improves both the spatial and temporal resolutions of the measurements, typically few microns and few microseconds, respectively. For example, a micro-hot-wire (called hot-point) has very small thermal inertia and the diaphragm of a micro-pressure-transducer has correspondingly fast dynamic response. Moreover, the microsensors' extreme miniaturization and low energy consumption make them ideal for monitoring the flow state without appreciably affecting it. Lastly, literally hundreds of microsensors can be fabricated on the same silicon chip at a reasonable cost, making them well suited for distributed measurements and control. The UCLA/Caltech team [see, for example, Ho and Tai, 1996; 1998, and references therein] has been very effective in developing many MEMS-based sensors and actuators for turbulence diagnosis and control.

13.9 Reactive Flow Control

13.9.1 Introductory Remarks

Targeted control implies sensing and reacting to particular quasi-periodic structures in a turbulent flow. For a boundary layer, the wall seems to be the logical place for such reactive control, because of the relative ease of placing something in there, the sensitivity of the flow in general to surface perturbations, and the proximity and therefore accessibility to the dynamically all important near-wall coherent events. According to Wilkinson (1990), there are very few actual experiments that use embedded wall sensors to initiate a surface actuator response [Alshamani et al., 1982; Wilkinson and Balasubramanian, 1985; Nosenchuck and Lynch, 1985; Breuer et al., 1989]. This ten-year-old assessment is fast changing, however, with the introduction of microfabrication technology that has the potential for producing small, inexpensive, programmable sensor/actuator chips. Witness the more recent reactive control attempts by Kwong and Dowling (1993), Reynolds (1993), Jacobs et al. (1993), Jacobson and Reynolds (1993a; 1993b; 1994; 1995; 1998), Fan et al. (1993), James et al. (1994), and Keefe (1996). Fan et al. and Jacobson and Reynolds even consider the use of self-learning neural networks for increased computational speeds and efficiency. Recent reviews of reactive flow control include those by Gad-el-Hak (1994; 1996), Lumley (1996), McMichael (1996), Mehregany et al. (1996), Ho and Tai (1996), and Bushnell (1998).

Numerous methods of flow control have already been successfully implemented in practical engineering devices. Yet, limitations exist for some familiar control techniques when applied to specific situations. For example, in attempting to reduce the drag or enhance the lift of a body having a turbulent boundary layer using global suction, global heating and cooling or global application of electromagnetic body forces, the actuator's energy expenditure often exceeds the savings derived from the predetermined active control strategy. What is needed is a way to reduce this penalty to achieve a more efficient control. Reactive control geared specifically toward manipulating the coherent structures in turbulent shear flows, though considerably more complicated than passive control or even predetermined active control, has the potential to do just that. As will be argued in this and the following sections, future systems for control of turbulent flows in general and turbulent boundary layers in particular could greatly benefit from the merging of the science of chaos control, the technology of microfabrication, and the newest computational tools collectively termed soft computing. Such systems are envisaged as consisting of a large number of intelligent, communicative wall sensors and actuators arranged in a checkerboard pattern and targeted toward controlling certain quasi-periodic, dynamically significant coherent structures present in the near-wall region.

13.9.2 Targeted Flow Control

Successful techniques to reduce the skin friction in a turbulent flow, such as polymers, particles, or riblets, appear to act indirectly through local interaction with discrete turbulent structures, particularly small-scale eddies, within the flow. Common characteristics of all these passive methods are increased losses in the near-wall region, thickening of the buffer layer, and lowered production of Reynolds shear stress [Bandyopadhyay, 1986]. Active control strategies that act directly on the mean flow, such as suction or lowering of near-wall viscosity, also lead to inhibition of Reynolds stress. However, skin friction is increased when any of these velocity-profile modifiers is applied globally.

Could these seemingly inefficient techniques, e.g. global suction, be used more sparingly and be optimized to reduce their associated penalty? It appears that the more successful drag-reducing methods, e.g. polymers, act selectively on particular scales of motion and are thought to be associated with stabilization of the secondary instabilities. It is also clear that energy is wasted when suction or heating/cooling is used to suppress the turbulence throughout the boundary layer when the main interest is to affect a near-wall phenomenon. One ponders, what would become of wall turbulence if specific coherent structures are to be targeted by the operator through a reactive control scheme, for modification? The myriad of organized structures present in all shear flows are instantaneously identifiable, quasi-periodic motions [Cantwell, 1981; Robinson, 1991]. Bursting events in wall-bounded flows, for example, are both intermittent and random in space as well as time. The random aspects of these events reduce the effectiveness of a predetermined active control strategy. If such structures are nonintrusively detected and altered, on the other hand, net performance gain might be achieved. It seems clear, however, that temporal phasing as well as spatial selectivity would be required to achieve proper control targeted toward random events.

A nonreactive version of the above idea is the *selective suction technique* that combines suction to achieve an asymptotic turbulent boundary layer and longitudinal riblets to fix the location of low-speed streaks. Although far from indicating net drag reduction, the available results are encouraging and further optimization is needed. When implemented via an array of reactive control loops, the selective suction method is potentially capable of skin-friction reduction that approaches 60%.

The genesis of the selective suction concept can be found in the papers by Gad-el-Hak and Blackwelder (1987b; 1989) and the patent by Blackwelder and Gad-el-Hak (1990). These researchers suggest that one possible means of optimizing the suction rate is to be able to identify where a low-speed streak is presently located and apply a small amount of suction under it. Assuming that the production of turbulence kinetic energy is due to the instability of an inflectional $U(y)$ velocity profile, one needs to remove only enough fluid so that the inflectional nature of the profile is alleviated. An alternative technique that could conceivably reduce the Reynolds stress is to inject fluid selectively under the high-speed regions. The immediate effect of normal injection would be to decrease the viscous shear at the wall resulting in less drag. In addition, the velocity profiles in the spanwise direction, $U(z)$, would have a smaller shear, $\partial U/\partial Z$, because the suction/injection would create a more uniform flow. Since Swearingen and Blackwelder (1984) and Blackwelder and Swearingen (1990) have found that inflectional $U(z)$ profiles occur as often as inflection points are observed in $U(y)$ profiles, suction under the low-speed streaks and/or injection under the high-speed regions would decrease this shear and hence the resulting instability.

The combination of selective suction and injection is sketched in Figure 13.7. In Figure 13.7a, the vortices are idealized by a periodic distribution in the spanwise direction. The instantaneous velocity profiles without transpiration at constant y and z locations are shown by the dashed lines in Figures 13.7b and 13.7c, respectively. Clearly, the $U(y_0, z)$ profile is inflectional, having two inflection points per wavelength. At z_1 and z_2, an inflectional $U(y)$ profile is also evident. The same profiles with suction at z_1 and z_3 and injection at z_2 are shown by the solid lines. In all cases, the shear associated with the inflection points would have been reduced. Since the inflectional profiles are all inviscidly unstable with growth rates proportional to the shear, the resulting instabilities would be weakened by the suction/injection process.

The feasibility of the selective suction as a drag-reducing concept has been demonstrated by Gad-el-Hak and Blackwelder (1989) and is indicated in Figure 13.8. Low-speed streaks were artificially generated in a laminar boundary layer using three spanwise suction holes as per the method proposed by Gad-el-Hak and Hussain (1986), and a hot-film probe was used to record the near-wall signature of the streaks. An open, feedforward control loop with a phase lag was used to activate a predetermined suction from a longitudinal slot located in between the spanwise holes and the downstream hot-film probe. An equivalent suction coefficient of $C_q = 0.0006$ was sufficient to eliminate the artificial events and prevent bursting. This rate is five times smaller than the asymptotic suction coefficient for a corresponding turbulent boundary layer. If this result is sustained in a naturally developing turbulent boundary layer, a skin-friction reduction of close to 60% would be attained.

Gad-el-Hak and Blackwelder (1989) propose to combine suction with non-planar surface modifications. Minute longitudinal roughness elements if properly spaced in the spanwise direction greatly reduce the spatial randomness of the low-speed streaks [Johansen and Smith, 1986]. By withdrawing the streaks forming

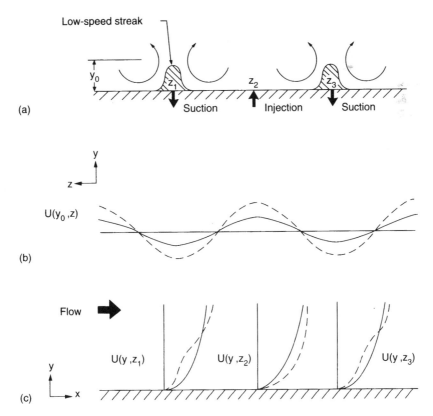

FIGURE 13.7 Effects of suction and injection on velocity profiles. Broken lines: reference profiles. Solid lines: profiles with transpiration applied. (a) Streamwise vortices in the y–z plane, suction/injection applied at z_1, z_2, and z_3. (b) Resulting spanwise velocity distribution at $y = y_0$. (c) Velocity profiles normal to the surface.

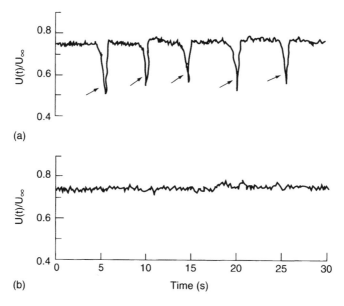

FIGURE 13.8 Effects of suction from a streamwise slot on five artificially induced burstlike events in a laminar boundary layer. (a) $C_q = 0$. (b) $C_q = 0.0006$. [After Gad-el-Hak and Blackwelder, 1989.]

near the peaks of the roughness elements, less suction should be required to achieve an asymptotic boundary layer. Experiments by Wilkinson and Lazos (1987) and Wilkinson (1988) combine suction and blowing with thin-element riblets. Although no net drag reduction is yet attained in these experiments, their results indicate some advantage of combining suction with riblets as proposed by Gad-el-Hak and Blackwelder (1987b; 1989).

The recent numerical experiments of Choi et al. (1994) also validate the concept of targeting suction/injection to specific near-wall events in a turbulent channel flow. Based on complete interior flow information and using the rather simple, heuristic control law proposed earlier by Gad-el-Hak and Blackwelder (1987b), Choi et al.'s direct numerical simulations indicate a 20% net drag reduction accompanied by significant suppression of the near-wall structures and the Reynolds stress throughout the entire wall-bounded flow. When only wall information was used, a drag reduction of 6% was observed; a rather disappointing result considering that sensing and actuation took place at *every* grid point along the computational wall. In a practical implementation of this technique, even fewer wall sensors would perhaps be available, measuring only a small subset of the accessible information and thus requiring even more sophisticated control algorithms to achieve the same degree of success. Low-dimensional models of the near-wall flow and soft computing tools can help in constructing more effective control algorithms. Both of these topics will be discussed in following paragraphs.

Time sequences of the numerical flowfield of Choi et al. (1994) indicate the presence of two distinct drag-reducing mechanisms when selective suction/injection is used. First, deterring the sweep motion, without modifying the primary streamwise vortices above the wall, and consequently moving the high-shear regions from the surface to the interior of the channel, thus directly reducing the skin friction. Secondly, changing the evolution of the wall vorticity layer by stabilizing and preventing lifting of the near-wall spanwise vorticity, thus suppressing a potential source of new streamwise vortices above the surface and interrupting a very important regeneration mechanism of turbulence.

Three modern developments have relevance to the issue at hand. Firstly, the recently demonstrated ability to revert a chaotic system to a periodic one may provide optimal nonlinear control strategies for further reduction in the amount of suction (or the energy expenditure of any other active wall-modulation technique) needed to attain a given degree of flow stabilization. This is important since, as can be seen from the Kármán integral momentum equation, net drag reduction achieved in a turbulent boundary layer increases as the suction coefficient decreases. Secondly, to selectively remove the randomly occurring low-speed streaks, for example, would ultimately require reactive control. In that case, an event is targeted, sensed, and subsequently modulated. Microfabrication technology provides opportunities for practical implementation of the required large array of inexpensive, programmable sensor/actuator chips. Thirdly, newly introduced soft computing tools include neural networks, fuzzy logic, and genetic algorithms are now more advanced as well as more widely used as compared to just few years ago. These tools could be very useful in constructing effective adaptive controllers.

13.9.3 Reactive Feedback Control

As was schematically depicted in Figure 13.3, a control device can be passive, requiring no auxiliary power, or active, requiring energy expenditure. Active control is further divided into predetermined or reactive. Predetermined control includes the application of steady or unsteady energy input without regard to the particular state of the flow. The control loop in this case is open as was shown in Figure 13.4a, and no sensors are required. Because no sensed information is being fed forward, this open control loop is not a feedforward one. Reactive control is a special class of active control where the control input is continuously adjusted based on measurements of some kind. The control loop in this case can either be an open, feedforward one (Figure 13.4b) or a closed, feedback loop (Figure 13.4c).

Moin and Bewley (1994) categorize reactive feedback control strategies by examining the extent to which they are based on the governing flow equations. Four categories are discerned: adaptive, physical model-based, dynamical systems-based, and optimal control (Figure 13.3). Note that except for adaptive control, the other three categories of reactive feedback control can also be used in the feedforward mode or the combined feedforward–feedback mode. Also, in a convective environment such as that for a boundary layer, a controller would perhaps combine feedforward and feedback information and may include elements from each of the four classifications. Each of the four categories is briefly described below.

Adaptive schemes attempt to develop models and controllers via some learning algorithm without regard to the details of the flow physics. System identification is performed independently of the flow dynamics or the Navier–Stokes equations that govern this dynamics. An adaptive controller tries to optimize a specified performance index by providing a control signal to an actuator. In order to update its parameters, the controller thus requires feedback information relating to the effects of its control. The most recent innovation in adaptive flow control schemes involves the use of neural networks that relate the sensor outputs to the actuator inputs through functions with variable coefficients and nonlinear, sigmoid saturation functions. The coefficients are updated using the so-called back-propagation algorithm, and complex control laws can be represented with a sufficient number of terms. Hand tuning is required, however, to achieve good convergence properties. The nonlinear adaptive technique has been used with different degrees of success by Fan et al. (1993) and Jacobson and Reynolds (1993b; 1995; 1998) to control, respectively, the transition process in laminar boundary layers and the bursting events in turbulent boundary layers.

Heuristic physical arguments can instead be used to establish effective control laws. That approach obviously will work only in situations in which the dominant physics are well understood. An example of this strategy is the active cancellation scheme, used by Gad-el-Hak and Blackwelder (1989) in a physical experiment and by Choi et al. (1994) in a numerical experiment, to reduce the drag by mitigating the effect of near-wall vortices. As mentioned earlier, the idea is to oppose the near-wall motion of the fluid, caused by the streamwise vortices, with an opposing wall control, thus lifting the high-shear region away from the surface and interrupting the turbulence regeneration mechanism.

Nonlinear dynamical systems theory allows turbulence to be decomposed into a small number of representative modes for which dynamics are examined to determine the best control law. The task is to stabilize the attractors of a low-dimensional approximation of a turbulent chaotic system. The best known strategy is the OGY method that, when applied to simpler, small-number of degrees of freedom systems, achieves stabilization with minute expenditure of energy. This and other chaos control strategies, especially as applied to the more complex turbulent flows, will be discussed later.

Finally, optimal control theory applied directly to the Navier–Stokes equations can, in principle, be used to minimize a cost function in the space of the control. This strategy provides perhaps the most rigorous theoretical framework for flow control. As compared to other reactive control strategies, optimal control applied to the full Navier–Stokes equations is also the most computer-time intensive. In this method, feedback control laws are derived systematically for the most efficient distribution of control effort to achieve a desired goal. Abergel and Temam (1990) developed such optimal control theory for suppressing turbulence in a numerically simulated, two-dimensional Navier–Stokes flow, but their method requires impractical full flow-field information. Choi et al. (1993) developed a more practical, wall-information-only, sub-optimal control strategy that they applied to the one-dimensional stochastic Burgers equation. Later application of the sub-optimal control theory to a numerically simulated turbulent channel flow has been reported by Moin and Bewley (1994) and Bewley et al. (1997; 1998). The book edited by Sritharan (1998) provides eight articles that focus on the mathematical aspects of optimal control of viscous flows. The more recent book by Aamo and Krstic (2003) focuses on feedback control of stabilization and mixing.

13.9.4   Required Characteristics

The randomness of the bursting events necessitates temporal phasing as well as spatial selectivity to effect selective control. Practical applications of methods targeted at controlling a particular turbulent structure to achieve a prescribed goal would therefore require implementing a large number of surface sensors/actuators together with appropriate control algorithms. That strategy for controlling wall-bounded turbulent flows has been advocated by, among others and in chronological order, Gad-el-Hak and Blackwelder (1987b; 1989), Lumley (1991; 1996), Choi et al. (1992), Reynolds (1993), Jacobson and Reynolds (1993a; 1993b; 1994; 1995; 1998), Gad-el-Hak (1993; 1994; 1996; 1998; 2000), Moin and Bewley (1994), McMichael (1996), Mehregany et al. (1996), Blackwelder (1998), Delville et al. (1998), and Perrier (1998).

It is instructive to estimate some representative characteristics of the required array of sensors/actuators. Consider a typical commercial aircraft cruising at a speed of $U_\infty = 300$ m/s and at an altitude of 10 km. The density and kinematic viscosity of air and the unit Reynolds number in this case are, respectively, $\rho = 0.4 \text{ kg/m}^3$, $v = 3 \times 10^{-5} \text{ m}^2/\text{s}$, and $Re = 10^7/\text{m}$. Assume further that the portion of fuselage to be controlled has turbulent boundary layer characteristics that are identical to those for a zero-pressure-gradient flat plate at a distance of 1 m from the leading edge. In this case, the skin-friction coefficient[4] and the friction velocity are, respectively, $C_f = 0.003$ and $u_\tau = 11.62$ m/s. At this location, one viscous wall unit is only $v/u_\tau = 2.6$ microns. In order for the surface array of sensors/actuators to be hydraulically smooth, it should not protrude beyond the viscous sublayer, or $5v/u_\tau = 13 \mu\text{m}$.

Wall-speed streaks are the most visible, reliable, and detectable indicators of the preburst turbulence production process. The detection criterion is simply low velocity near the wall, and the actuator response should be to accelerate (or to remove) the low-speed region before it breaks down. Local wall motion, tangential injection, suction, heating or electromagnetic body force, all triggered on sensed wall-pressure or wall-shear stress, could be used to cause local acceleration of near-wall fluid.

The numerical experiments of Berkooz et al. (1993) indicate that effective control of bursting pair of rolls may be achieved by using the equivalent of two wall-mounted shear sensors. If the goal is to stabilize or to eliminate *all* low-speed streaks in the boundary layer, a reasonable estimate for the spanwise and streamwise distances between individual elements of a checkerboard array is, respectively, 100 and 1000 wall units,[5] or 260 μm and 2600 μm, for this particular example. A reasonable size for each element is

---

[4]Note that the skin friction decreases as the distance from the leading increases. It is also strongly affected by such things as the externally imposed pressure gradient. Therefore, the estimates provided in here are for illustration purposes only.

[5]These are equal to, respectively, the average spanwise wavelength between two adjacent streaks and the average streamwise extent for a typical low-speed region. One can argue that those estimates are too conservative: once a region is *relaminarized*, it would perhaps stay as such for quite a while as the flow convects downstream. The next row of sensors/actuators may therefore be relegated to a downstream location well beyond 1000 wall units. Relatively simple physical or numerical experiments could settle this issue.

probably one-tenth of the spanwise separation, or 26 μm. A (1 m × 1 m) portion of the surface would have to be covered with about $n = 1.5$ million elements. This is a colossal number, but the density of sensors/actuators could be considerably reduced if we moderate our goal of targeting every single bursting event (and also if less conservative assumptions are made).

It is well known that not every low-speed streak leads to a burst. On the average, a particular sensor would detect an incipient bursting event every wall-unit interval of $P^+ = P u_\tau^2/\nu = 250$, or $P = 56$ μs. The corresponding dimensionless and dimensional frequencies are $f^+ = 0.004$ and $f = 18$ kHz, respectively. At different distances from the leading edge and in the presence of nonzero-pressure gradient, the sensors/actuators array would have different characteristics, but the corresponding numbers would still be in the same ballpark as estimated in here.

As a second example, consider an underwater vehicle moving at a speed of $U_\infty = 10$ m/s. Despite the relatively low speed, the unit Reynolds number is still the same as estimated above for the air case, $Re = 10^7$/m, due to the much lower kinematic viscosity of water. At one meter from the leading edge of an imaginary flat plate towed in water at the same speed, the friction velocity is only $u_\tau = 0.13$ m/s, but the wall unit is still the same as in the aircraft example, $\nu/u_\tau = 2.6$ μm. The density of required sensors/actuators array is the same as computed for the aircraft example, $n = 1.5 \times 10^6$ elements/m². The anticipated average frequency of sensing a bursting event is, however, much lower at $f = 600$ Hz.

Similar calculations have been recently made by Gad-el-Hak (1993; 1994; 1998; 2000), Reynolds (1993), and Wadsworth et al. (1993). Their results agree closely with the estimates made here for typical field requirements. In either the airplane or the submarine case, the actuator's response need not be too large. As will be shown later, wall displacement on the order of 10 wall units (26 μm in both examples), suction coefficient of about 0.0006, or surface cooling/heating on the order of 40°C/2°C (in the first and second example, respectively) should be sufficient to stabilize the turbulent flow.

As computed in the two examples above, both the required size for a sensor/actuator element and the average frequency at which an element would be activated are within the presently known capabilities of microfabrication technology. The number of elements needed per unit area is, however, alarmingly large. The unit cost of manufacturing a programmable sensor/actuator element would have to come down dramatically, perhaps matching the unit cost of a conventional transistor,[6] before the idea advocated in here would become practical.

An additional consideration to the size, amplitude and frequency response is the energy consumed by each sensor/actuator element. Total energy consumption by the entire control system obviously has to be low enough to achieve net savings. Consider the following calculations for the aircraft example. One meter from the leading edge, the skin-friction drag to be reduced is approximately 54 N/m². Engine power needed to overcome this retarding force per unit area is 16 kW/m², or 10^4 μW/sensor. If a 60% drag-reduction is achieved,[7] this energy consumption is reduced to 4320 μW/sensor. This number will increase by the amount of energy consumption of a sensor/actuator unit, but hopefully not back to the uncontrolled levels. The voltage across a sensor is typically in the range of $V = 0.1$–1 V, and its resistance in the range of $R = 0.1$–1 MΩ. This means a power consumption by a typical sensor in the range of $P = V^2/R = 0.1$–10 μW, well below the anticipated power savings due to reduced drag.

For a single actuator in the form of a spring-loaded diaphragm with a spring constant of $k = 100$ N/m, and oscillating up and down at the bursting frequency of $f = 18$ kHz, with an amplitude of $y = 26$ microns, the power consumption is $P = (1/2)ky^2f = 600$ μW/actuator. If suction is used instead, $C_q = 0.0006$, and assuming a pressure difference of $\Delta p = 10^4$ N/m² across the suction holes/slots, the corresponding power consumption for a single actuator is $P = C_q U_\infty \Delta p/n = 1200$ μW/actuator. It is clear then that when the power penalty for the sensor/actuator is added to the lower-level drag, a net savings is still achievable. The corresponding actuator power penalties for the submarine example are even

---

[6]The transistor was invented in 1947. In the mid 1960s, a single transistor sold for around $70. In 1997, Intel's Pentium II processor (microchip) contained 7.5×10^6 transistors and cost around $500, that is less than $0.00007 per transistor!

[7]A not-too-farfetched goal according to the selective suction results presented earlier.

smaller ($P = 20\,\mu\text{W/actuator}$ for the wall motion actuator, and $P = 40\,\mu\text{W/actuator}$ for the suction actuator), and larger savings are therefore possible.

13.10   Magnetohydrodynamic Control

13.10.1   Introductory Remarks

Magnetohydrodynamics (MHD) is the science underlying the interaction of an electrically conducting fluid with a magnetic field. Several decades ago a number of researchers from around the world were vigorously attempting to exploit the electric current and thus the electric power that results when a fluid conductor moves in the presence of a magnetic field. The fluid internal energy (or enthalpy) is directly converted into electrical energy, eliminating the traditional intermediate mechanical step. The usual turbine and generator in conventional power plants are therefore combined in a single unit, in essence an electromagnetic turbine with no moving parts. The inverse device is an electromagnetic pump where, in the presence of a magnetic field, a conducting fluid is caused to move when electric current passes through it. An excellent primer for the topic of MHD can be found in the book by Shercliff (1965).

Examples of conducting fluids (in descending order of electrical conductivities) include liquid metals, plasmas, molten glass, and sea water. The electrical conductivity of mercury is $\sigma \approx 10^6\,\text{mhos/m}$ and for sea water, $\sigma \approx 4\,\text{mhos/m}$. Possible useful applications include MHD power generation, propulsion, and liquid metals used as coolants for magnetically confined fusion reactors. Though fictitious, the Soviet submarine in Tom Clancy's best-seller "*The Hunt for Red October*" was endowed with an extremely quiet MHD propulsion system. With the collapse of communism, the interest in this field has shifted somewhat to more peaceful uses as for example flow control, i.e., the ability to manipulate a flowfield in order to achieve a beneficial goal such as drag reduction, lift enhancement, and mixing augmentation. Here the so-called Lorentz body forces that result when a conducting fluid moves in the presence of a magnetic field are exploited to effect desired changes in the flowfield, for example to suppress turbulence.

Of particular interest to this chapter is the development of efficient reactive control strategies that employ the Lorentz forces to enhance the performance of sea vessels. Methods to delay transition, prevent separation, and reduce skin-friction drag in turbulent boundary layers are sought. Control strategies targeted toward certain coherent structures in a turbulent flow are particularly sought. We start here by developing the von Kármán momentum integral equation and the instantaneous equations of motion at the wall both in the presence of Lorentz force. This should prove useful to make quick estimates of system behavior under different operating conditions. Results are usually not as accurate as solving the differential conservation equations themselves, but more accurate than dimensional analysis. This is followed by an outline of a suggested reactive control strategy that exploits electromagnetic body forces.

13.10.2   The von Kármán Equation

The Lorentz force (body force per unit volume) is given by the cross-product of the current density vector, $\vec{j}$, and the magnetic flux density vector, $\vec{B}$:

$$\vec{F} = \vec{j} \times \vec{B} \tag{13.13}$$

where the current is the sum of the applied and induced contributions:

$$\vec{j} = \sigma(\vec{E} + \vec{u} \times \vec{B}) \tag{13.14}$$

where σ is the electrical conductivity of the fluid, $\vec{E}$ is the applied electric field vector, and $\vec{u}$ is the fluid velocity vector. Thus the Lorentz force is in general given by:

$$\vec{F} = \sigma(\vec{E} \times \vec{B}) + \sigma(\vec{u} \times \vec{B}) \times \vec{B} \tag{13.15}$$

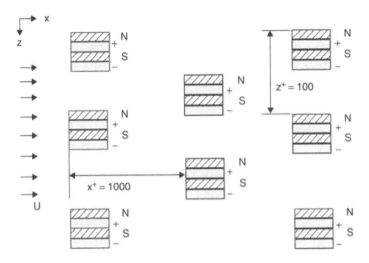

FIGURE 13.9   EMHD tiles for reactive control of turbulent boundary layers.

Due to the rather low electrical conductivity of sea water, the induced Lorentz force is typically very small in such application and the only way to effect a significant body force is to cross the magnetic field with an applied electric field. For such application, therefore, the term electromagnetic forcing (EMHD) is commonly used in place of MHD.

Assume a two-dimensional boundary-layer flow of a conducting fluid. Further assume that the magnetic and electric fields are generated using alternating electrodes and magnets parallel to the streamwise direction (see Figure 13.9). Basically both fields are in the plane (y, z). The resulting Lorentz force has components in the streamwise (x), normal (y), and spanwise (z) directions given by, respectively:

$$F_x = \sigma(E_y B_z - E_z B_y) - \sigma u(B_z^2 + B_y^2) \tag{13.16}$$

$$F_y = 0 - \sigma v B_z^2 \tag{13.17}$$

$$F_z = 0 - \sigma v B_z B_y \tag{13.18}$$

where u and v are respectively the streamwise and normal velocity components. Each of the first terms on the right hand sides of Equations 13.16–13.18 is due to the applied electric field. The second term is the induced Lorentz force and is negligible for low-conductivity fluids such as sea water. Therefore, for such application and particular arrangement of electrodes/magnets, the electromagnetic body force is predominately in the streamwise direction, $F_x \approx \sigma(E_y B_z - E_z B_y)$.

We now write the continuity and streamwise momentum equations. For an incompressible, two-dimensional flow and neglecting the gravitational body force:

$$\frac{\partial u}{\partial x} + \frac{\partial v}{\partial y} = 0 \tag{13.19}$$

$$\rho \frac{\partial u}{\partial t} + \rho u \frac{\partial u}{\partial x} + \rho v \frac{\partial u}{\partial y} = -\frac{\partial p}{\partial x} + \frac{\partial \tau}{\partial y} + \sigma(E_y B_z - E_z B_y) - \sigma u(B_z^2 + B_y^2) \tag{13.20}$$

where τ is the viscous shear stress. The pressure term can be written in terms of the velocity outside the boundary layer, U_o, which in general is a function of x and t. Note that the total drag should be the sum of the usual viscous and form drags plus (or minus) the Lorentz force. In Equation 13.20, the induced Lorentz force is always negative (for positive u). The applied force can be positive or negative depending on the direction of the applied electric field, $\vec{E}$, relative to the magnetic field, $\vec{B}$. In any case, when the Lorentz force is positive, it constitutes a thrust (negative drag).

At a given x location, the streamwise momentum equation can readily be integrated in y from the wall to the edge of the boundary layer. Invoking the continuity equation and using the usual definitions for the skin friction coefficient (C_f), and the displacement and momentum thicknesses (δ^* and δ_θ), the resulting integral equation reads:

$$C_f \equiv \frac{2\tau_w}{\rho U_o^2} = \frac{2}{U_o^2}\frac{\partial(U_o\delta^*)}{\partial t} + 2\frac{\partial\delta^*}{\partial x} - 2\frac{v_w}{U_o} + 2\delta_\theta\left(2 + \frac{\delta^*}{\delta_\theta}\right)\frac{1}{U_o}\frac{\partial U_o}{\partial x}$$

$$+ \frac{2\sigma}{\rho U_o^2}\int_0^\infty [E_y B_z - E_z B_y]\,dy - \frac{2\sigma}{\rho U_o^2}\int_0^\infty u[B_z^2 + B_y^2]\,dy \qquad (13.21)$$

where v_w is the injection (or suction) velocity normal to the wall. Note that upward/downward wall motion are equivalent to wall injection/suction. If σ is not constant, it has to be included in the two integrals on the right hand side of Equation 13.21.

It is clear that the following effects contribute to an *increase* in the skin friction: temporal acceleration of freestream, growth of momentum thickness, wall suction (or downward wall motion), favorable pressure gradient, and positive streamwise Lorentz force. When such force points opposite of the flow direction, the skin friction decreases and the velocity profile tends to become more inflectional. But once again remember to add (or subtract) the drag (or thrust) due to the applied as well as induced Lorentz forces.

Note that Equation 13.21 is valid for steady or unsteady flows, for Newtonian or non-Newtonian fluids, and for laminar or turbulent flows (mean quantities are used in the latter case). The two integrals in Equation 13.21 can be computed once the electric and magnetic fields are known and an approximate velocity profile is assumed. For sea water, the last integral in Equation 13.21 is neglected.

13.10.3 Vorticity Flux at the Wall

Another very useful equation is the differential momentum equation written at the wall ($y = 0$). Here we assume a Newtonian fluid but allow viscosity to vary spatially and the flow to be three-dimensional. For a non-moving wall,[8] the instantaneous streamwise, normal and spanwise momentum equations read respectively:

$$\rho v_w \frac{\partial u}{\partial y}\bigg|_{y=0} + \frac{\partial p}{\partial x}\bigg|_{y=0} - \frac{\partial \mu}{\partial y}\bigg|_{y=0}\frac{\partial \mu}{\partial y}\bigg|_{y=0} - \sigma\left[E_y B_z - E_z B_y\right]_{y=0} = \mu\frac{\partial^2 u}{\partial y^2}\bigg|_{y=0} \qquad (13.22)$$

$$0 + \frac{\partial p}{\partial y}\bigg|_{y=0} - 0 - F_y|_{y=0} = \mu\frac{\partial^2 v}{\partial y^2}\bigg|_{y=0} \qquad (13.23)$$

$$\rho v_w \frac{\partial w}{\partial y}\bigg|_{y=0} + \frac{\partial p}{\partial z}\bigg|_{y=0} - \frac{\partial \mu}{\partial y}\bigg|_{y=0}\frac{\partial w}{\partial y}\bigg|_{y=0} - F_z|_{y=0\cdot} = \mu\frac{\partial^2 w}{\partial y^2}\bigg|_{y=0} \qquad (13.24)$$

where $F_y|_{y=0}$ and $F_z|_{y=0}$ are respectively the wall values of the Lorentz force in the normal and spanwise directions. In Equation 13.22, the streamwise Lorentz force at the wall, $F_x|_{y=0}$, is due only to the applied electric field (the induced force is zero at the wall since u vanishes there for non-moving walls). The expression for F_x in Equation 13.22 is for an array of alternating electrodes and magnets parallel to the streamwise direction,[9]

---

[8]Wall is not moving in the streamwise and spanwise directions. Normal wall motions are equivalent to suction/injection, and hence are allowed in the present formulation.

[9]Note that for such an array, the applied part of both $F_y|_{y=0}$ and $F_z|_{y=0}$ is identically zero, but the induced part is non-zero if $v_w \neq 0$. In sea water applications, the induced Lorentz force is negligible regardless of any suction/injection or normal wall motion.

but the streamwise Lorentz force can readily be computed for any other configuration of magnets and electrodes. If the electric conductivity varies spatially, the value of σ at the wall should be used in Equation 13.22.

The right hand side of Equation 13.22 is the (negative of) wall flux of spanwise vorticity, while that of Equation 13.24 is the (negative of) wall flux of streamwise vorticity. Much like transpiration, pressure gradient, or viscosity variations, the wall value of the Lorentz force determines the sign and intensity of wall vorticity flux. Streamwise force contributes to spanwise vorticity flux, and spanwise force contributes to streamwise vorticity flux. Take for example a positive streamwise Lorentz force. This is a negative term on the left hand side of Equation 13.22, which makes the curvature of the streamwise velocity profile at the wall more negative (i.e., instantaneously fuller velocity profile). The wall is then a source of spanwise vorticity, and the flow is more resistant to transition and separation. In a turbulent flow, a positive streamwise Lorentz force leads to suppression of normal and tangential Reynolds stresses.

13.10.4 EMHD Tiles for Reactive Control

This subsection shall outline a reactive control strategy that exploits the Lorentz forces to modulate the flow of an electrically conducting fluid such as sea water. The idea is to target the low-speed streaks in the near-wall region of a turbulent boundary layer. The electric field is applied only when and where it is needed, and hence the power consumed by the reactive control system is kept far below that consumed by a predetermined active control system. According to recent numerical and experimental results, brute force application of steady or time-dependent Lorentz force — to reduce drag for example — do not achieve the breakeven point. The reason being the high energy expenditure by predetermined control systems [see, for example, the two meeting proceedings edited by Gerbeth, 1997, and Meng, 1998].

The top view in Figure 13.9 depicts the proposed distribution of EMHD tiles to achieve targeted control in a turbulent wall-bounded flow. Each tile consists of two streamwise strips constituting the north and south poles of a permanent magnet and another two strips constituting positive and negative electrodes.[10] The tiles are staggered in a checkerboard configuration and are separated by 100 wall units in the spanwise direction (z), and 1000 wall units in the streamwise direction (x). The length and width of each tile is about 50 wall units.[11]

A shear stress sensor having a spatial resolution of 2–5 wall units is placed just upstream and along the centerline of each tile. Each sensor and corresponding actuator is connected via a closed, feedback control loop. The tile is activated to give a positive/negative streamwise Lorentz force when a low- or high-speed region is detected. The control law used can be based on simple physical arguments or more complex self-learning neural network. Other possible (albeit more sophisticated) control laws can be based on nonlinear dynamical systems theory or optimal/suboptimal control theory.

For a simpler physical or numerical experiment that does not require sensors and complex closed-loop control, a single low-speed streak is artificially generated in a laminar boundary layer using two suction holes as depicted in the (x–z) view in Figure 13.10. The two holes are separated in the spanwise direction by 100 wall units, and each is about 0.5 wall unit in diameter. Such method was successfully used by Gad-el-Hak and Hussain (1986) to generate artificial low-speed streaks as well as simulated bursting events in a laminar environment. In the proposed experiment, high-speed regions and counter-rotating streamwise vortices are also generated and, if desired, one can readily target any of these simulated events for elimination. The necessary actuation would of course depend on the kind of coherent structure to be targeted.

A single EMHD tile is placed about 50 wall units directly downstream of the suction holes and is activated with a suitable time delay after a suction pulse is applied, in essence using a simple, predetermined, open-loop control. A single electric pulse[12] triggers a positive Lorentz force to eliminate the artificially

---

[10]The cathode and anode are interchangeable depending on the desired direction of the Lorentz force.

[11]The length and width of each magnet pole (or electrode) are thus 50 and 12.5 wall units, respectively.

[12]Note that in this simulated environment, the electric pulse would have the same polarity if only low-speed streaks are to be obliterated.

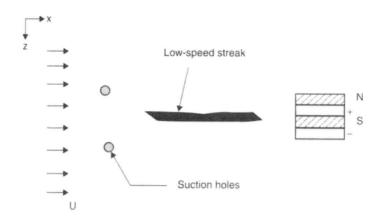

FIGURE 13.10   Control of a single artificial low-speed streak.

generated low-speed streak. The experiment would be repeated several times, and the actuator location, phase lag, strength, and duration could be optimized to achieve the desired goal.

The streamwise Lorentz force necessary to eliminate the resulting artificial low-speed streak could be computed from the measurements, and thus the energy expenditure necessary to eliminate all the low-speed streaks in a real turbulent boundary layer could readily be estimated.

13.11   Chaos Control

In the theory of dynamical systems, the so-called butterfly effect denotes sensitive dependence of nonlinear differential equations on initial conditions, with phase-space solutions initially very close together separating exponentially. The solution of nonlinear dynamical systems of three or more degrees of freedom may be in the form of a strange attractor that has an intrinsic structure that contains a well-defined mechanism to produce a chaotic behavior without requiring random forcing. Chaotic behavior is complex, aperiodic and, though deterministic, appears to be random.

A question arises naturally: just as small disturbances can radically grow within a deterministic system to yield rich, unpredictable behavior, can minute adjustments to a system parameter be used to reverse the process and control, i.e., regularize, the behavior of a chaotic system? Recently, that question was answered in the affirmative theoretically as well as experimentally, at least for system orbits, which reside on low-dimensional strange attractors [see the review by Lindner and Ditto, 1995]. Before describing such strategies for controlling chaotic systems, recent attempts to construct a low-dimensional dynamical systems representation of turbulent boundary layers will be summarized. Such construction is a necessary first step to be able to use chaos control strategies for turbulent flows. Additionally, as argued by Lumley (1996), a low-dimensional dynamical model of the near-wall region used in a Kalman filter [Banks, 1986; Petersen and Savkin, 1999] can make the most of the partial information assembled from a finite number of wall sensors. Such filter minimizes in a least square sense the errors caused by incomplete information, and thus globally optimizes the performance of the control system.

13.11.1   Nonlinear Dynamical Systems Theory

Boundary-layer turbulence is described by a set of nonlinear partial differential equations and is characterized by an infinite number of degrees of freedom. This makes it rather difficult to model the turbulence using a dynamical systems approximation. Complex, infinite-dimensional flow can be decomposed into several low-dimensional subunits, because quasi-periodic coherent structures dominate the dynamics of seemingly random turbulent shear flows. This implies that low-dimensional, localized dynamics can exist in formally infinite-dimensional extended systems such as open turbulent flows. Reducing the flow physics to finite-dimensional dynamical systems enables a study of this behavior by examining the fixed points and the topology of their stable and unstable manifolds. From the dynamical systems theory viewpoint, the meandering of low-speed streaks is interpreted as hovering of the flow state near an unstable fixed point in the low-dimensional state space. An intermittent event that produces high wall stress — a burst — is interpreted as a jump along a heteroclinic cycle to a different unstable fixed point that occurs when the state has wandered too far from the first unstable fixed point. Delaying this jump by holding the system near the first fixed point should lead to lower momentum transport in the wall region and, therefore, to lower skin-friction drag. Reactive control means sensing the current local state and through appropriate manipulation keeping the state close to a given unstable fixed point, thereby preventing further production of turbulence. Reducing the bursting frequency by say 50%, may lead to a comparable reduction in skin-friction drag. For a jet, relaminarization may lead to a quiet flow and very significant noise reduction.

In one significant attempt the proper orthogonal, or Karhunen-Loève, decomposition method has been used to extract a low-dimensional dynamical system from experimental data of the wall region [Aubry et al., 1988; Aubry, 1990]. Aubry et al. (1988) expanded the instantaneous velocity field of a turbulent boundary layer using experimentally determined eigenfunctions that are in the form of streamwise rolls. They expanded the Navier–Stokes equations using these optimally chosen, divergence-free, orthogonal functions. They then applied a Galerkin projection and truncated the infinite-dimensional representation to obtain a ten-dimensional set of ordinary differential equations. These equations represent the dynamical behavior of the rolls, and are shown to exhibit a chaotic regime as well as an intermittency due to a burst-like phenomenon. However, Aubry et al.'s ten-mode dynamical system displays a *regular* intermittency, in contrast to actual turbulence as well as to the chaotic intermittency encountered by Pomeau and Manneville (1980), in which event durations are distributed stochastically. Nevertheless, the major conclusion of Aubry et al.'s study is that the bursts appear to be produced autonomously by the wall region even without turbulence, but are triggered by turbulent pressure signals from the outer layer.

More recently, Berkooz et al. (1991) generalized the class of wall-layer models developed by Aubry et al. (1988) to permit uncoupled evolution of streamwise and cross-stream disturbances. Berkooz et al.'s results suggest that the intermittent events observed in Aubry et al.'s representation do not arise solely because of the effective closure assumption incorporated, but are rather rooted deeper in the dynamical phenomena of the wall region. The book by Holmes et al. (1996) details the Cornell University research group attempts at describing turbulence as a low-dimensional dynamical system.

In addition to the reductionist viewpoint exemplified by the work of Aubry et al. (1988) and Berkooz et al. (1991), attempts have been made to determine directly the dimension of the attractors underlying specific turbulent flows. Again, the central issue here is whether or not turbulent solutions to the infinite-dimensional Navier–Stokes equations can be asymptotically described by a finite number of degrees of freedom. Grappin and Léorat (1991) computed the Lyapunov exponents and the attractor dimensions of two- and three-dimensional periodic turbulent flows without shear. They found that the number of degrees of freedom contained in the large scales establishes an upper bound for the dimension of the attractor. Deane and Sirovich (1991) and Sirovich and Deane (1991) numerically determined the number of dimensions needed to specify chaotic Rayleigh-Bénard convection over a moderate range of Rayleigh numbers, Ra. They suggested that the *intrinsic* attractor dimension is $\mathcal{O}[Ra^{2/3}]$.

The corresponding dimension in wall-bounded flows appears to be dauntingly high. Keefe et al. (1992) determined the dimension of the attractor underlying turbulent Poiseuille flows with spatially periodic boundary conditions. Using a coarse-grained numerical simulation, they computed a lower bound on the Lyapunov dimension of the attractor to be approximately 352 at a pressure-gradient Reynolds number of 3200. Keefe et al. (1992) argue that the attractor dimension in fully-resolved turbulence is unlikely to be much larger than 780. This suggests that periodic turbulent shear flows are deterministic chaos and that a strange attractor underlies solutions to the Navier–Stokes equations. Temporal unpredictability in the turbulent Poiseuille flow is thus due to the exponential spreading property of such attractors. Although finite, the computed dimension invalidates the notion that the global turbulence can be attributed to the interaction of a *few* degrees of freedom. Moreover, in a physical channel or boundary layer, the flow is not periodic and is open. The attractor dimension in such case is not known but is believed to be even higher than the estimate provided by Keefe et al. for the periodic (*quasi-closed*) flow.

In contrast to closed, absolutely unstable flows, such as Taylor–Couette systems, where the number of degrees of freedom can be small, local measurements in open, convectively unstable flows, such as boundary layers, do not express the global dynamics; the attractor dimension in that case may inevitably be too large to be determined experimentally. According to the estimate provided by Keefe et al. (1992), the colossal data required (about 10^D, where D is the attractor dimension) for measuring the dimension simply exceeds current computer capabilities. Turbulence near transition or near a wall is an exception to that bleak picture. In those special cases, a relatively small number of modes is excited and the resulting *simple* turbulence can therefore be described by a dynamical system of a reasonable number of degrees of freedom.

13.11.2 Chaos Control

Another question of greater relevance is that, given a dynamical system in the chaotic regime, is it possible to stabilize its behavior through some kind of active control? While other alternatives have been devised [e.g., Fowler, 1989; Hübler and Lüscher, 1989; Huberman, 1990; Huberman and Lumer, 1990], the recent method proposed by workers at the University of Maryland [Ott et al., 1990a; 1990b; Shinbrot et al., 1990; 1992a; 1992b; 1992c; 1998; Romeiras et al., 1992] promises to be a significant breakthrough. Comprehensive reviews and bibliographies of the emerging field of chaos control can be found in the articles by Shinbrot et al. (1993), Shinbrot (1993; 1995; 1998), and Lindner and Ditto (1995).

Ott et al. (1990a) demonstrated, through numerical experiments with the Henon map, that it is possible to stabilize a chaotic motion about any pre-chosen, unstable orbit through the use of relatively small perturbations. The procedure consists of applying minute time-dependent perturbations to one of the system parameters to control the chaotic system around one of its many unstable periodic orbits. In this context, targeting refers to the process whereby an arbitrary initial condition on a chaotic attractor is steered toward a prescribed point (target) on this attractor. The goal is to reach the target as quickly as possible using a sequence of small perturbations (Kostelich et al., 1993a).

The success of the Ott–Grebogi–Yorke's (OGY) strategy for controlling chaos hinges on the fact that beneath the apparent unpredictability of a chaotic system lies an intricate but highly ordered structure. Left to its own recourse, such a system continually shifts from one periodic pattern to another, creating the appearance of randomness. An appropriately controlled system, on the other hand, is locked into one particular type of repeating motion. With such reactive control the dynamical system becomes one with a stable behavior.

The OGY-method can be simply illustrated by the schematic in Figure 13.11. The state of the system is represented as the intersection of a stable manifold and an unstable one. The control is applied intermittently whenever the system departs from the stable manifold by a prescribed tolerance, otherwise the control is shut off. The control attempts to put the system back onto the stable manifold so that the state converges toward the desired trajectory. Unmodeled dynamics cause noise in the system and a tendency for the state to wander off in the unstable direction. The intermittent control prevents this and the desired trajectory is achieved. This efficient control is not unlike trying to balance a ball in the center of a horse saddle [Moin and Bewley, 1994]. There is one stable direction (front and back) and one unstable direction (left and right). The restless horse is the unmodeled dynamics, intermittently causing the ball to move in the wrong direction. The OGY-control needs only be applied, in the most direct manner possible, whenever the ball wanders off in the left and right direction.

The OGY-method has been successfully applied in a relatively simple experiment in which reverse chaos was obtained in a parametrically driven, gravitationally buckled, amorphous magnetoelastic ribbon [Ditto et al. (1990); Ditto and Pecora (1993)]. Garfinkel et al. (1992) applied the same control strategy to stabilize drug-induced cardiac arrhythmias in sections of a rabbit ventricle. Other extensions, improvements, and applications of the OGY-method include higher-dimensional targeting [Auerbach et al., 1992; Kostelich et al., 1993b]; controlling chaotic scattering in Hamiltonian (i.e., nondissipative, area conservative) systems [Lai et al., 1993a; 1993b]; synchronization of identical chaotic systems that govern communication, neural, or biological processes [Lai and Grebogi, 1993]; use of chaos to transmit

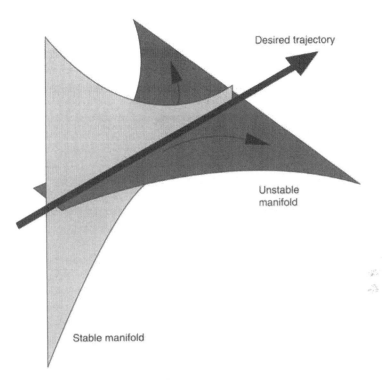

Desired trajectory

Unstable
manifold

Stable manifold

FIGURE 13.11 The OGY-method for controlling chaos.

information [Hayes et al., 1994a; 1994b]; control of transient chaos [Lai et al., 1994]; and taming spatio-temporal chaos using a sparse array of controllers [Chen et al., 1993; Qin et al., 1994; Auerbach, 1994].

In a more complex system, such as a turbulent boundary layer, there exist numerous interdependent modes and many stable as well as unstable manifolds (directions). The flow can then be modeled as coherent structures plus a parameterized turbulent background. The proper orthogonal decomposition (POD) is used to model the coherent part, because POD guarantees the minimum number of degrees of freedom for a given model accuracy. Factors that make turbulence control a challenging task are the potentially quite large perturbations caused by the unmodeled dynamics of the flow, the non-stationary nature of the desired dynamics, and the complexity of the saddle shape describing the dynamics of the different modes. Nevertheless, the OGY-control strategy has several advantages that are of special interest in the control of turbulence: (1) the mathematical model for the dynamical system need not be known; (2) only *small* changes in the control parameter are required; and (3) noise can be tolerated (with appropriate penalty).

Keefe (1993a; 1993b) made a useful comparison between two nonlinear control strategies as applied to fluid problems: the OGY feedback method (previously described) and the H-method, a model-based control strategy originated by Hübler and Lüscher (1989). Both novel control methods are essentially generalizations of the classical perturbation cancellation technique: apply a prescribed forcing to subtract the undesired dynamics and impose the desired one. The OGY-method exploits the sensitivity of chaotic systems to stabilize existing periodic orbits and steady states. Some feedback is needed to steer the trajectories toward the chosen fixed point, but the required control signal is minuscule. In contrast, Hübler's scheme does not explicitly make use of the system sensitivity. It produces general control response (periodic or aperiodic) and needs little or no feedback, but its control inputs are generally large. The OGY-method exploits the nonlinearity of a dynamical system; indeed the presence of a strange attractor and the extreme sensitivity of the dynamical system to initial conditions are essential to the success of this method. In contrast, the H-method works equally for both linear and nonlinear systems.

Keefe (1993a) first examined numerically the two schemes as applied to fully-developed and transitional solutions of the Ginzburg–Landau equation, an evolution equation that governs the initially weakly nonlinear stages of transition in several flows and that possesses both transitional and fully-chaotic solutions. The Ginzburg–Landau equation has solutions that display either absolute or convective instabilities, and is thus a reasonable model for both closed and open flows. Keefe's main conclusion is that control of nonlinear systems is best obtained by making maximum use possible of the underlying natural dynamics. If the goal dynamics is an unstable nonlinear solution of the equation and the flow is nearby at the instant control is applied, both methods perform reliably and at low-energy cost in reaching and maintaining this goal. Predictably, the performance of both control strategies degrades due to noise and the spatially discrete nature of realistic forcing.

Subsequently, Keefe (1993b) extended the numerical experiment in an attempt to reduce the drag in a channel flow with spatially periodic boundary conditions. The OGY-method reduces the skin friction to 60–80% of the uncontrolled value at a mass-flux Reynolds number of 4408. The H-method fails to achieve any drag reduction when starting from a fully-turbulent initial condition, but shows potential for suppressing or retarding laminar-to-turbulence transition. Keefe (1993a) suggests that the H-strategy might be more appropriate for boundary-layer control, while the OGY-method might best be used for channel flows.

It is also relevant to note the work of Bau and his colleagues at the University of Pennsylvania [Singer et al., 1991; Wang et al., 1992], who devised a feedback control to stabilize (*relaminarize*) the naturally occurring chaotic oscillations of a toroidal thermal convection loop heated from below and cooled from above. Based on a simple mathematical model for the thermosyphon, Bau and his colleagues constructed a reactive control system that was used to significantly alter the flow characteristics inside the convection loop. Their linear control strategy, perhaps a special version of the OGY's chaos control method, consists simply of sensing the deviation of fluid temperatures from desired values at a number of locations inside the thermosyphon loop and then altering the wall heating either to suppress or to enhance such deviations. Wang et al. (1992) also suggested extending their theoretical and experimental method to more complex situations such as those involving Bénard convection [Tang and Bau, 1993a; 1993b]. Hu and Bau (1994) used a similar feedback control strategy to demonstrate that the critical Reynolds number for the loss of stability of planar Poiseuille flow can be significantly increased or decreased.

Other attempts to use low-dimensional dynamical systems representation for flow control include the work of Berkooz et al. (1993), Corke et al. (1994), and Coller et al. (1994a; 1994b). Berkooz et al. (1993) applied techniques of modern control theory to estimate the phase-space location of dynamical models of the wall-layer coherent structures and used these estimates to control the model dynamics. Since discrete wall-sensors provide incomplete knowledge of phase-space location, Berkooz et al. maintain that a nonlinear observer, which incorporates past information and the equations of motion into the estimation procedure, is required. Using an extended Kalman filter, they achieved effective control of a bursting pair of rolls with the equivalent of two wall-mounted shear sensors. Corke et al. (1994) used a low-dimensional dynamical system based on the proper orthogonal decomposition to guide control experiments for an axisymmetric jet. By sensing the downstream velocity and actuating an array of miniature speakers located at the lip of the jet, their feedback control succeeded in converting the near-field instabilities from spatial-convective to temporal-global. Coller et al. (1994a; 1994b) developed a feedback control strategy for strongly nonlinear dynamical systems, such as turbulent flows, subject to small random perturbations that kick the system intermittently from one saddle point to another along heteroclinic cycles. In essence, their approach is to use local, weakly nonlinear feedback control to keep a solution near a saddle point as long as possible, but then to let the natural, global nonlinear dynamics run its course when *bursting* (in a low-dimensional model) occurs. Though conceptually related to the OGY-method, Coller et al.'s method does not actually stabilize the state but merely holds the system near the desired point longer than it would otherwise stay.

Shinbrot and Ottino (1993a; 1993b) offer yet another strategy presumably most suited for controlling coherent structures in area-preserving turbulent flows. Their geometric method exploits the premise that the dynamical mechanisms that produce the organized structures can be remarkably simple. By repeated stretching and folding of "horseshoes" that are present in chaotic systems, Shinbrot and Ottino have demonstrated numerically as well as experimentally the ability to create, destroy, and manipulate coherent structures in chaotic fluid systems. The key idea to create such structures is to intentionally place folds of horseshoes near low-order periodic points. In a dissipative dynamical system, volumes contract in state space and the co-location of a fold with a periodic point leads to an isolated region that contracts asymptotically to a point. Provided that the folding is done properly, it counteracts stretching. Shinbrot and Ottino (1993a) applied the technique to three prototypical problems: a one-dimensional chaotic map, a two-dimensional one, and a chaotically advected fluid. Shinbrot (1995; 1998) and Shinbrot et al. (1998) provide recent reviews of the stretching/folding as well as other chaos control strategies.

13.12   Soft Computing

The term soft computing was coined by Lotfi Zadeh of the University of California, Berkeley, to describe several ingenious modes of computations that exploit tolerance for imprecision and uncertainty in complex systems to achieve tractability, robustness, and low cost [Yager and Zadeh, 1992; Bouchon-Meunier et al., 1995a; 1995b; Jang et al., 1997]. The principle of complexity provides the impetus for soft computing: as the complexity of a system increases, the ability to predict its response diminishes until a threshold is reached beyond which precision and relevance become almost mutually exclusive [Noor and Jorgensen, 1996]. In other words, precision and certainty carry a cost. By employing modes of reasoning — probabilistic reasoning — that are approximate rather than exact, soft computing can help in searching for globally optimal design or achieving effectual control while taking into account system uncertainties and risks.

Soft computing refers to a domain of computational intelligence that loosely lies between purely numerical (hard) computing and purely symbolic computations. Alternatively, one can think about symbolic computations as a form of artificial intelligence lying between biological intelligence and computational intelligence (soft computing). The schematic in Figure 13.12 illustrates the general idea. Artificial intelligence relies on symbolic information processing techniques and uses logic as representation and inference mechanisms. It attempts to approach the high level of human cognition. In contrast, soft computing is based on modeling low-level cognitive processes and strongly emphasizes modeling of uncertainty as well as learning. Computational intelligence mimics the ability of the human brain to employ modes of reasoning that are approximate. Soft computing provides a machinery for the numeric representation of the types of constructs developed in the symbolic artificial intelligence. The boundaries between these paradigms are of course *fuzzy*.

The principal constituents of soft computing are neurocomputing, fuzzy logic, and genetic algorithms, as depicted in Figure 13.12. These elements, together with probabilistic reasoning, can be combined in hybrid arrangements resulting in better systems in terms of parallelism, fault tolerance, adaptivity, and uncertainty management. To my knowledge, only neurocomputing has been employed for fluid flow control, but the other tools of soft computing may be just as useful to construct powerful controllers and have been used as such in other fields such as large-scale subway controllers and video cameras. A brief description of those three constituents follows.

Neurocomputing is inspired by the neurons of the human brain and how they work. Neural networks are information processing devices that can learn by adapting synaptic weights to changes in the surrounding environment, can handle imprecise, fuzzy, noisy, and probabilistic information. They can also generalize from known tasks (examples) to unknown ones. Actual engineering oriented hardware is termed artificial neural networks (ANN) while algorithms are called computational neural networks (CNN). The nonlinear, highly parallel networks can perform any of the following tasks: classification, pattern matching, optimization, control, and noise removal. As modeling and optimization tools, neural networks are particularly useful when good analytic models are either unknown or extremely complex.

An artificial neural network consists of a large number of highly interconnected processing elements — essentially equations known as "transfer functions" — that are analogous to human neurons. These are tied together with weighted connections that are analogous to human synapses. A processing unit takes weighted signals from other units, possibly combines them, and gives a numeric result. The behavior of neural networks — how they map input data — is influenced primarily by the transfer functions of the processing elements, how the transfer functions are interconnected, and the weights of those interconnections.

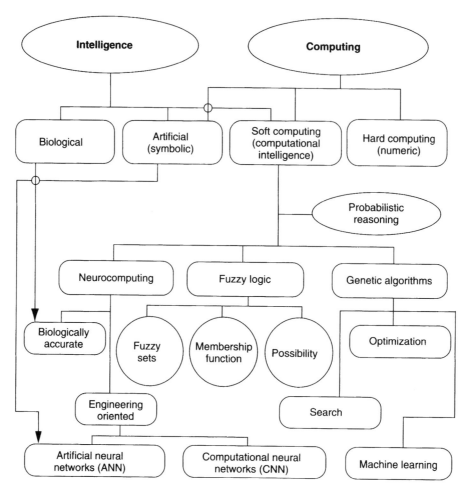

FIGURE 13.12   Tools for soft computing.

Learning typically occurs by example through exposure to a set of input–output data, where the training algorithm adjusts the connection weights (synapses). These connection weights store the knowledge necessary to solve specific problems. As an example, it is now possible to use neural networks to sense odors in many different applications [Ouellette, 1999]. The electronic noses (*e-noses*) are on the verge of finding commercial applications in medical diagnostics, environmental monitoring, and the processing and quality control of foods. Neural networks as used in fluid flow control will be covered in the following subsection.

Fuzzy logic was introduced by Lotfi Zadeh in 1965 as a mathematical tool to deal with uncertainty and imprecision. The book edited by Yager and Zadeh (1992) is an excellent primer to the field. For computing and reasoning, general concepts (such as size) are implemented into a computer algorithm by using mostly words (such as small, medium, or large). Fuzzy logic, therefore, provides a unique methodology for computing with words. Its rationalism is based on three mathematical concepts: fuzzy sets, membership function, and possibility. As dictated by a membership function, fuzzy sets allow a gradual transition from belonging to not belonging to a set. The concept of possibility provides a mechanism for interpreting factual statements involving fuzzy sets. Three processes are involved in solving a practical problem using fuzzy logic: fuzzification, analysis, and defuzzification. Given a complex, unsolvable problem in real space, those three steps involve enlarging the space and searching for a solution in the new superset, then specializing this solution to the original real constraints.

Genetic algorithms are search algorithms based loosely on the mechanics of natural selection and natural genetics. They combine survival of the fittest among string structures with structured yet randomized information exchange, and are used for search, optimization, and machine learning. For control, genetic algorithms aim at achieving minimum cost function and maximum performance measure while satisfying the problem constraints. The books by Goldberg (1989), Davis (1991), and Holland (1992) provide introduction to the field.

In the Darwinian principle of natural selection, the fittest members of a species are favored to produce offspring. Even biologists cannot help but be awed by the complexity of life observed to evolve in the relatively short time suggested by the fossil records. A living being is an amalgam of characteristics determined by the thousands of genes in its chromosomes. Each gene may have several forms or alternatives called alleles that produce differences in the set of characteristics associated with that gene. The chromosomes are therefore the organic devices through which the structure of a creature is encoded, and this living being is created partly through the process of decoding those chromosomes. Genes transmit hereditary characters and form specific parts of a self-perpetuated deoxyribonucleic acid (DNA) in a cell nucleus. Natural selection is the link between the genes and the performance of their decoded structures. Simply put, the process of natural selection causes those genes that encode successful structures to reproduce more often than those that do not.

In the early 1970s, John H. Holland of the University of Michigan, introduced a man-made version of the procedure of natural evolution in an attempt to solve difficult problems. The candidate solutions to a problem are ranked by the genetic algorithm according to how well they satisfy a certain criterion, and the fittest members are the most favored to combine amongst themselves to form the next generation of the members of the *species*. Fitter members presumably produce even fitter offspring and therefore better solutions to the problem at hand. Solutions are represented by binary strings, each trial solution is coded as a vector called chromosome. The elements of a chromosome are described as genes, and its varying values at specific positions are called alleles. Good solutions are selected for reproduction based on a fitness function using genetic recombination operators such as crossover and mutation. The main advantage of genetic algorithms is their global parallelism in which the search efforts to many regions of the search area are simultaneously allocated.

Genetic algorithms have been used for the control of different dynamical systems, as for example the optimization of robot trajectories. But to my knowledge and at the time of writing this chapter, reactive control of turbulent flows is yet to benefit from this powerful soft computing tool. In particular, when a finite number of sensors is used to gather information about the state of the flow, a genetic algorithm perhaps combined with a neural network can adapt and learn to use current information to eliminate the uncertainty created by insufficient a priori information.

13.12.1 Neural Networks for Flow Control

Biologically inspired neural networks are finding increased applications in many fields of science and technology. Modeling of complex dynamical systems, adaptive noise canceling in telephones and modems, bomb sniffers, mortgage-risk evaluators, sonar classifiers, and word recognizers are but a few of existing usage of neural nets. The book by Nelson and Illingworth (1991) provides a lucid introduction to the field, and the review article by Antsaklis (1993) focuses on the use of neural nets for the control of complex dynamical systems. For flow control applications, neural networks offer the possibility of adaptive controllers that are simpler and potentially less sensitive to parameter variations as compared to conventional controllers. Moreover, if a colossal number of sensors and actuators is to be used, the massively parallel computational power of neural nets will surely be needed for real-time control.

The basic elements of a neural network are schematically shown in Figure 13.13. Several inputs are connected to the nodes (neurons or processing elements) that form the input layer. There are one or more hidden layers, followed by an output layer. Note that the number of connections is higher than the total number of nodes. Both numbers are chosen based on the particular application and can be arbitrarily large for complex tasks. Simply put, the multi-task — albeit simple — job of each processing element is to evaluate each of the input signals to that particular element, calculate the weighted sum of the combined inputs, compare that total to some threshold level, and finally determine what the output should

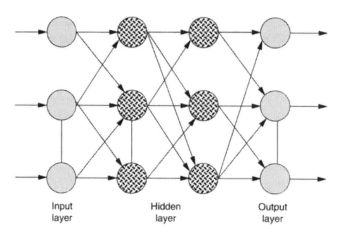

Input   Hidden   Output
layer   layer   layer

FIGURE 13.13   Elements of a neural network.

be. The various weights are the adaptive coefficients that vary dynamically as the network *learns* to perform its assigned task; some inputs are more important than others. The threshold, or transfer, function
is generally nonlinear; the most common one being the continuous sigmoid, or S-shaped, curve, which
approaches a minimum and maximum value at the asymptotes. If the sum of the weighted inputs is larger
than the threshold value, the neuron generates a signal; otherwise no signal is fired. Neural networks can
operate in feedforward or feedback mode.[13] Complex systems, for which dynamical equations may not be
known or may be too difficult to solve, can be modeled using neural nets.

For flow control, neural networks provide convenient, fast, nonlinear adaptive algorithms to relate sensor
outputs to actuator inputs via variable-coefficient functions and nonlinear, sigmoid saturation functions.
With no prior knowledge of the pertinent dynamics, a self-learning neural network develops a model for
that dynamics through observations of the applied control and sensed measurements. The network is by
nature nonlinear and can therefore better handle nonlinear dynamical systems, a difficult task when classical (linear or weakly nonlinear) control strategies are attempted. The feedforward type of neural network acts as a nonlinear filter forming an output from a set of input data. The output can then be
compared to some desired output, and the difference (error) is typically used in a back-propagation algorithm that updates the network parameters.

The number of researchers using neural networks to control fluid flows is growing rapidly; this chapter provides only a small sample. Using a pre-trained neural network, Fan et al. (1993) conducted a conceptual reactive flow control experiment to delay laminar-to-turbulence transition. Numerical
simulations of their flow control system demonstrate almost complete cancellation of single and multiple artificial wave disturbances. Their controller also successfully attenuated a natural disturbance signal
with developing wave packets from an actual wind-tunnel experiment.

Jacobson and Reynolds (1993b; 1995; 1998) used neural networks to minimize the boundary velocity
gradient of three model flows: the one-dimensional stochastic Burgers equation; a two-dimensional computational model of the near-wall region of a turbulent boundary layer; and a real-time turbulent flow
with a spanwise array of wall actuators together with upstream and downstream wall-sensors. For all
three problems, the neural network successfully learned about the flow and developed into proficient
controllers. For the laboratory experiments, however, Jacobson and Reynolds (1995) report that the
neural network training time was much longer and the performance was no better than a simpler ad hoc

---

[13]Note that this terminology refers to the direction of information through the network. When a neural net is used
as a controller, the overall control loop is, however, a feedback, closed loop: the self-learning network dynamically
updates its various parameters by comparing its output to a desired output, thus requiring feedback information
relating to the effect of its control.

controller that they developed. Jacobson and Reynolds emphasize that alternative neural net configurations and convergence algorithms may, however, greatly improve the network performance.

Using the angle of attack and angular velocity as inputs, Faller et al. (1994) trained a neural network to model the measured unsteady surface pressure over a pitching airfoil [see Schreck et al., 1995]. Following training and using the instantaneous angle of attack and pitch rate as the only inputs, their network was able to accurately predict the surface pressure topology as well as the time-dependent aerodynamic forces and moments. The model was then used to develop a neural network controller for wing-motion actuator signals that in turn provided direct control of the lift-to-drag ratio across a wide range of time-dependent motion histories.

As a final example, Kawthar-Ali and Acharya (1996) developed a neural network controller for use in suppressing the dynamic-stall vortex that periodically develops in the leading edge of a pitching airfoil. Based on the current state of the unsteady pressure field, their control system specified the optimum amount of leading-edge suction to achieve complete vortex suppression.

13.13   Use of MEMS for Reactive Control

Current usage for microelectromechanical systems (MEMS) includes accelerometers for airbags and guidance systems, pressure sensors for engine air intake and blood analysis, rate gyroscopes for antilock brakes, microrelays and microswitches for semiconductor automatic test equipment, and microgrippers for surgical procedures [Angell et al., 1983; Gabriel et al., 1988; 1992; O'Connor, 1992; Gravesen et al., 1993; Bryzek et al., 1994; Gabriel, 1995; Ashley, 1996; Ho and Tai, 1996; 1998; Hogan, 1996; Ouellette, 1996; 2003; Paula, 1996; Robinson et al., 1996a; 1996b; Tien, 1997; Amato, 1998; Busch-Vishniac, 1998; Kovacs, 1998; Knight, 1999; Epstein, 2000; O'Connor and Hutchinson, 2000; Goldin et al., 2000; Chalmers, 2001; Tang and Lee, 2001; Nguyen and Wereley, 2002; Karniadakis and Beskok, 2002; Madou, 2002; DeGaspari, 2003; Ehrenman, 2004; Sharke, 2004]. There is considerable work under way to include other applications, one example being the micro-steam engine described by Lipkin (1993), Garcia and Sniegowski (1993; 1995), and Sniegowski and Garcia (1996). A second example is the 3-cm $\times$ 1.5-cm digital light processor that contains 0.5–2 million individually addressable micromirrors each typically measuring 16 μm on a side. Texas Instruments, Inc., is currently producing such a device with a resolution of 2000 $\times$ 1000 pixels, for high-definition televisions and other display equipments. The company maintains that when mass produced, such a device would cost on the order of $100, i.e., less than $0.0001 per actuator!

MEMS would be ideal for the reactive flow control concept advocated in this chapter. Methods of flow control targeted toward specific coherent structures involve nonintrusive detection and subsequent modulation of events that occur randomly in space and time. To achieve proper targeted control of these quasi-periodic vortical events, temporal phasing as well as spatial selectivity are required. Practical implementation of such an idea necessitates the use of a large number of intelligent, communicative wall sensors and actuators arranged in a checkerboard pattern. This chapter has provided estimates for the number, characteristics, and energy consumption of such elements required to modulate the turbulent boundary layer that develops along a typical commercial aircraft or nuclear submarine. An upper-bound number to achieve total turbulence suppression is about one million sensors/actuators per square meter of the surface, although as argued earlier, the actual number needed to achieve effective control could perhaps be 1–2 orders of magnitude below that.

The sensors would be expected to measure the amplitude, location, and phase or frequency of the signals impressed upon the wall by incipient bursting events. Instantaneous wall-pressure or wall-shear stress can be sensed, for example. The normal or in-plane motion of a minute membrane is proportional to the respective point force of primary interest. For measuring wall pressure, microphone-like devices respond to the motion of a vibrating surface membrane or an internal elastomer. Several types are available including variable-capacitance (condenser or electret), ultrasonic, optical (e.g., optical-fiber and diode-laser), and piezoelectric devices [see, for example, Löfdahl et al., 1993; 1994; Löfdahl and Gad-el-Hak, 1999]. A potentially useful technique has been tried at MIT [Warkentin et al., 1987; Young et al., 1988; Haritonidis et al., 1990a; 1990b]. An array of extremely small (0.2 mm in diameter) laser-powered microphones (termed picophones) was machined in silicon using integrated circuit fabrication techniques, and was used for field measurement of the instantaneous surface pressure in a turbulent boundary layer. The wall-shear stress, though smaller and therefore more difficult to measure than pressure, provides a more reliable signature of the near-wall events.

Actuators are expected to produce a desired change in the targeted coherent structures. The local acceleration action needed to stabilize an incipient bursting event can be in the form of adaptive wall, transpiration, wall heat transfer, or electromagnetic body force. Traveling surface waves can be used to modify a locally convecting pressure gradient such that the wall motion follows that of the coherent event causing the pressure change. Surface motion in the form of a Gaussian hill with height $y^2 = \mathcal{O}[10]$ should be sufficient to suppress typical incipient bursts [Lumley, 1991; Carlson and Lumley, 1996]. Such time-dependent alteration in wall geometry can be generated by driving a flexible skin using an array of piezoelectric devices (dilate or contract depending on the polarity of current passing through them), electromagnetic actuators, magnetoelastic ribbons (made of nonlinear materials that change their stiffness in the presence of varying magnetic fields), or Terfenol-d rods (a novel metal composite, developed at Grumman Corporation, which changes its length when subjected to a magnetic field). Other exotic materials can also be used for actuation. For example, electrorheological fluids [Halsey and Martin, 1993] instantly solidify when exposed to an electric field, and may thus be useful for the present application. Recently constructed microactuators specifically designed for flow control include those by Wiltse and Glezer (1993), James et al. (1994), Jacobson and Reynolds (1995), Vargo and Muntz (1996), and Keefe (1996).

Suction/injection at many discrete points can be achieved by simply connecting a large number of minute streamwise slots, arranged in a checkerboard pattern, to a low-pressure or high-pressure reservoir located underneath the working surface. The transpiration through each individual slot is turned on and off using a corresponding number of independently controlled microvalves. Alternatively, positive-displacement or rotary micropumps [Sen et al., 1996; Sharatchandra et al., 1997] can be used for blowing or sucking fluid through small holes or slits. Based on the results of Gad-el-Hak and Blackwelder (1989), equivalent suction coefficients of about 0.0006 should be sufficient to stabilize the near-wall region. Assuming that the skin-friction coefficient in the uncontrolled boundary layer is $C_f = 0.003$, and assuming further that the suction used is sufficient to establish an asymptotic boundary layer ($d\delta_\theta/dx = 0$, where δ_θ is the momentum thickness), the skin friction in the reactively controlled case is then $C_f = 0 + 2C_q = 0.0012$, or 40% of the original value. The net benefit will, of course, be reduced by the energy expenditure of the suction pump (or micropumps) as well as the array of microsensors and microvalves.

Finally, if the bursting events are to be eliminated by lowering the near-wall viscosity, direct electric-resistance heating can be used in liquid flows, and thermoelectric devices based on the Peltier effect can be used for cooling in the case of gaseous boundary layers. The absolute viscosity of water at 20°C decreases by approximately 2% for each 1°C rise in temperature, while for room-temperature air, μ decreases by approximately 0.2% for each 1°C drop in temperature. The streamwise momentum equation written at the wall can be used to show that a suction coefficient of 0.0006 has approximately the same effect on the wall-curvature of the instantaneous velocity profile as a surface heating of 2°C in water or a surface cooling of 40°C in air [Liepmann and Nosenchuck, 1982; Liepmann et al., 1982].

Sensors and actuators of the types discussed in this section can be combined on individual electronic chips using microfabrication technology. The chips can be interconnected in a communication network that is controlled by a massively parallel computer or a self-learning neural network, perhaps each sensor/actuator unit communicating only with its immediate neighbors. In other words, it may not be necessary for one sensor/actuator to exchange signals with another far away unit. Factors to be considered in an eventual field application of chips produced using microfabrication processes include sensitivity of sensors, sufficiency and frequency response of actuators' action, fabrication of large arrays at affordable prices, survivability in the hostile field environment, and energy required to power the sensors/actuators. As argued by Gad-el-Hak (1994; 1996; 2000), sensor/actuator chips currently produced are small enough for typical field application, and they can be programmed to provide a sufficiently large/fast action in response to a certain sensor output [see also Jacobson and Reynolds, 1995]. Present prototypes are, however, still quite expensive as well as delicate. But so was the transistor when first introduced! It is hoped that the unit price of future sensor/actuator elements would follow the same dramatic trends witnessed in the case of the simple transistor and even the much more complex integrated circuit. The price anticipated by Texas Instruments for an array of one million mirrors hints that the technology is well on its way to mass-produce phenomenally inexpensive microsensors and microactuators. Additionally, current automotive applications are a rigorous proving ground for MEMS: under-the-hood sensors can already withstand harsh conditions such as intense heat, shock, continual vibration, corrosive gases, and electromagnetic fields.

13.14 Conclusions

This chapter has emphasized the frontiers of the field of flow control, reviewed the important advances that took place during the past few years, and provided a blueprint for future progress. In two words, the future of flow control is in taming turbulence by targeting its coherent structures: *reactive control*. Recent developments in chaos control, microfabrication, and soft computing tools are making it more feasible to perform reactive control of turbulent flows to achieve drag reduction, lift enhancement, mixing augmentation, and noise suppression. Field applications, however, have to await further progress in those three modern areas.

The outlook for reactive control is quite optimistic. Soft computing tools and nonlinear dynamical systems theory are developing at a fast pace. MEMS technology is improving even faster. The ability of Texas Instruments to produce an array of one million individually addressable mirrors for around 0.01 cent per actuator is a foreteller of the spectacular advances anticipated in the near future. Existing automotive applications of MEMS have already proven the ability of such devices to withstand the harsh environment under the hood. For the first time, targeted control of turbulent flows is now in the realm of possibility for future practical devices. What is needed now is a focused, well-funded research and development program to make it all come together for field application of reactive flow control systems.

It may be worth recalling that a mere 10% reduction in the total drag of an aircraft translates into a saving of one billion dollars in annual fuel cost (at 1999 prices) for the commercial fleet of aircraft in the United States alone. Contrast this benefit to the annual worldwide expenditure of perhaps a few million dollars for all basic research in the broad field of flow control. Taming turbulence, though arduous, will pay for itself in gold. Reactive control as difficult as it seems, is neither impossible nor a pie in the sky. Beside, lofty goals require strenuous efforts. Easy solutions to difficult problems are likely to be wrong as implied by the witty words of the famed journalist Henry Louis Mencken (1880–1956): *"There is always an easy solution to every human problem — neat, plausible, and wrong."*

References

Aamo, O.M., and Krstic, M. (2003) *Flow Control by Feedback: Stabilization and Mixing*, Springer-Verlag, New York.

Abergel, F., and Temam, R. (1990) "On Some Control Problems in Fluid Mechanics," *Theor. Comput. Fluid Dyn.*, 1, pp. 303–325.

Alshamani, K.M.M., Livesey, J.L., and Edwards, F.J. (1982) "Excitation of the Wall Region by Sound in Fully Developed Channel Flow," *AIAA J.*, 20, pp. 334–339.

Amato, I. (1998) "Formenting a Revolution, in Miniature," *Science*, 282, no. 5388, 16 October, pp. 402–405.

Angell, J.B., Terry, S.C., and Barth, P.W. (1983) "Silicon Micromechanical Devices," *Faraday Trans. I*, 68, pp. 744–748.

Antonia, R.A., Fulachier, L., Krishnamoorthy, L.V., Benabid, T., and Anselmet, F. (1988) "Influence of Wall Suction on the Organized Motion in a Turbulent Boundary Layer," *J. Fluid Mech.*, 190, pp. 217–240.

Antsaklis, P.J. (1993) "Control Theory Approach," in *Mathematical Approaches to Neural Networks*, J.G. Taylor, ed., pp. 1–23, Elsevier, Amsterdam.

Aronson, D., and Löfdahl, L. (1994) "The Plane Wake of a Cylinder: An Estimate of the Pressure Strain Rate Tensor," *Phys. Fluids*, 6, pp. 2716–2721.

Aronson, D., Johansson, A.V., and Löfdahl, L. (1997) "A Shear-Free Turbulent Boundary Layer — Experiments and Modeling," *J. Fluid Mech.*, **338**, pp. 363–385.

Ashley, S. (1996) "Getting a Microgrip in the Operating Room," *Mech. Eng.*, **118**, September, pp. 91–93.

Aubry, N. (1990) "Use of Experimental Data for an Efficient Description of Turbulent Flows," *Appl. Mech. Rev.*, **43**, pp. S240–S245.

Aubry, N., Holmes, P., Lumley, J.L., and Stone, E. (1988) "The Dynamics of Coherent Structures in the Wall Region of a Turbulent Boundary Layer," *J. Fluid Mech.*, **192**, pp. 115–173.

Auerbach, D. (1994) "Controlling Extended Systems of Chaotic Elements," *Phys. Rev. Lett.*, **72**, pp. 1184–1187.

Auerbach, D., Grebogi, C., Ott, E., and Yorke, J.A. (1992) "Controlling Chaos in High Dimensional Systems," *Phys. Rev. Lett.*, **69**, pp. 3479–3482.

Bakewell, H.P., and Lumley, J.L. (1967) "Viscous Sublayer and Adjacent Wall Region in Turbulent Pipe Flow," *Phys. Fluids*, **10**, pp. 1880–1889.

Bandyopadhyay, P.R. (1986) "Review—Mean Flow in Turbulent Boundary Layers Disturbed to Alter Skin Friction," *J. Fluids Eng.*, **108**, pp. 127–140.

Banks, S.P. (1986) *Control Systems Engineering*, Prentice-Hall International, Englewood Cliffs, New Jersey.

Barnwell, R.W., and Hussaini, M.Y., eds. (1992) *Natural Laminar Flow and Laminar Flow Control*, Springer-Verlag, New York.

Berkooz, G., Fisher, M., and Psiaki, M. (1993) "Estimation and Control of Models of the Turbulent Wall Layer," *Bul. Am. Phys. Soc.*, **38**, p. 2197.

Berkooz, G., Holmes, P., and Lumley, J.L. (1991) "Intermittent Dynamics in Simple Models of the Turbulent Boundary Layer," *J. Fluid Mech.*, **230**, pp. 75–95.

Bewley, T.R., Moin, P., and Temam, R. (1997) "Optimal and Robust Approaches for Linear and Nonlinear Regulation Problems in Fluid Mechanics," AIAA Paper No. 97–1872, Reston, Virginia.

Bewley, T.R., Temam, R., and Ziane, M. (1998) "A General Framework for Robust Control in Fluid Mechanics," Center for Turbulence Research No. CTR-Manuscript-169, Stanford University, Stanford, California.

Blackwelder, R.F. (1978) "The Bursting Process in Turbulent Boundary Layers," in *Workshop on Coherent Structure of Turbulent Boundary Layers*, C.R. Smith, and D.E. Abbott, eds., pp. 211–227, Lehigh University, Bethlehem, Pennsylvania.

Blackwelder, R.F. (1988) "Coherent Structures Associated with Turbulent Transport," in *Transport Phenomena in Turbulent Flows*, M. Hirata, and N. Kasagi, eds., pp. 69–88, Hemisphere, New York.

Blackwelder, R.F. (1998) "Some Notes on Drag Reduction in the Near-Wall Region," in *Flow Control: Fundamentals and Practices*, M. Gad-el-Hak, A. Pollard, and J.-P. Bonnet, eds., pp. 155–198, Springer-Verlag, Berlin.

Blackwelder, R.F., and Eckelmann, H. (1979) "Streamwise Vortices Associated with the Bursting Phenomenon," *J. Fluid Mech.*, **94**, pp. 577–594.

Blackwelder R.F., and Gad-el-Hak M. (1990) "Method and Apparatus for Reducing Turbulent Skin Friction," United States Patent No. 4,932,612.

Blackwelder, R.F., and Haritonidis, J.H. (1983) "Scaling of the Bursting Frequency in Turbulent Boundary Layers," *J. Fluid Mech.*, **132**, pp. 87–103.

Blackwelder, R.F., and Kovasznay, L.S.G. (1972) "Time-Scales and Correlations in a Turbulent Boundary Layer," *Phys. Fluids*, **15**, pp. 1545–1554.

Blackwelder, R.F., and Swearingen, J.D. (1990) "The Role of Inflectional Velocity Profiles in Wall Bounded Flows," in *Near-Wall Turbulence: 1988 Zoran Zaric Memorial Conference*, S.J. Kline, and N.H. Afgan, eds., pp. 268–288, Hemisphere, New York.

Bouchon-Meunier, B., Yager, R.R., and Zadeh, L.A., eds. (1995a) *Fuzzy Logic and Soft Computing*, World Scientific, Singapore.

Bouchon-Meunier, B., Yager, R.R., and Zadeh, L.A., eds. (1995b) *Advances in Intelligent Computing — IPMU '94*, Lecture Notes in Computer Science, vol. 945, Springer-Verlag, Berlin.

Breuer, K.S., Haritonidis, J.H., and Landahl, M.T. (1989) "The Control of Transient Disturbances in a Flat Plate Boundary Layer through Active Wall Motion," *Phys. Fluids A*, **1**, pp. 574–582.

Bryzek, J., Peterson, K., and McCulley, W. (1994) "Micromachines on the March," *IEEE Spectrum*, 31, May, pp. 20–31.

Busch-Vishniac, I.J. (1998) "Trends in Electromechanical Transduction," *Phys. Today*, 51, July, pp. 28–34.

Bushnell, D.M. (1983) "Turbulent Drag Reduction for External Flows," AIAA Paper No. 83-0227, New York.

Bushnell, D.M. (1994) "Viscous Drag Reduction in Aeronautics," *Proc. Nineteenth Congress of the International Council of the Aeronautical Sciences*, vol. 1, pp. XXXIII–LVI, paper no. ICAS-94-0.1, AIAA, Washington, D.C.

Bushnell, D.M. (1998) "Frontiers of the 'Responsibly Imaginable' in Aeronautics," Dryden Lecture, AIAA Paper No. 98-0001, Reston, Virginia.

Bushnell, D.M., and Hefner, J.N., eds. (1990) *Viscous Drag Reduction in Boundary Layers*, AIAA, Washington, D.C.

Bushnell, D.M., and McGinley, C.B. (1989) "Turbulence Control in Wall Flows," *Annu. Rev. Fluid Mech.*, 21, pp. 1–20.

Bussmann, K., and Münz, H. (1942) "Die Stabilität der laminaren Reibungsschicht mit Absaugung," *Jahrb. Dtsch. Luftfahrtforschung*, 1, pp. 36–39.

Butler, K.M., and Farrell, B.F. (1993) "Optimal Perturbations and Streak Spacing in Wall-Bounded Shear Flow," *Phys. Fluids A*, 5, pp. 774–777.

Cantwell, B.J. (1981) "Organized Motion in Turbulent Flow," *Ann. Rev. Fluid Mech.* 13, pp. 457–515.

Carlson, H.A., and Lumley, J.L. (1996) "Flow Over an Obstacle Emerging from the Wall of a Channel," *AIAA J.*, 34, pp. 924–931.

Chalmers, P. (2001) "Relay Races," *Mech. Eng.*, 123, January, pp. 66–68.

Chen, C.-C., Wolf, E.E., and Chang, H.-C. (1993) "Low-Dimensional Spatiotemporal Thermal Dynamics on Nonuniform Catalytic Surfaces," *J. Phys. Chem.*, 97, pp. 1055–1064.

Choi, H., Moin, P., and Kim, J. (1992) "Turbulent Drag Reduction: Studies of Feedback Control and Flow Over Riblets," Department of Mechanical Engineering Report No. TF-55, Stanford University, Stanford, California.

Choi, H., Moin, P., and Kim, J. (1994) "Active Turbulence Control for Drag Reduction in Wall-Bounded Flows," *J. Fluid Mech.*, 262, pp. 75–110.

Choi, H., Temam, R., Moin, P., and Kim, J. (1993) "Feedback Control for Unsteady Flow and Its Application to the Stochastic Burgers Equation," *J. Fluid Mech.*, 253, pp. 509–543.

Coller, B.D., Holmes, P., and Lumley, J.L. (1994a) "Control of Bursting in Boundary Layer Models," *Appl. Mech. Rev.*, 47, no. 6, part 2, pp. S139–S143.

Coller, B.D., Holmes, P., and Lumley, J.L. (1994b) "Control of Noisy Heteroclinic Cycles," *Physica D*, 72, pp. 135–160.

Corino, E.R., and Brodkey, R.S. (1969) "A Visual Investigation of the Wall Region in Turbulent Flow," *J. Fluid Mech.*, 37, pp. 1–30.

Corke, T.C., Glauser, M.N., and Berkooz, G. (1994) "Utilizing Low-Dimensional Dynamical Systems Models to Guide Control Experiments," *Appl. Mech. Rev.*, 47, no. 6, part 2, pp. S132–S138.

Corrsin, S. (1957) "Some Current Problems in Turbulent Shear Flow," in *Symp. on Naval Hydrodynamics*, F.S. Sherman, ed., pp. 373–400, National Academy of Sciences/National Research Council Publication No. 515, Washington, D.C.

Corrsin, S. (1959) "Outline of Some Topics in Homogenous Turbulent Flow," *J. Geophys. Res.*, 64, pp. 2134–2150.

Davis, L., ed. (1991) *Handbook of Genetic Algorithms*, Van Nostrand Reinhold, New York.

Deane, A.E., and Sirovich, L. (1991) "A Computational Study of Rayleigh–Bénard Convection. Part 1. Rayleigh-Number Scaling," *J. Fluid Mech.*, 222, pp. 231–250.

DeGaspari, J. (2003) "Mixing It Up," *Mech. Eng.*, 125, August, pp. 34–38.

Delville, J., Cordier L., Bonnet J.-P. (1998) "Large-Scale-Structure Identification and Control in Turbulent Shear Flows," in *Flow Control: Fundamentals and Practices*, M. Gad-el-Hak, A. Pollard, and J.-P. Bonnet, eds., pp. 199–273, Springer-Verlag, Berlin.

Ditto, W.L., and Pecora, L.M. (1993) "Mastering Chaos," *Sci. Am.*, 269, August, pp. 78–84.

Ditto, W.L., Rauseo, S.N., and Spano, M.L. (1990) "Experimental Control of Chaos," *Phys. Rev. Lett.*, 65, pp. 3211–3214.

Donohue, G.L., Tiederman, W.G., and Reischman, M.M. (1972) "Flow Visualization of the Near-Wall Region in a Drag-Reducing Channel Flow," *J. Fluid Mech.*, 56, pp. 559–575.

Ehrenman, G. (2004) "Shrinking the Lab Down to Size," *Mech. Eng.*, 126, May, pp. 26–29.

Eléna, M. (1975) "Etude des Champs Dynamiques et Thermiques d'un Ecoulement Turbulent en Conduit avec Aspiration á la Paroi," Thése de Doctorat des Sciences, Université d'Aix-Marseille, Marseille, France.

Eléna, M. (1984) "Suction Effects on Turbulence Statistics in a Heated Pipe Flow," *Phys. Fluids*, 27, pp. 861–866.

Epstein, A.H. (2000) "The Inevitability of Small," *Aerosp. Am.*, 38, March, pp. 30–37.

Falco, R.E. (1974) "Some Comments on Turbulent Boundary Layer Structure Inferred from the Movements of a Passive Contaminant," AIAA Paper No. 74-99, New York.

Falco, R.E. (1977) "Coherent Motions in the Outer Region of Turbulent Boundary Layers," *Phys. Fluids*, 20, no. 10, part II, pp. S124–S132.

Falco, R.E. (1980) "The Production of Turbulence Near a Wall," AIAA Paper No. 80-l356, New York.

Falco, R.E. (1983) "New Results, a Review and Synthesis of the Mechanism of Turbulence Production in Boundary Layers and Its Modification," AIAA Paper No. 83-0377, New York.

Falco, R.E. (1991) "A Coherent Structure Model of the Turbulent Boundary Layer and Its Ability to Predict Reynolds Number Dependence," *Phil. Trans. R. Soc. London A*, 336, pp. 103–129.

Faller, W.E., Schreck, S.J., and Luttges, M.W. (1994) "Real-Time Prediction and Control of Three-Dimensional Unsteady Separated Flow Fields Using Neural Networks," AIAA Paper No. 94-0532, Washington, D.C.

Fan, X., Hofmann, L., and Herbert, T. (1993) "Active Flow Control with Neural Networks," AIAA Paper No. 93-3273, Washington, D.C.

Favre, A., Dumas, R., Verollet, E., and Coantic, M. (1966) "Couche Limite Turbulente sur Paroi Poreuse avec Aspiration," *J. Méc.*, 5, pp. 3–28.

Fiedler, H.E. (1986) "Coherent Structures," in *Advances in Turbulence*, G. Comte-Bellot, and J. Mathieu, eds., pp. 320–336, Springer-Verlag, Berlin.

Fiedler, H.E. (1988) "Coherent Structures in Turbulent Flows," *Prog. Aerosp. Sci.*, 25, pp. 231–269.

Fiedler, H.E., and Fernholz, H.-H. (1990) "On Management and Control of Turbulent Shear Flows," *Prog. Aerosp. Sci.*, 27, pp. 305–387.

Fiedler, H.E., Glezer, A., and Wygnanski, I. (1988) "Control of Plane Mixing Layer: Some Novel Experiments," in *Current Trends in Turbulence Research*, H. Branover, M. Mond, and Y. Unger, eds., pp. 30–64, AIAA, Washington, D.C.

Fowler, T.B. (1989) "Application of Stochastic Control Techniques to Chaotic Nonlinear Systems," *IEEE Trans. Autom. Control*, 34, pp. 201–205.

Gabriel, K.J. (1995) "Engineering Microscopic Machines," *Sci. Am.*, 260, September, pp. 150–153.

Gabriel, K.J., Jarvis, J., and Trimmer, W., eds. (1988) *Small Machines, Large Opportunities: A Report on the Emerging Field of Microdynamics, National Science Foundation*, published by AT&T Bell Laboratories, Murray Hill, New Jersey.

Gabriel, K.J., Tabata, O., Shimaoka, K., Sugiyama, S., and Fujita, H. (1992) "Surface-Normal Electrostatic/Pneumatic Actuator," in *Proc. IEEE Micro Electro Mechanical Systems '92*, pp. 128–131, 4–7 February, Travemünde, Germany.

Gad-el-Hak, M. (1989) "Flow Control," *Appl. Mech. Rev.*, 42, pp. 261–293.

Gad-el-Hak, M. (1993) "Innovative Control of Turbulent Flows," AIAA Paper No. 93-3268, Washington, D.C.

Gad-el-Hak, M. (1994) "Interactive Control of Turbulent Boundary Layers: A Futuristic Overview," *AIAA J.*, 32, pp. 1753–1765.

Gad-el-Hak, M. (1996) "Modern Developments in Flow Control," *Appl. Mech. Rev.*, 49, pp. 365–379.

Gad-el-Hak, M. (1998) "Frontiers of Flow Control," in *Flow Control: Fundamentals and Practices*, M. Gad-el-Hak, A. Pollard, and J.-P. Bonnet, eds., pp. 109–153, Springer-Verlag, Berlin.

Gad-el-Hak, M. (2000) *Flow Control: Passive, Active, and Reactive Flow Management*, Cambridge University Press, London, United Kingdom.

Gad-el-Hak, M., and Bandyopadhyay, P.R. (1994) "Reynolds Number Effects in Wall-Bounded Flows," *Appl. Mech. Rev.*, **47**, pp. 307–365.

Gad-el-Hak, M., and Blackwelder, R.F. (1987a) "Simulation of Large-Eddy Structures in a Turbulent Boundary Layer," *AIAA J.*, **25**, pp. 1207–1215.

Gad-el-Hak, M., and Blackwelder, R.F. (1987b) "A Drag Reduction Method for Turbulent Boundary Layers," AIAA Paper No. 87-0358, New York.

Gad-el-Hak, M., and Blackwelder, R.F. (1989) "Selective Suction for Controlling Bursting Events in a Boundary Layer," *AIAA J.*, **27**, pp. 308–314.

Gad-el-Hak, M., and Bushnell, D.M. (1991a) "Status and Outlook of Flow Separation Control," AIAA Paper No. 91-0037, New York.

Gad-el-Hak, M., and Bushnell, D.M. (1991b) "Separation Control: Review," *J. Fluids Eng.*, **113**, pp. 5–30.

Gad-el-Hak, M., and Hussain, A.K.M.F. (1986) "Coherent Structures in a Turbulent Boundary Layer. Part 1. Generation of 'Artificial' Bursts," *Phys. Fluids*, **29**, pp. 2124–2139.

Gad-el-Hak, M., and Tsai, H.M. (2005) *Transition and Turbulence Control*, World Scientific Publishing, Singapore.

Gad-el-Hak, M., Blackwelder, R.F., and Riley, J.J. (1981) "On the Growth of Turbulent Regions in Laminar Boundary Layers," *J. Fluid Mech.*, **110**, pp. 73–95.

Gad-el-Hak, M., Pollard, A., Bonnet, J.-P. (1998) *Flow Control: Fundamentals and Practices*, Springer-Verlag, Berlin.

Garcia, E.J., and Sniegowski, J.J. (1993) "The Design and Modelling of a Comb-Drive-Based Microengine for Mechanism Drive Applications," in *Proc. Seventh International Conference on Solid-State Sensors and Actuators (Transducers '93)*, pp. 763–766, Yokohama, Japan, 7–10 June.

Garcia, E.J., and Sniegowski, J.J. (1995) "Surface Micromachined Microengine," *Sensors Actuators A*, **48**, pp. 203–214.

Garfinkel, A., Spano, M.L., Ditto, W.L., and Weiss, J.N. (1992) "Controlling Cardiac Chaos," *Science*, **257**, pp. 1230–1235.

Gerbeth, G., ed. (1997) *Proceedings of the International Workshop on Electromagnetic Boundary Layer Control (EBLC) for Saltwater Flows*, 7–8 July, Forschungszentrum Rossendorf, Dresden, Germany.

Goldberg, D.E. (1989) *Genetic Algorithms in Search, Optimization, and Machine Learning*, Addison-Wesley, Reading, Massachusetts.

Goldin, D.S., Venneri, S.L., and Noor, A.K. (2000) "The Great out of the Small," *Mech. Eng.*, **122**, November, pp. 70–79.

Grappin, R., and Léorat, J. (1991) "Lyapunov Exponents and the Dimension of Periodic Incompressible Navier–Stokes Flows: Numerical Measurements," *J. Fluid Mech.*, **222**, pp. 61–94.

Gravesen, P., Branebjerg, J., and Jensen, O.S. (1993) "Microfluidics — A Review," *J. Micromech. Microeng.*, **3**, pp. 168–182.

Gutmark, E.J., and Grinstein, F.F. (1999) "Flow Control with Noncircular Jets," *Annu. Rev. Fluid Mech.*, **31**, pp. 239–272.

Gutmark, E.J., and Ho, C.-M. (1986) "Visualization of a Forced Elliptical Jet," *AIAA J.*, **24**, pp. 684–685.

Gutmark, E.J., Schadow, K.C., and Yu, K.H. (1995) "Mixing Enhancement in Supersonic Free Shear Flows," *Annu. Rev. Fluid Mech.*, **27**, pp. 375–417.

Halsey, T.C., and Martin, J.E. (1993) "Electrorheological Fluids," *Sci. Am.* **269**, October, pp. 58–64.

Haritonidis, J.H., Senturia, S.D., Warkentin, D.J., and Mehregany, M. (1990a) "Optical Micropressure Transducer," U.S. Patent number 4,926,696.

Haritonidis, J.H., Senturia, S.D., Warkentin, D.J., and Mehregany, M. (1990b) "Pressure Transducer Apparatus," U.S. Patent number 4,942,767.

Hayes, S., Grebogi, C., and Ott, E. (1994a) "Communicating with Chaos," *Phys. Rev. Lett.*, **70**, pp. 3031–3040.

Hayes, S., Grebogi, C., Ott, E., and Mark, A. (1994b) "Experimental Control of Chaos for Communication," *Phys. Rev. Lett.*, **73**, pp. 1781–1784.

Hinze, J.O. (1975) *Turbulence*, second edition, McGraw-Hill, New York.

Ho, C.-M., and Tai, Y.-C. (1996) "Review: MEMS and Its Applications for Flow Control," *J. Fluids Eng.*, **118**, pp. 437–447.

Ho, C.-M., and Tai, Y.-C. (1998) "Micro-Electro-Mechanical Systems (MEMS) and Fluid Flows," *Annu. Rev. Fluid Mech.*, **30**, pp. 579–612.

Hogan, H. (1996) "Invasion of the Micromachines," *New Scientist*, **29**, June, pp. 28–33.

Holland, J.H. (1992) *Adaptation in Natural and Artificial Systems*, MIT Press, Cambridge, Massachusetts.

Holmes, P., Lumley, J.L., and Berkooz, G. (1996) *Turbulence, Coherent Structures, Dynamical Systems and Symmetry*, Cambridge University Press, Cambridge, Great Britain.

Hu, H.H., and Bau, H.H. (1994) "Feedback Control to Delay or Advance Linear Loss of Stability in Planar Poiseuille Flow," in *Proc. R. Soc. London A*, **447**, pp. 299–312.

Huberman, B. (1990) "The Control of Chaos," *Proc. Workshop on Applications of Chaos*, 4–7 December, San Francisco.

Huberman, B.A., and Lumer, E. (1990) "Dynamics of Adaptive Systems," *IEEE Trans. Circuits Syst.*, **37**, pp. 547–550.

Hübler, A., and Lüscher, E. (1989) "Resonant Stimulation and Control of Nonlinear Oscillators," *Naturwissenschaften*, **76**, pp. 67–69.

Huerre, P., and Monkewitz, P.A. (1990) "Local and Global Instabilities in Spatially Developing Flows," *Annu. Rev. Fluid Mech.*, **22**, pp. 473–537.

Hussain, A.K.M.F. (1986) "Coherent Structures and Turbulence," *J. Fluid Mech.*, **173**, pp. 303–356.

Iglisch, R. (1944) "Exakte Berechnung der laminaren Reibungsschicht an der längsangeströmten ebenen Platte mit homogener Absaugung," *Schriften Dtsch. Akad. Luftfahrtforschung B*, **8**, pp. 1–51.

Jacobs, J., James, R., Ratliff, C., and Glazer, A. (1993) "Turbulent Jets Induced by Surface Actuators," AIAA Paper No. 93-3243, Washington, D.C.

Jacobson, S.A., and Reynolds, W.C. (1993a) "Active Control of Boundary Layer Wall Shear Stress Using Self-Learning Neural Networks," AIAA Paper No. 93-3272, AIAA, Washington, D.C.

Jacobson, S.A., and Reynolds W.C. (1993b) "Active Boundary Layer Control Using Flush-Mounted Surface Actuators," *Bul. Am. Phys. Soc.*, **38**, p. 2197.

Jacobson, S.A., and Reynolds, W.C. (1994) "Active Control of Transition and Drag in Boundary Layers," *Bul. Am. Phys. Soc.*, **39**, p. 1894.

Jacobson, S.A., and Reynolds, W.C. (1995) "An Experimental Investigation Towards the Active Control of Turbulent Boundary Layers," Department of Mechanical Engineering Report No. TF-64, Stanford University, Stanford, California.

Jacobson, S.A., and Reynolds, W.C. (1998) "Active Control of Streamwise Vortices and Streaks in Boundary Layers," *J. Fluid Mech.*, **360**, pp. 179–211.

James, R.D., Jacobs, J.W., and Glezer, A. (1994) "Experimental Investigation of a Turbulent Jet Produced by an Oscillating Surface Actuator," *Appl. Mech. Rev.*, **47**, no. 6, part 2, pp. S127–S1131.

Jang, J.-S.R., Sun, C.-T., and Mizutani, E. (1997) *Neuro-Fuzzy and Soft Computing*, Prentice Hall, Upper Saddle River, New Jersey.

Johansen, J.B., and Smith, C.R. (1986) "The Effects of Cylindrical Surface Modifications on Turbulent Boundary Layers," *AIAA J.*, **24**, pp. 1081–1087.

Joslin, R.D., Erlebacher, G., Hussaini, M.Y. (1996) "Active Control of Instabilities in Laminar Boundary Layers — Overview and Concept Validation," *J. Fluids Eng.*, **118**, pp. 494–497.

Karniadakis, G.E., and Beskok A. (2002) *Microflows: Fundamentals and Simulation*, Springer-Verlag, New York.

Kawthar-Ali, M.H., and Acharya, M. (1996) "Artificial Neural Networks for Suppression of the Dynamic-Stall Vortex over Pitching Airfoils," AIAA Paper No. 96-0540, Washington, D.C.

Keefe, L.R. (1993a) "Two Nonlinear Control Schemes Contrasted in a Hydrodynamic Model," *Phys. Fluids A*, **5**, pp. 931–947.

Keefe, L.R. (1993b) "Drag Reduction in Channel Flow Using Nonlinear Control," AIAA Paper No. 93-3279, Washington, D.C.

Keefe, L.R. (1996) "A MEMS-Based Normal Vorticity Actuator for Near-Wall Modification of Turbulent Shear Flows," *Proc. Workshop on Flow Control: Fundamentals and Practices*, J.-P. Bonnet,

M. Gad-el-Hak, and A. Pollard, eds., pp. 1–21, 1–5 July, Institut d'Etudes Scientifiques des Cargèse, Corsica, France.

Keefe, L.R., Moin, P., and Kim, J. (1992) "The Dimension of Attractors Underlying Periodic Turbulent Poiseuille Flow," *J. Fluid Mech.*, **242**, pp. 1–29.

Keith, W.L., Hurdis, D.A., and Abraham, B.M. (1992) "A Comparison of Turbulent Boundary Layer Wall-Pressure Spectra," *J. Fluids Eng.*, **114**, pp. 338–347.

Kim, H.T., Kline, S.J., and Reynolds, W.C. (1971) "The Production of Turbulence Near a Smooth Wall in a Turbulent Boundary Layer," *J. Fluid Mech.*, **50**, pp. 133–160.

Kim, J., Moin, P., and Moser, R.D. (1987) "Turbulence Statistics in Fully-Developed Channel Flow at Low Reynolds Number," *J. Fluid Mech.*, **177**, pp. 133–166.

Klewicki, J.C., Murray, J.A., and Falco, R.E. (1994) "Vortical Motion Contributions to Stress Transport in Turbulent Boundary Layers," *Phys. Fluids*, **6**, pp. 277–286.

Kline, S.J., and Runstadler, P.W. (1959) "Some Preliminary Results of Visual Studies of the Flow Model of the Wall Layers of the Turbulent Boundary Layer," *J. Appl. Mech.*, **26**, pp. 166–170.

Kline, S.J., Reynolds, W.C., Schraub, F.A., and Runstadler, P.W. (1967) "The Structure of Turbulent Boundary Layers," *J. Fluid Mech.*, **30**, pp. 741–773.

Knight, J. (1999) "Dust Mite's Dilemma," *New Scientist*, **162**, no. 2180, 29 May, pp. 40–43.

Kolmogorov, A.N. (1941a) "The Local Structure of Turbulence in Incompressible Viscous Fluid for Very Large Reynolds Number," *Dokl. Akad. Nauk SSSR* **30**, pp. 301–305. (Reprinted in *Proc. R. Soc. London A*, **434**, pp. 9–13, 1991.)

Kolmogorov, A.N. (1941b) "On Degeneration of Isotropic Turbulence in an Incompressible Viscous Liquid," *Dokl. Akad. Nauk. SSSR*, **31**, pp. 538–540.

Kostelich, E.J., Grebogi, C., Ott, E., and Yorke, J.A. (1993a) "Targeting from Time Series," *Bul. Am. Phys. Soc.*, **38**, p. 2194.

Kostelich, E.J., Grebogi, C., Ott, E., and Yorke, J.A. (1993b) "Higher-Dimensional Targeting," *Phys. Rev. E*, **47**, pp. 305–310.

Kovacs, G.T.A. (1998) *Micromachined Transducers Sourcebook*, McGraw-Hill, New York.

Kovasznay, L.S.G. (1970) "The Turbulent Boundary Layer," *Annu. Rev. Fluid Mech.*, **2**, pp. 95–112.

Kovasznay, L.S.G., Kibens, V., and Blackwelder, R.F. (1970) "Large-Scale Motion in the Intermittent Region of a Turbulent Boundary Layer," *J. Fluid Mech.*, **41**, pp. 283–325.

Kwong, A., and Dowling, A. (1993) "Active Boundary Layer Control in Diffusers," AIAA Paper No. 93-3255, Washington, D.C.

Lachmann, G.V., ed. (1961) *Boundary Layer and Flow Control*, vols. 1 and 2, Pergamon Press, Oxford, Great Britain.

Lai, Y.-C., and Grebogi, C. (1993) "Synchronization of Chaotic Trajectories Using Control," *Phys. Rev. E*, **47**, pp. 2357–2360.

Lai, Y.-C., Deng, M., and Grebogi, C. (1993a) "Controlling Hamiltonian Chaos," *Phys. Rev. E*, **47**, pp. 86–92.

Lai, Y.-C., Grebogi, C., and Tél, T. (1994) "Controlling Transient Chaos in Dynamical Systems," in *Towards the Harnessing of Chaos*, M. Yamaguchi, ed., Elsevier, Amsterdam, the Netherlands.

Lai, Y.-C., Tél, T., and Grebogi, C. (1993b) "Stabilizing Chaotic-Scattering Trajectories Using Control," *Phys. Rev. E*, **48**, pp. 709–717.

Laufer, J. (1975) "New Trends in Experimental Turbulence Research," *Annu. Rev. Fluid Mech.*, **7**, pp. 307–326.

Liepmann, H.W., and Nosenchuck, D.M. (1982) "Active Control of Laminar–Turbulent Transition," *J. Fluid Mech.*, **118**, pp. 201–204.

Liepmann, H.W., Brown, G.L., and Nosenchuck, D.M. (1982) "Control of Laminar Instability Waves Using a New Technique," *J. Fluid Mech.*, **118**, pp. 187–200.

Ligrani, P.M., and Bradshaw, P. (1987) "Spatial Resolution and Measurements of Turbulence in the Viscous Sublayer Using Subminiature Hot-Wire Probes," *Exp. Fluids*, **5**, pp. 407–417.

Lindner, J.F., and Ditto, W.L. (1995) "Removal, Suppression and Control of Chaos by Nonlinear Design," *Appl. Mech. Rev.*, **48**, pp. 795–808.

Lipkin, R. (1993) "Micro Steam Engine Makes Forceful Debut," *Science News* **144**, September, p. 197.

Löfdahl, L., and Gad-el-Hak, M. (1999) "MEMS Applications in Turbulence and Flow Control," *Prog. Aerosp. Sci.*, **35**, pp. 101–203.

Löfdahl, L., Glavmo, M., Johansson, B., and Stemme, G. (1993) "A Silicon Transducer for the Determination of Wall-Pressure Fluctuations in Turbulent Boundary Layers," *Appl. Sci. Res.*, **51**, pp. 203–207.

Löfdahl, L., Kälvesten, E., and Stemme, G. (1994) "Small Silicon Based Pressure Transducers for Measurements in Turbulent Boundary Layers," *Exp. Fluids*, **17**, pp. 24–31.

Löfdahl, L., Kälvesten, E., and Stemme, G. (1996) "Small Silicon Pressure Transducers for Space-Time Correlation Measurements in a Flat Plate Boundary Layer," *J. Fluids Eng.*, **118**, pp. 457–463.

Lumley, J.L. (1991) "Control of the Wall Region of a Turbulent Boundary Layer," in *Turbulence: Structure and Control*, J.M. McMichael, ed., pp. 61–62, 1–3 April, Ohio State University, Columbus, Ohio.

Lumley, J.L. (1996) "Control of Turbulence," AIAA Paper No. 96-0001, Washington, D.C.

Lüscher, E., and Hübler, A. (1989) "Resonant Stimulation of Complex Systems," *Helv. Phys. Acta*, **62**, pp. 544–551.

Madou, M. (2002) *Fundamentals of Microfabrication*, 2nd edition, CRC Press, Boca Raton, Florida.

McMichael, J.M. (1996) "Progress and Prospects for Active Flow Control Using Microfabricated Electromechanical Systems (MEMS)," AIAA Paper No. 96-0306, Washington, D.C.

Mehregany, M., DeAnna, R.G., and Reshotko, E. (1996) "Microelectromechanical Systems for Aerodynamics Applications," AIAA Paper No. 96-0421, Washington, D.C.

Meng, J.C.S., ed. (1998) *Proceedings of the International Symposium on Sea Water Drag Reduction*, 22–23 July, Naval Undersea Warfare Center, Newport, Rhode Island.

Moin, P., and Bewley, T. (1994) "Feedback Control of Turbulence," *Appl. Mech. Rev.*, **47**, no. 6, part 2, pp. S3–S13.

Narasimha, R., and Sreenivasan, K.R. (1979) "Relaminarization of Fluid Flows," in *Advances in Applied Mechanics*, vol. 19, C.-S. Yih, ed., pp. 221–309, Academic Press, New York.

Nelson, M.M., and Illingworth, W.T. (1991) *A Practical Guide to Neural Nets*, Addison-Wesley, Reading, Massachusetts.

Nguyen, N.-T., and Wereley, S.T. (2002) *Fundamentals and Applications of Microfluidics*, Artech House, Norwood, Massachusetts.

Noor, A., and Jorgensen, C.C. (1996) "A Hard Look at Soft Computing," *Aerosp. Am.*, **34**, September, pp. 34–39.

Nosenchuck, D.M., and Lynch, M.K. (1985) "The Control of Low-Speed Streak Bursting in Turbulent Spots," AIAA Paper No. 85-0535, New York.

O'Connor, L. (1992) "MEMS: Micromechanical Systems," *Mech. Eng.*, **114**, February, pp. 40–47.

O'Connor, L., and Hutchinson, H. (2000) "Skyscrapers in a Microworld," *Mech. Eng.*, **122**, March, pp. 64–67.

Offen, G.R., and Kline, S.J. (1974) "Combined Dye-Streak and Hydrogen-Bubble Visual Observations of a Turbulent Boundary Layer," *J. Fluid Mech.*, **62**, pp. 223–239.

Offen, G.R., and Kline, S.J. (1975) "A Proposed Model of the Bursting Process in Turbulent Boundary Layers," *J. Fluid Mech.*, **70**, pp. 209–228.

Oldaker, D.K., and Tiederman, W.G. (1977) "Spatial Structure of the Viscous Sublayer in Drag-Reducing Channel Flows," *Phys. Fluids*, **20**, no. 10, part II, pp. S133–144.

Ott, E., Grebogi, C., and Yorke, J.A. (1990a) "Controlling Chaos," *Phys. Rev. Lett.*, **64**, pp. 1196–1199.

Ott, E., Grebogi, C., and Yorke, J.A. (1990b) "Controlling Chaotic Dynamical Systems," in *Chaos: Soviet–American Perspectives on Nonlinear Science*, D.K. Campbell, ed., pp. 153–172, American Institute of Physics, New York.

Ouellette, J. (1996) "MEMS: Mega Promise for Micro Devices," *Mech. Eng.*, **118**, October, pp. 64–68.

Ouellette, J. (1999) "Electronic Noses Sniff Out New Markets," *Ind. Physicist*, **5**, no. 1, pp. 26–29.

Ouellette, J. (2003) "A New Wave of Microfluidic Devices," *Ind. Physicist*, **9**, no. 4, pp. 14–17.

Panton, R.L., ed. (1997) *Self-Sustaining Mechanisms of Wall Turbulence*, Computational Mechanics Publications, Southampton, Great Britain.

Paula, G. (1996) "MEMS Sensors Branch Out," *Aerosp. Am.*, **34**, September, pp. 26–32.

Perrier, P. (1998) "Multiscale Active Flow Control," in *Flow Control: Fundamentals and Practices*, M. Gad-el-Hak, A. Pollard, and J.-P. Bonnet, eds., pp. 275–334, Springer-Verlag, Berlin.

Petersen, I.R., and Savkin, A.V. (1999) *Robust Kalman Filtering for Signals and Systems with Large Uncertainties*, Birkhäuser, Boston.

Pomeau, Y., and Manneville, P. (1980) "Intermittent Transition to Turbulence in Dissipative Dynamical Systems," *Commun. Math. Phys.*, 74, pp. 189–197.

Prandtl, L. (1904) "Über Flüssigkeitsbewegung bei sehr kleiner Reibung," *Proc. Third Int. Math. Cong.*, pp. 484–491, Heidelberg, Germany.

Pretsch, J. (1942) "Umschlagbeginn und Absaugung," *Jahrb. Dtsch. Luftfahrtforschung*, 1, pp. 54–71.

Qin, F., Wolf, E.E., and Chang, H.-C. (1994) "Controlling Spatiotemporal Patterns on a Catalytic Wafer," *Phys. Rev. Lett.*, 72, pp. 1459–1462.

Reischman, M.M., and Tiederman, W.G. (1975) "Laser-Doppler Anemometer Measurements in Drag-Reducing Channel Flows," *J. Fluid Mech.*, 70, pp. 369–392.

Reynolds, W.C. (1993) "Sensors, Actuators, and Strategies for Turbulent Shear-Flow Control," invited oral presentation at *AIAA Third Flow Control Conference*, 6–9 July, Orlando, Florida.

Riley, J.J., and Gad-el-Hak, M. (1985) "The Dynamics of Turbulent Spots," in *Frontiers in Fluid Mechanics*, S.H. Davis, and J.L. Lumley, eds., pp. 123–155, Springer-Verlag, Berlin.

Robinson, E.Y., Helvajian, H., and Jansen, S.W. (1996a) "Small and Smaller: The World of MNT," *Aerosp. Am.*, 34, September, pp. 26–32.

Robinson, E.Y., Helvajian, H., and Jansen, S.W. (1996b) "Big Benefits from Tiny Technologies," *Aerosp. Am.*, 34, October, pp. 38–43.

Robinson, S.K. (1991) "Coherent Motions in the Turbulent Boundary Layer," *Annu. Rev. Fluid Mech.*, 23, pp. 601–639.

Robinson, S.K., Kline, S.J., and Spalart, P.R. (1989) "A Review of Quasi-Coherent Structures in a Numerically Simulated Turbulent Boundary Layer," NASA Technical Memorandum No. TM-102191, Washington, D.C.

Romeiras, F.J., Grebogi, C., Ott, E., and Dayawansa, W.P. (1992) "Controlling Chaotic Dynamical Systems," *Physica D*, 58, pp. 165–192.

Rotta, J.C. (1970) "Control of Turbulent Boundary Layers by Uniform Injection and Suction of Fluid," in *Seventh Cong. of the International Council of the Aeronautical Sciences*, paper no. 70-10, ICAS, Rome, Italy.

Runstadler, P.G., Kline, S.J., and Reynolds, W.C. (1963) "An Experimental Investigation of Flow Structure of the Turbulent Boundary Layer," Department of Mechanical Engineering Report No. MD-8, Stanford University, Stanford, California.

Saffman, P.G. (1978) "Problems and Progress in the Theory of Turbulence," in *Structure and Mechanisms of Turbulence II*, H. Fiedler, ed., pp. 273–306, Springer-Verlag, Berlin.

Saffman, P.G., and Baker, G.R. (1979) "Vortex Interactions," *Annu. Rev. Fluid Mech.*, 11, pp. 95–122.

Schreck, S.J., Faller, W.E., and Luttges, M.W. (1995) "Neural Network Prediction of Three-Dimensional Unsteady Separated Flow Fields," *J. Aircraft*, 32, pp. 178–185.

Sen, M., Wajerski, D., and Gad-el-Hak, M. (1996) "A Novel Pump for MEMS Applications," *J. Fluids Eng.*, 118, pp. 624–627.

Sharatchandra, M.C., Sen, M., and Gad-el-Hak, M. (1997) "Navier–Stokes Simulations of a Novel Viscous Pump," *J. Fluids Eng.*, 119, pp. 372–382.

Sharke, P. (2004) "Water, Paper, Glass," *Mech. Eng.*, 126, May, pp. 30–32.

Shercliff, J.A. (1965) *A Textbook of Magnetohydrodynamics*, Pergamon Press, Oxford, Great Britain.

Shinbrot, T. (1993) "Chaos: Unpredictable Yet Controllable?" *Nonlinear Sci. Today*, 3, pp. 1–8.

Shinbrot, T. (1995) "Progress in the Control of Chaos," *Adv. Phys.*, 44, pp. 73–111.

Shinbrot, T. (1998) "Chaos, Coherence and Control," in *Flow Control: Fundamentals and Practices*, M. Gad-el-Hak, A. Pollard, and J.-P. Bonnet, eds., pp. 501–527, Springer-Verlag, Berlin.

Shinbrot, T., and Ottino, J.M. (1993a) "Geometric Method to Create Coherent Structures in Chaotic Flows," *Phys. Rev. Lett.*, 71, pp. 843–846.

Shinbrot, T., and Ottino, J.M. (1993b) "Using Horseshoes to Create Coherent Structures in Chaotic Fluid Flows," *Bul. Am. Phys. Soc.*, **38**, p. 2194.

Shinbrot, T., Bresler, L., and Ottino, J.M. (1998) "Manipulation of Isolated Structures in Experimental Chaotic Fluid Flows," *Exp. Therm. Fluid Sci.*, **16**, pp. 76–83.

Shinbrot, T., Ditto, W., Grebogi, C., Ott, E., Spano, M., and Yorke, J.A. (1992a) "Using the Sensitive Dependence of Chaos (the "Butterfly Effect") to Direct Trajectories in an Experimental Chaotic System," *Phys. Rev. Lett.*, **68**, pp. 2863–2866.

Shinbrot, T., Grebogi, C., Ott, E., and Yorke, J.A. (1992b) "Using Chaos to Target Stationary States of Flows," *Phys. Lett. A*, **169**, pp. 349–354.

Shinbrot, T., Grebogi, C., Ott, E., and Yorke, J.A. (1993) "Using Small Perturbations to Control Chaos," *Nature*, **363**, pp. 411–417.

Shinbrot, T., Ott, E., Grebogi, C., and Yorke, J.A. (1990) "Using Chaos to Direct Trajectories to Targets," *Phys. Rev. Lett.*, **65**, pp. 3215–3218.

Shinbrot, T., Ott, E., Grebogi, C., and Yorke, J.A. (1992c) "Using Chaos to Direct Orbits to Targets in Systems Describable by a One-Dimensional Map," *Phys. Rev. A*, **45**, pp. 4165–4168.

Singer, J., Wang, Y.-Z., and Bau, H.H. (1991) "Controlling a Chaotic System," *Phys. Rev. Lett.*, **66**, pp. 1123–1125.

Sirovich, L., and Deane, A.E. (1991) "A Computational Study of Rayleigh-Bénard Convection. Part 2. Dimension Considerations," *J. Fluid Mech.*, **222**, pp. 251–265.

Smith, C.R., and Metzler, S.P. (1982) "A Visual Study of the Characteristics, Formation, and Regeneration of Turbulent Boundary Layer Streaks," in *Developments in Theoretical and Applied Mechanics*, vol. XI, T.J. Chung and G.R. Karr, eds., pp. 533–543, University of Alabama, Huntsville, Alabama.

Smith, C.R., and Metzler, S.P. (1983) "The Characteristics of Low-Speed Streaks in the Near-Wall Region of a Turbulent Boundary Layer," *J. Fluid Mech.*, **129**, pp. 27–54.

Smith, C.R., and Schwartz, S.P. (1983) "Observation of Streamwise Rotation in the Near-Wall Region of a Turbulent Boundary Layer," *Phys. Fluids*, **26**, pp. 641–652.

Sniegowski, J.J., and Garcia, E.J. (1996) "Surface Micromachined Gear Trains Driven by an On-Chip Electrostatic Microengine," *IEEE Electron Device Lett.*, **17**, July, p. 366.

Sreenivasan, K.R. (1989) "The Turbulent Boundary Layer," in *Frontiers in Experimental Fluid Mechanics*, M. Gad-el-Hak, ed., pp. 159–29, Springer-Verlag, New York.

Sritharan, S.S., ed. (1998) *Optimal Control of Viscous Flow*, SIAM, Philadelphia.

Swearingen, J.D., and Blackwelder, R.F. (1984) "Instantaneous Streamwise Velocity Gradients in the Wall Region," *Bul. Am. Phys. Soc.*, **29**, p. 1528.

Tang, J., and Bau, H.H. (1993a) "Stabilization of the No-Motion State in Rayleigh–Bénard Convection through the Use of Feedback Control," *Phys. Rev. Lett.*, **70**, pp. 1795–1798.

Tang, J., and Bau, H.H. (1993b) "Feedback Control Stabilization of the No-Motion State of a Fluid Confined in a Horizontal Porous Layer Heated from Below," *J. Fluid Mech.*, **257**, pp. 485–505.

Tang, W.C., and Lee, A.P. (2001) "Military Applications of Microsystems," *The Ind. Physicist*, 7, February, pp. 26–29.

Taylor, G.I. (1935) "Statistical Theory of Turbulence," *Proc. R. Soci. London A*, **151**, pp. 421–478.

Tennekes, H., and Lumley, J.L. (1972) *A First Course in Turbulence*, MIT Press, Cambridge, Massachusetts.

Tiederman, W.G., Luchik, T.S., and Bogard, D.G. (1985) "Wall-Layer Structure and Drag Reduction," *J. Fluid Mech.*, **156**, pp. 419–437.

Tien, N.C. (1997) "Silicon Micromachined Thermal Sensors and Actuators," *Microscale Thermophys. Eng.*, **1**, pp. 275–292.

Townsend, A.A. (1961) "Equilibrium Layers and Wall Turbulence," *J. Fluid Mech.*, **11**, pp. 97–120.

Townsend, A.A. (1970) "Entrainment and the Structure of Turbulent Flow," *J. Fluid Mech.*, **41**, pp. 13–46.

Ulrich, A. (1944) "Theoretische Untersuchungenüber die Widerstandsersparnis durch Laminarhaltung mit Absaugung," *Schriften Dtsch. Akad. Luftfahrtforschung B*, **8**, p. 53.

Vargo, S.E., and Muntz, E.P. (1996) "A Simple Micromechanical Compressor and Vacuum Pump for Flow Control and Other Distributed Applications," AIAA Paper No. 96-0310, AIAA, Washington, D.C.

Verollet, E., Fulachier, L., Dumas, R., and Favre, A. (1972) "Turbulent Boundary Layer with Suction and Heating to the Wall," in *Heat and Mass Transfer in Boundary Layers*, N. Afgan, Z. Zaric and P. Anastasijevec, eds., vol. 1, pp. 157–168, Pergamon Presss, Oxford.

Viswanath, P.R. (1995) "Flow Management Techniques for Base and Afterbody Drag Reduction," *Prog. Aerosp. Sci.*, **32**, pp. 79–129.

Wadsworth, D.C., Muntz, E.P., Blackwelder, R.F., and Shiflett, G.R. (1993) "Transient Energy Release Pressure Driven Microactuators for Control of Wall-Bounded Turbulent Flows," AIAA Paper No. 93-3271, AIAA, Washington, D.C.

Wang, Y., Singer, J., and Bau, H.H. (1992) "Controlling Chaos in a Thermal Convection Loop," *J. Fluid Mech.*, **237**, pp. 479–498.

Warkentin, D.J., Haritonidis, J.H., Mehregany, M., and Senturia, S.D. (1987) "A Micromachined Microphone with Optical Interference Readout," in *Proc. Fourth Int. Conf. on Solid-State Sensors and Actuators (Transducers '87)*, June, Tokyo, Japan.

Wilkinson, S.P. (1988) "Direct Drag Measurements on Thin-Element Riblets with Suction and Blowing," AIAA Paper No. 88-3670-CP, Washington, D.C.

Wilkinson, S.P. (1990) "Interactive Wall Turbulence Control," in *Viscous Drag Reduction in Boundary Layers*, D.M. Bushnell and J.N. Hefner, eds., pp. 479–509, AIAA, Washington, D.C.

Wilkinson, S.P., and Balasubramanian, R. (1985) "Turbulent Burst Control through Phase-Locked Surface Depressions," AIAA Paper No. 85-0536, New York.

Wilkinson, S.P., and Lazos, B.S. (1987) "Direct Drag and Hot-Wire Measurements on Thin-Element Riblet Arrays," in *Turbulence Management and Relaminarization*, H.W. Liepmann and R. Narasimha, eds., pp. 121–131, Springer-Verlag, New York.

Wilkinson, S.P., Anders, J.B., Lazos, B.S., and Bushnell, D.M. (1988) "Turbulent Drag Reduction Research at NASA Langley: Progress and Plans," *Int. J. Heat Fluid Flow*, **9**, pp. 266–277.

Williams, D.R., and Amato, C.W. (1989) "Unsteady Pulsing of Cylinder Wakes," in *Frontiers in Experimental Fluid Mechanics*, M. Gad-el-Hak, ed., pp. 337–364, Springer-Verlag, New York.

Willmarth, W.W. (1975a) "Structure of Turbulence in Boundary Layers," *Adv. Appl. Mech.*, **15**, pp. 159–254.

Willmarth, W.W. (1975b) "Pressure Fluctuations beneath Turbulent Boundary Layers," *Annu. Rev. Fluid Mech.*, **7**, pp. 13–37.

Wiltse, J.M., and Glezer, A. (1993) "Manipulation of Free Shear Flows Using Piezoelectric Actuators," *J. Fluid Mech.*, **249**, pp. 261–285.

Yager, R.R., and Zadeh, L.A., eds. (1992) *An Introduction to Fuzzy Logic Applications in Intelligent Systems*, Kluwer Academic, Boston.

Young, A.M., Goldsberry, J.E., Haritonidis, J.H., Smith, R.I., and Senturia, S.D. (1988) "A Twin-Interferometer Fiber-Optic Readout for Diaphragm Pressure Transducers," in *IEEE Solid-State Sensor and Actuator Workshop*, 6–9 June, Hilton Head, South Carolina.

Zilberman, M., Wygnanski, I., and Kaplan, R.E. (1977) "Transitional Boundary Layer Spot in a Fully Turbulent Environment," *Phys. Fluids*, **20**, no. 10, part II, pp. S258–S271.

14

Reactive Control for Skin-Friction Reduction

Haecheon Choi
Seoul National University

14.1 Introduction

Recently, interest in how to control thermal and fluid phenomena has increased significantly due to an immense need for control technologies aimed at drag reduction, noise reduction, and mixing enhancement. In the case of large-scale vehicles, such as an airplane or oil tanker, the skin-friction drag is about 70 to 90% of the total drag. Therefore, reducing turbulent fluctuations to reduce skin-friction is one of the most important problems in fluid mechanics. On the other hand, a great benefit also can be obtained by increasing turbulence. For example, turbulence should be increased in order to mix air and fuel efficiently or to decrease the form drag of bluff bodies such as a golf ball.

In spite of its importance, research on flow control has been limited because most flows occurring in industrial machinery and transportation vehicles are turbulent. Before the1960s, turbulent flow was considered to be a completely irregular motion, but now we know that coherent structures exist in turbulent flow, which play an important role in momentum transport [Cantwell, 1981; Robinson, 1991]. One of the most important coherent structures in wall-bounded flows may be the near-wall streamwise vortex, which is responsible for the turbulence production and skin-friction increase near the wall [Kline et al., 1967; Kim et al., 1971; Kravchenko et al., 1993; Choi et al., 1994]. Therefore, attempts to control turbulent flows for engineering applications have been focused on the manipulation of these coherent structures.

Before the 1990s, most turbulence-control strategies for wall-bounded flows were based on passive approaches such as riblets or large-eddy breakup devices (LEBUs). Such devices are passive in that they have no feedback (or feedforward) loop to sense or manipulate flow structures; also, they require no energy input. So far, it is known that one of the most successful passive devices for skin-friction reduction is the riblet, which produces a maximum 8% drag reduction [Walsh, 1982]. However, there is a limit to reducing skin-friction drag by passive devices alone.

Since 1990, there have been several studies on the active feedback or feedforward (i.e., reactive) control of dynamically significant coherent structures to achieve skin-friction reduction using direct numerical simulation. The basic strategy of the most successful controls is to manipulate the coherent structures, especially the near-wall streamwise vortices. For example, in Choi et al. (1994), the blowing or suction velocity at the wall was given as exactly opposite to the wall-normal component of the velocity at $y^+ = yu_\tau/v = 10$, which is associated with the wall-normal velocity induced by the near-wall streamwise vortices, resulting in more than 25% drag reduction. Here, y is the wall-normal distance from the wall, u_τ is the wall-shear velocity, and v is the kinematic viscosity. However, sensing at $y^+ = 10$ is not an easy task, so controls based on sensing at the wall have been suggested in the literature [Lee et al., 1997, 1998; Koumoutsakos, 1999]. On the other hand, blowing or suction from a hole or slot is difficult in real life; for example, at high altitude, icing blocks the hole such that the actuation mechanism completely fails. Thus, wall movement instead of blowing and suction also has been suggested in the literature [Carlson and Lumley, 1996; Endo et al., 2000; Kang and Choi, 2000]. In spite of all these successful results, actual implementation of the reactive-control algorithm is still very difficult, mainly because the size of near-wall streamwise vortices is so small (submillimeter) and their occurrence is irregular in space and time. Therefore, detecting the vortices is very difficult, as is controlling them. Fortunately, the recent development of micromachining technology makes it possible to fabricate mechanical parts of micron size [Ho and Tai, 1996, 1998]. Such technology provides us with microsensors and microactuators that can be applied to reactive flow control.

Quite a few review papers on flow control exist in the literature. Therefore, we do not intend to cover all of the different aspects of flow control here but rather will limit our discussion to reactive flow controls based on sensing of coherent structures for skin-friction reduction. Those who are interested in the general aspects of flow control should refer to Bushnell and McGinley (1989), Moin and Bewley (1994), Gad-el-Hak (1994, 1996), Ho and Tai (1996, 1998), and Bewley (2001).

In this chapter, we will first discuss the near-wall streamwise vortices in a boundary layer and then review some successful reactive control strategies for skin-friction reduction using blowing and suction or active wall motion based on sensing of coherent structures. Finally, some issues on practical implementation will be reviewed in the last section.

14.2 Near-Wall Streamwise Vortices

The most well known structure in the turbulent boundary layer is the streaky structure as shown in Figure 14.1a. Here, the streaky structure denotes the alternating low- and high-speed fluid pattern in the spanwise direction. Kline et al. (1967) showed through flow visualization that turbulent production in the boundary layer is closely associated with lift-up, oscillation, and breakup of the streaky structures. Corino and Brodkey (1969) reported that turbulent production occurs locally when high-speed fluids approach the wall (called sweep). This streaky structure is closely related to the near-wall streamwise vortices that produce sweep and ejection and thus generate turbulence. Recently, Choi et al. (1994) and Kravchenko et al. (1993) showed that high skin friction on the wall has a high correlation with the near-wall streamwise vortices, and the sweep event of those vortices significantly increases the skin friction nearby (Figure 14.1b).

As a successful passive device, the riblets produce a maximum 8% drag reduction, in which viscous drag is reduced by restricting the location of streamwise vortices above the wetted surface such that only a limited area of the riblets is exposed to downwash of the high-speed fluid that the vortices induce [Choi et al., 1993a]. Therefore, it is logical to search for a reactive control method that is efficient and more effective in controlling the near-wall streamwise vortices to achieve greater skin-friction reduction than the riblets.

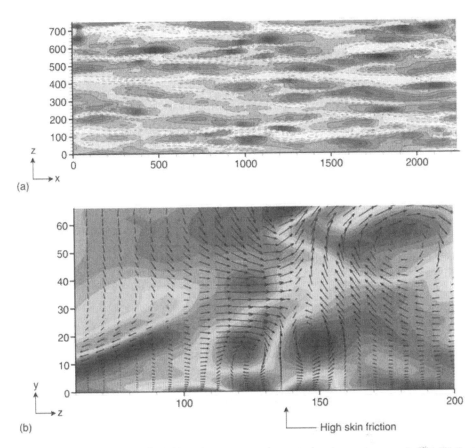

FIGURE 14.1   (a) Contours of the skin-friction fluctuations at the wall. Negative contours are dashed. (b) Contours of the streamwise vort icity and cross-flow vector (v, w) in a (y, z) plane. Contours in gray denote high magnitudes of the streamwise vorticity.

14.3   Opposition Control: Control Based on Physical Intuition

Gad-el-Hak and Blackwelder (1989) experimentally investigated the feasibility of removing some or all of the turbulence-producing eddy structures in a turbulent boundary layer by using selective suction relative to the spatial location of the eddy structure. They showed that the selective suction eliminates the low-speed streak and prevents bursting at much lower suction strength than that required by the conventional suction method, implying that selective suction may be a good control strategy for skin-friction reduction.

Choi et al. (1994) proposed two active control strategies for drag reduction using direct numerical simulation of turbulent channel flow: (1) control with the wall-normal velocity (blowing and suction) at the wall (v control; Figure 14.2a), and (2) control with the spanwise velocity at the wall (w control; Figure 14.2b). The control algorithm was that the control-input velocity at the wall is negatively proportional to the instantaneous velocity at a location near the wall. For instance, in the case of v control, the blowing and suction velocity at the wall was exactly opposite to the wall-normal component of the velocity at a prescribed y location, y_d; i.e., $v_w = -v|_{y=y_d}$ (Figure 14.2a).

Figure 14.3a shows the time histories of the pressure gradients that were required to drive a fixed mass flow rate for the unmanipulated fully developed channel flow and for manipulated (v control) channel flows. Substantial skin-friction reduction was obtained ($\approx$25% on each wall) with $y_d^+ \approx 10$. For other y_d^+ locations, either the drag was substantially increased ($y_d^+ \approx 26$) or the reduction was small ($y_d^+ \approx 5$). Noting that the streamwise vortices led to strong spanwise velocity as well as wall-normal velocity, the out-of-phase boundary condition was applied to the spanwise velocity at the surface; i.e., $w_w = -w|_{y=y_d}$ (Figure 14.2b). Several

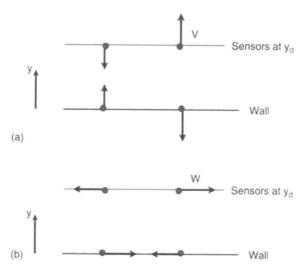

FIGURE 14.2 Schematic diagram of controls used by Choi et al. (1994): (a) *v* control; (b) *w* control. (Reprinted by permission of Cambridge University Press from Choi, H. et al. [1994] "Active Turbulence Control for Drag Reduction in Wall-Bounded Flows," *J. Fluid Mech.* **262**, p. 75.).

sensor locations ranging from $y_d^+ \approx 5$ to 26 were tested, and the best result was obtained with $y_d^+ \approx 10$, yielding about 30% drag reduction (Figure 14.3b), which is slightly better than the optimum *v* control. With $y_d^+ > 20$, the drag was increased.

The velocity, pressure, and vorticity fluctuations as well as the Reynolds shear stress were significantly reduced throughout the channel (see Figure 14.4 for the change in strength of the streamwise vorticity). Instantaneous flow fields showed that streaky structures below $y^+ \approx 5$ were clearly diminished by the active control, and the physical spacing of the streaky structure above $y^+ \approx 5$ was increased in the manipulated channels. The active blowing and suction or spanwise velocity at the wall significantly affected turbulence statistics above the wall.

A schematic diagram of the drag-reduction mechanism is shown in Figure 14.5. The high shear-rate regions on the wall are moved to the interior of the channel ($y^+ \approx 5$) by the control schemes. The sweep motion due to strong streamwise vortices is directly deterred by the *v* control. In the case of *w* control, a wall-normal velocity is induced very near the wall by the imposed spanwise velocity distribution at the wall; i.e., from the continuity equation at the wall, $\partial v / \partial y = -\partial w / \partial z \neq 0$, leading to higher values of *v* near the wall.

14.4   Reactive Controls Based on Sensing at the Wall

Even though Choi et al. (1994) showed that turbulent skin friction could be reduced by suppressing the dynamically significant coherent structures present in the wall region (i.e., near-wall streamwise vortices), practical implementation of their control scheme is nearly impossible at this time because one has to measure spatial distribution of the wall-normal velocity at $y^+ = 10$. They also considered control of the turbulent boundary layer by placing sensors only at the wall for practical implementation; using the Taylor series expansion, the wall-normal velocity at $y^+ = 10$ was approximated as $v\big|_{y^+_m} \sim -\frac{\partial}{\partial z}\frac{\partial w}{\partial y}\big|_w$. However, a control based on this variable yielded only about 6% drag reduction, which is smaller than the riblets' maximum drag reduction.

Motivated by the drag reduction techniques discussed in Choi et al. (1994), Lee et al. have presented two successful active control methods based on wall measurement for skin-friction reduction. One uses the neural network [Lee et al., 1997], and the other uses the suboptimal control theory [Lee et al., 1998]. Both approaches applied blowing and suction to turbulent channel flow. Koumoutsakos (1999) presented a feedback-control result in a turbulent channel flow, where the sensing variable is the vorticity flux at the

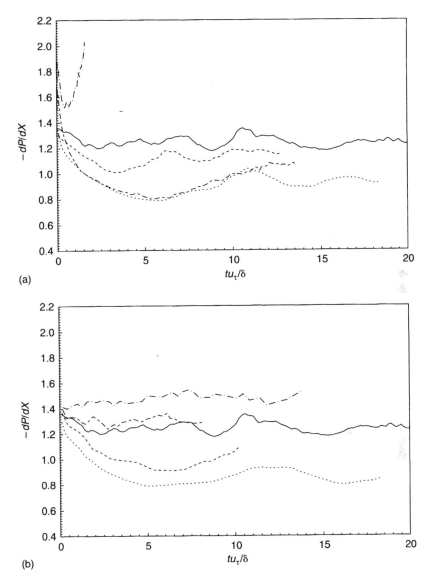

FIGURE 14.3 Time history of the pressure gradient required to drive a fixed-mass flow rate in the cases of (a) v control and (b) w control. Note: —, unmanipulated channel; ---, manipulated channel with sensors at $y_d^+ \approx 5$; . . ., $y_d^+ \approx 10$; – – –, $y_d^+ \approx 20$; – · –, $y_d^+ \approx 26$. (Reprinted by permission of Cambridge University Press from Choi, H. et al. [1994] "Active Turbulence Control for Drag Reduction in Wall-Bounded Flows," *J. Fluid Mech.* **262**, p. 75.)

wall and the actuation mechanism is blowing and suction at the wall. In this section, we summarize these three studies.

14.4.1 Neural Network

The objective of the work reported in Lee et al. (1997) was to seek wall actuations to obtain skin-friction reduction, (blowing and suction) based on sensing of the wall shear stress. This requires knowledge of how the wall shear stress responds to wall actuation. Because of the complexity of turbulent flow, however, it is not possible to find such a correlation in a closed form or to approximate it in a simple form. Thus, the authors used a neural network to approximate the correlation that predicts the optimal wall actuation to achieve minimum skin-friction drag.

FIGURE 14.4 Contours of the streamwise vorticity in a cross-flow plane. The contour levels are the same for the cases with and without control, and negative contours are dotted. (Reprinted by permission of Cambridge University Press from Choi, H. et al. [1994] "Active Turbulence Control for Drag Reduction in Wall-Bounded Flows," *J. Fluid Mech.* 262, p. 75.)

A standard two-layer feedforward network with hyperbolic tangent hidden units and a linear output unit was used. The functional form of the final neural network was

$$v_w(x, z, t) = W_a \tanh\left(\sum_{j=-(N-1)/2}^{(N-1)/2} W_j \left.\frac{\partial w}{\partial y}\right|_w (x, z + j\Delta z, t) - W_b \right) - W_c, \tag{14.1}$$

where W denotes the weights, N is the total number of input weights, and Δz is the grid spacing in the span-wise direction. Seven neighboring points ($N = 7$), including the point of interest, in the spanwise direction

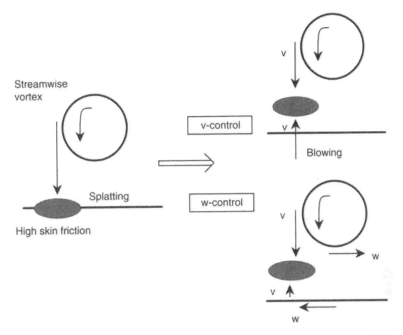

FIGURE 14.5 Schematic diagram of the drag-reduction mechanism by active controls [Choi et al., 1994]. (Reprinted by permission of Cambridge University Press from Choi, H. et al. [1994] "Active Turbulence Control for Drag Reduction in Wall-Bounded Flows," *J. Fluid Mech.* **262**, p. 75.)

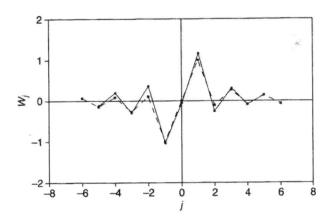

FIGURE 14.6 Weight distribution: —, from neural network; ---, suboptimal control.

(corresponding to approximately 90 wall units) were found to provide enough information to adequately train and control the near-wall structures responsible for high skin friction. The network was trained to minimize the sum of a weighted-squared error that is proportional to the square of the difference between the desired control input and the network control input. In its off-line training and control, the wall-normal velocity at $y^+ = 10$ was considered as a desired control input. After training was completed, the weight distribution in the spanwise direction was obtained, as shown in Figure 14.6. Applying this control scheme to a turbulent channel flow at low Reynolds number resulted in about 20% drag reduction. The computed flow fields were examined to determine the mechanism by which the drag reduction was achieved. The most salient feature of the controlled case was that the strength of the near-wall streamwise vortices was substantially reduced. This result further substantiates the notion that a successful suppression of the near-wall streamwise vortices leads to a significant reduction in drag.

Lee et al. (1997) further considered a neural network with continuous on-line training, from which a very simple control scheme was obtained:

$$\hat{v}_w = C \frac{ik_z}{|k_z|} \left. \frac{\partial \hat{w}}{\partial y} \right|_w \tag{14.2}$$

Here, the hat (^) denotes the Fourier component, k_z is the spanwise wavenumber, and C is a positive scale factor that determines the root-mean-square (rms) value of the actuation. This control worked equally well as Equation (14.1).

14.4.2 Suboptimal Control

A systematic approach based on an optimal control theory was proposed by Abergel and Temam (1990). Choi et al. (1993b) proposed a suboptimal control procedure in which the iterations required for a global optimal control were avoided by seeking an optimal condition over a short time. Lee et al. (1998) derived a simple control scheme by minimizing cost functions related to the near-wall streamwise vortices. They demonstrated that a wise choice of the cost function coupled with a variation of the formulation can lead to a more practical control law. They showed how to choose a cost function and how to minimize it to yield simple feedback control laws that require quantities measurable only at the wall. One of the laws requires spatial information on the wall pressure over the entire wall, and the other requires information on one component of the wall shear stress. They then derive more practical control schemes that require only local wall pressure or local wall shear-stress information and show that the schemes work equally well. Here, we will explain the suboptimal control approach based on sensing of the wall shear stress.

   In Lee et al. (1998), a cost function was chosen based on the observation of the successful controls by Choi et al. (1994). The blowing and suction by Choi et al. (1994), equal and opposite to the wall-normal velocity component at $y^+ = 10$, effectively suppresses a streamwise vortex by counteracting up-and-down motion induced by the vortex. This blowing and suction action creates locally high pressure in the near-wall region marked with + and low pressure in the region marked with – in Figure 14.7. A crucial aspect of their analysis is the observation that this blowing and suction action increases the pressure gradient in the spanwise direction under the streamwise vortex near the wall and subsequently increases the spanwise velocity gradient $\partial w/\partial y$ near the wall. The above argument suggests that we should seek the blowing and suction velocity that increases $\partial w/\partial y$ near the wall for a short time (i.e., in the suboptimal sense) in order to achieve a drag reduction similar to that achieved by Choi et al. (1994). The cost function $J(v_w)$ to be

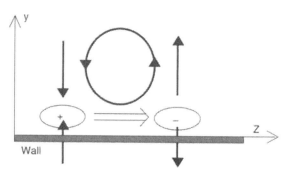

FIGURE 14.7   Effect of active blowing and suction on the pressure and spanwise-velocity gradient near the wall. (Reprinted by permission of Cambridge University Press from Lee, C. et al. [1998] "Suboptimal control of turbulent channel flow for drag reduction," *J. Fluid Mech.* **358**, p. 245.)

minimized is then:

$$J(v_w) = \frac{l}{2A\Delta t} \int_S \int_t^{t+\Delta t} v_w^2 \, dt \, dS - \frac{1}{2A\Delta t} \int_S \int_t^{t+\Delta t} \left(\frac{\partial w}{\partial y}\right)_w^2 \, dt \, dS, \tag{14.3}$$

where the integrations are over the wall, S, in space and over a short duration in time, Δt, which typically corresponds to the time step used in the numerical computation; l is the relative price of the control, as the first term on the right-hand side represents the cost of the actuation v_w. Note that there is a minus sign in front of the second term because we want to maximize $\partial w/\partial y$ at the wall. It should be noted that the spanwise velocity gradient at the wall will *eventually* be reduced when the strength of the near-wall streamwise vortices is reduced through successful control. Here, blowing and suction that increase the spanwise-velocity gradient for the next step are sought as a suboptimal control. After some mathematical procedure, the blowing and suction velocity of minimizing the cost function is obtained as:

$$\hat{v}_w = C \frac{ik_z}{k} \frac{\partial \hat{w}}{\partial y}\bigg|_w, \tag{14.4}$$

where the hat denotes the Fourier coefficient; $k^2 = k_x^2 + k_z^2$, k_x and k_z are the streamwise and spanwise wavenumbers respectively; and C is a positive scale factor that determines the cost of the actuation. The above equation indicates that the optimum wall actuation should be proportional to the spanwise derivative of the spanwise shear at the wall, $(v_w \sim \frac{\partial}{\partial z} \frac{\partial w}{\partial y}|_w)$, with the high wavenumber components reduced by $1/k$. Note that the scale factor C in Equation (14.4) is arbitrary: in the simulations, the rms value of v_w was taken to be equal to that of the wall-normal velocity at $y^+ = 10$, which gives the same rms value of wall actuations as that of Choi et al. (1994). Applying Equation (14.4) to turbulent channel flow produced about 22% drag reduction. Turbulence modification was also very similar to that shown in Choi et al. (1994) and Lee et al. (1997).

Note that the control law from neural network (Lee et al. 1997) shown in Equation (14.2) produced almost the same distribution of blowing and suction as Equation (14.4) because the near-wall structures have a relatively slow variation in the streamwise direction (the $k_x = 0$ component is dominant). This blowing and suction distribution is also very similar to the one based on $y^+ = 10$ data.

This control scheme, however, is still impractical to implement because it is expressed in terms of the Fourier components (i.e., in wavenumber space), which require information over the entire spatial domain. Therefore, the inverse transform of ik_z/k sought numerically so that the convolution integral could be used to express the control law in physical space. For example, the discrete representation of the control law, Equation (14.4) then became:

$$v_w(x, z, t) = C \sum_j W_j \frac{\partial w}{\partial y}\bigg|_w (x, z + j\Delta z, t). \tag{14.5}$$

The weights W_j decayed rapidly with distance from the point of interest, suggesting that the optimum actuation can be obtained by a local weighted average of $\partial w/\partial y|_w$. The result obtained using an 11-point average yielded about the same drag reduction as that obtained from full integration using Equation (14.4). It is remarkable that localized information can produce such a significant drag reduction. The weight distribution W_j was very similar to the one found by applying the neural network to the same turbulent flow (see Figure 14.6).

14.4.3 Vorticity Flux Control

For an incompressible viscous flow over a stationary wall, the vorticity fluxes at the wall are directly proportional to the pressure gradients:

$$v \frac{\partial \omega_x}{\partial y}\bigg|_w = \frac{1}{\rho} \frac{\partial p}{\partial z}\bigg|_w, \quad -v \frac{\partial \omega_z}{\partial y}\bigg|_w = \frac{1}{\rho} \frac{\partial p}{\partial x}\bigg|_w, \tag{14.6}$$

where p is the pressure and ω_x and ω_z are the streamwise and spanwise vorticity components. The flux of the wall-normal vorticity ω_y can be determined from the kinematic condition ($\nabla \cdot \omega = 0$).

A feedback-control vorticity-flux algorithm was presented and applied to a turbulent channel flow in Koumoutsakos (1999), where the wall pressure was sensed and its gradient (the wall vorticity flux) was calculated. Unsteady blowing and suction at the wall were the control inputs, and its strength was calculated explicitly by formulating Lighthill's mechanism of vorticity generation at a no-slip wall. The author considered a series of vorticity flux (or, equivalently, pressure gradient) sensors on the wall at locations (x_i, z_i), $i = 1, 2, 3, \ldots, M$, and determined the blowing and suction strengths necessary to achieve a desired vorticity flux profile at the wall at a time instant, k, by solving the linear set of:

$$Bu_k + X_{k-1} = D_k, \tag{14.7}$$

where D_k is an $M \times 1$ vector of the *desired* vorticity flux at the sensor locations, X_{k-1} is an $M \times 1$ vector of the *measured* vorticity flux at the sensor locations, and u_k is an $N \times 1$ vector of blowing and suction strengths at the actuator locations. B is an $M \times N$ matrix whose elements B_{ji} are a function of the relative locations of the sensors (j index) and actuators (i index) (see [Koumoutsakos, 1999] for more details). The unknown blowing and suction strengths were determined by solving Equation (14.7).

An additional equation that determines how the desired and measured vorticity flux components are related was given as:

$$\begin{pmatrix} v\dfrac{\partial \omega_x}{\partial y} \\[2mm] v\dfrac{\partial \omega_z}{\partial y} \end{pmatrix}_{\text{control}} = \begin{pmatrix} a & b \\ c & d \end{pmatrix} \begin{pmatrix} v\dfrac{\partial \omega_x}{\partial y} \\[2mm] v\dfrac{\partial \omega_z}{\partial y} \end{pmatrix}_{\text{measured}}, \tag{14.8}$$

where the coefficients, a, b, c, and d, can be chosen *a priori* and they may be constant or spatially varying. The simulations presented in Koumoutsakos (1999) have been conducted with the set of parameters $a = b = c = 0$ and $d = \pm 1$ equivalent to considering the *in-* and *out-of-phase* control respectively of the *spanwise* vorticity flux. Here, the *in-phase* control implies enhancement of the wall vorticity flux, whereas the *out-of-phase* control implies cancellation of the induced vorticity flux. The author presented the results for the case of *out-of-phase* control.

For the out-of-phase control, the streaks were eliminated, and highly spanwise correlated patterns were established for the spanwise vorticity flux, shear stresses, and actuation strengths (see, for example, Figure 14.8). These spanwise-correlated structures and further flow visualizations suggested the formation of unsteady spanwise vortical rollers in the inner layer of the wall. These spanwise vortical rollers resulted in the formation of positive and negative shear stresses at the wall. The elimination of streaks and the disruption of the near-wall processes by the establishment of the particular vortical rollers resulted in skin-friction

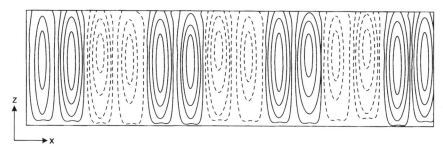

FIGURE 14.8 Highly spanwise correlated pattern of the blowing and suction as seen in Koumoutsakos (1999): —, blowing; ---, suction. (Reprinted by permission of the American Institute of Physics from Koumoutsakos, P. [1999] "Vorticity Flux Control for a Turbulent Channel Flow," *Phys. Fluids* 11, p. 248.)

reduction on the order of 40%. The regularity in the resulting actuation strength as shown in Figure 14.8 suggests that it may be possible to devise an open-loop control law, but further study in that direction was not conducted.

14.5 Active Wall Motion: Numerical Studies

Although the control schemes presented in the previous section (i.e., sensing at the wall) are a significant improvement over that of Choi et al. (1994), which requires velocity information above the wall, other issues must be resolved before these control schemes can be implemented in real situations (e.g., time delay between sensing and actuation). Other approaches, such as active wall motion using deformable actuators, may be more practical. So far, three studies involving active wall motion have been carried out using numerical simulation. A brief overview of these studies is given in this section.

14.5.1 Streak Control

Carlson and Lumley (1996) simulated turbulent flow in a minimal channel with time-dependent wall geometries in order to investigate the effect of wall movement on the skin friction. The control device consisted of *one* actuator (Gaussian shape and about 12 wall units in height) *on one of the* channel walls:

$$h(x, z, t) = \varepsilon(t)\exp\left\{-\sigma^{-2}\left[(x - x_0)^2 + \left(\frac{z - z_0}{\mu}\right)^2\right]\right\},\qquad(14.9)$$

where h is the time-varying (due to $\varepsilon(t)$) height of the actuator; (x_0, z_0) is the center location of the actuator; $\sigma = 0.18\delta$; $\mu = 1\delta$ or 2δ; and δ is the channel half-width. The control device was located underneath either a high-speed streak or a low-speed streak (note that the minimal channel investigated in Carlson and Lumley has only one vortical structure on one side of the channel, and thus only one pair of high- and low-speed streaks exists on the controlled wall).

Raising the actuator underneath a low-speed streak increased drag, but raising it underneath a high-speed streak reduced drag. The authors claimed that the actuator lifts fast-moving fluid in the high-speed region away from the wall, allowing the adjacent low-speed region to expand and thereby lowering the average wall shear stress. Conversely, raising an actuator underneath a low-speed streak allows the adjacent high-speed region to expand, which increases the skin friction (Figure 14.9). However, they could not present the amount of time-averaged drag reduction due to wall motion because only a short time control was conducted in their study.

Later, Endo et al. (2000) designed a realistic array of deformable actuators elongated in the streamwise direction. Each actuator moved up and down and induced the wall-normal velocity at the wall. The wall velocity of the actuator was determined by:

$$v_w(x, z, t) = v_m(t)f(x)\exp\left[-\frac{(z - z_0)^2}{\sigma_z^2}\right]\sin\left[\frac{2\pi(z - z_0)}{m_z}\right],\qquad(14.10)$$

$$v_m(t) = \begin{cases} \alpha\tanh\left(\frac{\partial}{\partial z}\frac{\partial u}{\partial y}\frac{1}{\beta}\right) - \gamma y_m & \text{if } \frac{\partial}{\partial z}\frac{\partial w}{\partial y} < 0, \\[2mm] -\gamma y_m & \text{otherwise} \end{cases}\qquad(14.11)$$

where α, β, γ, σ_z, m_z and y_m are the control parameters; $f(x)$ is a hyperbolic-tangent function (see Endo et al., 2000); and v_m is the wall velocity at the center of the actuator (x_0, z_0). Each sensor was located in front of the actuator and measured the spanwise distribution of the streamwise wall shear rate $\left(\frac{\partial}{\partial z}\frac{\partial u}{\partial y}\big|_w\right)$ (see Equation [14.11]).

With the control scheme above, Endo et al. (2000) obtained about 10% drag reduction. Because measurement of the streamwise shear rate was a concern in their approach, the main mechanism of the drag reduction is believed to attenuate and stabilize streak meandering and thus to hamper the regeneration cycle of the near-wall turbulence [Hamilton et al., 1995].

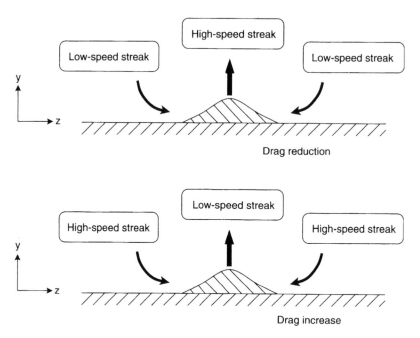

FIGURE 14.9   The mechanism for drag reduction or increase due to wall motion as seen in Carlson and Lumley (1996). (Reprinted by permission of Cambridge University Press from Carlson, A., and Lumley, J.L. [1996] "Active Control in the Turbulent Wall Layer of a Minimal Flow Unit," *J. Fluid Mech.* **329**, p. 341.)

14.5.2   Near-Wall Streamwise-Vortex Control

Kang and Choi (2000) conducted numerical simulations to investigate a possibility of reducing skin-friction drag in a turbulent channel flow with active wall motions. The wall was locally (and continuously) deformed according to the successful control strategy developed from the suboptimal control [Lee et al., 1998] and thus the induced velocity at the wall by wall motion was represented as:

$$\hat{v}_w = C \frac{ik_z}{k} \frac{\partial \hat{w}}{\partial y}\bigg|_w, \tag{14.12}$$

which is the same as Equation (14.4). Because the Fourier transformation is not realistic (one needs information on an entire wall), a more realistic wall-motion control strategy was devised by applying the 11-point weighted average of $\partial w/\partial y|_w$ in physical space (see Equation [14.5]).

As a result, about 13 to 15% drag reduction was obtained due to active wall motions. Active wall motions significantly weakened the strength of the streamwise vortices near the wall. An instantaneous wall shape during control is shown in Figure 14.10. The rms amplitude of wall deformation was about 3 wall units. The wall shapes were elongated in the streamwise direction and resembled riblets in appearance (a similar wall shape was also observed in [Endo et al., 2000]). The average peak-to-peak distance of the grooved wall shape was around 100 wall units. Riblets of this size hardly reduce the skin-friction drag; therefore, the mechanism of the drag reduction is essentially different from that of riblets. That is, the active wall motion directly suppresses the near-wall streamwise vortices by the induced blowing and suction, whereas passive riblets restrict the location of the streamwise vortices above the wetted surface such that only a limited area of the riblets is exposed to the downwash motion that the vortices induce.

14.6 Concluding Remarks

There have been many experimental studies on passive or open-loop active control in a turbulent boundary layer. Most work has focused on understanding the flow response to external disturbances, such as

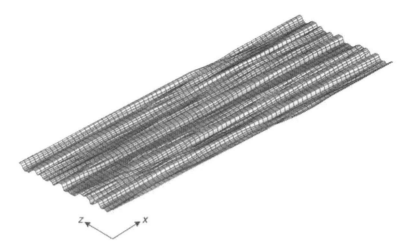

FIGURE 14.10 Instantaneous wall shape. (Reprinted by permission of the American Institute of Physics from Kang, S., and Choi, H. [2000] "Active Wall Motions for Skin-Friction Drag Reduction," Phys. Fluids 12, p. 3301.)

uniform or periodic blowing and suction, or to surface roughness. This work did not seriously consider drag reduction performance because measuring the skin friction in a perturbed turbulent boundary layer is not easy. To the best of author's knowledge, no study has shown an experimental evidence of drag reduction using an active open-loop control strategy with a constraint of zero-net mass flow rate. To date, only a very few preliminary experimental studies have considered reactive (feedback or feedforward) control for skin-friction reduction in a turbulent or laminar boundary layer. Jacobson and Reynolds (1998) implemented a springboard-type actuator in a laminar boundary layer, which produces a quasi-steady pair of counter rotating streamwise vortices. They showed that transition was delayed by 40 displacement thickness, and the mean and spanwise variation of wall shear stress was substantially reduced downstream of a single transverse array of actuators. Rathnasingham and Breuer (1997) constructed a linear feedforward control algorithm and applied it to a turbulent boundary layer for skin-friction reduction. Using three wall-based sensors and a single actuator, it was shown that the control produced a maximum of 31% reduction in the rms streamwise velocity fluctuations. However, they did not show the detailed skin-friction variation due to the measurement difficulty.

In this chapter, we have shown that the skin friction can be reduced by manipulating the near-wall streamwise vortices or streaks using either blowing and suction or wall motion. For the past 10 years, quite a few different successful control strategies have been presented using numerical simulations and have provided information about what to manipulate for skin-friction reduction. The maximum reduction of skin-friction drag obtained so far is relaminarization at low Reynolds number (see [Choi et al., 1994] for v and w control; [Bewley et al., 2001] for optimal control). The control methods based on wall sensing using neural network or suboptimal control produced a maximum of 20% drag reduction. However, in a practical implementation, sensing and controlling the vortices are difficult tasks. One of the main reasons for this is that the near-wall structures appear randomly in space and in time. Thus, one has to place the sensors and actuators over the entire surface in order to obtain a substantial decrease of drag. The other main reason is that the typical diameter of the near-wall streamwise vortices is about 30 wall units [Kim et al., 1987], which is about 100 μm in a laboratory experiment. In order to detect and control these vortices, the size of sensors and actuators should be comparable to that of the vortices.

Practical implementation of reactive control algorithms developed from numerical studies has been hampered by difficulties in manufacturing microsensors and actuators using traditional machining tools. However, recent development of micromachining technology now makes it possible to fabricate micron-size mechanical parts, and thus microsensors and microactuators can be mass produced. So far, micro-electromechanical system (MEMS) technology has been successfully applied to active open-loop control of flow over bluff bodies (e.g., [Ho and Tai, 1998], and the references therein). Also, microsensors and microactuators for boundary-layer flow have been developed using MEMS technology [Ho and Tai, 1998]. Spatial distribution of the skin friction has been measured using microsensors, and changes in boundary-layer flow due to microactuators have been thoroughly investigated. Researchers are now implementing the reactive control algorithms in laboratory experiments using MEMS technology and are expecting a significant amount of skin-friction reduction comparable to that obtained from numerical studies.

In most numerical studies, sensors and actuators are collocated and distributed all over the computational domain. In reality, sensors and actuators cannot be collocated, and actuators should be located downstream of the sensors. Therefore, how to efficiently distribute the sensors and actuators is an important issue. A second issue is how to develop a new control method that is well suited for experimental approach (note that most control methods investigated so far started from numerical studies). The third issue is the Reynolds number effect. Relaminarization of turbulent flow due to control, as has been observed in numerical studies, might happen because of low Reynolds numbers. At low Reynolds number, the dynamics of boundary-layer flow is mostly governed by near-wall phenomena, whereas it is affected not only by the near-wall behavior but also by the fluid motion in the buffer or outer layer. Therefore, various control algorithms may have to be developed depending on the Reynolds number.

Acknowledgments

This work has been supported by the Creative Research Initiatives of the Korean Ministry of Science and Technology.

References

Abergel, F., and Temam, R. (1990) "On Some Control Problems in Fluid Mechanics," *Theor. Comp. Fluid Dyn.*, 1, pp. 303–325.

Bewley, T.R. (2001) "Flow Control: New Challenges for a New Renaissance," *Prog. Aerosp. Sci.*, 37, pp. 21–58.

Bewley, T.R., Moin, P., and Temam, R. (2001) "DNS-based Predictive Control of Turbulence: An Optimal Benchmark for Feedback Algorithms," *J. Fluid Mech.*, 447, pp. 179–226.

Bushnell, D.M., and McGinley, C.B. (1989) "Turbulence Control in Wall Flows," *Annu. Rev. Fluid Mech.*, 21, pp. 1–20.

Cantwell, B.J. (1981) "Organized Motion in Turbulent Flow," *Annu. Rev. Fluid Mech.*, 13, pp. 457–515.

Carlson, A., and Lumley, J.L. (1996) "Active Control in the Turbulent Wall Layer of a Minimal Flow Unit," *J. Fluid Mech.*, 329, pp. 341–371.

Choi, H., Moin, P., and Kim, J. (1993a) "Direct Numerical Simulation of Turbulent Flow Over Riblets," *J. Fluid Mech.*, 255, pp. 503–539.

Choi, H., Temam, R., Moin, P., and Kim, J. (1993b) "Feedback Control for Unsteady Flow and Its Application to the Stochastic Burgers Equation," *J. Fluid Mech.*, 245, pp. 509–543.

Choi, H., Moin, P., and Kim, J. (1994) "Active Turbulence Control for Drag Reduction in Wall-Bounded Flows," *J. Fluid Mech.*, 262, pp. 75–110.

Corino, E.R., and Brodkey, R.S. (1969) "A Visual Investigation of the Wall Region in Turbulent Flow," *J. Fluid Mech.*, 37, pp. 1–30.

Endo, T., Kasagi, N., and Suzuki, Y. (2000) "Feedback Control of Wall Turbulence with Wall Deformation," *Int. J. Heat Fluid Flow*, 21, pp. 568–575.

Gad-el-Hak, M. (1994) "Interactive Control of Turbulent Boundary Layers: A Futuristic Overview," *AIAA J.*, 32, pp. 1753–1765.

Gad-el-Hak, M. (1996) "Modern Developments in Flow Control," *Appl. Mech. Rev.*, 49, pp. 365–380.

Gad-el-Hak, M., and Blackwelder, R.F. (1989) "Selective Suction for Controlling Bursting Events in a Boundary Layer," *AIAA J.*, 27, pp. 308–314.

Hamilton, J.M., Kim, J., and Waleffe, F. (1995) "Regeneration Mechanisms of Near-Wall Turbulence Structures," *J. Fluid Mech.*, 287, pp. 317–348.

Ho, C.-H., and Tai, Y.-C. (1996) "Review: MEMS and Its Applications for Flow Control," *J. Fluids Eng.*, **118**, pp. 437–446.

Ho, C.-M., and Tai, Y.-C. (1998) "Micro-Electro-Mechanical Systems (MEMS) and Fluid Flows," *Annu. Rev. Fluid Mech.*, **30**, pp. 579–612.

Jacobson, S.A., and Reynolds, W.C. (1998) "Active Control of Streamwise Vortices and Streaks in Boundary Layers," *J. Fluid Mech.*, **360**, pp. 179–212.

Kang, S., and Choi, H. (2000) "Active Wall Motions for Skin-Friction Drag Reduction," *Phys. Fluids*, **12**, pp. 3301–3304.

Kim, H.T., Kline, S.T., and Reynolds, W.C. (1971) "The Production of Turbulence Near a Smooth Wall in a Turbulent Boundary Layer," *J. Fluid Mech.*, **50**, pp. 133–160.

Kim, J., Moin, P., and Moser, R. (1987) "Turbulence Statistics in Fully Developed Channel Flow at Low Reynolds Number," *J. Fluid Mech.*, **177**, pp. 133–166.

Kline, S.J., Reynolds, W.C., Schraub, F.A., and Runstadler, P.W. (1967) "The Structure of Turbulent Boundary Layers," *J. Fluid Mech.*, **30**, pp. 741–774.

Koumoutsakos, P. (1999) "Vorticity Flux Control for a Turbulent Channel Flow," *Phys. Fluids*, **11**, pp. 248–250.

Kravchenko, A.G., Choi, H., and Moin, P. (1993) "On the Relation of Near-Wall Streamwise Vortices to Wall Skin Friction in Turbulent Boundary Layer," *Phys. Fluids* A, **5**, pp. 3307–3309.

Lee, C., Kim, J., Babcock, D., and Goodman, R. (1997) "Application of Neural Networks to Turbulence Control for Drag Reduction," *Phys. Fluids*, **9**, pp. 1740–1747.

Lee, C., Kim, J., and Choi, H. (1998) "Suboptimal Control of Turbulent Channel Flow for Drag Reduction," *J. Fluid Mech.*, **358**, pp. 245–258.

Moin, P., and Bewley, T.R. (1994) "Feedback Control of Turbulence," *Appl. Mech. Rev.*, **47**, pp. S3–S13.

Rathnasingham, R., and Breuer, K.S. (1997) "System Identification and Control of a Turbulent Boundary Layer," *Phys. Fluids*, **9**, pp. 1867–1869.

Robinson, S.K. (1991) "Coherent Motions in the Turbulent Boundary Layer," *Annu. Rev. Fluid Mech.*, **23**, pp. 601–639.

Walsh, M.J. (1982) "Turbulent Boundary Layer Drag Reduction Using Riblets," *AIAA Paper No.* 82-0169, AIAA, Washington, DC.

15

Toward MEMS Autonomous Control of Free-Shear Flows

Ahmed Naguib
Michigan State University

15.1 Introduction

Interest in the application of microelectromechanicalsystems (MEMS) technology to flow control and diagnostics started around the early 1990s. During this relatively short time, there have been a handful attempts aimed at using the new technology to develop and implement reactive control of various flow phenomena. The ultimate goal of these attempts has been to capitalize on the unique ability of MEMS to integrate sensors, actuators, driving circuitry, and control hardware in order to attain *autonomous* active flow management. To date, however, there remains to be a demonstration of a fully functioning MEMS flow control system whereby the multitude of information gathered from a distributed MEMS sensor array is successfully processed in real time using on-chip electronics to produce an effective response by a distributed MEMS actuator array.

Notwithstanding the inability of the research efforts in the 1990s to realize autonomous MEMS control systems, the lessons learned so far are valuable in understanding the strengths and limitations of MEMS in flow applications. In this chapter, the research work aimed at MEMS-based autonomous control of free-shear flows over the past decade is reviewed. The main intent is to use the outcome of these efforts as a telescope to peek through and project a vision of future MEMS systems for free-shear flow control.

The presentation of the material is organized as follows: first, important classifications of free-shear flows are introduced in order to facilitate subsequent discussions; second, a fundamental analysis concerning the usability of MEMS in different categories of free-shear flows is provided. This is followed by an outline of autonomous flow control system components, with a focus on MEMS in free-shear flow applications and related issues. Finally, the bulk of the chapter reviews prominent research efforts in the 1990s, leading to a vision of future systems.

15.2   Free-Shear Flows: A MEMS Control Perspective

Free-shear flows refer to the class of flows that develop without the influence imposed by direct contact with solid boundaries. However, in the absence of thermal gradients, nonuniform body-force fields, or similar effects, the vorticity in free-shear flows is actually acquired through contact with a solid boundary at one point in the history of development of the flow. The "free-shear state" of the flow is attained when it separates from the solid wall, carrying with it whatever vorticity was contained in the boundary layer at the point of separation. The mean velocity profile of the shear layer at the point of separation is inflectional, and hence is inviscidly unstable; that is, extremely sensitive to small perturbations if excited at the appropriate frequencies. This point is of fundamental importance to MEMS-based control.

Since MEMS devices are micron in scale they are only capable of delivering proportionally small energies when used as actuators. Therefore, if it were not for the high sensitivity of free-shear flows to disturbances at the point of separation, there would be no point in attempting to use MEMS actuators for shear-layer control. Moreover, active control of free-shear flows using MEMS as a disturbance source must be applied at or extremely close to the point of separation. The same statement can be extended to the more powerful conventional actuators, such as glow-discharge, large-scale piezoelectric devices, large-scale flaps, etc., if high-gain or efficient control is desired.

From a control point of view, it is useful to classify free-shear flows according to whether the separation line is stationary or moving. Stationary separation line (SSL) flows include jets, single- or two-stream shear layers, and backward facing step flows. In these flows, separation takes place at the sharply defined trailing edge of a solid boundary at the origin of the free-shear flow. On the other hand, moving separation line (MSL) flows include dynamic separation over pitching airfoils and wings, periodic flows through compressors, and forward facing step flows. The separation point in the latter, although steady on average, jitters due to the lack of a sharp definition of the geometry at separation (such as the nozzle lip for the shear layer surrounding a jet).

Whether the free-shear flow to be controlled is of the SSL or MSL type is extremely important, not only from the point of view of control feasibility and ease but also in terms of the need for MEMS versus conventional technology. In particular, in SSL flows, actuators — whether MEMS or conventional — can be located directly at the known point of separation where they would be effective. As the operating conditions change (for example, through a change in the speed of a jet) the same set of actuators can be used to affect the desired control. In contrast, in an MSL flow, the instantaneous location of the separation line has to be known and only actuators located along this line should be used for control.

To explain further, consider controlling dynamic separation over a pitching two-dimensional airfoil. As the airfoil is pitched from, say, zero to a sufficiently large angle of attack, a separated shear layer is formed. The separation line of this shear layer moves with the increasing angle of attack. To control this separating flow using MEMS or other conventional actuators, it is necessary to track the location of the separation point in real time in order to activate only those actuators that are located along the separation line. This would require distributed, or array, measurements on the surface of the airfoil. Furthermore, since the separation-line-locating algorithm is likely to employ spatial derivatives of the surface measurements, the surface sensors must have high spatial resolution and be packed densely. Therefore, in MSL problems, it seems inevitable to use MEMS sensor arrays if autonomous control is to be successful. This is believed to be true regardless of whether MEMS or conventional technology is used for actuation.

When considering MEMS control systems, another useful classification is that associated with the characteristic size of the flow. With the advent of MEMS, it is now common to observe flows confined to domains that are no larger than a few hundred microns in characteristic size. Such flows may be found in different microdevices including pumps, channels, nozzles, turbines, and others. Therefore, it is important to distinguish between microscale (MIS) flows and their macro counterparts (MAS). In particular, there should be no ambiguity that within a microdevice, only MEMS, or perhaps even NEMS (nano electro mechanical systems) are the only feasible method of control.

A good example of microdevices where free-shear flows may be encountered is the MIT microengine project [Epstein et al., 1997]. In the intricate, yet complex devices engineered in this project, shear layers exist in the microflows over the stator and rotor of the compressor and turbine and in the sudden expansion leading to the combustion chamber. Whether those shear layers and their susceptibility to excitation within the confines of the microdevice mimic that of macroscale shear layers is yet to be established. However, as indicated above, if control is to be exercised, the limitation imposed by the scale of the device dictates that any actuators and or sensors have to be as small as, if not smaller than, the device itself.

Active control of MIS flows is an area that is yet to be explored. This is in part because of the fairly short time since the interest in such flows started. Therefore, further discussion of the subject would be appropriately left until sufficient literature is available.

15.3 Shear-Layer MEMS Control System Components and Issues

To facilitate subsequent discussion of the research pertaining to MEMS-based control of free-shear flows, the different components of the control system will be discussed and analyzed here. Of particular importance are the issues specific to utilization of MEMS technology to realize the control system components. Figure 15.1 displays a general functional block diagram for feedback control systems in SSL and MSL flows. As seen from Figure 15.1, for both types of flows the information obtained from surface-mounted sensor arrays is fed to a flow-field estimator in order to predict the state of the flow field being controlled. If the desired flow field and deviations from it can be defined in terms of a signature measurable at the surface, the flow estimator may be bypassed altogether. Any difference between the measured and desired flow states is used to drive surface mounted actuators in a manner that would force the flow toward the desired state. In the case of the MSL flow, the current location of the separation line must also be identified and fed to the controller in order to operate the actuator set nearest to or at the position of separation (see Figure 15.1, bottom).

15.3.1 Sensors

When attempting to use MEMS sensors for implementing autonomous control of free-shear flows, several issues should be considered.

15.3.1.1 Sensor Types

From a practical point of view, it is very difficult if not impossible to use sensors that are embedded inside the flow to achieve the information feedback necessary for implementing closed-loop control. This is due primarily to the inability to use a sufficiently large number of sensors to provide the needed information without blocking and/or significantly altering the flow. Therefore, the deployment of the MEMS sensor arrays should be restricted to the surface. For flows where no heat transfer occurs, this limits the measurements primarily to the surface stresses: wall shear (τ_w), or tangential component; and wall pressure (p_w), or normal component. The former has one component in the streamwise (x) direction and the other in the spanwise (z) direction.

In addition to surface stresses, near-wall measurements of flow velocity may be achieved nonintrusively via optical means. An example of such system is currently being developed by Gharib et al. (1999), who are utilizing miniature diode lasers integrated with optics in a small package to develop mini-LDAs (laser doppler anemometers) for measurements of the wall shear and near-wall velocity. With the continued

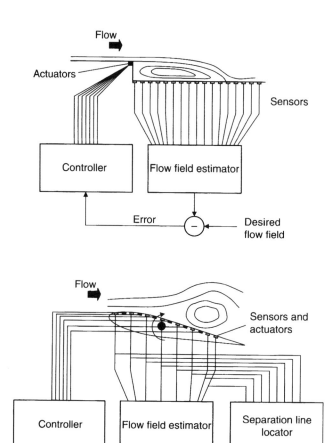

FIGURE 15.1   Conceptual block diagram of autonomous control systems for SSL (top) and MSL (bottom) flows.

miniaturization of components, it is not difficult to envision an array of MEMS-based LDA sensors deployed over surfaces of aero- and hydro-dynamic devices. Of course, with LDA sensors flow seeding is necessary. Hence, such optical techniques may be practical only for flows where natural contamination is present or when practical means for seeding the flow locally in the vicinity of the measurement volume can be devised.

Whereas surface sensors for MSL flows should cover the surface surrounding the flow to be controlled, those for SSL flows typically would be placed at the trailing edge. However, in certain instances, placement of the sensors downstream of the trailing edge may be feasible. Examples include sudden expansion and backward facing step flows (Figure 15.1, top) and jet flows, where a sting may be extended at the center of the jet for sensor mounting.

15.3.1.2   Sensor Characteristics

Properly designed and fabricated MEMS sensors should possess very high spatial and temporal resolution because of their extremely small size. Therefore, when considering the response of MEMS sensors for surface measurements, only the sensitivity and signal-to-noise ratio (SNR) are of concern. The need for high sensitivity and SNR is particularly important in detecting separation because the wall shear values are near zero in the vicinity of the separation line ($\tau_w = 0$ at separation for steady flows). Also, hydrodynamic surface pressure fluctuations are typically small for low-speed flows and require microphone-like sensitivities for measurements.

The concern regarding MEMS sensor sensitivity does not include indirect measurements of surface shear, such as that conducted using thermal anemometry, which can be achieved with higher sensitivity using MEMS sensors than using conventional sensors (e.g., [Liu et al., 1994; and Cain et al., 2000]). On the other hand, both direct measurements of the surface shear (using a floating element) and pressure measurements (through a deflecting diaphragm) rely on the force produced by those stresses acting on the area of the sensor. Since typical MEMS sensor dimensions are less than 500 μm on the side, the resulting force is extremely small. Because of this, no MEMS pressure sensor currently is known to have a sensitivity comparable to that of 1/8″ capacitive microphones [Naguib et al., 1999a]. Also, notwithstanding the creativity involved in developing a number of floating element designs and detection schemes [Padmanabhan et al., 1996; and Reshotko et al., 1996], the signal-to-noise ratio of such sensors remains significantly below that achievable with thermal sensors. Because of the direction reversing nature of separated flows, however, thermal shear sensors generally are not too useful in MSL flows. Direct or other-direction-sensitive sensors are required for conducting appropriate measurements in such flows.

15.3.1.3 Nature of the Measurements

Fundamentally, the information inferred from measurements of the wall-shear stress and near-wall velocity differs from that obtained from the surface pressure. The latter is known to be a global quantity that is influenced by both near-surface as well as remote flow structures. On the other hand, measurements of the shear stress and flow velocity provide information concerning flow structures that are in direct contact with the sensors.

Because of the global nature of pressure measurements, pressure sensors are probably the most suitable type of sensors for conducting measurements at the trailing edge in SSL flows. More specifically, in most if not all instances involving SSL flows, the control objective is aimed at controlling the flow downstream of the trailing edge. Therefore, local measurements of the wall shear and velocity would not be of great use in predicting the flow structure downstream of the trailing edge. A possible exception is when hydrodynamic feedback mechanisms are present, as is the case in a backward facing step where some structural influences downstream are naturally fed back to the trailing edge. In such flows, however, the instability of the shear layer may be *absolute* rather than *convective* [Huerre and Monkewitz, 1990]. Fiedler and Fernholz (1990) suggested that absolutely unstable flows are less susceptible to local periodic excitation such as that discussed here. Control provisions aimed at blocking the hydrodynamic feedback loop seem to be more effective in such cases.

Another important issue concerning the nature of surface pressure measurements is that they inevitably contain contributions from hydrodynamic as well as acoustic sources. The latter could be either a consequence of the flow itself (such as jet noise) or of the environment, emanating from other surrounding sources that are not related to the flow field. Typically, when the sound is produced by the flow, knowledge of the general characteristics of the acoustic field (direction of propagation, special symmetries, etc.) enables separation of the hydrodynamic and acoustic contributions to pressure measurements. In general, attention must be paid to ensure that the appropriate component of the pressure measurements is extracted.

15.3.1.4 Robustness and Packaging

Perhaps the primary concern for the use of MEMS in practical systems is whether the minute, fairly fragile devices can withstand the operating environment. One solution is to package the devices in isolation from their environment. This solution has been adopted, for example, in the commercially available MEMS accelerometers from Analog Devices, Inc. Such a solution, although possibly feasible for the mini-LDA systems discussed above, in general is not useful for flow applications. Most wall-shear and wall-pressure sensors and all actuators must interact directly with the flow. Therefore, the ability of the minute devices to withstand harsh, high-temperature, chemically reacting environments is of concern. Also, the possibility for mechanical failure during routine operation and maintenance must be accounted for.

15.3.1.5 Ability to Integrate with Actuators and Electronics

Although many types of MEMS sensors have been fabricated and characterized for use in flow applications, most of these sensors have been designed for isolated operation individually or in arrays. If autonomous control is to be achieved using the unique capability of MEMS for integration of components, the sensor fabrication technology must be compatible with that of the actuators and the circuitry. More specifically, if a particular fabrication sequence is successful in constructing a pressure sensor with certain desired characteristics, the same sequence may be unusable when the sensors are to be integrated with actuators on the same chip. Examples of the few integrated MEMS systems that were developed for autonomous control include the fully integrated system used by Tsao et al. (1997) for controlling the drag force in turbulent boundary layers and the actuator-sensor systems from Huang et al. (1996) aimed at controlling supersonic jet screech (discussed in detail later in this chapter).

Tsao et al. (1997) utilized a CMOS compatible technology to fabricate three magnetic flap-type actuators integrated with 18 wall-shear stress sensors and control electronics. In contrast, Huang et al. (1996 and 1998) developed hot-wire and pressure sensor arrays integrated with resonant electrostatic actuators. Typically two actuators were integrated with three to four sensors in the immediate vicinity of the actuators.

15.3.2   Actuators

15.3.2.1   Actuator Types and Receptivity

There seems to be a multitude of possible of ways to excite a shear layer that differ in their nature of excitation (mechanical, fluidic, acoustic, thermal, etc.), relative orientation to the flow (e.g., tangential versus lateral actuation), domain of influence (local versus global), and specific positioning with respect to the shear layer. Given the wide range of actuator types, it is confusing to choose the most appropriate type of actuation for a particular flow control application. Because of the micron-level disturbances introduced by MEMS actuators, however, it is not ambiguous that the actuation scheme providing "the most bang for the money" should be utilized; that is, the type of actuation that is most efficient or to which the flow has the largest *receptivity*. The ambiguity in picking and choosing from the list of actuators is for the most part due to our lack of understanding of the receptivity of flows to the different types of actuation.

When selecting a MEMS actuator, or a conventional actuator for that matter, for flow control, one is faced with a few fundamental questions. What type of actuator will achieve the control objective with minimum input energy? That is, what type of actuator produces the largest receptivity? Is it a mechanical, fluidic, acoustic, or other type of actuator? If mechanical, should it oscillate in the normal, streamwise, or spanwise direction? What actuator amplitude is needed to generate a certain flow-velocity disturbance magnitude? In the vicinity of the point of separation, is there an optimal location for the chosen type of actuator? All of these questions are fundamental not only to the design of the actuation system but also to the assessment of the feasibility of using MEMS actuators to accomplish the control goal.

15.3.2.2   Forcing Parameters

Perhaps one of the most limiting factors in the applicability of MEMS actuators in flow control is their micron-size forcing amplitude. Typical mechanical MEMS actuators are capable of delivering oscillation amplitudes ranging from a few microns to tens of microns. In shear flows evolving from the separation of a thin laminar boundary layer, such small amplitudes can be comparable to the momentum thickness of the separating boundary layer, resulting in significant flow disturbance. This is particularly true in SSL flows, where the actuator location can be maintained in the immediate vicinity of the separation point.

To appreciate the susceptibility of shear-layers to very low-level actuation, it is instructive to consider an order of magnitude analysis. For example, consider a situation where it is desired to attenuate or eliminate naturally existing two-dimensional disturbances in a laminar incompressible single-stream shear layer. One approach to attain this goal is through "cancellation" of the disturbance during its initial linear growth phase, where the streamwise development of the instability velocity amplitude, or *rms*, is given by:

$$u'(x) = u'_o \, e^{-\alpha_i (x - x_o)} \tag{15.1}$$

where, u' is an integral measure of the instability amplitude at streamwise location x from the point of separation of the shear layer (trailing edge in this example), u'_o is the *initial* instability amplitude, x_o is the *virtual origin* of the shear layer, and α_i is the imaginary wavenumber component, or spatial growth rate, of the instability wave.

To cancel the flow instability, a periodic disturbance at the same frequency but 180° out of phase may be introduced at some location $x = x^*$ near the point of separation of the shear layer. The amplitude of the disturbance introduced by the control should be of the same order of magnitude as that of the natural flow instability at the location of the actuator. This may be estimated from Equation (15.1) as follows:

$$u'(x^*) = u'_o e^{-\alpha_i(x^*-x_o)} \tag{15.2}$$

The location of the actuator should be less than one to two wavelengths (λ) of the flow instability downstream of the trailing edge to be within the linear region. This allows superposition of the control and natural instabilities, leading to cancellation of the latter. At the end of the linear range (approximately 2λ), the natural instability amplitude may be calculated from Equation (15.1) as:

$$u'(2\lambda) = u'_o e^{-\alpha_i(2\lambda-x_o)} \tag{15.3}$$

Dividing Equation (15.2) by Equation (15.3), one may obtain an estimate for the required actuator-induced disturbance amplitude in terms of the flow instability amplitude at the end of the linear growth zone:

$$u'(x^*) = u'(2\lambda)e^{-\alpha_i(x^*-2\lambda)} \tag{15.4}$$

The instability amplitude at the end of the linear region may be estimated from typical amplitude saturation levels (about 10% of the freestream velocity, U). Also, if the natural instability corresponds to the most unstable mode, its frequency would be given by (e.g., [Ho and Huerre, 1984]):

$$St = f\theta/U = 0.016 \tag{15.5}$$

where St is the Strouhal number and θ is the local shear layer thickness. For the most unstable mode, the instability convection speed is equal to $U/2$ and $-\alpha_i\theta = 0.1$ (see [Ho and Huerre, 1984], for instance). Using this information and writing the frequency in terms of the wavelength and convection velocity, Equation (15.5) reduces to:

$$-\alpha_i\lambda \approx (2 \times 0.16)^{-1} \tag{15.6}$$

for the most unstable mode.

Now, for the sake of the argument, assuming the actuator to be located at the shear layer separation point ($x^* = 0$), estimating $u'(2\lambda)$ as $0.1U$, and substituting from Equation (15.6) in Equation (15.4), one obtains:

$$u'(x^*) = 0.1Ue^{-\frac{2}{2\times0.16}} \approx 10^{-4}U \tag{15.7}$$

That is, the required actuator–generated disturbance velocity amplitude should be about 0.01% of the free stream velocity. Also, the actual forcing amplitude is not that given by Equation (15.7) but rather one that is related to it through an amplitude receptivity coefficient. That is:

$$u'_{act.} = u'(x^*) \times R \tag{15.8}$$

where R is the receptivity coefficient and $u'_{act.}$ is the actuation amplitude. Thus, if R is a number of order one, the required actuator velocity amplitude for exciting flow structures of comparable strength to the natural coherent structures in a high-speed shear layer (say, $U = 100$ m/s) is equal to 1 cm/s (i.e., of the order of a few cm/s). If the actuator can oscillate at a frequency of 10 kHz (easily achievable with MEMS), the corresponding actuation amplitude is only about a few microns!

In MSL flows, the ability of the sensor array and associated search algorithm to locate the instantaneous separation location of the shear layer and the actuator-to-actuator spacing may not permit flow excitation as close to the separation point as desired. In this case, even for extremely thin laminar shear layers, stronger actuators may be needed to compensate for the possible suboptimal actuation location. MEMS actuators that are capable of delivering hundreds of microns up to order of 1 mm excitation amplitude have been devised. These include the work of Miller et al. (1996) and Yang et al. (1997).

Although the more powerful, large-amplitude actuators provide much needed "muscle" to the miniature devices, they nullify one of the main advantages of MEMS technology: the ability to fabricate actuators that can oscillate mechanically at frequencies of hundreds of kHz. Traditionally, high-frequency (few to tens of kHz) excitation of flows was possible only via acoustic means. With the ability of MEMS to fabricate devices with "microinertia," it is now possible to excite high-speed flows using mechanical devices (e.g., see [Naguib et al., 1997]).

To estimate the order of magnitude of the required excitation frequency, the frequency of the linearly most unstable mode in two-dimensional shear layers may be used. This frequency may be estimated from Equation (15.5). Using such an estimate, it is straightforward to show that the most unstable mode frequency for typical high-speed MAS shear layers, such as that in transonic and supersonic jets, is of the order of tens to hundreds of kHz. On the other hand, if one considers a microscale (MIS) shear layer with a momentum thickness in the range of one to 10 microns and a modest speed of 1 m/s, the corresponding most unstable frequency is in the range of 1.6–16 kHz. Of course, this estimate assumes the MIS shear layer instability characteristics are similar to those of the MAS shear layer, an assumption that awaits verification.

An inherent characteristic of high-frequency MEMS actuators with tens of microns oscillation amplitudes is that they tend to be of the resonant type. Moreover, the Q factor of such large-amplitude microresonators tends to be large. Therefore, these high-frequency actuators are typically useful only at or very close to the resonant frequency of the actuator. This is a limiting factor not only from the perspective of shear layer control under different flow conditions, but also when multimode forcing is desired. For instance, Corke and Kusek (1993) have shown that resonant subharmonic forcing of the shear layer surrounding an axi-symmetric jet leads to a substantial enhancement in the growth of the shear layer. To implement this forcing technique it was necessary to force the flow at two different frequencies simultaneously using an array of miniature speakers. Although the modal shapes of the forcing employed by Corke and Kusek (1993) can be implemented easily using a MEMS actuator array distributed around the jet exit, only one frequency of forcing can be targeted with a single actuator design as discussed above. A possible remedy would use the MEMS capability of fabricating densely packed structures to develop an interleaved array of two different actuators with two different resonant frequencies.

15.3.2.3 Robustness, Packaging and Ability to Integrate With Sensors and Electronics

Similar to sensors, robustness of the MEMS actuators is essential if they are to be useful in practice. In fact, actuators tend to be more vulnerable to adverse effects of the flow environment than sensors are, as they typically protrude farther into the flow, exposing themselves to higher flow velocities, temperatures, forces, etc. Additionally, as discussed earlier, the fabrication processes of the actuators must be compatible with those of the sensors and circuitry if they are to be packaged together into an autonomous control system.

15.3.3 Closing the Loop: The Control Law

Perhaps one of the most challenging aspects of realizing MEMS autonomous systems in practice is one that is not related to the microfabrication technology itself. Given an integrated array of MEMS sensors and actuators that meets the characteristics described above, a fundamental question arises. How should the information from the sensors be processed to decide where, when, and how much actuation should be exercised to maintain a desired flow state? That is, what is the appropriate control law?

Of course, to arrive at such a control law, one needs to know the response of the flow to the range of possible actuation. This is far from being a straightforward task, however, given the nonlinearity of the system being controlled: the Navier–Stokes Equations. Moin and Bewley (1994) provide a good summary of various approaches that have been attempted to develop control laws for flow applications. Detailed discussion of the topic is not part of this chapter, and it is mentioned in passing here only to underline its significance for implementation of autonomous MEMS control systems.

15.4 Control of the Roll Moment on a Delta Wing

This section discusses a MEMS system aimed at realizing autonomous control of the roll moment acting on a delta wing. It is based on the work of fluid dynamics researchers from the University of California Los Angeles (UCLA) in collaboration with microfabrication investigators from the California Institute of Technology (Caltech). The premise of the control pursued by the UCLA/Caltech group is based on the dominating influence of the suction-side vortices of a delta wing on the lift force. In particular, when a delta wing is placed at a high angle of attack, the shear layer separating around the side edge of the wing rolls up into a persistent vortical structure. A pair of these vortices (one from each side of the wing) is known to be responsible for generating about 40% of the lift force. Plan- and end-view flow visualization images of such vortices are shown in Figure 15.2. Since the vortical structures evolve from a separating shear layer, their characteristics (e.g., location above the wing and strength) may be manipulated indirectly through alteration of the shear layer at or near the point of separation. Such a manipulation may be used to change the characteristics of only one of the vortices, thus breaking the symmetry of the flow structure and leading to a net rolling moment.

The shear layer separating from the edge of the delta wing is thin (order of 1 mm for the UCLA/Caltech work) and very sensitive to minute changes in the geometry. Therefore, as discussed earlier in this chapter, the use of microactuators to alter the characteristics of the shear layer and ultimately the vortical structure has a good potential for success. Furthermore, when the edge of the wing is rounded rather than sharp, the specific separation point location will vary with the distance from the wing apex, the flow velocity, and the position of the wing relative to the flow. Therefore, a distributed sensor–actuator array is needed to cover the area around the edge of the delta wing for detection of the separation line and actuation in its immediate vicinity.

15.4.1 Sensing

To detect the location of the separation line around the edge of the delta wing, the UCLA/Caltech group utilized an array of MEMS hot-wire shear sensors. The sensors, which are described in detail by Liu et al. (1994), consisted of 2 μm wide × 80 μm long polysilicon resistors that were micromachined on top of an evacuated cavity (a SEM view of one of the sensors is provided in Figure 15.3). The vacuum cavity provided thermal isolation against heat conduction to the substrate in order to maximize sensor cooling by the flow. The resulting sensitivity was about 15 mV/Pa, and the frequency response of the sensors was 10 kHz.

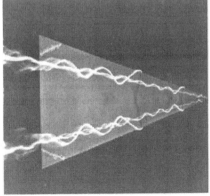

FIGURE 15.2   End (left) and plan (right) visualization of the flow around a delta wing at high angle of attack.

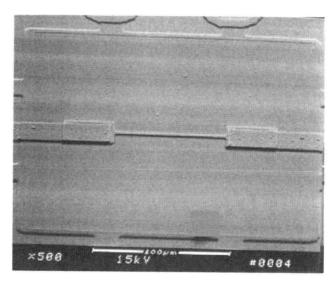

FIGURE 15.3   SEM image of the UCLA/Caltech shear stress sensor.

Because of directional ambiguity of hot-wire measurements and the three-dimensionality of the separation line, it was not possible to identify the location of separation from the instantaneous shear-stress values measured by the MEMS sensors. Instead, Lee et al. (1996) defined the location of the separation line as that separating the pressure- and suction-side flows in the vicinity of the edge of the wing. The distinction between the pressure and suction sides was based on the *rms* level of the wall-shear signal. This was possible because the unsteady separating flow on the suction side produced a highly fluctuating wall-shear signature in comparison to the more steady attached flow on the pressure side.

A typical variation in the *rms* value of the wall-shear sensor is shown as a function of the position around the leading edge of the wing in Figure 15.4. Note that the position around the edge is expressed in terms of the angle from the bottom side of the edge, as demonstrated by the inset in Figure 15.4. Because the *rms* is a time-integrated quantity, the detection criterion was primarily useful in identifying the average location of separation. In a more dynamic situation where, for example, the wing is undergoing a pitching motion, different criteria or sensor types should be used to track the instantaneous location of separation.

To map the separation position along the edge of the wing, Ho et al. (1998) utilized 64 hot-wire sensors that were integrated during the fabrication process on a flexible shear stress skin. This 80 μm thick skin, shown in Figure 15.5, covered a $1 \times 3 \, cm^2$ area and was wrapped around the curved leading edge of the delta wing. Using the *rms* criterion discussed in the previous paragraph, the separation line was identified for different flow speeds. A plot demonstrating the results is given in Figure 15.6, where the angle at which separation occurs (see inset in Figure 15.4 for definition of the angle) is plotted as a function of the distance from the wing apex. As seen from the figure, the separation line is curved, demonstrating the three-dimensional nature of the separation. Additionally, the separation point seems to move closer to the pressure side of the wing with increasing flow velocity (more details may be found in [Jiang et al., 2000]).

15.4.2   Actuation

Two different types of actuators were fabricated for use in the MEMS rolling-moment control system: (1) magnetic flap actuators, and (2) bubble actuators. The electromagnetic driving mechanism of the former was selected because of its ability to provide larger forces and displacements than the more common electrostatic type. Passive as well as active versions of the magnetic actuator were conceived. In the passive design, a $1 \times 1 \, mm^2$ ploysilicon flap (Figure 15.7) was supported by two flexible straight beams on a substrate. A 5 μm thick permalloy (80% Ni and 20% Fe) was electroplated on the surface of the 1.8 μm thick flap.

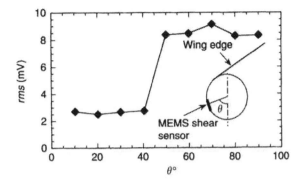

FIGURE 15.4   Distribution of surface-shear stress *rms* values around the edge of the delta wing.

FIGURE 15.5   UCLA/Caltech flexible shear-stress sensor array skin.

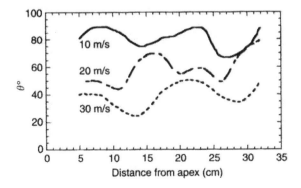

FIGURE 15.6   Separation line along the edge of the delta wing for different flow velocities.

FIGURE 15.7   SEM view of passive-type flap actuators.

The permalloy layer caused the flap to align itself with the magnetic field lines of a permanent magnet. Hence, it was possible to move the flap up or down by rotating a magnet embedded inside the edge of the wing as seen in Figure 15.8. The actuator is described in more detail in Liu et al. (1995).

A photograph of the active flap actuator is shown in Figure 15.9. The construction of this device is generally the same as that of the passive one, except for a copper coil deposited on the silicon nitride flap. A time varying current can be passed through the coil to modulate the flap motion around an average position (determined by the permalloy layer electroplated on the flap and the magnetic field imposed by an external permanent magnet). The actuator response was characterized by Tsao et al. (1994). The results demonstrated the ability of the actuator to produce tip displacements of more than $100\,\mu$m at frequencies of more than 1 kHz.

To develop more robust actuators that are not only useable in wind tunnels but also in practice, Grosjean et al. (1998) fabricated "balloon," or bubble, actuators. The basic principle of these actuators is based on inflating flush-mounted flexible silicon membranes using pressurized gas. As seen in Figure 15.10, the gas can be supplied through ports that are integrated under the membrane during the microfabrication process. When, inflated, the bubbles can extend to heights close to 1 mm. Figure 15.11 demonstrates bubble inflation with increasing pressure.

15.4.3   Flow Control

To test the ability of the actuators to produce a significant change in the roll moment on a delta wing, a 56.5° swept-angle model was used. The model, which was 30 cm long and 1.47 cm thick, was placed inside a $91\,\text{cm}^2$ test section of an open return wind tunnel. For different angles of attack and flow speeds, the moment and forces on the wing were measured using a six-component force/moment transducer.

Figures 15.12a through 15.12d display measurements of the change in the rolling moment as a function of the location of the actuator around the leading edge of the wing. Each of the four plots in Figure 15.12

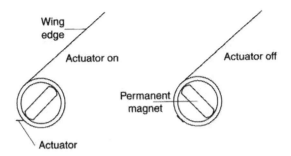

FIGURE 15.8   Permanent magnet actuation mechanism inside delta wing.

FIGURE 15.9   SEM view of active-type flap actuator.

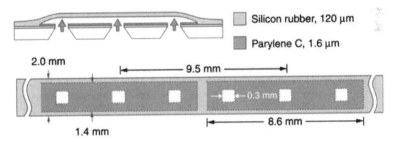

FIGURE 15.10   Schematic of bubble actuator details.

represents data acquired at a different angle of attack: $\alpha = 20°, 25°, 30°$, and $35°$ for plots a through d respectively. Also, different lines represent different Reynolds numbers.

The results shown in Figure 15.12 were obtained for the magnetic flap actuators. However, similar results were also produced using bubble actuators [Ho et al., 1998]. In all cases, the actuators were deflected by 2 mm only on one side of the wing. The results demonstrate a significant change in the rolling moment (up to 40% for $\alpha = 25$) for all angles of attack. The largest positive roll moment (rolling toward the actuation side) change is observed around an actuator location of approximately 50° from the pressure side of the wing. The location of maximum influence was found to be slightly upstream of the separation line (identified earlier using the MEMS sensor array; see Figure 15.4) as anticipated. Another important feature in Figure 15.12 is the apparent collapse of the different lines, which suggests that the actuation impact is affected very little, if any, by the changing Reynolds number.

In addition to producing a net positive roll moment, the miniature actuators are also capable of producing net negative moment at the lower angles of attack as implied by the negative peak depicted in

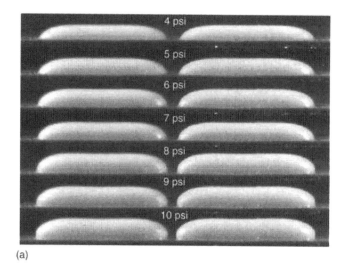

(a)

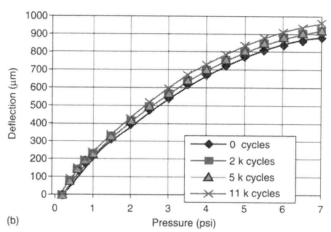

(b)

FIGURE 15.11   Bubble actuators (a) and their characteristics (b).

Figures 15.12a and 15.12b. Although, the largest negative roll value (found for an actuator location of approximately 100° from the pressure side) is not as strong as the largest positive moment, it can be used to augment the total roll moment control authority in a particular direction by using a two-sided actuation scheme. That is, if actuators at the appropriate locations on both edges of the delta wing were operated simultaneously to produce an upward motion on one side and a downward motion on the other, the two effects would superimpose to produce a larger net rolling moment. This type of moment augmentation was verified by Lee et al. (1996), as seen from the plot in Figure 15.13. An interesting aspect of the results shown in Figure 15.13 is that the net rolling moment produced by the two-sided actuation is equal to the sum of the moments produced by actuation on either side individually. This seems to suggest that the manipulation of each of the vortices is independent of the other.

From a physical point of view, the production of a net rolling moment by the upward motion of the actuator is caused by a resulting shift in the location of the nearby-vortex core. This was demonstrated clearly in the flow visualization images from Ho et al. (1998). Two of these images may be seen in Figure 15.14. In each image, a light-sheet cut was used to make the cores of the two edge vortices visible. The top image corresponds to the flow without actuation, whereas the bottom one was captured when the right-side actuators (of the bubble type in this case) were activated. A clear outward-shift of approximately one vortex

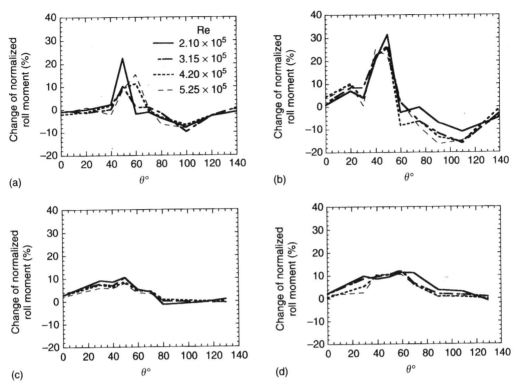

FIGURE 15.12 Actuation effect on delta wing rolling moment for different angles of attack: (a) 20°; (b) 25°; (c) 30°; (d) 35°.

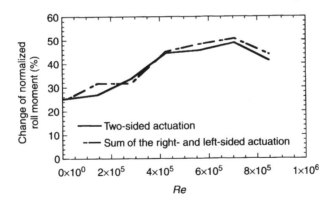

FIGURE 15.13 Roll moment augmentation using two-sided actuation.

diameter is observed for the right-side vortex core. A corresponding shift in the surface-pressure signature of the vortex presumably produces the observed net rolling moment.

The amount of vortex-core shift and resulting net moment appear to be proportional to the actuation stroke. This can be seen from the results in Figure 15.15, where the change in rolling moment due to actuation on one side is displayed for two different actuator displacements, 1 and 2 mm. As observed from the data, the peak moment produced by the 2 mm actuator is about five times larger than that produced by the 1 mm actuator. It is unclear, however, if the increase in the moment is a linear or nonlinear function of the actuator size, or stroke. In either case, the ability to adjust the amount of moment produced through the actuation scheme is essential if proportional control of the rolling moment is desired.

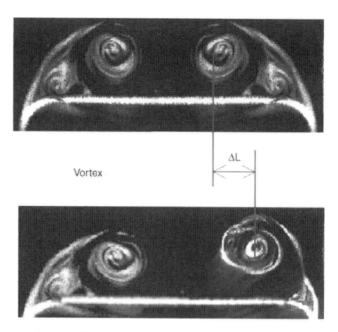

FIGURE 15.14   Flow visualization of the vortex structure on the suction side of the delta wing without (top) and with (bottom) actuation.

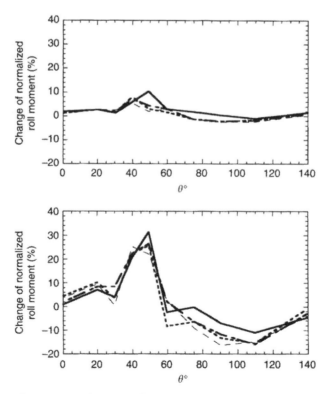

FIGURE 15.15   Effect of actuation amplitude on roll moment: 1 mm (top) and 2 mm (bottom).

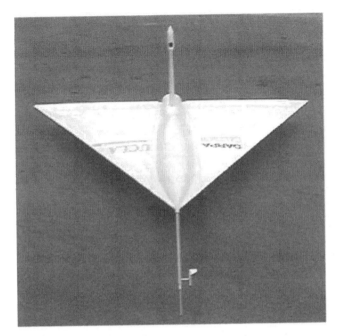

FIGURE 15.16   GRYPHON, the UCLA/Caltech MEMS flight research UAV.

Finally, Lee et al. (1996) also demonstrated that the utility of the delta wing control system is not limited to the roll component of the moment vector. The ability to control the pitch and yaw moments as well was demonstrated.

15.4.4   System Integration

Although sensing and actuation components of the MEMS roll-moment control system were fabricated and demonstrated successfully by the UCLA/Caltech group, a demonstration of the full integrated system in operation is yet to take place; that is, a demonstration of a real-time control action whereby the separation location is detected by the hot-wire sensor array and the output is fed to the appropriate algorithms to engage the actuators along the separation line — all done autonomously and for dynamic separation line conditions. Such a demonstration is perhaps more challenging from a fluid mechanics perspective than from the technology, or MEMS, point of view.

So far the work conducted by the UCLA/Caltech group has shown the ability to use MEMS to fabricate the sensors and actuators necessary to significantly modify the rolling moment on a delta wing (at least at the scale of the test model and under controlled wind tunnel conditions). This includes the compatibility of the fabrication processes of the sensor and actuators so that they can be integrated in arrays on the same substrate. However, from a fluid mechanics point of view, the use of hot wires to detect the location of dynamic three-dimensional separation is almost impossible. This is particularly true if real-time control is to be implemented where the separation position is to be located instantaneously rather than on average.

Finally, it should be added here that the UCLA/Caltech group has recently collaborated with AeroVironment to develop an instrumented research unmanned aerial vehicle (UAV) for testing of advanced MEMS control concepts for maneuvering of aircraft. The vehicle, which was called GRYPHON, is shown in the photograph in Figure 15.16.

15.5   Control of Supersonic Jet Screech

Reminiscent of several early efforts aimed at developing MEMS sensor–actuator systems for flow control, this study involved the collaboration of fluid dynamics researchers from the Illinois Institute of Technology (IIT) and microfabrication experts from the Center for Integrated Sensors and Circuits, Solid-State Electronics Laboratory at the University of Michigan (UM). The selection of supersonic screech noise for the scope of the MEMS control system was motivated by the desire to push and, therefore, identify the limits of MEMS technology rather than by the eventual practical implementation of such a system. That is, the jet screech environment provided harsh high-speed flow surroundings within which the devices not only had to operate but also produce a significant disturbance of the flow.

To explain further, it is useful to describe the general characteristics of jet screech. When a supersonic jet is operating at an off-design Mach number (i.e., the jet is under- or over-expanded), a set of shock cells forms. These cells interact with the axi-symmetric or helical flow structures that originate from the instability of the shear layer surrounding the jet at its exit to produce acoustic noise. The noise is primarily radiated in the upstream direction, where it excites the jet shear layer regenerating the flow structure. Thus, a self-sustained feedback cycle (first recognized by [Powel, 1953]) is established during screech.

The basic principle of the proposed IIT/UM screech control strategy is that of mode cancellation. Because the existence of axi-symmetric or helical shear layer modes is a necessary link in the screech generation chain of events, their elimination or weakening is anticipated to result in a corresponding cancellation or attenuation of screech noise. Such an elimination is to be achieved by the independent excitation of a flow structure of the same mode shape as, but out of phase with, that existing in the shear layer using MEMS actuators that are distributed around the nozzle lip. Since the MEMS forcing is achieved at the jet lip, where the shear layer stability growth is likely to be linear, the natural and forced disturbances will superpose destructively to weaken the shear layer structure. Also, the placement of the actuators in the *immediate* vicinity of the jet lip, or point of separation of the shear layer, renders the flow susceptible to the minute disturbances produced by the MEMS actuators.

The spatially-fixed location of the shear layer separation point in the jet flow alleviates the need for using sensor arrays to detect and locate the point of separation. Nevertheless, sensors are needed to provide reference information concerning the mode shape and phase of the existing shear layer structure. In this manner, the appropriate forcing phase of the actuators can be selected and adjusted with varying flow conditions.

15.5.1 Sensing

The IIT/UM group developed two types of sensors to provide the necessary measurements: hot wires and microphones. The hot wires were intended for direct measurements of the disturbance velocity in the shear layer near the nozzle lip, while the microphones were aimed at measuring the feedback acoustic waves that generate the disturbances during screech, as explained above. A close-up view of one of the MEMS hot wires overhanging a glass substrate is shown in Figure 15.17. Briefly, the hot wire is made from p^{++} polysilicon with typical dimensions of 12 μm thickness, 4 μm width, and 100 μm length. The corresponding resistance is in the range of 200–800 Ω. The glass substrate under the wire provides a good thermal-insulating base. Fabrication details are provided in Huang et al. (1996).

A sample calibration of one of the hot wires is shown in Figure 15.18 when the wire was operated at an overheat ratio of 1.05. Three different calibrations separated by time intervals of one hour and ten days are included in the figure. The calibration results display a change of about 2 volts of the hot-wire output for a velocity range of up to 35 m/s. A conventional hot wire at even larger overheats of more than 1.5 would exhibit a voltage output range of less than 1 volt. Hence, Figure 15.18 demonstrates the high sensitivity of MEMS (polysilicon) hot wires. Furthermore, considering the calibration results from the three different trials, very little change in the sensor characteristics is observed. Given the long time interval between some of the calibrations, the MEMS sensors are generally very stable. Further characterization of the sensors may be found in Naguib et al. (1999b).

On the other hand, the MEMS acoustic sensor, shown using a SEM view in Figure 15.19, consists of a stress-compensated PECVD silicon nitride/oxide, 0.4 μm thick diaphragm together with four monocrystalline ion-implanted p^{++} silicon piezoesistors. The coefficient π_{44} for this type of piezoresistor is about four times larger than that based on p-type polysilicon, thus leading to a higher transducer sensitivity. The piezoresistors are arranged in a full Wheatstone-bridge configuration for detection of the diaphragm deflection. For fabrication details, the reader is referred to Huang et al. (1998).

FIGURE 15.17 SEM view of the IIT/UM hot-wire sensor.

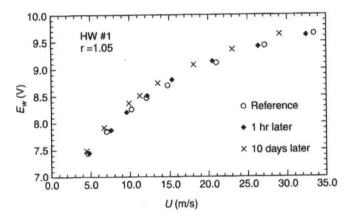

FIGURE 15.18 Static calibration of one of the IIT/UM hot-wire sensors.

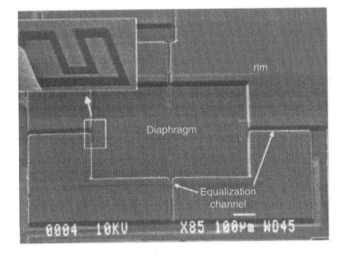

FIGURE 15.19 SEM view of IIT/UM acoustic sensor.

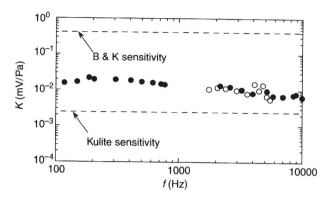

FIGURE 15.20   Frequency response of one of the IIT/UM acoustic sensors.

Figure 15.20 depicts the frequency characteristics of a $710 \times 710\,\mu m^2$ MEMS sensor obtained by Huang et al. (2002a). For reference, the sensitivities of a commercial silicon Kulite sensor and a 1/8″ B&K microphone are also provided in the figure. The open and closed circles represent calibration results for the same sensor obtained two-months apart. The results demonstrate a very good agreement between the open and closed circles. This agreement indicates that the MEMS transducers are stable and reliable. For frequencies below 1 kHz, the sensitivity of the MEMS sensors seems to be almost an order of magnitude larger than the sensitivity of the commercial Kulite sensor. At higher frequencies, however, the MEMS sensitivity is attenuated.

15.5.2   Actuation

The most common form of excitation of high-speed jets and shear layers has been via internal and external acoustic sources (e.g. [Moore, 1977]). The popularity of acoustic forcing has been due primarily to ease of implementation and the ability to generate the high-frequency disturbances required for effective forcing of high-speed shear flows. However, the receptivity of the jet to acoustic excitation requires matching of the wavenumbers of the acoustic and instability waves [Tam, 1986]. Although such a condition is achievable in supersonic jets, at low Mach numbers the instability wavelength is much smaller than the acoustic wavelength. In this case, conversion of acoustic energy into instability waves is dependent on the efficiency of the unsteady Kutta condition at the nozzle lip.

Historically, studies utilizing mechanical actuators have been limited to low flow speeds due to the inability of the mechanical actuators with their large inertia to operate at frequencies higher than tens to hundreds of herz. Therefore, the emergence of MEMS carried with it a hope to break this barrier through the fabrication of mechanical actuators with microinertia that can oscillate at high frequency. Ultimately, this may lead to the realization of more efficient actuators that can act directly on the flow without the need for a Kutta-like condition. For these reasons, micromechanical actuators were selected for the screech control problem.

Figure 15.21 shows a SEM photograph of one of the electrostatic actuators used in the jet study. The device consists of a free-floating section that is mounted to a glass substrate underneath using elastic folded beams. At the lower end of the floating section is a roughly $6\,\mu m \times 30\,\mu m \times 1300\,\mu m$ T-shaped actuator head. The actuator head is made to oscillate at an amplitude up to $70\,\mu m$ and a resonant frequency that can be tailored by adjusting the mass and stiffness of the actuator structure. Also, notice the porous actuator body, which results in a reduced dynamic load on the actuator head. For more details on the construction and manufacturing of the actuators, see Huang et al. (1996).

Two actuator designs were developed with a resonant frequency of 5 and 14 kHz. These values were selected to match the frequency of different screech modes in the High-Speed Jet Facility (HSJF) at IIT. The actuators were driven using a combination of 20 volt dc and a 40 volt peak-to-peak sinusoidal voltage at the actuator resonant frequency.

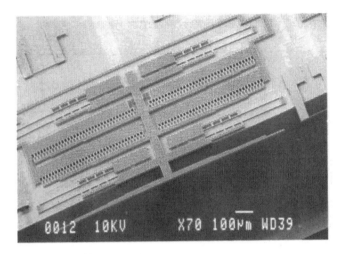

FIGURE 15.21 SEM view of IIT/UM electrostatic actuator.

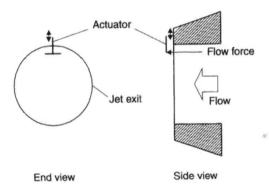

FIGURE 15.22 Schematic demonstrating direction of flow forces acting on the MEMS actuator.

The relative placement of one of the T-actuators with respect to the jet is illustrated in Figure 15.22. As depicted in the figure, the rigidity of the actuator with respect to out-of-plane deflection caused by flow forces is primarily dependent on the thickness of the device. For the device shown in Figure 15.21, the thickness was limited to a value of less than 12 μm because of the wet etching methodology employed in the fabrication sequence. An alternate fabrication approach, which utilized deep reactive ion etching (RIE) etching, was later utilized to fabricate thicker, more rigid actuators. Using this approach, actuators with a thickness of 50 μm were obtained. A SEM view of one of the thick actuators is provided in Figure 15.23.

15.5.3 Flow Control

The ability of the microactuators to disturb the shear layer surrounding the jet was tested for flow speeds of 70, 140, and 210 m/s. The corresponding Mach (M_j) and Reynolds numbers (Re; based on jet diameter, D) were 0.2, 0.4, and 0.6 and 1,18,533, 2,37,067 and 3,55,600 respectively. The corresponding linearly most unstable frequency of the jet shear layer was estimated to be 16, 44, and 91 kHz in order of increasing Mach number. Because the largest effect on the flow is achieved when forcing as close as possible to the most unstable frequency, the actuator frequency of 14 kHz was substantially lower than desired for the two largest jet speeds.

Using hot-wire measurements on the shear layer center line, velocity spectra of the natural and forced jet flows at the different Mach numbers were obtained. Those results are shown in Figure 15.24. The different

FIGURE 15.23   SEM view of a thick (50 μm) actuator.

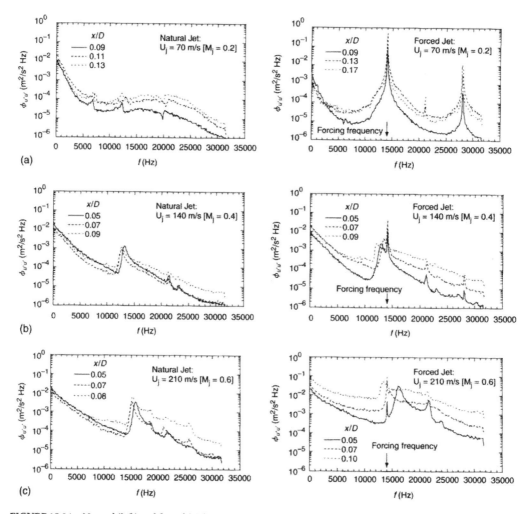

FIGURE 15.24   Natural (left) and forced (right) spectra in the IIT high-speed jet for a Mach number of: (a) 0.2; (b) 0.4; (c) 0.6.

lines in each of the plots represent spectra obtained at different streamwise (x) locations. For a Mach number of 0.2, the spectra obtained in the natural jet seem to be wide-band except for two fairly small peaks at approximately 7 and 13 kHz. When MEMS forcing is applied, a very strong peak is observed at the forcing frequency (Figure 15.24a, right). The magnitude of the peak seems to rise initially with downstream distance and then fall. In addition to the peak at the forcing frequency, a second strong peak at 28 kHz is observed at all streamwise locations. A third peak is also depicted at a frequency of 21 kHz at the second and third x locations. The existence of multiple peaks in the spectrum at frequencies that are multiples of the forcing frequency and its subharmonic suggests that the MEMS-introduced disturbance has a large enough amplitude to experience nonlinear effects of the flow.

The disturbance spectra for the natural jet at a Mach number of 0.4 are shown in Figure 15.24b (left). The corresponding spectra for the forced jet are shown in Figure 15.24b (right). As seen from Figure 15.24, the natural jet spectra possess a fairly large and broad peak at a frequency of about 13 kHz. Since the most unstable frequency of the jet shear layer at this Mach number is expected to be around 44 kHz, the 13 kHz peak does not seem to correspond to the natural mode of the shear layer or its subharmonic. When operating the MEMS actuator, a clear peak is depicted in the spectrum at the forcing frequency. The peak magnitude initially magnifies to reach a magnitude of more than an order of magnitude larger than the fairly strong peak depicted in the natural spectrum at 13 kHz. The MEMS-induced peak is also significantly sharper than the natural peak, presumably because of the more organized nature of forced modes.

Finally, the spectra obtained at a Mach number of 0.6 are displayed in Figure 15.24c. Similar to the results for 0.4 Mach number, a broad, fairly strong peak is depicted in the natural jet spectra. The frequency of this peak appears to be about 16 kHz at $x/D = 0.1$, which is considerably lower than the estimated most unstable frequency of 91 kHz. The peak frequency value decreases with increasing x. This decrease in the peak frequency with x may be symptomatic of probe feedback effects [Hussain and Zaman, 1978]. In addition to this strong peak, when forced using the MEMS actuator, a clearly observable peak at the forcing frequency is depicted in the spectrum of the forced shear layer. Unlike, the results for the forced jet at Mach numbers of 0.2 and 0.4, the forced jet spectrum for a Mach number of 0.6 does not contain spectral peaks at higher harmonics of the forcing frequency.

To evaluate the level of the disturbance introduced into the shear layer by the MEMS actuator, the energy content of the spectral peak at the forcing frequency was calculated from the phase-averaged spectra (to avoid inclusion of background turbulence energy). The forced disturbance energy ($<u_{rms,f}>$) dependence on the streamwise location is shown for all three Mach numbers in Figure 42.25. The disturbance energy is normalized by the jet velocity (U_j), and the streamwise coordinate is normalized by the jet diameter. Inspection of Figure 15.25 shows that for both Mach numbers of 0.2 and 0.4 no region of linear growth is detectable. For these two Mach numbers, ($<u_{rms,f}>$) only increases slightly before reaching a peak followed by a gradual decrease in value, a process that is reminiscent of nonlinear amplitude saturation.

On the other hand, the disturbance energy corresponding to a Mach number of 0.6 appears to experience linear growth over the first four streamwise positions before saturating. It appears also that only for 0.6 Mach number is the disturbance *rms* level appreciably lower than 1% of the jet velocity at the first streamwise location. In the work of Drubka et al. (1989), the fundamental and subharmonic modes in an acoustically excited incompressible axi-symmetric jet were seen to saturate when their *rms* value exceeded 1–2% of the jet velocity. Therefore, it seems that the IIT/UM MEMS actuator is capable of providing an excitation to the shear layer that is sufficient not only to disturb the flow but also to produce nonlinear forcing levels.

The magnitude of the MEMS forcing may also be appreciated further by comparison to other types of macroscale forcing. Thus, the disturbance *rms* value produced by internal acoustic [Lepicovsky et al., 1985] and glow discharge [Corke and Cavalieri, 1996] forcing is compared to the corresponding *rms* values produced by MEMS forcing in Figure 15.26. The results for MEMS forcing contained in the figure are those from Huang et al. (2002b) using the high-frequency MEMS actuators as well as those from the earlier study by Alnajjar et al. (1997) using the same type of MEMS actuators but at a forcing frequency of 5 kHz.

As seen from Figure 15.26, for most cases of MEMS forcing the MEMS-generated disturbance grows to a level that is similar to that produced by glow discharge and acoustic forcing. For a Mach number of 0.2,

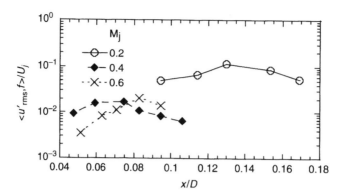

FIGURE 15.25   Streamwise development of the MEMS-induced disturbance *rms*.

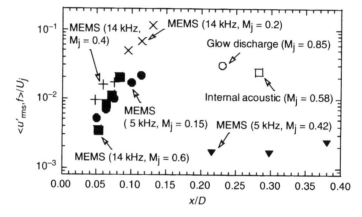

FIGURE 15.26   MEMS-induced disturbance *rms* compared to macroscale forcing schemes.

the forcing frequency of 14 kHz is almost equal to the most unstable frequency of the shear layer, and the resulting flow disturbance is at a level that is significantly higher than even that produced by the macroscale forcing methods. On the other hand, at the 0.42 Mach number, the 5-kHz MEMS actuator excites the flow at a frequency that is almost an order of magnitude lower than the most amplified frequency of the shear layer. The small amplification rate associated with a disturbance at a frequency that is significantly smaller than the natural frequency of the flow is believed to be responsible for the resulting small disturbance level at the Mach number of 0.42 when forcing with the 5 kHz actuator.

The ability of the MEMS devices to excite the flow on a par with other macro forcing devices is believed to be due to the ability to position the MEMS extremely close to the point of high-receptivity at the nozzle lip where the flow is sensitive to minute disturbances. To investigate this matter further, the radial position of the MEMS actuator with respect to the nozzle lip (y_{off}) was varied systematically. For all actuator positions, the flow was maintained at 70 m/s, while the actuator was traversed in the range from about 50 μm (outside the flow) to −150 μm (inside the flow) relative to the nozzle lip. The boundary layer at the exit of the jet at 70 m/s was laminar and had a momentum thickness of about 72 μm. Therefore, at its innermost location, the actuator penetrated into the flow a distance that was less than 20% of the boundary layer thickness, and hence, based on a Blasius profile, it was exposed to a velocity less than one fifth of the jet speed.

The energy content of the spectral peak at the forcing frequency was calculated from the disturbance spectra obtained at $x/D = 0.1$ for different radial actuator locations. The results are displayed in Figure 15.27. For reference, a dimension indicating the momentum thickness of the boundary layer emerging at the exit of the jet (θ_o) is included in the figure. As observed from the figure, the largest disturbance energy is produced when the actuator is closest to the nozzle lip ($y_{off} = 0$) and into the flow. If the actuator is placed a distance as small

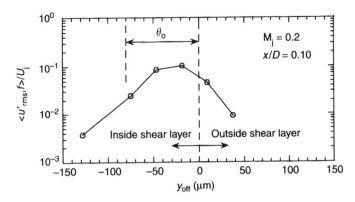

FIGURE 15.27   Effect of MEMS actuator radial position on the generated shear layer disturbance level.

as 75 μm (less than the size of a human hair) off the position corresponding to maximum response, an order of magnitude reduction in disturbance *rms* value is observed.

The reduction in actuator effectiveness as it is traversed radially outward is presumably because of the increased separation between the actuator and the shear layer. However, the reasons for observing the same trend when locating the actuator further into the flow are not equally clear. In fact, recently Alnajjar et al. (2000), using piezoelectric actuators, found the response of the same jet flow used for MEMS testing to be fairly insensitive to the radial location of the actuator inside the shear layer, with the largest response found near the center of the shear layer. This suggests that the reduced actuation impact seen in Figure 15.27 as the actuator is positioned farther into the flow is most likely due to a change in the behavior of the actuator itself. In particular, it is not unreasonable to expect that as the actuator head is exposed to larger flow speeds for positions farther into the flow, it may experience larger out-of-plane deflections. These deflections would cause the overlap area between the stationary and moving parts of the comb-drives to become smaller, leading to a smaller actuation force and amplitude.

15.5.4   System Integration

The fabrication sequence of the sensors for the screech control system was developed carefully to insure its compatibility with that of the actuators. As discussed at the beginning of the chapter, the compatibility of the fabrication sequence of the individual components is necessary if one is to capitalize on the MEMS technology's advantage of integrating component arrays to form full autonomous systems. For instance, to machine the MEMS acoustic sensors shown in Figure 15.19 integrated with the actuators on the same chip, a total of nine masks were required [Huang et al., 1998]. In contrast, fabrication of a similar sensor in isolation would probably require no more than three to four masks. SEM views of the electrostatic actuators integrated with hot wires and acoustic sensors are given in Figures 15.21 and 15.28 respectively.

To make external input–output connections to the sensors and actuators the glass substrate supporting the MEMS devices was fixed to a sector-shaped printed-circuit board using epoxy. Close-up and full views of the board may be seen in Figure 15.29. The full sensor–actuator array was then assembled by mounting the sixteen sector boards on a specially designed jet faceplate. Each of the individual boards was supported on a radial miniature traversing system to enable precise placement of the actuators at the location of maximum response (as dictated by the results in Figure 15.27). Figure 15.30 shows a schematic diagram of the jet faceplate, sector PC-boards, and miniature traverses. An image of the full MEMS array as mounted on the jet can be seen in Figure 15.31.

The distributed 16-actuator array provided the capability of exciting flow structures with azimuthal modes up to a mode number (m) of eight. This was verified using single hot-wire measurements at different locations around the perimeter of the jet. At each measurement location the wire was located approximately at the center of the shear layer. The amplitude and phase information of the measured disturbance was

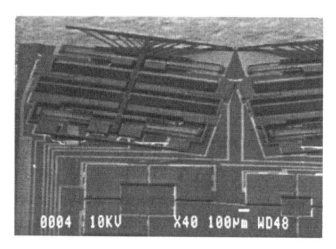

FIGURE 15.28   Two acoustic sensors integrated with two actuators.

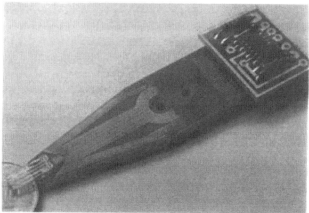

FIGURE 15.29   Packaging of the IIT/UM sensor–actuator system.

obtained from the cross-spectrum between the hot-wire data and the driving signal of one of the actuators. The results are demonstrated in Figure 15.32. As seen from the amplitude data, a fair amount of scatter is observed around an average value over the full circumference. The observed scatter is believed to be due to imprecise positioning of the hot wire at the center of the shear layer and variation in the amplitude of

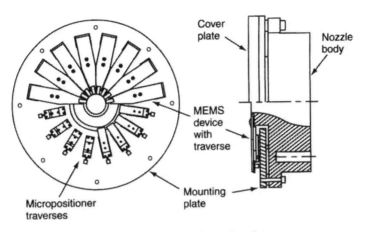

FIGURE 15.30 MEMS PC boards mounting provisions on the jet face plate.

FIGURE 15.31 Photograph of the MEMS sensor–actuator array mounted on the jet.

the different actuators. On the other hand, the phase results (Figure 15.32, bottom) indicate that the disturbance phase measured in the shear layer agrees with the excited flow modes.

The nonuniformity of the MEMS actuator amplitudes can be remedied easily by adjusting the individual actuator amplitudes to provide uniform forcing. A more significant problem encountered in forcing with the full array was slight deviations in the resonant frequency of the individual devices. For the most part, this deviation did not exceed ±2% of the average resonant frequency for a particular batch. This may not seem to be a substantial deviation. However, MEMS resonant devices tend to have very small damping ratios (high Q factor) that lead to very narrow resonant peaks. As a result, the devices cannot be operated at frequencies that deviate more than 0.5 to 1% from the resonance frequency while maintaining more than 20% of their resonance amplitude.

With the different actuators operating at slightly different frequencies it was not possible to sustain a specific phase relationship among the different actuators for extended periods. Therefore, the data shown in Figure 15.32 were acquired using a transient forcing scheme whereby the actuators were used to force a particular azimuthal mode only so long as the largest phase deviation did not exceed an acceptable tolerance. The best remedy to the problem with the current MEMS technology appears to be the fabrication of a large batch of devices, so that only those with matching resonant frequencies may be used.

In summary, the IIT/UM work has resulted in the realization of an integrated actuators and sensors system that can autonomously control shear layers in high-speed jets. The system was not actually used to control screech because the actuators operated successfully up to a Mach number of 0.8 but not in the high-unsteady-pressure environment encountered during screech. Nevertheless, the ability of the less-than-hair-width actuators to operate up to such a high speed while producing significant disturbance into the flow (even at frequencies less than 1/10 of the most unstable frequency) was quite an impressive demonstration of the potential of the technology.

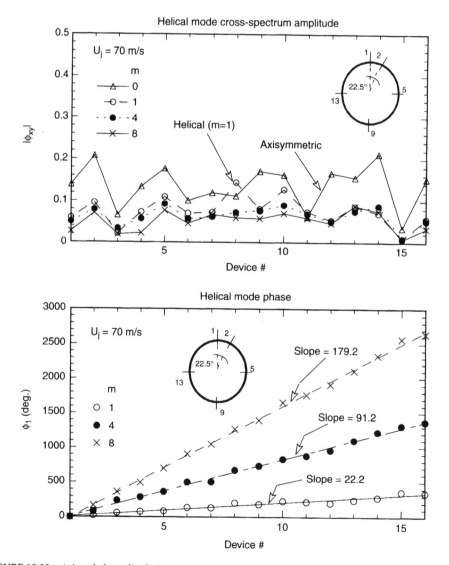

FIGURE 15.32   Azimuthal amplitude (top) and phase (bottom) distribution of the MEMS-induced disturbance.

15.6 Control of Separation over Low-Reynolds-Number Wings

Recently, researchers from the University of Florida have proposed a MEMS system for controlling separation at low Reynolds numbers. The primary motivation of the proposed system was to enhance the lift-to-drag ratio in the flight of micro-air-vehicles (MAVs). Because of their small size (a few centimeters characteristic size) and low speed, MAVs experience low Reynolds number flow phenomena during flight. One of these is an unsteady laminar separation that occurs near the leading edge of the wing and affects the aerodynamic efficiency of the wing adversely.

Figure 15.33 displays a schematic of the proposed control system components and test model geometry. The main idea is based on the deployment of integrated MEMS sensors and actuators near the leading edge of an airfoil, or wing section. Additional sensor arrays are to be used near the trailing edge of the wing. The leading edge sensors are intended for detection of the separation location in order to activate those actuators closest to that location for efficient control, as discussed previously. On the other hand, the trailing edge sensors are to be utilized to sense the location of flow reattachment. In this manner, it would be possible to adapt the magnitude and location of actuation in response to changes in the flow and thus, for instance, maintain the flow attached at a particular location on the wing. The ultimate benefit of such a control system is the manipulation of the aerodynamic forces on the wing for increased efficiency as well as maneuverability without the use of cumbersome mechanical systems.

In actual implementation, the University of Florida group adopted a hybrid approach whereby conventional-scale piezoelectric devices were used for actuation and MEMS sensors were used for measurements. Additionally, it appears that because of the difficulty in detecting the instantaneous separation location, as discussed in the delta wing control problem, a small step in the surface of the wing was introduced near the leading edge at the actuation location. Thus, the location of separation was fixed and there was no need to use leading edge sensors for initial testing of the controllability of the flow. The flow control test model is shown in Figure 15.34.

15.6.1 Sensing

To measure the unsteady wall shear stress, platinum-surface hot-wire sensors were microfabricated. The devices consisted of a 0.15 μm thick × 4 μm wide × 200 μm long platinum wire deposited on top of a 0.15 silicon nitride membrane. Beneath the membrane is a 10 μm deep vacuum cavity with a diameter of 200 μm. Similar to the UCLA/Caltech sensor the evacuated cavity was incorporated in the sensor design to maximize the thermal insulation to cooling effects other than that due to the flow. As a result the sensor

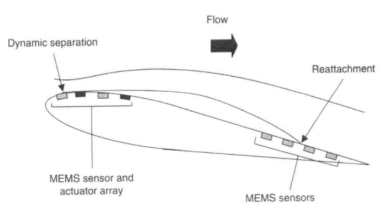

FIGURE 15.33 Control system components for University of Florida low Reynolds number wing control project.

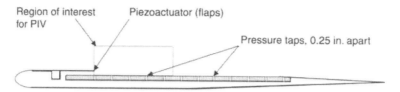

FIGURE 15.34   Test model for separation control experiments of University of Florida.

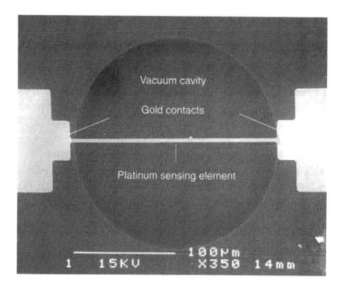

FIGURE 15.35   SEM view of University of Florida MEMS wall-shear sensor.

exhibited a static sensitivity as high as 11 mV/Pa when operating at an overheat ratio (operating resistance/cold resistance) of up to 2.0. The sensor details can be seen in the SEM image in Figure 15.35. For detailed characterization of the static and dynamic response of the sensor, refer to Chandrasekaran et al. (2000) and Cain et al. (2000).

15.6.2   Flow Control

Static surface pressure measurements and PIV images were used by Fuentes et al. (2000) to characterize the response of the reattaching flow to forcing with the piezoelectric actuators. The 51 mm wide × 16 m long flap-type actuators (see Figure 15.34) were operated at their resonance frequency of 200 Hz. The resulting static pressure (plotted as a coefficient of pressure, *CP*) distribution downstream of the 1.4 mm high step is given in Figure 15.36. Similar results without forcing are also provided in the figure for comparison. As seen from the figure, the minimum negative peak of *CP*, corresponding to the location of reattachment, shifts upstream with excitation. The extent of the shift is fairly significant, amounting to about 30% or so of the uncontrolled reattachment length.

The reduction in the reattachment length with forcing also can be depicted from the streamline plots obtained from PIV measurements (see Figure 15.37). However, the real benefit of the PIV data was to reveal the nature of the flow structure associated with actuation by capturing images that were phase-locked to different points of the forcing cycle. Those results are provided in Figure 15.38 for an approximately full cycle of the forcing. A convecting vortex structure is clearly seen in the sequence of streamline plots in Figures 15.38a through 15.38d. The observed vortex structures were periodic when an actuation amplitude of about 22 μm was used. For substantially smaller forcing amplitude, the generated vortices were found to be aperiodic.

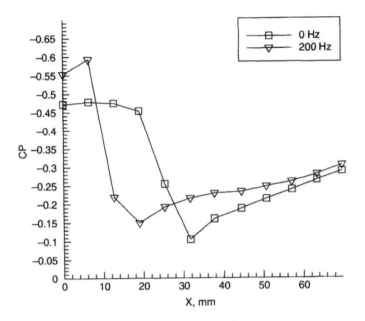

FIGURE 15.36 Static pressure distribution with and without control.

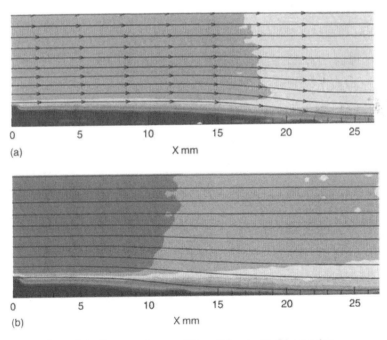

FIGURE 15.37 Streamlines of the flow over the step without (a) and with (b) actuation.

Similar to the UCLA/Caltech and IIT/UM efforts, the University of Florida work has demonstrated the ability to alter the flow significantly through low-level forcing. Additionally, high-sensitivity MEMS sensors were developed and tested. However, for all three efforts their remains to be a demonstration of a fully autonomous system in operation.

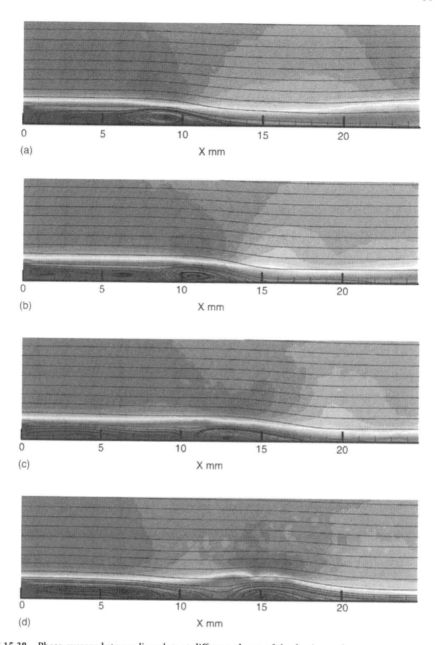

FIGURE 15.38 Phase-averaged stream line plots at different phases of the forcing cycle.

15.7 Reflections on the Future

When considering the potential use of MEMS for flow control, it is not difficult to find contradictory views within the fluid dynamics community. This is not surprising given the number of challenges facing the implementation and use of the fairly young technology. Challenges aside, however, there are certain capabilities that can be achieved only with MEMS technology. Examples include tens of kHz distributed mechanical actuators; sensor arrays that are capable of resolving the spatio-temporal character of the flow structure in high-Reynolds-number flows; integration of actuators, sensors, and electronics; and more. These are the kind of capabilities that seem to be needed if we are to have any hope of controlling such a difficult system as that governed by the Navier Stokes equations. Therefore, it is much more constructive to identify the challenges facing the use of MEMS and search for their solutions than to simply dismiss the technology along with its potential benefits. In this section, some of the leading challenges facing the attainment of autonomous MEMS control systems for shear layer control are highlighted. These are accompanied by the author's perspective on the hope of overcoming these challenges.

One of the main concerns regarding the implementation of MEMS devices is regarding their robustness, particularly if they have to be operated in harsh, high-temperature environments. For the most part, this concern stems from the micron size of the MEMS devices, which renders them vulnerable to large external forces. However, it is important to remember that as one shrinks a structure, the flow forces acting on it decrease along with its ability to sustain such forces. That is, to a certain extent the microscale devices may be as strong as, if not stronger than, their larger scale equivalents (at least if they are designed well). That is probably why the actuators from Naguib et al. (1997) operated properly while immersed in a Mach-0.8 shear layer, and the actuators and sensors of Huang et al. (2000) successfully completed a test flight while attached to the outside of an F-15 fighter jet. Furthermore, as new microfabrication techniques are devised for more resilient, chemically inert, harder materials than silicon, it will be possible to construct microdevices for harsh, high-temperature, chemically reacting environments. Some of the current notable efforts in this area are those concerned with micromachining of silicon carbide and diamond.

The robustness question is probably more critical from a practical point of view. That is, whereas a MEMS array of surface stress sensors deployed over an airplane wing may survive during flight, it may easily be crushed by a person during routine maintenance. However, such issues should, and could, be addressed at the design stage where, for instance, the sensor array might be designed to be normally hidden away and deploy only during flight. Additionally, the inherent array-fabricating ability of MEMS could be used to increase system robustness through redundancy. If a few sensors broke, other on-chip sensors could be used instead. If the number of malfunctioning sensors became unacceptable, the entire chip could be replaced with a new one. The economics of replacing MEMS system modules will likely be justified, as it seems natural that MEMS will eventually follow in the path of the IC chip with its low-cost bulk-fabrication technologies.

Beyond robustness, there will be a need to develop innovative approaches to enhance the signal-to-noise ratio of MEMS sensors. As discussed earlier, when shrinking sensors, their sensitivity often, but not always, decreases proportionally. Because for the most part traditional transduction approaches have been used with the smaller sensors, the overall signal-to-noise ratio cannot be maintained at desired levels. Hence, there is a need to identify ultrasensitive transduction methods. An example of such methods is the intragrain poly-diamond piezoresistive technology developed recently by Salhi and Aslam (1998). This technology promises the ability to integrate inexpensive poly-diamond piezoresitive gauges with a gauge factor of up to 4000 (20 times more sensitive than the best silicon sensors) into microsensors.

Finally, when it comes to actuation, one of the most challenging issues that need to be addressed is the sufficiency of MEMS actuation amplitudes. Notwithstanding the successful demonstrations of the IIT/UM and UCLA/Caltech groups discussed earlier in this chapter, boundary layers in practice tend to be significantly thicker and turbulent at separation than encountered in those experiments. Therefore, it is most likely that the use of MEMS actuators will be confined to controlled experiments in the laboratory (where they may be used, for example, for proof of concept experiments) and flows in microdevices. For large-scale flows, successful autonomous control systems will most probably be hybrids consolidating macroactuators with MEMS sensor arrays as in the University of Florida work. This will require developing clever techniques for integrating the fabrication processes of MEMS to those of large-scale devices in order to capitalize on the full advantage of MEMS.

Acknowledgment

The author greatly appreciates the help of Prof. Chih-Ming Ho at UCLA and Prof. Carol Bruce at the University of Florida for providing images and electronic copies of their publications for composition of this chapter.

References

Alnajjar, E., Naguib, A.M., Nagib, H.M., and Christophorou, C. (1997) "Receptivity of High-Speed Jets to Excitation Using an Array of MEMS-Based Mechanical Actuators," *Proceedings of ASME Fluids Engineering Division Summer Meeting*, paper FEDSM97-3224, 22–26 June, Vancouver, BC, Canada.

Alnajjar, E., Naguib, A., and Nagib, H. (2000) "Receptivity of an Axi-Symmetric Jet to Mechanical Excitation Using a Piezoelectric Actuator," *AIAA Fluids 2000*, AIAA paper number 2000-2557, 19–22 June, Denver, Colorado.

Cain, A., Chandrasekaran, V., Nishida, T., and Sheplak, M. (2000) "Development of a Wafer-Bonded, Silicon-Nitride, Membrane Thermal Shear-Stress Sensor with Platinum Sensing Element," *Solid-State Sensor and Actuator Workshop*, 4–8 June, Hilton Head Island, South Carolina.

Chandrasekaran, V., Cain, A., Nishida, T., and Sheplak, M. (2000) "Dynamic Calibration Technique for Thermal Shear Stress Sensors with Variable Mean Flow," *38th Aerospace Sciences Meeting & Exhibit*, AIAA paper number 2000-0508, 10–13 January, Reno, NV.

Corke, T.C., and Kusek, S.M. (1993) "Resonance in Axisymmetric Jets with Controlled Helical-Mode Input," *J. Fluid Mech.*, 249, pp. 307–36.

Corke, T.C., and Cavalieri, D. (1996) "Mode Excitation in a Jet at Mach 0.85," Bul. Am. Phys. Soc. 41, p. 1700, 49th American Physical Society Meeting, DFD, 24–26 Nov., Syracuse, NY.

Drubka, R.E., Reisenthel, P., and Nagib, H.M. (1989) "The Dynamics of Low Initial Disturbance Turbulent Jets," *Phys. Fluids A*, 1, pp. 1723–1735.

Epstein, A.H., Senturia, S.D., Al-Midani, O., Anathasuresh, G., Ayon, A., Breuer, K., Chen, K-S, Ehrich, F.E., Esteve, E., Frechette, L., Gauba, G., Ghodssi, R., Groshenry, C., Jacobson, S., Kerrebrok, J.L., Lang, J.H., Lin, C-C, London, A., Lopata, J., Mehra, A., Mur Miranda, J.O., Nagle, S., Orr, D.J., Piekos, E., Schmidt, M.A., Shirley, G., Spearing, S.M., Tan, C.S., Tzeng, Y-S., and Waitz, I.A. (1997) "Micro-Heat Engines, Gas Turbines, and Rocket Engines: The MIT Microengine Project," *AIAA Fluid Mechanics Summer Meeting*, AIAA Paper 97-1773, 29 June–2 July, Snowmass, CO.

Fiedler, H.E., and Fernholz, H.-H. (1990) "On Management and Control of Turbulent Shear Flows," *Prog. Aerosp. Sci.*, 27, pp. 305–87.

Fuentes, C., He, X., Carroll, B., Lian, Y., and Shyy, W. (2000) "Low Reynolds Number Flows Around an Airfoil with a Movable Flap: Part 1. Experiments," *AIAA Fluids 200 & Exhibit*, AIAA paper number 2000-2239, 19–22 June, Denver, CO.

Gharib, M., Modarress, D., Fourguette, D., and Taugwalder, F. (1999) "Design, Fabrication and Integration of Mini-LDA and Mini-Surface Stress Sensors for High Reynolds Number Boundary Layer Studies," *ONR Turbulence and Wakes Program Review*, 9–10 September, Palo Alto, CA.

Grosjean, C., Lee, G., Hong, W., Tai, Y.C., and Ho, C.M. (1998) "Micro Balloon Actuators for Aerodynamic Control," *Eleventh Annual International Workshop on Micro Electro Mechanical Systems (MEMS '98)*, pp. 166–71, 25–29 January, Heidelberg, Germany.

Ho, C.M., and Huerre, P. (1984) "Perturbed Free Shear Layers," *Annu. Rev. Fluid Mech.*, 16, pp. 365–424.

Ho, C.M., Huang, P.H., Lew, J., Mai, J., Lee, G.B., and Tai, Y.C. (1998) "MEMS: An Intelligent System Capable of Sensing-Computing-Actuating," *Proceedings of 4th International Conference on Intelligent Materials*, Society of Non-Traditional Technology, Tokyo, Japan, pp. 300–3.

Huang, A., Ho, C.M., Jiang, F., and Tai, Y.C. (2000) "MEMS Transducers for Aerodynamics: A Paradigm Shift," *AIAA 38th Aerospace Sciences Meeting & Exhibit*, AIAA paper number 00-0249, 10–13 January, Reno, NV.

Huang, C.C., Najafi, K., Alnajjar, E., Christophorou, C., Naguib, A., and Nagib, H.M. (1998) "Operation and Testing of Electrostatic Microactuators and Micromachined Sound Detectors for Active Control of High Speed Flows," *Eleventh Annual International Workshop on Micro Electro Mechanical Systems (MEMS '98)*, pp. 81–86, 25–29 January, Heidelberg, Germany.

Huang, C.C., Papp, J., Najafi, K., and Nagib, H.M. (1996) "A Microactuator System for the Study and Control of Screech in High-Speed Jets," *Nineth Annual International Workshop on Micro Electro Mechanical Systems (MEMS '96)*, pp. 19–24, 11–15 February, IEEE, San Diego, California.

Huang, C.C., Naguib, A., Soupos, E., and Najafi, K. "A Silicon Micromachined Microphone for Fluid Mechanics Research," (2002a) *J. Micromech. Microeng.*, **12**, pp. 1–8.

Huang, C., Christophorou, C., Najafi, K., Naguib, A., and Nagib, H.M. (2002b) "An Electrostatic Micro-actuator System for Application in High-Speed Jets," *J. Microelectromech. Syst.*, **11**, pp. 222–35.

Huerre, P., and Monkewitz, P.A. (1990) "Local and Global Instabilities in Spatially Developing Flows," *Annu. Rev. Fluid Mech.*, **22**, pp. 473–537.

Hussain, A.K.M.F., and Zaman, K.B.M.Q. (1978) "The Free Shear Layer Tone Phenomenon and Probe Interference," *J. Fluid Mech.*, **87**, pp. 349–84.

Jiang, F., Lee, G.B., Tai, Y.C., and HO, C.M. (2000) "A Flexible Micromaching-Based Shear-Stress Sensor Array and its Application to Separation-Point Detection," *Sensors Actuators*, **79**, pp. 194–203.

Lee, G.B., Ho, C.M., Jiang, F., Liu, C., Tsao, T., Tai, Y.C., and Scheuer, F. (1996) "Control of Roll Moment by MEMS," in *ASME MEMS*, Ng, W., ed.

Lepicovsky, J., Ahuja, K.K., and Burrin, R.H. (1985) "Tone Excited Jets: Part 3. Flow Measurements," *J. Sound Vib.*, **102**, pp. 71–91.

Liu, C., Tai, Y.C., Huang, J.B., and Ho, C.M. (1994) "Surface-Micromachined Thermal Shear Stress Sensor," *Appl. Microfab. Fluid Mech.* **FED-Vol. 197**, ASME, pp. 9–16.

Liu, C., Tsao, T., Tai, Y.C., Leu, J., Ho, C.M., Tang, W.L., and Miu, D. (1995) "Out-of-Plane Permanent Magnetic Actuators for Delta Wing Control," *Eighth Annual International Workshop on Micro Electro Mechanical Systems (MEMS '95)*, pp. 7–12, 29 January–2 February, IEEE Amsterdam, Netherlands.

Miller, R., Burr, G., Tai, Y.C., Psaltis, D., Ho, C.M., and Katti, R. (1996) "Electromagnetic MEMS Scanning Mirrors for Holographic Data Storage," *Technical Digest, Solid-State Sensor and Actuator Workshop*, pp. 183–86.

Moin, P., and Bewley, T. (1994) "Feedback Control of Turbulence," *Appl. Mech. Rev.* **47**, part 2, pp. S3–S13.

Moore, C.J. (1977) "The Role of Shear-Layer Instability Waves in Jet Exhaust Noise," *J. Fluid Mech.*, **80**, pp. 321–67.

Naguib, A., Christophorou, C., Alnajjar, E., and Nagib, H. (1997) "Arrays of MEMS-Based Actuators for Control of Supersonic Jet Screech," AIAA Summer Fluid Mechanics Meeting, AIAA paper number 97-1963, 29 June–2 July, Snowmass, CO.

Naguib, A., Soupos, E., Nagib, H., Huang, C., and Najafi, K. (1999a) "A Piezoresistive MEMS Sensor for Acoustic Noise Measurements," *5th AIAA/CEAS Aeroacoustics Conference*, AIAA paper number 99-1992, 10–12 May, Bellevue, WA.

Naguib, A., Benson, D., Nagib, H., Huang, C., and Najafi, K. (1999b) "Assessment of New MEMS-Based Hot Wires," *Proceedings of the 3rd ASME/JSME Joint Fluids Engineering Conference*, 18–22 July, San Francisco.

Padmanabhan, A., Goldberg, H.D., Breuer, K.S., and Schmidt, M.A. (1996) "A Wafer-Bonded Floating-Element Shear-Stress Microsensor with Optical Position Sensing by Photodiodes," *J. Microelectromech. Syst.*, **5**, pp. 307–15.

Powel, A. (1953) "On the Mechanism of Choked Jet Noise," *Proc. Phys. Soc. (London)*, **66**, no. 408B, pp. 1039–56.

Reshotko, E., Pan, T., Hyman, D., and Mehregany, M. (1996) "Characterization of Microfabricated Shear Stress Sensors," *Eighth Beer-Sheva International Seminar on MHD Flows and Turbulence*, 25–29 February, Jerusalem, Israel.

Salhi, S., and Aslam, D.M. (1998) "Ultra-High Sensitivity Intra-Grain Poly-Diamond Piezoresistors," *Sensors Actuators A*, **71**, pp. 193–97.

Tam, C.K.W. (1986) "Excitation of Instability Waves by Sound: A Physical Interpretation," *J. Sound Vib.*, **105**, pp. 169–72.

Tsao, F., Jiang, R., Miller, A., Tai, Y.C., Gupta, B., Goodman, R., Tung, S., and Ho, C.M. (1997) "An Integrated MEMS System for Turbulent Boundary Layer Control," *Technical Digest, Transducers '97* 1, pp. 315–18.

Tsao, T., Liu, C., Tai, Y.C., and Ho, C.M. (1994) "Micromachined Magnetic Actuator for Active Fluid Control," *Appl. Microfab. Fluid Mech.* **FED-Vol. 197**, ASME, pp. 31–38.

Yang, X., Tai, Y.C., and Ho, C.M. (1997) "Micro Bellow Actuators," *Technical Digest, International Conference on Solid-State Sensors and Actuators, Transducers '97* 1, pp. 45–48, 16–19 June, Chicago.

Index

C